Pub. No. 9

AMERICAN PRACTICAL NAVIGATOR

AN EPITOME OF NAVIGATION

ORIGINALLY BY

NATHANIEL BOWDITCH, LL.D.

Volume I

1984 Edition

PUBLISHED BY THE

DEFENSE MAPPING AGENCY HYDROGRAPHIC/TOPOGRAPHIC CENTER

For sale by authorized Sales Agents of the Defense Mapping Agency Office of Distribution Services

DMA STOCK NO. NVPUB9V1

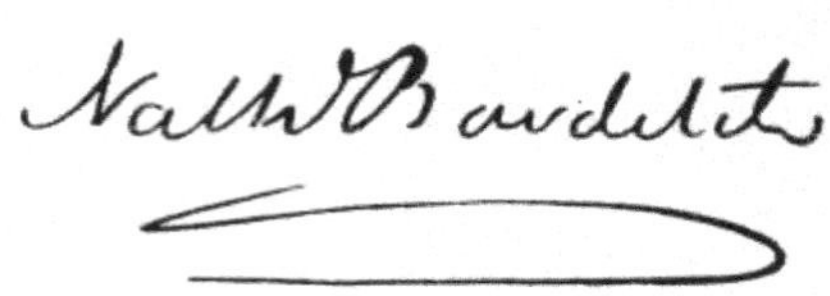

Last painting by Gilbert Stuart (1828). Considered by the family of Bowditch to be the best of various paintings made, although it was unfinished when the artist died.

NATHANIEL BOWDITCH

(1773–1838)

Nathaniel Bowditch was born on March 26, 1773, at Salem, Mass., fourth of the seven children of shipmaster Habakkuk Bowditch and his wife, Mary.

Since the migration of William Bowditch from England to the Colonies in the 17th century, the family had resided at Salem. Most of its sons, like those of other families in this New England seaport, had gone to sea, and many of them became shipmasters. Nathaniel Bowditch himself sailed as master on his last voyage, and two of his brothers met untimely deaths while pursuing careers at sea.

It is reported that Nathaniel Bowditch's father lost two ships at sea, and by late Revolutionary days he returned to the trade of cooper, which he had learned in his youth. This provided insufficient income to properly supply the needs of his growing family, and hunger and cold were often experienced. For many years the nearly destitute family received an annual grant of 15 to 20 dollars from the Salem Marine Society. By the time Nathaniel had reached the age of 10, the family's poverty necessitated his leaving school and joining his father in the cooper's trade.

Nathaniel was unsuccessful as a cooper, and when he was about 12 years of age, he entered the first of two ship-chandlery firms by which he was employed. It was during the nearly 10 years he was so employed that his great mind first attracted public attention. From the time he began school Bowditch had an all-consuming interest in learning, particularly mathematics. By his middle teens he was recognized in Salem as an authority on that subject. Salem being primarily a shipping town, most of the inhabitants sooner or later found their way to the ship chandler, and news of the brilliant young clerk spread until eventually it came to the attention of the learned men of his day. Impressed by his desire to educate himself, they supplied him with books that he might learn of the discoveries of other men. Since many of the best books were written by Europeans, Bowditch first taught himself their languages. French, Spanish, Latin, Greek, and German were among the two dozen or more languages and dialects he studied during his life. At the age of 16 he began the study of Newton's *Principia,* translating parts of it from the Latin. He even found an error in that classic, and though lacking the confidence to announce it at the time, he later published his findings and had them accepted.

During the Revolutionary War a privateer out of Beverly, a neighboring town to Salem, had taken as one of its prizes an English vessel which was carrying the philosophical library of a famed Irish scholar, Dr. Richard Kirwan. The books were brought to the Colonies and there bought by a group of educated Salem men who used them to found the Philosophical Library Company, reputed to have been the best library north of Philadelphia at the time. In 1791, when Bowditch was 18, two Harvard-educated ministers, Rev. John Prince and Rev. William Bentley, persuaded the Company to allow Bowditch the use of its library. Encouraged by these two men and a third—Nathan Read, an apothecary and also a Harvard man—Bowditch studied the works of the great men who had preceded him, especially the mathematicians and the astronomers. By the time he became of age, this knowledge, acquired before and after his long working hours and in his spare time, had made young Bowditch the outstanding mathematician in the Commonwealth, and perhaps in the country.

In the seafaring town of Salem, Bowditch was drawn to navigation early, learning the subject at the age of 13 from an old British sailor. A year later he began studying surveying, and in 1794 he assisted in a survey of the town. At 15 he devised an almanac reputed to have been of great accuracy. His other youthful accomplishments included the construction of a crude barometer and a sundial.

When Bowditch went to sea at the age of 21, it was as captain's writer and nominal second mate, the officer's berth being offered him because of his reputation as a scholar. Under Captain Henry Prince, the ship *Henry* sailed from Salem in the winter of 1795 on what was to be a year-long voyage to the Ile de Bourbon (now called Réunion) in the Indian Ocean.

Bowditch began his seagoing career when accurate time was not available to the average naval or merchant ship. A reliable marine chronometer had been invented some 60 years before, but the prohibitive cost, plus the long voyages without opportunity to check the error of the timepiece, made the large investment an impractical one. A system of determining longitude by "lunar distance," a method which did not require an accurate timepiece, was known, but this product of the minds of mathematicians and astronomers was so involved as to be beyond the capabilities of the uneducated seamen of that day. Consequently, ships navigated by a combination of dead reckoning and parallel sailing (a system of sailing north or south to the latitude of the destination and then east or west to the destination).

To Bowditch, the mathematical genius, computation of lunar distances was no mystery, of course, but he recognized the need for an easier method of working them in order to navigate ships more safely and efficiently. Through analysis and observation, he derived a new and simplified formula during his first trip.

John Hamilton Moore's *The Practical Navigator* was the leading navigational text when Bowditch first went to sea, and had been for many years. Early in his first voyage, however, the captain's writer-second mate began turning up errors in Moore's book, and before long he found it necessary to recompute some of the tables he most often used in working his sights. Bowditch recorded the errors he found, and by the end of his second voyage, made in the higher capacity of supercargo, the news of his findings in *The New Practical Navigator* had reached Edmund Blunt, a printer at Newburyport, Mass. At Blunt's request, Bowditch agreed to participate with other learned men in the preparation of an American edition of the thirteenth (1798) edition of Moore's work. The first American edition was published at Newburyport by Blunt in 1799. This edition corrected many of the errors that Moore had failed to correct. Although most of the errors were of little significance to practical navigation as they were errors in the fifth and sixth places of logarithm tables, some errors were significant. The most significant error was listing the year 1800 as a leap year in the table of the sun's declination. The consequence was that Moore gave the declination for March 1, 1800, as 7°11′. Since the actual value was 7°33′, the calculation of a meridian altitude would be in error by 22 minutes of latitude.

Bowditch's principal contribution to the first American edition was his chapter "The Method of finding the Longitude at Sea," which was his new method for computing the lunar distance. Following publication of the first American edition, Blunt obtained Bowditch's services in checking the American and English editions for further errors. Blunt then published a second American edition of Moore's thirteenth edition in 1800. When preparing a third American edition for the press, Blunt decided that Bowditch had revised Moore's work to such an extent that Bowditch should be named as author. The title was changed to *The New American Practical Navigator* and the book was published in 1802 as a first edition.

Bowditch made a total of five trips to sea, over a period of about 9 years, his last as master and part owner of the three-masted *Putnam.* Homeward bound from a 13-month voyage to Sumatra and the Ile de France (now called Mauritius) the *Putnam* approached Salem harbor on December 25, 1803, during a thick fog without having had a celestial observation since noon on the 24th. Relying upon his dead reckoning, Bowditch conned his wooden-hulled ship to the entrance of the rocky harbor, where he had the good fortune to get a momentary glimpse of Eastern Point, Cape Ann, enough to confirm his position. The *Putnam* proceeded in, past such hazards as "Bowditch's Ledge" (named after a great-grandfather who had wrecked his ship on the rock more than a century before) and anchored safely at 1900 that evening. Word of the daring feat, performed when other masters were hove-to outside the harbor, spread along the coast and added greatly to Bowditch's reputation. He was, indeed, the "practical navigator."

His standing as a mathematician and successful shipmaster earned him a lucrative (for those times) position ashore within a matter of weeks after his last voyage. He was installed as president of a Salem fire and marine insurance company, at the age of 30, and during the 20 years he held that position the company prospered. In 1823 he left Salem to take a similar position with a Boston insurance firm, serving that company with equal success until his death.

From the time he finished the "*Navigator*" until 1814, Bowditch's mathematical and scientific pursuits consisted of studies and papers on the orbits of comets, applications of Napier's rules, magnetic variation, eclipses, calculations on tides, and the charting of Salem harbor. In that year, however, he turned to what he considered the greatest work of his life, the translation into English of *Mécanique Céleste,* by Pierre Laplace. *Mécanique Céleste* was a summary of all the then known facts about the workings of the heavens. Bowditch translated four of the five volumes before his death, and published them at his own expense. He gave many formula derivations which Laplace had not shown, and also included further discoveries following the time of publication. His work made this information available to American astronomers and enabled them to pursue their studies on the basis of that which was already known. Continuing his style of writing for the learner, Bowditch presented his English version of *Mécanique Céleste* in such a manner that the student of mathematics could easily trace the steps involved in reaching the most complicated conclusions.

Shortly after the publication of *The New American Practical Navigator,* Harvard College honored its author with the presentation of the honorary degree of Master of Arts, and in 1816 the college made him an honorary Doctor of Laws. From the time the Harvard graduates of Salem first assisted him in his studies, Bowditch had a great interest in that college, and in 1810 he was elected one of its Overseers, a position he held until 1826, when he was elected to the Corporation. During 1826–27 he was the leader of a small group of men who saved the school from financial disaster by forcing necessary economies on the college's reluctant president. At one time Bowditch was offered a Professorship in Mathematics at Harvard but this, as well as similar offers from West Point and the University of

Virginia, he declined. In all his life he was never known to have made a public speech or to have addressed any large group of people.

Many other honors came to Bowditch in recognition of his astronomical, mathematical, and marine accomplishments. He became a member of the American Academy of Arts and Sciences, the East India Marine Society, the Royal Academy of Edinburgh, the Royal Society of London, the Royal Irish Academy, the American Philosophical Society, the Connecticut Academy of Arts and Sciences, the Boston Marine Society, the Royal Astronomical Society, the Palermo Academy of Science, and the Royal Academy of Berlin.

Nathaniel Bowditch outlived all of his brothers and sisters by nearly 30 years. Death came to him on March 16, 1838, in his sixty-fifth year. The following eulogy by the Salem Marine Society indicates the regard in which this distinguished American was held by his contemporaries:

"In his death a public, a national, a human benefactor has departed. Not this community, nor our country only, but the whole world, has reason to do honor to his memory. When the voice of Eulogy shall be still, when the tear of Sorrow shall cease to flow, no monument will be needed to keep alive his memory among men; but as long as ships shall sail, the needle point to the north, and the stars go through their wonted courses in the heavens, the name of Dr. Bowditch will be revered as of one who helped his fellow-men in a time of need, who was and is a guide to them over the pathless ocean, and of one who forwarded the great interests of mankind."

The New American Practical Navigator was revised by Nathaniel Bowditch several times after 1802 for subsequent editions of the book. After his death, Jonathan Ingersoll Bowditch, a son who made several voyages, took up the work and his name appeared on the title page from the eleventh edition through the thirty-fifth, in 1867. In 1868 the newly organized U.S. Navy Hydrographic Office bought the copyright. Revisions have been made from time to time to keep the work in step with navigational improvements. The name has been altered to the *American Practical Navigator,* but the book is still commonly known as "Bowditch." A total of more than 900,000 copies has been printed in about 70 editions during the more than a century and a half since the book was first published in 1802. It has lived because it has combined the best thoughts of each generation of navigators, who have looked to it as their final authority.

PREFACE

This 1984 edition has afforded the opportunity to incorporate changes to clarify certain parts of the text. In addition minor corrections and modifications have been made including a change to the name of the Agency responsible for this publication to Defense Mapping Agency Hydrographic/Topographic Center (DMAHTC).

Appendix Y, Buoyage Systems, has been expanded to include the *Modified U.S. Aids to Navigation System* and the *International Association of Lighthouse Authorities (IALA) Maritime Buoyage System,* Region A and Region B. In addition the metric side of Port Maury, the fictitious nautical chart in chapter V, has been revised to include the IALA Buoyage System, Region B.

A description of the NAVSTAR Global Positioning System (GPS) has been added to the Satellite Navigation chapter.

A new piloting method, called the slide bar technique, has been added to Chapter X, Piloting.

The description of Loran-A in Appendix R has been replaced by a table for the Greenwich Hour Angle (GHA) and Declination of the Sun for the years 1981–2016.

Appendix Z, Extracts from Chart No. 1, has been deleted from this edition.

The text of this 1984 edition was produced on magnetic tape which was then used to photocompose each page.

Users should refer corrections, additions, and comments for improving this product to DIRECTOR, DEFENSE MAPPING AGENCY HYDROGRAPHIC/TOPOGRAPHIC CENTER, ATTN: PR, Washington, D.C. 20315.

CONTENTS

CONTENTS

PART FOUR

THE PRACTICE OF NAVIGATION

PART FIVE

NAVIGATIONAL SAFETY

PART SIX

OCEANOGRAPHY

PART SEVEN

WEATHER

PART EIGHT

ELECTRONICS AND NAVIGATION

APPENDICES

PART ONE
FUNDAMENTALS

PART ONE

FUNDAMENTALS

PART ONE
FUNDAMENTALS

PART ONE

FUNDAMENTALS

CHAPTER I

HISTORY OF NAVIGATION

Introduction

101. Background.—Navigation began with the first man. One of his first conscious acts probably was to home on some object that caught his eye, and thus **land navigation** was undoubtedly the earliest form. His first venture upon the waters may have come shortly after he observed that some objects float, and through curiosity or an attempt at self-preservation he learned that a larger object, perhaps a log, would support him. **Marine navigation** was born when he attempted to guide his craft.

The earliest marine navigation was a form of **piloting,** which came into being as man became familiar with landmarks and used them as guides. **Dead reckoning** probably came next as he sought to predict his future positions, or perhaps as he bravely ventured farther from landmarks. **Celestial navigation,** as it is known today, had to await acquisition of information regarding the motions of the heavenly bodies, although these bodies were used to steer by almost from the beginning.

102. From art to science.—Navigation is the process of directing the movements of a craft from one point to another. To do this safely is an *art.* In perhaps 6,000 years—some writers make it 8,000—man has transformed this art almost into a *science,* and navigation today is so nearly a science that the inclination is to forget that it was ever anything else. It is commonly thought that to navigate a ship one must have a chart to determine the course and distance, a compass to steer by, and a means of determining the positions of the ship during the passage. *Must* have? The word "must" betrays how dependent the modern navigator has become upon the tools now in his hands. Many of the great voyages of history—voyages that made known much of the world—were made without one or more of these "essentials."

103. Epic voyages.—History records a number of great voyages of varying navigational significance. Little or nothing is known of the navigational accomplishments of the ancient mariners, but the record of the knowledge and equipment used during later voyages serves to illustrate periodic developments in the field.

104. Pre-Christian navigation.—Down through the stream of time a number of voyages have occurred without navigational significance. Noah's experience in the ark is of little interest navigationally, except for his use of a dove to locate land. There is evidence to support the view that at least some American Indians reached these shores by sea, the earliest of several groups probably having come about 2200 BC, the approximate time that a general exodus seems to have occurred from a center in southwestern Asia. This is about the time the Tower of Babel is believed to have been built. It is noteworthy that almost every land reached by the great European explorers was already inhabited.

It is not difficult to understand how a people not accustomed to the sea might make a single great voyage without contributing anything of significance to the advancement of navigation. Not so clear, however, is the fact that the Norsemen and the Polynesians, great seafaring people, left nothing more than conflicting traditions of their methods. The reputed length of the voyages made by these people suggests more advanced navigational methods than their records indicate, although the explanation may be that they left few written accounts of any kind. Or perhaps they developed their powers of perception to such an extent that navigation to them, was a highly advanced art. In this respect their navigation may not have differed greatly from that of some birds, insects, fishes, and animals.

One of the earliest well-recorded voyages is known today through the book of observations written by Pytheas of Massalia, a Greek astronomer and navigator. Sometime between the years 350 BC and 300 BC he sailed from a Mediterranean port and followed an established trade route to England. From there he ventured north to Scotland and Thule, the legendary land of the midnight sun. He went on to explore Norwegian fiords, and rivers in northwest Germany. He may have made his way into the Baltic.

Pytheas' voyage, and others of his time, were significant in that they were the work of men who had no compasses, no sextants, no chronometers, no electronic devices such as are commonplace today. The explanation of how they did it is not what some historians have said, that before seafaring men had adequate equipment, the compass especially, they hugged the shore and sailed only by daylight in fair weather. Many undoubtedly did use this practice. But the more intrepid did not creep along the coast, venturing nothing more daring than sailing from headland to headland. They were often out of sight of land, and yet knew sufficiently well where they were and how to get home again. They were able to use the sun, the stars, and the winds without the aid of mechanical devices.

Pytheas had none of the equipment considered essential by the modern navigator—none, at least, as it is thought of today. It would be incorrect, however, to say that he had no navigational aids whatever. He was not the first to venture upon the sea, and even in his time man was the inheritor of his predecessors' knowledge.

He must have known what the mariners of his time, Phoenician and Greek, knew about navigation. There was a fair store of knowledge about the movements of the stars, for example, which all seafaring men shared. They had a practical grasp of some part of what is now called celestial navigation, for the moving celestial bodies were their compasses. Pytheas may not have been acquainted with the *Periplus* of Scylax, the earliest known sailing directions, but it is reasonable to suppose that he had similar information.

If there were sailing directions, there may well have been charts of a sort, even though no record of them exists.

Even if Pytheas and his contemporaries had sailing directions and charts, these must have been far from comprehensive, and they undoubtedly did not cover the areas north of Britain. But these early seamen knew direction by day or night if the sky was clear, and they could judge it reasonably well when the sky was overcast, using the wind and the sea. They knew the hot Libyan wind from the desert—today called the **sirocco**—and the northern wind, the **mistral.**

They could estimate distance. Their ships must have carried some means of measuring time—the sand glass was known to the ancients—and they could estimate speed by counting the strokes of the oars, a common practice from galley to modern college racing shell. Mariners who spent their lives traveling the Mediterranean knew what their ships could do, even if today it is not known what they meant by "a day's sail"—whether 35 miles, or 50, or 100.

105. Sixteenth century navigation.—Progress in the art of navigation came slowly during the early centuries of the Christian era, all but stopped during the

Dark Ages, and then spurted forward when Europe entered a golden age of discovery. The circumnavigation of the globe by the expedition organized by Ferdinand Magellan, a disgraced Portuguese nobleman who sailed under the flag of Spain, was a voyage which illustrates the advances made during the 1,800 years following Pytheas.

Magellan was able to find justification for his belief that a navigable pass to the Pacific Ocean existed in high southern latitudes, in Martin Behaim's globe or chart of the world, in the globe constructed by Johann Schoner of Nuremberg in 1515, and in Leonardo da Vinci's map of the world drawn in the same year. He obtained further information for his voyage from Ruy Faleiro, an astronomer and cartographer whose charts, sailing directions, nautical tables, and instructions for use of the astrolabe and cross-staff were considered to be among the best available. Faleiro was also an advocate of the fallacious methods of determining longitude by variation.

When Magellan sailed in 1519, his equipment included sea charts, parchment skins to be made into charts en route, a terrestrial globe, wooden and metal theodolites, wooden and wood-and-bronze quadrants, compasses, magnetic needles, hour glasses and "timepieces," and a log to be towed astern.

So the 16th century navigator had crude charts of the known world, a compass to steer by, instruments with which he could determine his latitude, a log to estimate speed, certain sailing directions, and solar and traverse tables. The huge obstacle yet to be overcome was an accurate method of determining longitude.

106. Eighteenth century navigation.—Little is known today of the "timepieces" carried by Magellan, but surely they were not used to determine longitude. Two hundred years later, however, the chronometer began to emerge. With it, the navigator, for the first time, was able to determine his longitude accurately and fix his position at sea.

The three voyages of discovery made by James Cook of the Royal Navy in the Pacific Ocean between 1768 and 1779 may be said to mark the dawn of modern navigation. Cook's expedition had the full backing of England's scientific organizations, and he was the first captain to undertake extended explorations at sea with navigational equipment, techniques, and knowledge that might be considered modern.

On his first voyage Cook was provided with an astronomical clock, a "journeyman" clock, and a watch lent by the Astronomer Royal. With these he could determine longitude, using the long and tedious lunar distance method. On his second voyage four chronometers were provided. These instruments, added to those already possessed by the mariner, enabled Cook to navigate his vessels with a precision undreamed of by Pytheas and Magellan.

By the time Cook began his explorations, astronomers had made great contributions to navigational advancement, and the acceptance of the heliocentric theory of the universe had led to the publication of the first official nautical almanac. Charts had progressed steadily, and adequate projections were available. With increased understanding of variation, the compass had become reliable. Good schools of navigation existed, and textbooks which reduced the mathematics of navigation to the essentials had been published. Speed through the water could be determined with reasonable accuracy by the logs then in use. Most important, the first chronometers were being produced.

107. Twentieth century navigation.—The maiden voyage of the SS *United States* in July 1952 served to illustrate the progress made in navigation during the

175 years since Cook's voyages. Outstanding because of its record trans-Atlantic passage, the vessel is of interest navigationally in that it carried the most modern equipment then available and exemplified the fact that navigation had become nearly a science.

Each of the deck officers owned a sextant with which he could make observations more accurately than did Cook. Reliable chronometers, the product of hundreds of years of experimental work, were available to determine the time of each observation. The gyrocompass indicated true north regardless of variation and deviation.

Modern, convenient almanacs were used to obtain the coordinates of various celestial bodies, to an accuracy greater than needed. Easily used altitude and azimuth tables gave the navigator data for determining his Sumner (celestial) line of position by the method of Marcq St.-Hilaire. Accurate charts were available for the waters plied, sailing directions for coasts and ports visited, light lists giving the characteristics of the various aids to navigation along these coasts, and pilot charts and navigational texts for reference purposes.

Electronics served the navigator in a number of ways. Radio time signals and weather reports enabled him to check his chronometers and avoid foul weather. A radio direction finder was available to obtain bearings, and a radio telephone was used to communicate with persons on land and sea. The electrically operated echo sounder indicated the depth of water under the keel, radar the distances and bearings of objects within range, even in the densest fog. Using Loran, the navigator could fix the position of his ship a thousand miles and more from transmitting stations.

Piloting and Dead Reckoning

108. Background.—The history of piloting and dead reckoning extends from man's earliest use of landmarks to the latest model of the gyrocompass. In the thousands of years between, navigation by these methods has progressed from short passages along known coastlines to transoceanic voyages during which celestial observations cannot be, or are not, made.

109. Charts.—A form of sailing directions was written several hundred years before Christ. Although charts cannot be traced back that far, they may have existed during the same time. From earliest times men have undoubtedly known that it is more difficult to explain how to get to a place than it is to draw a diagram, and since the first charts known are comparatively accurate and cover large areas, it seems logical that earlier charts served as guides for the cartographers.

Undoubtedly, the first charts were not made on any "projection" (ch. III) but were simple diagrams which took no notice of the shape of the earth. In fact, these "plane" charts were used for many centuries after chart projections were available.

The **gnomonic projection** (art. 317) is believed to have been developed by Thales of Miletus (640–546 BC), who was chief of the Seven Wise Men of ancient Greece; founder of Greek geometry, astronomy, and philosophy; and a navigator and cartographer.

The size of the earth was measured at least as early as the third century BC, by Eratosthenes. He observed that at noon on the day of the summer solstice, a certain well at Syene (Assuan) on the Tropic of Cancer was lighted throughout its depth by the light of the sun as it crossed the meridian; but that at Alexandria, about 500 miles to the north, shadows were cast by the sun at high noon. He reasoned that this was due to curvature of the earth, which must be spherical. By double measurement of the arc of the meridian between the two places in degrees and stadia, Eratosthenes determined the circumference of the earth to be 252,000 stadia (art. 113).

Eratosthenes is believed to have been the first person to measure latitude, using the degree for this purpose. He constructed a 16-point wind rose, prepared a table of winds, and recognized local and prevailing winds. From his own discoveries and from information gleaned from the manuscripts of mariners, explorers, land travelers, historians, and philosophers, he wrote an outstanding description of the known world, which helped elevate geography to the status of a science.

Stereographic (art. 318) and **orthographic** (art. 319) **projections** were originated by Hipparchus in the second century BC.

Ptolemy's World Map. The Egyptian Claudius Ptolemy was a second century AD astronomer, writer, geographer, and mathematician who had no equal in astronomy until the arrival of Copernicus in the 16th century. An outstanding cartographer, for his time, Ptolemy constructed many charts and listed the latitudes and longitudes, as determined by celestial observations, of the places shown. As a geographer, however, he made his most serious mistake. Though Eratosthenes' calculations on the circumference of the earth were available to him, he took the estimate of the Stoic philosopher, Posidonius (circa 130–51 BC), who calculated the earth to be 180,000 stadia in circumference. The result was that those who accepted his work—and for many hundreds of years few thought to question it—had to deal with a concept that was far too small. In 1409 the Greek original of Ptolemy's *Cosmographia,* a book in which he declared this doctrine, was discovered and translated into Latin. It served as the basis for future cartographic work, and so it was that Columbus died convinced that he had found a shorter route to the East Indies. Not until 1669, when Jean Picard computed the circumference of the earth to be 24,500 miles, was a more accurate figure generally used.

Ptolemy's map of the world (fig. 109a) was a great achievement, however. It was the original conic projection, and on it he located some 8,000 places by latitude and longitude. It was he who fixed the convention that the top of the map is north.

Asian Charts. Through the Dark Ages some progress was made. Moslem cartographers as well as astronomers took inspiration from Ptolemy. However, they knew that Ptolemy had overestimated the length of the Mediterranean by some 20°. Charts of the Indian Ocean, bearing horizontal lines indicating parallels of latitude, and vertical lines dividing the seas according to the direction of the wind, were drawn by Persian and Arabian navigators. The prime meridian separated a windward from a leeward region and other meridians were drawn at intervals indicating "three hours sail." This information, though far from exact, was helpful to the sailing ship masters.

Portolan Charts. The mariners of Venezia (Venice), Livorno (Leghorn), and Genova (Genoa) must have had charts when they competed for Mediterranean trade before, during, and after the Crusades. Venice at one time had 300 ships, a navy of 45 galleys, and 11,000 men engaged in her maritime industry. But perhaps the rivalry was too keen for masters carelessly to leave charts lying about. At any rate, the earliest useful charts of the Middle Ages that are known today were drawn by seamen of Catalonia (now part of Spain).

The Portolan charts were constructed from the knowledge acquired by seamen during their voyages about the Mediterranean. The actual courses and dead reckoning distances between land points were used as a skeleton for the charts, and the

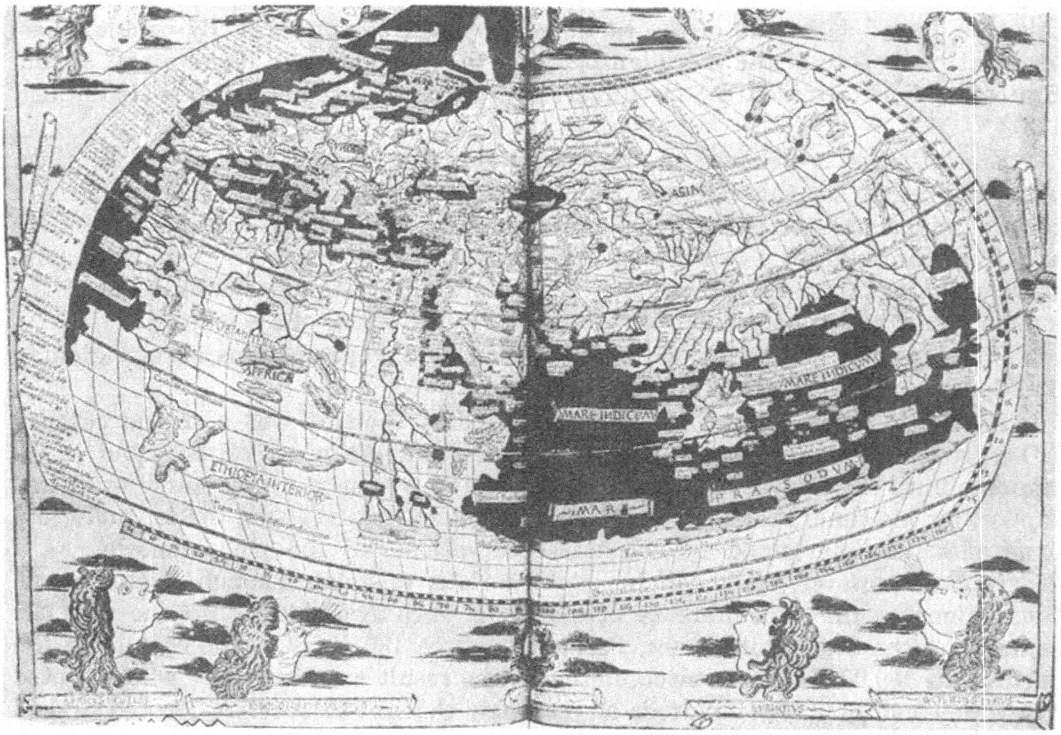

Courtesy of the Map Division of the Library of Congress.

FIGURE 109a.—The world, as envisioned by Ptolemy about AD 150. This chart was prepared in 1482 by Nicolaus Germanus for a translation of *Cosmographia*.

coasts between were usually filled in from data obtained in land surveys. After the compass came into use, these charts became quite accurate. Some, for example, indicated the distance between Gibraltar and Bayrut (Beirut) to be 3,000 Portolan miles, or 40°.5 of longitude. The actual difference of longitude is 40°.8.

These charts were distinguished by a group of long rhumb lines intersecting at a common point, surrounded by eight or 16 similar groups of shorter lines. Later *Portolanis* had a *rose dei venti* (rose of the winds), the forerunner of the compass rose, superimposed over the center (fig. 109b). They carried a scale of miles, located nearly all the known hazards to navigation, and had numerous notes of interest to the pilot. They were not marked with parallels of latitude or meridians of longitude, but present-day harbor and coastal charts trace their ancestry directly to them.

Padrón Real. The growing habit of assembling information for charts took concrete form in the *Padrón Real.* This was the pattern, or master, map kept after 1508 by the *Casa de Contratación* at Seville. It was intended to contain everything known about the world, and it was constructed from facts brought back by mariners from voyages to newly discovered lands. From it were drawn the charts upon which the explorers of the Age of Discovery most depended.

World maps of the Middle Ages. In 1515 Leonardo da Vinci drew his famous map of the world. On it, America is represented as extending more to the east and west than to the north and south, with only a chain of islands, the largest named Florida, between it and South America. A wide stretch of ocean is shown between South America and *Terra Australis Nondum Cognita,* the mythical south-seas continent whose existence in the position shown was not disproved until 250 years later.

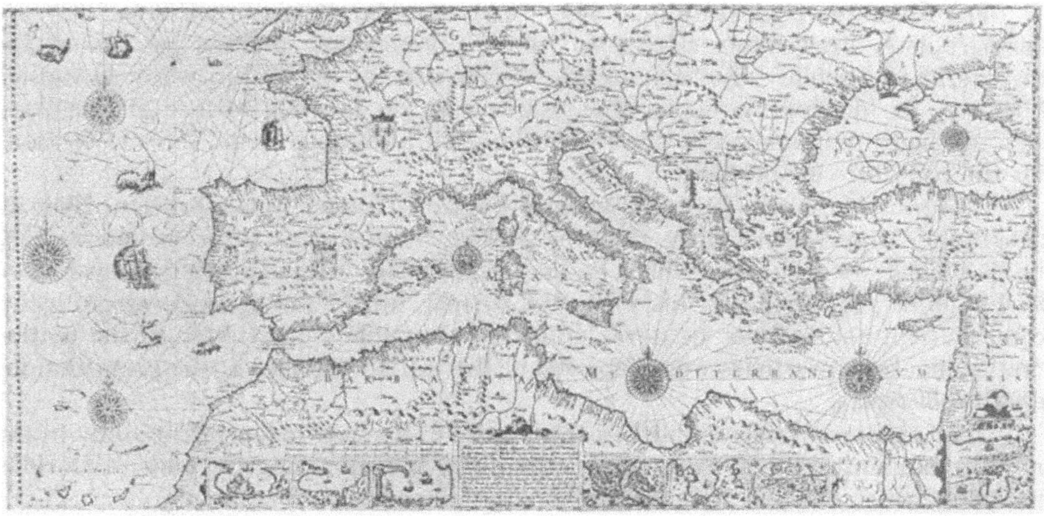

Courtesy of the Map Division of the Library of Congress.

FIGURE 109b.—A 14th-century Portolan chart.

Ortelius' atlas *Theatrum Orbis Terra* was published at Antwerp in 1570. One of the most magnificent ever produced, it illustrates Europe, Africa, and Asia with comparative accuracy. North and South America are poorly depicted, but Magellan's Strait is shown. All land to the south of it, as well as Australia, is considered part of *Terra Australis Nondum Cognita* (fig. 109c).

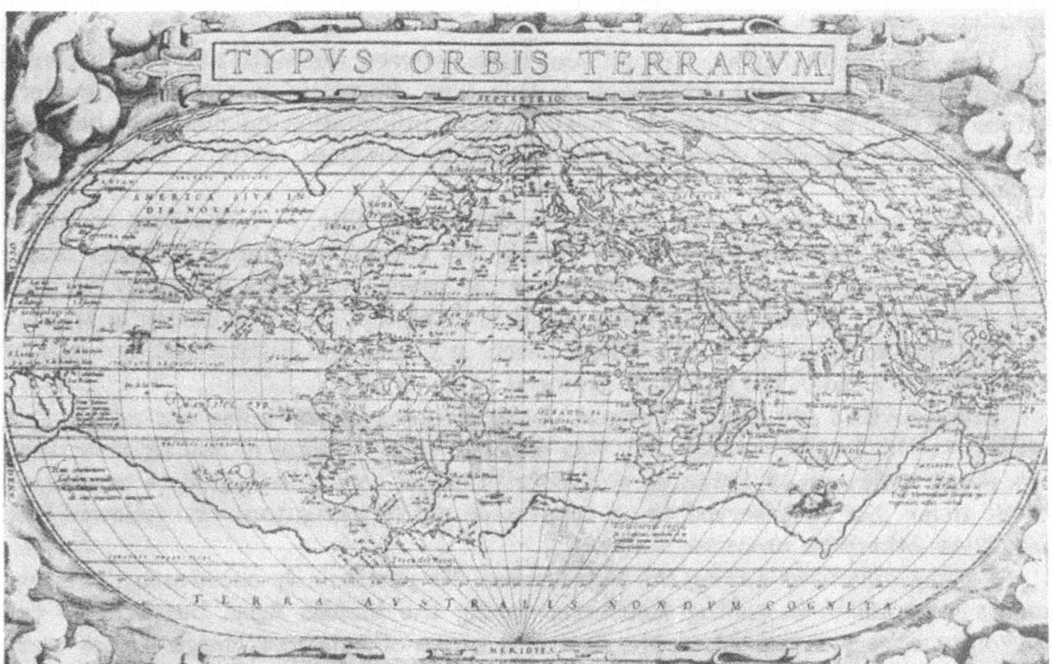

Courtesy of the New York Public Library.

FIGURE 109c.—Ortelius' world map, from his atlas *Theatrum Orbis Terra,* published at Antwerp in 1570.

The Mercator projection (art. 305). For hundreds, perhaps thousands, of years cartographers drew their charts as "plane" projections, making no use of the discoveries of Ptolemy and Hipparchus. As the area of the known world increased, however, the attempt to depict that larger area on the flat surface of the plane chart brought map makers to the realization that allowance would have to be made for the curvature of the earth.

Gerardus Mercator (Latinized form of Gerhard Kremer) was a brilliant Flemish geographer who recognized the need for a better method of chart projection. In 1569 he published a world chart which he had constructed on the principle since known by his name. The theory of his work was correct, but Mercator made errors in his computation, and because he never published a complete description of the mathematics involved, mariners were deprived of the full advantages of the projection for another 30 years.

Then Edward Wright published the results of his own independent study in the matter, explaining the Mercator projection fully and providing the table of meridional parts which enabled all cartographers to make use of the principle.

Wright was a mathematician at Caius College who developed the method and table and gave them to certain navigators for testing. After these proved their usefulness, Wright decided upon publication, and in 1599 *Certaine Errors in Navigation Detected and Corrected* was printed.

The Lambert projections. Johann Heinrich Lambert, 1728–1777, self-educated son of an Alsace tailor, designed a number of map projections. Some of these are still widely used, the most renowned being the **Lambert conformal** (art. 314).

110. Sailing directions.—From earliest times there has been a demand for knowledge of what lay ahead, and this gave rise to the early development of sailing directions (art. 1301).

The *Periplus* of Scylax, written sometime between the sixth and fourth centuries BC, is the earliest known book of this type. Surprisingly similar to modern sailing directions, it provided the mariner with information on distances between ports, aids and dangers, port facilities, and other pertinent matters. The following excerpt is typical:

"Libya begins beyond the Canopic mouth of the Nile. . . . The first people in Libya are the Adrymachidae. From Thonis the voyage to Pharos, a desert island (good harbourage but no drinking water), is 150 stadia. In Pharos are many harbors. But ships water at the Marian Mere, for it is drinkable. . . . The mouth of the bay of Plinthine to Leuce Acte (the white beach) is a day and night's sail; but sailing round by the head of the bay of Plinthine is twice as long. . . ."

Ports Around the World, Pytheas' book of observations made during his epic voyage in the fourth century BC, was another early volume of sailing directions. His rough estimates of distances and descriptions of coastlines would be considered crude today, but they served as an invaluable aid to navigators who followed him into these otherwise unknown waters.

Sailing directions during the Renaissance. No particularly noteworthy improvements were made in sailing directions during the Middle Ages, but in 1490 the *Portolano Rizo* was published, the first of a series of improved design. Other early volumes of this kind appeared in France and were called "routiers"—the **rutters** of the English sailor. In 1557 the Italian pilot Battista Testa Rossa published *Brieve Compendio del Arte del Navigar,* which was designed to serve the mariner on soundings and off. It forecast the single, all-inclusive volume that was soon to come, the **Waggoner.**

About 1584 the Dutch pilot Lucas Janszoon Waghenaer published a volume of navigational principles, tables, charts, and sailing directions which served as a guide for such books for the next 200 years. In *Spieghel der Zeevaerdt* (The Mariner's Mirror), Waghenaer gave directions and charts for sailing the waters of the Low Countries and later a second volume was published covering waters of the North and Baltic seas.

These "Waggoners" met with great success and in 1588 an English translation of the original book was made by Anthony Ashley. During the next 30 years, 24 editions of the book were published in Dutch, German, Latin, and English. Other authors followed the profitable example set by Waghenaer, and American, British, and French navigators soon had "Waggoners" for most of the waters they sailed.

The success of these books and the resulting competition among authors were responsible for their eventual discontinuance. Each writer attempted to make his work more inclusive than any other (the 1780 *Atlantic Neptune* contained 257 charts of North America alone) and the result was a tremendous book difficult to handle. They were too bulky, the sailing directions were unnecessarily detailed, and the charts too large. In 1795 the British Hydrographic Department was established, and charts and sailing directions were issued separately. The latter, issued for specific waters, were returned to the form of the original *Periplus.*

Modern sailing directions. The publication of modern sailing directions by the Defense Mapping Agency Hydrographic/Topographic Center is one of the achievements properly attributed to Matthew Fontaine Maury. During the two decades he headed the Depot of Charts and Instruments (renamed U.S. Naval Observatory and Hydrographical Office in 1854), Maury gathered data that led to the publication of eight volumes of sailing directions.

111. The compass.—Early in the history of navigation man noted that the pole star (it may have been *α Draconis* then) remained close to one point in the northern sky. This served as his compass. When it was not visible, he used other stars, the sun and moon, winds, clouds, and waves. The development of the magnetic compass, perhaps a thousand years ago, and the 20th century development of the gyrocompass, offer today's navigator a method of steering his course with an accuracy as great as he is capable of using.

The **magnetic compass** (art. 623) is one of the oldest of the navigator's instruments. Its origin is not known. In 203 BC, when Hannibal set sail from Italy, his pilot was said to be one *Pelorus.* Perhaps the compass was in use then; no one can say for certain that it was not. There is little to substantiate the story that the Chinese invented it, and the legend that Marco Polo introduced it into Italy in the 13th century is almost certainly false. It is sometimes stated that the Arabs brought it to Europe, but this, too, is unlikely. Probably it was known first in the west. The Norsemen of the 11th century were familiar with it, and about 1200 a compass used by mariners when the pole star was hidden was described by a French poet, Guyot de Provins.

A needle thrust through a straw and floated in water in a container comprised the earliest compass known. A 1248 writer, Hugo de Bercy, spoke of a new compass construction, the needle "now" being supported on two floats. Petrus Peregrinus de Maricourt, in his *Epistola de Magnete* of 1269, wrote of a pivoted floating compass with a lubber's line, and said that it was equipped with sights for taking bearings.

The reliability of the magnetic compass of today is a comparatively recent achievement. It was not until the 1870's that Sir William Thomson (Lord Kelvin) was able to successfully combine all of the requirements for a good dry-card compass, and mount it in a well-designed binnacle. The dry-card compass was the standard compass in the Royal Navy until 1906 when the Board of Admiralty adopted the liquid compass as the standard compass.

The **compass card,** according to tradition, originated about the beginning of the 14th century, when Flavio Gioja of Amalfi attached a sliver of lodestone or a magnetized needle to a card. But the rose on the compass card is probably older than the needle. It is the wind rose of the ancients. Primitive man naturally named directions by the winds. The prophet Jeremiah speaks of the winds from the four quarters of heaven (Jer. 49:36) and Homer named four winds—Boreas, Eurus, Notus, and Lephyrus. Aristotle is said to have suggested a circle of 12 winds, and Eratosthenes, who measured the world correctly, reduced the number to eight about 200 BC. The "Tower of Winds" at Athens, built about 100 BC, had eight sides. The Latin rose of 12 points was common on most compasses used in the Middle Ages.

Variation (art. 706) was well understood 200 years ago, and navigators made allowance for it, but earliest recognition of its existence is not known. Columbus and even the 11th century Chinese have been given credit for its discovery, but little proof can be offered for either claim.

The secular change in variation was determined by a series of magnetic observations made at Limehouse, England. In 1580 William Borough fixed the variation in that area at approximately 11°25′ east. Thirty-two years later Edmund Gunter, professor of astronomy at Gresham College, determined it to be 6°13′ east. At first it was believed that Borough had made an error in his work, but in 1633 a further decrease was found, and the earth's changing magnetic field was established.

A South Atlantic expedition was led by Edmond Halley at the close of the 17th century to gather data and to map, for the first time, lines of variation. In 1724 George Graham published his observations in proof of the diurnal change in variation. Canton determined that the change was considerably less in winter than in summer, and about 1785 the strength of the magnetic force was shown by Paul de Lamanon to vary in different places.

The existence of **deviation** (art. 709) was known to John Smith in 1627 when he wrote of the "bittacle" as being a "square box nailed together with wooden pinnes, because iron nails would attract the Compasse." But no one knew how to correct a compass for deviation until Captain Matthew Flinders, while on a voyage to Australia in HMS *Investigator* in 1801–02, discovered a method of doing so. Flinders did not understand deviation completely, but the vertical bar he erected to correct for it was part of the solution, and the **Flinders bar** (art. 720) used today is a memorial to its discoverer. Between 1839 and 1855 Sir George Airy, then Astronomer Royal, studied the matter further and developed combinations of permanent magnets and soft iron masses for adjusting the compass. The introduction, by Lord Kelvin, of short needles as compass magnets made adjustment more precise.

The gyrocompass (art. 631). The age of iron ships demanded a compass which could be relied upon to indicate true north at all times, free from disturbing forces of variation and deviation.

In 1851, at the Pantheon in Paris, Leon Foucault performed his famous pendulum experiment to demonstrate the rotation of the earth. Foucault's realization that the swinging pendulum would maintain the plane of its motion led him, the following year, to develop and name the first gyroscope, using the principle of a common toy called a "rotascope." Handicapped by the lack of a source of power to maintain the spin of his gyroscope, Foucault used a microscope to observe the indication of the earth's rotation during the short period in which his manually operated gyroscope remained in rotation. A gyrocompass was not practical until electric power became available, more than 50 years later, to maintain the spin of the gyroscope.

Elmer A. Sperry, an American, and Anschutz-Kampfe, a German, independently invented gyrocompasses during the first decade of the 20th century. Tested first in 1911 on a freighter operating off the East Coast of the United States and then on American warships, Sperry's compass was found adequate, and in the years following World War I gyrocompasses became standard equipment on all large naval and merchant ships.

Gyrocompass auxiliaries commonly used today were added later. These include gyro repeaters, to indicate the vessel's heading at various locations; gyro pilots, to steer vessels automatically; course recorders, to provide a graphic record of courses steered; gyro-magnetic compasses, to repeat headings of magnetic compasses so located as to be least affected by deviation; and others in the fields of fire control, aviation, and guided missiles.

112. The log.—Since virtually the beginning of navigation, the mariner has attempted to determine his speed in traveling from one point to another. The earliest method was probably by estimate.

The oldest speed measuring device known is the **Dutchman's log.** Originally, any object which would float was thrown overboard on the lee side, from a point well forward, and the time required for it to pass between two points on the deck was noted. The time, as determined by sand glass, was compared with the known distance along the deck between the two points to determine the speed.

Near the end of the 16th century a line was attached to the log, and as the line was paid out a sailor recited certain sentences. The length of line which was paid out during the recitation was used to determine the speed. There is record of this method having been used as recently as the early 17th century. In its final form this **chip log, ship log,** or **common log** consisted of the *log chip* (or *log ship*), *log line, log reel,* and *log glass.* The chip was a quadrant-shaped piece of wood weighted along its circumference to keep it upright in the water (fig. 112). The log line was made fast to the log chip by means of a bridle, in such manner that a sharp pull on the log line dislodged a wooden peg and permitted the log chip to be towed horizontally through the water, and hauled aboard. Sometimes a *stray line* was attached to the log to veer it clear of the ship's wake. In determining speed, the observer counted the knots in the log line which was paid out during a certain time. The length of line between knots and the number of seconds required for the sand to run out were changed from time to time as the accepted length of the mile was altered.

Courtesy of "Motor Boating & Sailing."

FIGURE 112.—The common or chip log, showing the log reel, the log line, the log chip, and the log glass.

The chip log has been superseded by patent logs that register on dials. However, the common log has left its mark on modern navigation, as the use of the term **knot** to indicate a speed of 1 nautical mile per hour dates from this device. There is evidence to support the opinion that the expression "dead reckoning" had its origin in this same device, or perhaps in the earlier Dutchman's log. There is logic in attributing "dead" reckoning to a reckoning relative to an object "dead" in the water.

Mechanical logs first appeared about the middle of the 17th century. By the beginning of the 19th century, the forerunners of modern mechanical logs were used by some navigators, although many years were to pass before they became generally accepted.

In 1773 logs on which the distance run was recorded on dials secured to the taffrail were tested on board a British warship and found reasonably adequate, although the comparative delicateness of the mechanism led to speculation about their long-term worth. Another type in existence at the time consisted of a wheel arrangement made fast on the underside of the keel, which transmitted readings to a dial inside the vessel as the wheel rotated.

An improved log was introduced by Edward Massey in 1802. This log gave considerably greater accuracy by means of a more sensitive rotator attached by a short length of line to a geared recording instrument. The difficulty with this log was that it has to be hauled aboard to take each reading. Various improvements were made, notably by Alexander Bain in 1846 and Thomas Walker in 1861, but it was not until 1878 that a log was developed in which the rotator could be used in conjunction with a dial secured to the after rail of the ship, and although refinements and improvements have been made, the patent log used today is essentially the same as that developed in 1878.

Engine revolution counters (art. 616) had their origin with the observations of the captains of the first paddle steamers, who discovered that by counting the paddle revolutions, they could, with practice, estimate their runs in thick weather as accurately as they could by streaming the log. Later developments led to the modern revolution counter on screw-type vessels, which can be used with reasonable accuracy if the propeller is submerged and an accurate estimate of slip is made.

Pitot-static and **impeller logs** (arts. 614, 615) are mechanical developments in the field of speed measurement. Each utilizes a retractable "rodmeter" which projects through the hull of the ship into the water. In the Pitot-static log, static and dynamic pressures on the rodmeter transmit readings to the master speed indicator. In the impeller log an electrical means of transmitting speed indications is used.

113. Units of distance and depth.—The modern navigator is concerned principally with four units of linear measure: the **nautical mile**, the **fathom**, the **foot** and the **meter**. Primitive man, however, used such natural units as the width of a finger, the **span** of his hand, the length of his foot, the distance from his elbow to the tip of the middle finger (the **cubit** of biblical renown), or the **pace** (sometimes one but usually a double step) to measure short distances.

Although the **Roman mile** had a value of about 1,488 meters or about 0.9248 of our statute mile of 5,280 feet, several standards were in use among the cities of ancient Greece at the same time. The Greek stadia being variable, there is uncertainty as to the accuracy of the measurement of the earth by Eratosthenes.

The **nautical mile** bears little relation to these land measures, which were not associated with the size of the earth. With the emergence of the nautical chart, it became customary to show a scale of miles on the chart, and the accepted value of this unit varied over the centuries with the changing estimates of the size of the earth. These estimates varied widely, ranging from about 44.5 to 87.5 modern nautical miles per degree of latitude, although generally they were too small. Columbus and Magellan used the value 45.3. Actually, the earth is about 32 percent larger. The *Almagest* of Ptolemy considered 62 Roman miles equivalent to 1 degree, but a 1466 edition of this book contained a chart of southern Asia drawn by Nicolaus Germanus on which 60 miles were shown to a degree. Whether the change was considered a correction or an adaptation to provide a more convenient relationship between the mile and the degree is not clear, but this is the earliest known use of this ratio.

Later, when the size of the earth was determined by measurement, the relationship of 60 Roman miles of 4,858.60 U. S. feet to a degree of latitude was seen to be in error. Both possible solutions to the problem—changing the ratio of miles to a degree, or changing the length of the mile—had their supporters, and neither group was able to convince the other. As a result, the shorter mile remained as the **land** or **statute mile** (now established as 5,280 feet in the United States), and the longer **nautical mile** gradually became established at sea. The earliest known reference to it by this name occurred in 1730.

Finer instruments and new methods make increasingly more accurate determinations of the size of the earth an ever-present possibility. Hence, a unit of length defined in terms of the size of the earth is undesirable. Recognition of this led, in 1875, to a change in the definition of the **meter** from one ten-millionth of the distance from the pole to the equator of the earth to the distance between two marks (approximately 39.37 U. S. inches) on a standard platinum-iridium bar kept at the Pavillon de Breteuill at Sevres, near Paris, France, by the International Commission of Weights and Measures. In further recognition of this principle, the International Hydrographic Bureau in 1929 recommended adoption of a standard value for the nautical mile, and proposed 1,852 international meters. This International Nautical Mile of 1,852 meters exactly has been adopted by nearly all maritime nations. The U. S. Departments of Defense and Commerce adopted this value in July 1954. With the yard-meter relationship then in use, the International Nautical Mile was equivalent to 6,076.10333 feet, approximately. Using the yard-meter exact relationship of 1 yard equal to 0.9144 meter adopted by the United States on July 1, 1959, the International Nautical Mile is equivalent to 6,076.11549 feet, approximately. In October 1960, the Eleventh General (International) Conference on Weights and Measures redefined the meter as equal to 1,650,763.73 wavelengths of the orange-red radiation in vacuum of krypton 86.

The **meter** as a unit of depth and height on U. S. nautical charts is of recent origin. The current policy of the Defense Mapping Agency Hydrographic/Topographic Center of converting new compilations of nautical and special purpose charts to the metric system was implemented on January 2, 1970.

The **fathom** as a unit of length or depth is of obscure origin, but primitive man considered it a measure of the outstretched arms, and the modern seaman still estimates the length of a line in this manner. That the unit was used in early times is indicated by reference to it in the detailed account given of the Apostle Paul's voyage to Rome, as recorded in the 27th chapter of the *Acts of the Apostles.* Posidonius reported a sounding of more than 1,000 fathoms in the second century BC. How old the unit was at that time is unknown.

114. Soundings.—Probably the most dangerous phase of navigation occurs when the vessel is "on soundings." Since man first began navigating the waters, the possibility of grounding his vessel has been a major concern, and frequent soundings have been the most highly valued safeguard against that experience. Undoubtedly used long before the Christian era, the lead line is perhaps the oldest instrument of navigation.

The lead line. The **hand lead** (art. 618), consisting of a lead weight attached to a line usually marked in fathoms, has been known since antiquity and, with the exception of the markings, is probably the same today as it was 2,000 or more years ago. The **deep sea lead,** a heavier weight with a longer line, was a natural outgrowth of the hand lead. A 1585 navigator speaks of soundings of 330 fathoms, and in 1773, in the Norwegian Sea, Captain Phipps had all the sounding lines on board spliced together to obtain a sounding of 683 fathoms. Matthew Fontaine Maury made his deep sea soundings by securing a cannon shot to a ball of strong twine. The heavy weight caused the twine to run out rapidly, and when bottom was reached, the twine was cut and the depth deduced from the amount remaining on the ball.

The sounding machine. The biggest disadvantage of the deep sea lead is that the vessel must be stopped if depths are to be measured accurately. This led to the development of the sounding machine.

Early in the 19th century a sounding machine similar to one of the earlier patent logs was invented. A wheel was secured just above the lead and the cast made in such a way that all the line required ran out freely and the lead sank directly to the bottom. The motion through the water during the descent set the wheel revolving, and this in turn caused the depth to be indicated on a dial. Ships sailing at perhaps 12 knots required 20 or 30 men to heave aboard the heavy line with its weight of 50 or more pounds after each cast. A somewhat similar device was the **buoy sounder.** The lead was passed through a buoy in which a spring catch was fitted and both were cast over the side. The lead ran freely until bottom was reached, when the catch locked, preventing further running out of the line. The whole assembly was then brought on board, the depth from the buoy to the lead being read.

The first use of the pressure principle to determine the depth of water occurred early in the 19th century when the "Self-acting Sounder" was introduced. A hollow glass tube open at its lower end contained an index which moved up in the tube as greater water pressure compressed the air inside. The index retained its highest position when hauled aboard the vessel, and its height was proportional to the depth of the water.

The British scientist, Sir William Thomson (Lord Kelvin) in 1878 perfected the sounding machine after repeated tests at sea. Prior to his invention, fibre line was used exclusively in soundings. His introduction of piano wire solved the problem of rapid descent of the lead and also that of hauling it back aboard quickly. The chemically coated glass tube which he used to determine depth was an improvement of earlier methods, and the worth of the entire machine is evidenced by the fact that it is still used in essentially the same form.

Echo sounding. Based upon the principle that sound travels through sea water at a nearly uniform rate, automatic depth-registering devices (art. 619) have been invented to indicate the depth of water under a vessel, regardless of its speed. In 1911 an account was published of an experiment performed by Alexander Behm of Kiel, who timed the echo of an underwater explosion, testing this theory. High frequency sounds in water were produced by Pierre Langevin, and in 1918 he used the principle for echo depth finding. The first practical echo sounder was developed by the United States Navy in 1922.

The actual time between emission of a sonic or ultrasonic signal and return of its echo from the bottom, the angle at which the signal is beamed downward in order that its echo will be received at another part of the vessel, and the phase difference between signal and echo have all been used in the development of the modern echo sounder.

115. Aids to navigation.—The Cushites and Libyans constructed towers along the Mediterranean coast of Egypt, and priests maintained beacon fires in them. These were the earliest known lighthouses. At Sigeum in the Troad (part of Troy) a lighthouse was built before 660 BC. One of the seven wonders of the ancient world was the lighthouse called the Pharos of Alexandria, which may have been more than 200 feet tall. It was built by Sostratus of Cnidus (Asia Minor) in the third century BC, during the reign of Ptolemy Philadelphus. The word "pharos" has since been a general term for lighthouses. Some time between 1584 and 1611 the light of Cordouan, the earliest waveswept lighthouse, was erected at the entrance to the Gironde river in western France. An oak log fire illuminated this structure until the 18th century.

Wood or coal fires were used in the many lighthouses built along the European and British coasts in the 17th and 18th centuries. One of these, the oak pile structure erected by Henry Whiteside in 1776 to warn shipmasters of Small's Rocks, subsequently played a major role in navigational history, as it was this light which figured in the discovery of the celestial line of position by Captain Thomas Sumner some 60 years later (art. 131).

In England such structures were privately maintained by interested organizations. One of the most famous of these groups, popularly known as "Trinity House," was organized in the 16th century, perhaps earlier, when a "beaconage and buoyage" fee was levied on English vessels. This prompted the establishment of Trinity House "to make, erect, and set up beacons, marks, and signs for the sea" and to provide vessels with pilots. The organization is now in its fifth century of operation, and its chief duties are to serve as a general lighthouse and pilotage authority, and to supply pilots.

The first lightship was a small vessel with lanterns hung from its yardarms. It was stationed at the Nore in the Thames estuary, in 1732.

The pilot's profession is not much younger than that of the mariner. The Bible relates (1 Kings 9:27) that Hiram of Tyre provided pilots for King Solomon. The duties of these pilots are not specified. In the first century AD, fishermen of the Gulf of Cambay, India, met seagoing vessels and guided them into port. It is probable that pilots were established in Delaware Bay earlier than 1756.

Seafaring people of the United States had erected lighthouses and buoys before the Revolutionary War, and in 1789 Congress passed legislation providing for federal expansion of the work. About 1767 the first buoys were placed in the Delaware River. These were logs or barrels, but about 1820 they were replaced with spar buoys. In that same year, the first lightship was established in Chesapeake Bay.

As the maritime interests of various countries grew, more and better aids to navigation were made available. In 1850 Congress prescribed the present system of coloring and numbering United States buoys (app. Y). Conformity as to shape resulted from the recommendations of the International Marine Conference of 1889. The second half of the 19th century saw the development of bell, whistle, and lighted buoys, and in 1910 the first lighted buoy in the United States utilizing high pressure acetylene apparatus was placed in service. Stationed at the entrance to Ambrose Channel in New York, it provided the basis for the high degree of perfection which has been achieved in the lighted buoy since that time. The complete buoyage system maintained by the U.S. Coast Guard today is chiefly a product of the 20th century. In 1900 there were approximately 5,000 buoys of all types in use in the United States, while today there are more than 20,000.

116. The sailings.—The various methods of mathematically determining course, distance, and position arrived at have a history almost as old as mathematics itself. Thales, Hipparchus, Napier, Wright, and others contributed the formulas that led to the tables permitting computation of course and distance by plane, traverse, parallel, middle-latitude, Mercator, and great-circle sailings.

Plane sailing (art. 901). Based upon the assumption that the surface of the earth is plane, or flat, this method was used by navigators for many centuries. The navigator solved problems by laying down his course relative to his meridian, and stepping off the distance run to the new position. This system is used with accuracy today in measuring short runs on a Mercator chart, which compensates for the convergence of the meridians, but on the plane chart, serious errors resulted. Early navigators might have obtained mathematical solutions to this problem, with no greater accuracy, but the graphical method was commonly used.

Traverse sailing (art. 901). Because sailing vessels were subject to the winds, navigators of old were seldom able to sail one course for great distances, and consequently a series of small triangles had to be solved. Equipment was designed to help seamen in maintaining their dead reckoning positions. The modern **rough log** evolved from the *log board,* hinged wooden boards that folded like a book and on which courses and distances were marked in chalk. Each day the position was determined from this data and entered in the ship's journal, today's **smooth log.**

The log board was succeeded by the **travas,** a board with lines radiating from the center in 32 compass directions. Regularly spaced along the lines were small holes into which pegs were fitted to indicate time run on the particular course. In 1627 John Smith described the travas as a "little round board full of holes upon lines like the compasse, upon which by the removing of a little sticke they (seamen) keepe an account, how many glasses (which are but halfe houres) they steare upon every point of the compasse."

These devices were of great value to the navigator in keeping a record of the courses and distances sailed, but still left him the long mathematical solutions necessary to determine the new position. In 1436 what appears to have been the first **traverse table** was prepared by Andrea Biancho. Using this table of solutions of right-angled plane triangles, the navigator was able to determine his course and distance made good after sailing a number of distances in different directions.

Parallel sailing (art. 901) was an outgrowth of the navigator's inability to determine his longitude. Not a mathematical solution in the sense that the other sailings are, it involved converting the distance sailed along a parallel (departure), as determined by dead reckoning, into longitude.

Middle-latitude sailing (art. 901). The inaccuracies involved in plane sailing led to the improved method of middle-latitude sailing early in the 17th century. A mathematician named Ralph Handson is believed to have been its inventor.

Middle-latitude sailing is based upon the assumption that the use of a parallel midway between those of departure and arrival will eliminate the errors inherent in plane sailing due to the convergence of the meridians. The assumption is reasonably accurate and although the use of Mercator sailing usually results in greater accuracy, middle-latitude sailing still serves a useful purpose.

Mercator sailing (art. 901). Included in Edward Wright's *Certaine Errors in Navigation Detected and Corrected,* of 1599, was the first published table of meridional parts which provided the basis for the most accurate of rhumb-line sailings—Mercator sailing.

Great-circle sailing (art. 903). For many hundreds of years mathematicians have known that a great circle is the shortest distance between two points on the surface of a sphere, but it was not until the 19th century that navigators began to regularly make use of this information.

The first printed description of great-circle sailing appeared in Pedro Nunes' 1537 *Tratado da Sphera.* The method had previously been proposed by Sebastian Cabot in 1498, and in 1524 Verrazano sailed a great-circle course to America. But the sailing ships could not regularly expect the steady winds necessary to sail such a course, and their lack of knowledge concerning longitude, plus the necessity of stopping at islands along their routes to take supplies, made it impractical for most voyages at that time.

The gradual accumulation of knowledge concerning seasonal and prevailing winds, weather conditions, and ocean currents eventually made it possible for the navigator to plan his voyage with more assurance. Nineteenth century writers of navigational texts recommended the use of great-circle sailing, and toward the close of that century such sailing became increasingly popular, particularly in the Pacific.

117. Hydrographic offices.—The practice of recording hydrographic data was centuries old before the establishment of the first official hydrographic office, in 1720. In that year the **Depot des Cartes, Plans, Journaux et Memoirs Relatifs a la Navigation** was formed in France with the Chevalier de Luynes in charge. The Hydrographic Department of the British Admiralty, though not established until 1795, has played a major part in European hydrographic work.

The **National Ocean Service** was originally founded when Congress, in 1807, passed a resolution authorizing a survey of the coast, harbors, outlying islands, and fishing banks of the United States. On the recommendation of the American Philosophical Society, President Jefferson appointed Ferdinand Hassler, a Swiss immigrant who had founded the Geodetic Survey of his native land, the first Director of the "Survey of the Coast." The survey was renamed "Coast Survey" in 1836.

The approaches to New York were the first sections of the coast charted, and from there the work spread northward and southward along the eastern seaboard. In 1844 the work was expanded and arrangements made to chart simultaneously the gulf and east coasts. Investigation of tidal conditions began, and in 1855 the first tables of tide predictions were published. The California gold rush gave impetus to the survey of the west coast, which began in 1850, the year California became a State. The survey ship *Washington* undertook investigations of the Gulf Stream. Coast pilots, or sailing directions, for the Atlantic coast of the United States were privately published in the first half of the 19th century, but about 1850 the Survey began accumulating data that led to federally produced coast pilots. The 1889 *Pacific Coast Pilot* was an outstanding contribution to the safety of west coast shipping.

In 1878 the survey was renamed "Coast and Geodetic Survey"; in 1970 the survey became the "National Ocean Survey," and in 1983 it became the "National Ocean Service."

Today the National Ocean Service provides the mariner with the charts and coast pilots of all waters of the United States and its possessions, and tide and tidal current tables for much of the world.

Defense Mapping Agency Hydrographic/Topographic Center. In 1830 the U.S. Navy established a "Depot of Charts and Instruments" in Washington, D.C. Primarily, it was to serve as a storehouse where such charts and sailing directions as were available, together with navigational instruments, could be assembled for issue to Navy ships which required them. Lieutenant L. M. Goldsborough and one assistant, Passed Midshipman R. B. Hitchcock, constituted the entire staff.

The first chart published by the Depot was produced from data obtained in a survey made by Lieutenant Charles Wilkes, who had succeeded Goldsborough in 1834, and who later earned fame as the leader of a United States exploring expedition to Antarctica.

From 1842 until 1861 Lieutenant Matthew Fontaine Maury served as Officer-in-Charge. Under his command the Depot rose to international prominence. Maury decided upon an ambitious plan to increase the mariner's knowledge of existing winds, weather, and currents. He began by making a detailed record of pertinent matter included in old log books stored at the Depot. He then inaugurated a hydrographic reporting program among shipmasters, and the thousands of answers received, along with the log book data, were first utilized to publish the *Wind and Current Chart of the North Atlantic* of 1847. The United States instigated an international conference in 1853 to interest other nations in a system of exchanging nautical information. The plan, which was Maury's, was enthusiastically adopted by other maritime nations, and is the basis upon which hydrographic offices operate today.

In 1854 the Depot was redesignated the "U.S. Naval Observatory and Hydrographical Office," and in 1866 Congress separated the two, broadly increasing the functions of the latter. The Office was authorized to carry out surveys, collect information, and print every kind of nautical chart and publication, all "for the benefit and use of navigators generally."

One of the first acts of the new Office was to purchase the copyright of *The New American Practical Navigator.* Several volumes of sailing directions had already been published. The first *Notice to Mariners* appeared in 1869. Daily broadcast of navigational warnings was inaugurated in 1907, and in 1912, following the sinking of the SS *Titanic,* Hydrographic Office action led to the establishment of the International Ice Patrol. The development by the U.S. Navy of an improved depth finder in 1922 made possible the acquisition of additional information concerning bottom topography. During the same year aerial photography was first employed as an aid in chart making. The Hydrographic Office published the first chart for lighter-than-air craft in 1923.

In 1962 the U.S. Navy Hydrographic Office was redesignated the U.S. Naval Oceanographic Office. In 1972 certain hydrographic functions of the latter office were transferred to the Defense Mapping Agency Hydrographic Center. In 1978 the Defense Mapping Agency Hydrographic/Topographic Center assumed hydrographic and topographic chart production functions.

The **International Hydrographic Organization (IHO)** was originally established in 1921 as the **International Hydrographic Bureau (IHB).** The present name was adopted in 1970 as a result of a revised international agreement among member nations. However, the former name, International Hydrographic Bureau, was retained for the IHO's administrative body of three Directors and a small Staff at the Organization's headquarters in Monaco.

The IHO (as did the former IHB) sets forth hydrographic standards as they are agreed upon by the member nations. All member States are urged and encouraged to follow these standards in their surveys, nautical charts, and publications. As these standards are uniformly adopted, the products of the world's hydrographic and oceanographic offices become more uniform. Much has been done in the field of standardization since the Bureau was founded.

The principal work undertaken by the IHO is:

1. to bring about a close and permanent association between national hydrographic offices;

2. to study matters relating to hydrography and allied sciences and techniques;

3. to further the exchange of nautical charts and documents between hydrographic offices of Member Governments;

4. to circulate the appropriate documents;

5. to tender guidance and advice upon request, in particular to countries engaged in setting up or expanding their hydrographic service;

6. to encourage coordination of hydrographic surveys with relevant oceanographic activities;

7. to extend and facilitate the application of oceanographic knowledge for the benefit of navigators;

8. to cooperate with international organizations and scientific institutions which have related objectives.

During the 19th century, many maritime nations established hydrographic offices to provide means for improving the navigation of naval and merchant vessels by providing nautical publications, nautical charts, and other navigational services. Nonuniformity of hydrographic procedures, charts, and publications was much in evidence. In 1889, an International Marine Conference was held at Washington, D.C., and it was proposed to establish a "permanent international commission." Similar proposals were made at the sessions of the International Congress of Navigation held at St. Petersburg in 1908 and again in 1912.

In 1919 the hydrographers of Great Britain and France cooperated in taking the necessary steps to convene an international conference of hydrographers. London was selected as the most suitable place for this conference, and on July 24, 1919, the First International Conference opened, attended by the hydrographers of 24 nations. The object of the conference was clearly stated in the invitation to attend. It read, "To consider the advisability of all maritime nations adopting similar methods in the preparation, construction, and production of their charts and all hydrographic publications; of rendering the results in the most convenient form to enable them to be readily used; of instituting a prompt system of mutual exchange of hydrographic information between all countries; and of providing an opportunity to consultations and discussions to be carried out on hydrographic subjects generally by the hydrographic experts of the world." In general, this is still the purpose of the International Hydrographic Organization. As a result of the conference, a permanent organization was formed and statutes for its operations were prepared. The International Hydrographic Bureau, now the International Hydrographic Organization, began its activities in 1921 with 18 nations as members. The Principality of Monaco was selected because of its easy communication with the rest of the world and also because of the generous offer of Prince Albert I of Monaco to provide suitable accommodations for the Bureau in the Principality. The

IHO, including the three Directors and their staff, is housed in its own headquarters which were built and are maintained by the Government of Monaco.

The works of the IHO are published in both French and English and are distributed through various media. Many of the publications are available to the general public, and a discount of 30 percent is offered to naval and merchant marine officers of any of the member nations. Inquiries as to the availability of the publications should be made directly to the "International Hydrographic Bureau, Avenue President J. F. Kennedy, Monte-Carlo, Monaco."

118. Navigation manuals.—Although navigation is as old as man himself, navigation textbooks, as they are thought of today, are a product of the last several centuries. Until the end of the Dark Ages such books, or manuscripts, as were available were written by astronomers for other astronomers. The navigator was forced to make use of these, gleaning what little was directly applicable to his profession. After 1500, however, the need for books on navigation resulted in the publication of a series of manuals of increasing value to the mariner.

Sixteenth century manuals. Frequently a command of Latin was required to study navigation during the 16th century. *Regimento do estrolabio e do quadrante* (fig. 130a), which was published at Lisbon in 1509, or earlier, explained the method of finding latitude by meridian observations of the sun and the pole star, contained a traverse table for finding the longitude by dead reckoning, and listed the longitudes of a number of places. Unfortunately, the author made several errors in transcribing the declination tables published by Abraham Zacuto in 1474, and this resulted in errors being made for many years in determining latitude. Nevertheless, the nameless writer of the *Regimento* performed a great service for all mariners. His "Handbook for the Astrolabe and Quadrant"—to translate the title—had many editions and many emulators.

In 1519 Fernandez de Encisco published his *Suma de Geographia,* the first Spanish manual. The book was largely a translation of the *Regimento,* but new information was included, and revisions were printed in 1530 and 1546.

The Flemish mathematician and astronomer R. Gemma Frisius published a book on navigation in 1530. This manual, entitled *De Principiis Astronomiae,* gave an excellent description of the sphere, although the astronomy was that of Ptolemy, and discussed at length the use of the globe in navigation. Gemma gave courses in terms of the principal winds, proposed that longitude be reckoned from the Fortunate Islands (Canary Islands), and gave rules for finding the dead reckoning position by courses and distances sailed.

Tratado da Sphera, Pedro Nunes' great work, appeared in 1537. In addition to the first printed description of great-circle sailing, Nunes' book included a section on determining the latitude by two altitudes of the sun (taken when the azimuths differed by not less than 40°) and solving the problem on a globe. The method was first proposed by Gemma. *Tratado da Sphera* contained the conclusion of a study of the "plane chart" which Nunes had made. He exposed its errors, but was unable to develop a satisfactory substitute.

During the years that followed, an extensive navigational literature became available. The Spaniards Pedro de Medina and Martin Cortes published successful manuals in 1545 and 1551, respectively. Medina's *Arte de Navegar* passed through 13 editions in several languages and *Breve de la Spera y de la Arte de Navegar,* Cortes' book, was eventually translated into English and became the favorite of the British navigator. Cortes discussed the principle which Mercator used 18 years later in constructing his famous chart, and he also listed accurately the distance between meridians at all latitudes.

The first western hemisphere navigation manual was published by Diego Garcia de Palacio at Mexico City in 1587. His *Instrucion Nauthica* included a partial glossary of nautical terms and certain data on ship construction.

John Davis' *The Seaman's Secrets* of 1594 was the first of the "practical" books. Davis was a celebrated navigator who asserted that it was the purpose of his book to give "all that is necessary for sailors, not for scholars on shore." Davis' book discussed at length the navigator's instruments, and went into detail on the "sailings." He explained the method of dividing a great circle into a number of rhumb lines, and the work he had done with Edward Wright qualified him to report on the method and advantages of Mercator sailing. He endorsed the system of determining latitude by two observations of the sun and the intermediate bearing.

Although best known for the presentation of the theory of Mercator sailing, Edward Wright's *Certaine Errors in Navigation Detected and Corrected* (1599) was a sound navigation manual in its own right. Particularly, he advocated correcting sights for dip, refraction, and parallax (ch. XVI).

Later manuals. The next 200 years saw a succession of navigation manuals made available to the navigator; so many that only a few can be mentioned. Among those which enjoyed the greatest success were Blundeville's *Exercises,* John Napier's *Mirifici Logarithmorum Canonis Descriptio* (which introduced the use of logarithms at sea), the tables and rules of Edmund Gunter, *Arithmetical Navigation* by Thomas Addison, and Richard Norwood's *The Sea-mans Practice* (which gave the length of the nautical mile as 6,120 feet). Robert Dudley filled four volumes in writing the *Arcano del Mare* (1646–47) as did John Robertson with *Elements of Navigation.* Jonas and John Moore, William Jones, and Samuel Dunns were others who contributed navigation books before Nathaniel Bowditch in America and J. W. Norie in England wrote the manuals which navigators found best suited to their needs.

Bowditch's *The New American Practical Navigator* was first published in 1802 (fig. 118), and Norie's *Epitome of Navigation* appeared the following year. Both were outstanding books which enabled the mariner of little formal education to grasp the essentials of his profession. The Englishman's book passed through 22 editions in that country before losing its popularity to Captain Lecky's famous *"Wrinkles" in Practical Navigation* of 1881. The *American Practical Navigator* is still read widely, more than a century-and-a-half after its original printing.

A number of worthy navigation manuals have appeared in recent years.

Celestial Navigation

119. Astronomy is sometimes called the oldest of sciences. The movements of the sun, moon, stars, and planets were used by the earliest men as guides in hunting, fishing, and farming. The first maps were probably of the heavens.

Babylonian priests studied celestial mechanics at a very early date, possibly as early as 3800 BC, more probably about 1500 years later. These ancient astronomers predicted lunar and solar eclipses, constructed tables of the moon's hour angle, and are believed to have invented the zodiac. The week and month as known today originated with their calendar. They grouped the stars by constellations. It is probable that they were arranged in essentially their present order as early as 2000 BC. The five planets easily identified by the unaided eye were known to the Babylonians, who were apparently the first to divide the sun's apparent motion

THE NEW AMERICAN

PRACTICAL NAVIGATOR;

BEING AN

EPITOME OF NAVIGATION;

CONTAINING ALL THE TABLES NECESSARY TO BE USED WITH THE

NAUTICAL ALMANAC,

IN DETERMINING THE

LATITUDE;

AND THE

LONGITUDE BY LUNAR OBSERVATIONS;

AND

KEEPING A COMPLETE RECKONING AT SEA:

ILLUSTRATED BY

PROPER RULES AND EXAMPLES:

THE WHOLE EXEMPLIFIED IN A

JOURNAL,

KEPT FROM

BOSTON TO MADEIRA,

IN WHICH ALL THE RULES OF NAVIGATION ARE INTRODUCED:

ALSO

The Demonstration of the most useful Rules of TRIGONOMETRY: With many useful Problems in MENSURATION, SURVEYING, and GAUGING: And a Dictionary of SEA-TERMS; with the Manner of performing the most common EVOLUTIONS at Sea.

TO WHICH ARE ADDED,

Some GENERAL INSTRUCTIONS and INFORMATION to MERCHANTS, MASTERS of VESSELS, and others concerned in NAVIGATION, relative to MARITIME LAWS and MERCANTILE CUSTOMS.

FROM THE BEST AUTHORITIES.

ENRICHED WITH A NUMBER OF

NEW TABLES,

WITH ORIGINAL IMPROVEMENTS AND ADDITIONS, AND A LARGE VARIETY OF NEW AND IMPORTANT MATTER:

ALSO,

MANY THOUSAND ERRORS ARE CORRECTED,

WHICH HAVE APPEARED IN THE BEST SYSTEMS OF NAVIGATION YET PUBLISHED.

BY **NATHANIEL BOWDITCH,**

FELLOW OF THE AMERICAN ACADEMY OF ARTS AND SCIENCES.

ILLUSTRATED WITH COPPERPLATES.

First Edition.

PRINTED AT NEWBURYPORT, (MASS.) 1802,

BY

EDMUND M. BLUNT, (Proprietor)

FOR CUSHING & APPLETON, SALEM.

SOLD BY EVERY BOOKSELLER, SHIP-CHANDLER, AND MATHEMATICAL-INSTRUMENT-MAKER, IN THE UNITED STATES AND WEST-INDIES.

FIGURE 118.—Original title page of *The New American Practical Navigator,* written by Nathaniel Bowditch and published in 1802.

about the earth into 24 equal parts. They published this and other astronomical data in ephemerides. There is evidence that the prophet Abraham had an excellent knowledge of astronomy.

The Chinese, too, made outstanding contributions to the science of the heavens. They may have fixed the solstices and equinoxes before 2000 BC. They had quadrants and armillary spheres, used water clocks, and observed meridian transits. These ancient Chinese determined that the sun made its annual apparent revolution about the earth in 365¼ days, and divided circles into that many parts, rather than 360. About 1100 BC the astronomer Chou Kung determined the sun's maximum declination within about 15′.

Astronomy was used by the Egyptians in fixing the dates of their religious festivals almost as early as the Babylonian studies. By 2000 BC or earlier the new year began with the heliacal rising of Sirius; that is, the first reappearance of this star in the eastern sky during morning twilight after having last been seen just after sunset in the western sky. The heliacal rising of Sirius coincided approximately with the annual Nile flood. The famous Pyramid of Cheops, which was probably built in the 17th century BC, was so constructed that the light of Sirius shone down a southerly shaft when at upper transit, and the light of the Pole Star shone down a northerly shaft at lower transit, the axes of the two shafts intersecting in the royal burial chamber. When the pyramid was constructed, *α Draconis,* not Polaris, was the Pole Star.

The Greeks learned of navigational astronomy from the Phoenicians. The earliest Greek astronomer, Thales, was of Phoenician ancestry. He is given credit for dividing the year of the western world into 365 days, and he discovered that the sun does not move uniformly between solstices. Thales is most popularly known, however, for predicting the solar eclipse of 585 BC, which ended a battle between the Medes and the Lydians. He was the first of the great men whose work during the next 700 years was the controlling force in navigation, astronomy, and cartography until the Renaissance.

120. Shape of the earth.—Advanced as the Babylonians were, they apparently considered the earth to be flat. Land surveys of about 2300 BC show a "salt water river" encircling the country (fig. 120).

But seafarers knew that the last to be seen of a ship as it disappeared over the horizon was the masthead. They recognized the longer summer days in England when they sailed to the tin mines of Cornwall, as early as 900 BC. In that "northland" the Mediterranean sailors noticed that the Pole Star was higher in the sky and the lower southern constellations were no longer visible. When Thales invented the gnomonic projection about 600 BC, he must have believed the earth to be a sphere. Two centuries later Aristotle wrote that the earth's shadow on the moon during an eclipse was always circular. Archimedes (287–212 BC) used a glass celestial globe with a smaller terrestrial globe inside it. Although the average man has understood the spherical nature of the earth for only a comparatively short period, learned astronomers have accepted the fact for more than 25 centuries.

121. Celestial mechanics.—Among astronomers the principal question for 2,000 years was not the shape of the earth, but whether it or the sun was the center of the universe. A stationary earth seemed logical to the early Greeks, who calculated that daily rotation would produce a wind of several hundred miles per hour at the equator. Failing to realize that the earth's atmosphere turns with it, they considered the absence of such a wind proof that the earth was stationary.

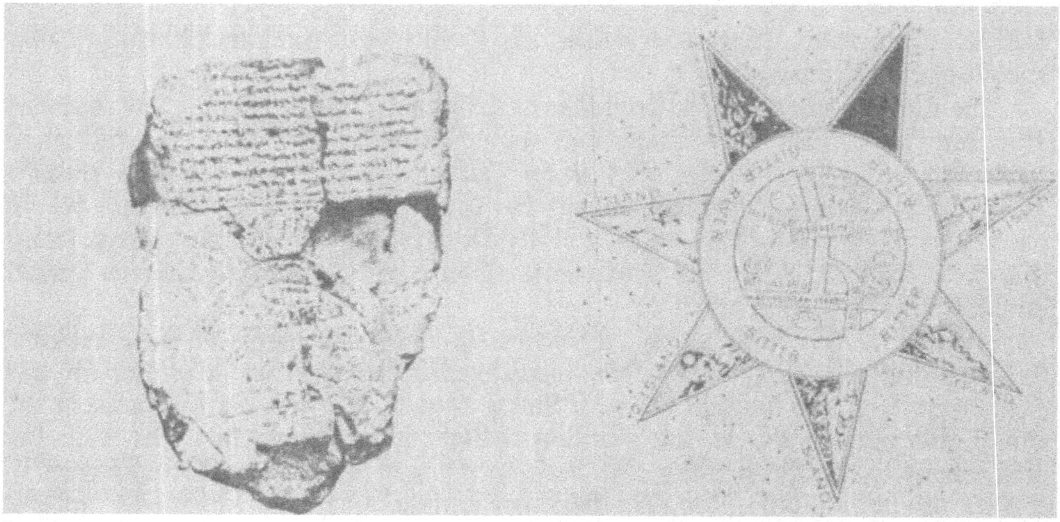

Courtesy of the Map Division of the Library of Congress.

FIGURE 120.—The original and reconstruction of a Babylonian map of about 500 BC. The Babylonians believed the earth to be a flat disk encircled by a salt water river.

The belief among the ancients was that all celestial bodies moved in circles about the earth. However, the planets—the "wanderers," as they were called—contradicted this theory by their irregular motion. In the fourth century BC Eudoxus of Cnidus attempted to account for this by suggesting that planets were attached to concentric spheres which rotated about the earth at varying speeds. The plan of **epicycles**, the theory of the universe which was commonly accepted for 2,000 years, was first proposed by Appolonius of Perga in the third century BC. Ptolemy accepted and amplified the plan, explaining it in his famous books, the *Almagest* and *Cosmographia.* According to Ptolemy, the planets moved at uniform speeds in small circles, the centers of which moved at uniform speeds in circles about the earth (fig. 121).

At first the Ptolemaic theory was accepted without question, but as the years passed, forecasts based upon it proved to be inaccurate. By the time the *Alfonsine Tables* were published in the 13th century AD, a growing number of astronomers considered the Ptolemaic doctrine unacceptable. However, Purbach, Regiomontanus, Bernhard Walther of Nuremberg, and even Tycho Brahe in the latter part of the 16th century, were among those who tried to reconcile the earth-centered epicyclic plan to the observed phenomena of the heavens.

As early as the sixth century BC, a brotherhood founded by Pythagoras, a Greek philosopher, proposed that the earth was round and self-supported in space, and that it, the other planets, the sun, and the moon revolved about a central fire which they called *Hestia,* the hearth of the universe. The sun and the moon, they said, shone by reflected light from Hestia.

The central fire was never located, however, and a few hundred years later Aristarchus of Samos advanced a genuine heliocentric theory. He denied the existence of Hestia and placed the sun at the center of the universe, correctly considering it to be a star which shone by itself. The Hebrews apparently understood the correct relationship at least as early as Abraham (about 2000 BC), and the early inhabitants of the Western Hemisphere probably knew of it before the Europeans did.

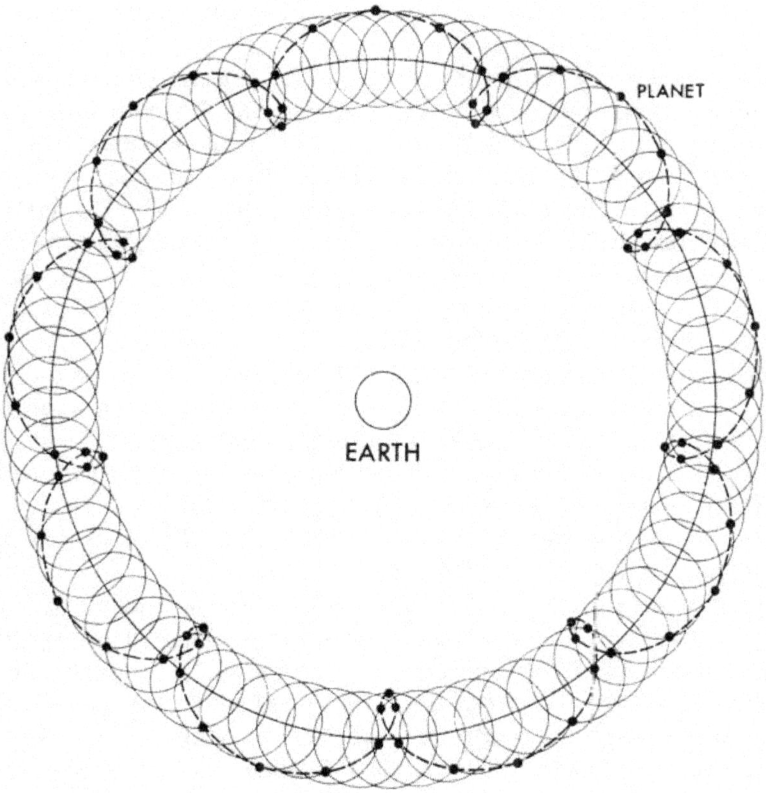

FIGURE 121.—The plan of epicycles, by which the ancients explained the retrograde motion of the planets. The planets were believed to rotate in small circles whose centers moved about the earth in a large circle.

The Ptolemaic theory was generally accepted until its inability to predict future positions of the planets could no longer be reconciled. Its replacement by the heliocentric theory is credited principally to Nicolaus Copernicus (or Koppernigk). After studying mathematics at the University of Cracow, Copernicus went to Bologna, where he attended the astronomical lectures of Domenicao Maria Novara, an advocate of the Pythagorean theory. Further study in Martianus Copella's *Satyricon,* which includes a discussion of the heliocentric doctrine, convinced him that the sun was truly the center of the universe.

Until the year of his death Copernicus tested his belief by continual observations, and in that year, 1543, he published *De Revolutionibus Orbium Coelestium.* In it he said that the earth rotated on its axis daily and revolved in a circle about the sun once each year. He placed the other planets in circular orbits about the sun also, recognizing that Mercury and Venus were closer than the earth, and the others farther out. He concluded that the stars were motionless in space and that the moon moved circularly about the earth. His conclusions did not become widely known until nearly a century later, when Galileo publicized them. Today, "heliocentric" and "Copernican" are synonymous terms used in describing the character of the solar system.

122. Other early discoveries.—A knowledge of the principal motions of the planets permitted reasonably accurate predictions of future positions. Other, less spectacular data, however, were being established to help round out the knowledge astronomers needed before they could produce the highly accurate almanacs known today.

More than a century before the birth of Christ, Hipparchus discovered the **precession of the equinoxes** (art. 1419) by comparing his own observations of the stars with those recorded by Timocharis and Aristyllus about 300 BC. Hipparchus cataloged more than a thousand stars, and compiled an additional list of time-keeping stars which differed in sidereal hour angle by 15° (1 hour), accurate to 15′. A spherical star map, or planisphere, and a celestial globe were among the equipment he designed. However, his instruments did not permit measurements of such precision that stellar parallax could be detected, and, consequently, he advocated the geocentric theory of the universe.

Three centuries later Ptolemy examined and confirmed Hipparchus' discovery of precession. He published a catalog in which he arranged the stars by constellations and gave the magnitude, declination, and right ascension (art. 1426) of each. Following Hipparchus, Ptolemy determined longitudes by eclipses. In the *Almagest* he included the plane and spherical trigonometry tables which Hipparchus had developed, mathematical tables, and an explanation of the circumstances upon which the equation of time (art. 1809) depends.

The next thousand years saw little progress in the science of astronomy. Alexandria continued as a center of learning for several hundred years after Ptolemy, but succeeding astronomers at the observatory confined their work to comments on his great books. The long twilight of the Dark Ages had begun.

Alexandria was captured and destroyed by the Arabs in AD 640, and for the next 500 years Moslems exerted the primary influence in astronomy. Observatories were erected at Baghdād and Damascus during the ninth century. Ibn Yunis' observatory near Cairo gathered the data for the Hakimite tables in the 11th century. Earlier, the Spanish, under Moorish tutelage, set up schools of astronomy at Cordova and Toledo.

123. Modern astronomy may be said to date from Copernicus, although it was not until the invention of the telescope, about 1608, that precise measurement of the positions and motions of celestial bodies was possible.

Galileo Galilei, an Italian, made outstanding contributions to the cause of astronomy, and these served as a basis for the work of later men, particularly Isaac Newton. He discovered Jupiter's satellites, providing additional opportunities for determining longitude on land. He maintained that it is natural for motion to be uniform and in a straight line and that a force is required only when direction or speed is changing. Galileo's support of the heliocentric theory, his use and improvement of the telescope, and particularly the clarity and completeness of his records provided firm footing for succeeding astronomers.

Early in the 17th century, before the invention of the telescope, Tycho Brahe found that planet Mars to be in a position differing by as much as 8′ from that required by the geocentric theory. When the telescope became available, astronomers learned that the apparent diameter of the sun varied during the year, indicating that the earth's distance from the sun varies, and that its orbit is not circular.

Johannes Kepler, a German who had succeeded Brahe and who was attempting to account for his 8′ discrepancy, published in 1609 two of astronomy's most important doctrines, the **law of equal areas,** and the **law of elliptical orbits.** Nine years later he announced his third law, relating the periods of revolution of any two planets to their respective distances from the sun (art. 1407).

Kepler's discoveries provided a mathematical basis by which more accurate tables of astronomical data were computed for the maritime explorers of the age. His realization that the sun is the controlling power of the system and that the orbital planes of the planets pass through its center almost led him to the discovery of the law of gravitation.

Sir Isaac Newton reduced Kepler's conclusions to the **universal law of gravitation** (art. 1407) when he published his three laws of motions in 1687. Because the planets exert forces one upon the other, their orbits do not agree exactly with Kepler's laws. Newton's work compensated for this and, as a result, the astronomer was able to forecast with greater accuracy the positions of the celestial bodies. The navigator benefited through more exact tables of astronomical data.

Between the years 1764 and 1784, the Frenchmen Lagrange and Laplace conclusively proved the solar system's mechanical stability. Early in the 19th century, Nathaniel Bowditch translated and commented upon Laplace's *Mécanique Céleste,* bringing it up-to-date. Prior to their work this stability had been questioned due to apparent inconsistencies in the motions of some of the planets. After their demonstrations, men were convinced and could turn to other important work necessary to refine and improve the navigator's almanac.

But there were real, as well as apparent, irregularities of motion which could not be explained by the law of gravitation alone. By this law the planets describe ellipses about the sun, and these orbits are repeated indefinitely, except as the other planets influence the orbits of each by their own gravitational pull. Urbain Leverrier, one-time Director of the Paris Observatory, found that the line of apsides of Mercury was advancing 43″ per century faster than it should, according to the law of gravitation and the positions of other known planets. In an attempt to compensate for the resulting errors in the predicted positions of the planet, he suggested that there must be a mass of circulating matter between the sun and Mercury. No such circulating matter has been found, however, and Leverrier's discovery is attributed to a shortcoming of Newton's law, as explained by Albert Einstein.

In Einstein's hands, Leverrier's 43″ became a fact as powerful as Brahe's 8′ had been in the hands of Kepler. Early in the 20th century, Einstein announced the general theory of relativity. He stated that for the planets to revolve about the sun is natural, and gravitational force is unnecessary for this, and he asserted that there need be no circulating matter to account for the motion of the perihelion of Mercury as this, too, is in the natural order of things. Calculated from his theory, the correction to the previously computed motion of the perihelion in 100 years is 42″.9.

Prior to Einstein's work, other discoveries had helped round out man's knowledge of the universe.

Aberration (art. 1417), discovered by James Bradley about 1726, accounted for the apparent shifting of the stars throughout the year, due to the combined orbital speed of the earth and the speed of light. Twenty years later Bradley described the periodic wobbling of the earth's axis, called **nutation** (art. 1417), and its effect upon precession of the equinoxes.

Meanwhile, in 1718 Edmond Halley, England's second Astronomer Royal, detected a motion of the stars, other than that caused by precession, that led him to conclude that they, too, were moving. By studying the works of the Alexandrian astronomers, he found that some of the most prominent stars had changed their positions by as much as 32′. Jacques Cassini gave Halley's discovery further support when he found, a few years later, that the declination of Arcturus had changed 5′ in the 100 years since Brahe made his observations. This **proper motion** (art. 1414) is motion in addition to that caused by precession, nutation, and aberration.

Sir William Herschel, the great astronomer who discovered the planet Uranus in 1781, proved that the solar system is moving toward the constellation *Hercules*. As early as 1828 Herschel advocated the establishment of a standard time system. Neptune was discovered in 1846 after its position had been predicted by the Frenchman Urbain Leverrier. Based upon the work of Percival Lowell, an American, Pluto was identified in 1930. Uranus, Neptune, and Pluto are of little concern to the navigator.

A more recent discovery may well have greater navigational significance. This is the existence of sources of electromagnetic energy in the sky in the form of **radio stars** (art. 1414). The sun has been found to transmit energy of radio frequency, and instruments have been built which are capable of tracking it across the sky regardless of weather conditions.

124. Sextant.—Prior to the development of the magnetic compass, the navigator used the heavenly bodies chiefly as guides by which to steer. The compass, however, led to more frequent long voyages on the open sea, and the need for a vertical-angle measuring device which could be used for determining altitude, so that latitude could be found.

Probably the first such device used at sea was the **common quadrant**, the simplest form of all such instruments. Made of wood, it was a fourth part of a circle, held vertical by means of a plumb bob. An observation made with this instrument at sea was a two- or three-man job. This device was probably used ashore for centuries before it went to sea, although its earliest use by the mariner is unknown.

Invented perhaps by Apollonius of Perga in the third century BC, the **astrolabe** (fig. 124a)—from the Greek for *star* and *to take*—had been made portable by the Arabs possibly as early as AD 700. It was in the hands of Christian pilots by the end of the 13th century, often as an elaborate and beautiful creation wrought of precious metals. Some astrolabes could be used as **star finders** (art. 2210) by fitting an engraved plate to one side. Large astrolabes were among the chief instruments of 15th and 16th century observatories, but the value of this instrument at sea was limited.

The principle of the astrolabe was similar to that of the common quadrant, but the astrolabe consisted of a metal disk, graduated in degrees, to which a movable sight vane was attached. In using the astrolabe, which may be likened to a pelorus held on its side, the navigator adjusted the sight vane until it was in line with the star, and then read the zenith distance from the scale. As with the common quadrant, the vertical was established by plumb bob.

Three men were needed to make an observation with the astrolabe (one held the instrument by a ring at its top, another aligned the sight vane with the body, a third made the reading) and even then the least rolling or pitching of a vessel caused large acceleration errors in observations. Therefore, navigators were forced to abandon the plumb bob and make the horizon their reference.

The **cross-staff** (fig. 124b) was the first instrument which utilized the visible horizon in making celestial observations. The instrument consisted of a long, wooden shaft upon which one of several cross-pieces was mounted perpendicularly. The cross-pieces were of various lengths, the one being used depending upon the angle to be measured. The navigator fitted the appropriate cross-piece on the shaft

Courtesy of the John Carter Brown Library, Brown University.

FIGURE 124a.—An ancient astrolabe, one of the earliest kinds of altitude-measuring instruments.

and, holding one end of the shaft beside his eye, adjusted the cross until its lower end was in line with the horizon and its upper end with the body. The shaft was calibrated to indicate the altitude of the body observed.

In using the cross-staff, the navigator was forced to look at the horizon and the celestial body at the same time. In 1590 John Davis, author of *The Seaman's Secrets,* invented the **backstaff** (fig. 124c) or **sea quadrant.** He was one of the few practical seamen (Davis Strait is named for him, in honor of his attempt to find the Northwest Passage) to invent a navigational device. The backstaff marked a long advance and was particularly popular among American colonial navigators.

In using this instrument, the navigator turned his back to the sun and aligned its shadow with the horizon. The backstaff had two arcs, and the sum of the values shown on each was the zenith distance of the sun. Later, this instrument was fitted with a mirror to permit observations of bodies other than the sun.

Another instrument developed about the same time was the **nocturnal** (fig. 124d). Its purpose was to provide the mariner with the appropriate correction to be made to the altitude of Polaris to determine latitude. By sighting on Polaris

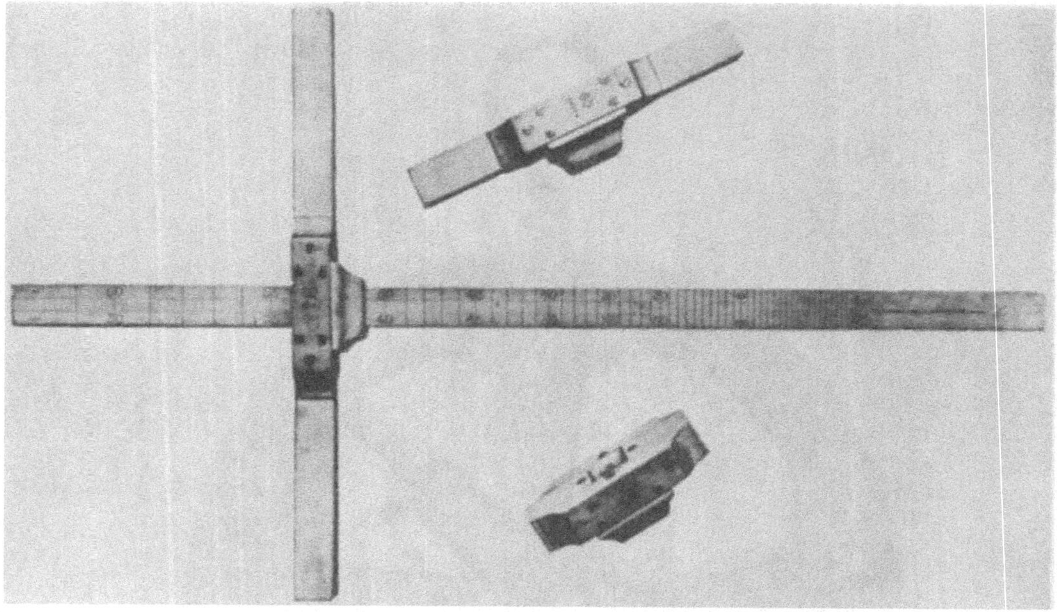

Courtesy of Peabody Museum of Salem.

FIGURE 124b.—The cross-staff, the first instrument to utilize the visible horizon in making celestial observations.

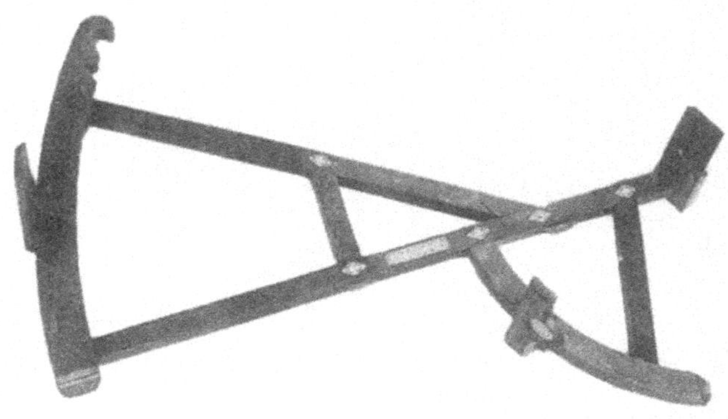

Courtesy of Peabody Museum of Salem.

FIGURE 124c.—The backstaff, or sea quadrant, a favorite instrument of American colonial navigators.

through the hole in the center of the instrument and adjusting the movable arm so that it pointed at Kochab, the navigator could read the correction from the instrument. Most nocturnals had an additional outer disk graduated for the months and days of the year and by adjusting this the navigator could also determine solar time.

Tycho Brahe designed several instruments with arcs of 60°, having one fixed sight and another movable one. He called the instruments **sextants** and the name is now commonly applied to all altitude-measuring devices used by the navigator (ch. XV). In 1700 Sir Isaac Newton sent to Edmond Halley, the Astronomer-Royal, a description of a device having double-reflecting mirrors, the principle of the modern marine sextant. However, this was not made public until after somewhat similar

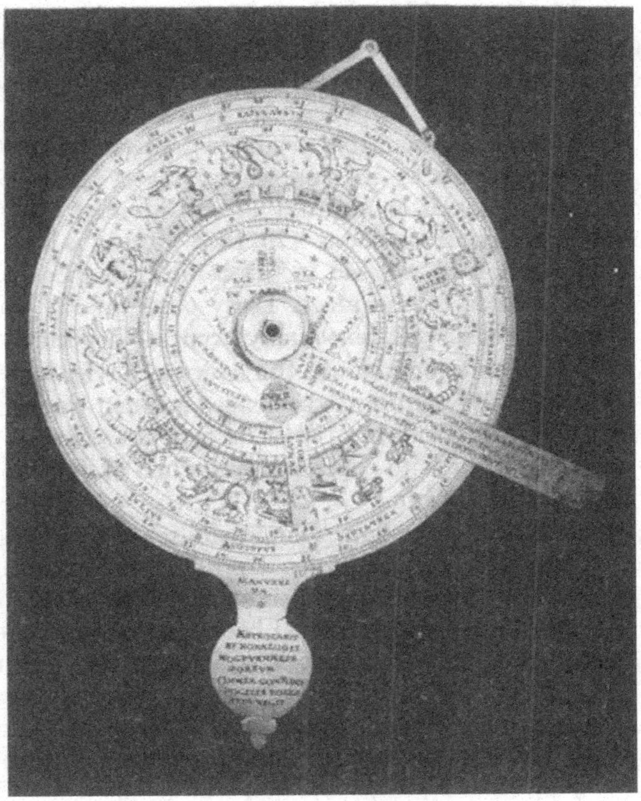

British Crown copyright. Science Museum, London, England.

FIGURE 124d.—The nocturnal, an instrument used to determine latitude by an observation of Polaris.

instruments had been made in 1730 by the Englishman John Hadley, and the American Thomas Godfrey.

The original instrument constructed by Hadley was, in fact, an *octant,* but due to the double-reflection principle it measured angles up to one-fourth of a circle, or 90°. Godfrey's instrument is reported to have been a *quadrant,* and so could measure angles through 180°. The two men received equal awards from England's Royal Society, as their work was considered to be a case of simultaneous independent invention, although Hadley probably preceded Godfrey by a few months in the actual construction of his sextant.

In the next few years both instruments were successfully tested at sea, but 20 years or more passed before the navigator gave up his backstaff or sea quadrant for the new device. In 1733 Hadley attached a spirit level to a quadrant, and with it was able to measure altitudes without reference to the horizon. Some years later the first bubble sextant (art. 1513) was developed.

Pierre Vernier, in 1631, had attached to the limb of a quadrant a second, smaller graduated arc, thereby permitting angles to be measured more accurately, and this device was incorporated in all later angle-measuring instruments.

The sextant has remained practically unchanged since its invention more than two centuries ago. The only notable improvements have been the addition of an endless tangent screw and a micrometer drum, both having been added during the 20th century.

125. Determining latitude.—The ability to determine longitude at sea is comparatively modern, but latitude has been available for thousands of years.

Meridian transit of the sun. Long before the Christian era, astronomers had determined the sun's declination for each day of the year, and prepared tables listing the data. This was a comparatively simple matter, for the zenith distance obtained by use of a shadow cast by the sun on the day of the winter solstice could be subtracted from that obtained on the day of the summer solstice to determine the range of the sun's declination, about 47°. Half of this is the sun's maximum declination, which could then be applied to the zenith distance recorded on either day to determine the latitude of the place. Daily observations thereafter enabled the ancient astronomers to construct reasonably accurate declination tables.

Such tables were available long before the average navigator was ready to use them, but certainly by the 15th century experienced seamen were determining their latitude at sea to within 1° or 2°. In his 1594 *The Seaman's Secrets,* Davis made use of his experience in high latitudes to explain the method of determining latitude by lower transit observations of the sun.

Ex-meridian observation of the sun. The possibility of overcast skies at the one time each day when the navigator could get a reliable observation for latitude led to the development of the "ex-meridian" sight. Another method, involving two sights taken with a considerable time interval between, had previously been known, but the mathematics were so involved that it is doubtful that many seamen made use of it.

There are two methods by which ex-meridian observations can be solved. The direct process was the more accurate, although it required a trigonometrical solution. By the latter part of the 19th century, tables were introduced which made the method of reduction to the meridian more practical and, when occasion demands such an observation, this is the method generally used today. However, with the development of line of position methods and the modern inspection table, ex-meridian observations have lost much of their popularity.

Latitude by Polaris. First use of the Pole Star to determine latitude is not known, but many centuries ago seamen who used it as a guide by which to steer were known to comment upon its change of altitude as they sailed north or south.

By Columbus' time some navigators were using Polaris to determine latitude, and with the invention of the nocturnal late in the 16th century, providing corrections to the observed altitude, the method came into more general use. The development of the chronometer in the 18th century permitted exact corrections, and this made determination of latitude by Polaris a common practice. Even today, more than a century after discovery of the celestial line of position, the method is still in use. The modern inspection table has eliminated the need for meridian observations as a special method for determining latitude. Perhaps when the almanacs and sight reduction tables make the same provision for solution of Polaris sights as they do for any other navigational star, this last of the special methods will cease to be used for general navigation. But customs die slowly, and one as well established as that of position finding in terms of separate latitude and longitude observations—instead of lines of position—is not likely to disappear completely for many years to come.

126. The search for a method of "discovering" longitude at sea.—A statement once quite common was, "The navigator always knows his latitude." A more accurate statement would have been, "The navigator never knows his longitude." In 1594 Davis wrote: "Now there be some that are very inquisitive to have a way to get the longitude, but that is too tedious for seamen, since it requireth the deep knowledge of astronomy, wherefore I would not have any man think that the longitude is to be found at sea by any instrument, so let no seamen trouble themselves with any such rule, but let them keep a perfect account and reckoning of the way of their ship." In speaking of conditions of his day, he was correct, for it was not until the 19th century that the average navigator was able to determine his longitude with accuracy.

Parallel sailing. Without knowledge of his longitude, the navigator of old found it necessary on an ocean crossing to sail northward or southward to the latitude of his destination, and then to follow that parallel of latitude until the destination was reached, even though this might take him far out of his way. Because of this practice, parallel sailing was an important part of the navigator's store of knowledge. The method was a crude one, however, and the time of landfall was often in error by a matter of days, and, in extreme cases, even weeks.

Eclipses. Almost as early as the rotation of the earth was established, astronomers recognized that longitude could be determined by comparing local time with that at the reference meridian. The problem was the determination of time at the reference meridian.

One of the first methods proposed was that of observing the disappearance of Jupiter's satellites as they were eclipsed by their planet. This method, originally proposed by Galileo for use on land, required the ability to observe and identify the satellites by using a powerful telescope, knowledge of the times at which the eclipses would take place, and the skill to keep the instrument directed at the bodies while aboard a small vessel on the high seas. Although used in isolated cases for many years, the method was not satisfactory at sea, due largely to the difficulty of observation (some authorities recommended use of a telescope as long as 18 or 19 feet) and the lack of sufficiently accurate predictions.

Variation of the compass was seriously considered as a method of determining longitude for 200 years or more. Faleiro, Magellan's advisor, believed it could be so utilized, and, until development of the chronometer, work was carried on to perfect the theory. Although there is no simple relationship between variation and longitude, those who advocated the method felt certain that research and investigation would eventually provide the answer. Many others were convinced that such a solution did not exist. In 1676, Henry Bond published *The Longitude Found,* in which he stated that the latitude of a place and its variation could be referred to the prime meridian to determine longitude. Two years later Peter Blackborrow rebutted with *The Longitude Not Found.*

Variation was put to good use in determining the nearness to land by shipmasters familiar with the waters they plied, but as the solution to the longitude problem it was a failure, and with the improvement of lunar distance methods and the invention of the chronometer, interest in the method waned. If it had been possible to provide the mariner with an accurate chart of variation, and to keep it up-to-date, a means of establishing an approximate line of position in areas where the gradient is large would have resulted; in many cases this would have established longitude if latitude were known.

Lunar distances. The first method widely used at sea to determine longitude with some accuracy was that of lunar distances (art. 131), by which the navigator determined GMT by noting the position of the relatively fast-moving moon among the stars. Both Regiomontanus, in 1472, and John Werner, in 1514, have been credited with being the first to propose the use of the lunar distance method. At least one source states that Amerigo Vespucci, in 1497, determined longitude using the moon's position relative to that of another body. One of the principal reasons for establishing the Royal Observatory at Greenwich was to conduct the observations necessary to provide more accurate predictions of the future positions of the moon. Astronomers, including the Astronomers Royal, favored this method, and half a century after the invention of the chronometer it was still being perfected. In 1802 Nathaniel Bowditch simplified the method and its explanation, thus eliminating much of the mystery surrounding it and making it understandable to the average mariner. By using Bowditch's method, the navigator was able to head more or less directly toward his destination, rather than travel the many additional miles often required in "running down the latitude" and then using parallel sailing. An explanation of the lunar distance method, and tables for its use, were carried in the *American Practical Navigator* until 1914.

The Board of Longitude. The lunar distance method, using the data and equipment available early in the 18th century, was far from satisfactory. Ships, cargoes, and lives were lost because of inaccurately determined longitudes. During the Age of Discovery, Spain and Holland posted rewards for solution to the problem, but in vain. When 2,000 men were lost as a squadron of British men-of-war ran aground on a foggy night in 1707, officers of the Royal Navy and Merchant Navy petitioned Parliament for action. As a result, the Board of Longitude was established in 1714, empowered to reward the person who could solve the problem of "discovering" longitude at sea. A voyage of 6 weeks duration was to be the test of proposed methods which were deemed worthy. The discoverer of a system which could determine the longitude within 60 miles by the end of the voyage was to receive £10,000; within 40 miles, £15,000; and within 30 miles, £20,000. These would be handsome sums today. In the 18th century they were fortunes.

127. Evolution of the chronometer.—Many and varied were the solutions proposed for finding longitude, and as the different methods were found unsatisfactory, it became increasingly apparent that the problem was one of keeping the time of the prime meridian. But the development of a device that would keep accurate time during a long voyage seemed to most men to be beyond the realm of possibility. Astronomers were flatly opposed to the idea and felt that the problem was properly theirs. There is even some evidence to indicate that the astronomers of the Board of Longitude made unfair tests of chronometers submitted to them.

Christian Huygens (1629–95), a Dutch scientist and mathematician, made a number of contributions of great value in the field of astronomy, but his most memorable work, to the navigator, was his attempt at constructing a perfect timepiece. It was probably Galileo who first suggested using a pendulum in keeping time. Huygens realized that an error would result from the use of a simple pendulum, however, and he devised one in which the bob hung from a double cord that passed between two plates in such a way that it traced a cycloidal path.

In 1660 Huygens built his first chronometer. The instrument utilized his cycloidal pendulum, actuated by a spring. To compensate for rolling and pitching, Huygens mounted the clock in gimbals. Two years later the instrument was tested at sea, with promising results. The loss of tension in the spring as it ran down was the major weakness in this clock. Huygens compensated for this by attaching oppositely tapered cones and a chain to the spring. A 1665 sea test of the new timepiece showed greater accuracy, but still not enough for determination of longitude. In 1674 he constructed a chronometer with a special balance and long balance-spring. Although it was the best marine timepiece then known, Huygens' last clock was also unsuited for use at sea due to the error caused by temperature changes.

John Harrison was a carpenter's son, born in Yorkshire in 1693. He followed his father's trade during his youth, but soon became interested in the repair and construction of clocks. At the age of 20 he completed his first timekeeper, a pendulum-type clock with wooden wheels and pinions. Harrison's gridiron pendulum, one which maintained its length despite temperature changes, was designed about 1720, and contained alternate iron and brass rods to eliminate distortion. Until the time that metal alloys having small coefficients of temperature expansion were developed, Harrison's invention was the type pendulum used by almost all clockmakers.

By 1728 Harrison felt ready to take his pendulum, an escapement he had invented, and plans for his own marine timepiece before the Board of Longitude. In London, however, Edmund Halley, Astronomer-Royal, advised him to first construct the timekeeper. Harrison did, and in 1735 he submitted his No. 1 chronometer (fig. 127). The Board authorized a sea trial aboard HMS *Centurion.* The following year, that vessel sailed for Lisbon with Harrison's clock on board, and upon her return, the error was found to be 3 minutes of longitude, a performance which astounded members of the Board. But the chronometer was awkward and heavy, being mounted on springs and enclosed in a gimballed wooden case and weighing some 72 pounds. The Board advanced Harrison a total of £1,250 between 1736 and 1760 to develop timekeepers No. 2 and No. 3.

British Crown copyright. From the original in the National Maritime Museum, London, England. Reproduced by permission of the Admiralty.

FIGURE 127.—Harrison's No. 1 chronometer. The first of four timekeepers constructed by Harrison, this clock weighs 65 pounds.

During the next few years he constructed two other chronometers, which were stronger and less complicated, although there is no record of their being tested by the Board of Longitude. Harrison continued to devote his life to the construction of an accurate clock to be used in determining longitude, and finally, as he approached old age, he developed his No. 4. Again he went before the Board, and again a test was arranged. In November of 1761, HMS *Deptford* sailed for Jamaica with No. 4 aboard, in the custody of Harrison's son, William. On arrival, after a passage lasting 2 months, the watch was only 9 seconds slow (2¼ minutes of longitude). In January of 1762 it was placed aboard HMS *Merlin* for the return voyage to England. When the *Merlin* anchored in English waters in April of that year, the total error shown by the chronometer was 1 minute 54.5 seconds. This is equal to less than a half degree of longitude, or less than the minimum error prescribed by the Board for the largest prize. Harrison applied for the full £20,000, but the Board, led by the Astronomer-Royal, allowed him only one-eighth of that, and insisted on another test.

William Harrison sailed again with No. 4 for Barbados in March of 1764, and throughout the almost four-months-long voyage the chronometer showed an error of only 54 seconds, or 13.5 minutes of longitude. The astronomers of the Board reluctantly joined in a unanimous declaration that Harrison's timepiece had exceeded all expectations, but they still would not pay him the full reward. Harrison was paid £7,500 in 1765. In addition to the plans, he was required to turn over all four of his timepieces. Even when this was done, the Board delayed payment further by having one of its members construct a timepiece from the plans. Not until 1773, Harrison's 80th year, was the rest of the reward paid, and only then because of intervention by the king himself.

Pierre LeRoy, a great French clockmaker, constructed a chronometer in 1766 which has since been the basis for all such instruments. LeRoy's several inventions made his chronometer a timepiece which has been described as a "masterpiece of simplicity, combined with efficiency." Others to contribute to the art of watchmaking included Ferdinand Berthoud of France and Thomas Mudge of England, each of whom developed new escapements. The balance wheel was improved by John Arnold, who invented the escapement acting in one direction only, substantially that used today. Acting independently, Thomas Earnshaw invented a similar escapement. He built the first reliable chronometer at a relatively low price. The chronometer the Board of Longitude had made from Harrison's plans cost £450; Earnshaw's cost £45.

Timepieces designed to provide the navigator with information other than time were popular a century or more ago. One showed the times of high and low water, the state of the tide at any time, and the phases of the moon; another gave the equation of time and the apparent motions of the stars and planets; a third offered the position of the sun and both mean and sidereal times. But the chronometers produced by LeRoy and Earnshaw were the ones of greatest value to the navigator; they gave him a simple and reliable method of determining his longitude.

Time signals, which permit the mariner at sea to check the error in his chronometer, are essentially a 20th century development. Telegraphic time signals were inaugurated in the United States at the end of the Civil War, and enabled ships to check their chronometers in port by **time ball** signals. Previously, the Navy's "standard" chronometer had been carried from port to port to allow such comparison. In their most advanced form, time balls were dropped by telegraphic action. In 1904 the first official "wireless" transmission of time signals began from a naval station at Navesink, N.J. These were low-power signals which could be heard for a distance of about 50 miles. Five years later the range had been doubled, and, as other nations began sending time signals, the navigator was soon able to check his chronometer around the world.

The search for longitude was ended.

128. Establishment of the prime meridian.—Until the beginning of the 19th century, there was little uniformity among cartographers as to the meridian from which longitude was measured. The navigator was not particularly concerned, as he could not determine his longitude, anyway.

Ptolemy, in the second century AD, had measured longitude eastward from a reference meridian 2 degrees west of the Canary Islands. In 1493 Pope Alexander VI drew a line in the Atlantic west of the Azores to divide the territories of Spain and Portugal, and for many years this meridian was used by chart makers of the two countries. In 1570 the Dutch cartographer Ortelius used the easternmost of the Cape Verde Islands. John Davis, in his 1594 *The Seaman's Secrets,* said the Isle of Fez in the Canaries was used because there the variation was zero. Mariners paid little attention, however, and often reckoned their longitude from several different capes and ports during a voyage, depending upon their last reliable fix.

The meridian of London was used as early as 1676, and over the years its popularity grew as England's maritime interests increased. The system of measuring longitude both east and west through 180° may have first appeared in the middle of the 18th century. Toward the end of that century, as the Greenwich Observatory increased in prominence, English map makers began using the meridian of that observatory as a reference. The publication by the Observatory of the first British *Nautical Almanac* in 1767 further entrenched Greenwich as the prime meridian. A later and unsuccessful attempt was made in 1810 to establish Washington as the prime meridian for American navigators and cartographers. At an international conference held in Washington in 1884 the meridian of Greenwich was officially established, by the 25 nations in attendance, as the prime meridian. Today all maritime nations have designated the Greenwich meridian the prime meridian, except in a few cases where local references are used for certain harbor charts.

129. Astronomical observatories.—Thousands of years before the birth of Christ, crude observatories existed, and astronomers constructed primitive tables which were the forerunners of modern almanacs. The famous observatory at Alexandria, the first "true" observatory, was constructed in the third century BC, but the Egyptians, as well as the Babylonians and Chinese, had already studied the heavens for many centuries. The **armillary sphere** (fig. 129a) was the principal instrument used by the early astronomers. It consisted of a skeleton sphere with several movable rings which could be adjusted to indicate the orbits of the various celestial bodies. One source attributes the invention of the armillary sphere to Eratosthenes in the third century BC; another says the Chinese knew it 2,000 years earlier, as well as the water clock and a form of astrolabe. The Alexandrian observatory was the seat of astronomical learning in the western world for several centuries, and there Hipparchus discovered the precession of the equinoxes, and Ptolemy did the work which led to his *Almagest.*

Astronomical study did not cease entirely during the Dark Ages. The Arabians erected observatories at Baghdād and Damascus in the ninth century AD, and observatories in Cairo and northwestern Persia followed. The Moors brought the astronomical knowledge of the Arabs into Spain, and the *Toledan Tables* of 1080

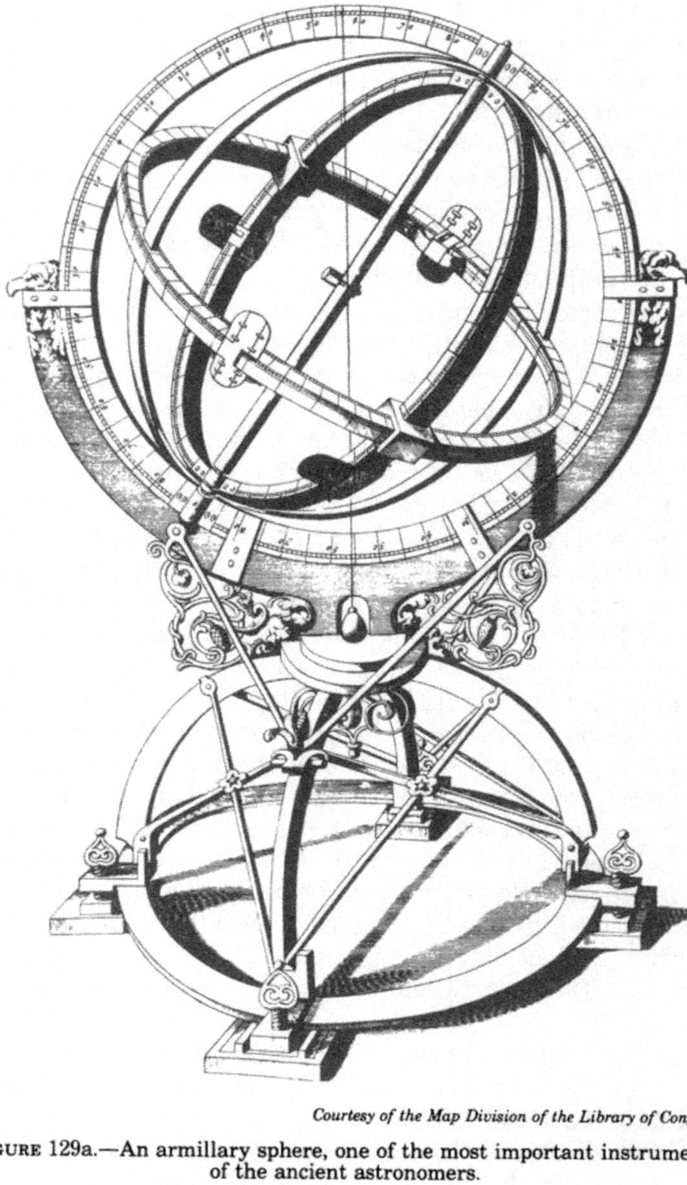

Courtesy of the Map Division of the Library of Congress.

FIGURE 129a.—An armillary sphere, one of the most important instruments of the ancient astronomers.

resulted from an awakening of scientific interest that brought about the establishment of schools of astronomy at Cordova and Toledo in the 10th century.

The great voyages of western discovery began early in the 15th century, and chief among those who recognized the need for greater precision in navigation was Prince Henry "The Navigator" of Portugal. About 1420 he had an observatory constructed at Sagres, on the southern tip of Portugal, so that more accurate information might be available to his captains. Henry's hydrographic expeditions added to the geographical knowledge of the mariner, and he was responsible for the simplification of many navigational instruments.

The Sagres observatory was rudimentary, however, and not until 1472 was the first complete observatory built in Europe. In that year Bernard Walther, a wealthy astronomer, constructed the Nuremberg Observatory, and placed Regiomontanus in charge. Regiomontanus, born Johann Müller, contributed a wealth of astronomical data of the greatest importance to the navigator.

The observatory at Cassel, built in 1561, had a revolving dome and an instrument capable of measuring altitude and azimuth at the same time. Tycho Brahe's Uraniburgum Observatory, located on the Danish island Hveen, was opened in 1576, and the results of his observations contributed greatly to the navigator's knowledge. Prior to the discovery of the telescope, the astronomer could increase the accuracy of his observations only by using larger instruments. Brahe used a quadrant with a radius of 19 feet, with which he could measure altitudes to 0.′6, an unprecedented degree of precision at that time. He also had an instrument with which he could determine altitude and azimuth simultaneously (fig. 129b). After Brahe, Kepler made use of the observatory and his predecessor's records in determining the laws which bear his name.

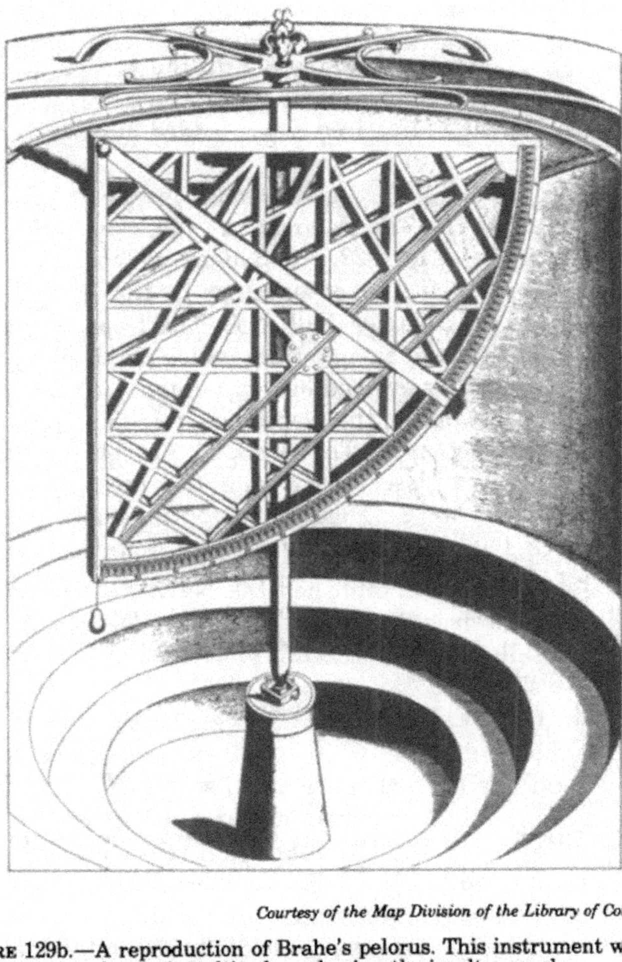

Courtesy of the Map Division of the Library of Congress.

FIGURE 129b.—A reproduction of Brahe's pelorus. This instrument was used to determine altitude and azimuth simultaneously.

The **telescope,** the modern astronomer's most important tool, was invented by Hans Lippershey about 1608. Galileo heard of Lippershey's invention, and soon improved upon it. In 1610 he discovered the four great moons of Jupiter, which led to the "longitude by eclipse" method successfully used ashore for many years and experimented with at sea. With the 32-power telescope he eventually built, Galileo was able to observe clearly the motions of sun spots, by which he proved that the sun rotates on its axis. In Paris, in 1671, the French National Observatory was established.

Greenwich Royal Observatory. England had no early privately supported observatories such as those on the continent. The need for navigational advancement was ignored by Henry VIII and Elizabeth I, but in 1675 Charles II, at the urging of John Flamsteed, Jonas Moore, Le Sieur de Saint-Pierre, and Christopher Wren, established the Greenwich Royal Observatory. Charles limited construction costs to £500, and appointed Flamsteed the first Astronomer-Royal, at an annual salary of £100. The equipment available in the early years of the observatory consisted of two clocks, a "sextant" of 7-foot radius, a quadrant of 3-foot radius, two telescopes, and the star catalog published almost a century before by Tycho Brahe. Thirteen years passed before Flamsteed had an instrument with which he could determine his latitude accurately. In 1690 a transit instrument equipped with a telescope and vernier was invented by Romer, and he later added a vertical circle to the device. This enabled the astronomer to determine declination and right ascension at the same time. One of these instruments was added to the equipment at Greenwich in 1721, replacing the huge quadrant previously used. The development and perfection of the chronometer in the next hundred years added further to the accuracy of observations.

Other national observatories were constructed in the years that followed: at Berlin in 1705, St. Petersburg in 1725, Palermo in 1790, Cape of Good Hope in 1820, Parramatta in New South Wales in 1822, and Sydney in 1855.

U.S. Naval Observatory. The first observatory in the United States is said to have been built in 1831–1832 at Chapel Hill, N.C. The Depot of Charts and Instruments, established in 1830, was the agency from which the U.S. Navy Hydrographic Office and the Naval Observatory evolved 36 years later. Under Lieutenant Charles Wilkes, the second Officer-in-Charge, the Depot about 1835 installed a small transit instrument for rating chronometers. The Mallory Act of 1842 provided for the establishment of a permanent observatory, and the director was authorized to purchase all such supplies as were necessary to continue astronomical study. The observatory was completed in 1844 and the results of its first observations were published 2 years later. Congress established the Naval Observatory as a separate agency in 1866. In 1873 a refracting telescope with a 26-inch aperture, then the world's largest, was installed. The observatory, located at Washington, D.C., has occupied its present site since 1893.

The **Mount Wilson Observatory** of the Carnegie Institution of Washington was built in 1904–05. The observatory's 100-inch reflector telescope opened wider the view of the heavens, and enabled astronomers to study the movements of celestial bodies with greater accuracy than ever before. But a still finer tool was needed, and in 1934 the 200-inch reflector for the **Palomar Mountain Observatory** was cast. The $6 million observatory was built by the Rockefeller General Education Board for the California Institute of Technology, which also operates the Mount Wilson Observatory. The 200-inch telescope makes it possible to see individual stars 20 million light-years away and galaxies at least 1.6 billion light-years away.

As with earlier instruments, the telescope has about reached the limit of practical size. Present efforts are being directed toward application of the electron microscope to the telescope, to increase the range of present instruments.

130. Almanacs.—From the beginning, astronomers have undoubtedly recorded the results of their observations. Tables computed from such results have been known for centuries. The work of Hipparchus, in the second century BC, and Ptolemy, in his famous *Almagest,* are examples. Then the *Toledan Tables* appeared in AD 1080, and the *Alfonsine Tables* in 1252. Even with these later tables, however, few copies were made, for printing had not yet been invented, and those that were available were kept in the hands of astronomers. Not until the 15th century were the first almanacs printed and made available to the navigator. In Vienna, in 1457, George Purbach issued the first almanac. Fifteen years later the Nuremberg Observatory, under Regiomontanus, issued the first of the ephemerides it published until 1506. These tables gave the great maritime explorers of the age the most accurate information available. In 1474 Abraham Zacuto introduced his *Almanach Perpetuum* (fig. 130a) which contained tables of the sun's declination in the most useful form yet available to the mariner. *Tabulae Prutenicae,* the first tables to be calculated on Copernican principles, were published by Erasmus Reinhold in 1551 and gave the mariner a clearer picture of celestial movements than anything previously available. The work of Brahe and Kepler at the Uraniburgum Observatory provided the basis for the publication of the *Rudolphine Tables* in 1627.

KL Março tem dias .xxxi. ha lũa .xxx. ho. dia .xij. ho. noite .xij.

		Di:	Lugar. sol	Declinaçã: sol.	
		Domes	Pisces	Graã	Minu.
d	Adrian martyr e simplici	1	20	3	59
e	Donado martyr e basilci	2	21	3	35
f	Marciniano martyr pécia	3	22	3	11
g	Lucio papa e martyre	4	23	2	48
A	Sam focas martyre.	5	24	2	24
b	Victor e victorino marti	6	25	2	0
c	Thome d'aquino confessor	7	26	1	36
d	Appollonio martyre.	8	72	1	12
e	Os qrenta martyres.	9	28	0	48
f	Alexandre e gaio martyr	10	29	0	24
g	Quarenta mil martyres	11	1 ♈	0	0
A	Sam gregorio papa e do	12	2	0	24
b	Sam leandre bispo bispo	13	3	0	48
c	Leco papa e martyr e do	14	4	1	12
d	longinus martyr.	15	5	1	36
e	Gertrudis virgem e mart	16	6	2	0
f	Patricio bispo e confesso	17	6	2	24
g	Anselmo bispo e doutor	18	7	2	48
A	Joseph d nossa senhora	19	8	3	11
b	Gutberto confessor	20	9	3	35
c	Bento abade fundador	21	10	3	59
d	Paulo bispo e confessor	22	11	4	22
e	Pigmene preste e martyr	23	12	4	46
f	Theodoro psbi	24	13	5	9
g	Anunciaçã de nossa senhora	25	14	5	33
A	Sam castorio martyre	26	15	5	56
b	Joam yrmitam confessor	27	16	6	19
c	Marcello papa e grondan	28	17	6	43
d	Quirino martyr e castalio	29	18	7	6
e	Sam segundo martyre	30	19	7	29
f	Felice pappa e babina vi	31	20	7	51

Courtesy of the John Carter Brown Library, Brown University.

FIGURE 130a.—An excerpt from the Portuguese *Regimento do estrolabio e do quadrante* of about 1509, giving the sun's declination and other data based upon Zacuto's calculations for the month of March. The first day of spring, the 11th by the Julian calendar then in use, is marked by the symbol of Aries, the ram (♈).

Still, the information contained in these books was intended primarily for the use of the astronomer, and the navigator carried the various tables only that he might make use of the portions applicable to his work. The first official almanac, *Connaissance des Temps,* was issued by the French National Observatory in 1696. The French Observatory rose to its greatest prominence during the 20 years that Urbain Leverrier held the position of director.

In 1767 the British *Nautical Almanac* was first published. Nevil Maskelyne was then Astronomer-Royal, and he provided the navigator with the best information available. The book contained tables of the sun's declination, and corrections to the observed altitude of Polaris. The moon's position relative to other celestial bodies was included at 12-hour intervals, and lunar distance tables gave the angular distance between the moon and certain other bodies at 3-hour intervals.

For almost a hundred years the British *Nautical Almanac* was the one used by American navigators, but in 1852 the Depot of Charts and Instruments published the first *American Ephemeris and Nautical Almanac,* for the year 1855.

Early American almanacs were distinguished by their excessive detail in some cases and shortage of data of importance to the navigator in others. Declination was given to the nearest $0''.1$ and the equation of time to the nearest $0^s.01$. Most figures were given only for noon at Greenwich, and a tedious interpolation was involved in converting the information to that at a given time at the longitude of the observer. Lunar distances were given at 3-hour intervals. Few star data were listed (fig. 130b).

Since 1858 the *American Nautical Almanac* has been printed without the ephemeris section, that part of value chiefly to astronomers. Until 1908 the positions of the brighter stars were given only for January 1st, and in relation to the meridian of Washington. Beginning in that year, the apparent places of 55 major stars were given for the first of each month. In 1912, the tables of lunar distances were omitted. In 1919, sunrise and sunset tables were added.

One of the greatest inconveniences involved in using the old almanacs was the astronomical day, which began at noon of the civil day of the same date. This system was abolished in 1925, and the United States adopted the expression "civil time" to designate time by the new system. Greenwich hour angle was first published for the moon in the *Lunar Ephemeris for Aviators* for the last 4 months of 1929. This publication became a supplement to the *Nautical Almanac* in 1931, and for 1932 the two were merged.

The *Air Almanac,* designed by Captain P. V. H. Weems, USN (Ret.), was published for 1933, giving Greenwich hour angle for all bodies included. For 1934 this information was given in the *Nautical Almanac,* and the *Air Almanac* was discontinued. The first British air almanac was published for the last quarter of 1937, and modified for 1939 with features followed closely in the first *American Air Almanac,* for 1941. In 1950 a revised *Nautical Almanac* appeared, patterned after the popular *American Air Almanac.* Starting with the 1953 edition, the British and American air almanacs were combined in a single publication. In that year the United States reverted to the expression "mean time" in place of "civil time." In 1958, the British and American nautical almanacs were combined, and in 1960, the name was standardized.

131. The navigational triangle.—It is customary for modern navigators to reduce their celestial observations by solving the triangle whose points are the elevated pole, the celestial body, and the zenith of the observer. The sides of this triangle are the polar distance of the body (codeclination), its zenith distance (coaltitude), and the polar distance of the zenith (colatitude of the observer).

FIXED STARS, 1855.

MEAN PLACES OF 100 PRINCIPAL FIXED STARS, FOR JANUARY 1, 1855.

Star's Name.	Mag.	Right Ascension.	An. Variation.	Declination.	An. Variation.
		h m s	s	° ′ ″	″
α ANDROMEDÆ .	2	0 0 53.97	+ 3.087	+28 17 23.3	+19.93
γ PEGASI (*Algenib*)	3.2	0 5 46.37	3.085	+14 22 38.1	20.05
β Hydri . . .	3	0 18 3.62	3.292	−78 4 23.1	20.23
α CASSIOPEÆ .	var.	0 32 18.36	3.356	+55 44 29.2	19.83
β Ceti . . .	2	0 36 18.45	3.016	−18 47 0.1	19.86
α URS. MIN. (*Polaris*)	2	1 6 29.82	+18.117	+88 32 11.3	+19.23
θ[1] Ceti . . .	3	1 16 46.57	3.000	− 8 55 58.6	18.74
α Eridani (*Achernar*)	1	1 32 18.42	2.238	−57 58 28.2	18.59
α ARIETIS . . .	2	1 59 0.44	3.365	+22 46 28.4	17.29
γ Ceti . . .	3.4	2 35 47.42	3.102	2 37 19.4	15.44
α CETI . . .	2.3	2 54 42.21	+ 3.129	+ 3 31 4.7	+14.40
α PERSEI . . .	2	3 13 59.52	4.243	49 20 26.8	13.25
η Tauri . . .	3	3 38 52.31	3.553	+23 39 11.0	11.54
γ[1] Eridani . . .	3	3 51 15.91	2.796	−13 55 26.7	10.59
α TAURI (*Aldebaran*)	1	4 27 36.26	3.436	+16 12 49.4	7.72
α AURIGÆ (*Capella*) .	1	5 5 59.03	+ 4.423	+45 50 41.8	+ 4.27
β ORIONIS (*Rigel*)	1	5 7 34.23	2.884	− 8 22 22.5	4.54
β TAURI . .	2	5 17 7.72	3.791	+28 28 48.3	3.55
δ ORIONIS . .	2	5 24 36.06	3.066	− 0 24 37.8	3.05
α Leporis . .	3	5 26 20.19	2.648	−17 55 46.0	2.94
ε ORIONIS . .	2	5 28 51.43	+ 3.044	− 1 17 54.6	+ 2.71
α Columbæ . .	2	5 34 24.05	2.177	−34 9 13.3	2.23
α ORIONIS . .	var.	5 47 19.35	3.249	+ 7 22 32.6	+ 1.11
μ Geminorum .	3	6 14 11.30	3.636	+22 34 59.9	− 1.37
α Argus (*Canopus*) .	1	6 20 44.13	1.330	−52 37 4.7	− 1.81
51 (Hev.) Cephei .	5	6 31 6.10	+30.650	+87 15 7.9	− 2.80
α CANIS MAJ. (*Sirius*)	1	6 38 45.60	2.646	−16 31 12.8	4.52
ε Canis Majoris . .	2.1	6 52 55.69	2.360	−28 46 40.3	4.58
δ Geminorum .	3.4	7 11 27.65	3.597	+22 14 41.7	6.16
α[2] GEMINOR. (*Castor*)	2.1	7 25 20.49	3.841	32 12 6.2	7.37
α CAN. MIN. (*Procyon*)	1	7 31 42.52	+ 3.145	+ 5 35 35.7	− 8.79
β GEMINOR. (*Pollux*)	1.2	7 36 26.23	3.681	+28 22 19.9	8.26
15 Argus . . .	3	8 1 22.22	2.557	−23 53 20.5	10.06
ε Hydræ . .	3.4	8 39 5.74	3.189	+ 6 56 52.2	12.86
ι Ursæ Majoris . .	3	8 49 15.44	4.123	+48 36 26.7	13.78
ι Argus . .	2	9 13 12.52	+ 1.602	−58 40 3.3	−14.91
α HYDRÆ . . .	2	9 20 27.65	2.951	− 8 1 56.8	15.36
θ Ursæ Majoris .	3	9 23 7.85	4.048	+52 20 6.3	16.13
ε Leonis . . .	3	9 37 36.82	3.424	24 26 22.0	16.34
α LEONIS (*Regulus*)	1.2	10 0 38.72	3.205	+12 40 26.4	17.40
η Argus . .	2	10 39 26.75	+ 2.306	−58 55 21.5	−18.74

FIGURE 130b.—Star data from the 1855 *Nautical Almanac*. The annual corrections in declination and right ascension can be used to obtain reasonably correct values today.

Lunar distances. A spherical triangle was first used at sea in solving lunar distance problems. Simultaneous or nearly simultaneous observations were made of the altitudes of the moon and the sun or a star near the ecliptic, and the angular distance between the moon and the other body. The zenith of the observer and the two celestial bodies formed the vertices of the triangle, whose sides were the two coaltitudes and the angular distance between the bodies. By means of a mathematical calculation the navigator "cleared" this distance of the effects of refraction and parallax applicable to each altitude, and other errors. The corrected value was then used as an argument for entering the almanac, which gave the true lunar distance from the sun and several stars at 3-hour intervals.

Previously, the navigator had set his watch or checked its error and rate, which could be relied upon for short periods, with the local mean time determined by celestial observations. The local mean time of the watch, properly corrected, applied to the Greenwich mean time obtained from the lunar distance observation, gave the longitude.

The mathematics involved was tedious, and few mariners were capable of solving the triangle until Nathaniel Bowditch published his simplified method in 1802 in *The New American Practical Navigator.* Chronometers were reliable by that time, but their high cost prevented their general use aboard the majority of naval and merchant ships. Using Bowditch's method, however, most navigators, for the first time, could determine their longitude, and so eliminate the need for parallel sailing and the lost time associated with it. The popularity of the lunar distance method is indicated by the fact that tables for its solution were carried in the *American Nautical Almanac* until the second decade of the 20th century.

The determination of latitude was considered a separate problem, usually solved by means of a meridian altitude or an observation of Polaris.

The time sight. The theory of the time sight (art. 2106) had been known to mathematicians since the dawn of spherical trigonometry, but not until the chronometer was developed could it be used by mariners.

The time sight made use of the modern navigational triangle. The codeclination, or polar distance, of the body could be determined from the almanac. The zenith distance (coaltitude) was determined by observation. If the colatitude were known, three sides of the triangle were available. From these the meridian angle was computed. The comparison of this with the Greenwich hour angle from the almanac yielded the longitude.

The time sight was mathematically sound, but the navigator was not always aware that the longitude determined was only as accurate as the latitude, and together they merely formed a point on what is known today as a line of position. If the observed body was on the prime vertical, the line of position ran north and south and a small error in latitude generally had little effect on the longitude. But when the body was close to the meridian, a small error in latitude produced a large error in longitude.

The line of position by celestial observation (art. 1703) was unknown until discovered in 1837 by 30-year-old Captain Thomas H. Sumner, a Harvard graduate and son of a United States Congressman from Massachusetts. The discovery of the "Sumner line," as it is sometimes called, was considered by Maury "the commencement of a new era in practical navigation." In Sumner's own words, the discovery took place in this manner:

"Having sailed from Charleston, S.C., 25th November, 1837, bound to Greenock, a series of heavy gales from the Westward promised a quick passage; after passing the Azores, the wind prevailed from the Southward, with thick weather; after passing Longitude 21° W., no observation was had until near the land; but soundings were had not far, as was supposed, from the edge of the Bank. The weather was now more boisterous, and very thick; and the wind still Southerly; arriving about midnight, 17th December, within 40 miles, by dead reckoning, of Tusker light; the wind hauled S.E., true, making the Irish coast a lee shore; the ship was then kept close to the wind, and several tacks made to preserve her position as nearly as possible until daylight; when nothing being in sight, she was kept on E.N.E. under short sail, with heavy gales; at about 10 A.M. an altitude of the sun was observed, and the Chronometer time noted; but, having run so far without any observation, it was plain the Latitude by dead reckoning was liable to error, and could not be entirely relied on.

"Using, however, this Latitude, in finding the Longitude by Chronometer, it was found to put the ship 15′ of Longitude, E. from her position by dead reckoning; which in Latitude 52° N. is 9 nautical miles; this seemed to agree tolerably well with the dead reckoning; but feeling doubtful of the Latitude, the observation was tried with a Latitude 10′ further N., finding this placed the ship E.N.E. 27 *nautical* miles, of the former position, it was tried again with a Latitude 20′ N. of the dead reckoning; this also placed the ship still further E.N.E., and still 27 *nautical miles* further; these three positions were then seen to lie in the direction of *Small's light.* It then at once appeared that the observed altitude must have happened at *all the three* points, and at *Small's light,* and at the ship, at the *same instant of time;* and it followed, that *Small's light* must bear E.N.E., if the Chronometer was right. Having been convinced of this truth, the ship was kept on her course, E.N.E., the wind being still S.E., and in less than an hour, *Small's light* was made bearing E.N.E. ½ E., and close aboard."

In 1843 Sumner published his book, *A New and Accurate Method of Finding a Ship's Position at Sea by Projection on Mercator's Chart,* which met with great acclaim. In it he proposed that a single time sight be solved twice, as he had done (fig. 131), using latitudes somewhat greater and somewhat less than that arrived at by dead reckoning, and joining the two positions obtained to form the line of position. It is significant that Sumner was able to introduce this revolutionary principle without seriously upsetting the method by which mariners had been navigating for years. Perhaps he realized that a better method could be derived, but almost certainly navigators would not have accepted the line of position so readily had he recommended that they abandon altogether the familiar time sight.

The Sumner method required the solution of two time sights to obtain each line of position. Many older navigators preferred not to draw the lines on their charts, but to fix their position mathematically by a method which Sumner had also devised and included in his book. This was a tedious procedure, but a popular one. Lecky recommended the method, and it was still in use early in the 20th century.

The alternative to working two time sights in the Summer method was to determine the azimuth of the body and to draw a line perpendicular to it through the point obtained by working a single time sight. Several decades after the appearance of Sumner's book, this method was made available to navigators through the publication of accurate azimuth tables, and the system was widely used until comparatively recent times. The 1943 edition of the *American Practical Navigator* included examples of its use. The 2-minute azimuth tables still found on many ships were designed principally for this purpose. The mathematical solution for azimuth was not at first a part of the time sight.

132. Modern methods of celestial navigation.—Sumner gave the mariner the line of position; St.-Hilaire the altitude difference or intercept method. Others who followed these men applied their principles to provide the navigator with rapid means for determining his position. The new navigational methods developed by these men, although based upon work done earlier, are largely a product of the 20th century.

Four hundred years ago Pedro Nunes used a globe to obtain a fix by two altitudes of the sun, and the azimuth angles. Fifty years later Robert Hues determined his latitude on a globe by using two observations and the time interval between them. G. W. Littlehales, of the U.S. Navy Hydrographic Office, advocated using a stereographic projection to obtain computed altitude and azimuth in his *Altitude, Azimuth, and Geographical Position,* published in 1906.

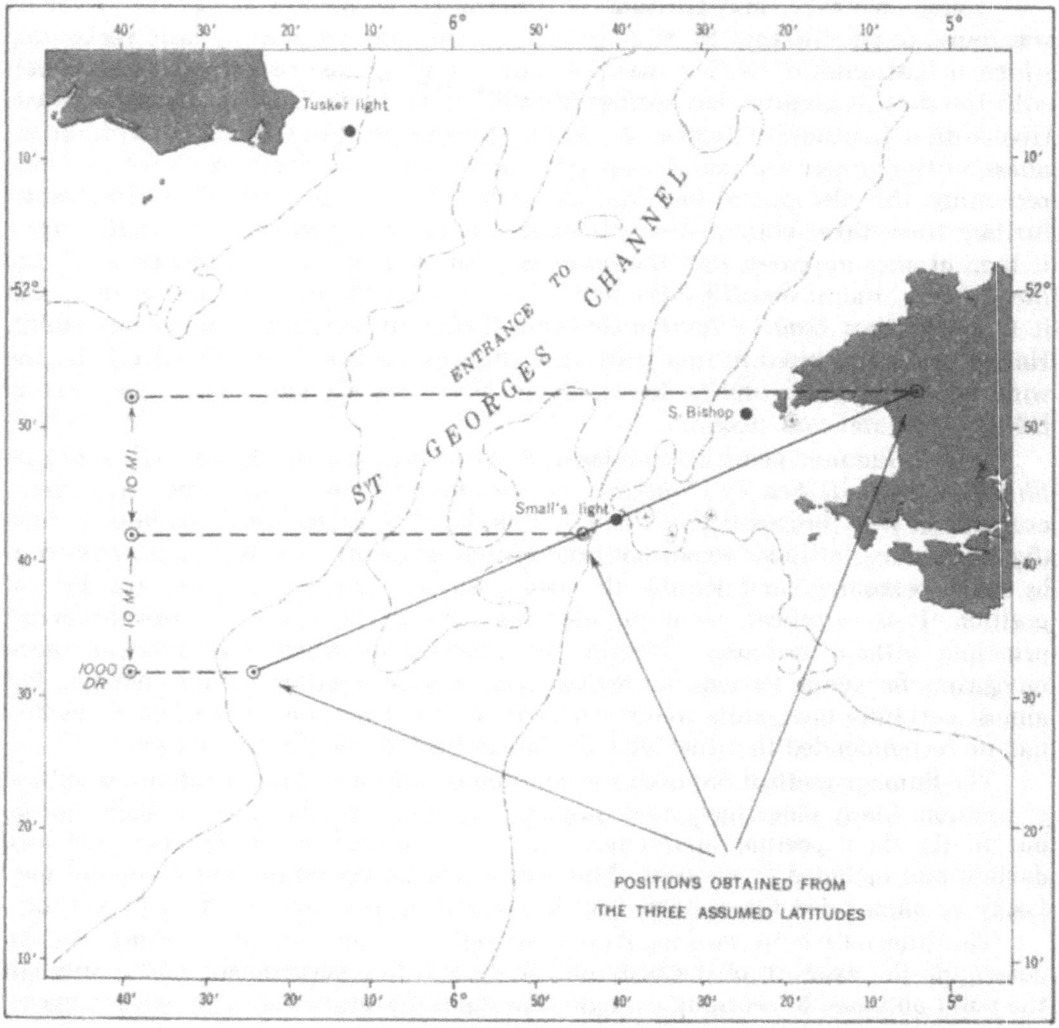

FIGURE 131.—The first celestial line of position, obtained by Captain Thomas Sumner in 1837.

Various graphic and mechanical methods have also been proposed. Of these, only one, the *Star Altitude Curves* of Captain P. V. H. Weems, USN (Ret.), has had wide usage, almost entirely among aviators. During World War II, some aircraft were fitted with a device called an "astrograph," which projected star altitude curves from film upon a special plotting sheet. The curves could be moved to allow for the earth's rotation. When they were properly oriented, part of the line of position could be traced on the plotting sheet. More generally, however, the navigational triangle has been solved mathematically or by the use of tables.

Spherical trigonometry is the basis for solving every navigational triangle, and until about 80 years ago the navigator had no choice but to completely solve each triangle himself. The cosine formula is a fundamental spherical trigonometry formula by which the navigational triangle can be conveniently solved. This formula was commonly used in lunar distance solutions when they were first introduced, but, because ambiguous results are obtained when the azimuth is close to 90° or 270°, mathematicians turned to the haversine, which has the advantage of increasing numerically from 0° to 180°. The **cosine-haversine formula** (art. 2109) was used by navigators until recent years.

Toward the end of the 19th century the "short" methods began to appear. About 1875, A. C. Johnson of the British Royal Navy published his book *On Finding the Latitude and Longitude in Cloudy Weather.* No plotting was involved in Johnson's method, but he made use of the principle that a single time sight be worked, rather than the two that Sumner proposed, and the line of position drawn through the point thus determined.

In 1879 Percy L. H. Davis, of the British Nautical Almanac Office, and Captain J. E. Davis collaborated on a *Sun's True Bearing or Azimuth Table,* which enabled the navigator to lay down a line of position using a computed azimuth. *Chronometer Tables,* published by Percy Davis 20 years later, covered latitudes up to 50° and gave local hour angle values for selected altitudes to 1 minute of arc. In 1905 his *Requisite Tables* were issued, enabling the mariner to "solve spherical triangles with three variable errors."

These were the first of a large number of "short" solutions which followed the work of Marcq St.-Hilaire. Generally, they consist of adaptations of the formulas of spherical trigonometry, and tables of logarithms in a convenient arrangement. It is customary for such methods to divide the navigational triangle into two right spherical triangles by dropping a perpendicular from one vertex to the side opposite. In some methods, partial solutions are made and the results tabulated. Aquino and Braga of Brazil; Ball, Comrie, Davis, and Smart of England; Bertin, Hugon, and Souillagouet of France; Fuss of Germany; Ogura and Yonemura of Japan; Blackburne of New Zealand; Pinto of Portugal: Garcia of Spain; and Ageton, Driesonstok, Gingrich, Rust, and Weems of the United States are but a few of those providing such solutions. Although "inspection tables" have largely superseded them, many of these "short" methods are still in use, kept alive largely by the compactness of their tables and the universality of their application. They are an intermediate step between the tedious earlier solutions and the fast tabulated ones, and they encouraged the navigator to work to a practical precision. The earlier custom of working to a precision not justified by the accuracy of the information used created a false sense of security in the mind of some navigators, especially those of little experience.

A book of tabulated solutions, from which an answer can be extracted by inspection, is not a new idea. Lord Kelvin, generally considered the father of modern navigational methods, expressed interest in such a method. However, solution of the many thousands of triangles involved would have made the project too costly if done by hand. Electronic computers have provided a practical means of preparing tables. In 1936 the first published volume of Pub. No. 214 was made available, and later Pub. No. 249 was provided for air navigators. British Admiralty editions of both these sets of tables have been published. Editions of Pub. No. 214 have also been published by the Instituto Hidrographico de la Marina, Cadiz, Spain, and by the Istituto Idrografico della Marina, Genova, Italy.

Electronics and Navigation

133. Electricity.—Twenty-five hundred years ago Thales of Miletus commented on basic electrical phenomena, but more than two millenniums passed before men first approached an understanding of electricity and the uses to which it could be put.

Until about 1682 the only known method of creating electricity was by rubbing glass with silk or amber with wool. Then Otto von Guericke of Magdeburg invented an "electric machine" and made possible the creation of electricity for experimental work. The Leyden jar, the electrical condenser (or machine) commonly used today, had its origin in 1745 when its principle was accidentally discovered independently by P. van Musschenbroek, of the University of Leyden, and von Kleist.

Stephen Gray, about 1729, demonstrated the difference between conductors and nonconductors, or insulators, and 10 years later Hawkesbee and DuFay, working independently, each discovered the positive and negative qualities of electricity.

In the middle of the 18th century Sir William Watson of England, developer of the Leyden jar in essentially its present form, sent electricity more than 2 miles by wire. Whether Watson was aware of the tremendous possibilities his experiment demonstrated is not known. Twenty-five years later, about 1774, Lesage devised what is believed to have been the first method of electrical communication. He had a separate wire for each letter of the alphabet and momentarily charged the appropriate wire to send each letter.

A German scholar, Francis Aepinus (1728–1802), was the first to recognize the reciprocal relationship of electricity and magnetism. In 1837 Karl Gauss and Wilhelm Weber collaborated in inventing a reflecting galvanometer for use in telegraphic work, which was the forerunner of the galvanometer at one time employed in submarine signaling. Michael Faraday (1791–1867), in a lifetime of experimental work, contributed most of what is known today in the field of electromagnetic induction. In 1864 James Clerk Maxwell of Edinburgh made public his electromagnetic theory of light. Many consider it the greatest single advancement in man's knowledge of electricity.

134. Electronics.—In 1887 Heinrich Hertz provided the proof of Maxwell's theory by producing electromagnetic waves and showing that they could be reflected. A decade later Joseph J. Thomson discovered the electron and so provided the basis for the development of the vacuum tube by Fleming and DeForest. In 1899 R. A. Fessenden pointed out that directional reception of radio signals was possible if a single coil or frame aerial was used as the receiving antenna. In 1895 Guglielmo Marconi transmitted a "wireless" message a distance of about 1 mile. By 1901 he was able to communicate between stations more than 2,000 miles apart. The following year Arthur Edwin Kennelly and Oliver Heaviside introduced the theory of an ionized layer in the atmosphere and its ability to reflect radio waves. Pulse ranging had its origin in 1925 when Gregory Breit and Merle A. Tuve used this principle to measure the height of the ionosphere.

135. Application of electronics to navigation.—Perhaps the first application of electronics to navigation was the transmission of radio **time signals** (art. 1826) in 1904, thus permitting the mariner to check his chronometer at sea. Telegraphic time signals had been sent since 1865, providing a means of checking the chronometer in various ports.

Next, radio broadcasts providing navigational warnings, begun in 1907 by the U.S. Navy Hydrographic Office, helped increase the safety of navigation at sea.

By the latter part of World War I the directional properties of a loop antenna were successfully utilized in the **radio direction finder** (art. 4101). The first radiobeacon was installed in 1921.

Early 20th century experiments by Behm and Langevin led to the development, by the U.S. Navy, of the first practical **echo sounder** (art. 619) in 1922.

As early as 1904, Christian Hulsmeyer, a German engineer, obtained patents in several countries on a proposed method of utilizing the reflection of radio waves as an obstacle detector and a navigational aid to ships. Apparently, the device was never constructed. In 1922 Marconi said, "It seems to me that it should be possible to design apparatus by means of which a ship could radiate or project a divergent beam of these rays (electromagnetic waves) in any desired direction, which rays if coming across a metallic object, such as another ship, would be reflected back to a receiver screened from the local transmitter on the sending ship, and thereby immediately reveal the presence and bearing of the other ship in fog or thick weather."

In the same year of 1922 two scientists, Dr. A. Hoyt Taylor and Leo C. Young, testing a communication system at the Naval Aircraft Radio Laboratory at Anacostia, D.C., noted fluctuations in the signals when ships passed between stations on opposite sides of the Potomac River. Although the potential value of the discovery was recognized, work on its exploitation did not begin until March 1934, when Young suggested to Dr. Robert M. Page, an assistant, that this might bear further investigation. By December, Page had constructed a pulse-signal device that determined the positions of aircraft. This was the first **radar** (art. 4201). In the spring of 1935 the British, unaware of American efforts, began work in this field, and developed radar independently. In 1937 the USS *Leary* tested the first seagoing radar. In 1940 United States and British scientists combined their efforts, resulting in more rapid progress. The British revealed the principle of the multicavity magnetron developed by J. T. Randall and H. A. H. Boot at the University of Birmingham in 1939. This magnetron made microwave radar practical. Probably no scientific or industrial development in history expanded so rapidly in all phases—research, development, design, production, trials, and training—and on such a scale. In 1945, at the close of hostilities of World War II, radar was made available for commercial use.

136. Development of hyperbolic radio aids.—The work on television and cosmic-ray counting devices in the decade prior to World War II provided the electronic techniques needed for the practical development of radio aids to navigation based upon the time of transmission of radio signals. Because the frequency stability of oscillators used in those early days was very poor—about a million times less than is available in 1984—it was obvious that only the *difference* in transmission times of two or more signals from different places could be measured. But this quantity would become useful only if the various signals could be kept in close synchronism by some control mechanism. Using this method, with the assumption of a constant velocity of propagation, it was clear that two signals would define a family of hyperbolic lines of position having the transmitting antennas as foci, and signals from either three or four different stations would establish a position fix.

It was also obvious, since the velocity of light is about 300 meters or nearly 1000 feet per microsecond, that time-difference measurements would have to be made within a very few microseconds if positional accuracy comparable to other methods of navigation were to be achieved. This precision generally exceeded that attained in ionospheric pulse-sounding techniques available at the time, but not by a very substantial margin. The only potentially difficult problem was the irregular variations of transmission times of most radio signals. These variations could be reduced or practically eliminated only by operating at very high frequencies where line-of-sight transmission could be achieved without interference from waves reflected from the ionosphere. It was natural, therefore, that the first operational hyperbolic radio aid was arranged to use these principles in the very high frequency part of the radio spectrum.

This first aid to navigation of the new kind was the British **Gee** system, proposed by Robert J. Dippy in 1937 and brought into operation by a team headed by Dippy in early 1942. This system was designed in accordance with the principles given above. Gee operated with a pulse length of about 5 microseconds at frequencies from 30 to 80 megahertz with separations between transmitters of the order of 100 miles. For high-flying aircraft the system could be used at distances up to 350 or 400 miles. Even though it was heavily jammed over western Europe, Gee was of the greatest importance to the night flying of the Bomber Command of the Royal Air Force, as it made return to bases in the British Isles relatively easy and accurate even under very poor flying conditions.

In 1940 the Microwave Committee of the U.S. National Defense Research Committee was assigned a project to develop a long-range, precision aircraft navigation system. Operational specifications for the system included an accuracy of about 1,000 feet at a range of 200 miles. To meet these requirements it was planned to use synchronized pairs of pulse-type transmitting stations separated by distances of several hundred miles. Transmitters radiating a peak power of about 1½ megawatts in the 30 to 40 megahertz band were contemplated. Except for instrumentation, the system would have been very similar to Gee.

The original system concepts used groundwave signals only. However during the course of system developments, measurements were made of the timing stability of pulsed radio waves having frequencies of from about 2 to 8 megahertz received via reflections from the ionosphere. Contrary to what was generally believed at the time, the stability of the signal reflected from the E-layer of the ionosphere was found to be quite good. Computations based on these measurements indicated that a long-range system using a combination of groundwaves and skywaves would provide a fixed accuracy of better than 5 miles at a range of 1,500 miles. The possibilities of a navigational system with this range and accuracy were so great that the original concept was abandoned and all efforts were concentrated toward this new goal. The revised project was assigned to the Radiation Laboratory of the Massachusetts Institute of Technology in the summer of 1941. Experimental transmitting stations were located at U.S. Coast Guard facilities near Montauk Point, New York, and Fenwick Island, Delaware.

In January 1942 the first skywave accuracy tests were made; a radio-frequency band was selected. Trials in moving vehicles were undertaken in June. By October 1942 a four-station chain was inaugurated for extended field trials by the U.S. Navy. About 40 receiver-indicators were installed in naval vessels during the next 4 or 5 months. Data were rapidly taken that defined the necessary skywave correction to reduce nighttime E-layer signals to the equivalent groundwave readings given on the charts.

On January 1, 1943, the administration of the new Loran program was assigned to the U.S. Navy. The U.S. Coast Guard and the Royal Canadian Navy were assigned responsibility for operation of the transmitting stations. The Loran system became fully operational in the spring of 1943 when charts for the four-station North Atlantic chain became available. The first chain comprised the two test stations at Montauk Point and Fenwick Island plus two new stations at Baccaro and Deming, Nova Scotia. The Fenwick station was first moved to Bodie Island, North Carolina, and later to Cape Hatteras, North Carolina. The Montauk Point station was moved to Nantucket Island. Installations in the Aleutians and the South Pacific soon followed.

This first version of Loran which operated on channels in the 1800 to 2000 kilohertz band was originally called **Standard Loran** to distinguish it from other experimental versions then being evaluated. Standard Loran later became known as **Loran-A.**

The most successful variation of Standard Loran during World War II was known as **Skywave Synchronized (SS) Loran.** This SS Loran operated at 2 megahertz, but, as its name implies, the stations maintained synchronization by using skywaves rather than the groundwave. This system was usable only during nighttime because of radio propagation conditions. SS Loran was first tested on the night of April 10, 1943, between Fenwick Island, Delaware, and Bonavista, Newfoundland, 1,100 miles away. Observations at the Radiation Laboratory, Cambridge, Massachusetts, revealed a line of position probable error of about 0.5 mile. This demonstrated the important fact that the errors of a few microseconds in the skywave transmission would not prevent position fixing to a useful accuracy when a sufficiently long baseline could be used. By the fall of 1943 two SS Loran pairs were in operation with transmitting stations at East Brewster, Massachusetts, Gooseberry Falls, Montana, Montauk Point, New York, and Key West, Florida. Extensive evaluation flights by U.S. and Allied Forces revealed an average position-fixing error of 1 to 2 miles.

In the early spring of 1944, the four SS Loran stations in the U.S. were dismantled, and the equipment was installed in Europe and North Africa. Stations were located in Scotland, Tunisia, Algeria, and Libya. This system became operational in October 1944 and was used extensively for night bombing operations. The combination of very long baselines (approximately 950 miles) and favorable baseline orientation provided nighttime service over a large part of Europe with an accuracy of 1 to 2 miles. SS Loran systems were also operated successfully in Southeast Asia. Lack of daytime coverage was the major limitation of SS Loran.

Skywave Long Baseline Loran was tested by the U.S. Coast Guard shortly after World War II. This system was similar to SS Loran but operated at 10.585 megahertz during the day and at 2 megahertz during the night for synchronization purposes. In order to provide normal 2 megahertz service, transmitters were operated at 2 megahertz during the day as well as at night, being controlled by the synchronization on 10.585 megahertz in daytime.

Preliminary tests were conducted between Chatham, Massachusetts, and Fernandina, Florida, in May 1944. These tests were followed by additional tests between Hobe Sound, Florida, and Point Chinato, Puerto Rico, in December 1945 and January 1946. Results of these tests demonstrated the basic concepts to be sound, but the difficulty in obtaining a suitable frequency allocation ended development.

It was recognized early that a low frequency Loran system would provide improved accuracy and greatly extended navigational coverage during the day and night with fewer transmitting stations. The first experimental low frequency Loran system, operating at 180 kilohertz and called **LF Loran,** was placed in operation in 1945 with transmitting stations at Cape Cod, Massachusetts, Cape Fear, North Carolina, and Key Largo, Florida. Monitor stations for overwater observations were installed at Bermuda, the Azores, Puerto Rico, and Trinidad. Overland signals were observed at monitor stations in Ohio, Minnesota, and aboard specially equipped aircraft.

The LF Loran system was basically an extension of the techniques of 2 megahertz Loran to the lower frequency. However, the LF stations operated in synchronized triplets instead of pairs, and the individual radio-frequency cycles of the master and slave pulses were displayed on the user's receiver-indicator. The receivers were designed to provide for visual matching of pulse and cycles. A rough match was made first using the envelopes of the two pulses, as in standard Loran, and then a fine measurement was made by matching selected radio-frequency cycles within each pulse.

In 1946 all equipment installed in the experimental east coast LF Loran system was transferred to northwest Canada where it served the requirements of special Arctic maneuvers in the area. Upon completion of the maneuvers, a joint Canadian-United States project was initiated to evaluate the system. Nine fixed-monitor stations and a number of specially equipped aircraft were placed in operation and comprehensive tests were made over a period of months. These operational tests, together with results of the east coast tests, showed that the LF Loran system could operate with substantially longer baselines than was feasible with the 2 megahertz system and that a 24-hour service coverage over land would be of the order of two-thirds of that of sea water as opposed to the almost negligible overland coverage provided by 2 megahertz Loran. The accuracy achieved was equivalent to an average line of position error of 160 feet at 750 miles. Beyond 750 miles, accuracy deteriorated rapidly due to skywave interference.

However, operators found that they could not select the correct pair of cycles more than 75 percent of the time without prior knowledge of the correct pulse envelope delay. The resulting positional ambiguities were operationally unacceptable; the system was judged unsatisfactory for general purpose navigation. Work was begun in 1946 on the development of cycle-identification and phase-measuring techniques to correct these positional ambiguities. This work by government and industry culminated in the field tests of a low frequency, cycle matching Loran system called **Cyclan.** This name was derived from **Cycl**e matching Lor**an.**

Cyclan was the first fully automatic Loran system. The cyclic ambiguity problem was solved through the use of pulse transmissions on two frequencies 20 kilohertz apart. At first 180 and 200 kilohertz were used, followed by operation on 160 and 180 kilohertz. Slope matching on the first 50 microseconds of the pulse was followed by cycle matching within the pulse envelope for precise determination of arrival time-differences. Incorrect cycle matching at one frequency was readily apparent by an obvious mismatch at the second frequency utilized. Cyclan coverage was limited to the groundwave region and, depending on local noise, gave a range of about 1,000 to 1,500 miles. Operational tests with Cyclan were complicated by serious interference problems involving broadcast stations and aeronautical radiobeacons on adjacent frequencies. The tests did show, however, that the radiofrequency cycle identification problem could be solved. Significant progress was also made in instrumentation. It became necessary to seek another solution when the 1947 Atlantic City Radio Conference designated the 90 to 110 kilohertz band (20 kilohertz bandwidth) for the development of long range navigational systems; Cyclan required a total bandwidth of approximately 40 kilohertz.

Navaglobe was an early system investigated as a potential low frequency system operating within the 90 to 110 kilohertz band. Work on this system started in 1945. The directional characteristics were obtained from a configuration of three vertical antennas placed at the corners of an equilateral triangle. The antennas were excited alternately in pairs so that three overlapping figure-eight patterns were obtained. Measurement of the relative amplitudes of the received signals determined the navigator's bearing from the transmitting station. Cross bearings were required to establish position. To obtain range information, parallel development of a distance measuring system called **Facom** was carried out. This system also operated in the 90 to 110 kilohertz band. Coarse distance data were developed by comparing the phase of a low frequency modulating tone on a local oscillator with the phase of a similar tone on the continuous wave signal from the Facom ground station. Fine distance measurements were made on the radio frequency cycles in the carrier.

Navarho, the combined Navaglobe-Facom system, was extensively evaluated during 1957. The project was discontinued because the overall system performance was unsatisfactory.

Navarho was the first system to attempt the distance-difference measurement from observation of the change of phase of the received signal relative to a very stable, locally generated reference signal of the same frequency. To obtain useful navigational accuracy the frequencies of the transmitted signals and of the receiver's reference signals had to be synchronized with an accuracy of a part in a billion or better. One of the first commercial cesium beam frequency standards was used to control the frequency of the signals radiated from three towers at Camden, New York. Although the airborne crystal oscillators were awkward to operate, requiring close attention to attain the necessary stability, the results of numerous flights out to ranges of 2,000 miles demonstrated acceptable range and accuracy in the distance measurement; the bearing measurements at long range, however, were relatively poor.

In 1952 work began under government contract on a long range, automatic, ground-reference tactical bombing system known as **Cytac.** A pulsed, hyperbolic navigation system operating in the 90 to 110 kilohertz band was an integral part of the Cytac system. Equipment development was completed in 1955, and three transmitting stations were constructed at Forestport, New York; Carolina Beach, North Carolina; and Carrabelle, Florida. Tests with the navigational component of the system throughout 1956 showed that automatic instrumentation could solve the radio frequency cycle identification problem and could measure time-difference in a hyperbolic system with an average error of a few tenths of a microsecond. The coverage area extended from the Atlantic Ocean to the Mississippi River, and from the Great Lakes to the Gulf of Mexico. Monitor stations installed at widely separated locations collected data during a year of testing. The results of the tests demonstrated that the system was not only capable of a high degree of precision but also that the laws controlling its accuracy were sufficiently well known to permit sound predictions of accuracy prior to installation. For operational reasons, the Cytac concept was abandoned. Its use as a navigational aid was immediately apparent.

In 1957 an operational requirement for a highly accurate long range maritime radionavigation aid was developed. The stated accuracy and range requirements were considerably in excess of the capabilities of existing Loran-A equipment. On the basis of the results of the Cytac tests, it was believed that this requirement could be satisfied by implementing the Cytac concepts as well as some of the Cytac equipment. Consequently, equipment from stations at Forestport, New York, and Carrabelle, Florida, was transferred to Martha's Vineyard, Massachusetts, and Jupiter, Florida, respectively. These stations operating in conjunction with the existing station at Carolina Beach, North Carolina, were placed in operation in 1957. The U. S. Coast Guard in accordance with the U. S. Federal Laws assumed responsibility for operation of the stations in August 1958. Comprehensive tests by both ships and aircraft showed that the original concepts were sound. The new system, designated **Loran-C**, was at that time placed in operational status.

Following the closing of the Radiation Laboratory at the Massachusetts Institute of Technology, a small group of the scientists moved to the Cruft Laboratory of Harvard University, intending to apply some of the new techniques the war had brought forward to various investigations of radio wave propagation. A theoretical study was made of the stability of the phase of a modulated signal that might be expected under the conditions of interference between modes of propagation in long-distance ionospheric transmission. This study indicated that a relatively long-period modulation might be measureable with nearly the accuracy that had been achieved in the LF Loran trials. Refinement of this concept led to the proposal in 1947 of a system called **Radux.**

This new system was to be very similar to LF Loran, except that the arrival times would be measured in terms of the phase of 200 hertz modulation instead of the time of a pulse having a duration of about 300 microseconds. Because a 200 hertz modulation could be radiated from a good low frequency antenna at a frequency as low as 40 or 50 kilohertz, it was hoped that a 1,000-mile baseline could be used and that useful service could be provided as far as 3,000 miles. However, little was known about propagational characteristics in this frequency region.

After 2 to 3 years of study of the proposal, the U. S. Navy assigned the Navy Electronics Laboratory (NEL) the task of making the necessary tests to develop Radux as an aid to navigation. With the advice of an informal steering committee, representing several naval technical bureaus and laboratories and commercial contractors, NEL proceeded to build and install transmitters where suitable antennas could be found, initially in Hawaii and San Diego, and to procure and operate receivers. This work occupied nearly the entire decade of the 1950's. The results were very much like those of the work of the Radiation Laboratory on LF Loran; the range and accuracy were less than desired.

Going back to 1953, Dr. Louis Essen of the National Physical Laboratory (NPL) in Teddington, England, who had designed by far the best crystal oscillator then available (the Essen ring) and who was developing the first practical cesium frequency standard, visited Harvard University. He called attention to the fact that the British Post Office transmitter at Rugby had begun transmitting a standard frequency of 60 kilohertz. This transmission was made for only 1 hour per day, but it was derived from an Essen ring oscillator and the frequency was accurately monitored by both the Post Office and NPL. The group at Harvard University immediately began observing this transmission, and made the pleasing discovery that the frequency could be measured to a part in 10^{10} during the 1-hour transmission. This accuracy exceeded by a factor of 10^2 to 10^3 that available through high-frequency transmission of standard frequencies. This link from Rugby to Cambridge, Massachusetts became for a time the primary intercomparison mechanism between the British and American systems of standard time, which differed greatly in those days. The first international intercomparison of cesium-controlled clocks was made in the same way.

The Rugby (GBR) frequency was stabilized in 1954. This started an era in which the frequencies of most VLF transmissions have been stabilized so that they can be used for frequency intercomparisons and for new kinds of propagational research. It was soon found that daily measurement yielded an accuracy of about 2 parts in 10^{11} in frequency, even at distances of thousands of miles.

The Radux work was being done under conditions of military classification. These new VLF discoveries were therefore published only through their bearing on frequency comparison. In April, 1955, however, a letter report to the Office of Naval Research and verbal and other communications with the Naval Laboratories recommended extension of the navigational efforts to the very low frequencies; in particular to those below 14 kilohertz, where circuit bandwidths are so low that the frequencies are not of general interest for communication.

It had been found that the accuracy of Radux, while inadequate to resolve cyclic ambiguities at 40 kilohertz (periods of 25 microseconds), would resolve the four times larger phase ambiguities at a frequency of 10 kilohertz. A composite system was therefore proposed which could operate as Radux, probably at 40 kilohertz, while coherent bursts of 10 kilohertz carrier, radiated from the same antenna at different times, would permit measurements of much greater precision but with 8-mile ambiguities resolvable, at least in theory, by measurements of the 200 hertz modulation of Radux. This composite system was called **Radux-Omega,** and was investigated for a few years. It rapidly became obvious that this marriage was an unfortunate one, because the useful range of the 10 kilohertz component greatly exceeded that of Radux, while the relatively short baselines of Radux spoiled the geometrical accuracy of the Omega (10 kilohertz) component at long distances, and kept the ultimate accuracy far less than it might otherwise be.

One of the primary reasons for the suggestion of such a low modulation frequency as 200 hertz was to guard against the cyclic ambiguities that are a feature of any phase measuring system. When measuring at 200 hertz in a hyperbolic system, any possible ambiguities would be separated by 400 miles or more and could be disregarded as an operational problem.

During the decade beginning in 1950, forces were at work that made the design of Radux less attractive than it had seemed in 1947. The most obvious change was a general increase in the desired accuracy of a navigational aid. When Loran was developed the only standard of excellence in the deep-water environment was celestial navigation, with perhaps 3 nautical miles as a typical error. Loran, especially Loran-C and a number of short-distance aids, gave navigators a desire for higher accuracy. Two other factors became important: (1) a general improvement in dead reckoning, including the inertial systems, and (2) a very great increase in the reliability of electronic devices. The first of these made the recovery of a lost cycle count much more probable, while the second made the loss of a count itself a relatively unlikely event.

The overall result of these external forces was to reduce by a large factor the fear of lane ambiguity. It therefore became possible to think of satisfying the need for accuracy beyond that of Radux by using a much more ambiguous system, with real hope that the possibility of ambiguities would not become a serious operational defect.

These considerations gradually led to the abandoning of Radux, leaving Omega to stand by itself. During the hybrid period, enough data had been taken to confirm the phenomenal range available at VLF, with timing errors that did not increase markedly with increasing distance. It was also realized that long baselines are especially effective on a spherical earth. For example, with a baseline subtending 60 degrees of arc the divergence of the hyperbolic lines of position is limited to a factor of 2, instead of increasing infinitely as it does with a baseline negligibly long in comparison with the curvature of the earth. And, of course, if a pair can link the opposite ends of a diameter of the earth there is no divergence at all, and a measurement accuracy of 12 microseconds in time-difference corresponds to a positional accuracy of a nautical mile, anywhere on earth.

The Omega experiments began with the pair linking Hawaii and California, operating at first at 12.5 kilohertz, although experiments were rapidly extended throughout the 9 to 15 kilohertz region, and even higher at times. One of the earliest experiments gave an exciting and convincing demonstration of the merits of VLF cycle matching. Commander Lyle C. Read, USN, who was the Radux Program Officer at the Navy Electronics Laboratory, was traversing the baseline with an early Omega receiver on a naval vessel, essentially counting the number of wavelengths between Hawaii and California as a check on the then somewhat nebulous ideas about phase velocity at very low frequencies. Fortunately, the ship got a little ahead of schedule and Commander Read was able to induce the captain to make a standard 360° turn midway in the passage. Although only a single pair was being tracked, the double amplitude of the sinusoidal variation on the phase record gave the diameter of the ship's turning circle within 50 yards.

This and many other demonstrations of the sensitivity of the Omega technique accelerated the decision to concentrate on VLF because of its range and potential accuracy, and to accept the best that could be done to solve the ambiguity problem.

From this point, the work went rapidly for a time. A station in the Panama Canal Zone was borrowed from Naval Communications, and one in Wales from the British Post Office. With these, and San Diego and Hawaii, the network was large enough to permit monitoring from Alaska to South America and from Hawaii to Europe. Early ideas of the velocity of propagation and of its variation with time of day were refined (a process that is still being carried forward) and innumerable trials and demonstrations were conducted.

By 1966 Omega signals were being transmitted on a regular basis from stations located in New York, Hawaii, Trinidad, and Norway. But since these stations utilized existing facilities and developmental equipment, none of these stations was capable of transmitting the power required for an operational system. However, signals were being transmitted full time from a four station complex providing the vital ingredients necessary to further system development.

By 1976 seven of the eight stations of the fully implemented system were in normal operation. The developmental station at Trinidad remained in operation pending implementation of the South Pacific station.

Other developments include **Decca** (art. 4344), a short to medium range hyperbolic system, which was first used under the code name "QM" during the landings in Normandy in 1944. Another World War II development was the rotating electronic beam utilized in the German navigation system called **sonne,** later further perfected by the British under the name **Consol** (art. 4354).

In the late 1950's the Decca Navigator Company Ltd., developed an experimental VLF radionavigation system known as **Delrac,** a name derived from **Decca long-range area coverage.** In principle, this system was similar to Omega.

137. Development of satellite navigation.—The Navy Navigation Satellite System (NAVSAT) was developed within the Navy to fulfill a requirement established by the Chief of Naval Operations for an accurate worldwide navigation system for all naval surface vessels, aircraft, and submarines. The system was conceived and developed by the Applied Physics Laboratory of The Johns Hopkins University under Navy contract.

The underlying concept that led to development of the system dates back to 1957 and the first launch of an artificial satellite into orbit—Russia's **Sputnik I.** Dr.

William H. Guier and Dr. George C. Wieffenbach at the Applied Physics Laboratory of the Johns Hopkins University were monitoring the famous "beeps" transmitted by the passing satellite. They plotted the received signals at precise intervals, and noticed that a characteristic doppler curve emerged. Since celestial bodies followed fixed orbits, they reasoned that this curve could be used to describe the satellite orbit. Later, they demonstrated that they could determine all of the orbital parameters for a passing satellite by doppler observation of a single pass from a single fixed station. The doppler shift apparent while receiving a transmission from a passing satellite was proven to be an effective measuring device for establishing the satellite orbit.

Dr. Frank T. McClure, also of the Applied Physics Laboratory concluded inversely that if the satellite-orbit were known, doppler shift measurement could be used to determine one's position on earth—thereby suggesting a new method for navigation—a more precise method than any yet known, available anywhere on earth without regard for weather conditions. His studies earned for him the first National Aeronautics and Space Administration award for important contributions to space study development.

In 1958, on the strength of Dr. McClure's studies, the Applied Physics Laboratory proposed to the Bureau of Naval Weapons that possibilities be explored for establishing a satellite doppler navigation system. The Chief of Naval Operations set forth requirements for such a system to provide accurate all-weather worldwide navigation, recommending to the Advanced Research Projects Agency that funds be made available for the purpose. Although this was only one of a number of proposals to utilize satellites for navigation, it was accepted, and, until 1960, all work on the system was performed by the Navy with that Agency's backing. An experimental satellite that failed to achieve orbit in September 1959 indicated the feasibility of tracking by doppler; the first successful launching of a prototype system satellite in April 1960 demonstrated its operational usefulness for navigation.

138. Development of inertial navigation.—The first inertial navigation system was developed in 1942 for use in the V-2 missile by the Peenemunde group under the leadership of Dr. Wernher von Braun. This system used two 2-degree-of-freedom gyroscopes and an integrating accelerometer to determine the missile velocity. By the end of World War II, the Peenemunde group had developed a stable platform with three single-degree-of-freedom gyroscopes and an integrating accelerometer.

Following World War II inertial navigation development in the United States was conducted by four groups, one sponsored by the Army and three by the Air Force. The Army group included the Peenemunde group under Dr. Wernher von Braun. This group later became the inertial group for the National Aeronautics and Space Administration. The Air Force-sponsored groups were Northrup Aircraft, Autonetics Division of North American Aviation, and the Massachusetts Institute of Technology Instrumentation Laboratory, which was later to become The Charles Stark Draper Laboratory.

At first the systems developed for the Air Force were combinations of stellar and inertial systems. As the state-of-the-art improved, purely inertial systems were designed. During this development the principal proponent of purely inertial systems was Dr. Charles Stark Draper of the Massachusetts Institute of Technology.

The Autonetics Division of North American Aviation ingeniously adapted one of its systems for shipboard use. In 1958 this system was used to navigate the USS *Nautilus* under the ice to the North Pole.

The development of purely inertial systems for air and marine applications proceeded along parallel lines. Missile and space applications followed.

The development of the **Ship Inertial Navigation System (SINS)** began in 1951 and was completed in 1954. The initial system test on the highway provided realistic operating conditions under close monitoring and control. The results of the shipboard test of SINS in 1955 indicated that it provided what was needed for the fleet ballistic missile submarine.

Conclusion

139. Navigation has come a long way, but there is no evidence that it is nearing the end of its development. Progress will continue as long as man remains unsatisfied with the means at his disposal.

References

Collinder, Per. *A History of Marine Navigation.* Tr. Maurice Michael. New York, St. Martin's, 1955.

Hewson, J. B. *A History of the Practice of Navigation.* Glasgow, Brown, 1951.

May, W. E. *A History of Marine Navigation.* Oxfordshire, G. T. Foulis and Company Limited, 1973.

Petze, C. L., Jr. *The Evolution of Celestial Navigation.* Vol. 26, Ideal Series. New York, Motor Boating, 1948.

Pierce, J. A., and R. H. Woodward. "The Development of Long-Range Hyperbolic Navigation in the United States." Harvard University, Cambridge, Massachusetts, Office of Naval Research Technical Report No. 620, February 1971.

Stewart, J. Q., and N. L. Pierce. "The History of Navigation." *Marine and Air Navigation* (Boston, Ginn, 1944). Chap. 29.

Taylor, E. G. R. *The Mathematical Practitioners of Tudor and Stuart England.* London, Cambridge University Press, 1955.

Taylor, E. G. R. *The Haven-Finding Art.* London, Hollis and Carter, 1956.

Taylor, E. G. R. *The Geometrical Seaman.* London, Hollis and Carter, 1962.

Waters, D. W. *The Art of Navigation in England in Elizabethan and Early Stuart Times.* New Haven, Yale University Press, 1958.

Wroth, L. C. *Some American Contributions to the Art of Navigation,* 1519–1802. Providence, John Carter Brown Library, 1947.

In addition, articles pertaining to the history of navigation are frequently carried in certain periodicals, including:

The American Neptune. (Salem)

The Journal of Navigation. (London)

The Nautical Magazine. (Glasgow)

NAVIGATION: *Journal of the Institute of Navigation.*

NAVIGATION: *Revue Technique de Navigation Maritime,* Aerienne *et Spatiale.* (Paris)

United States Naval Institute Proceedings. (Annapolis)

CHAPTER II

BASIC DEFINITIONS

201. Navigation is the process of directing the movements of a craft, expeditiously and safely, from one point to another. The word *navigate* is from the Latin *navigatus,* the past participle of the verb *navigere,* which is derived from the words *navis,* meaning "ship," and *agere,* meaning "to move" or "to direct." Navigation of water craft is called **marine navigation** to distinguish it from navigation of aircraft, called **air navigation.** Navigation of a vessel on the surface is sometimes called **surface navigation** to distinguish it from **underwater navigation** of a submerged vessel. Navigation of vehicles across land or ice is called **land navigation.** The expression **lifeboat navigation** is used to refer to navigation of lifeboats or life rafts, generally involving rather crude methods. The expression **polar navigation** refers to navigation in the regions near the geographical poles of the earth, where special techniques are employed.

The principal divisions of navigation are as follows:

Dead reckoning is the determination of position by advancing a known position for courses and distances. A position so determined is called a **dead reckoning position.** It is generally accepted that the course *steered* and the speed *through the water* should be used, but the expression is also used to refer to the determination of position by use of the course and speed expected to be made good over the ground, thus making an estimated allowance for disturbing elements such as current and wind. A position so determined is better called an **estimated position.** The expression "dead reckoning" probably originated from use of the Dutchman's log a buoyant object thrown overboard, to determine the speed of the vessel relative to the object, which was assumed to be *dead* in the water. Apparently, the expression **deduced reckoning** was used when allowance was made for current and wind. It was often shortened to *ded reckoning* and the similarity of this expression to *dead reckoning* was undoubtedly the source of the confusion that is still associated with these expressions.

Piloting (or **pilotage**) is navigation involving frequent or continuous determination of position or a line of position relative to geographic points, and usually requiring need for close attention to the vessel's draft with respect to the depth of water. It is practiced in the vicinity of land, dangers, etc., and requires good judgment and almost constant attention and alertness on the part of the navigator. **Celestial navigation** is navigation using information obtained from celestial bodies.

Radionavigation is navigation using radio waves for determination of position or of a line of position. Radar navigation and satellite navigation are parts of the radionavigation division. **Radar navigation** involves the use of radio waves, usually in the centimeter band, to determine the distance and direction of an object reflecting the waves to the sender. **Satellite navigation** involves the use of artificial earth satellites for determination of position.

The term **electronic navigation** is used to refer to navigation involving the use of electronics in any way. Thus, the term includes the use of the gyrocompass for steering and the echo sounder when piloting. Because of the wide use of electronics in navigation equipment, the term electronic navigation has limited value as a term for a division of navigation.

Electronics is the science and technology relating to the emission, flow, and effects of electrons in a vacuum or through a semiconductor, and to systems using devices in which this action takes place.

202. The earth is approximately an **oblate spheroid** (a sphere flattened at the poles). Approximations of its dimensions and the amount of flattening are given in appendix D. However, for many navigational purposes, the earth is assumed to be a sphere, without intolerable error.

The **axis of rotation** or **polar axis** of the earth is the line connecting the North Pole and the South Pole.

203. Circles of the earth.—A **great circle** is the line of intersection of a sphere and a plane through the center of the sphere (fig. 203a). This is the largest circle that can be drawn on a sphere. The shortest line on the surface of a sphere between two points on that surface is part of a great circle. On the spheroidal earth the shortest line is called a **geodesic.** A great circle is a near enough approximation to a geodesic for most problems of navigation.

A **small circle** is the line of intersection of a sphere and a plane which does not pass through the center of the sphere (fig. 203a).

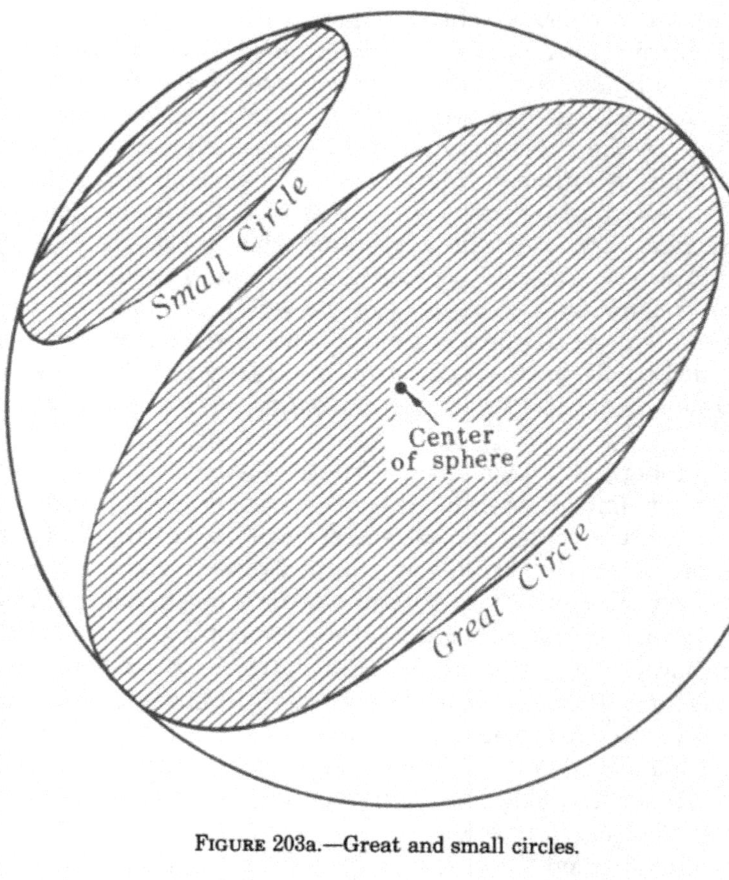

FIGURE 203a.—Great and small circles.

A **meridian** is a great circle through the geographical poles of the earth. Hence, all meridians meet at the poles, and their planes intersect each other in a line, the **polar axis** (fig. 203b). The term **meridian** is usually applied to the **upper branch** only, that half from pole to pole which passes through a given point. The other half is called the **lower branch.**

The **prime meridian** is that meridian used as the origin for measurement of longitude (fig. 203c). The prime meridian used almost universally is that through the original position of the British Royal Observatory at Greenwich, near London.

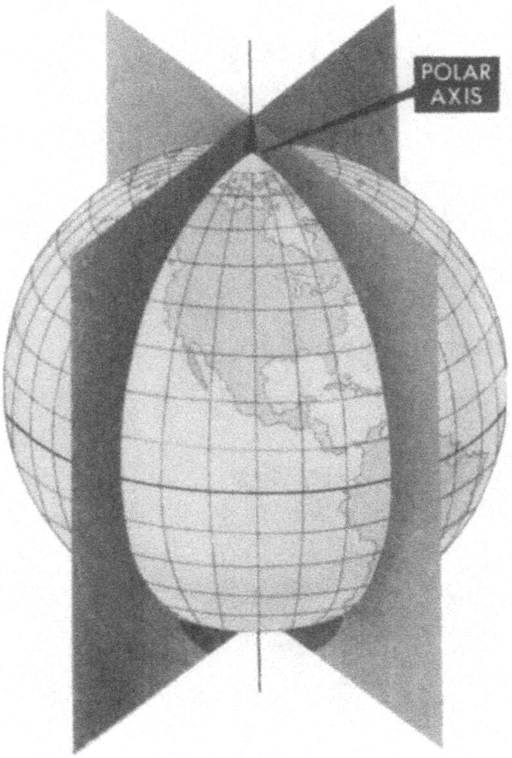

FIGURE 203b.—The planes of the meridians meet at the polar axis.

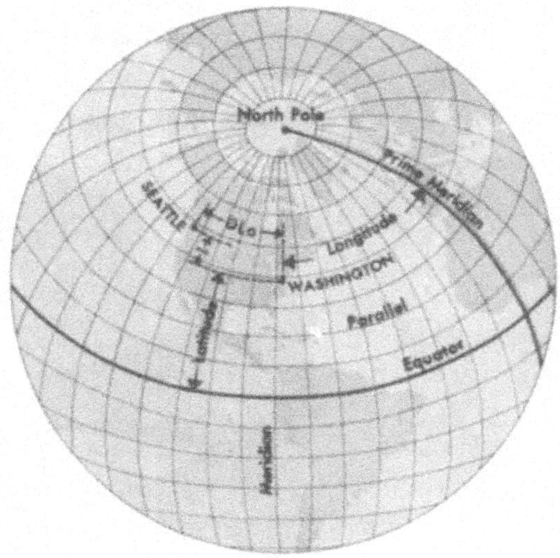

FIGURE 203c.—Circles and coordinates of the earth. All parallels except the equator are small circles; the equator and meridians are great circles.

The **equator** is the terrestrial great circle whose plane is perpendicular to the polar axis (fig. 203d). It is midway between the poles.

A **parallel** or **parallel of latitude** is a circle on the surface of the earth, parallel to the plane of the equator (fig. 203e). It connects all points of equal latitude. The equator, a great circle, is a limiting case connecting points of 0° latitude. The poles, single points at latitude 90°, are the other limiting case. All other parallels are small circles.

204. Position on the earth.—A position on the surface of the earth (except at either of the poles) may be defined by two magnitudes called **coordinates.** Those customarily used are *latitude* and *longitude.* A position may also be expressed in relation to known geographical positions.

Latitude (L, lat.) is angular distance from the equator, measured northward or southward along a meridian from 0° at the equator to 90° at the poles (fig. 203c). It is designated *north* (N) or *south* (S) to indicate the direction of measurement.

The **difference of latitude (*l*, D. Lat.)** between two places is the angular length of arc of any meridian between their parallels (fig. 203c). It is the numerical difference of the latitudes if the places are on the same side of the equator, and the

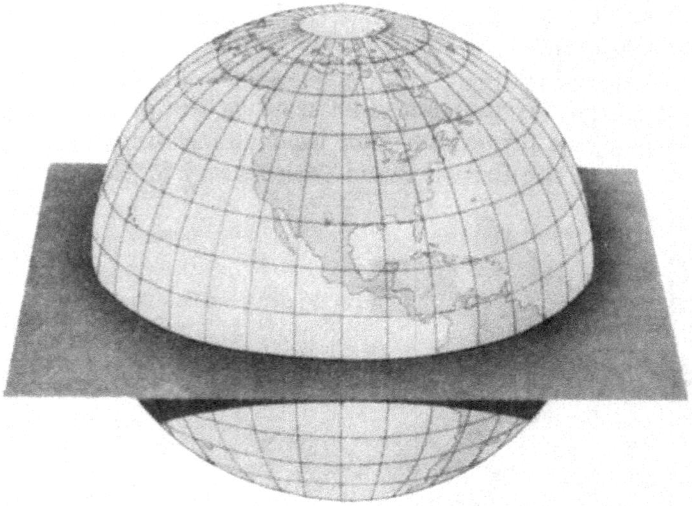

FIGURE 203d.—The equator is a great circle midway between the poles.

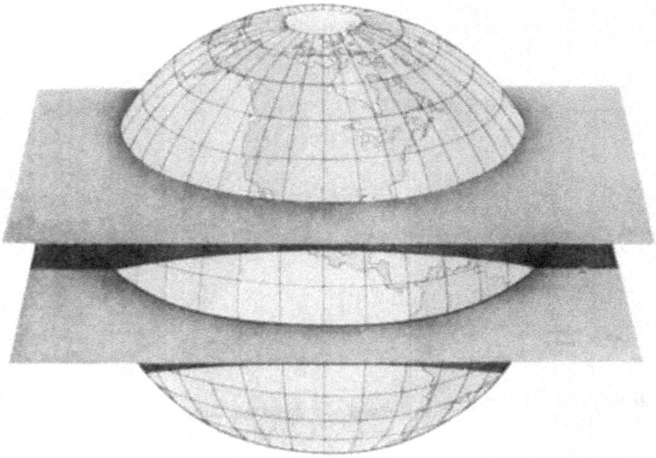

FIGURE 203e.—A parallel of latitude is parallel to the equator.

sum if they are on opposite sides. It may be designated *north* (N) or *south* (S) when appropriate.

The **middle** or **mid latitude (Lm)** between two places on the same side of the equator is half the sum of their latitudes. Mid latitude is labeled N or S to indicate whether it is north or south of the equator. The expression is occasionally used with reference to two places on opposite sides of the equator, when it is equal to half the *difference* between the two latitudes, and takes the name of the place farthest from the equator. However, this usage is misleading, as it lacks the significance usually associated with the expression. When the places are on opposite sides of the equator, two mid latitudes are generally used, the average of each latitude and 0°.

Longitude (λ, **long.**) is the arc of a parallel or the angle at the pole between the prime meridian and the meridian of a point on the earth, measured eastward or westward from the prime meridian through 180° (fig. 203c). It is designated *east* (E) or *west* (W) to indicate the direction of measurement.

The **difference of longitude (DLo)** between two places is the shorter arc of the parallel or the smaller angle at the pole between the meridians of the two places (fig. 203c). If both places are on the same side (east or west) of Greenwich, DLo is the numerical difference of the longitudes of the two places; if on opposite sides, DLo is the numerical sum unless this exceeds 180°, when it is 360° minus the sum. The distance between two meridians at any parallel of latitude, expressed in distance units, usually nautical miles, is called **departure (p, Dep.)**. It represents distance made good to the east or west as a craft proceeds from one point to another. Its numerical value between any two meridians decreases with increased latitude, while DLo is numerically the same at any latitude. Either DLo or p may be designated *east* (E) or *west* (W) when appropriate.

205. Distance on the earth.—Distance (D, Dist.) is the spatial separation of two points, and is expressed as the length of a line joining them. On the surface of the earth it is usually stated in miles. Navigators customarily use the **nautical mile (mi., M)** of 1,852 meters exactly. This is the value suggested by the International Hydrographic Bureau in 1929, and since adopted by most maritime nations. It is often called the **International Nautical Mile** to distinguish it from slightly different values used by some countries. On July 1, 1959, the United States adopted the exact relationship of 1 yard=0.9144 meter. The length of the International Nautical Mile is consequently equal to 6,076.11549 feet (approximately).

For most navigational purposes the nautical mile is considered the length of 1 minute of latitude, or of any great circle of the earth, regardless of location. On the Clarke spheroid of 1866, used for mapping North America, the length of 1 minute of latitude varies from about 6,046 feet at the equator to approximately 6,108 feet at the poles. The length of 1 minute of a great circle of a sphere having an area equal to that of the earth, as represented by this spheroid, is 6,080.2 United States feet. This was the standard value of the nautical mile in the United States prior to adoption of the international value. A **geographical mile** is the length of 1 minute of the equator, or about 6,087 feet.

The **land** or **statute mile (mi., St M)** of 5,280 feet is commonly used for navigation on rivers and lakes, notably the Great Lakes of North America.

The nautical mile is about 38/33 or approximately 1.15 statute miles. A conversion table for nautical and statute miles is given in table 20.

Distance, as customarily used by the navigator, refers to the length of the **rhumb line** connecting two places. This is a line making the same oblique angle with all meridians. Meridians and parallels (including the equator) which also maintain constant true directions, may be considered special cases of the rhumb line. Any other rhumb line spirals toward the pole, forming a **loxodromic curve** or **loxodrome** (fig. 205). Distance along the great circle connecting two points is customarily designated **great-circle distance.**

206. Speed (S) is rate of motion, or distance per unit of time.

A **knot (kn.)**, the unit of speed commonly used in navigation, is a rate of 1 nautical mile per hour. The expression "knots per hour" refers to acceleration, not speed.

The expression **speed of advance (SOA)** is used to indicate the speed intended to be made along the track (art. 207), and **speed over ground (SOG)** the speed along the path actually followed. **Speed made good (SMG)** is the speed along the course made good.

207. Direction on the earth.—Direction is the position of one point relative to another, without reference to the distance between them. In navigation, direction is

FIGURE 205.—A rhumb line or loxodrome.

customarily expressed as the angular difference in degrees from a reference direction, usually north or the ship's head. Compass directions (east, south by west, etc.) or points (of 11¼° or 1/32 of a circle) are seldom used by modern navigators for precise directions.

Course (C, Cn) is the horizontal direction in which a vessel is steered or intended to be steered, expressed as angular distance from north, usually from 000° at north, clockwise through 360°. Strictly, the term applies to direction *through the water,* not the direction intended to be made good *over the ground.* The course is often designated as **true, magnetic, compass,** or **grid** as the reference direction is true, magnetic, compass, or grid north, respectively. **Track made good (TMG)** is the single resultant direction from the point of departure to point of arrival at any given time. Sometimes the expression **course of advance (COA)** is used to indicate the direction intended to be made good over the ground, and **course over ground (COG)** the direction of the path actually followed, usually a somewhat irregular line. **Course line** is a line, as drawn on a chart, extending in the direction of a course.

In making computations it is sometimes convenient to express a course as an angle from *either* north or south, through 90° or 180°. In this case it is designated **course angle (C)** and should be properly labeled to indicate the origin (prefix) and direction of measurement (suffix). Thus, C N35°E=Cn 035° (000°+35°), C N155°W=Cn 205° (360°−155°), C S47°E=Cn 133° (180°−47°). But Cn 260° may be either C N100°W or C S80°W, depending upon the conditions of the problem.

The symbol C is always used for *course angle,* and is usually used for *course* where there is little or no possibility of confusion.

Track (TR) is the intended or desired horizontal direction of travel with respect to the earth and also the path of intended travel. The terms **intended track** and **trackline** are also used to indicate the path of intended travel (fig. 207a). The path actually followed is usually a somewhat irregular line. The track consists of one or a series of course lines from the point of departure to the destination, along which it is intended the vessel will proceed. A great circle which a vessel intends to follow approximately is called a **great-circle track.**

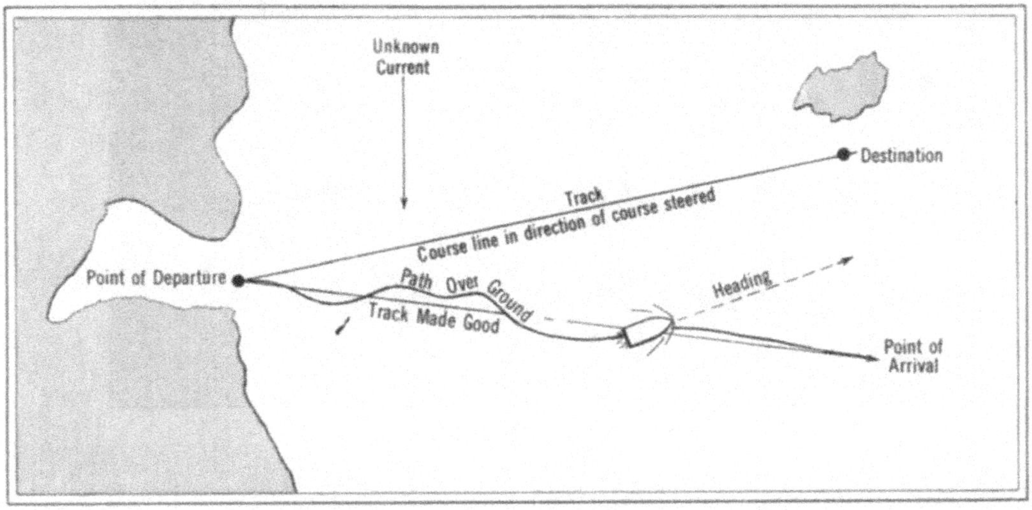

FIGURE 207a.—Course line, track, track made good, and heading.

Heading (Hdg., SH) is the direction in which a vessel is pointed, expressed as angular distance from north, usually from 000° at north, clockwise through 360°. *Heading* should not be confused with *course. Heading* is a constantly changing value as a vessel oscillates or yaws back and forth across the course due to the effects of sea, wind, and steering error.

Bearing (B, Brg.) is the direction of one terrestrial point from another, expressed as angular distance from a reference direction, usually from 000° at the reference direction, clockwise through 360°. When measured through 90° or 180° from *either* north or south, it is called **bearing angle (B)**, which bears the same relationship to bearing as *course angle* does to *course. Bearing* and *azimuth* are sometimes used interchangeably, but the latter is better reserved exclusively for reference to horizontal direction of a point on the celestial sphere from a point on the earth.

A relative bearing is one relative to the heading, or to the vessel itself. It is usually measured from 000° at the heading, clockwise through 360°. However, it is sometimes conveniently measured right or left from 0° at the ship's head through 180°. This is particularly true when using table 7. Older methods, such as indicating the number of degrees or points from some part of the vessel (10° forward of the starboard beam, two points on the port quarter, etc.) are seldom used by modern navigators to indicate precise directions, except for bearings dead ahead or astern, or broad on the bow, beam, or quarter.

To convert a relative bearing to a bearing from north (fig. 207b), express the relative bearing in terms of the 0°–360° system and add the heading:

True Bearing = Relative Bearing + True Heading.

Thus, if another vessel bears 127° relative from a ship whose heading is 150°, the bearing from north is 127° + 150° = 277°. If the total exceeds 360°, subtract this amount. To convert a bearing from north to a relative bearing, subtract the heading:

Relative Bearing = True Bearing − True Heading.

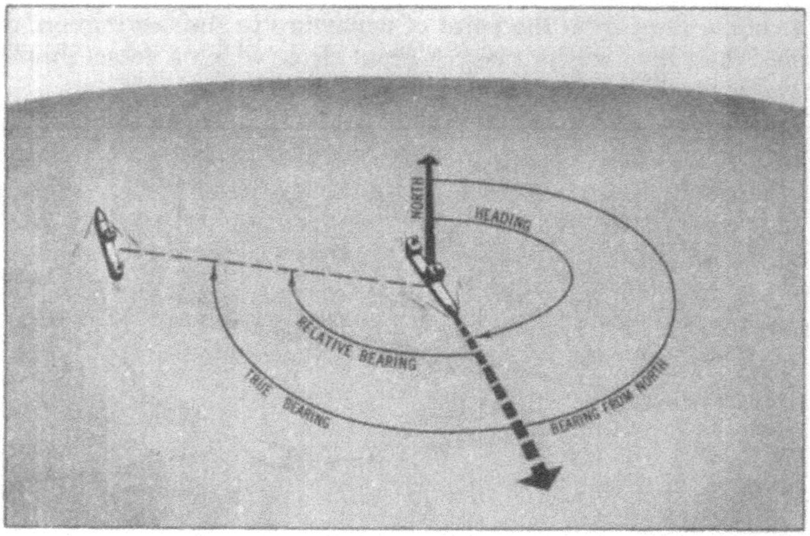

FIGURE 207b.—Relative bearing.

Thus, a lighthouse which bears 241° from north bears 241° − 137° = 104° relative from a ship whose heading is 137°. If the heading is larger than the true bearing, add 360° to the true bearing before subtracting.

CHAPTER III

CHART PROJECTIONS

General

301. The navigator's chart.—A **map** is a conventional representation, usually on a plane surface, of all or part of the physical features of the earth's surface or any part of it. A **chart** is such a representation intended primarily for navigation. A **nautical** or **marine chart** is one intended primarily for marine navigation. It generally shows depths of water (by soundings and sometimes also by depth curves), aids to navigation, dangers, and the outline of adjacent land and such land features as are useful to the navigator.

Chart making presents the problem of representing the surface of a spheroid upon a plane surface. The surface of a sphere or spheroid is said to be **undevelopable** because no part of it can be flattened without distortion. A **map projection** or **chart projection** is a method of representing all or part of the surface of a sphere or spheroid upon a plane surface. The process is one of transferring points on the surface of the sphere or spheroid onto a plane, or onto a **developable** surface (one that can be flattened to form a plane) such as a cylinder or cone. If points on the surface of the sphere or spheroid are projected from a single point (including infinity), the projection is said to be **perspective** or **geometric.** Most map projections are not perspective.

302. Selecting a projection.—Each projection has distinctive features which make it preferable for certain uses, no one projection being best for all conditions. These distinctive features are most apparent on charts of large areas. As the area becomes smaller, the differences between various projections become less noticeable until on the largest scale chart, such as of a harbor, all projections become practically identical. Some of the desirable properties are:

1. *True shape* of physical features.
2. *Correct angular relationship.* A projection with this characteristic is said to be **conformal** or **orthomorphic.**
3. *Equal area,* or the representation of areas in their correct relative proportions.
4. *Constant scale* values for measuring distances.
5. *Great circles* represented as straight lines.
6. *Rhumb lines* represented as straight lines.

It is possible to preserve any one and sometimes more than one property in any one projection, but it is impossible to preserve all of them. For instance, a projection cannot be both conformal and equal area, nor can both great circles and rhumb lines be represented as straight lines.

303. Types of projection.—Projections are usually classified primarily as to the type of developable surface to which the spherical or spheroidal surface is transferred. They are sometimes further classified as to whether the projection (but not necessarily the charts made by it) is centered on the equator (**equatorial**), a pole (**polar**), or some point or line between (**oblique**). The name of a projection often indicates its type and sometimes, in addition, its principal feature.

The projection used most frequently by mariners is commonly called **Mercator,** after its inventor (art. 109). Classified according to type this is a cylindrical projection upon a plane, the cylinder conceived as being tangent along the equator. A similar projection based upon a cylinder tangent along a meridian is called **transverse Mercator** or **transverse orthomorphic.** It is sometimes called **inverse Mercator** or **inverse orthomorphic.** If the cylinder is tangent along a great circle other than the equator or a meridian, the projection is called **oblique Mercator** or **oblique orthomorphic.**

In a **simple conic** projection points on the surface of the earth are conceived as transferred to a tangent cone. In a **Lambert conformal** projection the cone intersects the earth (a **secant** cone) at two small circles. In a **polyconic** projection, a series of tangent cones is used.

An **azimuthal** or **zenithal** projection is one in which points on the earth are transferred directly to a plane. If the origin of the projecting rays is the center of the earth, a **gnomonic** projection results; if it is the point opposite the plane's point of tangency, a **stereographic** projection; and if at infinity (the projecting lines being parallel to each other), an **orthographic** projection (fig. 303). The gnomonic, stereographic, and orthographic are perspective projections. In an **azimuthal equidistant** projection, which is not perspective, the scale of distances is constant along any radial line from the point of tangency.

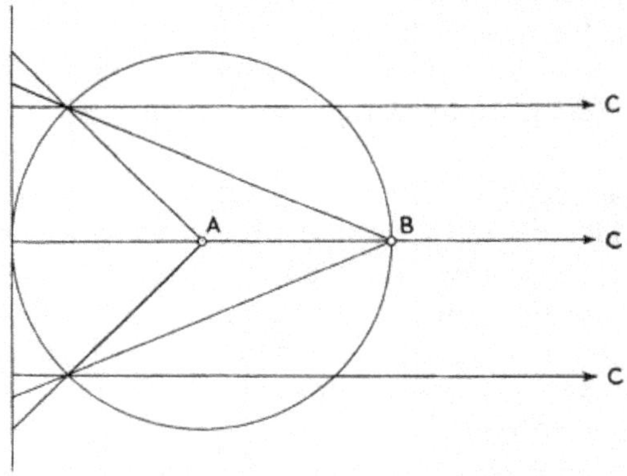

FIGURE 303.—Azimuthal projections: *A*, gnomonic; *B*, stereographic; *C* (at infinity), orthographic.

Cylindrical and plane projections can be considered special cases of conical projections with the heights infinity and zero, respectively.

A **graticule** is the network of latitude and longitude lines laid out in accordance with the principles of any projection.

Cylindrical Projections

304. Features.—If a cylinder is placed around the earth, tangent along the equator, and the planes of the meridians are extended, they intersect the cylinder in a number of vertical lines (fig. 304). These lines, all being vertical, are parallel, or everywhere equidistant from each other, unlike the terrestrial meridians, which become closer together as the latitude increases. On the earth the parallels of latitude are perpendicular to the meridians, forming circles of progressively smaller diameter as the latitude increases. On the cylinder they are shown perpendicular to the projected meridians, but because a cylinder is everywhere of the same diameter, the projected parallels are all the same size.

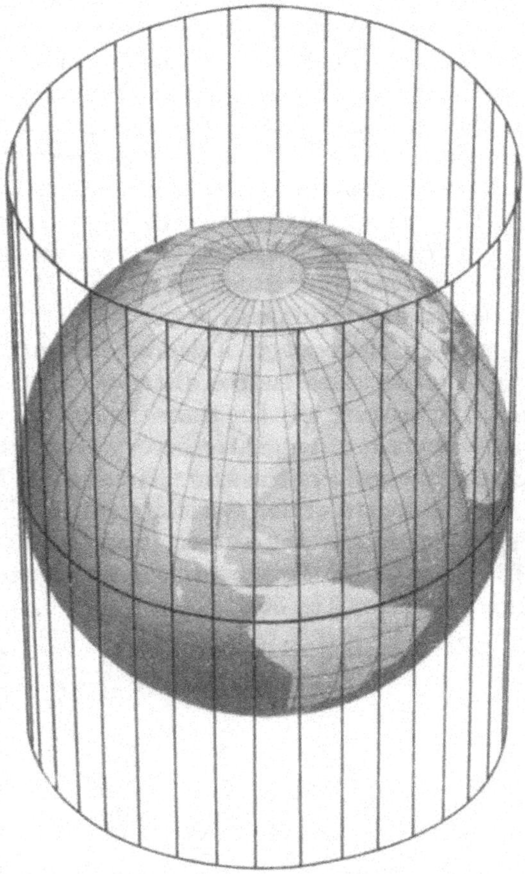

FIGURE 304.—A cylindrical projection.

If the cylinder is cut along a vertical line (a meridian) and spread out flat, the meridians appear as equally spaced, vertical lines, and the parallels as horizontal lines. The spacing of the parallels relative to each other differs in the various types of cylindrical projections.

The cylinder may be tangent along some great circle other than the equator, forming an oblique or transverse cylindrical projection, on which the pattern of latitude and longitude lines appears quite different, since the line of tangency and the equator no longer coincide.

305. Mercator projection.—A conformal projection upon a plane used for navigation is the **Mercator**, named for its inventor Gerhard Kremer (Mercator), a Flemish geographer. It is not perspective and the parallels cannot be located by geometrical projection, the spacing being derived mathematically. The use of a tangent cylinder to explain the development of the projection has been used, but the relationship of the terrestrial latitude and longitude lines to those on the cylinder is often carried beyond justification, resulting in misleading statements and illustrations.

The distinguishing feature of the Mercator projection (fig. 305) among other projections is that both the meridians and parallels are expanded at the same ratio with increased latitude. The expansion is equal to the secant of the latitude, with a small correction for the ellipticity of the earth. Since the secant of 90° is infinity, the projection cannot include the poles. Expansion is the same in all directions and angles are correctly shown, the projection being conformal. Rhumb lines appear as straight lines, the directions of which can be measured directly on the chart. Distances can also be measured directly, to practical accuracy, but not by a single distance scale over the entire chart, unless the spread of latitude is small. The latitude scale is customarily used for measuring distances, the expansion of the scale being the same as that of distances at the same latitude. Great circles, except meridians and the equator, appear as curved lines concave to the equator (fig. 310a). Small areas appear in their correct shape but of increased size unless they are near the equator. Plotting of positions by latitude and longitude is done by means of rectangular coordinates, as on any cylindrical projection.

306. Meridional parts.—At the equator a degree of longitude is approximately equal in length to a degree of latitude. As the distance from the equator increases, degrees of latitude remain approximately the same (not exactly because the earth is not quite a sphere), while degrees of longitude become progressively shorter. Since degrees of longitude appear everywhere the same length in the Mercator projection, it is necessary to increase the length of the meridians if the expansion is to be equal in all directions. Thus, to maintain the correct proportions between degrees of latitude and degrees of longitude, the former are shown progressively longer as the distance from the equator increases (fig. 305).

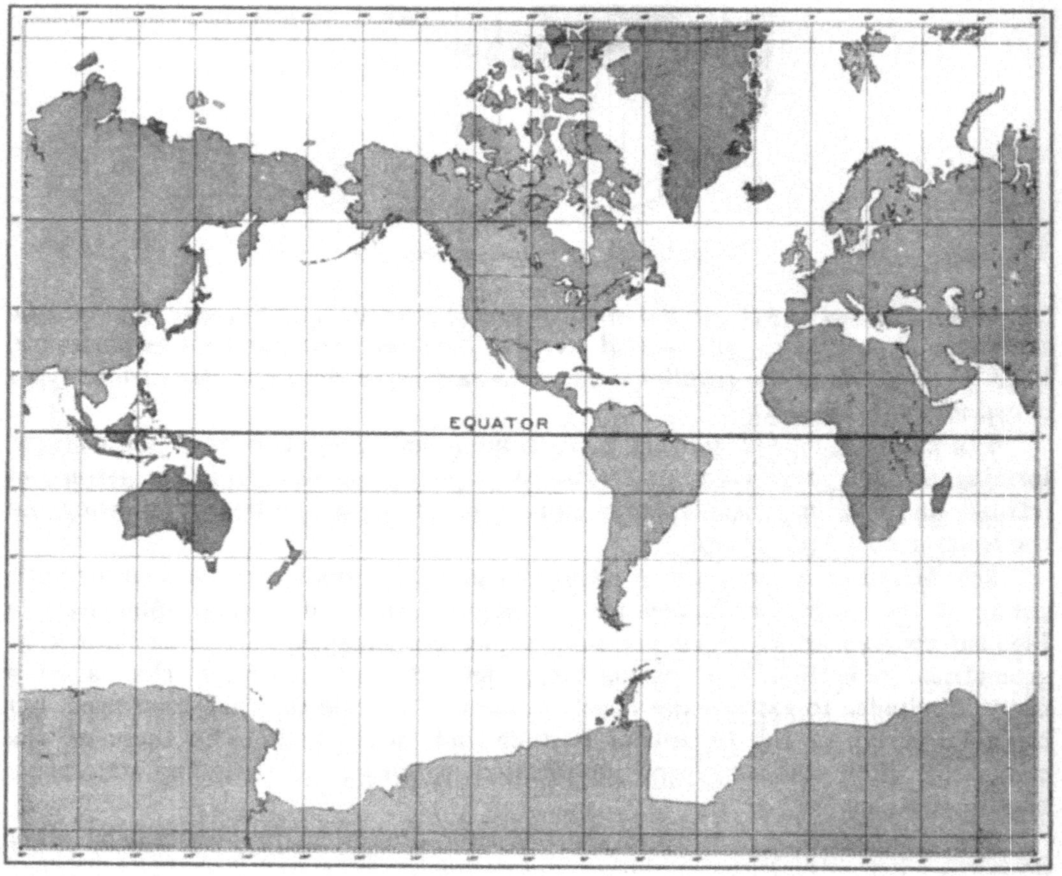

FIGURE 305.—A Mercator map of the world.

The length of the meridian, as thus increased between the equator and any given latitude, expressed in minutes of the equator as a unit, constitutes the number of **meridional parts (M)** corresponding to that latitude. Meridional parts, given in table 5 for every minute of latitude from the equator to the pole, afford facilities for constructing a Mercator chart and for solving problems in Mercator sailing. These values are for the WGS ellipsoid of 1972. The formula for meridional parts, given in the explanation to table 5, is derived from an integral representing the exact relationship.

307. Mercator chart construction.—To construct a Mercator chart, first select the scale and then proceed as follows:

Draw a series of vertical lines to represent the meridians, spacing them in accordance with the scale selected. If the chart is to include the equator, the distances of the various parallels from the equator are given directly in table 5, although it may be desirable to convert the tabulated values to more convenient units. Thus, if 1°(60′) of longitude is to be shown as 1 inch, each meridional part will be $\frac{1}{60}$ or 0.01667 inch in length. The distance, in inches, of any parallel from the equator is then determined by dividing its meridional parts by 60 or multiplying them by 0.01667.

If the equator is not to be included, the **meridional difference (m)** is used. This is the difference between the meridional parts of the various latitudes and that of the lowest parallel (the one nearest the equator) to be shown. Distances so determined are measured from the lowest parallel.

It is often desired to show a minimum area on a chart of limited size, to the largest possible scale. The scale is then dictated by the limitations.

When the graticule has been completed, the features to be shown are located by means of the latitude and longitude scales.

Example.—A Mercator chart is to be constructed at the maximum scale on a sheet of paper 35 × 46 inches, with a minimum 2-inch margin outside the **neatline** limiting the charted area. The minimum area to be covered is lat. 44°–50° north and long. 56°–68° west.

Solution.—Step one: Determine which dimension to place horizontal. From table 5 the meridional difference is:

$$\begin{array}{lr} M_{50^\circ} & 3456.6 \\ M_{44^\circ} & \underline{2929.6} \\ m & 527.0 \end{array}$$

The chart is to cover at least 12° (68°–56°) of longitude. The longitude is therefore to cover a distance of 12×60=720 meridional parts. Since there are a greater number of meridional parts of longitude to be shown than of latitude, the long dimension is placed horizontal.

Step two: Determine whether the latitude or longitude is the limiting scale factor. The number of inches available for latitude coverage is 31 (35 inches minus a 2-inch margin top and bottom). If 527 meridional parts are to be shown in 31 inches, each meridional part will be

$$\frac{31}{527}=0.05882 \text{ inch.}$$

There are 46–4=42 inches available for longitude, and therefore the length of each meridional part will be

$$\frac{42}{720}=0.05833 \text{ inch.}$$

Thus, the longitude is the limiting scale factor, for all of the desired area could not be shown in the available space if the larger scale were to be used. Using the smaller scale, it is found that 30.74 inches (0.05833×527) will be needed to show the desired latitude coverage. The top and bottom margins can be increased slightly, or additional latitude coverage can be shown. If it is desired to include the additional coverage, the amount can be determined by dividing the available space, 31 inches, by the scale, 0.05833. This is 531.5 meridional parts, or 4.5 more than the minimum. By inspection of table 5, it is seen that the latitude can be extended either 3′.3 below 44° or 2′.9 above 50°. Suppose it is decided that the margin will be increased slightly and only the desired minimum coverage shown.

Step three: Determine the spacing of the meridians and parallels. Meridians 1° or 60′ apart will be placed 60×0.05833=3.50 inches apart. Next, determine each degree of latitude separately. First, compute the meridional difference between the lowest parallel and the various parallels to be shown:

$M_{45°}$	3013.5	$M_{46°}$	3098.8	$M_{47°}$	3185.7	$M_{48°}$	3274.2	$M_{49°}$	3364.5	$M_{50°}$	3456.6
$M_{44°}$	2929.6	$M_{44°}$	2929.6	$M_{44°}$	2929.6	$M_{44°}$	2929.6	$M_{44°}$	2929.6	$M_{44°}$	2929.6
m	83.9	m	169.2	m	256.1	m	344.6	m	434.9	m	527.0

Next, determine the distance of each parallel from that of L 44°N by multiplying its meridional difference by the scale, 0.05833:

L 44° to L 45°=0.05833× 83.9= 4.89 in.
L 44° to L 46°=0.05833×169.2= 9.87 in.
L 44° to L 47°=0.05833×256.1=14.94 in.
L 44° to L 48°=0.05833×344.6=20.10 in.
L 44° to L 49°=0.05833×434.9=25.37 in.
L 44° to L 50°=0.05833×527.0=30.74 in.

Step four: Draw the graticule. Draw a horizontal line 2.13 inches

$$\left(\frac{35-30.74}{2}\right)$$

from the bottom. This is the lower neatline. Label it "44°N." Draw the right-hand neatline 2 inches from the edge. Label it "56°W." Along the lower parallel measure off distances in units of 3.50 inches from λ 56°W at the right to λ 68°W at the left. Through the points thus located draw vertical lines to represent the meridians. Along any meridian measure upward from the horizontal line a series of distances as determined by the calculations above. Through these points draw horizontal lines to represent the parallels. Label the meridians and parallels as shown in figure 307.

Step five: Mark off the latitude and longitude scales around the neatline. The scales can be graduated in units as small as desired. Determine the longitude scale by dividing the degrees into equal parts. Establish the latitude scale by computing each subdivision of a degree in the same manner as described above for whole degrees. In low latitudes degrees of latitude can be divided into equal parts without serious loss of accuracy.

Step six: Fill in the desired detail.

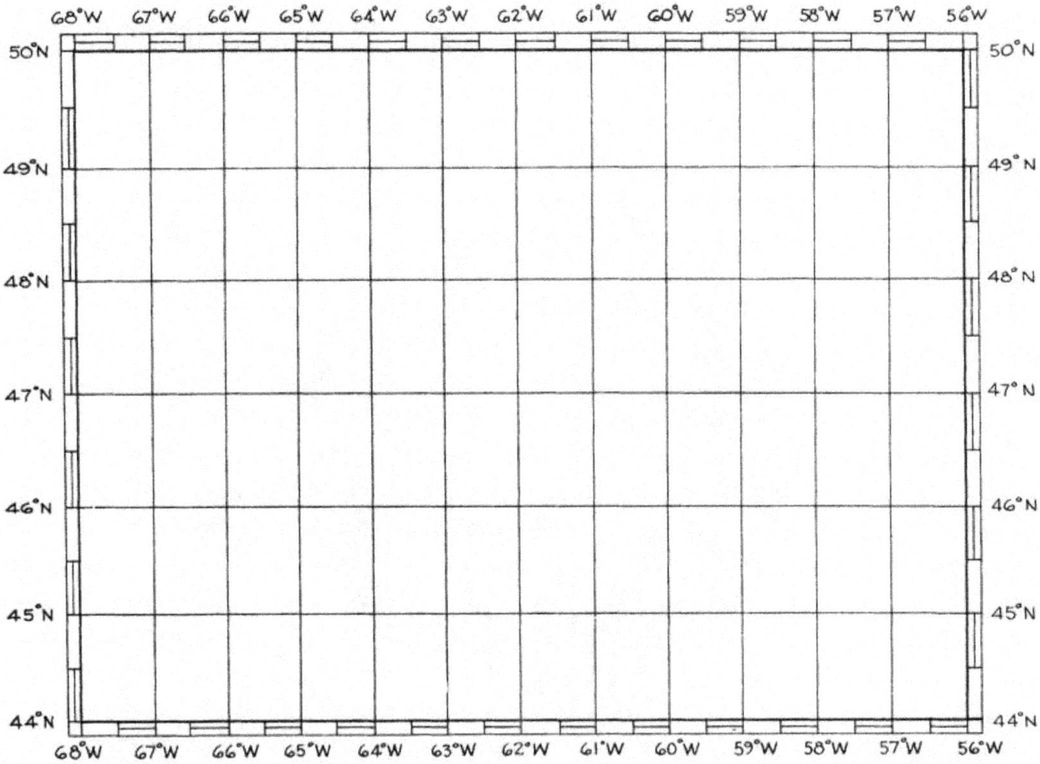

FIGURE 307.—The graticule of a Mercator chart from L 44°N to L 50°N and from λ 56°W to λ 68°W.

In *south* latitude the distance between consecutive parallels increases toward the *south*. The top parallel is drawn first and distances measured downward from it. Latitude labels increase toward the *south* (down).

In *east* longitude the longitude labels increase toward the *east* (right).

308. Transverse and oblique Mercator projections.—If Mercator principles are used to construct a chart, but with the cylinder tangent along a meridian, a **transverse Mercator** or **transverse orthomorphic** projection results. The word "inverse" is sometimes used in place of "transverse" with the same meaning. If the cylinder is tangent at some great circle other than the equator or a meridian (fig. 308a), the projection is called **oblique Mercator** or **oblique orthomorphic.** These projections utilize a **fictitious graticule** similar to but offset from the familiar network of meridians and parallels (fig. 308b). The tangent great circle is the **fictitious equator.** Ninety degrees from it are two **fictitious poles.** A group of great circles through these poles and perpendicular to the tangent great circle are the **fictitious meridians,** while a series of circles parallel to the plane of the tangent great circle form the **fictitious parallels.**

The actual meridians and parallels appear as curved lines (figs. 309 and 310b).

A straight line on the transverse or oblique Mercator projection makes the same angle with all fictitious meridians, but not with the terrestrial meridians. It is therefore a **fictitious rhumb line.** Near the tangent great circle a straight line closely approximates a great circle. It is in this area that the chart is most useful.

The **Universal Transverse Mercator (UTM) grid** is a military grid superimposed upon a transverse Mercator graticule, or the representation of these grid lines upon any graticule.

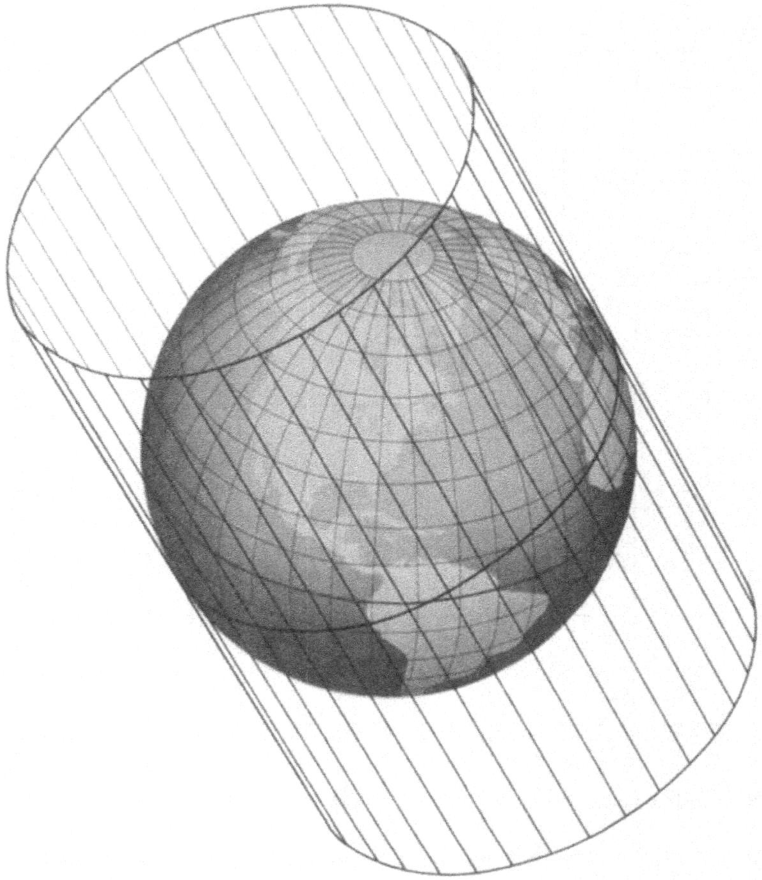

FIGURE 308a.—An oblique Mercator projection.

This grid system and these projections are often used for large-scale (harbor) nautical charts and military charts.

309. Transverse Mercator projection.—A special case of the Mercator projection in which the cylinder is tangent along a meridian is called a **transverse (inverse) Mercator** or **transverse (inverse) orthomorphic** projection. Since the area of minimum distortion is near a meridian, this projection is useful for charts covering a large band of latitude and extending a relatively short distance on each side of the tangent meridian (fig. 309). It is sometimes used for star charts showing the evening sky at various seasons of the year (figs. 2205–2208).

310. Oblique Mercator projection.—The Mercator projection in which the cylinder is tangent along a great circle other than the equator or a meridian is called an **oblique Mercator** or **oblique orthomorphic** projection. This projection is used principally where it is desired to depict an area in the near vicinity of an oblique great circle, as, for instance, along the great-circle route between two important, widely separated centers. Figure 310a is a Mercator map showing Washington and Moscow and the great circle joining them. Figure 310b is an oblique Mercator map with the great circle between these two centers as the tangent great circle or fictitious equator (as in fig. 308b). The limits of the chart of figure 310b are indicated in figure 310a. Note the large variation in scale as the latitude changes.

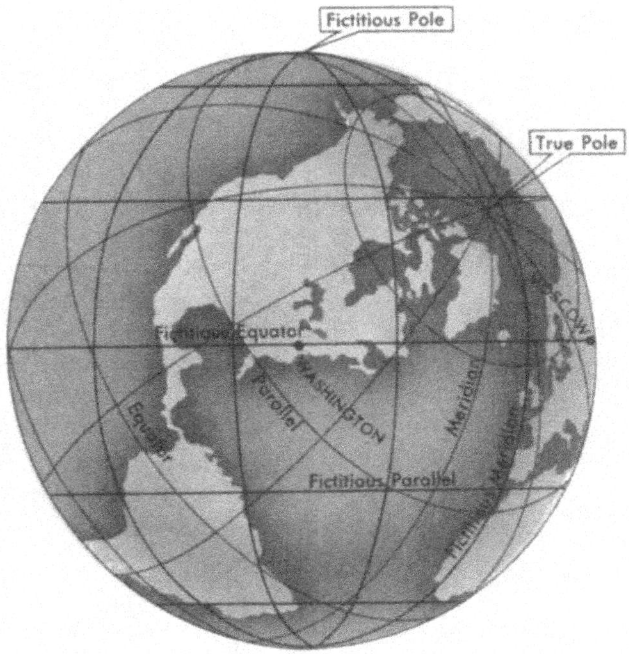

FIGURE 308b.—The fictitious graticule of an oblique Mercator projection.

311. Rectangular projection.—A cylindrical projection similar to the Mercator but with uniform spacing of the parallels is called a **rectangular** projection (fig. 311). It is convenient for graphically depicting information where distortion is not important. The principal navigational use of this projection is for the star chart of the *Air Almanac* (art. 2204), where positions of stars are plotted by rectangular coordinates representing declination (ordinate) and sidereal hour angle (abscissa). Since the meridians are parallel, the parallels of latitude (including the equator and the poles) are all represented by lines of equal length.

Conic Projections

312. Features.—A conic projection is produced by transferring points from the surface of the earth to a cone or series of cones which are then cut along an element and spread out flat to form the chart. If the axis of the cone coincides with the axis of the earth, the usual situation, the parallels appear as arcs of circles and the meridians as either straight or curved lines converging toward the nearer pole. Excessive distortion is usually avoided by limiting the area covered to that part of the cone near the surface of the earth. A parallel along which there is no distortion is called a **standard parallel.** Neither the **transverse conic** projection, in which the axis of the cone is in the equatorial plane, nor the **oblique conic** projection, in which the axis of the cone is oblique to the plane of the equator, are ordinarily used for navigation, their chief use being for illustrative maps.

The appearance and features of conic projections are varied by using cones tangent at various parallels, using a secant (intersecting) cone, or by using a series of cones.

313. Simple conic projection.—A conic projection using a single tangent cone is called a **simple conic** projection (fig. 313a). The height of the cone increases as the latitude of the tangent parallel decreases. At the equator the height reaches infinity

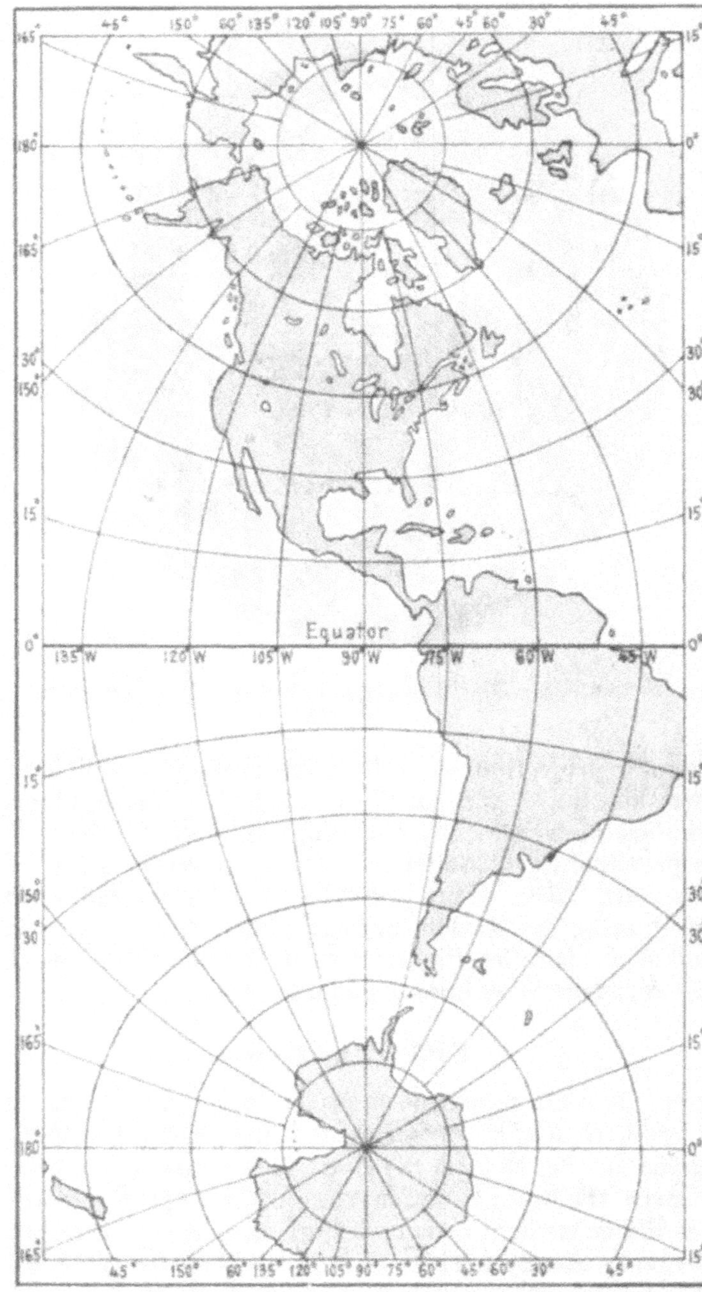

FIGURE 309.—A transverse Mercator map of the Western Hemisphere.

and the cone becomes a cylinder. At the pole its height is zero and it becomes a plane. As in the Mercator projection, the simple conic projection is not perspective, as only the meridians are projected geometrically, each becoming an element of the cone. When this is spread out flat to form a map, the meridians appear as straight lines converging at the apex of the cone. The standard parallel, or that at which the cone is tangent to the earth, appears as the arc of a circle with its center at the apex of the cone, or the common point of intersection of all the meridians. The other parallels are concentric circles, the distance along any meridian between

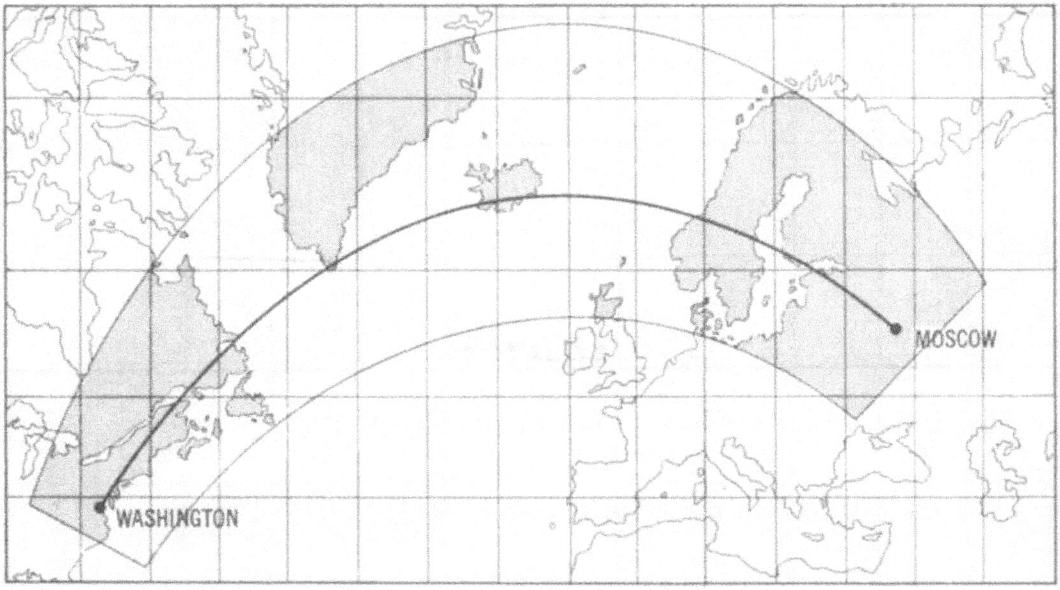

FIGURE 310a.—The great circle between Washington and Moscow as it appears on a Mercator map. See figures 308b and 310b.

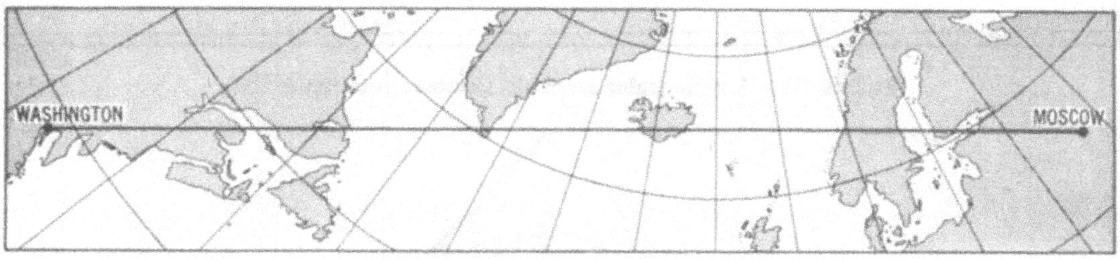

FIGURE 310b.—An oblique Mercator map based upon a cylinder tangent along the great circle through Washington and Moscow. The map includes an area 500 miles on each side of the great circle. The limits of this map are indicated on the Mercator map of figure 310a.

consecutive parallels being in correct relation to the distance on the earth, and hence derived mathematically. The pole is represented by a circle (fig. 313b). The scale is correct along any meridian and along the standard parallel. All other parallels are too great in length, the error increasing with increased distance from the standard parallel. Since the scale is not the same in all directions about every point, the projection is not conformal, its principal disadvantage for navigation. Neither is it an equal-area projection.

Since the scale is correct along the standard parallel and varies uniformly on each side, with comparatively little distortion near the standard parallel, this projection is useful for mapping an area covering a large spread of longitude and a comparatively narrow band of latitude. It was developed by Claudius Ptolemy in the second century after Christ to map just such an area, the Mediterranean.

314. Lambert conformal projection.—The useful latitude range of the simple conic projection can be increased by using a secant cone intersecting the earth at two standard parallels (fig. 314). The area between the two standard parallels is compressed, and that beyond is expanded. Such a projection is called a **secant conic** or **conic projection with two standard parallels.**

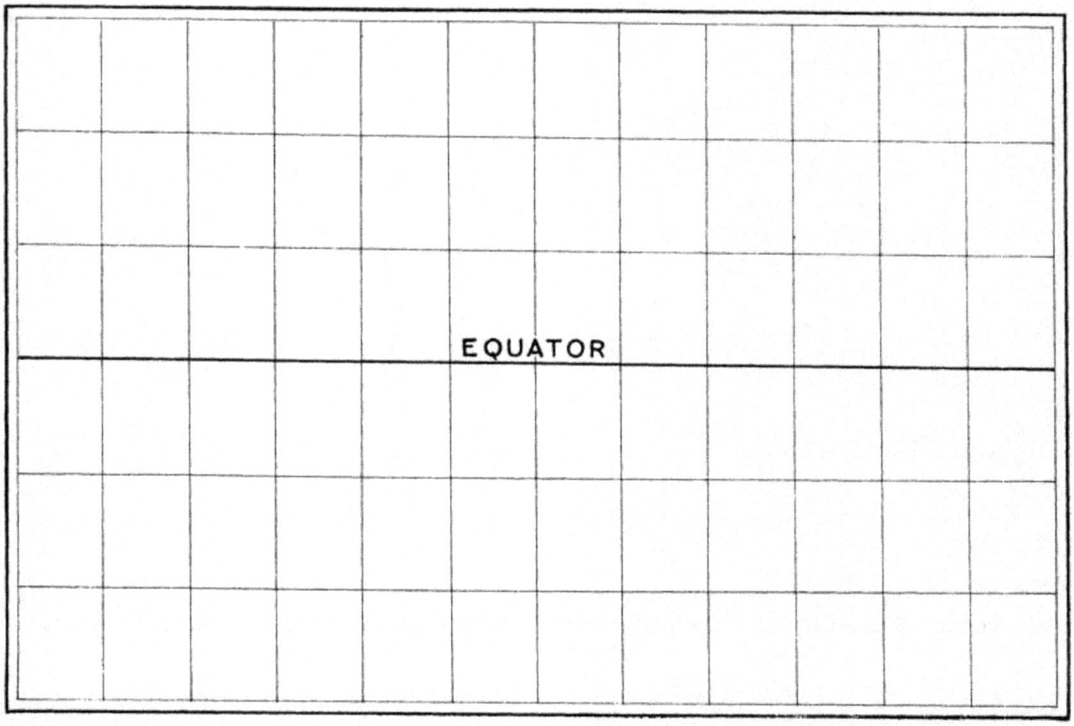

FIGURE 311.—A rectangular graticule. Compare with figure 305.

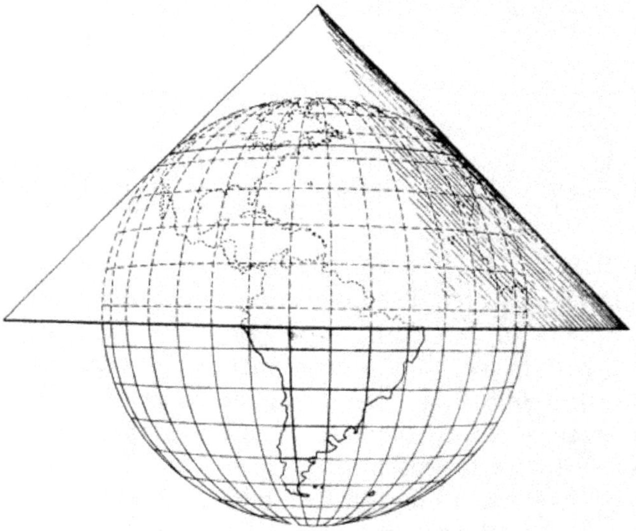

FIGURE 313a.—A simple conic projection.

If, in such a projection, the spacing of the parallels is altered so that the distortion is the same along them as along the meridians, the projection becomes conformal. This is known as the **Lambert conformal** projection, after its eighteenth century Alsatian inventor, Johann Heinrich Lambert. It is the most widely used conic projection for navigation, though its use is more common among aviators than mariners. Its appearance is very much the same as that of the simple conic projection. If the chart is not carried far beyond the standard parallels, and if these

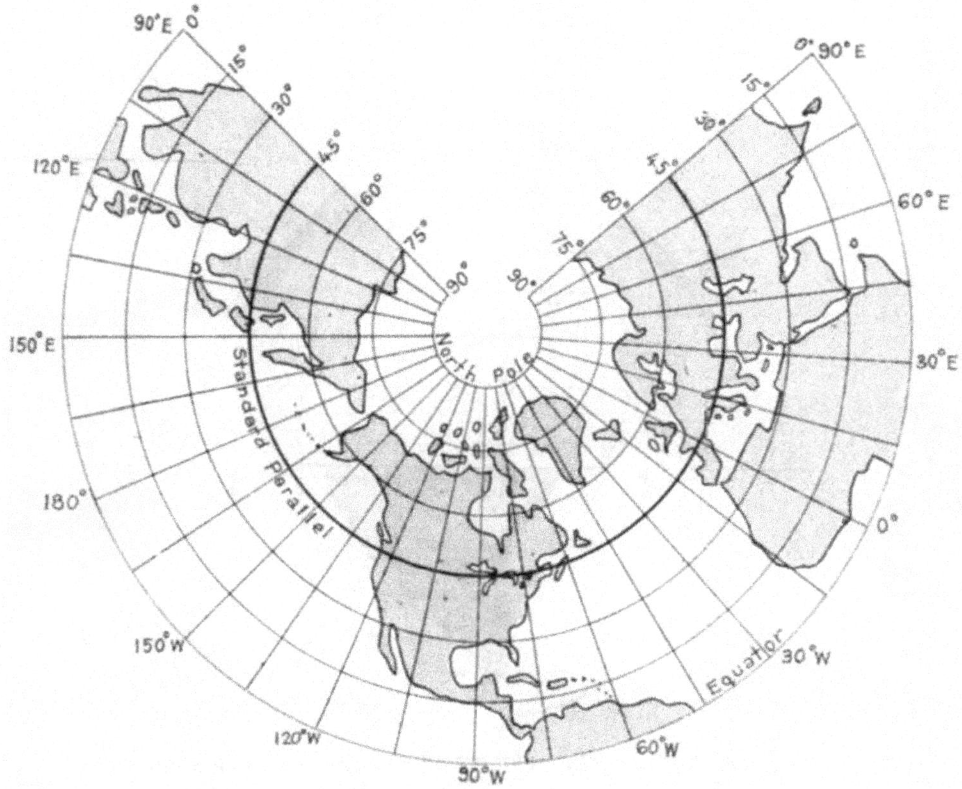

FIGURE 313b.—A simple conic map of the Northern Hemisphere.

are not a great distance apart, the distortion over the entire chart is small. A straight line on this projection so nearly approximates a great circle that the two can be considered identical for many purposes of navigation. Radio bearings, from signals which are considered to travel great circles, can be plotted on this projection without the correction needed when they are plotted on a Mercator chart. This feature, gained without sacrificing conformality, has made this projection popular for aeronautical charts, since aircraft make wide use of radio aids to navigation. It has made little progress in replacing the Mercator projection for marine navigation, except in high latitudes. In a slightly modified form this projection has been used for polar charts (art. 321).

315. Polyconic projection.—The latitude limitations of the secant conic projection can be essentially eliminated by the use of a series of cones, resulting in a **polyconic** projection. In this projection each parallel is the base of a tangent cone. At the edges of the chart the area between parallels is expanded to eliminate gaps. The scale is correct along any parallel and along the central meridian of the projection. Along other meridians the scale increases with increased difference of longitude from the central meridian. Parallels appear as nonconcentric circles and meridians as curved lines converging toward the pole and concave to the central meridian.

The polyconic projection is widely used in atlases, particularly for areas of large range in latitude and reasonably large range in longitude, as for a continent such as North America. However, since it is not conformal, this projection is not customarily used in navigation.

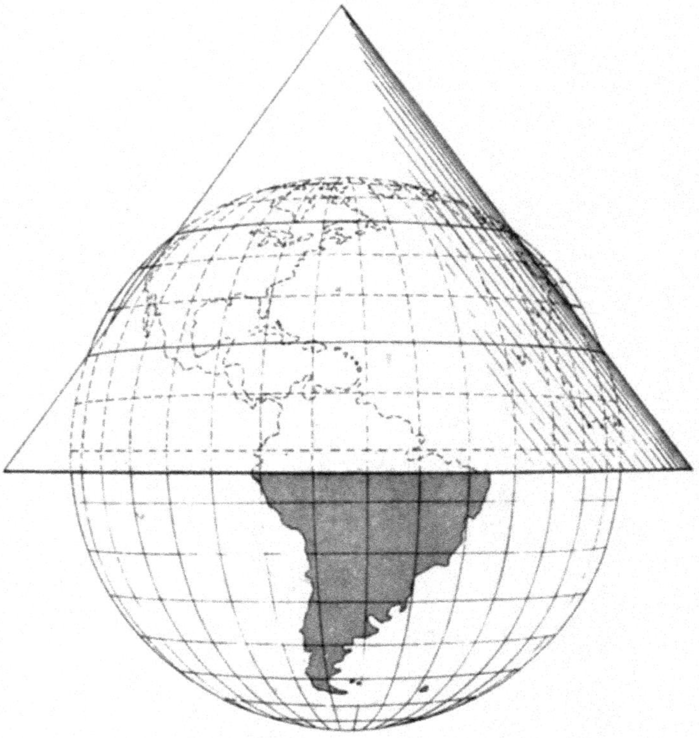

FIGURE 314.—A secant cone for a conic projection with two standard parallels.

Azimuthal Projections

316. Features.—If points on the earth are projected directly to a plane surface, a map is formed at once, without cutting and flattening, or "developing." This can be considered a special case of a conic projection in which the cone has zero height.

The simplest case of the azimuthal projection is one in which the plane is tangent at one of the poles. The meridians are straight lines intersecting at the pole, and the parallels are concentric circles with their common center at the pole. Their spacing depends upon the method of transferring points from the earth to the plane.

If the plane is tangent at some point other than a pole, straight lines through the point of tangency are great circles, and concentric circles with their common center at the point of tangency connect points of equal distance from that point. Distortion, which is zero at the point of tangency, increases along any great circle through this point. Along any circle whose center is the point of tangency, the distortion is constant. The bearing of any point from the point of tangency is correctly represented. It is for this reason that these projections are called **azimuthal.** They are also called **zenithal.** Several of the common azimuthal projections are perspective.

317. Gnomonic projection.—If a plane is tangent to the earth, and points are projected geometrically from the center of the earth, the result is a **gnomonic** projection (fig. 317a). This is probably the oldest of the projections, believed to have been developed by Thales about 600 BC. Since the projection is perspective, it can be demonstrated by placing a light at the center of a transparent terrestrial globe and holding a flat surface tangent to the sphere.

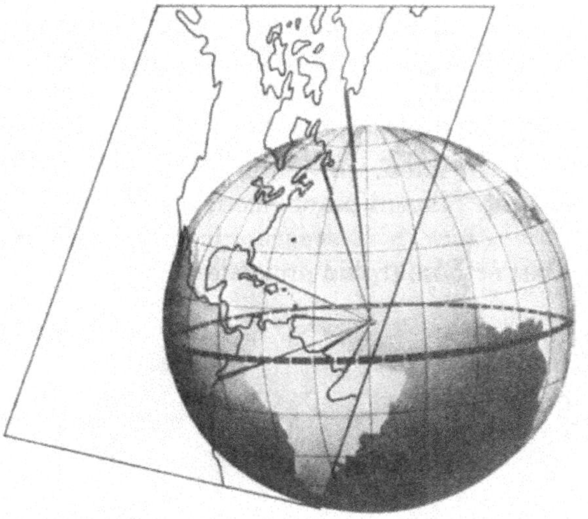

FIGURE 317a.—An oblique gnomonic projection.

For the oblique case the meridians appear as straight lines converging toward the nearer pole. The parallels, except the equator, appear as curves (fig. 317b). As in all azimuthal projections, bearings from the point of tangency are correctly represented. The distance scale, however, changes rapidly. The projection is neither conformal nor equal area. Distortion is so great that shapes, as well as distances and areas, are very poorly represented, except near the point of tangency.

FIGURE 317b.—An oblique gnomonic map with point of tangency at latitude 30°N, longitude 90°W.

The usefulness of the projection rests upon the one feature that *any* great circle appears on the map as a straight line. This is apparent when it is realized that a great circle is the line of intersection of a sphere and a plane through the center of the sphere, this center being the origin of the projecting rays for the map. This plane intersects any other nonparallel plane, including the tangent plane, in a straight line. It is this one useful feature that gives charts made on this projection the common name **great-circle charts.**

Gnomonic charts published by DMAHTC bear instructions for determining direction and distance on the charts. The principal navigational use of such charts is for plotting the great-circle track between points, for planning purposes. Points along the track are then transferred, by latitude and longitude, to the navigational chart, usually one on the Mercator projection. The great circle is then followed approximately by following the rhumb line from one point to the next (art. 903).

318. Stereographic projection.—If points on the surface of the earth are projected geometrically onto a tangent plane, from a point on the surface of the earth opposite the point of tangency, a **stereographic** projection results (fig. 318a). It is also called an **azimuthal orthomorphic** projection.

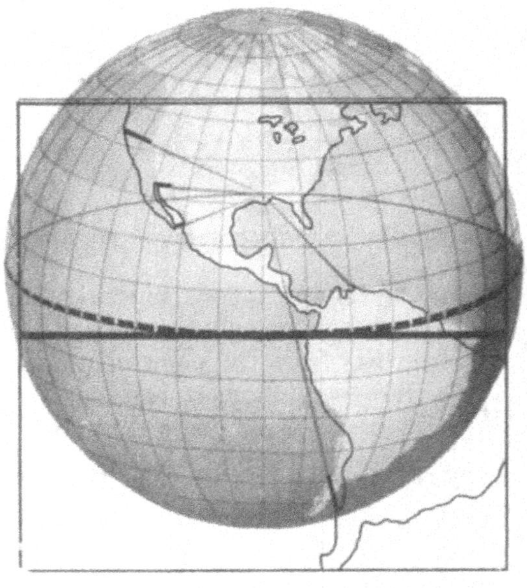

FIGURE 318a.—An equatorial stereographic projection.

The scale of the stereographic projection increases with distance from the point of tangency, but more slowly than in the gnomonic projection. An entire hemisphere can be shown on the stereographic projection without excessive distortion (fig. 318b). As in other azimuthal projections, great circles through the point of tangency appear as straight lines. All other circles, including meridians and parallels, appear as circles or arcs of circles.

The principal navigational use of the stereographic projection is for charts of the polar regions and devices for mechanical or graphical solution of the navigational triangle (art. 2122). A **Universal Polar Stereographic (UPS)** grid, mathematically adjusted to the graticule is used as a reference system.

319. Orthographic projection.—If terrestrial points are projected geometrically from infinity (projecting lines parallel) to a tangent plane, an **orthographic** projection results (fig. 319a). This projection is neither conformal nor equal area and has no advantages as a map projection. Its principal navigational use is in the field of navigational astronomy, where it is useful for illustrating or graphically solving the navigational triangle and for illustrating celestial coordinates. If the plane is tangent at a point on the equator, the usual case, the parallels (including the equator) appear as straight lines and the meridians as ellipses, except that the meridian through the point of tangency appears as a straight line and the one 90° away as a circle (fig. 319b).

FIGURE 318b.—A stereographic map of the Western Hemisphere.

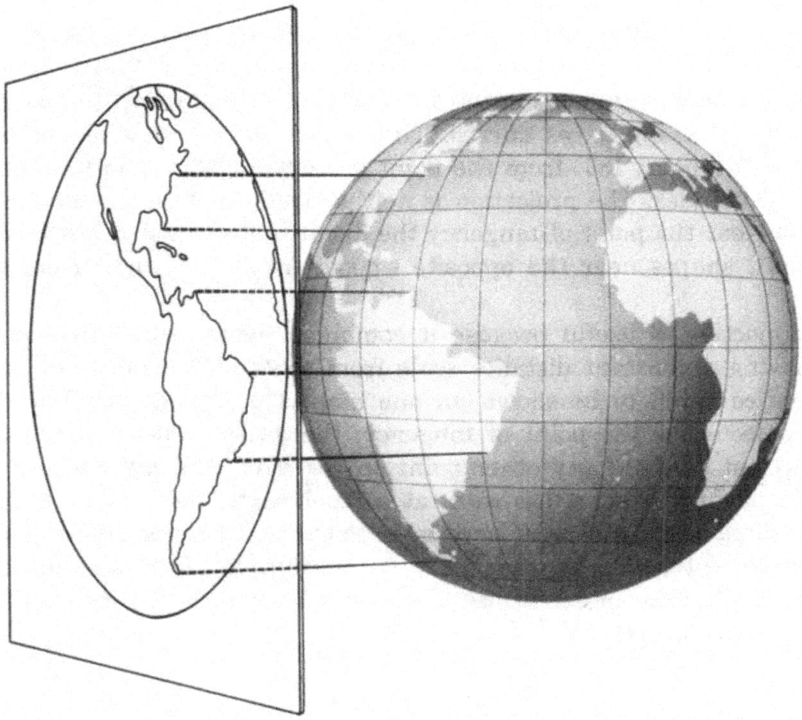

FIGURE 319a.—An equatorial orthographic projection.

320. Azimuthal equidistant projection.—An azimuthal projection in which the distance scale along any great circle through the point of tangency is constant is called an **azimuthal equidistant** projection. If a pole is the point of tangency, the

FIGURE 319b.—An orthographic map of the Western Hemisphere.

meridians appear as straight radial lines and the parallels as concentric circles, equally spaced. If the plane is tangent at some point other than a pole, the concentric circles represent distance from the point of tangency. In this case meridians and parallels appear as curves. The projection can be used to portray the entire earth, the point 180° from the point of tangency appearing as the largest of the concentric circles. The projection is neither conformal nor equal area, nor is it perspective. Near the point of tangency the distortion is small, but it increases with distance until shapes near the opposite side of the earth are unrecognizable (fig. 320).

The projection is useful because it combines the three features of being azimuthal, having a constant distance scale from the point of tangency, and permitting the entire earth to be shown on one map. Thus, if an important harbor or airport is selected as the point of tangency, the great-circle course, distance, and track from that point to any other point on the earth are quickly and accurately determined. For communication work at a fixed point, the point of tangency, the path of an incoming signal is at once apparent if the direction of arrival has been determined. The direction to train a directional antenna for desired results can be determined easily. The projection is also used for polar charts and for the familiar star finder and identifier, No. 2102-D (art. 2210).

Polar Charts

321. Polar projections.—Special consideration is given to the selection of projections for polar charts, principally because the familiar projections become special cases with unique features.

In the case of cylindrical projections in which the axis of the cylinder is parallel to the polar axis of the earth, distortion becomes excessive and the scale

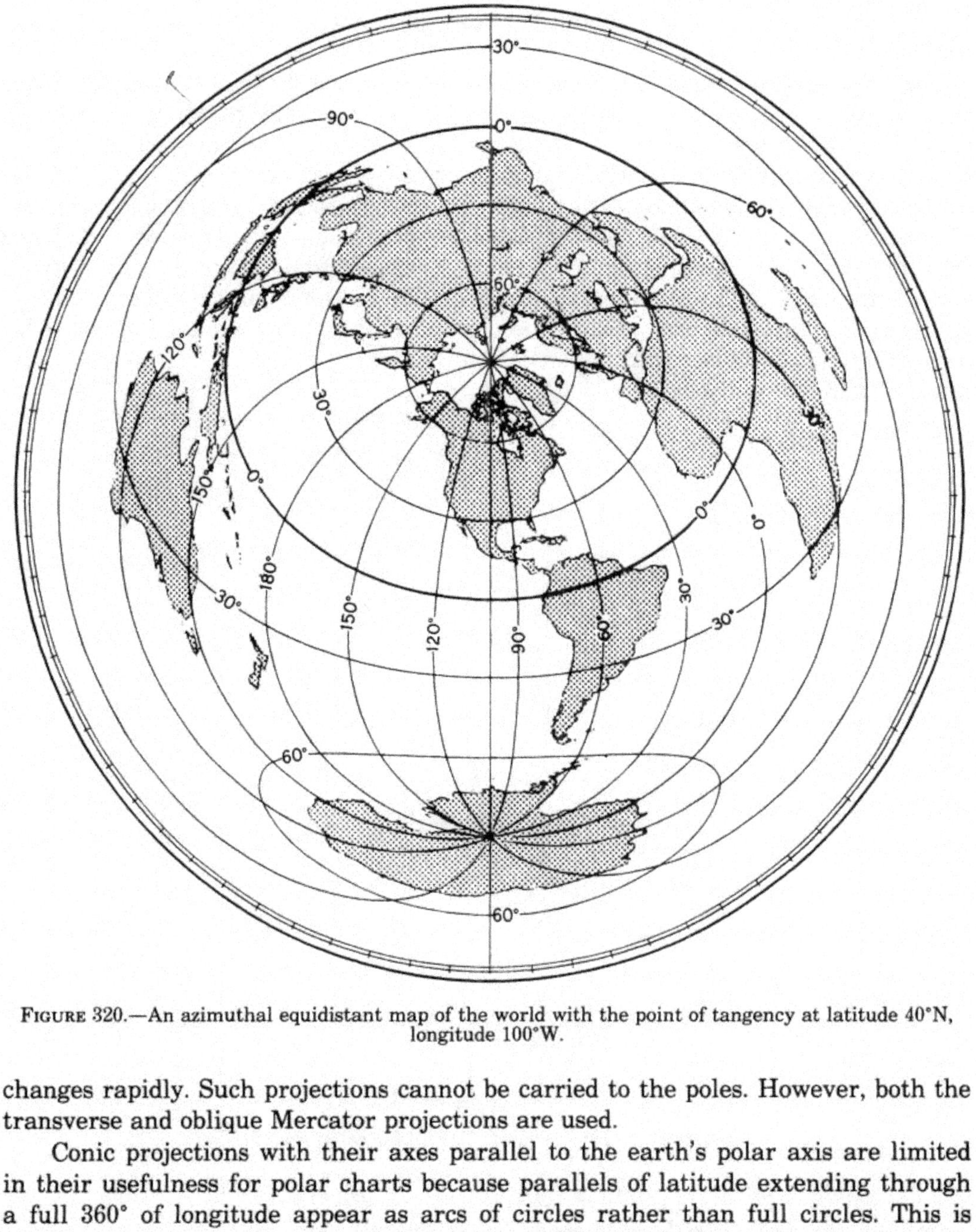

FIGURE 320.—An azimuthal equidistant map of the world with the point of tangency at latitude 40°N, longitude 100°W.

changes rapidly. Such projections cannot be carried to the poles. However, both the transverse and oblique Mercator projections are used.

Conic projections with their axes parallel to the earth's polar axis are limited in their usefulness for polar charts because parallels of latitude extending through a full 360° of longitude appear as arcs of circles rather than full circles. This is because a cone, when cut along an element and flattened, does not extend through a full 360° without stretching or resuming its former conical shape. The usefulness of such projections is also limited by the fact that the pole appears as an arc of a circle instead of a point. However, by using a parallel very near the pole as the higher standard parallel, a conic projection with two standard parallels can be made which requires little stretching to complete the circles of the parallels and eliminate that of the pole. Such a projection, called the **modified Lambert conformal** or **Ney's** projection, is useful for polar charts. It is particularly acceptable to those accustomed to using the ordinary Lambert conformal charts in lower latitudes.

Azimuthal projections are in their simplest form when tangent at a pole, since the meridians are straight lines intersecting at the pole, and parallels are concentric circles with their common center at the pole. Within a few degrees of latitude of the pole they all look essentially alike, but as the distance becomes greater, the spacing of the parallels becomes distinctive in each projection. In the polar azimuthal equidistant it is uniform; in the polar stereographic it increases with distance from the pole until the equator is shown at a distance from the pole equal to twice the length of the radius of the earth, or about 27% too much; in the polar gnomonic the increase is considerably greater, becoming infinity at the equator; in the polar orthographic it decreases with distance from the pole (fig. 321). All of these but the last are used for polar charts.

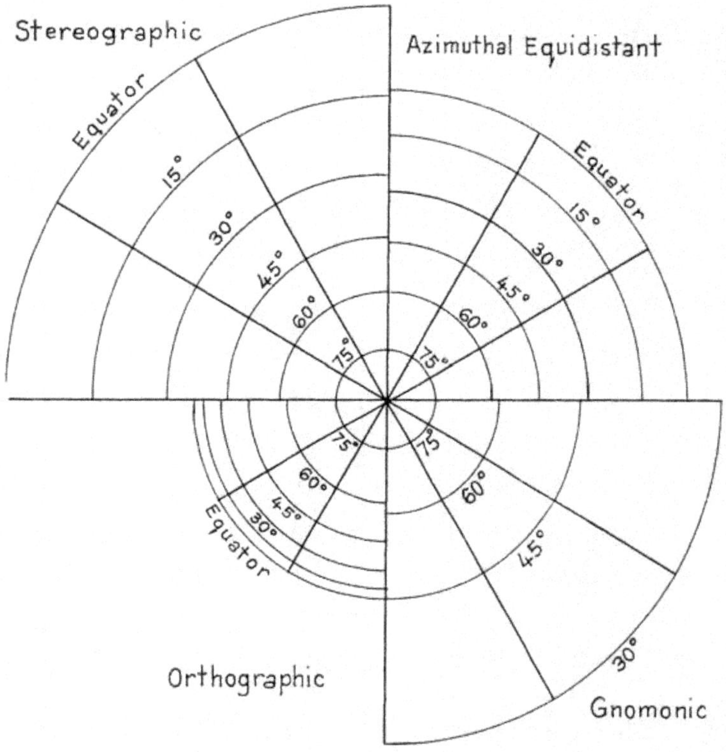

FIGURE 321.—Expansion of polar azimuthal projections.

322. Selection of a polar projection.—The principal considerations in the choice of a suitable projection for polar navigation are:

1. *Conformality.* It is desirable that angles be correctly represented so that plotting can be done directly on the chart, without annoying corrections.

2. *Great-circle representation.* Since great circles are more useful than rhumb lines in high latitudes, it is desirable that great circles be represented by straight lines.

3. *Scale variation.* Constant scale over an entire chart is desirable.

4. *Meridian representation.* Straight meridians are desirable for convenience and accuracy of plotting, and for grid navigation (art. 2510).

5. *Limits of utility.* Wide limits are desirable to reduce to a minimum the number of projections needed. The ideal would be a single projection for world coverage.

The projections commonly used for polar charts are the modified Lambert conformal, gnomonic, stereographic, and azimuthal equidistant. Near the pole there is little to choose between them. Within the limits of practical navigation all are essentially conformal and on all a great circle is nearly a straight line.

As the distance from the pole increases, however, the distinctive features of each projection become a consideration. The modified Lambert conformal projection is virtually conformal over its entire extent, and the amount of its scale distortion is comparatively little if it is carried only to about 25° or 30° from the pole. Beyond this, the distortion increases rapidly. A great circle is very nearly a straight line anywhere on the chart. Distances and directions can be measured directly on the chart in the same manner as on a Lambert conformal chart. However, for highly accurate work this projection is not suitable, for it is not strictly conformal, and great circles are not exactly straight lines.

The polar gnomonic projection is the one polar projection on which great circles are exactly straight lines. A complete hemisphere cannot be represented upon a plane, because the radius of 90° from the center would become infinity.

The polar stereographic projection (fig. 322) is conformal over its entire extent, and a great circle differs but little from a straight line. The scale distortion is not excessive for a considerable distance from the pole, but is greater than that of the modified Lambert conformal projection.

The polar azimuthal equidistant projection is useful for showing a large area such as a hemisphere, because there is no expansion along the meridians. However, the projection is not conformal, and distances cannot be measured accurately in any but a north-south direction. Great circles other than the meridians differ somewhat from straight lines. The equator is a circle centered at the pole.

The two projections most commonly used for charts for ordinary navigation near the poles are the modified Lambert conformal and the polar stereographic. When a directional gyro is used as a directional reference, the track of the craft is approximately a great circle. A desirable chart is one on which a great circle is represented as a straight line with a constant scale and with angles correctly represented. These requirements are not met entirely by any single projection, but they are approximated by both the modified Lambert conformal and the polar stereographic. The scale is more nearly constant on the former, but the projection is not strictly conformal. The polar stereographic is conformal, and its maximum scale variation can be reduced by using a plane which intersects the earth at some parallel intermediate between the pole and the lowest parallel, so that that portion within this standard parallel is compressed, and that portion outside is expanded.

The selection of a suitable projection for use in polar regions, as in other areas, depends upon the requirements, which establish relative importance of the various features. For a relatively small area, any of several projections is suitable. For a large area, however, the choice is more critical. If grid directions (art. 2510) are to be used, it is important that all units in related operations use charts on the same projection, with the same standard parallels, so that a single grid direction exists between any two points. Nuclear powered submarine operations under the polar icecap have increased the need for grid directions in marine navigation.

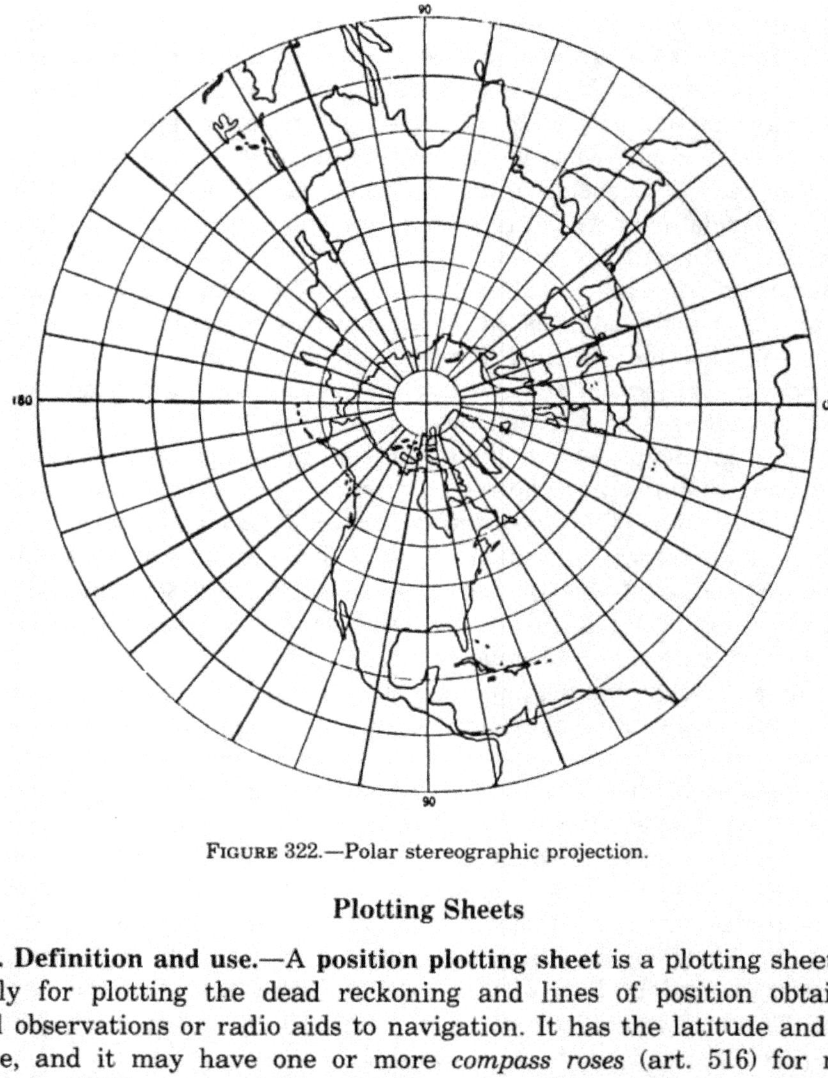

FIGURE 322.—Polar stereographic projection.

Plotting Sheets

323. Definition and use.—A **position plotting sheet** is a plotting sheet designed primarily for plotting the dead reckoning and lines of position obtained from celestial observations or radio aids to navigation. It has the latitude and longitude graticule, and it may have one or more *compass roses* (art. 516) for measuring direction, but little or no additional information. The meridians are usually unlabeled by the publisher so the plotting sheet can be used for any longitude.

Plotting sheets are less expensive to produce than charts and are equally suitable or superior for some purposes. They are used primarily when land, visual aids to navigation, and depth of water are not important.

Any projection can be used for constructing a plotting sheet, but that used for the navigator's charts is customarily employed also for his plotting sheets.

324. Small area plotting sheets.—A Mercator plotting sheet can be constructed by the method explained in article 307. For a relatively small area a good approximation can be more quickly constructed by the navigator by either of two alternative methods based upon a graphical solution of the secant of the latitude, which approximates the expansion.

First method (fig. 324a). *Step one.* Draw a series of equally spaced, vertical lines at any spacing desired. These are the meridians; label them at any desired interval, as 1′, 2′, 5′, 10′, 30′, 1°, etc.

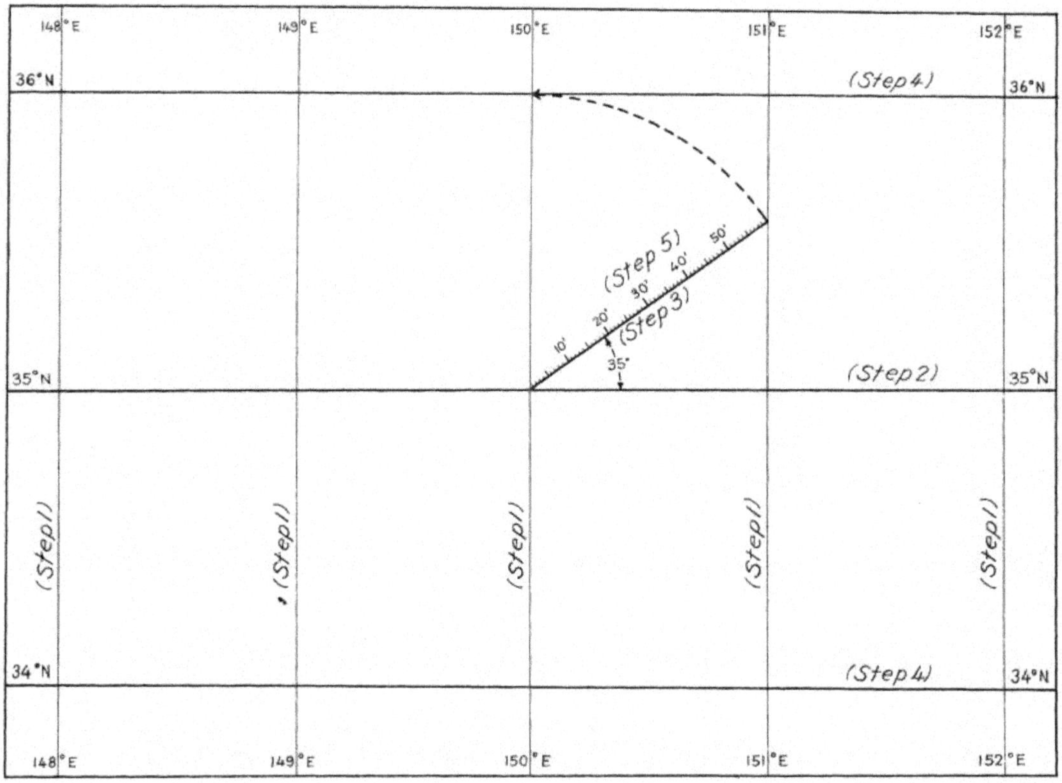

FIGURE 324a.—Small area plotting sheet with selected longitude scale.

Step two. Through the center of the sheet draw a horizontal line to represent the parallel of the mid-latitude of the area to be covered, and label it.

Step three. Through any convenient point, such as the intersection of the central meridian and the parallel of the mid-latitude, draw a line making an angle with the *horizontal* equal to the mid-latitude. In figure 324a this angle is 35°.

Step four. Draw in and label additional parallels. The length of the oblique line between consecutive meridians is the perpendicular distance between consecutive parallels, as shown by the broken arc. The number of minutes of arc between consecutive parallels thus drawn is the same as that between the meridians shown.

Step five. Graduate the oblique line into convenient units. If 1′ is selected, this scale serves as both a latitude and mile scale. It can also be used as a longitude scale by measuring horizontally from a meridian instead of obliquely along the line.

Second method (fig. 324b). *Step one.* At the center of the sheet draw a circle with a radius equal to 1° (or any other convenient unit) of latitude at the desired scale. If a sheet with a compass rose is available, as in figure 324b, the compass rose can be used as the circle and will prove useful for measuring directions. It need not limit the scale of the chart, as an additional concentric circle can be drawn and desired graduations extended to it.

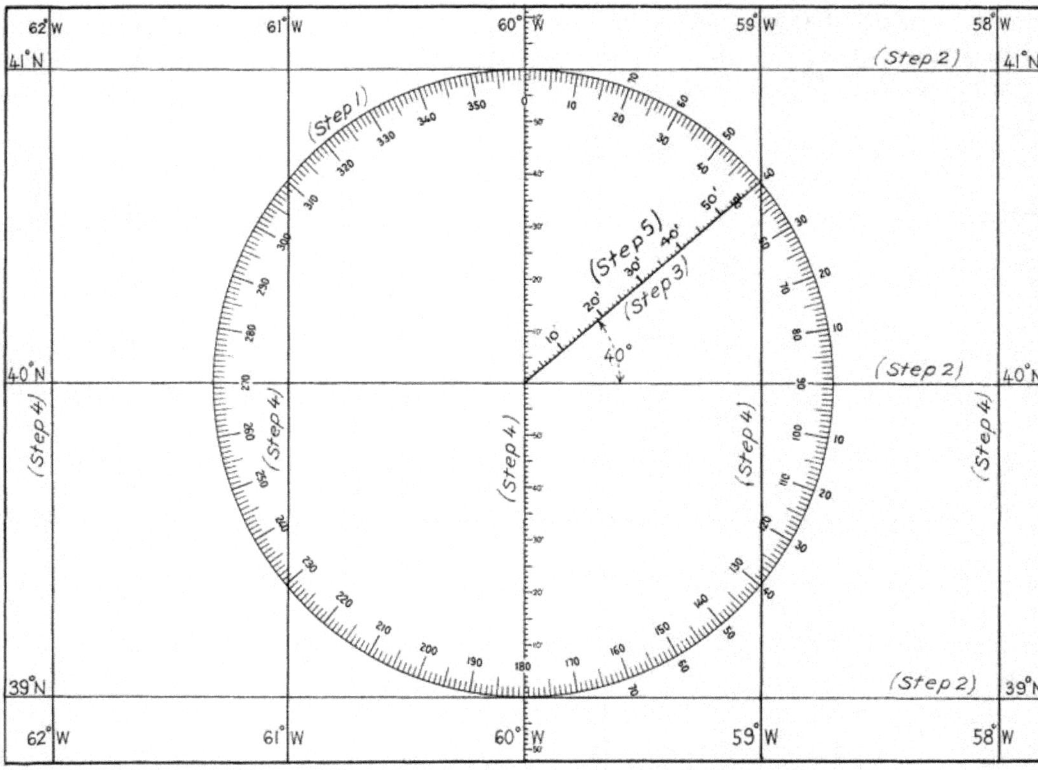

FIGURE 324b.—Small area plotting sheet with selected latitude scale.

Step two. Draw horizontal lines through the center of the circle and tangent at the top and bottom. These are parallels of latitude; label them accordingly, at the selected interval (as every 1°, 30′, etc.).

Step three. Through the center of the circle draw a line making an angle with the *horizontal* equal to the mid-latitude. In figure 324b this angle is 40°.

Step four. Draw in and label the meridians. The first is a vertical line through the center of the circle. The second is a vertical line through the intersection of the oblique line and the circle. Additional meridians are drawn the same distance apart as the first two.

Step five. Graduate the oblique line into convenient units. If 1′ is selected, this scale serves as a latitude and mile scale. It can also be used as a longitude scale by measuring horizontally from a meridian instead of obliquely along the line.

The same end result is produced by either method. The first method, starting with the selection of the longitude scale, is particularly useful when the longitude limits of the plotting sheet determine the scale. When the latitude coverage is more important, the second method may be preferable. If a standard size is desired, part of the sheet can be printed in advance, forming what is called a **universal plotting sheet.** This is done by the Defense Mapping Agency Hydrographic/Topographic Center. In either method a central compass rose might be printed. In the first method the meridians may be shown at the desired interval and the mid parallel may be printed and graduated in units of longitude. In using the sheet it is necessary only to label the meridians and draw the oblique line and from it determine the interval and draw in and label additional parallels. If the central meridian is graduated, the oblique line need not be. In the second method the parallels may be shown at the desired interval, and the central meridian may be printed and graduated in units of latitude. In using the sheet it is necessary only to label the parallels, draw the oblique line and from it determine the interval and draw in and label additional meridians. If the central meridian is graduated, as shown in figure 324b, the oblique line need not be.

Both methods use a constant relationship of latitude to longitude over the entire sheet and both fail to allow for the ellipticity of the earth. For practical navigation these are not important considerations for a small area. If a larger area is to be shown or if more precise results are desired, the method of article 307 should be used.

Grids

325. Purpose and definition of grid.—No system has been devised for showing the surface of the earth *on a flat surface,* without distortion. Moreover, the appearance of any portion of the surface varies with the projection and, in many cases, with the location of the portion with respect to the point or line of tangency. For some purposes (particularly military) it is desirable to be able to identify a location or area by rectangular coordinates, using numbers or letters, or a combination of numbers and letters, without the necessity of indicating the units used or assigning a name (north, south, east, or west), thus reducing the possibility of a mistake. This is accomplished by means of a **grid.** In its usual form this consists of two series of lines which are mutually perpendicular *on the chart,* with suitable designators.

326. Types of grids.—A grid may use the rectangular graticule of the Mercator projection, or a set of arbitrary lines on a particular projection. The most widely used system of the first is called the **World Geographic Referencing System (Georef).** It is merely a method of designating latitude and longitude by a system of letters and numbers instead of by angular measure, and therefore is not strictly a grid, except on a Mercator projection. It is particularly useful for operations extending over a wide area. Examples of the second type of grid are the **Universal Transverse Mercator (UTM) grid,** the **Universal Polar Stereographic (UPS) grid,** and the **Temporary Geographic Grid (TGG).** Since these systems are used primarily by military forces, they are sometimes called **military grids.**

References

Chamberlin, Wellman. *The Round Earth on Flat Paper.* Washington, National Geographic Society, 1947.

Deetz, C. H., and O. S. Adams. *Elements of Map Projection.* 5th ed. U. S. Coast and Geodetic Survey Special Publication No. 68. Washington, U. S. Govt. Print. Off., 1945.

Greenhood, David. *Down to Earth: Mapping for Everybody.* New York, Holiday House, 1944.

Hinks, A. R. *Map Projections.* 2nd ed. London, Cambridge University Press, 1921.

Jameson, A. H., and M. T. M. Ormsby. *Mathematical Geography.* Vol. I, *Elementary Surveying and Map Projection.* London, Pitman, [1942?].

Jervis, W. W. *The World in Maps.* New York, Oxford, 1937.

Mainwaring, James. *An Introduction to the Study of Map Projection.* London, Macmillan, 1942.

Raisz, Erwin. *General Cartography.* New York, McGraw-Hill, 1938.

Steers, J. A. *An Introduction to the Study of Map Projections.* Rev. ed. London, U. of London Press, 1929.

CHAPTER IV

VISUAL AND AUDIBLE AIDS TO NAVIGATION

401. Introduction.—The term **aid to navigation,** as used herein, means any device external to a vessel intended to be of assistance to a navigator in his determination of position or safe course, or to provide him with a warning to dangers or obstructions to navigation. This term includes lighthouses, beacons, sound signals, buoys, marine radiobeacons, racons, and the medium and long range radionavigation systems. The discussion of the various aids to navigation in this chapter is limited to the visual and audible aids established in the navigable waters of the United States and its possessions.

Aids to navigation are placed at various points along the coast and navigable waterways as markers and guides to mark safe water and to provide navigators with means to determine their position with relation to the land and to hidden dangers. Within the bounds of actual necessity, each aid is designed to be seen or heard so that it provides the necessary system coverage to enable safe transit of a waterway.

As all aids to navigation serve the same general purpose, structural differences are solely for the purpose of meeting the conditions and requirements of the particular location at which the aid is established.

The maintenance of marine aids to navigation is a function of the United States Coast Guard. This responsibility includes the maintenance of lighthouses, radiobeacons, racons, Loran, sound signals, buoys, and beacons upon all navigable waters of the United States and its possessions, including the Atlantic and Pacific coasts of the continental United States, the Great Lakes, the Mississippi River and its tributaries, Puerto Rico, the U.S. Virgin Islands, the Hawaiian Islands, Alaska, Trust Territory of the Pacific Islands, and such other places where aids to navigation are required to serve the needs of the armed forces.

Lights on Fixed Structures

402. Lights on fixed structures vary from the tallest lighthouse on the coast, flashing with an intensity of millions of candlepower, to a simple battery-powered lantern on a wooden pile in a small creek. Being in fixed positions enabling accurate charting, lights provide navigators with reliable means to determine their positions with relation to land and hidden dangers during daylight and darkness. The structures are often distinctively colored to facilitate their observation during daylight.

A **major light** is a light of high intensity and reliability exhibited from a fixed structure or on a marine site (except range lights). Major lights include primary seacoast lights and secondary lights. **Primary seacoast lights** are those major lights established for the purpose of making landfalls and coastwise passages from headland to headland. **Secondary lights** are those major lights, other than primary seacoast lights, established at harbor entrances and other locations where high intensity and reliability are required. Major lights are usually located at manned or monitored automated stations.

A **minor light** is an automatic unmanned (unwatched) light on a fixed structure showing usually low to moderate intensity. Minor lights are established in harbors, along channels, rivers, and isolated locations. They usually have the same numbering, coloring, and light and sound characteristics as the lateral system of buoyage (art. 411).

Lighthouses (fig. 402a), all of which exhibit major lights, are placed where they will be of most use: on prominent headlands, at entrances, on isolated dangers, or at other points where it is necessary that mariners be warned or guided. Their principal purpose is to support a light at a considerable height above the water. In many instances, sound signals, radiobeacon equipment, and operating personnel are housed in separate buildings located near the tower. Such a group of facilities is called a **light station.**

Many of the lighthouses which were originally tended by resident keepers are now operated automatically. There are also many automatic lights on smaller structures maintained through periodic visits of Coast Guard cutters or of attendants in charge of a group of such aids. The introduction of new automatic apparatus means that the relative importance of lights cannot be judged on the basis of whether or not they have resident keepers.

Offshore light stations and large navigational buoys (art. 407) have replaced lightships. The offshore light stations in U.S. waters, such as the one shown in figure 402b have helicopter landing surfaces. In the 1983 *Light List,* the CHESAPEAKE LIGHT station is described as a blue tower on a white square superstructure on four black piles. "CHESAPEAKE" is on sides; the piles are floodlighted sunset to sunrise.

Range lights (fig. 402c) are pairs of lights so located as to form a range in line with the center of channels or entrance to a harbor. The rear light is higher than the front light and a considerable distance in back of it, thus enabling the mariner to use the range by keeping the lights in line as he progresses up the channel. Range lights are sometimes used during daylight hours through the use of high intensity lights. Otherwise, the range light structures are equipped with daymarks (art. 412) for ordinary daytime use.

Range lights are usually white, red, or green, and display various characteristics to differentiate them from surrounding lights.

A **directional light** is a single light which projects a beam of high intensity, separate color, or special characteristic in a given direction. It has limited use for those cases where a two-light range may not be practicable or necessary, and for special applications. The directional light is essentially a narrow sector light with or without adjacent sectors which give information as to the direction of and relative displacement from the narrow sector.

Aeronautical lights, which are lights of high intensity, may be the first lights observed at night from vessels approaching the coast. Those situated near the coast are accordingly listed in the *List of Lights* (art. 1306) in order that the navigator may be able to obtain more complete information concerning their description. These lights are not listed in the U. S. Coast Guard *Light List.*

Aeronautical lights are placed in geographic sequence in the body of the text of the *List of Lights* along with lights for marine navigation. It should be borne in mind, however, that these lights are not designed or maintained for marine navigation, and that they are *subject to changes of which neither lighthouse authorities nor the marine navigator may receive prompt notification.*

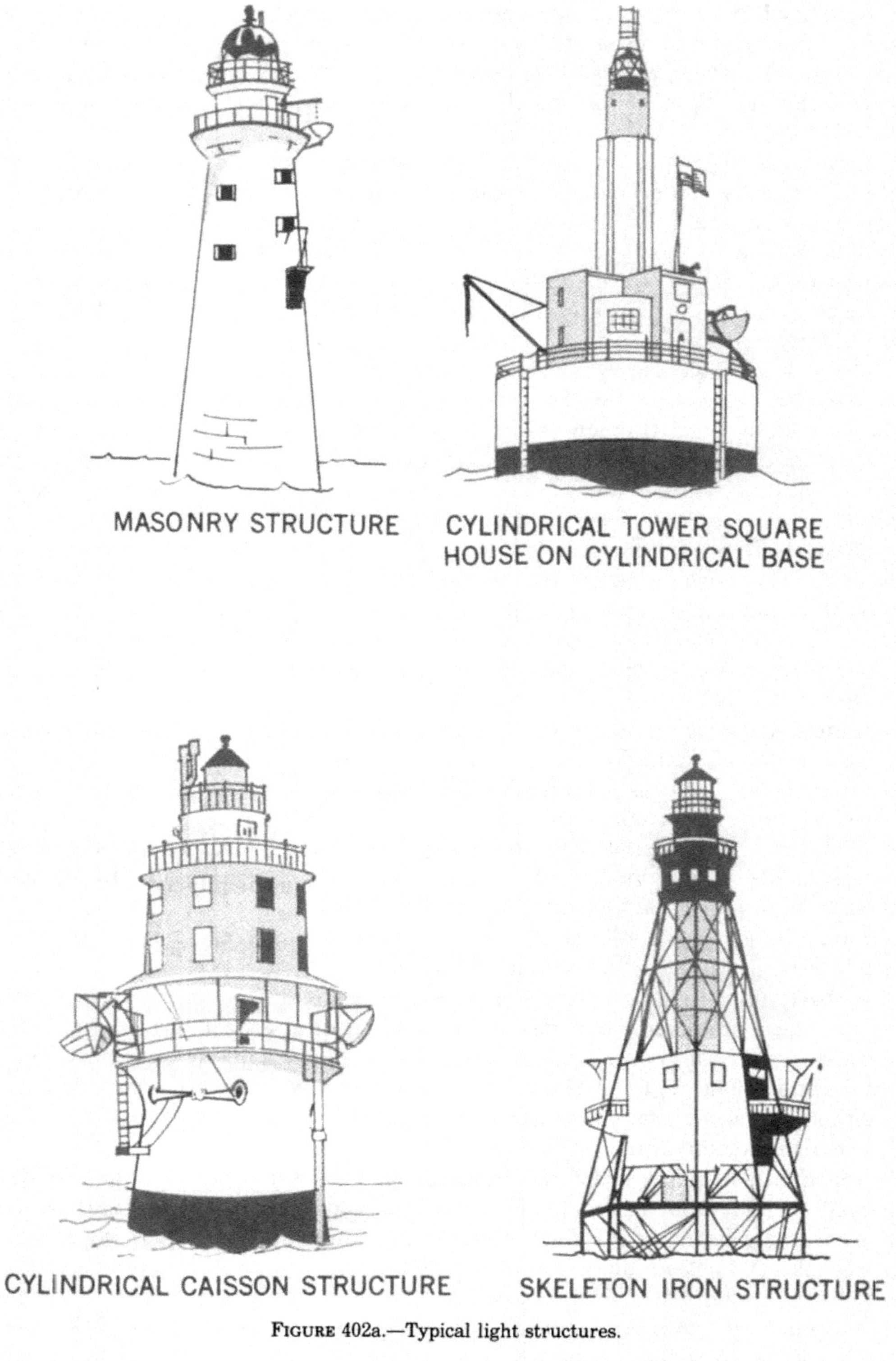

FIGURE 402a.—Typical light structures.

Bridges across navigable waters of the United States are generally marked with red, green, and white lights for nighttime navigation. Red lights mark piers and other parts of the bridge. Red lights are also used on drawbridges to show when

FIGURE 402b.—Typical offshore light station.

they are in the closed position. Green lights are used to mark the centerline of navigable channels through fixed bridges. The preferred channel, if there are two or more channels through the bridge, is marked by three white lights in a vertical line above the green light.

Green lights are also used on drawbridges to show when they are in the open position. Because of the variety of drawbridges, the position of the green lights on the bridge will vary according to the type of structure. Navigational lights on bridges are prescribed by regulation.

Bridges infrequently used may be unlighted. In unusual cases the type and method of lighting may be different than normally found.

FIGURE 402c.—Range lights.

Drawbridges required to be operated for passage of vessels operate upon sound and light signals given by the vessel and acknowledged by the bridge. These signals are prescribed by regulation.

In addition to lighting, certain bridges may be equipped with sound signals and radar reflectors where unusual geographic or weather conditions require them.

Light Characteristics

403. Characteristics.—Lights are given distinctive characteristics so that one light may be distinguished from another navigational light or from the general background of shore lights or as a means of conveying certain definite information. This distinctiveness may be obtained by giving each light a distinctive sequence of light and dark intervals, having lights that burn steadily and others that flash or occult, or by giving each light a distinctive color, or color sequence. In the light lists, the dark intervals are referred to as **eclipses.** An **occulting light** is a light totally eclipsed at regular intervals, the duration of light always being greater than the duration of darkness. A **flashing light** is a light which flashes at regular intervals, the duration of light always being less than the duration of darkness. An **equal interval light** is a light which flashes at regular intervals, the duration of light always being equal to the duration of darkness. This light is also called an **isophase light.**

404. Light phase characteristics (figs. 404a and 404b) are the distinctive sequences of light and dark intervals or distinctive sequences in the variations of the luminous intensity of a light. The light phase characteristics of lights which change color do not differ from those of lights which do not change color. A light showing different colors alternately is described as an **alternating light.** The alternating characteristic may be used with other light phase characteristics.

A *Light List* entry for an **alternating fixed and flashing light** may be given as:

Al. F.W., F.R. and Fl. W.
90^s (F.W., 59^s, F.R., 14^s,
Fl. W. 3^s (high intensity),
F.R. 14^s).

CHARACTERS OF LIGHTS

CLASS	ABBREVIATION	GENERAL DESCRIPTION	ILLUSTRATION
Fixed light	F	A continuous and steady light.	
Occulting light	Oc, Occ	The total duration of light in a period is longer than the total duration of darkness and the intervals of darkness (eclipses) are usually of equal duration. Eclipse regularly repeated.	
Group - occulting light	Oc (2), Gp Occ (2)	An occulting light in which a group of eclipses, specified in number, is regularly repeated.	
Composite group - occulting light	Oc (2 + 1), Gp Occ (2 + 1)	A light similar to a group-occulting light except that successive groups in a period have different numbers of eclipses.	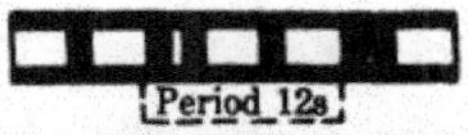
Isophase light	Iso	A light in which all the durations of light and darkness are clearly equal.	
Flashing light	Fl	A light in which the total duration of light in a period is shorter than the total duration of darkness and the appearances of light (flashes) are usually of equal duration (at a rate of less than 50 flashes per minute).	
Long - flashing light	L Fl	A single-flashing light in which an appearance of light of not less than 2 sec. duration (long flash) is regularly repeated.	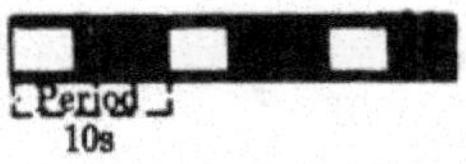
Group - flashing light	Fl (3), Gp Fl (3)	A flashing light in which a group of flashes, specified in number, is regularly repeated.	
Composite group - flashing light	Fl (2 + 1), Gp Fl (2 + 1)	A light similar to a group-flashing light except that successive groups in a period have different numbers of flashes.	
Quick light	Q, Qk Fl	A light in which a flash is regularly repeated. Flashes are repeated at a rate of not less than 50 flashes per minute but less than 80 flashes per minute.	
Group quick light	Q (3), Gp Fl (3)	A light in which a specified group of flashes is regularly repeated. Flashes are repeated at a rate of not less than 50 flashes per minute but less than 80 flashes per minute.	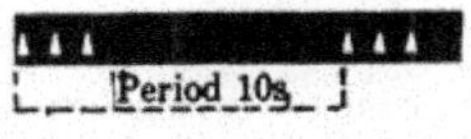
	Q (9), Gp Fl (9)		
	Q (6) + LFl, Gp Fl (6)		

FIGURE 404a.—Light phase characteristics.

CHARACTERS OF LIGHTS

CLASS	ABBREVIATION	GENERAL DESCRIPTION	ILLUSTRATION
Interrupted quick light	IQ, Int Qk Fl	A light in which the sequence of quick flashes is interrupted by regularly repeated eclipses of constant and long duration.	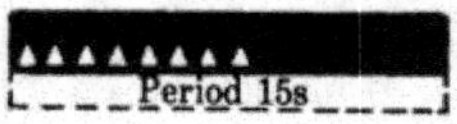
Continuous very quick light	VQ, Qk Fl	A very quick light in which a flash is regularly repeated. Flashes are repeated at a rate of not less than 80 flashes per minute but less than 160 flashes per minute.	
Group very quick light	VQ (3), Gp Fl (3)	A very quick light in which a specified group of flashes is regularly repeated.	
	VQ (9), Gp Fl (9)		
	VQ (6) + LFl, Gp Fl 6 + LFl		
Interrupted very quick light	IVQ, Int Qk Fl	A light in which the sequence of quick flashes is interrupted by regularly repeated eclipses of constant and long duration.	
Continuous ultra quick light	UQ, Qk Fl	An ultra quick light in which a flash is regularly repeated. Flashes are repeated at a rate of not less than 160 flashes per minute.	
Interrupted ultra quick light	IUQ	An ultra quick light in which the sequence of flashes is interrupted by eclipses of long duration.	
Morse code light	Mo (U)	A light in which appearances of light of two clearly different durations are grouped to represent a character or characters in the Morse Code.	
Fixed and flashing light	FFl	A light in which a fixed light is combined with a flashing light of higher luminous intensity.	
Alternating light	Al, Alt	A light showing different colors alternately.	

NOTES: 1. Alternating lights may be used in combined form with most of the previous classes of lights.

2. The second abbreviation shown for a light, if any, is alternate U.S. usage and is being superceded as new editions of charts are produced.

FIGURE 404b.—Light phase characteristics.

With each 90^s period the light is first fixed white for 59^s, then fixed red for 14^s, then there is a flash of white for 3^s, and finally the light is fixed red for 14^s.

A *Light List* entry for a **group flashing light** may be given as:

Fl. (2) W., 15^s
0.2^s fl., 3.0^s ec.
0.2^s fl., 11.6^s ec.
(2 flashes).

Within each 15^s period, there is first a white flash of $0^s.2$ duration, the light is eclipsed (extinguished) for 3^s, then there is a white flash of $0^s.2$ duration, and then the light is eclipsed for $11^s.6$ before the sequence begins again.

A *Light List* entry for a **composite group flashing light** may be given as:

Fl. (1+2) W., 15^s
0.2^s fl., 5.8^s ec.
0.2^s fl., 2.8^s ec.
0.2^s fl., 5.8^s ec.
(3 flashes).

Within each 15^s period, there is first a white flash of $0^s.2$ duration, the light is eclipsed for $5^s.8$, then there is a white flash of $0^s.2$ duration, the light is eclipsed for $2^s.8$, and then there is a $0^s.2$ duration white flash followed by a $5^s.8$ eclipse. Thus, the first group consists of a single flash; the second group consists of two flashes. This is indicated by the (1+2) notation.

Most lighted aids to navigation are automatically extinguished during daylight hours by switches activated by daylight. These switches are not of equal sensitivity. Therefore, all lights do not come on or go off at the same time. Mariners should take this fact into account when identifying aids to navigation during twilight periods when some lighted aids are on while others are not.

405. Sectors of colored glass or plastic are placed in the lanterns of certain lights to mark shoals or to warn mariners off the nearby land. Lights so equipped show one color from most directions and a different color or colors over definite arcs of the horizon as indicated in the light lists and upon the charts. A sector changes the color of a light, when viewed from certain directions, but not the characteristic. For example, a flashing white light having a red sector, when viewed from within the sector, will appear flashing red.

Sectors may be but a few degrees in width, marking an isolated rock or shoal, or of such width as to extend from the direction of the deep water toward shore. Bearings referring to sectors are expressed in degrees as observed from a vessel toward the light.

In the majority of cases, water areas covered by red sectors should be avoided, the exact extent of the danger being determined from an examination of the charts. In some cases a narrow sector may mark the best water across a shoal. A narrow sector may also mark a turning point in a channel.

The transition from one color to the other is not abrupt, but changes through an arc of uncertainty of about 2° or less, which depends upon the optical design of the components of the lighting apparatus.

406. Factors affecting visual range and apparent characteristics.—The condition of the atmosphere has a considerable effect upon the distance at which lights can be seen. Sometimes lights are obscured by fog, haze, dust, smoke, or precipitation which may be present at the light or between it and the observer, but not at the observer and possibly unknown to him. On the other hand, refraction may often cause a light to be seen farther than under ordinary circumstances. A light of low intensity will be easily obscured by unfavorable conditions of the atmosphere and less dependence can be placed on its being seen. For this reason, the intensity of a light should always be considered when expecting to sight it in thick weather. Haze and distance may reduce the apparent duration of the flash of a flashing light. In some conditions of the atmosphere white lights may have a reddish hue. In clear weather green lights may have a whitish hue.

It should be remembered that lights placed at great elevations are more frequently obscured by clouds, mist, and fog than those near sea level.

In regions where ice conditions prevail in the winter, the lantern panes of unattended lights may become covered with ice or snow, which will greatly reduce the luminous ranges of the lights and may also cause lights to appear of different color.

The increasing use of brilliant shore lights for advertising, illuminating bridges, and other purposes, may cause navigational lights, particularly those in densely inhabited areas, to be outshone and difficult to distinguish from the background lighting. Mariners are requested by the U. S. Coast Guard to report such cases as outlined above in order that steps may be taken to attempt to improve the conditions.

The "loom" of a powerful light is often seen beyond the geographic range of the light. The loom may sometimes appear sufficiently sharp to obtain a bearing.

At short distances, some of the *brighter* flashing lights may show a faint continuous light between flashes.

It should be borne in mind that, when attempting to sight a light at night, the geographic range is considerably increased from aloft. By noting a star immediately over the light an accurate compass bearing may be indirectly obtained on the light from the navigating bridge although the light is not yet visible from that level.

The distance of an observer from a light cannot be estimated by its apparent intensity. *Always check the characteristics of lights* in order that powerful lights visible in the distance shall not be mistaken for nearby lights showing similar characteristics at lower intensity (such as those on lighted buoys).

If lights are not sighted within a reasonable time after prediction, a dangerous situation may exist requiring prompt resolution or action to insure the safety of the vessel.

The apparent characteristic of a complex light may change with the distance of the observer. For example, a light which actually displays a characteristic of fixed white varied by flashes of alternating white and red (the phases having a decreasing range of detection in the order: flashing white, flashing red, fixed white) may, when first sighted in clear weather, show as a simple flashing white light. As the vessel draws nearer, the red flash will become visible and the characteristic will apparently be alternating flashing white and red. Later, the fixed white light will be seen between the flashes and the true characteristic of the light finally recognized—fixed white, alternating flashing white and red (F.W.Alt.Fl. W. and R.).

There is always the possibility of a light being extinguished. In the case of unattended lights, this condition might not be immediately detected and corrected. The mariner should immediately report this condition. During periods of armed conflict, certain lights may be deliberately extinguished without notice if the situation warrants such action.

Large Navigational Buoys

407. Large navigational buoys and offshore light stations have replaced lightships. These 40-foot diameter buoys (fig. 407) may show secondary lights (art. 402) from heights of about 36 feet above the water. In addition to the light, these buoys may mount a radiobeacon and provide sound signals. A station buoy may be moored nearby.

FIGURE 407.—Large navigational buoy.

Station buoys, often called **watch buoys,** are sometimes moored near navigational buoys to mark the approximate station should the buoy be carried away or temporarily removed. Since these buoys are always unlighted and, in some cases, moored as much as a mile from the navigational buoys, the danger of a closely passing vessel colliding with them is always present—particularly so during darkness or periods of reduced visibility.

Experience shows that offshore light stations cannot be safely used as leading marks to be passed close aboard, but should always be left broad off the course, whenever searoom permits.

Buoyage and Beaconage

408. Buoys are used to delineate channels, indicate shoals, mark obstructions, and warn the mariner of dangers where the use of fixed aids for such purposes would be uneconomical or impracticable. By their color, shape, number, and light or sound characteristics, buoys provide indications to the mariner as to how he may avoid navigational hazards.

There are many different sizes and types of buoys to meet the wide range of environmental conditions and user requirements. The principal types of buoys used by the United States are lighted, lighted sound, unlighted sound, and unlighted. Some examples of these types are illustrated in Chart No. 1.

A **lighted buoy** consists of a floating hull with a tower on which a lantern is mounted. Batteries to power the light are contained in special pockets in the buoy hull. To keep the buoy in an upright stable position a large counterweight (fig. 408a) sometimes is extended from a tube attached to the base of the hull below the water surface. The radar reflector (art. 4207), on those buoys so equipped, forms a part of the buoy tower.

Lighted sound buoys have the same general configuration as lighted buoys but are equipped with either a gong, bell, whistle, or electronic horn. Bells and gongs on buoys are sounded by tappers that hang from the tower and swing as the buoys roll in the sea. Bell buoys produce sound of only one tone; gong buoys produce several tones.

Whistle buoys make a loud moaning sound caused by the rising and falling motions of the buoy in the sea. A sound buoy equipped with an electronic horn will produce a pure tone at regular intervals and will operate continually regardless of the sea state.

Unlighted sound buoys have the same general appearance as lighted buoys (except for old whistle buoys) but are not equipped with any light apparatus.

Unlighted buoys have either a can or nun shape. **Can buoys** have a cylindrical shape whereas **nun buoys** have a conical shape usually located on top of a cylindrical shape. Since these buoys are unlighted there is no requirement for battery pockets, and the hull of the buoy forms part of the shape.

Buoys are *floating aids* and therefore require moorings to hold them in position. Typically the mooring consists of chain and a large concrete sinker (fig. 408b). Because buoys are subjected to waves, wind, tides, and other conditions, the moorings must be deployed in lengths greater than the water depth. The scope of chain can be as much as 5 times the depth of water or more but normally will be about 3 times the water depth. For this reason the buoy can be expected to swing in a circle as the current, wind, and wave conditions change.

409. Fallibility of buoys.—It is *imprudent* for a navigator to rely on floating aids to navigation to always maintain their charted positions and to constantly and unerringly display their advertised characteristics.

The buoy symbol shown on charts indicates the approximate position of the buoy body and the sinker which secures the buoy to the seabed. The approximate position is used because of practical limitations in placing and keeping buoys and

FIGURE 408a.—Buoy showing counterweight.

FIGURE 408b.—Sinkers used to anchor buoys.

their sinkers in exact geographical locations. These limitations include, but are not limited to, inherent inaccuracies in position fixing methods, prevailing atmospheric and sea conditions, the slope of and the material making up the seabed, the fact that buoys are moored to sinkers with more chain than the water depth, and the fact that the positions of the buoys and the sinkers are not under continuous surveillance but are normally checked only during periodic maintenance visits which often occur more than a year apart. The position of the buoy can be expected to shift inside and outside the area shown by the chart symbol due to the forces of nature. The mariner is also cautioned that buoys are liable to be missing, shifted, overturned, etc. Lighted buoys may be extinguished or sound signals may not function because of ice, running ice, natural causes, collisions, or other accidents.

For these reasons, a prudent mariner must not rely completely upon the position or operation of buoys, but will also navigate using bearings of charted features, structures, and aids to navigation on shore. Further, a vessel attempting to pass too close always risks a collision with a yawing buoy or with the obstruction which the buoy marks.

The concept that a wreck buoy always occupies a position directly over the wreck it is intended to mark is erroneous. Buoys must be placed in position by a vessel. It is usually physically impossible for these vessels to maneuver directly over a wreck to place the sinker without incurring serious underwater damage. For this reason, a wreck buoy is usually placed on the seaward or channelward side of a wreck, the proximity thereto being governed by existing conditions. To avoid confusion in some situations, two buoys may be used to mark the wreck. Both may not be located on the seaward or channelward side of the wreck, but the wreck will lie between them. Obviously, the mariner should not attempt to pass between buoys so placed.

Sunken wrecks are not always static. They are sometimes moved away from their buoys by severe sea conditions or other causes. Just as shoals may shift away from the buoys placed to mark them, wrecks may shift away from wreck buoys.

All buoys should, therefore, be regarded as warnings, guides, or aids but not as infallible navigation marks, especially those located in exposed positions. Whenever possible, a mariner should navigate by bearings or angles of reliable and identifiable fixed charted features or landmarks and by soundings rather than by sole reliance on buoys.

410. Buoyage Systems.—The International Association of Lighthouse Authorities (IALA) Maritime Buoyage System (combined cardinal and lateral) is being implemented by nearly every maritime jurisdiction in the world as either Region A buoyage (red to port) or Region B buoyage (red to starboard). For a description of Region A and Region B buoyage systems and an illustration showing an outline of the regions see appendix Y.

Conversion in Region A began in 1977 and will continue until 1987 and possibly later. Conversion in Region B, which includes North, Central, and South America, Japan, South Korea and the Philippines has begun (1983) and it can be anticipated that several years will be required to complete its transformation.

In 1982, the United States agreed to make *modifications* to incorporate that IALA Maritime Buoyage System for Region B. The Modified U.S. Aid System is described below. The U.S. lateral system is not materially changed.

In the lateral system, used on all navigable waters of the United States, the coloring, shape, numbering, and lighting of buoys indicate the direction to a danger relative to the course which should be followed. The color, shape, lights, and numbers of buoys in the lateral system as used by the United States are determined relative to a direction *from* seaward. Along the *coasts* of the United States, the *clockwise* direction around the country is arbitrarily considered to be the direction "from seaward." Proceeding in a westerly and northerly direction on the Great Lakes (except Lake Michigan), and in a southerly direction on Lake Michigan, is proceeding "from seaward." On the Intracoastal Waterway proceeding in a general southerly direction along the Atlantic coast, and in a general westerly direction along the Gulf coast, is considered as proceeding "from seaward." On the Mississippi and Ohio Rivers and their tributaries the aids to navigation characteristics are determined as proceeding from sea toward the head of navigation although local terminology describes "left bank" and "right bank" as proceeding with the flow of the river. Some countries using the lateral system have methods of coloring their buoys and lights opposite to that of the United States. Appendix Y treats this subject in greater detail.

Modified U.S. Aid System

The description that follows is that of the new Modified U.S. Aids to Navigation System. Significant changes include: black lateral buoys to be made green; red and black horizontally banded preferred channel buoys to be made red and green horizontally banded; white lights to be replaced by lateral colors (red or green) on lateral aids; black and white vertically striped mid-channel aids to be made red and white vertically striped; special purpose and other non-lateral aids to be made yellow, those which are lighted are to be fitted with yellow lights. *Within the text the old characteristic will be described in parenthesis.* Until the conversion program is completed mariners should be familiar with both systems and alert to the fact that changes may not be immediately reflected in published charts. All changes in aids to navigation will be published in the U.S. Coast Guard's *Local Notice to Mariners* and, where appropriate, the DMAHTC weekly *Notice to Mariners.* Illustrations of the Modified U.S. Aids to Navigation System is shown in appendix Y.

Colors. When proceeding from seaward:

(a) Green(black) buoys mark the port side of channels, or the location of wrecks or obstructions which must be passed by keeping the buoy on the left hand.

(b) Red buoys mark the starboard side of channels, or the location of wrecks or obstructions which must be passed by keeping the buoy on the right hand.

(c) Red and green(black) horizontally banded preferred channel buoys mark junctions or bifurcations in the channel, or wrecks or obstructions which may be passed on either side. If the topmost band is green(black), the preferred channel will be followed by keeping the buoy on the port hand. If the topmost band is red, the preferred channel will be followed by keeping the buoy on the starboard hand.

(d) Red(black) and white vertically striped safe water buoys mark the fairway or mid-channel.

Shapes. In order to provide ready identification certain unlighted buoys are differentiated by shape.

(a) Red buoys, or red and green(black) horizontally banded buoys with the topmost band red are conical shaped and called nun buoys.

(b) Green(black) buoys, or green(black) and red horizontally banded buoys with the topmost band black are cylindrical shaped and called can buoys.

(c) Red(black) and white vertically striped buoys are spherical shaped buoys. Lighted buoys, sound buoys, and spar buoys are not differentiated by shape to indicate the side on which they should be passed. No special significance is attached to the shape of these buoys, their purpose being indicated only by the color, number, or light characteristics. However, safe water buoys will eventually be fitted with a spherical red and white vertical striped topmark to aid in the buoy identification.

Numbers. (a) All solid colored buoys are numbered, the red buoys bearing even numbers and the green(black) buoys bearing odd numbers, the numbers for each increasing from numbers where required.

(b) No other color buoys are numbered; however, any color buoys may be lettered for the purpose of identification.

Light colors. Red(or white) lights on buoys are used only on red buoys or red and green(black) horizontally banded buoys with the topmost band red. Green(or white) lights on buoys are used only on the green(black) buoys or green(black) and red horizontally banded buoys with the topmost band green(black). White lights may be used only on "safe water" aids which show a Morse characteristic.

Light rhythmics. (See figs. 404a and 404b). (a) Lights on red buoys or green(black) buoys, if not occulting or isophase, will generally be regularly flashing. For ordinary purposes, the frequency of flashes will be not more than 50 flashes per minute (single flashing). For purposes when it is desired that lights have a distinct cautionary significance, as at sharp turns or sudden constrictions in the channel, or to mark wrecks or dangerous obstructions, the frequency of flashes will be not less than 50 flashes but not more than 80 flashes per minute (continuous quick).

(b) Lights on preferred channel buoys will show a series of grouped flashes, specified in number with successive groups in a period having different number of flashes—composite group flashing (or a quick light in which the sequence of flashes is interrupted by regularly repeated eclipses of constant and long duration—interrupted quick flashing).

(c) Lights on safe water buoys will always show a white Morse Code "A" (Short-Long) flash, this combination recurring at the rate of about eight times per minute.

Daylight controls. Lighted buoys are equipped with a special device which automatically controls the electric current to the light. This device causes the light to operate during hours of darkness and to be extinguished during daylight hours. These devices are not of equal sensitivity, therefore all lights do not come on or go off at the same time. (Mariners should ensure correct identification of aids during twilight periods when some lighted aids to navigation are on while others are not.)

Reflective material is placed on buoys to assist in their detection at night by use of a searchlight. The color of the reflective material agrees with the buoy color. Green reflective material is placed only on green(black) buoys, red reflective material is placed only on red buoys. Red or green reflective material will be placed on preferred channel (junction) buoys; red if topmost band is red or green if topmost band is green(black). White reflective material on buoys with lateral significance is limited to use on safe water (fairway) buoys. Special purpose buoys (which have no lateral significance) display yellow(white) reflective material. Warning or regulatory buoys display orange reflective horizontal bands and warning symbol, and intracoastal waterway display a yellow reflective horizontal strip in addition to red or green reflective material coincidental to the buoy color.

Special purpose buoys. (a) Buoys for special purposes are colored yellow (white buoys mark anchorage areas; white buoys with green tops are used in connection with dredging and survey operations; white and black horizontally banded buoys mark fish net areas). White and international orange buoys either horizontally banded or vertically striped, are for warning or regulatory purposes to which neither the lateral system nor the other special purpose colors apply.

(b) The shape of special purpose buoys has no significance. They are not numbered, but may be lettered. They display a yellow(amber or white) color light with fixed or slow flash characteristics preferred.

Buoys marking wrecks. Buoys established by the Coast Guard to mark wrecks are generally placed on the seaward or channel side of the wreck and as near to the wreck as conditions will permit. Caution must be exercised when navigating in the vicinity of buoys marking wrecks because, due to sea action, the wreck may shift in location between times that the buoy is established and later checked and serviced.

Station buoys. Buoys are sometimes placed in close proximity to a floating aid to mark the station in case the regular aid is accidentally shifted from station. Station buoys are colored and numbered the same as the regular aid to navigation.

Certain aids to navigation may be fitted with, or have incorporated in their design, radar reflectors designed to enhance their ability to reflect radar energy. In general, these reflectors will materially improve the aids for use by vessels equipped with radar.

411. Beacons are fixed aids to navigation placed on shore or on marine sites. If unlighted, the beacon is referred to as a **daybeacon.** A daybeacon is identified by its color and the color, shape, and number of its daymark. The simplest form of daybeacon consists of a single pile with a daymark affixed at or near its top (fig. 411).

Daybeacons may be used instead of range lights (art. 402) to form a **range** (art. 1004).

Daymarks serve to make aids to navigation readily visible and easily identifiable against daylight viewing backgrounds. For example, the distinctive color pattern and shape of a lighthouse aid identification during the daytime as does the color and shape of a buoy. The size of the daymark that is required to make the aid conspicuous depends upon how far the aid must be seen. On those structures which do not by themselves present an adequate viewing area to be seen at the required distance, the aid is made more visible by affixing a daymark to the structure. These daymarks have a distinctive shape and color depending upon the purpose of the aid.

FIGURE 411.—Daybeacon.

Most daymarks also display numbers or letters so that the daymark can be more readily identified as a particular aid. The numbers and letters, as well as portions of most daymarks (and portions of unlighted buoys) are made to be retro-reflective to enhance their illumination by the mariner.

Increasing amounts of information are conveyed by a daymark as the mariner approaches. At the detection distance, the daymark will convey only the information of its existence; it will be just detectable from its background. At the recognition distance, the daymark can be recognized as an aid to navigation. At this distance the distinctive shape or color pattern is recognizable. At the identification distance, when the number or letter can be read, the daymark can be identified as a particular aid.

The detection, recognition, and identification distances vary widely for any particular daymark depending upon the viewing conditions. This is an inherent limitation of any visual signal but is especially true for passive visual signals which utilize the sun as the source for their signal energy. The reflectivity of the daymark surface varies with the angle of the sun relative to the daymark. This causes the luminance of the daymark to vary. The detection, recognition, and identification distances depend upon the relative difference between the luminance of the daymark and that of the background, the position of the sun relative to the observer, and the meteorological visibility.

Beginning in 1975, a revised system of daymarks was gradually being implemented in the United States. The significant changes include the following:

1. On port side daymarks, green is used in lieu of the colors black or white; green numbers and letters are used.

2. On starboard side daymarks, red numbers and letters are used in lieu of white numbers and letters.

3. On ICW daymarks (art. 414), a yellow horizontal reflective strip is used in lieu of a yellow reflective border as the marking.

4. On junction daymarks, green is used in lieu of black in the color pattern.

Sound Signals

412. Sound signals.—Most lighthouses, light platforms, and some minor light structures and buoys are equipped with sound-producing instruments to aid the mariner in periods of low visibility.

Charts and light lists of the particular area should be consulted for positive identification. Caution: buoys fitted with a bell, gong, or whistle and actuated by wave motion may produce no sound when the sea is calm. Their positive identification is not always possible.

Any sound-producing instrument operated in time of fog from a definite point shown on the charts, such as a lighthouse, or buoy, serves as a useful fog signal. To be effective as an aid to navigation, a mariner must be able to identify it and to know from what point it is sounded.

At all lighthouses equipped with sound signals, these signals are operated by mechanical or electrical means and are sounded during periods of low visibility, providing the desirable feature of positive identification.

The characteristics of mechanized signals are varied blasts and silent periods. A definite time is required for each signal to perform a complete cycle of changes. Where the number of blasts and the total time for a signal to complete a cycle is not sufficient for positive identification, reference may be made to details in the *Light List* regarding the exact length of each blast and silent interval. The various types of sound signals also differ in tone, and this facilitates recognition of the respective stations.

Diaphones produce sound by means of a slotted piston moved back and forth by compressed air. Blasts may consist of two tones of different pitch, in which case the first part of the blast is high and the last of a low pitch. These alternate-pitch signals are called "two-tone."

Diaphragm horns produce sound by means of a disc diaphragm vibrated by compressed air or electricity. Duplex or triplex horn units of differing pitch produce a chime signal.

Sirens produce sound by means of either a disc or a cup-shaped rotor actuated by compressed air, steam, or electricity.

Whistles produce sound by compressed air emitted through a circumferential slot into a cylindrical bell chamber.

Bells are sounded by means of a hammer actuated by a descending weight, compressed gas or electricity.

413. Limitations of sound signals.—Sound signals depend upon the transmission of sound through air. As aids to navigation, they have limitations that should be considered. Sound travels through the air in a variable and frequently unpredictable manner.

It has been clearly established that:

1. Sound signals are heard at greatly varying distances and that the distance at which a sound signal can be heard may vary with the bearing of the signal and may be different on occasion.

2. Under certain conditions of the atmosphere, when a sound signal has a combination high and low tone, it is not unusual for one of the tones to be inaudible. In the case of sirens, which produce a varying tone, portions of the blast may not be heard.

3. There are occasionally areas close to the signal in which it is wholly inaudi-

ble. This is particularly true when the sound signal is screened by intervening land or other obstruction, or the signal is on a high cliff.

4. A fog may exist a short distance from a station and not be observable from it, so that the signal may not be in operation.

5. Some sound signals cannot be started at a moment's notice.

6. Even though a sound signal may not be heard from the deck or bridge of a ship when the engines are in motion, it may be heard when the ship is stopped, or from a quiet position. Sometimes it may be heard from aloft though not on deck.

7. The loudness of the sound emitted by a sound signal may be greater at a distance than in the immediate proximity.

All these considerations point to the necessity for the utmost caution when navigating near land in a fog. Mariners are therefore warned that sound signals can never be implicitly relied upon, and that the practice of taking soundings of the depth of water should never be neglected. Particular attention should be given to placing lookouts in positions in which the noises in the ship are least likely to interfere with hearing a sound signal. Sound signals are valuable as warnings but the mariner should not place implicit reliance upon them in navigating his vessel. They should be considered solely as warning devices.

Emergency sound signals are sounded at some of the light and fog signal stations when the main and stand-by sound signal is inoperative. Some of these emergency sound signals are of a different type and characteristic than the main sound signal. The characteristics of the emergency sound signals are listed in the *Light List.*

The mariner must not assume:

1. That he is out of ordinary hearing distance because he fails to hear the sound signal.

2. That, because he hears a sound signal faintly, he is at a great distance from it.

3. That he is near to it because he hears the sound plainly.

4. That the distance from and the intensity of a sound on any one occasion is a guide to him for any future occasion.

5. That the sound signal is not sounding because he does not hear it, even when in close proximity.

6. That the sound signal is in the direction the sound appears to come from.

414. Intracoastal Waterway aids to navigation.—The Intracoastal Waterway (ICW) runs parallel to the Atlantic and gulf coasts from Manasquan Inlet on the New Jersey shore to the Mexican border. Aids marking these waters have some portion of them marked with yellow as shown in Chart No. 1. Otherwise, the coloring and numbering of buoys and beacons follow the same system as that in other U. S. waterways.

In order that vessels may readily follow the Intracoastal Waterway route where it coincides with another marked waterway such as an important river, special markings are employed. These special markings are applied to the buoys or other aids which already mark the river or waterway for other traffic. These aids are then referred to as "Dual Purpose" aids. The marks consist of a yellow square or a yellow triangle, placed on a conspicuous part of the dual purpose aid. The yellow square, in outline similar to a can buoy, indicates that the aid on which it is placed should be kept on the left hand when following the Intracoastal Waterway down the coast. The yellow triangle has the same meaning as a nun; it should be kept on the right side. Where such dual purpose marking is employed, the mariner following the Intracoastal Waterway disregards the color and shape of the aid on which the mark is placed, being guided solely by the shape of the yellow mark.

415. Mississippi River system.—Aids to navigation on the Mississippi River and its tributaries in the Second Coast Guard District and parts of the Eighth Coast Guard District generally conform to the lateral system of buoyage. The following differences are significant:

1. Buoys are not numbered.

2. The numbers on lights and daybeacons do not have lateral significance; they indicate the mileage from a designated point downstream, normally the river mouth.

3. Flashing lights on the left side proceeding upstream show single green or white flashes while those on the right side show double (group flashing) red or white flashes.

4. "Crossing daymarks" are used to indicate where the channel crosses from one side of the river to the other.

416. The Uniform State Waterway Marking System (USWMS) was developed jointly by the U. S. Coast Guard and state boating administrators to assist the small craft operator in those state waters marked by participating states. The USWMS consists of two categories of aids to navigation. One is a system of aids to navigation, generally compatible with the Federal lateral system of buoyage, to supplement the federal system in state waters. The other is a system of regulatory markers to warn the small craft operator of dangers or to provide general information and directions.

On a well-defined channel, including a river or other relatively narrow, natural or improved waterway, solid colored red and black buoys are established in pairs (called "gates"), one on each side of the navigable channel which they mark, and opposite to each other to inform the user that the channel lies between the buoys and that he should pass between the buoys. The buoy which marks the left side of the channel viewed looking upstream or toward the head of navigation is colored all black; the buoy which marks the right side of the channel is colored all red.

On an irregularly-defined channel, solid colored buoys may be staggered on alternate sides of the channel but spaced at sufficiently close intervals to inform the user that the channel lies between the buoys and that he should pass between the buoys.

When there is no well-defined channel or when a body of water is obstructed by objects whose nature or location is such that the obstruction can be approached by a vessel from more than one direction, aids to navigation having cardinal meaning may be used. The aids conforming to the cardinal system consist of three distinctly colored buoys:

1. A white buoy with a red top is used to indicate to a vessel operator that he must pass to the south or west of the buoy.

2. A white buoy with a black top is used to indicate to a vessel operator that he must pass to the north or east of the buoy.

3. A buoy showing alternate vertical red and white stripes is used to indicate to a vessel operator that an obstruction to navigation extends from the nearest shore to the buoy and that he must not pass between the buoy and the nearest shore.

The shape of buoys has no significance in the USWMS.

Regulatory buoys are colored white with international orange horizontal bands completely around the buoy circumference. One band is at the top of the buoy with a second band just above the waterline of the buoy so that both orange bands are clearly visible from approaching vessels.

Geometric shapes are placed on the white portion of the buoy body and are colored international orange. The authorized geometric shapes and meanings associated with them are as follows:

1. A vertical open faced diamond shape means danger.

2. A vertical open faced diamond shape having a cross centered in the diamond means that vessels are excluded from the marked area.

3. A circular shape means that vessels in the marked area are subject to certain operating restrictions.

4. A square or rectangular shape indicates that directions or information is contained inside.

Regulatory markers consist of square and rectangular shaped signs displayed from a fixed structure. Each sign is white with an international orange border. Geometric shapes with the same meanings as those displayed on buoys are centered on the sign boards. The geometric shape displayed on a regulatory marker is intended to convey specific meaning to a vessel operator—whether or not he should stay well clear of the marker or may safely approach the marker in order to read any wording on the marker.

417. Private aids to navigation are those aids not established and maintained by the U. S. Coast Guard. Private aids include those established by other federal agencies with prior U. S. Coast Guard approval, those aids to navigation on marine structures or other works which the owners are legally obligated to establish, maintain, and operate as prescribed by the U. S. Coast Guard, and those aids which are merely desired, for one reason or another, by the individual, corporation, state or local government, or other body that has established the aid with U. S. Coast Guard approval.

Before any private aid to navigation consisting of a fixed structure is placed in the navigable waters of the United States, authorization to erect such structure shall first be obtained from the District Engineer, U. S. Army Corps of Engineers, in whose district the aid will be located.

Private aids to navigation are similar to the aids established and maintained by the U. S. Coast Guard, but are specially designated on the chart and *Light List*.

Although private aids to navigation are inspected periodically by the U. S. Coast Guard, the mariner should exercise special caution when using them for general navigation.

418. Protection by law.—All aids to navigation, including private aids, are protected by law. The *Code of Federal Regulations* (33 CFR 70) refers.

It is unlawful to take possession of or make use of for any purpose, or build upon, alter, deface, destroy, move, injure, obstruct by fastening vessels thereto or otherwise, or in any manner whatever impair the usefulness of any aid to navigation established and maintained by the United States.

Whenever any vessel collides with an aid to navigation established and maintained by the United States or any private aid established or maintained in accordance with 33 CFR 64, 66, 67, or 68, or is connected with any such collision, it shall be the duty of the person in charge of such vessel to report the accident to the nearest Officer in Charge, Office of Marine Inspection, U. S. Coast Guard.

CHAPTER V

THE NAUTICAL CHART

General Information

501. Introduction.—A nautical chart is a conventional graphic representation, on a plane surface, of a navigable portion of the surface of the earth. It shows the depth of water by numerous soundings, and sometimes by soundings and depth contours, the shoreline of adjacent land, topographic features that may serve as landmarks, aids to navigation, dangers, and other information of interest to navigators. It is designed as a work sheet on which courses may be plotted, and positions ascertained. It assists the navigator in avoiding dangers and arriving safely at his destination. The nautical chart is one of the most essential and reliable aids available to the navigator.

502. Projections.—Nearly all nautical charts used for ordinary purposes of navigation are constructed on the Mercator projection (art. 305). Large-scale harbor charts are sometimes constructed on the transverse Mercator projection. Charts for special purposes, such as great-circle sailing or polar navigation, are on appropriate projections; great-circle sailing charts are usually on the gnomonic projection (art. 317); polar charts are often on the polar stereographic projection (art. 322). The principal projections, with their navigational uses, are discussed in chapter III.

503. Scale.—The *scale* of a chart is the ratio of a given distance on the chart to the actual distance which it represents on the earth. It may be expressed in various ways. The most common are:

A simple ratio or fraction known as the **representative fraction.** For example, 1:80,000 or

$$\frac{1}{80,000}$$

means that one unit (such as an inch) on the chart represents 80,000 of the same unit on the surface of the earth. This scale is sometimes called the **natural** or **fractional scale.**

A statement of that distance on the earth shown in one unit (usually an inch) on the chart, or vice versa. For example, "30 miles to the inch" means that 1 inch on the chart represents 30 miles of the earth's surface. Similarly, "2 inches to a mile" indicates that 2 inches on the chart represent 1 mile on the earth. This is sometimes called the **numerical scale.**

Graphic scale. A line or bar may be drawn at a convenient place on the chart and subdivided into nautical miles, yards, etc. All charts vary somewhat in scale from point to point, and in some projections the scale is not the same in all directions about a single point. A single subdivided line or bar for use over an entire chart is shown only when the chart is of such scale and projection that the scale varies a negligible amount over the chart, usually one of about 1:75,000 or larger. Since 1 minute of latitude is very nearly equal to 1 nautical mile, the latitude scale serves as an approximate graphical scale. On most nautical charts the east and west borders are subdivided to facilitate distance measurements.

On a Mercator chart the scale varies with the latitude. This is noticeable on a chart covering a relatively large distance in a north-south direction. On such a chart the scale at the latitude in question should be used for measuring distances.

Of the various methods of indicating scale, the graphical method is normally available in some form on the chart. In addition, the scale is customarily stated on charts on which the scale does not change appreciably over the chart.

The ways of expressing the scale of a chart are readily interchangeable. For instance, in a nautical mile there are about 6,076.11549 feet or $6{,}076.11549 \times 12 = 72{,}913.39$ inches. If the natural scale of a chart is 1:80,000, one inch of the chart represents 80,000 inches of the earth, or a little more than a mile. To find the exact amount, divide the scale by the number of inches in a mile, or

$$\frac{80{,}000}{72{,}913.39} = 1.097.$$

Thus, a scale of 1:80,000 is the same as a scale of 1.097 (or approximately 1.1) miles to an inch. Stated another way, there are:

$$\frac{72{,}913.39}{80{,}000} = 0.911$$

(approximately 0.9) inch to a mile. Similarly, if the scale is 60 nautical miles to an inch, the representative fraction is $1{:}(60 \times 72{,}913.39) = 1{:}4{,}374{,}803$. Table 37 provides the scale equivalents.

A chart covering a relatively large area is called a *small-scale* chart and one covering a relatively small area is called a *large-scale* chart. Since the terms are relative, there is no sharp division between the two. Thus, a chart of scale 1:100,000 is large scale when compared with a chart of 1:1,000,000 but small scale when compared with one of 1:25,000.

504. Chart classification by scale.—Charts are constructed on many different scales, ranging from about 1:2,500 to 1:14,000,000 (and even smaller for some world charts). Small-scale charts covering large areas are used for planning and for offshore navigation. Charts of larger scale, covering smaller areas, should be used as the vessel approaches pilot waters. Several methods of classifying charts according to scale are in use in various nations. The following classifications of nautical charts are those used by the National Ocean Service.

Sailing charts are the smallest scale charts used for planning, fixing position at sea, and for plotting the dead reckoning while proceeding on a long voyage. The scale is generally smaller than 1:600,000. The shoreline and topography are generalized and only offshore soundings, the principal navigational lights, outer buoys, and landmarks visible at considerable distances are shown.

General charts are intended for coastwise navigation outside of outlying reefs and shoals. The scales range from about 1:150,000 to 1:600,000.

Coast charts are intended for inshore coastwise navigation where the course may lie inside outlying reefs and shoals, for entering or leaving bays and harbors of considerable width, and for navigating large inland waterways. The scales range from about 1:50,000 to 1:150,000.

Harbor charts are intended for navigation and anchorage in harbors and small waterways. The scale is generally larger than 1:50,000.

In the classification system used by the Defense Mapping Agency Hydrographic/Topographic Center, the sailing charts are incorporated in the general charts classification (smaller than about 1:150,000); those coast charts especially useful for approaching more confined waters (bays, harbors) are classified as **approach charts.**

505. Accuracy.—The accuracy of a chart depends upon:

1. *Thoroughness and up-to-dateness of the survey and other navigational information.* Some estimate of the accuracy of the survey can be formed by an examination of the source notes given in the title of the chart. If the chart is based upon very old surveys, it should be used with caution. Many of the earlier surveys were made under conditions that were not conducive to great accuracy. It is safest to question every chart based upon surveys of doubtful accuracy.

The number of soundings and their spacing is some indication of the completeness of the survey. Only a small fraction of the soundings taken in a thorough survey are shown on the chart, but sparse or unevenly distributed soundings indicate that the survey was probably not made in detail. Large or irregular blank areas, or absence of depth contours (commonly called **depth curves**), generally indicate lack of soundings in the area. If the water surrounding such a blank area is deep, there is generally considerable depth in the blank; conversely, shallow water surrounding such an area indicates the strong possibility of shoal water. If neighboring areas abound in rocks or are particularly uneven, the blank area should be regarded with additional suspicion. However, it should be kept in mind that relatively few soundings are shown when there is a large number of depth contours or where the bottom is flat or gently and evenly sloping. Additional soundings are shown when they are helpful in indicating the uneven character of a rough bottom (figs. 505a and 505b).

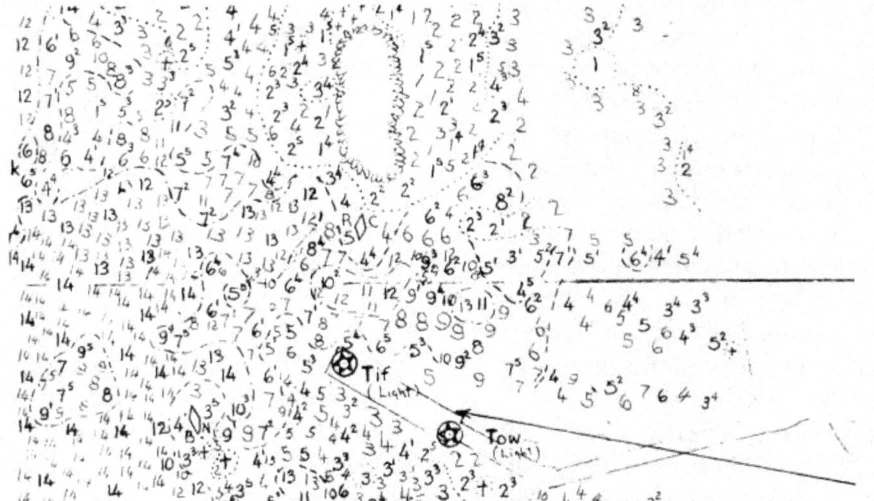

FIGURE 505a.—Part of a boat sheet, showing the soundings obtained in a survey.

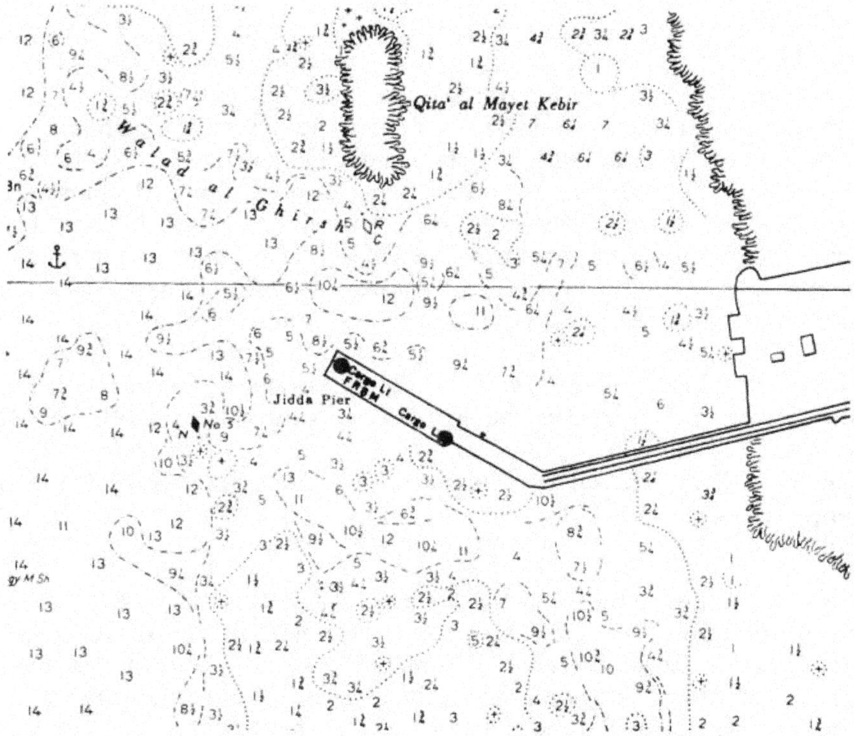

FIGURE 505b.—Part of a nautical chart made from the boat sheet of figure 505a. Compare the number of soundings in the two figures.

Even a detailed survey may fail to locate every rock or pinnacle, and in waters where their existence is suspected, the best methods for determining their presence are wire drag surveys. Areas that have been dragged may be indicated on the chart and a note added to show the effective depth at which the drag was operated.

Changes in the contour of the bottom are relatively rapid in areas where there are strong currents or heavy surf, particularly when the bottom is composed principally of soft mud or sand. The entrances to bar harbors are especially to be regarded with suspicion. Similarly, there is sometimes a strong tendency for dredged channels to shoal, especially if they are surrounded by sand or mud, and cross currents exist. Notes are sometimes shown on the chart when the bottom contours are known to change rapidly. However, the absence of such a note should not be regarded as evidence that rapid change does not occur.

Changes in aids to navigation, structures, etc., are more easily determined, and charts are generally corrected in this regard to the date of printing. However, there is always the possibility of a change having occurred since the chart was printed. All issues of *Notice to Mariners* printed after that date (art. 506) should be checked to insure accuracy in this respect.

2. *Suitability of the scale for the design and intended navigational use.* The same detail cannot be shown on a small-scale chart as on one of a larger scale. On small-scale charts detailed information, including minor aids to navigation, is omitted or generalized in the areas covered by larger scale charts. Therefore, it is good practice to use the largest scale chart available when in the vicinity of shoals or other dangers.

3. *Presentation and adequacy of data.* The amount and kind of detail to be shown, and the method of presentation, are continually under study by charting

agencies. Development of a new navigational aid may render many previous charts inadequate. An example is radar. Many of the charts produced before radar became available lack the detail needed for reliable identification of targets.

Part of the responsibility for the continuing accuracy of charts lies with the user. If charts are to remain reliable, they must be corrected as indicated by the *Notice to Mariners.* In addition, the user's reports of errors and changes and his suggestions often are useful to the publishing agencies in correcting and improving their charts. Navigators and maritime activities have contributed much to the reliability and usefulness of the modern nautical chart. If a chart becomes wet, the expansion and subsequent shrinkage when the chart dries are likely to cause distortion.

506. Dates on charts.—The system of dates now used on charts published by the Defense Mapping Agency Hydrographic/Topographic Center and the National Ocean Service is as follows:

First edition. The original date of issue of a new chart is shown at the top center margin, thus:

1st Ed., Sept. 1970

New edition. A new edition is made when, at the time of printing, the corrections are too numerous or too extensive to be reported in *Notice to Mariners,* making previous printings obsolete. The date of the first edition is retained at the top margin. At the lower left-hand corner it is replaced by the number and date of the new edition. The latter date is the same as that of the latest *Notice to Mariners* to which the chart has been corrected, thus:

5th Ed., July 8, 1978

Revised print. A revised print published by the National Ocean Service may contain corrections which have been published in *Notice to Mariners* but does not supersede a current edition. The date of the revision is shown to the right of the edition date, thus:

5th Ed., July 8, 1978; Revised 4/16/83

Reprint. A reprint is initiated by a low stock situation and is a reprint of the chart with a limited number of corrections from *Notice to Mariners.* The magnetic variation data on a reprint published by the Defense Mapping Agency Hydrographic/Topographic Center is updated to the latest epoch at the time of printing.

Chart Reading

507. Chart symbols.—Much of the information contained on charts is shown by conventional symbols which make no attempt at accuracy in scale or detail, but are shown at the correct location and make possible the showing of a large amount of information without congestion or confusion. The standard symbols and abbreviations which have been approved for use on regular nautical charts published by the United States of America are shown in Chart No. 1, *Nautical Chart Symbols and Abbreviations.* A knowledge of the meanings of these symbols is essential to a full understanding of charts. Fictitious sample charts (figs. 507a and 507b) show some of these symbols.

Most of the symbols and abbreviations shown in Chart No. 1 are in agreement with those recommended by the International Hydrographic Organization (IHO). Symbol and abbreviation status is indicated by alphanumeric style differences in the first column of Chart No. 1. The status is explained in the general remarks section of Chart No. 1.

The symbols and abbreviations on any given chart may differ somewhat from those shown in Chart No. 1 because of a change in the standards since printing of the chart or because the chart was published by an agency having a different set of standards.

508. Lettering.—Certain standards regarding lettering have been adopted, except on charts made from reproducibles furnished by foreign nations.

Vertical type is used for features which are dry at high water and not affected by movement of the water, except for heights above water.

Slanting type is used for water, underwater, and floating features, except soundings.

The type of lettering used may be the only means of determining whether a feature may be visible at high tide. For instance, a rock might bear the title "............. Rock" whether or not it extends above the surface. If the name is given in vertical letters, the rock constitutes a small islet; if in slanting type, the rock constitutes a reef.

509. The shoreline shown on nautical charts represents the line of contact between the land and a selected water elevation. In areas affected by tidal fluctuations, this line of contact is usually the mean high-water line. In confined coastal waters of diminished tidal influence, a mean water level line may be used. The shoreline of interior waters (rivers, lakes) is usually a line representing a specified elevation above a selected datum. A shoreline is symbolized by a heavy line. A broken line indicates that the charted position is approximate only. The nature of the shore may be indicated, as shown by the symbols in part A of Chart No. 1.

Where the low-water line differs considerably from the high-water line, the low-water line may be indicated by dots in the case of mud, sand, gravel, or stones, with the kind of material indicated, and by a characteristic symbol in the case of rock or coral. The area alternately covered and uncovered may be shown by a tint which is usually a combination of the land tint and a blue water tint as shown in figures 507a and 507b.

The apparent shoreline is used on charts to show the outer edge of marine vegetation where that limit would reasonably appear as the shoreline to the mariner, or where it prevents the shoreline from being clearly defined. The apparent shoreline is symbolized by a light line. The inner edge is marked by a broken line when no other symbol (such as a cliff, levee, etc.) furnishes such a limit. The area between inner and outer limits may be given the combined land-water tint or the land tint.

510. Water areas.—Soundings or depths of water are shown in several ways. Individual soundings are shown by numbers. These do not follow the general rule for lettering. They may be either vertical or slanting, or both may be used on the same chart to distinguish between the data based upon different surveys, different datums, smaller scale charts, or furnished by different authorities.

The unit of measurement used for soundings on each chart is shown in large block letters at the top and bottom of the chart. When the unit of measurement is meters or meters and decimeters, SOUNDINGS IN METERS is shown. When soundings in fathoms or fathoms and fractions are used, SOUNDINGS IN FATHOMS is shown, and when the soundings are in fathoms and feet, SOUNDINGS IN FATHOMS AND FEET is shown.

A depth conversion scale is placed outside the neatline on the chart for use in converting charted depths to feet, meters, or fathoms.

"No bottom" soundings are indicated by a number with a line over the top and a dot over the line, thus: $\dot{\overline{45}}$. This indicates that the spot was sounded to the depth indicated without reaching the bottom. Areas which have been wire dragged (fig. 510a) are shown by a broken limiting line, and the clear effective depth is indicated, with a characteristic symbol under the numbers. On charts of the Defense Mapping Agency Hydrographic/Topographic Center, a purple tint is shown within the limits of the swept area unless such tinting would result in excessive use of purple, in which case a green tint is shown within the limits of the swept area.

The soundings are supplemented by a series of *depth contours* (fig. 510b) connecting points of equal depth. These lines present a graphic indication of the configuration of the bottom. The types of lines used for various depths are shown in part R of Chart No. 1. On some charts depth contours are shown in solid lines, the depth represented by each being shown by numbers placed in breaks in the lines, as with land contours. Solid line depth contours are derived from intensively developed hydrographic surveys. *A broken or indefinite contour is substituted for a solid depth contour whenever the reliability of the contour is questionable.* Depth contours are labeled with numerals in the unit of measurement of the soundings. This type chart, presenting a more detailed indication of the bottom configuration with fewer numerical soundings, is particularly useful to the vessel equipped with an echo sounder permitting continuous determination of a profile of the bottom. Such a chart, to be reliable, can be made only for areas which have been surveyed in great detail.

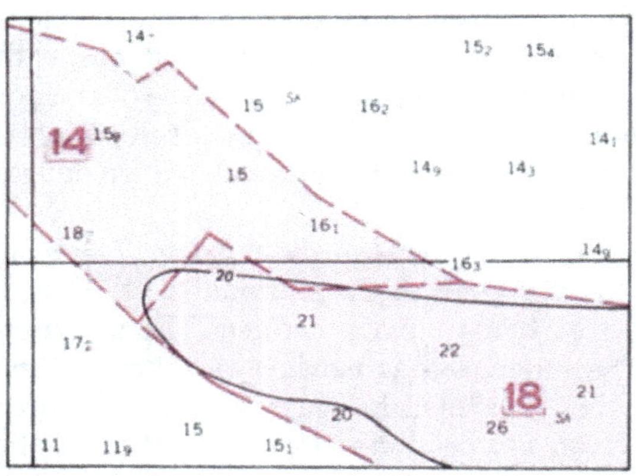

FIGURE 510a.—Swept area.

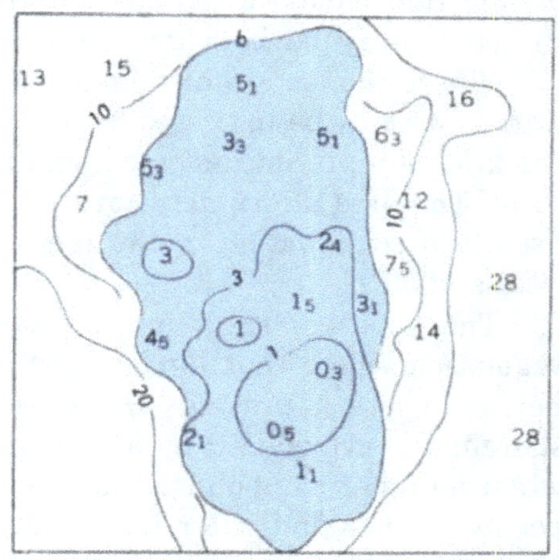

FIGURE 510b.—Depth contours.

Areas which uncover at low tide are tinted as indicated in article 509. Those areas out to a given depth often are given a blue tint, and occasionally a lighter blue is carried to some greater depth. On older charts the one-, two-, and three-fathom curves have stippled edges. Charts designed to give maximum emphasis to the configuration of the bottom show depths, beyond the 100-fathom curve, over the entire chart by depth contours similar to the contours shown on land areas to indicate graduations in height. These are called *bottom contour* or *bathymetric* charts.

The side limits of dredged channels are indicated by broken lines. The *project depth* (art. 2715) and the date of dredging, if known, are shown by a statement in or along the channel. The possibility of silting should be considered. Local authorities should be consulted for the *controlling depth* (art. 2715).

The chart scale is generally too small to permit all soundings to be shown. In the selection of soundings to be shown, *least* depths are generally chosen first and a sounding pattern worked out to provide safety, a practical presentation of the bottom configuration, and a neat appearance. Depths greater than those indicated may be found close to charted depths, but steep changes in depth are given every consideration in sounding selection. Also, the state of the tide affects the depth at any given moment. An isolated shoal sounding should be approached with caution, or avoided, unless it is known that the area has been wire dragged, for there is always the possibility that a depth less than the least shown may have escaped detection. Also, the shoal area near a coast little frequented by vessels is sometimes not surveyed with the same thoroughness as other areas. Such areas and those where rocks, coral, etc., are known to exist should be entered with caution, or avoided.

The substance forming the bottom is shown by abbreviations, as listed in part S of Chart No. 1. The meaning of some of the less-well-known terms is given below:

Ooze is a soft, slimy, organic sediment composed principally of shells or other hard parts of minute organisms.

Marl is a crumbling, earthy deposit, particularly one of clay mixed with sand, lime, decomposed shells, etc. A layer of marl may become quite compact.

Shingle consists of small, rounded, waterworn stones. It is similar to gravel but with the average size of stone generally larger.

Schist is crystalline rock of a finely laminated nature.

Madrepore is a stony coral which often forms an important building material for reefs.

Lava is rock in the fluid state, or such material after it has solidified. It is formed at very high temperature and issues from the earth through volcanoes.

Pumice is cooled volcanic glass with a great number of minute cavities caused by the expulsion of water vapor at high temperature, resulting in a very light material.

Tufa is a porous rocky deposit sometimes formed in streams and in the ocean near the mouths of rivers.

Scoria (plural *scoriae*) is rough, cinderlike lava.

Sea tangle is any of several species of seaweed, especially those of large size.

Spicules are the small skeletons of various marine animals such as sponges.

Foraminifera (plural) are small marine animals with hard shells of from one to many chambers.

Globigerina is a very small marine animal of the foraminifera order, with a chambered shell, or the shell of such an animal. In large areas of the ocean the calcareous shells of these animals are very numerous, being the principal constituent of a soft mud or **globigerina ooze**, forming part of the ocean bed.

Diatom is a microscopic animal with external skeletons of silica, often found in both fresh and salt water. Part of the ocean bed is composed of a sedimentary ooze consisting principally of large collections of the skeletal remains of diatoms.

Radiolaria (plural) are minute sea animals with a siliceous outer shell. The skeletons of these animals are very numerous, especially in the tropics.

Pteropod is a small marine animal with or without a shell and having two thin, winglike feet. These animals are often so numerous they cover the surface of the sea for miles. In some areas their shells cover the bottom.

Polyzoa (plural) are very small marine animals which reproduce by budding, many generations often being permanently connected by branchlike structures.

Cirripeda (plural) are barnacles and certain other parasitic marine animals.

Fucus is a coarse seaweed growing attached to rocks.

Matte is a dense, twisted growth of a sea plant such as grass.

"Calcareous" is an adjective meaning "containing or composed of calcium or one of its compounds."

511. Chart sounding datum.—*Depths.* All depths indicated on charts are reckoned from some selected level of the water, called the *chart sounding datum.* The various chart datums are explained in chapter XXXI. On charts made from surveys conducted by the United States the chart datum is selected with regard to the tides of the region, so that depths might be shown in their least favorable aspect. On charts based upon those of other nations the datum is that of the original authority. When it is known, the datum used is stated on the chart. In some cases where the chart is based upon old surveys, particularly in areas where the range of tide is not great, the actual chart datum may not be known.

For National Ocean Service charts of the Atlantic and gulf coasts of the United States and Puerto Rico the chart datum is *mean low water.* For charts of the Pacific coast of the United States, including Alaska, it is *mean lower low water.* Most Defense Mapping Agency Hydrographic/Topographic Center charts are based upon *mean low water, mean lower low water,* or *mean low water springs.* The chart datum for charts published by other countries varies greatly, but is usually lower than mean low water. On charts of the Baltic Sea, Black Sea, the Great Lakes, and other areas where tidal effects are small or without significance, the datum adopted is an arbitrary height approximating the mean water level.

The chart datum of the largest-scale charts of an area is generally the same as the reference level from which height of tide is tabulated in the tide tables.

The height of a chart datum is usually only an approximation of the actual mean value specified, for determination of the actual mean height usually requires a longer series of tidal observations than is available to the cartographer, and the height changes somewhat over a period of time.

Since the chart datum is generally a computed mean or average height at some state of the tide, the depth of water at any particular moment may be less than shown on the chart. For example, if the chart datum is *mean lower low water,* the depth of water at *lower low water* will be less than the charted depth about as often as it is greater. A lower depth is indicated in the tide tables by a minus sign (−).

Heights. The shoreline shown on charts is the high-water line, generally the level of mean high water. The heights of lights, rocks, islets, etc., are generally reckoned from this level. However, heights of islands, especially those at some distance from the coast, are often taken from sources other than hydrographic surveys, and may be reckoned from some other level, often mean sea level. The plane of reference for topographic detail is frequently not stated on the chart.

Since heights are usually reckoned from high water and depths from some form of low water, the reference levels are seldom the same. This is generally of little practical significance, but it might be of interest under some conditions, particularly where the range of tide is large.

512. Dangers are shown by appropriate symbols, as indicated in part O of Chart No. 1.

A rock that uncovers at mean high water may be shown as an islet. If an isolated, offlying rock is known to uncover at the chart datum but to be covered at high water, the appropriate symbol is shown and the height above the chart datum, if known, is usually given, either by statement such as "*Uncov 2 ft*" or by the figure indicating the number of feet above the chart datum underlined and usually enclosed in parentheses, thus: (2). This is illustrated in figure 512a. A rock which does not uncover is shown by the appropriate symbol. If it is considered a danger to surface vessels, the symbol is enclosed by a dotted curve for emphasis.

A distinctive symbol is used to show a detached coral reef which uncovers at the chart datum. For a coral or rocky reef which is submerged at chart datum, the sunken rock symbol or an appropriate statement is used, enclosed by a dotted or broken line if the limits have been determined.

Several different symbols are used for wrecks, depending upon the nature of the wreck or scale of the chart. The usual symbol for a visible wreck is shown in figure 512b. A sunken wreck with less than 11 fathoms of water over it is considered dangerous and its symbol is surrounded by a dotted curve. The safe clearance depth found over a wreck is indicated by a standard sounding number placed at the wreck, (fig. 512c). If this depth is determined by a wire drag, the sounding is underscored by the wire drag symbol (art. 510). An unsurveyed wreck over which the exact depth is unknown, but is considered to have a safe clearance to the depth shown is depicted as shown in figure 512c.

Tide rips, eddies, and kelp are shown by symbol or lettering.

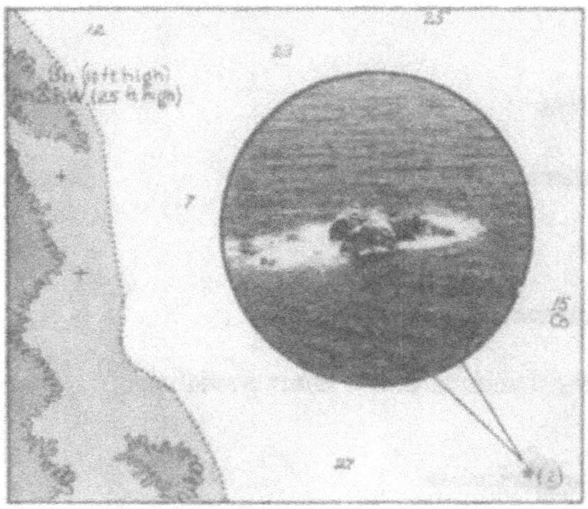

FIGURE 512a.—A rock awash.

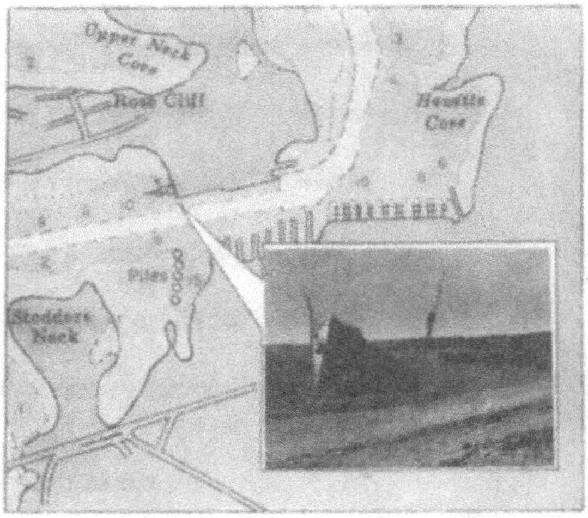

FIGURE 512b.—A visible wreck.

Piles, dolphins (clusters of piles), snags, stumps, etc., are shown by small circles and a label identifying the type of obstruction. If such dangers are submerged, the letters "Subm" precede the label.

Fish stakes and traps are shown when known to be permanent or hazardous to navigation.

The importance of knowing the chart symbols for dangers to navigation cannot be emphasized strongly enough. Most dangers are emphasized with a blue tint and dotted line surrounding the danger. Some of the danger symbols are shown in figure 512c.

513. Aids to navigation are shown by symbol, as given in Chart No. 1, usually supplemented by abbreviations and sometimes by additional descriptive text. In order to render the symbols conspicuous it is necessary to show them in greatly exaggerated size relative to the scale of the chart. It is therefore important that the

	Sunken wreck dangerous to surface navigation (less than 11 fathoms over wreck).
	Sunken rock dangerous to navigation.
(3) Wk	Wreck over which depth is known.
21 Wk	Wreck with depth cleared by wire drag.
(8) Wk	Unsurveved wreck over which the exact depth is unknown, but is considered to have a safe clearance to the depth shown.
(5) Rk	Shoal sounding on isolated rock.
(Co 8)	Coral reef covered at sounding datum.
(Foul)	Foul ground, Foul bottom.
*(2) or ✲(2)	Rock which covers and uncovers with height above chart sounding datum.
⁞ Subm Piles	Submerged pilings.
(3) Rep (1974)	Depth reported in 1974.
Reef	Reef of unknown extent.
(4) Obstr	Obstruction
■ ■ Platform (lighted) HORN	Offshore platform (unnamed)
	Drying (or uncovering) heights, above chart sounding datum.

FIGURE 512c.—Danger symbols.

navigator know which part of the symbol represents the actual position of the aid. For floating aids (lightships and buoys), the position part of the symbol marks the *approximate* location of the anchor or sinker, the aid swinging in an orbit around this approximate position.

The principal charted aids to navigation are lighthouses, other lights on fixed structures, beacons, lightships, radiobeacons, and buoys. The number of aids shown and the amount of information concerning them varies with the scale of the chart. Unless otherwise indicated, lights which do not alternate in color are white, and alternating lights are red and white. Light lists give complete navigational information concerning them.

Lighthouses and *other lights on fixed structures* are shown as black dots surrounded by nautical purple disks or as black dots with purple flare symbols. The center of the black dot is the position of the light.

On large-scale charts the characteristics of lights are shown in the following order:

Characteristic	*Example*	*Meaning*
1. Character	Fl (2)	group flashing (2 flashes)
2. Color	R	red
3. Period	10s	10 seconds
4. Height	160 ft	160 feet
5. Range	19M	19 nautical miles (See article 1307)
6. Number	"6"	light number 6

The legend for this light would appear on the chart:

Fl (2) R 10s 160 ft 19 M "6"

On older charts this form is varied slightly. As the chart scale becomes smaller the six items listed above are omitted in the following order: first, height; second, period (seconds); third, number (of flashes, etc.) in group; fourth, light number; fifth, visibility. Names of unnumbered lights are shown when space permits.

Daybeacons (unlighted beacons) are shown as depicted in Chart No. 1. When daybeacons are shown by small triangles, the center of the triangle marks the position of the aid. Except on Intracoastal Waterway charts and charts of state waterways the abbreviation Bn is shown beside the symbol, with the appropriate abbreviation for color if known. For black beacons the triangle is solid black and there is no color abbreviation. All beacon abbreviations are in vertical lettering, as appropriate for fixed aids (fig. 513a).

Lightships are shown by ship symbol, the center of the small circle at the base of the symbol indicating the approximate position of the lightship's anchor. The circle is overprinted by a small purple disk as shown in figure 513b or a purple flare emanating from the top of the symbol. As a floating aid, the light characteristics and the name of the lightship are given in leaning letters.

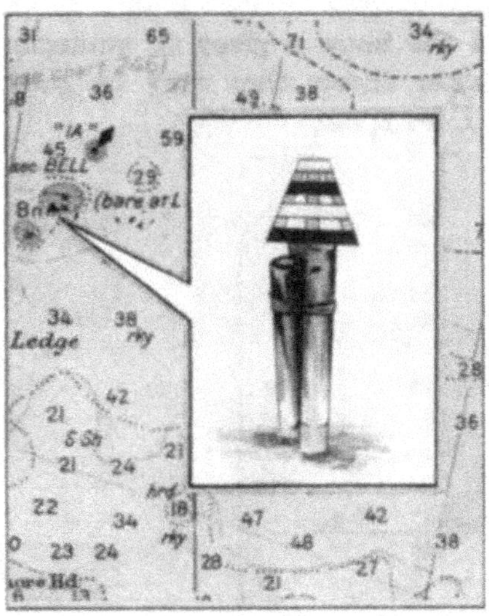

FIGURE 513a.—A daybeacon.

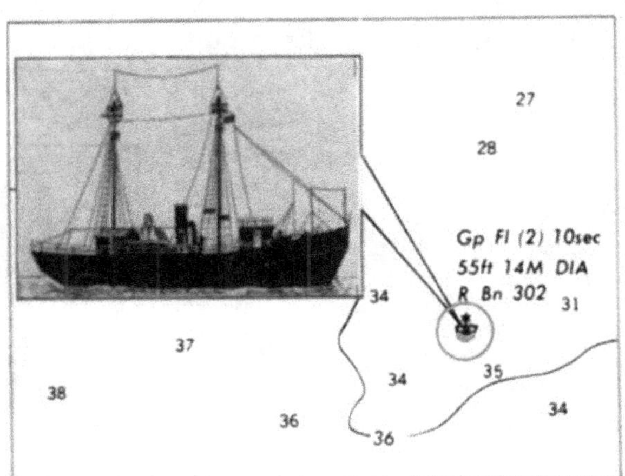

FIGURE 513b.—A lightship with a radiobeacon.

Radiobeacons are indicated on the chart by a small purple circle, as shown in figure 513b, accompanied by the appropriate abbreviation to indicate whether an ordinary radiobeacon (R Bn) or a radar beacon (Racon). The same symbol is used for a radio direction finder station with the abbreviation "RDF" and a coast radar station with the abbreviation Ra. Other radio stations are indicated by a small black circle with a dot in the center, or a smaller circle without a dot, and the appropriate abbreviation. In every case the center of the circle marks the position of the aid.

Buoys, except mooring buoys, are usually shown by a diamond-shaped symbol and a small dot or small circle in conjunction with one of its points (at one of its acute angles). The dot or small circle indicates the approximate position of the buoy's sinker. A mooring buoy is shown by a distinctive symbol as indicated in part L of Chart No. 1. The small circle interrupting the symbol's base line indicates the approximate position of the sinker.

A black buoy is shown by a solid black diamond symbol, without abbreviation. For all other buoys, color is indicated by an abbreviation, or in full by a note on the chart. In addition, the diamond-shaped symbols of red buoys often are colored purple. A buoy symbol with a line connecting the side points (shorter axis), half of the symbol being purple or open and the other half black, indicates horizontal bands. A line connecting the upper and lower points (longer axis) represents vertical stripes. Two lines connecting the opposite *sides* of the symbol indicate a checkered buoy.

There is no significance to the angle at which the diamond-shape appears on the chart. The symbol is placed so as to avoid interference with other features of the chart.

Lighted buoys are indicated by a purple flare emanating from the buoy symbol or by a small purple disk centered on the dot or small circle indicating the approximate position of the buoy's sinker, as shown in figure 513c.

Abbreviations for light characteristics, type and color of buoy, number of the buoy, and any other pertinent information given near the symbol are in slanting letters. The letter *C, N,* or *S,* indicates a can, nun, or spar, respectively (art. 409). The words "bell," "gong," and "whistle," are shown as *BELL, GONG,* and *WHIS,* respectively. The number or letter designation of the buoy is given in quotation marks on National Ocean Service charts. On other charts they may be given without quotation marks or punctuation, thus: No 1, No 2, etc.

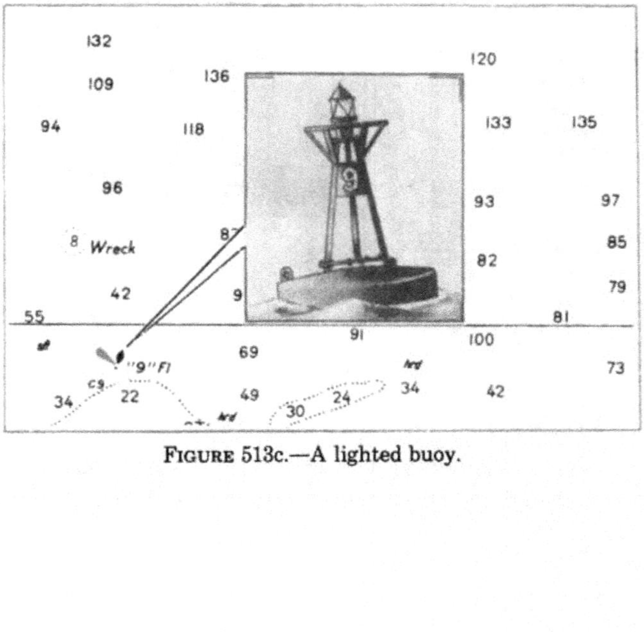

FIGURE 513c.—A lighted buoy.

Station buoys are not shown on small-scale charts, but are given on some large-scale charts.

Aeronautical lights included in the light lists are shown by the lighthouse symbol, accompanied by the abbreviation "AERO." The completeness to which the characteristics are shown depends principally upon the effective range of other navigational lights in the vicinity, and the usefulness of the light for marine navigation.

Ranges are indicated by a broken or solid line. The solid line, which indicates that part of the range intended for navigation, may be broken at irregular intervals to avoid being drawn through soundings. That part of the range line drawn only to guide the eye to the objects to be kept in range is broken at regular intervals. If the direction is given, it is expressed in degrees clockwise from true north.

Sound signal apparatus is indicated by the appropriate word in capital letters (*HORN, BELL, GONG,* etc.) or an abbreviation indicating the type of sound. Sound signals of all types other than submarine sound signals are represented by three arcs of concentric circles within an angle of 45°, oriented and placed as necessary for clarity. The letters "DFS" indicate a **distance finding station** having synchronized sound and radio signals. The location of a sound signal which does not accompany a visual aid, either lighted or unlighted, is shown by a small circle and the appropriate word in vertical block letters.

Private aids, when shown, are marked "Priv maintd." Some privately maintained unlighted aids are indicated by a small circle accompanied by the word "Marker," or a larger circle with a dot in the center and the word "MARKER." The center of the circle indicates the position of the aid. A privately maintained lighted aid has the light symbol and is accompanied by the characteristics and the usual indication of its private nature. Private aids should be used with caution.

A *light sector* is the sector or area bounded by two radii and the arc of a circle in which a light is visible or in which it has a distinctive color different from that of adjoining sectors. The limiting radii are indicated on the chart by dotted lines.

Colors of the sectors are indicated by words spelled out if space permits, or by abbreviation (W, R, etc.) if it does not.

Limits of light sectors and arcs of visibility *as observed from a vessel* are given in the light lists, in clockwise order.

514. Land areas.—The amount of detail shown on the land areas of nautical charts depends upon the scale and the intended purpose of the chart.

Relief is shown by contours and form lines.

Contours are lines connecting points of equal elevation. The heights represented by the contours are indicated in slanting figures at suitable places along the lines. Heights are usually expressed in feet (or in meters with means for conversion to feet). The interval between contours is uniform over any one chart, except that certain intermediate contours are sometimes shown by broken line. When contours are broken, their locations are approximate.

Form lines are approximations of contours used for the purpose of indicating relative elevations. They are used in areas where accurate information is not available in sufficient detail to permit exact location of contours. Elevations of individual form lines are not indicated on the chart.

Spot elevations are generally given only for summits or for tops of conspicuous landmarks. The heights of spot elevations and contours are given with reference to mean high water when this information is available.

When there is insufficient space to show the heights of islets or rocks, they are indicated by slanting figures enclosed in parentheses in the water area nearby.

Cities and roads. Cities are shown in a generalized pattern that approximates their extent and shape. Street names are generally not charted except those along the waterfront on the largest scale charts. In general, only the main arteries and thoroughfares or major coastal highways are shown on smaller scale charts. Occasionally, highway numbers are given. When shown, trails are indicated by a light broken line. Buildings along the waterfront or individual ones back from the waterfront but of special interest to the mariner are shown on large-scale charts. Special symbols are used for certain kinds of buildings, as indicated in part I of Chart No. 1. Both single and double track railroads are indicated by a single line with cross marks. In general, city electric railways are not charted. A fence or sewer extending into the water is shown by a broken line, usually labeled. Airports are shown on small-scale charts by symbol and on large-scale charts by shape and extent of runways. Breakwaters and jetties are shown by single or double lines depending upon the scale of the chart. A submerged portion and the limits of the submerged base are shown by broken lines.

515. Landmarks are shown by symbols, as given in Chart No. 1.

A large circle with a dot at its center is used for selected landmarks that have been accurately located. Capital letters are used to identify the landmark: HOUSE, FLAGPOLE, STACK, sometimes followed by "(conspic)."

A small circle without a dot is used for landmarks not accurately located. Capital and lower case letters are used to identify the landmark: Mon, Cup, Dome. The abbreviation "PA," for position approximate, is used when necessary as a safety feature.

When only one object of a group is charted, its name is followed by a descriptive legend in parenthesis, including the number of objects in the group, for example (TALLEST OF FOUR) or (NORTHEAST OF THREE).

Some of the accompanying labels on a chart are interpreted as follows:

Building or **house.** One of these terms, as appropriate, is used when the entire structure is the landmark, rather than an individual feature of it.

A **spire** is a slender pointed structure extending above a building. It is seldom less than two-thirds of the entire height of the structure, and its lines are rarely broken by stages or other features. The term is not applied to a short pyramid-shaped structure rising from a tower or belfry.

A **cupola** (kū'pō·lâ) is a small dome-shaped tower or turret rising from a building (fig. 515).

A **dome** is a large, rounded, hemispherical structure rising above a building, or a roof of the same shape. A prominent example is that of the Capitol of the United States, in Washington, D.C.

A **chimney** is a relatively small, upright structure projecting above a building for the conveyance of smoke.

A **stack** is a tall smokestack or chimney. The term is used when the stack is more prominent as a landmark than accompanying buildings.

A **flagpole** is a single staff from which flags are displayed. The term is used when the pole is not attached to a building.

The term **flagstaff** is used for a flagpole rising from a building.

A **flag tower** is a scaffold-like tower from which flags are displayed.

A **radio tower** is a tall pole or structure for elevating radio antennas.

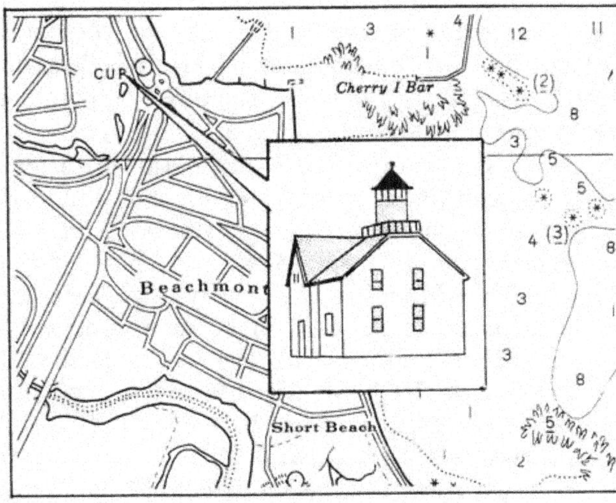

FIGURE 515.—A cupola.

A **radio mast** is a relatively short pole or slender structure for elevating radio antennas, usually found in groups.

A **tower** is any structure with its base on the ground and high in proportion to its base, or that part of a structure higher than the rest, but having essentially vertical sides for the greater part of its height.

A **lookout station** or **watch tower** is a tower surmounted by a small house from which a watch is kept regularly.

A **water tower** is a structure enclosing a tank or standpipe so that the presence of the tank or standpipe may not be apparent.

A **standpipe** is a tall cylindrical structure, in a waterworks system, the height of which is several times the diameter.

The term **tank** is used for a water tank elevated high above the ground by a tall skeleton framework.

The expression **gas tank** or **oil tank** is used for the distinctive structures described by these words.

516. Miscellaneous.—*Measured mile.* A measured nautical mile indicated on a chart is accurate to within 6 feet of the correct length. Most measurements in the United States were made before 1959, when the United States adopted the International Nautical Mile. The new value is within 6 feet of the previous standard length of 6,080.20 feet, adjustments not having been made. If the measured distance differs from the standard value by more than 6 feet, the actual measured distance is stated and the words "measured mile" are omitted.

Periods after abbreviations in water areas are omitted, as these might be mistaken for rocks. However, a lower case *i* or *j* is dotted.

Courses shown on charts are given in true directions, to the nearest minute of arc.

Bearings shown are in true directions *toward* (not from) the objects.

Commercial radio broadcasting stations are shown on charts when they are of value to the mariner either for obtaining radio bearings or as landmarks.

Rules of the road. Lines of demarcation between the areas in which international and inland navigation rules apply are shown only when they cannot be adequately described in notes on the chart.

Compass roses are placed at convenient locations on Mercator charts to facilitate the plotting of bearings and courses. The outer circle is graduated in degrees with zero at true north. The inner circle is graduated in points and degrees with the arrow indicating magnetic north.

Magnetic information. On many charts magnetic variation is given to the nearest 15′ by notes in the centers of compass roses; the annual change is given to the nearest 1′ to permit correction of the given value at a later date. When this is done, the magnetic information is updated when a new edition is issued. The current practice of the Defense Mapping Agency Hydrographic/Topographic Center is to give the magnetic variation to the nearest 1′, but the magnetic information on new editions is only updated to conform with the latest epoch (1980.0, 1985.0, etc.). Whenever a chart is reprinted, the magnetic information is updated to the latest epoch. On other charts the variation is given by a series of **isogonic lines** connecting points of equal variation, usually a separate line being given for each degree of variation. The line of zero variation is called the **agonic line.** Many plans and insets show neither compass roses nor isogonic lines, but indicate magnetic information by note. A local magnetic disturbance of sufficient force to cause noticeable deflection of the magnetic compass, called **local attraction,** is indicated by a note on the chart.

Currents are sometimes shown on charts by means of arrows giving the directions, and figures giving the speeds. The information thus given refers to the usual or average conditions, sometimes based upon very few observations. It is not safe to assume that conditions at any given time will not differ considerably from those shown.

Longitudes are reckoned eastward and westward from the meridian of Greenwich, England, unless otherwise stated.

Notes on charts should be read with care, as they may give important information not graphically presented. Several types of notes are used. Those in the margin give such information as the chart number and (sometimes) publication and edition notes, identification of adjoining charts, etc. Notes in connection with the chart title include such information as scale, sources of charted data, tidal information, the unit in which soundings are given, cautions, etc. Another class of notes is that given in proximity to the detail to which it refers. Examples of this type of note are those referring to local magnetic disturbance, controlling depths of channels, measured miles, dangers, dumping grounds, anchorages, etc.

Overlapping charts constructed on different horizontal geodetic datums (app. X) may carry the following note:

CAUTION

Differences in latitude and longitude may exist between this and other charts of the area; therefore, the transfer of positions from one chart to another should be done by bearings and distances from common features.

Horizontal geodetic datum shifts may be given to provide the corrections necessary to shift to a different datum (app. X). It is the practice of the Defense Mapping Agency Hydrographic/Topographic Center to provide, if plottable, the corrections to new charts and new editions of charts that are necessary to shift the geodetic datum to the World Geodetic System.

Anchorage areas are shown within purple broken lines and labeled as such. Anchorage berths are shown as purple circles with the number or letter assigned to the berth inscribed within the circle. Caution notes are sometimes shown when there are specific anchoring regulations.

Spoil areas are shown within short broken black lines. The area is tinted blue (National Ocean Service charts only) and labeled SPOIL AREA.

Firing and bombing practice areas in the United States territorial and adjacent waters are shown on National Ocean Service charts and Defense Mapping Agency Hydrographic/Topographic Center charts of the same area and comparable scale. Danger areas established for short periods of time are not charted, but are announced locally. Danger areas in effect for longer periods are published in the *Notice to Mariners.* Any aid to navigation established to mark a danger area or a fixed or floating target is shown on charts.

Traffic separation schemes show routes to increase safety of navigation, particularly in areas of high density shipping. Traffic separation schemes are shown on standard nautical charts of scale 1:600,000 and larger and are printed in purple. The arrows printed on charts to indicate tracks are intended to give the general direction of traffic only, ships need not set their courses strictly by the arrows. At points where several recommended routes meet, circular or triangular separation zones with traffic direction arrows are shown.

Recommended tracklines, portrayed in black, are used to indicate suggested courses through particular passages and are selected according to their value for oceangoing ships.

A *logarithmic time-speed-distance nomogram* with an explanation of its application is shown on harbor charts at scales of 1:40,000 and larger.

Tidal boxes (fig. 516a) are shown on charts of scales 1:75,000 and larger.

TIDAL INFORMATION

Place	Position		Height above datum of soundings			
			Mean High Water		Mean Low Water	
	N. Lat.	E. Long.	Higher	Lower	Lower	Higher
			meters	meters	meters	meters
Olongapo	14°49′	120°17′	...0.9...	... 0.4...	...0.0...	... 0.3....

FIGURE 516a.—Tidal box.

Tabulations of controlling depths (fig. 516b) are shown on National Ocean Service harbor charts.

Title. The chart title may be at any convenient location, usually in some area not important to navigation. It is composed of several distinctive parts as shown in figure 516c.

Reproductions of Foreign Charts

517. Modified facsimile charts are modified reproductions of foreign charts produced in accordance with bilateral agreements. Such agreements serve to provide the mariner with more up-to-date charts.

Modified facsimile charts published by the Defense Mapping Agency Hydrographic/Topographic Center are, in general, reproduced with minimal changes. Such changes may include all or part of the following:

1. The original name of the chart is removed and replaced by an anglicized version.

NANTUCKET HARBOR Tabulated from surveys by the Corps of Engineers - report of June 1972 and surveys of Nov. 1971							
Controlling depths in channels entering from seaward in feet at Mean Low Water					Project Dimensions		
Name of Channel	Left outside quarter	Middle half of channel	Right outside quarter	Date of Survey	Width (feet)	Length (naut. miles)	Depth M.L.W. (feet)
Entrance Channel	**11.1**	**15.0**	**15.0**	11-71	300	1.2	15
Note.-The Corps of Engineers should be consulted for changing conditions subsequent to the above.							

FIGURE 516b.—Tabulations of controlling depths.

GENERAL GEOGRAPHIC AREA → WEST INDIES
CUBA—SOUTH COAST

SPECIFIC LOCALITY → GUANTANAMO BAY

SOURCE AND DATE(S) OF SURVEY → From U.S. Navy surveys to 1951
with additions and corrections to 1975

CHART SOUNDING DATUM → SOUNDINGS IN METERS
(Under 31 in meters and half meters)
reduced to the approximate level of Mean Sea Level
HEIGHTS IN METERS ABOVE MEAN SEA LEVEL
Contour interval 10 meters

LEGEND → For Symbols and Abbreviations, see Chart No. 1
Names and boundaries are not necessarily authoritative

PROJECTION → MERCATOR PROJECTION
HORIZONTAL DATUM → WORLD GEODETIC SYSTEM—1972 DATUM
SCALE → SCALE 1:200,000 AT LAT. 19°50′

FIGURE 516c.—A chart title.

2. English language equivalents of names and terms on the original chart are printed in a suitable glossary on the reproduction, as appropriate.

3. All hydrographic information, except bottom characteristics, is shown as depicted on the original chart.

4. Bottom characteristics are shown as depicted in Chart No. 1.

5. The unit of measurement used for soundings is shown in block letters outside the upper and lower neatlines.

6. A scale for converting charted depth to feet, meters, or fathoms is added.

7. A blue tint is shown from a significant depth curve to the shoreline.

8. A blue tint is added to all dangers enclosed by a dotted danger curve.

9. A blue tint is added to dangerous wrecks, foul areas, obstructions, rocks awash, sunken rocks, and swept wrecks.

10. Aids to navigation, landmarks, and special area symbols and abbreviations on the original chart are changed to conform with Chart No. 1.

11. Caution notes are shown in purple and enclosed in a box.

12. Restricted, danger, and prohibited areas are usually outlined in purple and labeled "RESTRICTED AREA," "DANGER AREA," etc.

13. Traffic separation schemes are shown in purple.

14. A note on traffic separation schemes, printed in purple, is added to the chart.

15. Wire dragged (swept) areas are shown in purple or green.

16. If plottable, suitable corrections are provided to shift the horizontal datum to the World Geodetic System (1972).

518. International charts.—The need for mariners and chartmakers to understand and use nautical charts of different nations became increasingly apparent during the late 19th and 20th centuries as the maritime nations of the world developed their own establishments for the compilation and publication of nautical charts from hydrographic surveys. There followed a growing awareness that international standardization of symbols and presentation was desirable, which led to 22 maritime nations sending their representatives to a Hydrographic Conference in London in 1919. That conference resulted in the establishment of the International Hydrographic Bureau (IHB) in Monaco in 1921, where the seat of the International Hydrographic Organization (IHO), with a membership of over 40 States remains today.

Recognizing that there was considerable duplication of effort by various Member States when each was charting the same parts of the ocean, and being conscious of the significant level of standardization in chart symbolization which had been reached, a move was made by the IHO in 1967 to introduce the first international chart. A Committee of representatives from six Member States was organized which reported in 1970. The Committee drew up plans and specifications for two series of international charts of the oceans on scales 1:10,000,000 and 1:3,500,000, respectively. The limits of each of some 83 of these charts, giving worldwide small scale navigational cover, were agreed, and responsibility for compiling each of these has subsequently been accepted by Member States' Hydrographic Offices.

Once a Member State publishes an international chart, reproduction material is made available to any other Member State which may wish to print the chart for its own purposes.

International charts can be identified by the letters INT before the chart number and the International Hydrographic Organization seal in addition to what other seals may appear on the chart.

Chart Numbering System

519. Chart numbering system.—The numbering of nautical charts produced and issued by the Defense Mapping Agency Hydrographic/Topographic Center and the National Ocean Service is based on a system in which numbers are assigned in accordance with the scale range and geographical area of coverage of a chart. With the exception of certain charts produced for military use only, one- to five-digit numbers are used. And with the exception of one-digit numbers, the first digit identifies the area; the number of digits establishes the scale range (fig. 519a). The *one-digit numbers* are used for products in the chart system which are not actually charts, such as Chart No. 1, *Nautical Chart Symbols and Abbreviations,* chart 5, *National Flags and Ensigns,* and foreign symbols and abbreviations sheets for military use.

Number of Digits	Scale
1	No Scale
2	1:9,000,000 and smaller.
3	1:2,000,000 to 1:9,000,000.
4	Nonnavigational and special purpose.
5	1:2,000,000 and larger.

FIGURE 519a.—Scale ranges for number of digits in chart number.

Two- and three-digit numbers are assigned to those small-scale charts which depict the major portion of an ocean basin or a large area, with the first digit identifying the ocean basin (fig. 519b). Two-digit numbers are used for charts of scale 1:9,000,000 and smaller. Three-digit numbers are used for charts of scale 1:2,000,000 to 1:9,000,000.

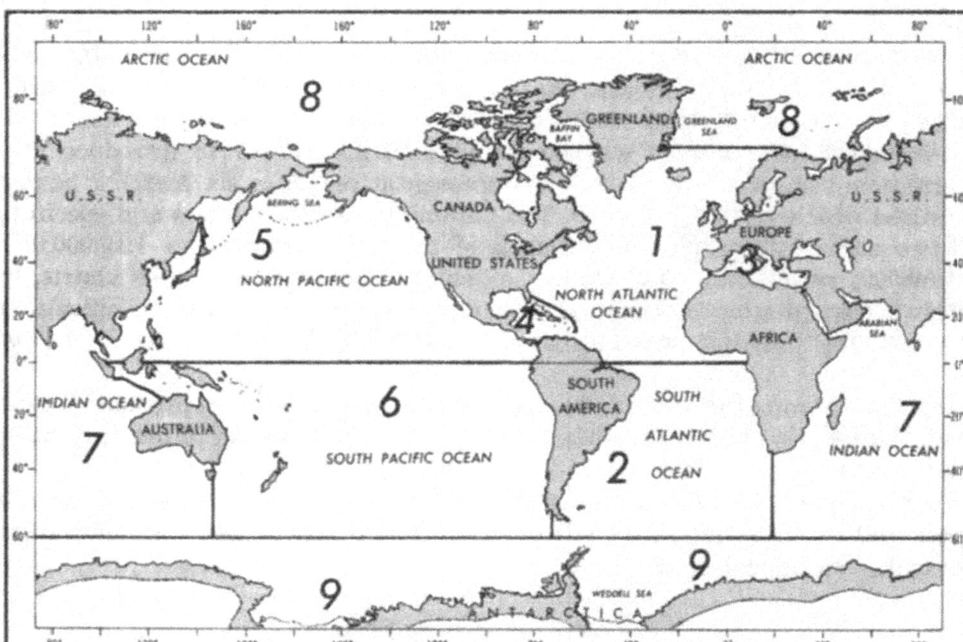

FIGURE 519b.—Ocean basins.

Due to the limited sizes of certain ocean basins, no charts for navigational use at scales of 1:9,000,000 and smaller are published to cover these basins. The otherwise unused two-digit numbers (30 to 49 and 70 to 79) are assigned to special world charts, such as chart 33, *Horizontal Intensity of the Earth's Magnetic Field,* chart 42, *Magnetic Variation,* and chart 76, *Standard Time Zone Chart of the World.*

One exception to the scale range criteria for three-digit numbers is the use of three-digit numbers for a series of position plotting sheets which are of larger scale than 1:2,000,000 because they have application in ocean basins and can be used in all longitudes.

Four-digit numbers are used for nonnavigational and special purpose charts, such as chart 5090, *Maneuvering Board,* chart 5101, *Gnomonic Plotting Chart North Atlantic,* and chart 7707, *Omega Plotting Chart.*

Five-digit numbers are assigned to those charts of scale 1:2,000,000 and larger that cover portions of the coastline rather than significant portions of ocean basins. These charts are based on the regions of the nautical chart index (fig. 519c).

The first of the five digits indicates the region; the second digit indicates the subregion; the last three digits indicate the geographical sequence of the chart within the subregion. Many numbers have been left unused in order that future charts may be placed in their proper geographical sequence as they are produced.

In order to establish a logical numbering system within the geographical subregions (for the 1:2,000,000 and larger-scale charts), a worldwide skeleton framework of coastal charts was laid out at a scale 1:250,000. This skeleton series was used as basic coverage for the numbering except in areas where a coordinated series at about this scale already existed. An example of an exception is the coast of Norway were a coordinated series of 1:200,000 coast charts is in existence. Within each region, the geographical subregions are numbered counterclockwise around the continents, and within each subregion the basic (1:250,000 skeleton) series also is numbered counterclockwise around the continents. The skeleton coverage is assigned generally every 20th digit, except that the first 40 numbers in each subregion are reserved for smaller-scale coverage. Charts with scales larger than the skeleton coverage are assigned one of the 19 numbers following the number assigned to the skeleton sheet within which it falls. Thus, charts on the west coast of the Iberian Peninsula and the northwest coast of Africa are numbered as shown in figure 519d.

As shown in figure 519d, five-digit numbers are assigned to the charts produced by other hydrographic offices. This numbering system is applied to foreign charts so that they can be filed in logical sequence with the charts produced by the Defense Mapping Agency Hydrographic/Topographic Center and the National Ocean Service.

Exceptions to the numbering system to satisfy military needs are as follows:

1. Bottom contour and non-submarine contact charts at a scale larger than 1:2,000,000 do not portray portions of a coastline but chart parts of the ocean basins. In view of the characteristics of these charts, they are identified with an alphabetical character plus four digits. All bottom contour charts will eventually be Omega versions only. The letter *B* denotes an interim version. The letter *C* denotes bottom contour charts with Loran-C information. The letter *E* denotes bottom contour charts with Omega information.

2. Combat charts at a scale of 1:50,000, which would otherwise be assigned five-digit numbers, are assigned four digits separated by a letter of the alphabet. The first two digits indicate the region and subregion; the third character is a letter of the alphabet; the last two digits indicate the geographical sequence of the chart within the subregion.

The Defense Mapping Agency stock number is shown in the lower left-hand corner of the chart directly under the chart number. The letters and numbers have chart significance as indicated in figure 519e.

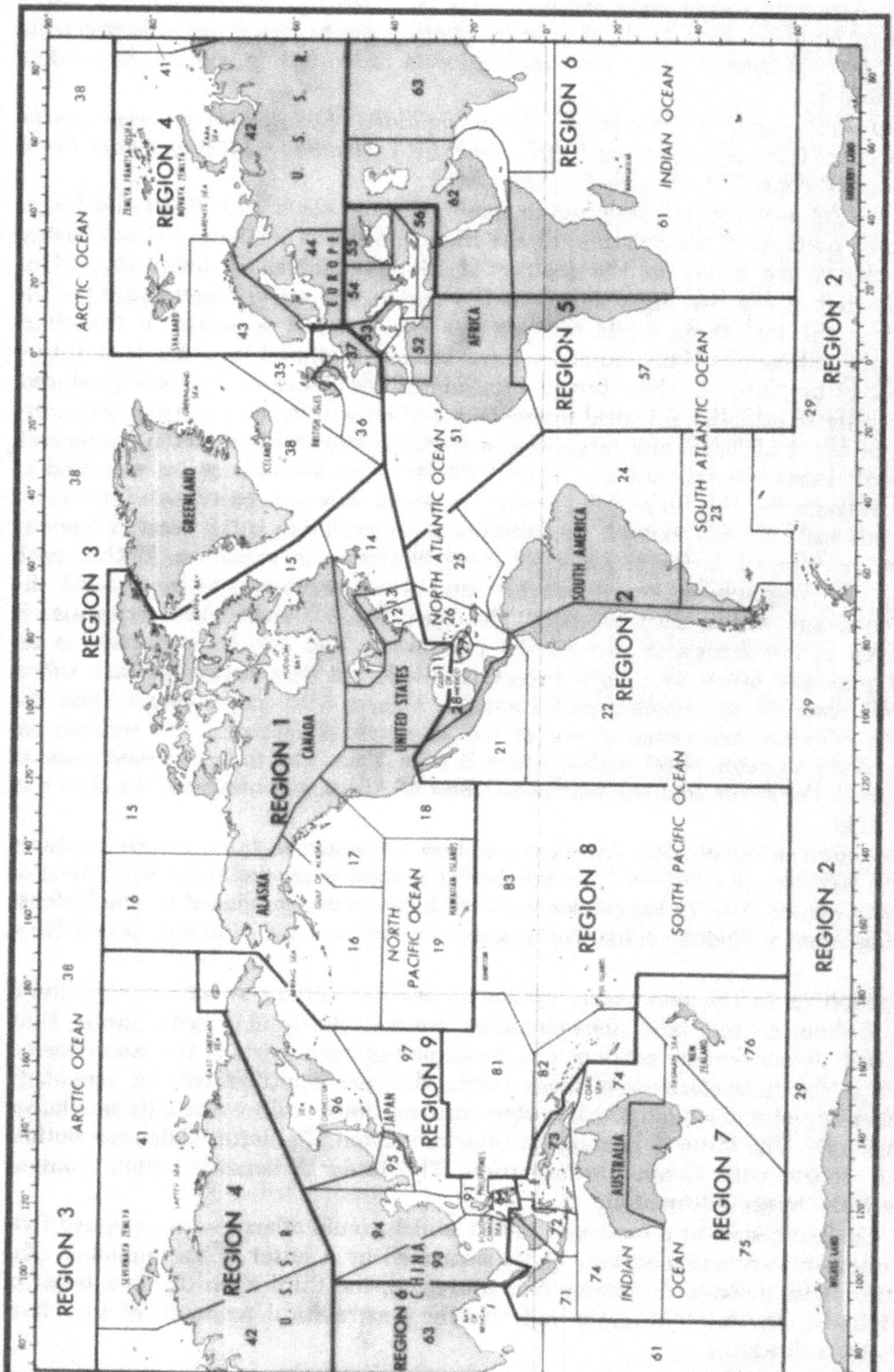

FIGURE 519c.—Regions and subregions of the nautical chart index.

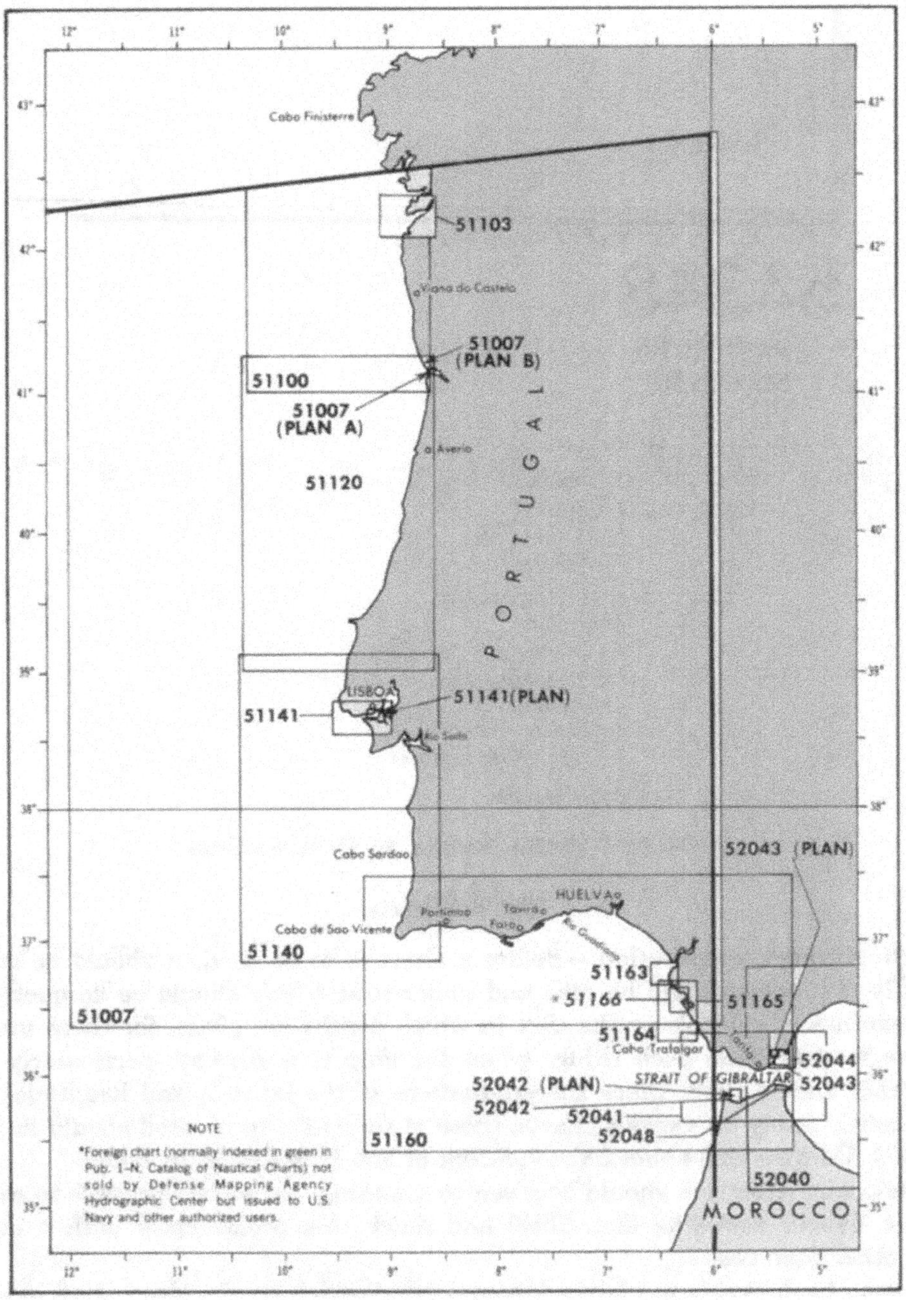

FIGURE 519d.—Area of subregion 51 illustrating the numerical sequence of larger-scale charts along a coast.

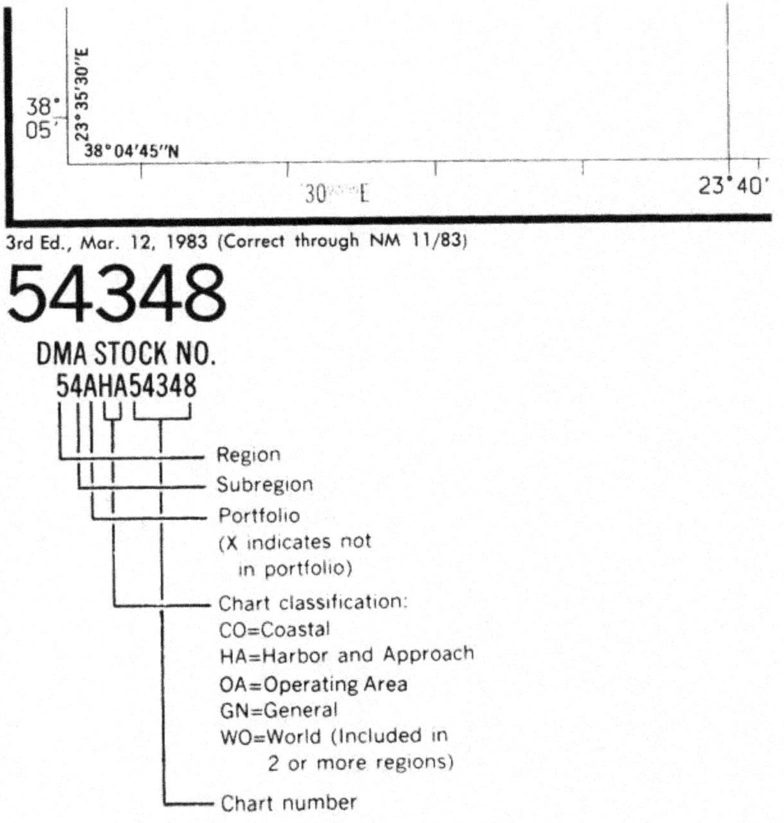

FIGURE 519e.—Defense Mapping Agency stock number.

Use of Charts

520. Advance preparation.—Before a chart is to be used, it should be studied carefully. All notes should be read and understood. There should be no question of the meanings of symbols or the unit in which depths are given, for there may not be time to determine such things when the ship is underway, particularly if an emergency should arise. Since the graduations of the latitude and longitude scales differ considerably on various charts, those of the chart to be used should be noted carefully. Dangers and abnormal conditions of any kind should be noted.

Particular attention should be given to soundings. It is good practice to select a realistic danger sounding (art. 1013) and mark this prominently with a colored pencil other than red.

It may be desirable to place additional information on the chart. Arcs of circles might be drawn around navigational lights to indicate the limit of visibility at the height of eye of an observer on the bridge. Notes regarding the appearance of light structures, tidal information, prominent ranges, or other information from the light lists, tide tables, tidal current tables, and sailing directions might prove helpful.

The particular preparation to be made depends upon the requirements and the personal preferences and experience of the individual navigator. The specific information selected is not important. But it *is* important that the navigator familiarize himself with his chart so that in an emergency the information needed will be available and there will be no question of its meaning.

521. Maintaining charts.—The print date in the lower left-hand corner of the chart is the date of the latest *Notice to Mariners* used to update the chart. Responsibility for maintaining it after this date lies with the user. *An uncorrected chart is a menace.* The various issues of *Notice to Mariners* subsequent to the print date contain information needed for maintaining charts. The more urgent items are also given in advance in the *Daily Memorandum* or by radio broadcast. A convenient way of keeping a record of the *Notice to Mariners* corrections made to each chart on hand is by means of the 5x8-inch *Chart/Publication Correction Record Card* (DMAHTC 8660/9).

Periodically DMAHTC publishes a *Summary of Corrections* containing previously published *Notice to Mariners* corrections.

When a new edition of a chart is published, it should be obtained and the old one retired from use. The very fact that a new edition has been prepared generally indicates that there have been changes that cannot adequately be shown by hand correction.

522. Use and stowage of charts.—Charts are among the most important aids of the navigator, and should be treated as such. When in use they should be spread out flat on a suitable chart table or desk, and properly secured to prevent loss or damage. Every effort should be made to keep charts dry, for a wet chart stretches and may not return to the original dimensions after drying. The distortion thus introduced may cause inaccurate results when measurements are made on the chart.

Permanent corrections to charts should be made in ink so that they will not be inadvertently erased. All other lines should be drawn lightly in pencil so that they can be easily erased without removing permanent information or otherwise damaging the chart. To avoid possible confusion, lines should be drawn no longer than necessary, and adequately labeled. When a voyage is completed, the charts should be carefully and thoroughly erased unless there has been an unusual incident such as a grounding or collision, when they should be preserved without change, as they will undoubtedly be requested by the investigating authority. After a chart has been erased, it should be inspected carefully for possible damage and for incompletely erased or overlooked marks that might prove confusing when the chart is next used.

When not in use charts should be stowed flat in their proper drawers or portfolios, with a minimum of folding. The stowed charts should be properly indexed so that any desired one can be found when needed. In removing or replacing a chart, care should be exercised to avoid damage to it or other charts.

A chart that is given proper care in use and stowage can have a long and useful life.

523. Chart lighting.—In the use of charts it is important that adequate lighting be provided. However, the light on the bridge of a ship underway at night should be such as to cause the least interference with the darkness-adaptation of the eyes of bridge personnel who watch for navigational lights, running lights, dangers, etc. Experiments by the Department of the Navy have indicated that red light is least disturbing to eyes which have been adapted to maximum vision during darkness. In some instances red lights, filters, or goggles have been provided on the bridges or in chartrooms of vessels. However, the use of such light seriously affects the appearance of a chart. Red, orange, and buff disappear. Other colors may appear changed. This has led to the substitution of nautical purple for red and orange, and gray for buff on some charts. However, before a chart is used in any light except white, a preliminary test should be made and the effect noted carefully. If a glass or plastic top is provided for the chart table or desk, a dim white light *below* the chart may provide sufficient illumination to permit chart reading, without objectionable disturbance of night vision.

524. Use of small-craft charts.—Although the small-craft charts published by the National Ocean Service are designed primarily for boatmen, these charts at scales of 1:80,000 and larger are in some cases the only charts available of inland waters transited by large vessels. In other cases the small-craft charts may provide a better presentation of navigational hazards than the standard nautical chart because of scale and detail. *Therefore, it behooves navigators of large vessels transiting inland waters not to ignore the small-craft charts.*

PART TWO
PILOTING AND DEAD RECKONING

PART TWO

PILOTING AND DEAD RECKONING

CHAPTER VI

INSTRUMENTS FOR PILOTING AND DEAD RECKONING

Introduction

601. Kinds of instruments.—The word "instrument" has several meanings, at least two of which apply to navigation: (1) an implement or tool, and (2) a device by which the present value of a quantity is measured. Thus, a straightedge and a mechanical log are both instruments, the first serving as a tool, and the second as a measuring device. This chapter is concerned with the navigational instruments used for plotting, and those for measuring distance or speed, depth, and direction. Instruments for measuring time are discussed in chapter XV. These quantities are the basic data in dead reckoning (ch. VIII) and piloting (ch. X). Other instruments are discussed in chapters XV and XXXVII.

In addition to the instruments discussed, several others are important to the navigator. Binoculars are helpful in observing landmarks. A flashlight has many uses, the principal one being to illuminate the scales of instruments when they are to be read at night. Erasers should be soft, and pencils should not be so hard that they damage the surface of the chart. The navigator's chart is discussed in chapter V.

Plotting Instruments

602. Dividers and compasses.—**Dividers,** or "pair of dividers," is an instrument originally used for dividing a line into equal segments. The instrument consists essentially of two hinged legs with pointed ends which can be separated to any distance from zero to the maximum imposed by physical limitations. The setting is retained either by friction at the hinge, as in the usual navigational dividers, or by means of a screw acting against a spring.

If one of the legs carries a pencil or ruling pen, the instrument is called **compasses.** The two legs may be attached to a bar of metal or wood instead of being hinged, thus permitting greater separation of the points. Such an instrument is called **beam compasses** or **beam dividers.**

The principal use of dividers in navigation is to measure or transfer distances on a chart, as described in article 804. Compasses are used for drawing distance circles or any plotting requiring an arc of a circle.

The friction at the hinge of most dividers and compasses can be varied, and should be adjusted so that the instrument can be manipulated easily with one hand, but will retain the separation of the points in normal handling. A drop of oil on the hinge may be required occasionally. The points should be sharp, and should have equal length, permitting them to be brought close together for the measurement of very short distances.

For navigation, it is desirable to have dividers and compasses with comparatively long legs, to provide adequate range for most requirements. It is desirable to learn to manipulate dividers or compasses with one hand.

603. Parallel rulers are an instrument for transferring a line parallel to itself. In its most common form it consists of two parallel bars or rulers, connected in such a manner that when one is held in place on a flat surface, the other can be moved, remaining parallel to its original direction. Firm pressure is required on one ruler while the other is being moved, to prevent slippage. The principal use of parallel rulers in navigation is to transfer the direction of a charted line to a compass rose, and vice versa.

The edges of the rulers should be truly straight; and in the case of double-edged rulers, should be parallel to each other in order that either edge can be used. Parallelism can be tested by comparison of all edges with the same straight line, as a meridian or parallel of a Mercator chart. The linkage can be tested for looseness and lack of parallelism by "walking" the rulers between parallel lines on opposite sides of the chart and back again.

Some metal parallel rulers have a protractor engraved on the upper surface to permit orientation of the ruler at any convenient meridian.

In one type of instrument, parallelism during transfer is obtained by supporting a single ruler on two knurled rollers. Both rollers have the same diameter, and the motion of one is transmitted to the other by an axle having a cover which provides a convenient handle. This type of ruler is convenient and accurate, and is less likely to slip than the linked double-ruler type. However, care is necessary to prevent its rolling off the chart table when the vessel is rolling or pitching.

Directions can also be transferred by means of two triangles such as are used in drafting, or by one triangle and a straightedge. One edge of a triangle is aligned in the desired direction and the triangle is then moved along a straightedge held firmly against one of its other edges until the first edge is at the desired place on the chart. Some triangles have protractors (art. 604) engraved on them to assist in transferring lines. Such a triangle becomes a form of plotter (art. 605).

604. Protractor.—A **protractor** is a device for measuring angles on a chart or other surface. It consists essentially of a graduated arc, usually of 180°, on suitable material such as metal or plastic.

A **three-arm protractor** consists essentially of a circular protractor with three radial arms attached. This instrument is discussed in greater detail in article 1102.

605. Plotters.—The increased popularity of graphical methods in practical navigation during recent decades has resulted in the development of a wide variety of devices to facilitate plotting. In its most common form, such a device consists essentially of a protractor combined with a straightedge. There are two general types, one having no movable parts, and the other having a pivot at the center of the arc of the protractor to permit rotation of the straightedge around the protractor. Examples of the fixed type are illustrated in figure 605. Those shown were designed for air navigation, but are applicable to many processes of marine navigation. The direction of the straightedge is controlled by placing the center of the protractor arc and the desired scale graduation on the same reference line. If the reference line is a meridian, the directions shown by the straightedge are true geographic directions. If, as in some processes of celestial navigation, it is desired to plot a line perpendicular to another line, the direction may be measured from a parallel of latitude or its equivalent, instead of adding or subtracting 90° from the value and measuring from a meridian. Some fixed-type plotters have auxiliary scales labeled to indicate true direction if a parallel is used as the reference.

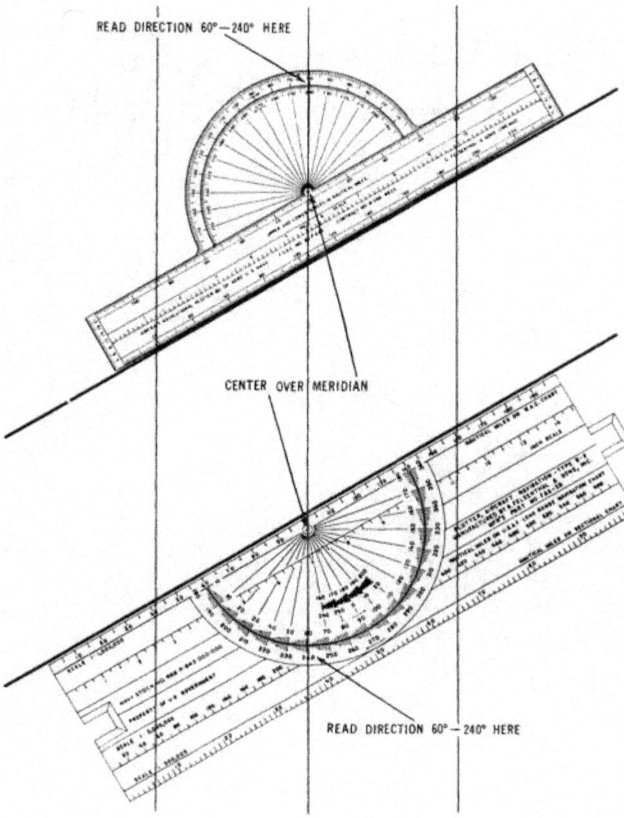

FIGURE 605.—Two plotters having no movable parts.

Most plotters also provide linear distance scales, as shown in figure 605. In the movable-arm type of plotter, a protractor is aligned with a meridian, and the movable arm is rotated until it is in the desired direction.

606. Drafting machine.—If a chart table of sufficient size is available, a **drafting machine** (fig. 606) is probably the most desirable plotting instrument. The straightedge of this instrument can be clamped so as to retain its direction during movement over the entire plotting area. Straightedges of various lengths and linear scales are interchangeable. Some models make provision for mounting two straightedges perpendicular to each other. However, for most purposes of navigation, the perpendicular is more conveniently obtained by the use of a triangle with a single straightedge. The movable protractor also retains its orientation, and can be adjusted to conform to the compass rose of a chart secured in any position on the chart table. Directions of the straightedge can then be read or set on the protractor without reference to charted compass roses. Use of the clamped protractor requires that charted meridians be straight and parallel, as on a Mercator chart (art. 305). Its use is restricted with projections such as the Lambert conformal (art. 314), on which meridians converge.

When a drafting machine is used, the chart or plotting sheet is first secured to the chart table. The straightedge is aligned with a meridian (or parallel) and clamped in position. The protractor is then adjusted so that 000° and 180° (090° and 270° if a parallel is used) are at the ruler indices, and clamped. With this setting, any subsequent position of the ruler is indicated as a true direction. If the protrac-

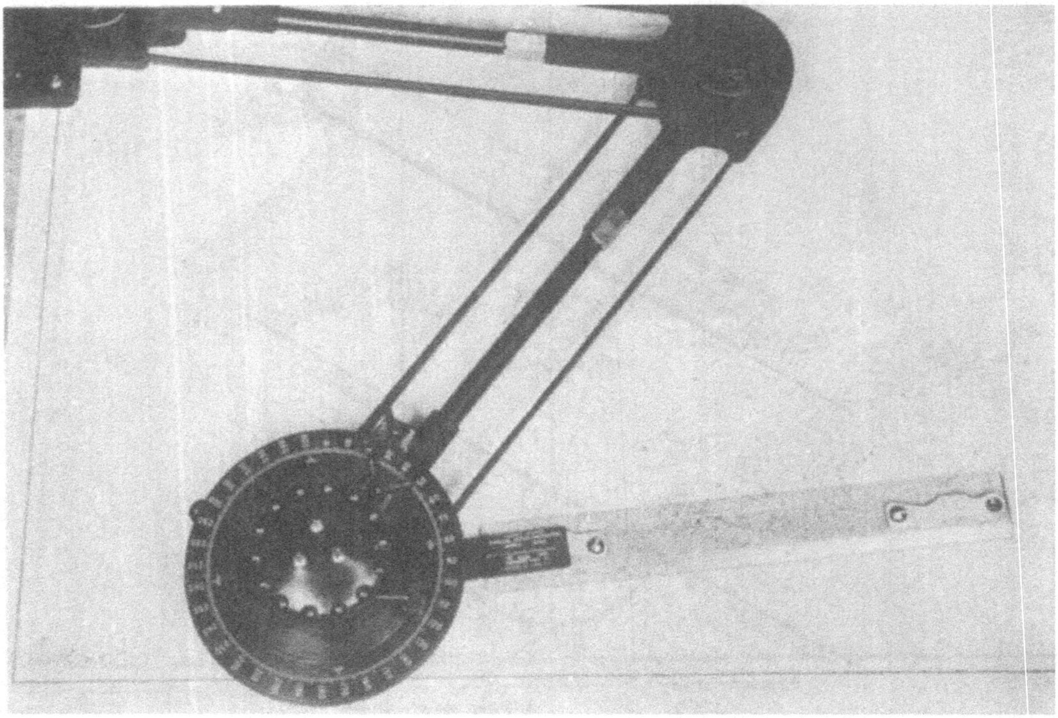

FIGURE 606.—Drafting machine.

tor is offset by the amount of the compass error (ch. VII), true directions can be plotted by setting the straightedge at the compass direction on the protractor, without need for applying compass error arithmetically. However, it is generally preferable to keep it set to true directions.

If accurate results are to be obtained, the anchor base must be rigidly fastened to the chart table. This should be checked from time to time, as the base may be loosened by vibration or normal use. The pivots in the anchor base should be firm without binding. The endless belts of the parallel motion mechanism should be taut if rigidity of the ruler is to be preserved. Provision is usually made for adjusting each of the various rulers to uniformity of alignment so that any other ruler can be substituted without changing the setting. As with parallel rulers, the device can be checked for parallelism by means of meridians or parallels on opposite sides of a Mercator chart.

Distance and Speed Measurement

607. Units of measurement.—Mariners generally measure horizontal distances in nautical miles (art. 205), but occasionally in yards or feet. Feet, meters, and fathoms are used for measuring depth of water, and feet or meters for measuring height above water. Nations which have adopted the metric system use meters in place of yards, feet, and fathoms, and for some purposes they use kilometers in place of nautical miles. Conversion factors for these and other units are given in appendix D. Nautical miles of 6,076.11549 feet (approximately) and land or statute miles of 5,280 feet can be interconverted by means of table 20. Meters, feet, and fathoms can be interconverted by means of table 21.

Speed is customarily expressed in knots (art. 206), or for some purposes, in kilometers per hour, or yards or feet per minute. For short distances, a nautical mile can be considered equal to 2,000 yards or 6,000 feet. This is a useful relationship because

$$\frac{6{,}000 \text{ feet}}{60 \text{ minutes}} = 100 \text{ feet per minute.}$$

Thus, speed in knots is equal approximately to hundreds of feet per minute or, hundreds of yards per 3-minute interval.

608. Distance, speed, and time are related by the formula

$$\text{distance} = \text{speed} \times \text{time}.$$

Therefore, if any two of the three quantities are known, the third can be found. The units, of course, must be consistent. Thus, if speed is measured in knots, and time in hours, the answer is in nautical miles. Similarly, if distance is measured in yards, and time in minutes, the answer is in yards per minute.

Table 19 is a speed, time, and distance table which supplies one of the three values if the other two are known. It is intended primarily for use in finding the distance steamed in a given time at a known speed. Table 18 is for use in determining speed by measuring the time needed to steam exactly one mile.

The solution of problems involving distance, speed, and time can easily be accomplished by means of a slide rule (art. 115 vol. II). If the index of scale *C* is set opposite speed in knots on scale *D*, the distance in nautical miles appears on scale *D* opposite time in hours on scale *C*. If 60 of scale *C* is set opposite speed in knots on scale *D*, the distance covered in any number of *minutes* is shown on scale *D* opposite the minutes on scale *C*. Several circular slide rules particularly adapted for solution of distance, speed, and time problems have been devised. One of these, called the "Nautical Slide Rule," is shown in figure 608.

609. Measurement of distance to an object can be made in a variety of ways, as by radar (art. 4202), vertical angle (table 9), range finder, or by several indirect methods. Another method used principally for measuring distance between ships in formation, but useful in measuring other distances, is by means of a small, hand-held instrument called a **stadimeter.**

Two types of stadimeters are illustrated in figure 609a. Both the Brandon or sextant type and the Fisk type operate on the principle used in table 9:

In a plane right triangle, *ABC*, having opposite sides *a*, *b*, and *c*,

$$\tan A = \frac{a}{b}, \text{ and } b = a \cot A.$$

This is applied to the stadimeter as shown in figure 609b. The height of the object is set on the height scale of the instrument, and the measured subtended angle is expressed in yards on the distançe (range) scale. To measure the angle, one directs the line of sight through the instrument to the water line of the object observed, and adjusts the range index until the reflection of the top of the object is seen in coincidence with the water line. If the readings are not within the scale of the instrument, some fraction or multiple of the height can be used and a corresponding adjustment made to the answer. Thus, if *half* the height is set on the instrument, the distance indicated is *half* the correct value.

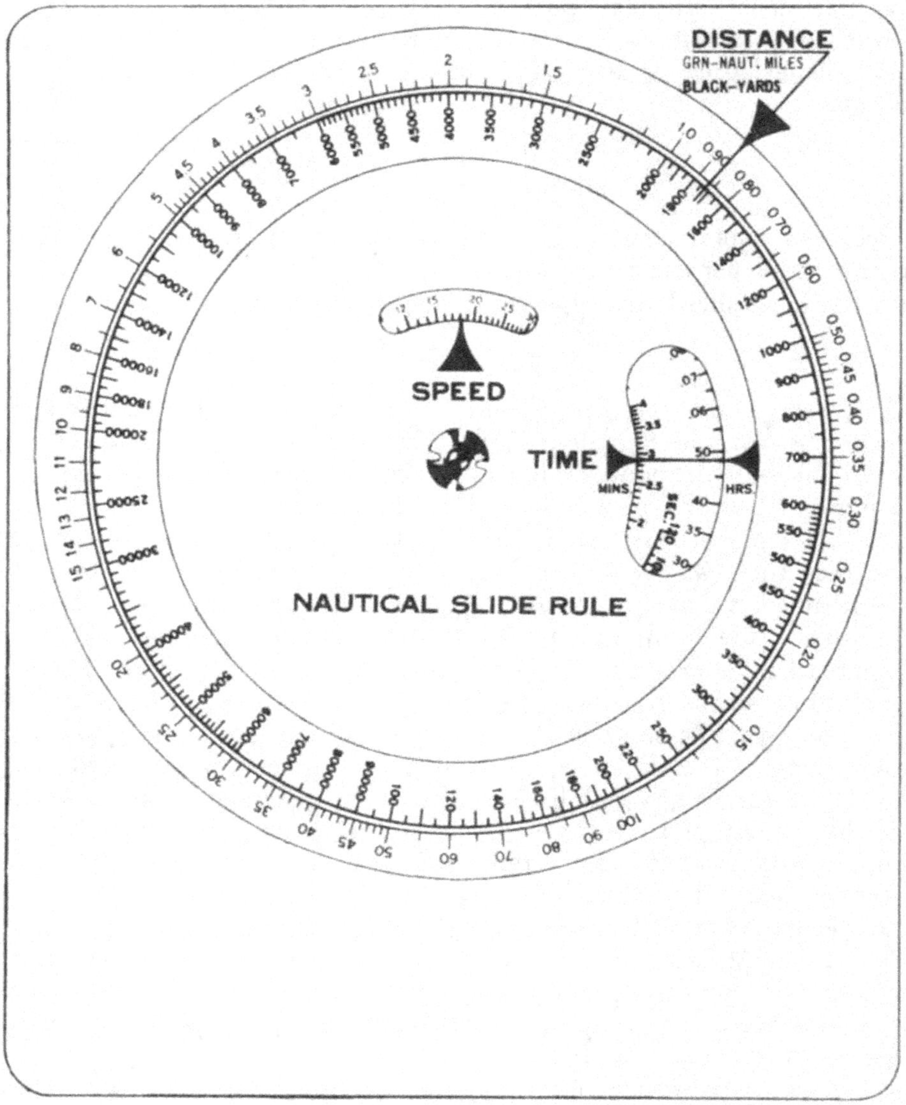

FIGURE 608.—The Nautical Slide Rule.

Since the observer's eye is not at the water level, a right angle is not necessarily formed between the line of sight and the top of the observed object. However, the resulting error is so small that it can be neglected under ordinary circumstances.

The aspect of a ship observed should be considered in stadimeter ranges. Thus, little error is introduced if the observer is broad on the beam of the other vessel, as in figure 609b, but less accuracy is obtained if the other vessel presents an end-on view, unless the water line directly below the masthead is correctly estimated.

A stadimeter can be used to indicate that a *change* in distance has occurred, even when the height of the object is not known. Similar indication of a change in distance can be obtained by a sextant (art. 1005), or the actual distance can be determined by the measured angle and table 9 if the height is known.

610. Measurement of distance traveled may be made directly, or the distance can be determined indirectly by means of the speed and time, using the relationship given in article 608.

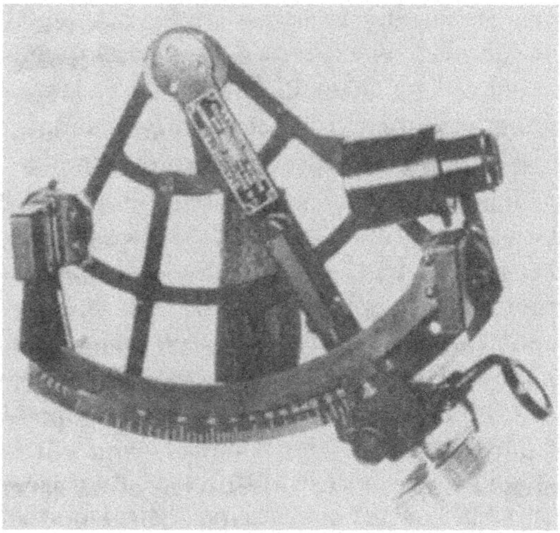

FIGURE 609a.—Stadimeters. Brandon (sextant) type at left; Fisk type at right.

FIGURE 609b.—Geometry of a stadimeter measurement. The distance $b=a \cot A$.

One of the simplest mechanical distance-measuring devices is the **taffrail log,** consisting of (1) a *rotator* which turns like a screw propeller when it is towed through the water; (2) a braided *log line,* up to 100 fathoms in length, which tows the rotator and transmits its rotation to an indicator on the vessel; and (3) a dial and pointer mechanism which registers the distance traveled through the water. In some installations, the readings of the register are transferred electrically to a dial on or near the bridge.

The taffrail log is usually streamed from the ship's quarter, although it may be carried at the end of a short boom extending outboard from the vessel. The log line should be sufficiently long, and attached in such position, that the rotator is clear of the disturbed water of the wake of the vessel; otherwise an error is introduced. Errors may also be introduced by a head or following sea; by mechanical wear or damage, such as a bent fin; or by fouling of the rotator, as by seaweed or refuse.

An accurately calibrated taffrail log in good working order provides information of sufficient reliability for most purposes of navigation. Its readings should be checked at various speeds by towing it over a known distance in an area free from currents. Usually, the average of several runs, preferably in opposite directions, is more accurate than a single one. If an error is found, it is expressed as a percentage and applied to later readings. The calibration should be checked from time to time.

Although a taffrail log is included in the equipment carried by many oceangoing vessels, the convenience and reliability of other methods of determining distance or speed have reduced the dependence formerly placed upon this instrument

611. Measurement of speed.—Speed can be determined indirectly by means of distance and time, or it can be measured directly. All instruments now in common use for measuring speed determine rate of motion *through the water.* This is done (1) by electromagnetic induction, (2) by differential pressure or measurement of the water pressure due solely to the forward motion of the vessel, (3) by measuring the resistance to the motion of the vessel, (4) by means of a small screw propeller having a speed of rotation proportional to the speed of the vessel, and (5) by determining the relationship between vessel speed and speed of rotation of its screw or screws. Instruments for measuring speed, like those for measuring distance, are called **logs.**

Before the development of modern logs, speed was determined in a number of ways. Perhaps the most common primitive device is the **chip log** (art. 112), although a **ground log** (a weight, with line attached, which was thrown overboard and rested on the bottom in shallow water) and a **Dutchman's log** (art. 112) have also been used. These devices are rarely used by modern navigators.

Speed over the bottom can be determined (1) by direct measurement as by doppler sonar speed log (ch. XLV) and sensing accelerations (ch. XLVI); (2) by measuring on the chart or plotting sheet the distance made good between fixes, and dividing this by the time; or (3) by finding the vector sum of velocity through the water and velocity of the current.

612. Problems of water-speed measurement.—Speed measured relative to water is not a stable well-defined quantity because of the motion of the water itself. Most speed logs now used to measure speed through the water measure speed relative to water within the hydrodynamic influence of the vessel's hull and in the immediate vicinity of the motion sensor itself. Speed measured with respect to a small volume of the water disturbed by a vessel's hull may vary significantly from speed with respect to a nearby volume of water. In addition, the motions of a vessel, such as yaw and pitch, introduce variations in the speed over the ground. These speed variations combined with sensor response characteristics can generate appreciable errors in the speed measurement. Many of the uncertainties and errors in the measurement of a vessel's speed are functions of the ocean environment and of the characteristics of the vessel carrying the speed sensor. These causes of measurement uncertainty limit the ultimate accuracy of a speed log installation, irrespective of the accuracy of the instrumentation itself.

The **potential flow field** represents the changes in the water velocity and pressure distributions caused by the shape, size, and orientation to the flow of the vessel carrying the speed sensor. As water changes its flow direction to pass around the underwater body of a vessel of a given configuration, the resulting accelerations and decelerations generate water velocities near the hull that are significantly different from the velocity of water relative to the vessel far from the influence of the vessel's hull.

Since each hull configuration experiences different local velocities, each system (hull and sensor) must be calibrated to remove the *system errors* inherent in the combination of hull and sensor. The usefulness of measured-mile calibrations in calibrating out system errors is limited by the water depth. Many of the available measured-mile courses are too shallow for accurate calibration. The viscosity of the water results in a friction **boundary layer** or layer of water carried along with the hull. The thickness of this boundary layer varies from fractions of inches at the bow to the order of feet near the stern. In this layer, the water velocity changes from zero at the hull to a value within several percent of free-stream velocity at the outer edge of the boundary layer. As a consequence, sensing elements usually have to extend beyond the boundary layer.

Appendages, such as sonar domes, create wakes that cause error in speed sensors located downstream. For optimum operation, speed sensors should not be mounted near wakes of appendages. However, it is not always possible to place a sensor entirely out of the wake of an appendage, and the resulting error must be accepted.

Shallow water effect is a particular aspect of the potential-flow problem that occurs when a vessel is in shallow water. The closeness of the bottom changes the potential-flow velocity distribution by restricting the region in which the water can flow around the hull, causing an increase in the speed reading of a bottom-mounted sensor. The speed problem is compounded because shallow water increases the drag of a vessel causing a decrease of actual speed. It is possible that the speed-reading error coupled with the actual speed decrease may result in no change in indicated speed, while, in fact, the vessel has actually slowed. The precise effects of shallow water on the speed reading are difficult to determine for any specific vessel, and there is no rule of thumb that is applicable to all types and sizes of hulls.

The speed sensor is usually rigidly attached to and undergoes the motions of the vessel carrying it. The vessel may be considered a rigid body with six degrees of freedom, three translation degrees of the vessel's center of gravity and three rotational degrees about the center of gravity. The speed sensor or sensing region is seldom located at the vessel's center of gravity; therefore, the sensor undergoes additional linear displacements proportional to the distance from the center of gravity and the rotation rate.

Average motion errors result from static or dynamic orientation difference from the designed sensor orientation. Speed sensors are usually oriented to measure the water velocity component parallel to the longitudinal axis of the vessel. Neglecting the effects of vessel trim on the water flow near the hull, a change in the static trim from the design trim could reduce the sensed velocity by the cosine of the trim angle through the geometric effects alone. Also, the vessel is normally pitching, yawing, and rolling. The instantaneous pitch and yaw angles cause similar cosine function reductions because the vessel's actual velocity vector is not parallel to the longitudinal axis, and the sensor signal is always reduced by these effects. A more significant error results when the vessel has a drift component due to wind. The speed sensor will not measure this leeway. Most of the average motion errors are quite small (less than a few tenths of 1 percent) and can be neglected. Leeway due to wind may be several knots and cannot be neglected if accurate velocity sensing is required.

Oscillatory errors are the differences between the instantaneous speed reading and the vessel's speed caused by its motion. The oscillatory errors tend to average out over a period of several minutes but may cause appreciable errors in applications requiring a continuous speed input. While almost all of the vessel's motions affect the instantaneous speed reading, the primary cause of the oscillatory error is the pitching motion. For example, for a vessel pitching with an amplitude of only

2½ degrees and a period of 5 seconds with the sensor mounted 10 feet below the center of gravity, the oscillatory error is ±⅓ knot.

Maneuvering errors are those speed-sensing errors caused by controlled motions of the vessel. For example, when a vessel turns, a drift angle is developed between the heading and the actual water velocity vector at the sensor. The speed sensor only measures the longitudinal velocity component, and the speed is reduced by the cosine of the drift angle at the sensor.

Variability is one of the primary characteristics of water movement in the ocean. This allows prediction of only average values, with the possibility that at any particular time the actual current may be quite different in both direction and magnitude. For example, in the Gulf Stream off Florida, there is 72 percent probability that the current will be toward the northeast with a strength greater than 2 knots. However, there is a 4 percent probability of no current, and a 7 percent probability that the current will be toward the southwest with a strength of ⅔ to 1 knot. In the Sargasso Sea there is almost equal probability of any current direction with magnitudes of ⅓ to ⅔ knot. Thus, it becomes obvious that in correcting for current, average values can be predicted, but large errors are possible unless currents can be accurately measured.

Problems of water-speed measurement by doppler sonar speed logs are discussed in chapter XLV.

613. The electromagnetic underwater log, commonly called the **EM log,** consists essentially of a rodmeter, sea valve, indicator-transmitter, and remote control unit (fig. 613a). The rodmeter is a strut of streamlined cross section. A sensing device near its tip develops a signal voltage proportional to the speed of the water past it. Of the two general types of rodmeters, one is fixed hull mounted and the other is retractable through a sea valve. The sea valve is mounted in the hull of the vessel and provides a watertight support through which the retractable rodmeter protrudes. It also seals the vessel's hull when the rodmeter is removed. The indicator-transmitter houses all moving parts of the equipment and performs the following functions: (1) It indicates the vessel's speed in knots on a dial, (2) it operates synchro transmitters to generate corresponding synchro signals for transmission of speed signals to receivers, (3) it registers on a counter the distance in nautical miles that the vessel has traveled through the water, and (4) it develops a synchro signal representing distance for transmission to receivers. The remote control unit may be used to set speed into the indicator-transmitter. In this mode the underwater log is being used as a **dummy log.**

The fixed rodmeter is designed for submarines, mounts on the exterior of the hull, and does not require a sea valve. The hull penetration is small since only the connector end of the rodmeter passes into the hull. There may be two or three rods installed, depending on the submarine. One is usually mounted topside. A fixed rodmeter has also been developed for use on surface vessels.

The principle of the electromagnetic log is that any conductor will produce a voltage when it is moved across a magnetic field or when a magnetic field is moved with respect to the conductor. Figure 613b illustrates this in elementary form. Note that the direction of the field, the direction of motion, and the direction of the induced voltage are all at right angles to each other. If the magnetic field remains constant, the magnitude of the voltage will be proportional to the speed of movement in the direction indicated.

In the electromagnetic log system, a magnetic field produced by a coil in the sensing unit at the outer end of the rodmeter is set up in the water in which the

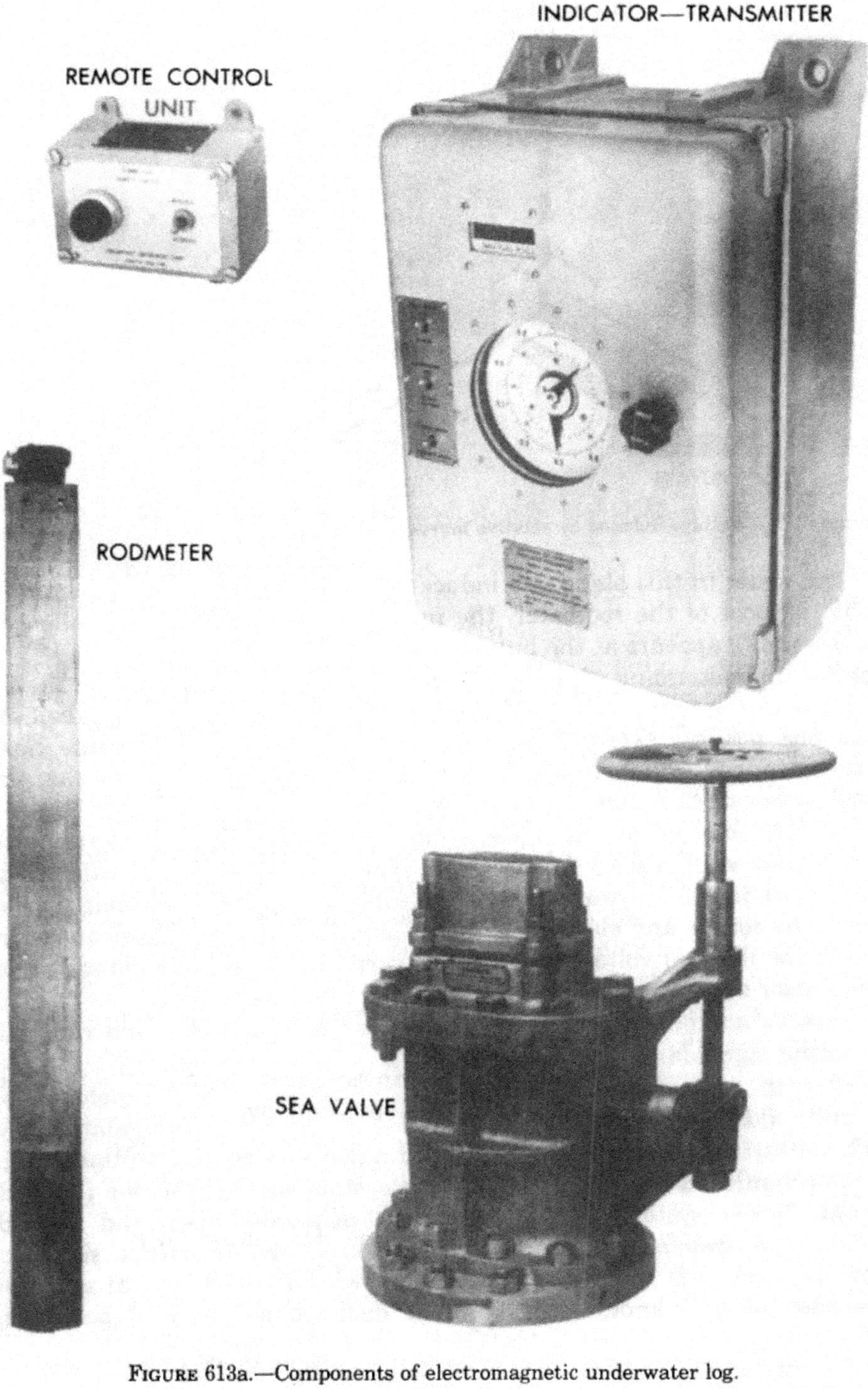

FIGURE 613a.—Components of electromagnetic underwater log.

vessel is floating. The sensing unit's outer surface is an insulating layer or boot, except for two Monel (nickel-copper alloy) buttons, one on each side of the rodmeter. As shown in the inset at the upper left of figure 613c, the horizontal plane in which the buttons are located is in the water. The axis of the coil in the sensing unit is perpendicular to this plane, and so are its magnetic flux lines where they cut the plane. If the vessel is moving in the direction indicated by the white arrow, the flux

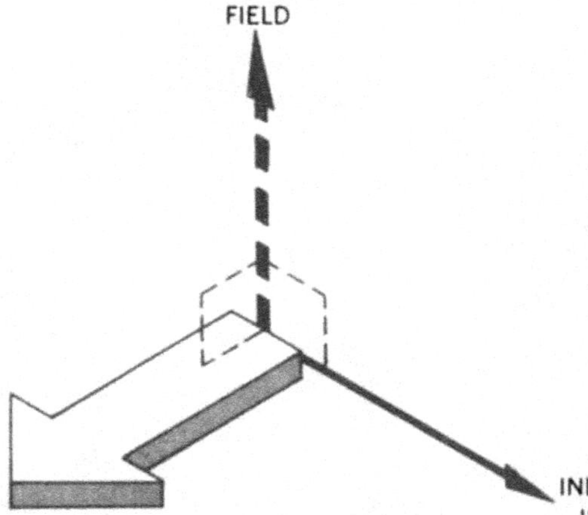

FIGURE 613b.—Voltage induced by relative movement between magnetic field and conductor.

lines cut the water in this plane and induce a voltage in it. Since the plane is cut by the insulating boot of the rodmeter, the induced voltage, sometimes called electromotive force (emf), appears at the buttons in the boot.

A better understanding of this effect can be obtained by comparing the inset of figure 613c with the main part of the figure. The main part shows the coil, its flux field, and one possible symmetrical "water circuit" (the broken white line from button to button) in which the voltage is induced. The inset shows the plane in which the "water circuit" (one of the many that are possible) exists.

The voltage induced by the vessel's motion is, in general, proportional to the rodmeter's speed with respect to the water. The induced voltage is affected by the flow characteristics of the water past the rodmeter (whether laminar (smooth) or turbulent). The sensor and electronics are designed so that no significant current is drawn from the induced voltage and normal variations of water conductivity do not affect the sensor accuracy.

The vessel's motion other than forward speed, such as pitch and roll, will also produce output signals from the rodmeter.

In order to accurately measure the precise speed signal generated by the sensing unit, most electromagnetic log systems employ a "null-balance" type of electronic voltmeter. For systems currently in naval service, this voltmeter employs an electromechanical servo and dial indicator system such as the one illustrated in figure 613d. Newer systems use all electronic instrumentation and have digital displays. Typical system characteristics include: (1) sensor output signal of 325 microvolts per knot; (2) instrument speed accuracy of 0.05 knot; (3) sensitivity to speed changes of 0.01 knot; (4) instrument distance accuracy of better than 1 percent of water distance traveled; and (5) capability to compensate for different hull flow characteristics. Several commercial electromagnetic logs provide slightly less precision.

614. Speed measurement by dynamic water pressure.—When an object is moving through a fluid such as water or air, its forward side is exposed to a *dynamic* pressure which is proportional to the speed at which the object is moving, in addition to the *static* pressure due to depth and density of the fluid above the object. Therefore, if dynamic pressure can be measured, this principle can be used for determining speed.

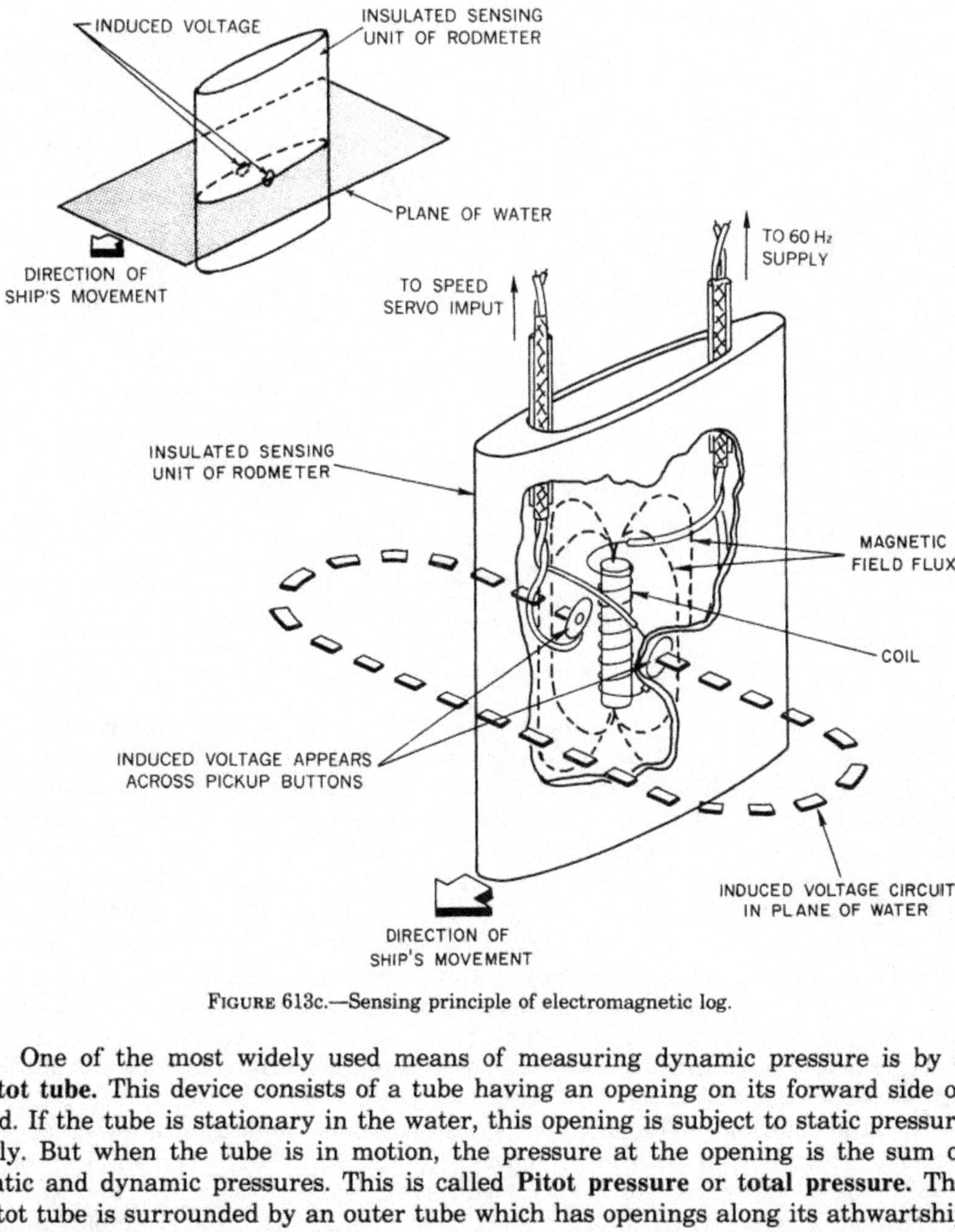

FIGURE 613c.—Sensing principle of electromagnetic log.

One of the most widely used means of measuring dynamic pressure is by a **Pitot tube.** This device consists of a tube having an opening on its forward side or end. If the tube is stationary in the water, this opening is subject to static pressure only. But when the tube is in motion, the pressure at the opening is the sum of static and dynamic pressures. This is called **Pitot pressure** or **total pressure.** The Pitot tube is surrounded by an outer tube which has openings along its athwartship sides. Whether the tube is stationary or in motion, these openings are subject to static pressure only.

In the **Pitot-static log,** the Pitot tube is in the form of a vertical "rodmeter" which extends through and is supported by a sea valve in the vessel's bottom. The tube extends 24 to 30 inches below the bottom of the vessel into water relatively undisturbed by motion of the hull. The two pressures, Pitot and static, are led to separate bellows attached to opposite ends of a centrally pivoted lever. This lever is electrically connected to a mechanism which controls the speed of a pump. When the vessel is dead in the water, the pressures are equal, and the pump is stopped. When the ship is moving, the pump speed is regulated so that the pressures in the two bellows are equalized. Thus, the pump speed is proportional to the ship speed.

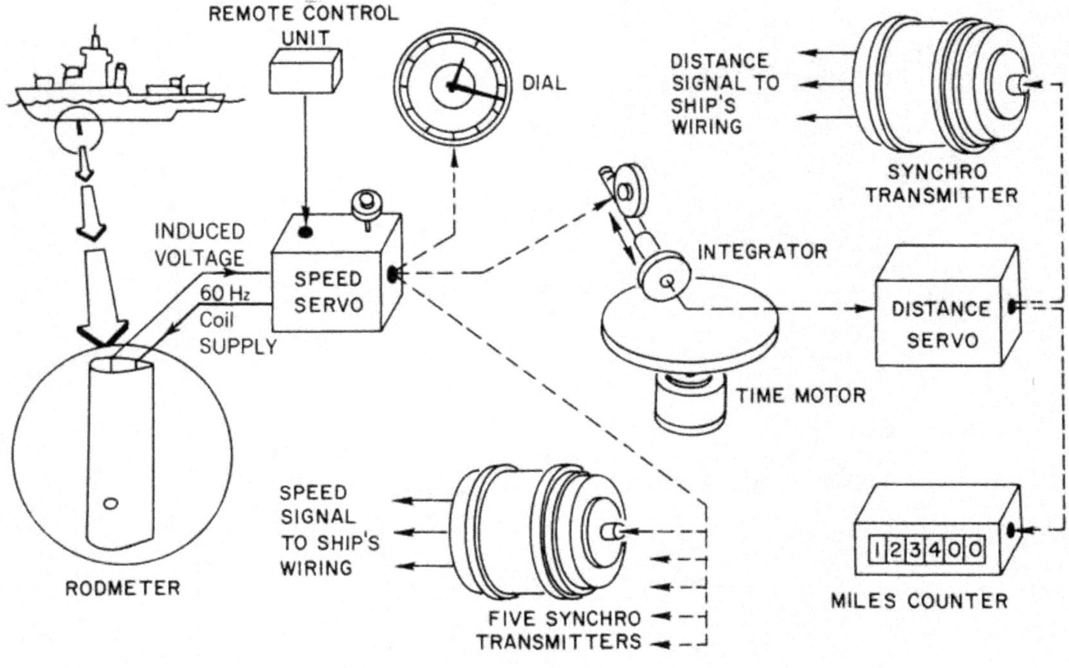

FIGURE 613d.—Electromagnetic log speed measuring system.

Various less accurate instruments have been devised for determining speed by measuring water pressure due to forward motion of the vessel. These are relatively simple, inexpensive instruments intended primarily for use by small craft. One instrument, the **force log**, has a strut which the water pressure forces aft against a calibrated spring. A flexible hydraulic cable transmits the motion to a speed indicator. The force log is probably the oldest of the speed logs used today. The principle of this log is that the resistance (drag) force on a drag strut is proportional to the square of the vessel's speed. Another instrument uses a small scoop attached to the hull of the vessel. The pressure of the water scooped up is transmitted by tubing to the speed indicator, which is essentially a pressure gauge graduated in knots. A third type measures the drag of a small towed object. The accuracy of such devices depends to a large extent upon the refinements of design, manufacture, installation, maintenance, and calibration.

615. Impeller log.—The impeller log may be a hull-mounted or a towed log. The impeller (propeller) rotates as it moves through the water. The number of its revolutions is proportional to the distance traveled through the water, and its speed of rotation is proportional to the vessel's speed. These logs usually employ a magnetic-induction type of pulse frequency generator so that no physical contact, other than bearing surface, is required between the rotor and the body of the instrument. This design permits the use of simple and accurate instrumentation. The characteristic curve of output frequency versus speed for this log is quite linear, except at very low speeds. The nonlinear curve at low speeds is the result of bearing drag on the otherwise freely rotating impeller.

616. Speed by engine revolution counter.—The number of turns of a propeller shaft is proportional to the distance traveled. If the element of time is added, speed can be determined. If the screw were advancing through a solid substance, the distance it would advance in one revolution would be the **pitch** of the screw. Thus, if a propeller having a pitch of 10 feet turns at 200 revolutions per minute, it advances 2,000 feet in one minute, equivalent to a speed of 19.75 knots. It does not do so in water because of **slip,** the difference between the distance it would advance in a solid substance and actual distance traveled, expressed as a percentage of the former. For example, if slip is 18 percent, both the ship's speed and distance covered are reduced by this percentage. Thus, instead of 19.75 knots, the speed is only $19.75 \times 0.82 = 16.2$ knots.

Slip depends upon the type and speed of rotation of the propeller, the type of ship, the condition of loading and ship's bottom, the state of the sea and the ship's course relative to it, and the apparent wind. Despite the many variables, slip can be determined with sufficient accuracy for practical navigation. This is usually accomplished by steaming a known distance and noting the time of passage. The speed corresponding to the number of revolutions being used can then be determined by means of the formula of article 608, in the form

$$\text{speed} = \frac{\text{distance}}{\text{time}}$$

or by reference to table 18 (if the distance is exactly one mile). Thus, speed can be determined directly, without computing slip, and a table or curve of ship speed for various engine revolution speeds can be made. Appendix U provides guidelines for determining speed in this manner. Any suitable distance can be used, but a distance of one nautical mile has been measured at various convenient locations. Each such **measured mile** is suitably marked on the beach, and shown on the chart, with the course to steer.

This method of determining speed is widely used in the merchant marine. By means of an **engine revolution counter** the number of revolutions during any suitable time interval can be measured. If a **tachometer** is available, the *rate* of shaft revolution is determined, usually in revolutions per minute. For best results, allowance should be made for condition of the bottom, draft and trim of the vessel, and the state of the sea.

Depth Measurement

617. Importance.—Accurate knowledge of the depth of water under a vessel is of such navigational importance that there is a legal requirement that American merchant vessels of 500 gross tons or more engaged in ocean and coastwise service "shall be fitted with an efficient mechanical deep-sea sounding apparatus in addition to the deep-sea hand leads."

618. The lead (lĕd) is a device consisting of a suitably marked line having a weight attached to one of its ends. It is used for measuring depth of water. Although the lead is probably the oldest of all navigational aids, it is still a highly useful device, particularly in periods of reduced visibility. Although its greatest service is generally in the shoal water near the shore, it sometimes can provide valuable information when the vessel is out of sight of land.

Two types of lead are in common use: the **hand lead,** weighing from 7 to 14 pounds and having a line marked to about 25 fathoms; and the **deep-sea (dipsey) lead,** weighing from 30 to 100 pounds and having a line marked to 100 fathoms or more in length. The markings commonly used on lead lines are as follows:

Distance from lead in fathoms	*Marking*	*Distance from lead in fathoms*	*Marking*
2	two strips of leather	20	short line with two knots
3	three strips of leather	25	short line with one knot
5	white rag (usually cotton)	30	short line with three knots
7	red rag (usually wool)	35	short line with one knot
10	leather with hole	40	short line with four knots
13	same as three fathoms	45	short line with one knot
15	same as five fathoms	50	short line with five knots
17	same as seven fathoms		etc.

Fathoms marked on the lead line are called **marks.** The intermediate whole fathoms are called **deeps.** In reporting depths it is customary to use these terms, as "by the mark five," "deep six," etc. The only fractions of a fathom usually reported are halves and quarters, the customary expressions being "and a half, eight," "less a quarter, four," etc. A practice sometimes followed is to place distinctive markings on the hand lead line at each foot near the critical depths of the vessel with which it is to be used. The markings should be placed on the lead line when it is wet, and the accuracy of the marking should be checked from time to time to detect any changes in the length of the line. The distance from the hand of the leadsman to the surface of the water under various conditions of loading should be determined so that correct allowance can be made when the marking nearest the surface cannot be observed.

The lead itself has a recess in its bottom. If this recess is filled with tallow or other suitable substance, a sample of the bottom can sometimes be obtained. This information can prove helpful in establishing the position of the vessel. If tallow is not available, some other substance can be used. Soap is suitable if it is replaced from time to time. When the recess is filled for obtaining a sample, the lead is said to be **armed** with the substance used.

619. Echo sounder.—Most soundings are made by means of an **echo sounder.** In this instrument a pulse of electrical energy is converted periodically to sound energy and transmitted downward by a transducer. When the energy strikes the bottom (or any other object having acoustic properties different from those of water), a portion is reflected back to the transducer as an echo. This energy is reconverted to electrical energy for presentation. Because the speed of sound in water is nearly constant, the amount of time which elapses between the transmission of a pulse and the reception of its echo is a measure of the distance traveled, or in this case, depth.

Depth information is presented in either of two ways; namely, an indicator consisting of a cathode-ray tube or a recorder which records depth on calibrated paper.

There are many forms of echo sounder. A typical installation (fig. 619a) consists essentially of a receiver-transmitter, transducer, and interconnecting cables. The receiver-transmitter includes all components and subassemblies of the system, except the transducer and interconnecting cables. The recorder performs the function of recording depth versus time on a paper roll visible through a window in the cabinet's front door. When the recorder is used, it keys the transmitter at a predetermined rate. The cathode-ray tube indicator, mounted below the recorder, indicates depth on a cathode-ray tube; a calibrated circular dial overlays the cathode-ray tube providing a means for reading the depth through an opening in the cabinet's front door. When the indicator is used, it keys the transmitter at a predetermined rate. The transducer converts the electrical energy to sound energy, transmits the sound into the water, receives returned echoes, and converts the returned energy to electrical energy.

When the recorder of the typical installation (fig. 619a) is energized by placing the depth range switch in one of the three recorder positions (600 feet, 600 fathoms, 6000 fathoms), a specially treated recording paper, held between two rollers, is moved at a uniform speed horizontally in front of a grounded plate. At the same time a stylus assembly is moved at a uniform speed vertically across the face of the paper. The rate of movement depends on the range selected, and is so fixed that one of the two stylii appears at the top of the recording paper (0 depth) when the transmitter is keyed and reaches the bottom of the paper at the same time as an echo would be received from the maximum depth of the selected depth range.

The recording paper is marked by the application of a voltage between the stylus and the grounded plate when the echo is received.

In the same installation, depth indication on the 100 feet and 100 fathom scales is given on the face of a cathode-ray tube by radial modulation of a circular sweep. The illuminated trace follows a circular course at a constant angular velocity. The time to complete one revolution is the time required for an echo to return from 100 feet or 100 fathoms at the assumed speed of sound in water. The transmitted pulse and the returned echoes cause radial modulations. Beams of light thus appear behind a calibrated screen covering the indicator tube face, one at position zero (transmitted pulse) and the others at positions corresponding to the echoes (fig. 619b).

The transmitter is keyed mechanically by the recorder or electronically by the cathode-ray tube indicator circuit, depending on the depth range scale in use. Keying may be automatic or manual.

The operator must observe certain precautions in his use of this typical installation. He must change depth range scales when conditions warrant. For example, a depth of 300 feet can be recorded on the 6000-fathom scale, better on the 600-fathom scale, but best on the 600-foot scale. If a range is selected which is less than the water depth, the echo will return either after the stylus that keyed the transmitter has left the paper (resulting in no indication) or after the other stylus has reached the paper (resulting in a false indication). Similarly, if an indicator range is selected which is less than the water depth, the result will be no indication or a false indication.

Examples of how false echoes can be produced follow. Suppose the water depth beneath the vessel is 1,300 feet and the 600-foot recorder scale is in use. Refer to figure 619c. Stylus A marks the paper at zero and travels downward while stylus B travels upward. Stylus B must travel a distance corresponding to 1,200 feet before it appears at the zero line. If the water depth is 1,300 feet, stylus B will mark the paper at 100 feet. It is obvious that if the water depth is from 600 to 1,200 feet, no echo will appear on the paper.

If the water depth is 120 feet and the 100-foot indicator scale is in use, a false echo will appear at 20 feet.

In either case, the false indications can be avoided by starting on the highest scale and then switching to the scale best suited to give optimum presentation.

The receiver gain must be set for optimum response. Too high a gain will result in reverberation which manifests itself as an elongation of the transmitted pulse. Too high a gain may also result in multiple echoes. Multiple echoes are caused by

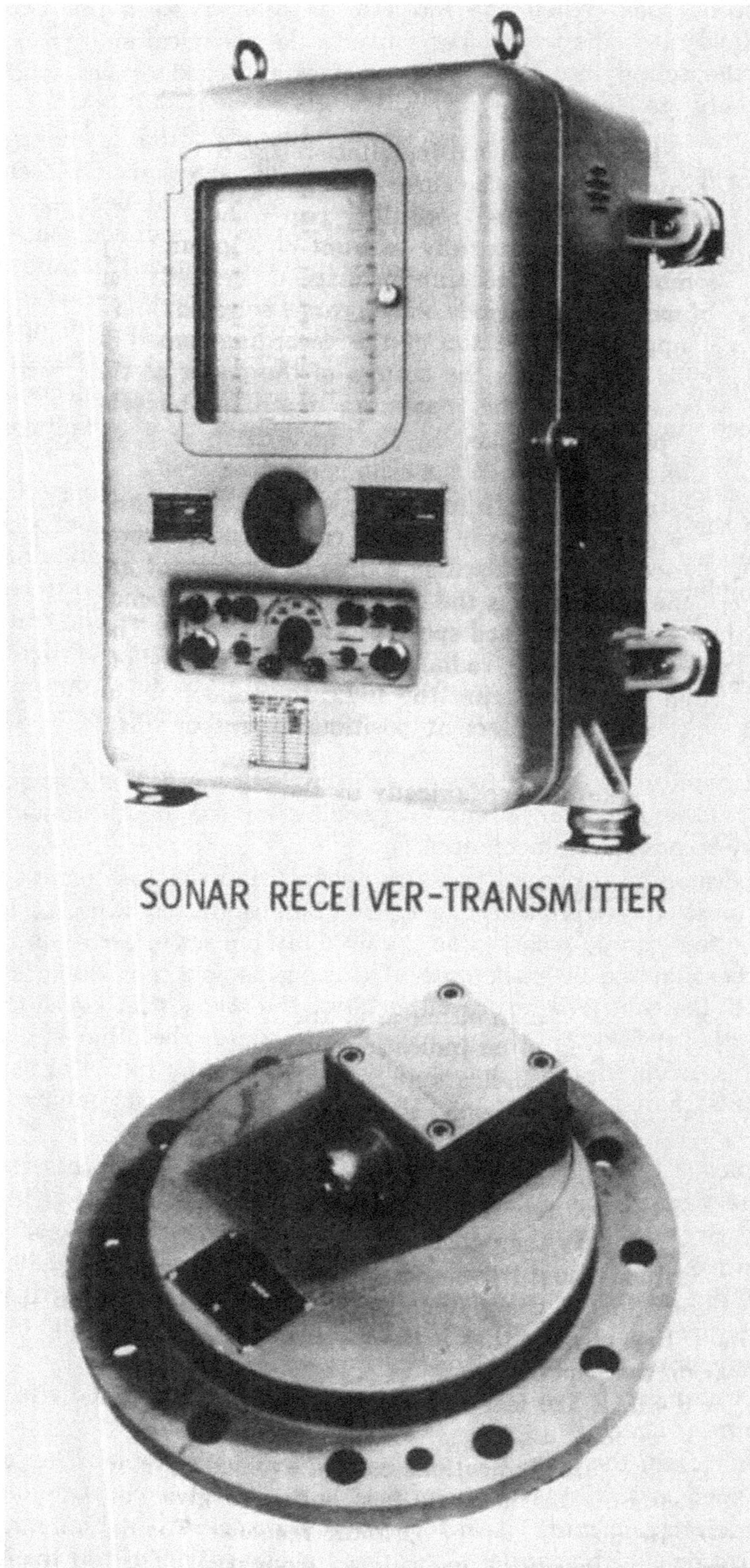

FIGURE 619a.—Echo sounder.

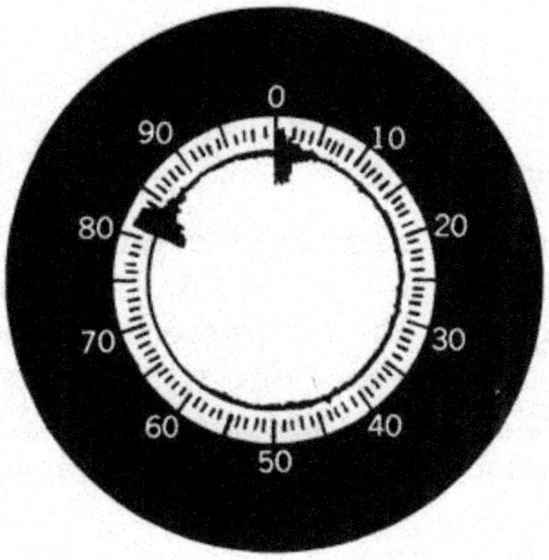

Figure 619b.—Cathode-ray tube indicator.

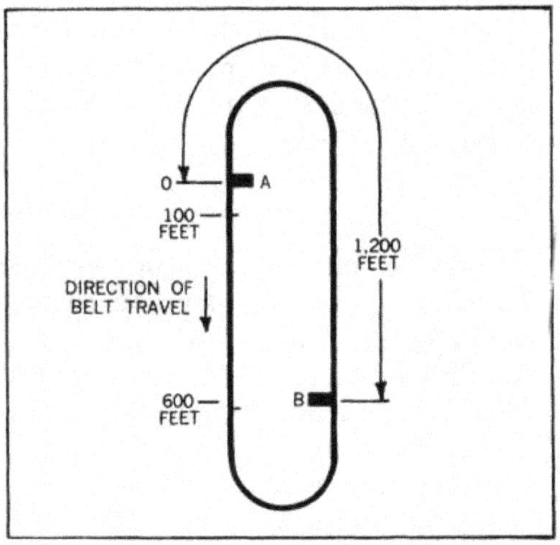

Figure 619c.—Side view of stylus belt.

the returning echo striking the vessel's keel or the water surface, reflecting back to the bottom, and again returning to the transducer. Very often, several of these multiple echoes can be seen. In shallow water, multiple echoes may produce a straight line the full length of the recorder paper or, if operating on the indicator ranges, produce a solid mass of echoes which merges with the initial pulse. Since multiple echoes are considerably attenuated with respect to the original echo, they may be eliminated by operating with a lower gain setting. Too low a gain, on the other hand, will not develop enough voltage from an otherwise suitable echo to mark the paper.

Echo sounders of American manufacture are calibrated for a speed of sound of 4,800 feet per second. The actual speed varies primarily with the temperature, pressure, and salinity, as discussed in article 3503, but in the ocean is nearly always faster than the speed of calibration. The error thus introduced is on the side of safety unless the water is fresh or very cold. Soundings shown on charts are those obtained by an echo sounder without correction, and can therefore be compared directly with the readings obtained aboard ship since the variation in speed from mean conditions is not great. Only in precise scientific work should it be necessary to correct the readings for actual sound speed under prevailing conditions. Accurate adjustment can be made only if information is available on conditions at various depths.

Errors are sometimes introduced by false bottoms. If soft mud covers the ocean floor, some of the sound-wave energy may penetrate to a harder layer beneath, resulting in indication of two bottoms. It is not unusual in deep water to receive a strong return at a depth of about 200 fathoms during the day, and somewhat nearer the surface at night. This is called the **phantom bottom** or **deep scattering layer.** It is believed to be due to large numbers of tiny marine animals. Schools of fish return an echo sufficiently strong to make the echo sounder a valuable aid to commercial fishermen.

In modern equipment the sound waves, whether sonic or ultrasonic, are produced electrically by means of a **transducer,** a device for converting electrical energy to sound waves, or vice versa. The transducer utilizes either the piezoelectric properties of quartz or the magnetostriction properties of nickel and its alloys.

Early models produced sound signals by striking the ship's hull with a mechanical hammer in the forward part of the vessel. The echo was received by a microphone in the after part of the vessel, depth being determined by the angle at which the signal returned.

Direction Measurement

620. Reference directions.—A horizontal direction is generally expressed as an angle between a line extending in some **reference direction** and a line extending in the given direction. The angle is numerically equal to the difference between the two directions, called the **angular distance** from the reference direction. Unless the reference direction is stated or otherwise understood, the intended direction is in doubt. Thus, to a navigator, direction 135° is southeast. To an astronomer or surveyor, it may be northwest.

A number of reference directions are used in navigation. If a direction is stated in three figures, without designation of reference direction, it is generally understood that the direction is related to true (geographical) north. When grid navigation (art. 2510) is being used, particularly in high latitudes, grid north is generally used as the reference direction. The reference direction for magnetic directions is magnetic north, and that for compass directions is compass north. For relative bearings it is the heading of the ship. For amplitudes, the reference direction is east or west, usually 090° or 270° true, but magnetic, compass, or even grid east or west may be used. In maneuvering situations, the heading of another vessel might be used as the reference direction.

The primary function of an instrument used for measuring direction is to determine the reference direction. This having been done, other directions can be indicated by a compass rose oriented in the reference direction. North is established by some form of compass. A compass rose is attached to the north-seeking element so that other directions can be determined directly. However, if one always keeps in mind that the primary function of the instrument is to indicate a reference direction, he should be able to avoid some of the mistakes commonly made in the application of compass errors.

621. Desirable characteristics of a navigational compass.—To adequately serve its purpose, a navigational compass needs to have certain characteristics to permit it to meet requirements of accuracy, reliability, and convenience.

The most important characteristic is accuracy. No other quality, however important or to whatever extent it may be possessed, compensates for the lack of accuracy. This does not mean that the compass need be without error, but that such errors as it may possess can be readily determined. Provisions should be made for removing deviation or reducing it to a minimum (ch. VII). If accurate horizontal directions are to be determined, the compass needs to be provided with some type of compass rose maintained in a horizontal position. Adequate sighting equipment is needed if bearings are to be observed, and an index is needed to mark the forward direction parallel to the keel if heading is to be measured. Accurate readings cannot be expected from a compass that **hunts** (oscillates) excessively. A characteristic closely related to accuracy is precision. The amount of precision required varies somewhat with the use and depends as much upon the steadiness of the compass and its design as upon its inherent qualities.

A compass is reliable when its operation is not often interrupted; when its indications are relatively free from unknown or unsuspected disturbances; when it is little affected by extremes of temperature, moisture, vibration, or the shock of gunfire; and when it is not so sensitive that large errors are introduced by ordinary changes in conditions or equipment near the compass.

The value of a compass is dependent somewhat upon the convenience with which it can be used. Accuracy, too, may be involved. Thus, a compass should not be installed in such a position that one must be in an unnatural or uncomfortable position to use it. A compass intended for use in obtaining bearings is of reduced value if it is installed at a location that does not permit an unobstructed view in most directions. The compass graduations and index should be clean, adequately lighted if the instrument is to be used at night, and clearly marked.

622. Kinds of compasses.—The compasses commonly used by the mariner are (1) *magnetic* and (2) *gyroscopic*. The magnetic compass tends to align itself with the magnetic lines of force of the earth, while the gyrocompass seeks the true (geographic) meridian. The word "compass" is also applied to instruments which do not continuously indicate some form of north. Thus, the free gyro (art. 630) tends to remain approximately aligned with any great circle to which it is set.

A compass may be designated to indicate its principal use, as a **standard, steering,** or **boat compass.** The compass designated as standard is usually a magnetic compass installed in an exposed position having an unobstructed view in most directions, permitting accurate determination of error. Preferably, it is located at a magnetically favorable position near the bridge. Before the development of a reliable gyrocompass, the standard compass was used for navigation of the vessel and for determining the error of the steering compass.

Although the modern, reliable gyrocompass has largely superseded the magnetic compass for most purposes, directional information is so important to a vessel that the availability of a second method is considered justified. It is wise to understand both types, keep a record of errors and the performance of all compasses, and to compare the indications of magnetic and gyrocompasses at frequent intervals, as every half hour when underway.

623. Magnetic compasses.—If a small magnet is pivoted at its center of gravity in such manner that it is free to turn and dip, it will tend to align itself with the magnetic field of the earth (art. 706). It thus provides a directional reference and becomes a simple compass. However, such a compass would not be adequate for use aboard ship. For this purpose a compass should have a stronger directive element than that provided by a single, pivoted magnet, should have provision for measuring various directions, should have some means of damping the oscillations of the directive element, should be approximately horizontal, and should have some means of neutralizing local magnetic influences.

In a mariner's compass, several magnets are mounted parallel to each other. To them is attached a **compass card** having a compass rose to indicate various directions (art. 624). Both magnets and compass card are enclosed in a bowl having a glass top through which the card can be seen. The bowl is weighted at the bottom and is suspended in gimbals in such manner that it remains nearly horizontal as the vessel rolls and pitches. In nearly all modern compasses the bowl is filled with a liquid that supplies a buoyant force almost equal to the force of gravity acting upon the directive element and card. This reduces the friction on the pivot (a metal point in a jeweled bearing), and provides a means of damping the oscillations of the compass card. The card is mounted in such manner as to remain in an essentially horizontal position. A mark called a **lubber's line** is placed on the inner surface of the bowl, adjacent to the compass card, to indicate the forward direction parallel to the keel when the bowl is correctly installed. The gimbals used for mounting the compass bowl are attached to a stand called a **binnacle**, which in most installations is permanently and rigidly attached to the deck of the vessel, usually on its longitudinal center line. Most binnacles provide means for neutralization of local magnetic influences due to magnetism within the vessel. A cover or "hood" is provided to protect the compass from the elements, dust, etc.

Directional information is of such importance that selection and installation of a suitable compass should be made carefully, seeking such guidance as may be needed. In the U.S. Navy this is covered by systems command directives. For merchant vessels and yachts, one would do well to consult a dependable compass adjustor before selecting and installing a compass or making any alteration in the vicinity of the compass. Common errors are the use of a compass designed for a different type craft (as an aircraft compass in a boat), permitting chrome plating of a binnacle by someone who does not know how to do this without creating a magnetic field, authorizing electric welding of steel near the compass, improper installation of magnetic equipment or electric appliances near the compass, allowing short circuits to occur in the vicinity of the compass, etc.

After the compass has been selected and installed, proper adjustment and compensation (ch. VII) are important, and future care of the instrument should not be neglected. It should be checked and overhauled at regular intervals, and any indication of malfunctioning or deterioration, however slight, should not be overlooked. Discoloration of the liquid or the presence of a bubble, for instance, indicates a condition that should be investigated and corrected at once. If it becomes necessary to add liquid, one should be certain that he has the correct substance, and should attempt to determine the source of the leak. Except as a temporary expedient, this is best done by a professional. Some compasses should be protected from prolonged exposure to sunlight, to prevent discoloration of the card and liquid.

624. The compass card is composed of light, nonmagnetic material. In nearly all modern compasses the card is graduated in 360°, increasing clockwise from north through east, south, and west. An older system still used somewhat is to graduate the card through 90° in each quadrant, increasing from both north and south. Some compass cards are graduated in "points," usually in addition to the degree graduations. There are 32 **points of the compass**, 11¼° apart. The four **cardinal points** are north, east, south, and west. Midway between these are four **intercardinal points** at northeast, southeast, southwest, and northwest. These eight points are the only ones appearing on the cards of compasses used by the U.S. Navy. The eight points between cardinal and intercardinal points are named for the two directions between which they lie, the cardinal name being given first, as north northeast, east northeast, east southeast, etc. The remaining 16 points are named for the nearest cardinal or intercardinal point "by" the next cardinal point in the direction of measurement, as north by east, northeast by north, etc. Smaller graduations are provided by dividing each point into four "quarter points," thus producing 128 graduations altogether. There are several systems of naming the quarter points. That used in the U.S. Navy when quarter points were used is given in table 2.

The naming of the various graduations of the compass card in order is called **boxing the compass**, an important attainment by the student mariner of earlier generations. The point system of indicating relative bearings (art. 1004) survived long after degrees became almost universally used for compass and true directions. Except for the cardinal and intercardinal points, and occasionally the two-point graduations, all of which are used to indicate directions generally (as "northwest winds," meaning winds from a general northwesterly direction), the point system has become largely historical.

625. The U.S. Navy 7½-inch compass has a liquid-filled bowl in which a 7½-inch aluminum card is pivoted. There is provision for either one or two pairs of magnets, symmetrically placed. The card and magnet assembly is provided with a central float or air chamber to reduce the weight on the pivot to between 60 and 90 grains (0.14 and 0.21 oz.) at 60°F when the correct compass fluid is used. Older compasses use a fluid consisting of 45 percent ethyl alcohol and 55 percent distilled water. Newer compasses use a highly refined petroleum distillate similar to varsol. Use of this oil increases the stability and efficiency of the compass. A hollow cone extends into the underside of the float. The bottom of this cone is open. The pointed top has a jewel bearing of synthetic sapphire. The card-float-magnet assembly rests on an osmiumiridium tipped pivot at the jewel center. This pivot extends upward from the bottom of the bowl. This compass is illustrated in figure 625.

The compass bowl is made of cast bronze, and has a tightly gasketed glass top cover to prevent leakage of the liquid. A bellows-type expansion chamber is provided to allow for changes in volume of the liquid as the temperature changes. The top rim or bezel of the bowl is accurately machined so that an azimuth or bearing circle can be placed over it. The compass is equipped with a gimbal ring for keeping the compass level when mounted in a binnacle. In addition to providing support for the compass, the binnacle has provision for housing the correctors used to partially neutralize local magnetic effects within the vessel.

FIGURE 625.—U. S. Navy 7½-inch compass.

626. The U. S. Navy 5-inch compass is lighter in weight and requires less space than the 7½-inch compass. This U. S. Navy No. 3 compass has a brass compass card with photo-etched perforations which permit underlighting with red light to meet darkness adaptation requirements. When such a card is used with both transmitted and reflected light in all combinations, there is a "twilight zone" in which the intensity of the light transmitted through the perforations is equal to that of the reflected light from the surrounding area. In the U.S. Navy No. 3 compass this problem is overcome by the installation of a shaded lamp inside the binnacle hood. The light from this lamp is directed at the lubber's line and adjacent compass card area. This light enables daylight viewing of the compass card. The red illumination is required only when practically complete darkness prevails.

In addition to providing support for the compass, the binnacle illustrated in figure 626a has provision for housing the correctors used to partially neutralize local magnetic effects within the vessel. The correctors consist of a tube assembly with heeling magnet, quadrantal correctors, fixed fore-and-aft and athwartship permanent magnetic correctors, and permanent magnet correctors rotatable about fore-and-aft and athwartship axes.

Courtesy of John E. Hand & Sons Company.

FIGURE 626a.—Binnacle for U. S. Navy 5-inch magnetic compass.

The fixed permanent magnets are in the form of wire magnet bundles of up to seven magnets each and are contained in three magnet tubes as shown in figure 626a. The athwartship tube is fixed to the after side of the binnacle. One fore-and-aft tube is on the port side of the binnacle; the other is on the starboard side.

The fixed fore-and-aft and athwartship magnets are used to obtain a "coarse" correction; the rotatable correctors are used to obtain a "fine" correction. Normally the coarse correction reduces the deviation from its original value to about 5°, the fine correction reduces the deviation to its residual value.

Except for the heeling magnet the various correctors are shown in schematic form in figure 626b. When the rotatable magnets on either axis lie in a plane parallel to the plane of the compass card, the associated magnetic field has no effect on the compass card magnets. This magnetic field is shown schematically in figure 626c for the magnets rotatable about the athwartship axis. When the rotatable magnets are inclined to the plane of the compass card, the effect on the compass card magnets varies with the sine function of the angle of inclination, the maximum effect being when the plane of the rotatable magnets is inclined 90° to the plane of the compass card.

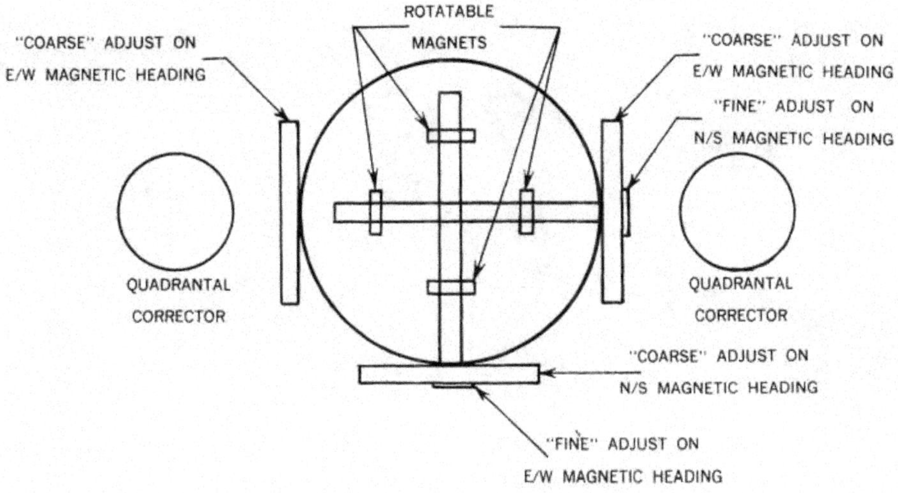

Figure 626b.—Correcting system of U. S. Navy 5-inch magnetic compass.

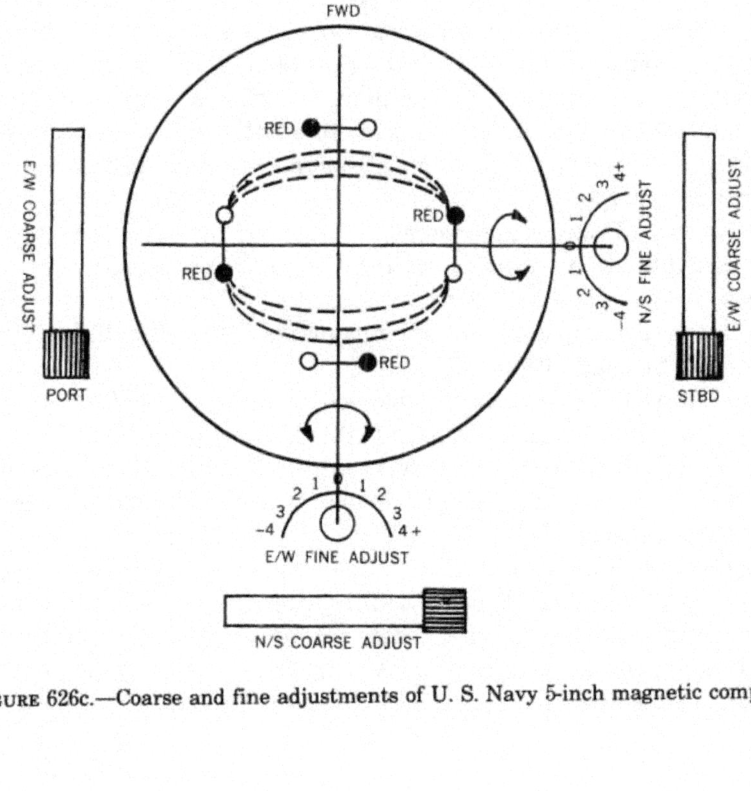

Figure 626c.—Coarse and fine adjustments of U. S. Navy 5-inch magnetic compass.

Figure 626d illustrates the fine adjustment control used to rotate the magnets. The fine adjustment should be limited to approximately the last 5° of deviation correction or to the range in which the sine function is more nearly linear.

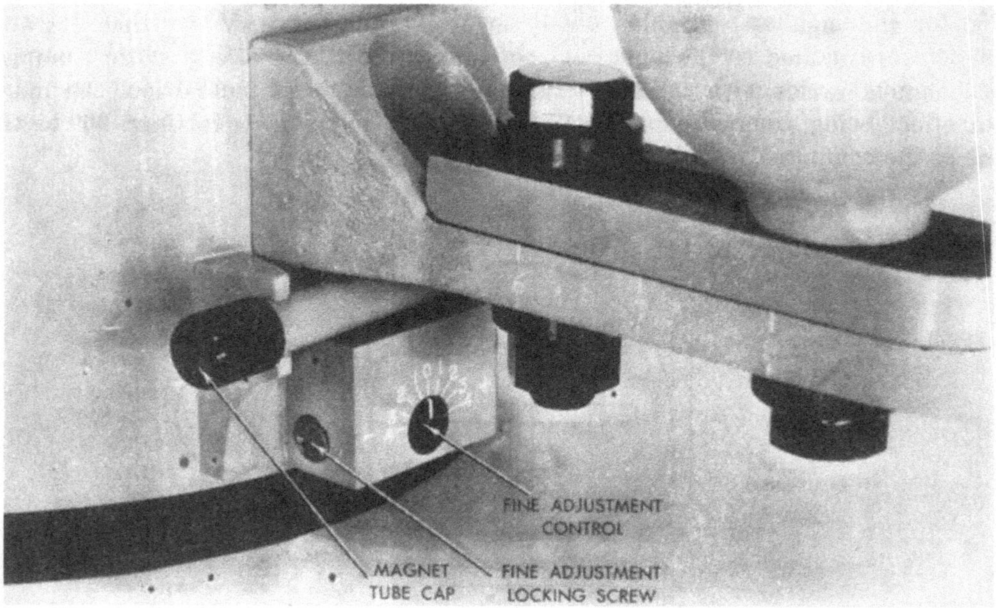

Courtesy of John E. Hand & Sons Company.

FIGURE 626d.—Fine adjustment control.

627. Other magnetic compasses.—The U. S. Navy No. 5 magnetic boat compass (fig. 627a) is a top reading, flat glass topped unit. The 3-inch compass card is made of sheet aluminum and incorporates an annular float to obtain a degree of buoyancy in the Varsol compass fluid. The card is supported on a jewel post without the use of gimbals. The compass bowl employs a bellows expansion chamber which permits volumetric changes of the compass fluid without bubble formation.

The correctors consist of quadrantal correctors, fore-and-aft and athwartship permanent magnet correctors.

A wide variety of magnetic compasses are used in merchant ships and yachts. The basic principles of operation of all magnetic compasses are the same, the various types differing only in details of construction. A feature which is widely used in commercial compasses is a hemispherical top (fig. 627b) which provides magnification of the graduations.

Reflection binnacles providing a periscopic readout in the wheelhouse enable mounting of the compass where it is usually less subject to the vessel's magnetic field and the installed electrical and electronic equipment than a wheelhouse installation. Location on the flying bridge also serves to facilitate compass adjustment and bearing observations.

628. Magnetic compass limitations.—Because of its essential simplicity, a magnetic compass does not easily become totally inoperative. Being independent of any power supply or other service, a magnetic compass may survive major damage to its ship without losing its utility. Small boat compasses often remain serviceable under the most rigorous conditions.

FIGURE 627a.—U. S. Navy No. 5 magnetic boat compass.

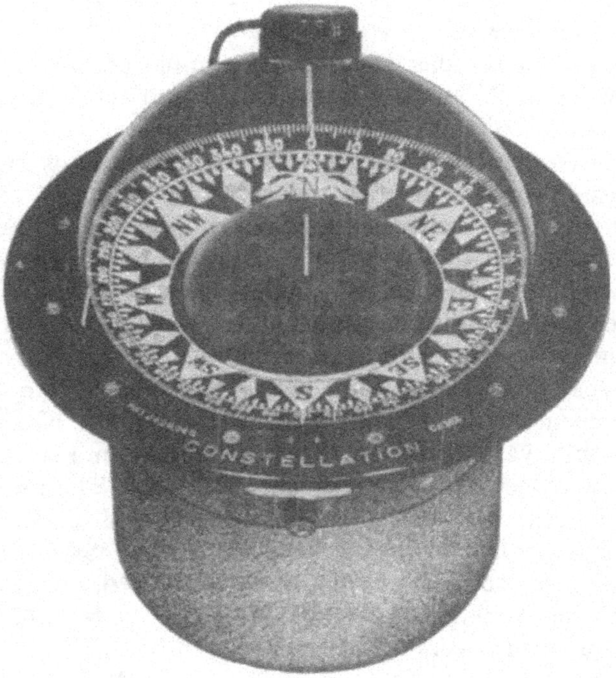

Courtesy of Danforth Division of The Eastern Company, a Connecticut Corporation.

FIGURE 627b.—A compass with a hemispherical top.

Despite its great reliability, however, a magnetic compass is subject to some limitations. Since it responds to *any* magnetic field, it is affected by any change in the local magnetic situation. Hence, the undetected presence or change of position of magnetic material near the compass may introduce an unknown error. Thus, an error might be introduced by a steel wrench or paint can left near the compass, or by a change in position of a steel boom or gun in the vicinity of the compass. Even such small amounts of magnetic material as might be included in a pocketknife or steel keys are sufficient to affect the compass if brought as close as they are when on the person of an individual standing by a compass. Nylon clothing may also introduce error in a magnetic compass. As distance from the compass increases, the strength of the magnetic field needed to introduce an error increases. A cargo of large amounts of iron or steel may be sufficient to affect the compass. The compass may also be affected by changes of the magnetic characteristics of the vessel itself. Such changes may occur during a protracted docking period, during a long sea voyage on substantially the same course, when repairs or changes of equipment are made, if the ship sustains heavy shock as by gunfire or riding out a heavy sea, if the vessel is struck by lightning, or if a short circuit occurs near the compass.

The directive force acting upon a magnetic compass is the horizontal component of the earth's magnetic field. This component is strongest at or near the magnetic equator, decreasing to zero at the magnetic poles (ch. VII). Near the magnetic poles, therefore, the magnetic compass is useless (art. 2513), and in a wider area its indications are of questionable reliability. The magnetic field of the earth has a number of local anomalies due to the presence of magnetic material within the earth. During magnetic storms (art. 2513) it may be altered considerably. Changes in the magnetic field surrounding a vessel, due either to changes of the field itself or to change of position of the vessel within the field, affect the magnetism of the vessel and the correctors used to neutralize this effect, with a possible disturbance of the balance set up between them.

For these and other reasons, frequent determination of compass error is necessary for safe navigation. Methods of determining and correcting compass error are discussed in chapter VII.

629. Magnetic compass accessories.—Compass heading is indicated by the lubber's line. Compass bearings may be measured by sighting across the compass, bringing the object and the vertical axis of the compass in line. Accuracy in making this alignment is increased by the use of a device to direct the line of sight across the center of the compass. Perhaps the simplest device of this kind is a **bearing bar,** consisting of two vertical **sighting vanes** mounted at opposite ends of a horizontal bar having a small pivot which fits into a hole drilled part way through the glass cover of the compass, at its center. The "near" vane (nearer the eye of the observer) has a very thin, open, vertical slot through which the line of sight is directed; the "far" vane has a thin, vertical wire or thread mounted on a suitable frame. The bar is rotated until the object is in line with the two vanes. The bearing is the reading of the compass in line with the vanes, on the far side from the observer. If a reflecting surface is pivoted to the far vane to permit observation of the azimuth (art. 1430) of a celestial body, the device is called an **azimuth instrument.** Bearing bars and azimuth instruments are usually used only with smaller compasses, and never with an after-reading compass.

Larger compasses or repeaters (art. 643) are usually provided with a **bearing circle** or **azimuth circle** (fig. 629). These devices take a variety of forms, but consist essentially of two parts: (1) a pair of sighting vanes attached to a ring which fits snugly over the compass, and (2) a mirror to reflect the compass graduation into the line of sight. The use of these devices is similar to that of the bearing bar and azimuth instrument. The azimuth circle has a pivoted reflecting surface attached to the far vane, to permit observation of celestial bodies. In most cases it also has a reflecting mirror and prism mounted on opposite sides of the ring, midway between the vanes. The prism is covered with opaque material except for a thin, vertical slot at its center. The surface of the mirror is curved so that reflection of sunlight falling upon it is in the form of a slender vertical line (at the distance of the prism) of about the same width as the slot. When the azimuth circle is adjusted so that this line of light falls upon the slot, a thin, bright line appears on the compass card graduations at the bearing of the sun. Most bearing and azimuth circles are provided with reverse compass rose graduations to permit reading of relative bearings or azimuths (by the vanes) at a mark on top of the compass bowl, in line with the lubber's line; bubbles for indicating the level position during observation; means for adjusting the snugness of the fit over the compass bowl; and handles for turning the device.

If a bearing or azimuth circle does not fit snugly over the compass bowl, an error might be introduced. Inaccuracy may also result from tilting of the reflecting surface of an azimuth circle with respect to the vertical plane through the line of sight. This can be checked by comparing an azimuth of the sun observed by means of the prism with one observed with the sighting vanes (with suitable protection being provided for the eyes). If the prism attachment is not available, a check can be made by comparing observed (compass) azimuths at different altitudes with computed (true) values at the time of observation. If both observed and computed azimuths are correct, the difference between them will be constant (if the compass error remains constant throughout the observation).

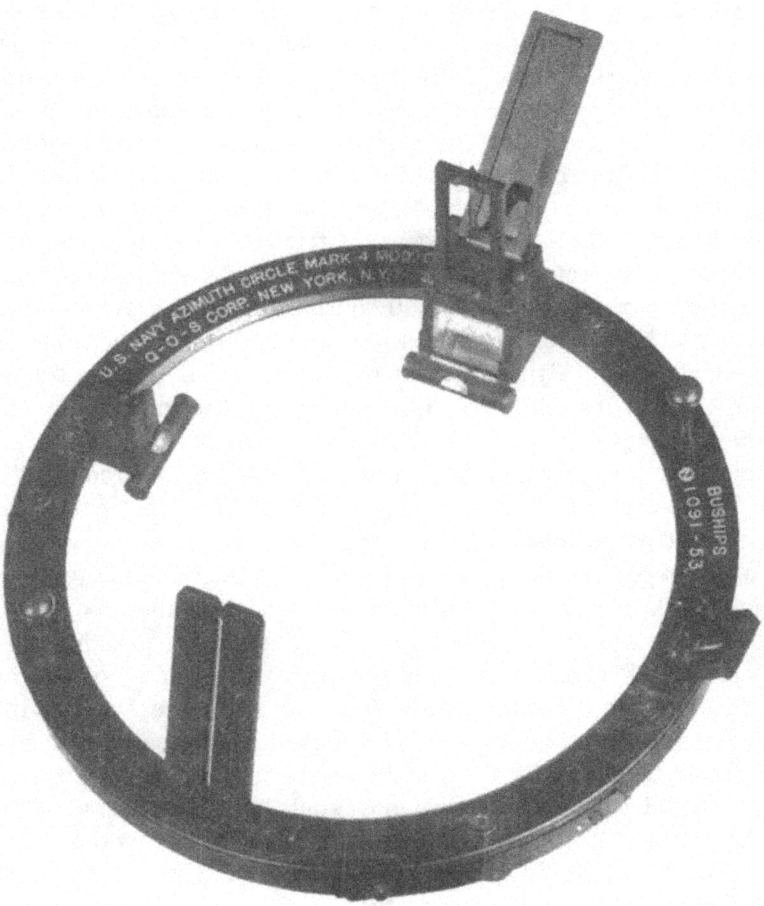

FIGURE 629.—An azimuth circle.

None of the bearing or azimuth instruments described above can be used with a compass not designed for it, as one having a hemispherical top, or an after-reading compass.

Some modern magnetic compasses are provided with electrical pick-offs of sufficient sensitivity that the instrument can be used to control such devices as remote indicators, automatic steering equipment, course recorders, and dead reckoning equipment without disturbing the reliability of the compass. However, these devices are more commonly controlled by a gyrocompass and hence are considered later in the chapter, after a discussion of this type compass.

630. The gyroscope.—Leon Foucault, a French physicist, first demonstrated the rotation of the earth by means of a pendulum. However, the pendulum was not entirely acceptable as proof of rotation because it required the earth's gravity for operation. In 1852, he gave the name **gyroscope** to a toy top which had been known for a quarter of a century as a "rotascope." By means of the gyroscope, Foucault illustrated the earth's rotation without the use of gravity.

A conventional gyroscope consists of a comparatively massive, wheel-like rotor balanced in gimbals which permit rotation in any direction about three mutually perpendicular axes through the center of gravity. The three axes are called the **spin axis,** the **horizontal axis,** and the **vertical axis,** as shown in figure 630a.

Since the rapidly spinning rotor is balanced at its center of gravity, it is in a state of neutral rotational equilibrium. If the gimbal bearings were completely frictionless, the spin axis would retain its direction in space despite any motion applied to the system as a whole, as by the rotation of the earth. This property is called **gyroscopic inertia.** Thus, if the spin axis were directed toward a star, the axis would continue to point toward the star during its apparent motion across the sky. To an observer on the earth, the spin axis would appear to change direction as the earth rotated eastward. This phenomenon, also known as **rigidity in space,** can be demonstrated by slowly tilting the base of the gyroscope as shown in figure 630b. If the gyroscope rotor is stationary, bearing friction will cause the rotor to tilt as the base is tilted. If the rotor is spinning, the rotor maintains the original plane of rotation as the gyroscope is tilted. It will continue to maintain the original plane of rotation no matter how much the gyroscope is tilted, as long as it continues to spin with sufficient velocity. Although bearing friction still affects the gyroscope, it affects it to a lesser degree than when the rotor was stationary.

Gyroscopic inertia depends upon angular velocity, mass, and the radius at which the mass is concentrated. For a given mass, maximum effect is obtained, therefore, from a mass rotating at high speed with the principal part of the mass concentrated near the periphery of the wheel.

Gyroscopic precession is that property of a gyroscope exhibited when a force is applied which tends to change the *direction* in space of the spin axis. The motion resulting from such a force is not in line with the force, as might be expected, but *perpendicular* to it. Precession can be demonstrated by applying a torque to the spinning gyroscope about its horizontal axis. This is done by applying a force at point A as illustrated in figure 630c. The gyroscope instead of turning about the horizontal axis as it would if it were not spinning, turns or precesses about its vertical axis. The direction of precession is such that it appears as though a force applied to the rotor at A is, instead, applied at a point 90° away in the direction of spin from point A. Similarly, if a torque is applied about the vertical axis, the gyroscope will precess about its horizontal axis in a direction such that it appears

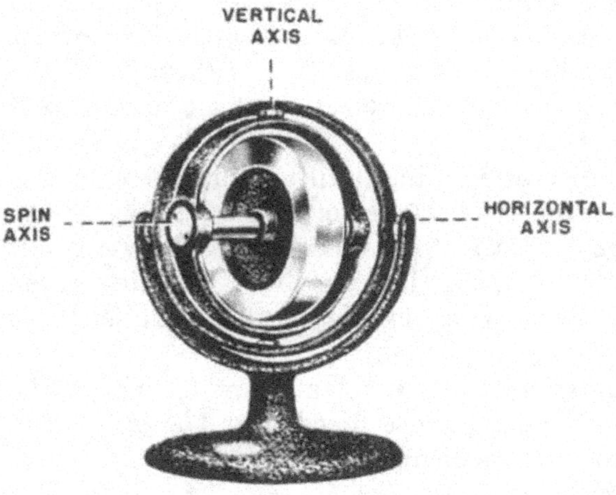

FIGURE 630a.—Gyroscope.

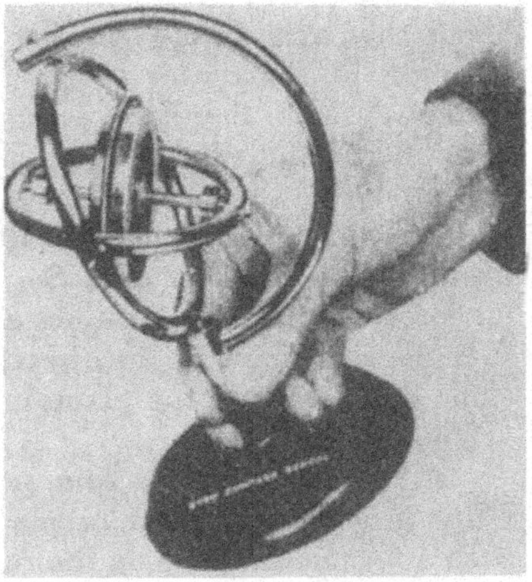

FIGURE 630b.—Demonstration of gyroscopic inertia.

as though the force is applied at a point 90° away in the direction of spin from the point where the force is applied on the rotor.

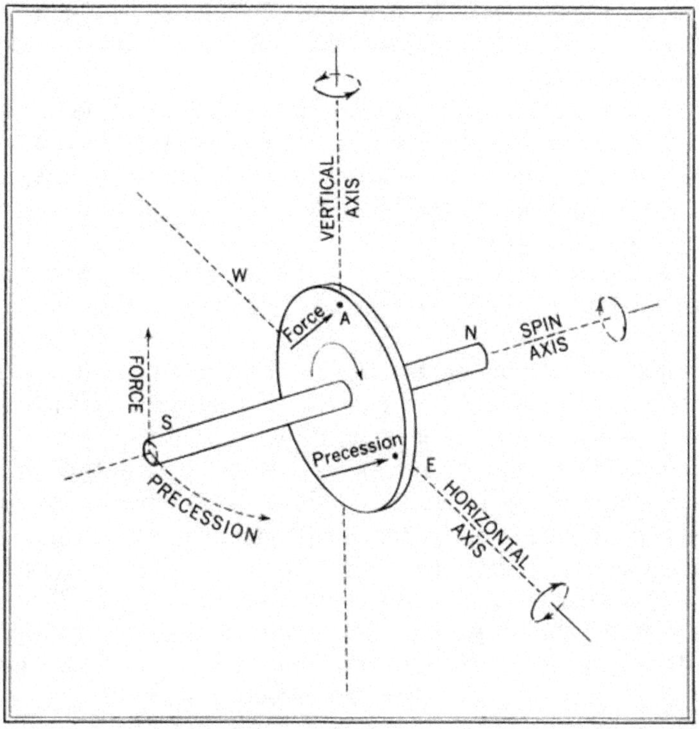

FIGURE 630c.—Axes of a gyroscope, and the direction of precession.

A **torque** is defined as that which effects or tends to effect rotation or torsion and which is measured by the product of the applied force and the perpendicular distance from the line of action of the force to the axis of rotation. It is obvious that a force acting through or parallel to an axis cannot produce any turning effect about that axis. In a gyroscope the three axes about which rotation is possible all intersect at the center of gravity of the entire system (excluding the supporting frame). A force, therefore, acting through the center of gravity acts through all the axes and cannot exert torque about any axis. But a force acting at any other point will produce a torque about one or more axes.

Precession can be caused only by a force attempting to tilt or turn the spin axis about one of the other axes. So, a force through the center of gravity of a gyroscope (or force of translation) cannot cause the gyroscope to precess, but can only cause it to move as a whole in the direction of the force, with its axle always pointing in the same direction and the plane in which the rotor is spinning always parallel to its original plane of spin.

A torque about the spin axis of a gyroscope does not attempt to change the plane in which the rotor is spinning, so it cannot cause precession. The only effect such a torque can have on the rotor is to increase or decrease its speed.

Any torque about either the horizontal or vertical axis of a gyroscope will cause it to precess about an axis at right angles to that about which the torque acts. This precession will continue as long as the torque acts, but will cease when the torque is removed. If the plane in which the torque is acting remains unchanged, the gyroscope will precess until the plane of the spin of the rotor is in the plane of the torque. When this position is reached, the torque will be about the spin axis and can cause no further precession. If, however, the plane in which the torque acts moves at the same rate and in the same direction as the precession it causes, the precession will be continuous.

The rotor of the conventional gyroscope previously described has three degrees of freedom: (1) freedom to spin on its axle, (2) freedom to tilt about its horizontal axis, and (3) freedom to turn about its vertical axis. These three degrees of freedom permit the rotor to assume any position with respect to the supporting frame. This gyroscope is called a **free gyroscope.**

The term **degree-of-freedom** refers to the number of orthogonal axes about which the spin axis is free to rotate, the spin axis not being counted in one convention.

The reason for gyroscopic precession may be explained simply by considering what happens to a single particle on the rim of the gyroscope wheel as shown in figure 630d. Assume that the wheel is spinning in the direction of arrow *R*. Also assume that a force *F* is applied against the wheel at point *B* on the particular particle *P*, which happens to be at the position shown at any particular instant.

Force *F* exerts a force upon this small particle along the vector *BL* and therefore accelerates it in that direction. During a short interval of time, the acceleration will give the particle a component of velocity *BL*. This velocity vector and the velocity of the particle along the vector *BJ*, due to rotation of the wheel, have as a resultant the vector *BK*, different in direction from *BJ*. This is equivalent to a rotation about axis, *YY*. Therefore, the effect of a torque acting about the *XX* axis is to cause a rotation of the gyroscope wheel about the *YY* axis. This rotation about the *YY* axis is gyroscopic precession.

If the gyroscope, or **gyro** as it is commonly called, is mounted at the equator with its spin axis pointing east and west, figure 630e illustrates how it would appear from a point in space beyond the South Pole. From the observation point in space, the earth is seen turning from west to east at a rate of 15° per hour carrying the gyro with it. However, the spin axis of the gyro, because of gyroscopic inertia, remains in rigid space just as it did when the base was tilted in figure 630b. To an observer on earth the same gyro appears to rotate about its horizontal axis with an angular velocity equal but opposite in direction to the rate of rotation of the earth. This effect, commonly referred to as **horizontal earth rate**, is equal to the rate of rotation of the earth (**earth rate**) times the cosine function of the latitude. It is therefore zero at the poles and increases to earth rate at the equator.

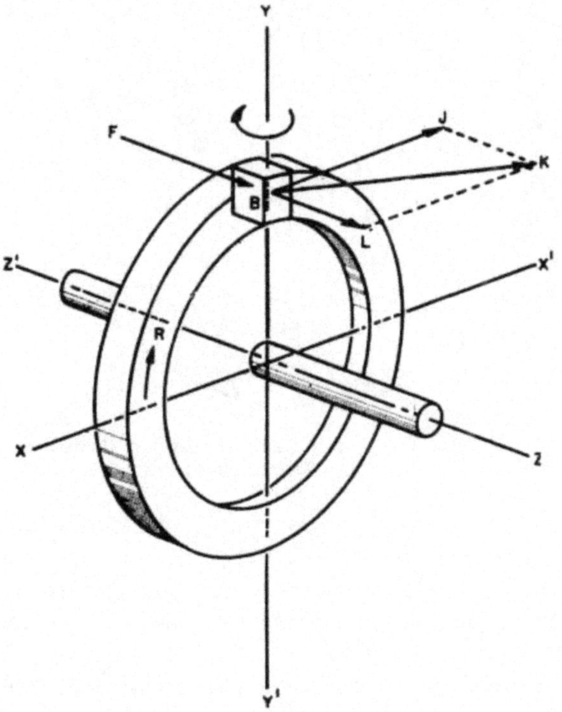

FIGURE 630d.—Forces causing precession of gyroscope rotor.

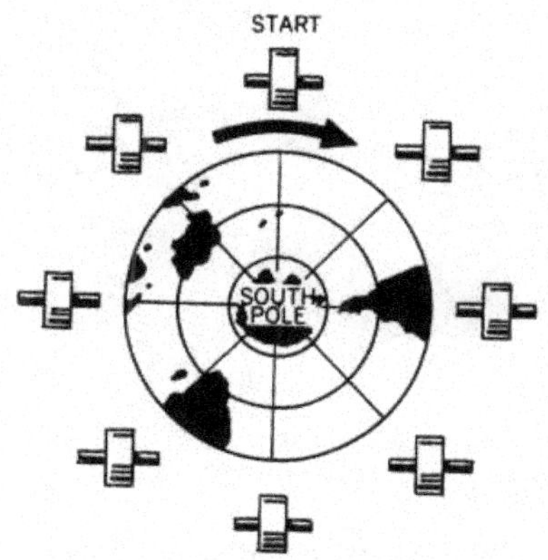

FIGURE 630e.—Demonstration of horizontal earth rate of gyro.

Similarly, if the gyro is mounted at the North or South Pole with its spin axis horizontal, as shown in figure 630f, the gyro will appear to rotate about its vertical axis. This effect is commonly referred to as **vertical earth rate**. At points between the poles and the equator, the gyro appears to turn partly about the horizontal axis and partly about the vertical axis as shown in figure 630g, because it is affected by both horizontal and vertical earth rates.

In general, horizontal earth rate causes the spin axis of the gyro to appear to tilt about its horizontal axis; vertical earth rate causes the gyro to appear to rotate about its vertical axis. The apparent motion of stars (art. 1416) can be used as a convenient reminder of the effect of the earth's rotation on a **free gyro**. Being rigid in space, the spin axis remains pointing at the same fixed star as the earth rotates. Thus, the spin axis describes a circle about Polaris in a counterclockwise direction as the earth rotates.

With reference to space the direction of the spin axis of the free gyro remains the same as the earth rotates. With respect to the earth, however, the spin axis rotates as just described. It is this rotation with respect to the earth which makes it possible to apply the force of gravity so as to convert the free gyro into a north-seeking gyrocompass.

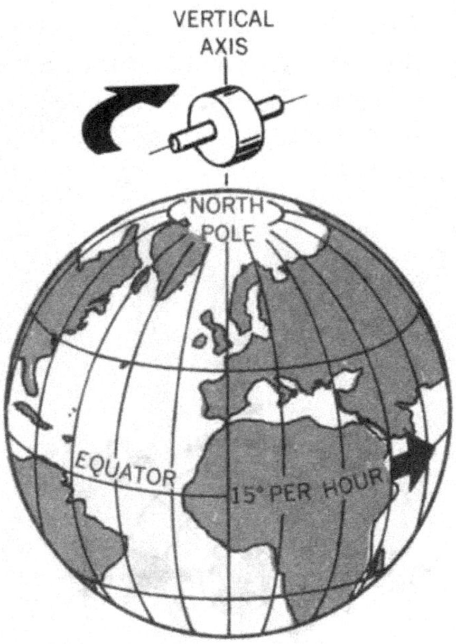

FIGURE 630f.—Demonstration of vertical earth rate of gyro.

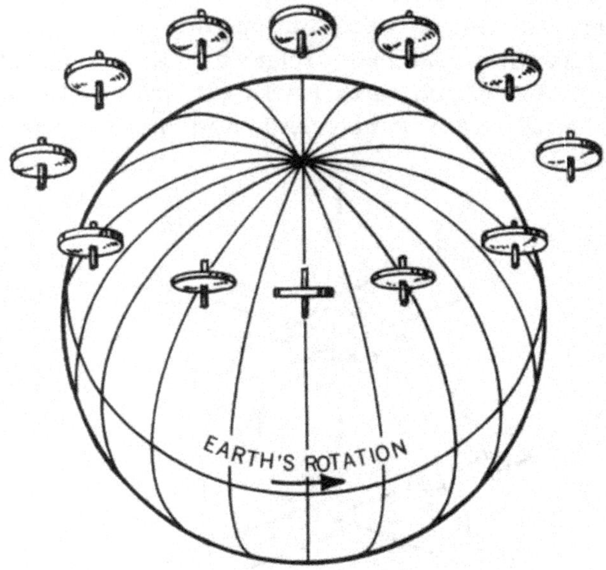

FIGURE 630g.—Combined effects of horizontal and vertical earth rates.

631. The gyrocompass depends upon four natural phenomena for its operation. It is only the methods whereby these phenomena are utilized that distinguish one type of gyrocompass from another. Of the four natural phenomena, two are inherent properties of the gyroscope, namely gyroscopic inertia and gyroscopic precession; the other two are the earth's rotation and gravity.

Before a free gyro can be converted into a gyrocompass, the mounting structure must be changed slightly. As shown in figure 631a, the rotor is mounted in a sphere (**gyrosphere**) and the sphere is supported in what is called the **vertical ring.** The sphere and vertical ring are, in turn, mounted in a base called the **phantom.** Means are provided for the vertical ring and phantom to follow the gyro as it turns about its vertical axis.

With no further additions, the gyro shown in figure 631a will, neglecting friction, maintain its direction in space so long as no outside forces are exerted on it. To make the gyro into a gyrocompass, the gyro has to be made *to seek and maintain true north.* Since north is the direction represented by a horizontal line in the plane of the meridian, some means have to be provided to: (1) make the gyro spin axis seek the meridian plane, (2) make the spin axis horizontal, and (3) make it maintain its position once reached.

The first step in making a gyro a gyrocompass is to make the gyro seek the meridian. To do this, a weight *W* is added to the bottom of the vertical ring, as shown in figure 631b. This causes the vertical ring to be pendulous about the horizontal axis.

If as at *A* of figure 631c the gyro is at the *equator,* the spin axis is horizontal pointing east-west, and the rotor is spinning clockwise as viewed from the west, the rotor and vertical ring are vertical and no torque is created by the added weight. At this point both properties of the gyroscope, gyroscopic inertia and precession, are brought into play. As the earth rotates, the spin axis and, therefore, the vertical ring become inclined to the horizontal as shown at *B.* The weight *W* is raised against the pull of gravity and consequently causes a torque about the horizontal

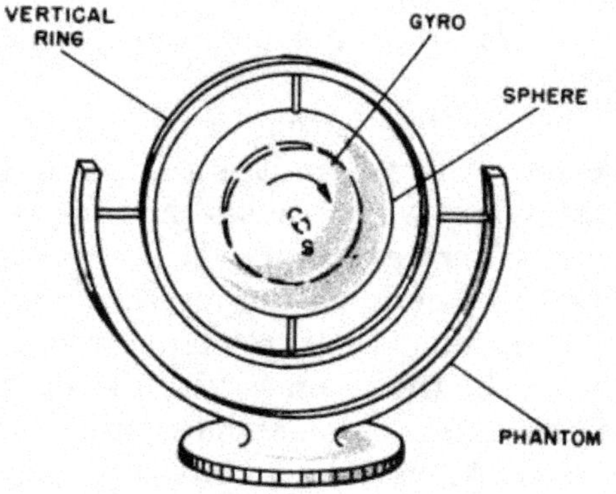

FIGURE 631a.—Modified model gyroscope.

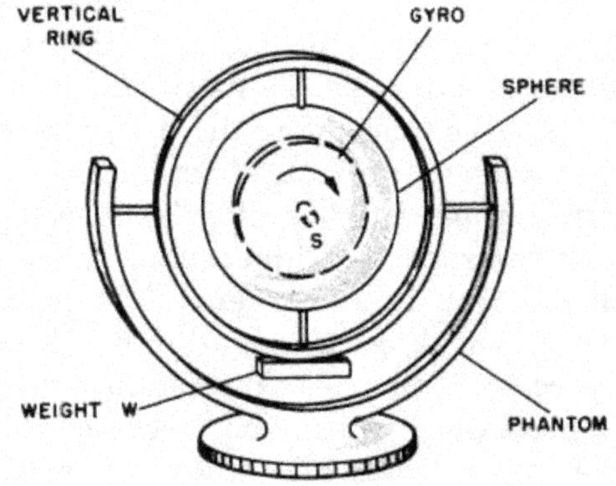

FIGURE 631b.—Making the free gyro seek north by the addition of a weight to the vertical ring.

axis of the gyro. This torque causes precession about the vertical axis in the direction indicated at *C*. The spin axis then has moved out of its original east-west direction.

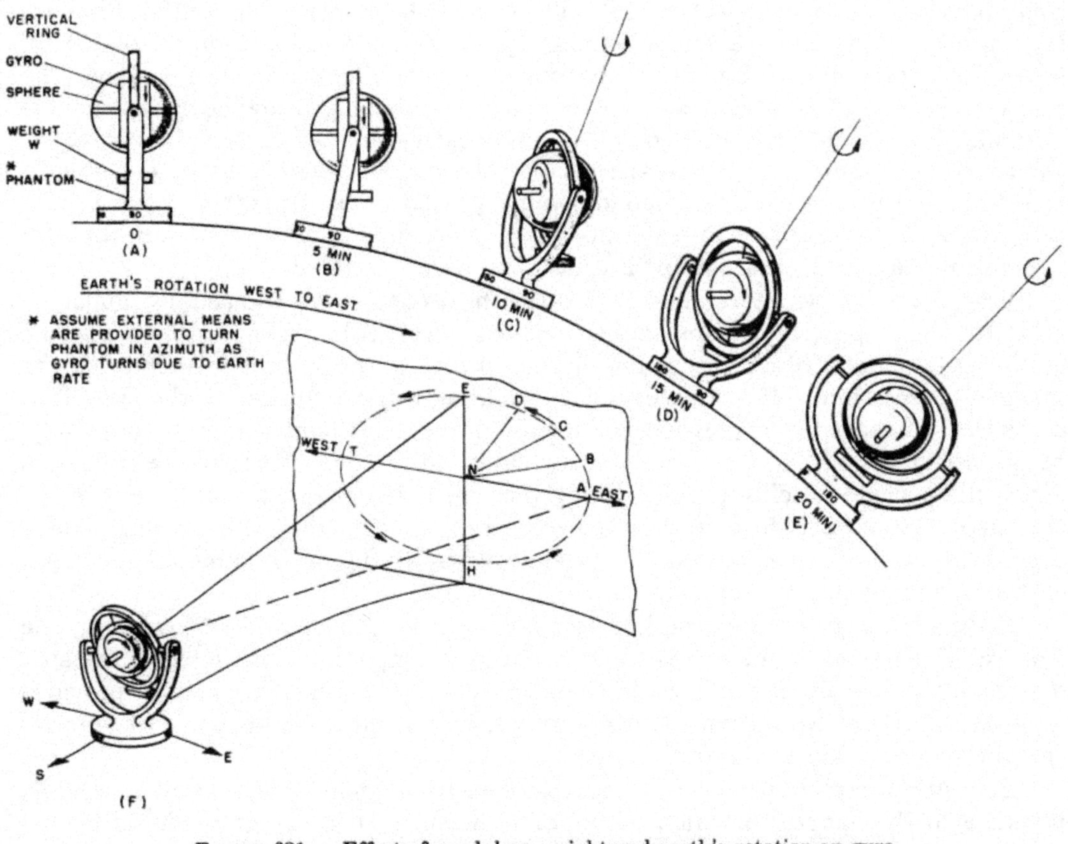

FIGURE 631c.—Effect of pendulous weight and earth's rotation on gyro.

As the end of the spin axis which was first pointing east (which will now be referred to as the north end) continues to rise, the torque on the gyro caused by the weight becomes greater since the moment arm through which the weight acts gets longer due to the greater tilt. Since the speed of precession is closely proportional to the tilt, the gyro turns about its vertical axis as shown at D at an increasing speed until the axis is on the meridian.

At the meridian, the tilt, the torque caused by the weight, and speed of precession are all at a maximum. It should be noted here that it is the righting couple applied to the tilted axis by the pendulous weight which causes the compass to precess past the meridian, the kinetic energy of precession having negligible effect. After the north end of the spin axis crosses the meridian the higher (north) end of the tilted axle is to the west of the meridian. As a result, referring back to figure 630e, the earth's rotation reduces the tilt. As the tilt becomes less, the speed of precession in azimuth decreases. Finally the spin axis becomes horizontal and precession to the west stops; the weight on the vertical ring causes no torque about the horizontal axis since its force acts in the same plane as the gyro rotor and vertical ring. At this point, the axle has precessed as far west of the meridian as it was to the east originally.

As the earth continues to rotate, the north end of the spin axis starts to dip. The weight W is raised on the opposite side of the horizontal axis in the opposite direction carrying the spin axis back across the meridian to its original position. At this point, the cycle is repeated and will go on indefinitely. The oscillation about the meridian may be clearly understood by referring to F in figure 631c which shows the movement of the end of the spin axis projected on a vertical plane. The ellipse is the result of a displacement of the spin axis only a few degrees from the meridian. If the spin axis were pointing east-west at the beginning of the cycle as shown at A in figure 631c, precession would take place through 180 degrees in each direction, and at one extreme the axle would point east, at the other, west. In any case, the gyro never comes to rest since there is no force tending to restore the spin axis to the horizontal position until it has passed the meridian.

The ratio of the movement about the horizontal axis (caused by apparent rotation) to the precessional movement about the vertical axis caused by the swing of the weight determines the shape of the ellipse. If the weight is increased, the speed of precession will increase and the ellipse will be flatter. If the weight is decreased, the speed of precession will decrease and the ellipse would, theoretically, be almost circular. The time, in minutes, required for one complete oscillation, is called the period of oscillation. For any given wheel and speed at a certain point on the earth, the period will be nearly the same regardless of the angle through which the wheel oscillates. The period can be changed by changing the amount of weight on the bottom of the vertical ring.

With such a gyroscope modified by hanging a weight on the vertical ring, the first requirement to make a gyroscope into a gyrocompass, that of making the spin axis seek the meridian, has been fulfilled. However, some means must be provided for damping these oscillations so the gyro wheel will quickly come to rest with its spin axis level in the north-south position.

To damp the oscillations of the spin axis about the meridian a small weight W_1 is added to the sphere in which the rotor is housed. This weight is placed on the east side of the sphere in a position shown in figure 631d. With the spin axis level, the torque produced by gravity acting upon the weight W_1 is restrained by the vertical axis bearings. When the spin axis tilts due to earth rate, the vertical axis is no longer vertical; the force of gravity, however, still pulls straight down on the weight. This allows the torque to act about the vertical axis.

Now, with both weights, the spin axis will begin to tilt due to earth rate and as soon as it tilts, the spin axis precesses toward the meridian and downward toward

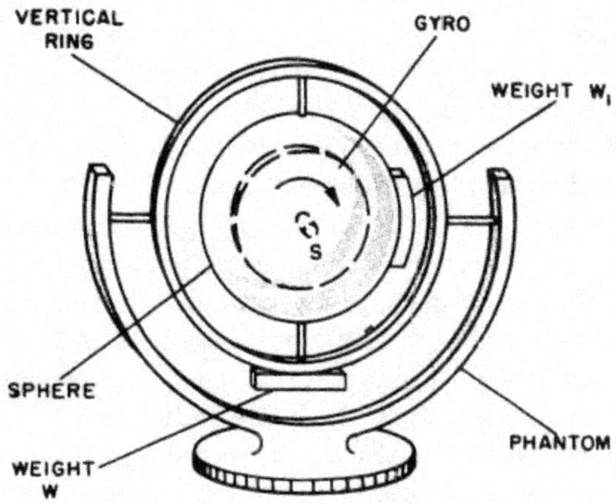

FIGURE 631d.—Modified model gyroscope with weights on the vertical ring and sphere.

the level position. As a result of the leveling action of weight W_1, the spin axis is not tilted up as much when it reaches the meridian as it was with only weight W. Since the spin axis is not tilted as much, the torque produced by weight W is not as great. Therefore, the spin axis will not precess as far to the west of the meridian as it was east of the meridian when it was started.

After reaching a point where the spin axis is level and as far west of the meridian as it is going due to the action of weight W, earth rate is still causing the spin axis to tilt downward. As a result, the forces due to the weights are reversed and torques are created which precess the gyro to the east and up. The same action takes place in the reverse direction. The gyro is not precessed as far to the east as it was to the west. Thus, the added weight W_1 causes the ellipse to be reduced each successive oscillation; the north end of the gyro axle will follow a spiral path as shown in figure 631e instead of an elliptical path as previously.

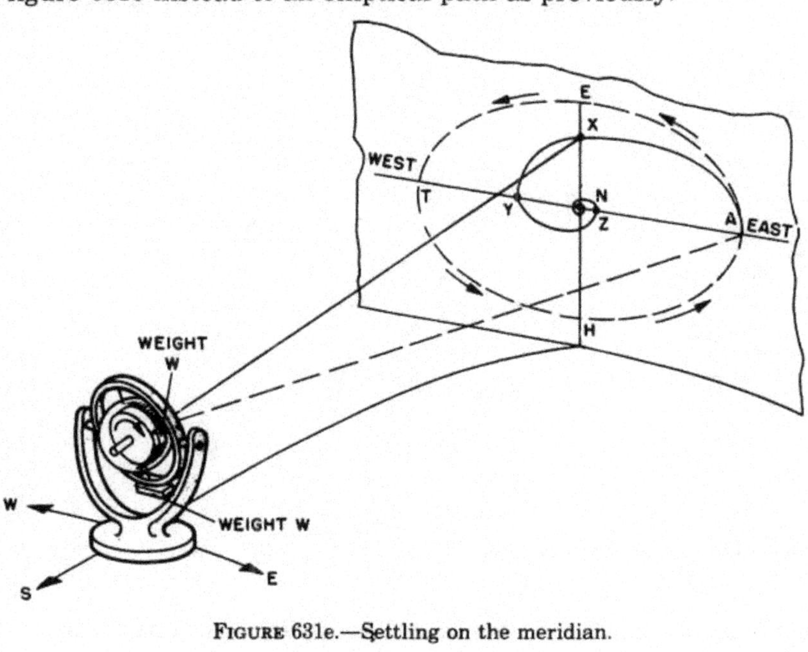

FIGURE 631e.—Settling on the meridian.

A careful consideration of the action of the two weights will make it apparent that the only position of rest that the gyro can find will be with the spin axis horizontal and on the meridian. The period of the compass can be changed by varying the weight W. With a given period the speed with which it settles to a level position can be changed by varying the weight W_1.

With the second modification to the model gyroscope (adding a weight to the sphere in which the gyro is housed), the second requirement to make a gyroscope into a gyrocompass, that of making the spin axis horizontal, has been fulfilled. However, when this modified gyroscope is moved, accelerations on the weights, because they are pendulous, will cause torques on the gyroscope. Also any change in latitude from the equator will result in false indications. The model gyroscope, as so far modified, does not fulfill the third requirement, that of *maintaining* the level position in the plane of the meridian once reached. Therefore, some means must be provided to eliminate the effect of unbalanced weights hanging on the modified gyroscope; means must be provided to compensate for false indications resulting from change in latitude from the equator.

Practical gyrocompasses employ both pendulous and nonpendulous gyroscopes. In addition these compasses have means of compensating for influences that might introduce errors into their indications.

One method of utilizing precession to cause the gyroscope of a practical gyrocompass to seek north is illustrated in figure 631f. Two reservoirs connected by a tube are attached to the bottom of the case enclosing the gyro rotor, with one reservoir north of the rotor and the other south of it. The reservoirs are filled with mercury to such a level that the weight below the spin axis is equal to the weight above it, so that the gyroscope is nonpendulous. The system of reservoirs and connecting tubes is called a **mercury ballistic.** In practice, there are usually four symmetrically placed reservoirs.

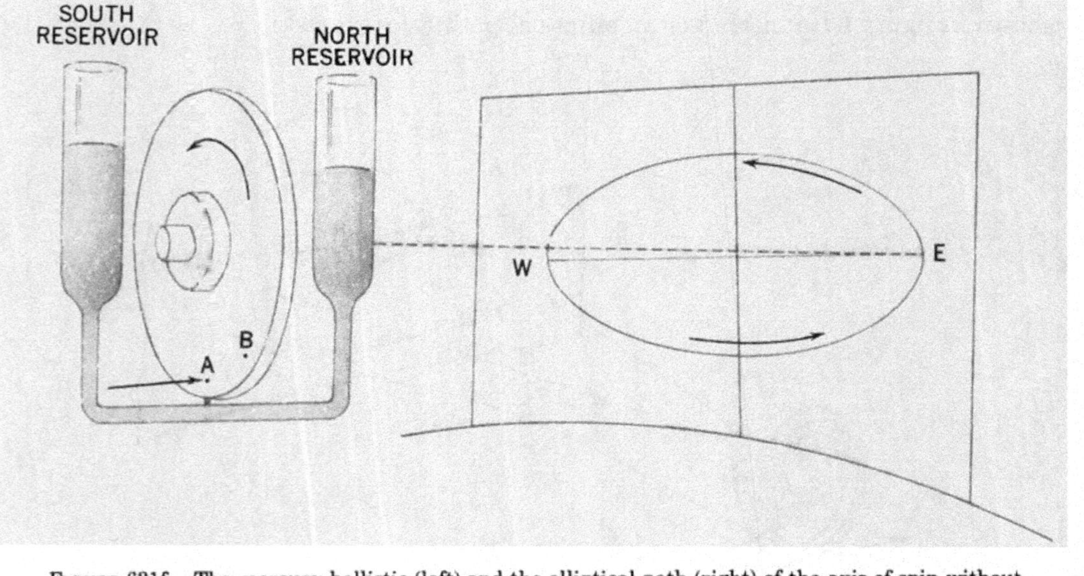

FIGURE 631f.—The mercury ballistic (left) and the elliptical path (right) of the axis of spin without damping.

Suppose that the spin axis is horizontal but is directed to the eastward of north. As the earth rotates eastward on its axis, the spin axis tends to maintain its direction in space; that is, it appears to follow a point, such as a star rising in the northeastern sky. With respect to the earth, the north reservoir rises and some of the mercury flows under the force of gravity into the south reservoir. The south side becomes heavier than the north side, and a force is applied to the bottom of the rotor case at point *A*. If the gyro rotor is spinning in the direction shown, the north end of the spin axis precesses slowly to the westward, following an elliptical path. When it reaches the meridian, upward tilt reaches a maximum. Precession continues, so that the axis is carried past the meridian and commences to sink as the earth continues to rotate. When the sinking has continued to the point where the axis is horizontal again, the excess mercury has returned to the north reservoir and precession stops. As sinking continues, due to continued rotation of the earth, an excess of mercury accumulates in the north reservoir, thus reversing the direction of precession and causing the spin axis to return slowly to its original position with respect to the earth, following the path shown at the right of figure 631f. One circuit of the ellipse requires about 84 minutes.

The elliptical path is symmetrical with respect to the meridian, and, neglecting friction, would be retraced indefinitely, unless some method of damping the oscillation were found. One method is by offsetting the point of application of the force from the mercury ballistic. Thus, if the force is applied not in the vertical plane, but at a point to the eastward of it, as at *B* in figure 631f, the resulting precession causes the spin axis to trace a spiral path as shown in figure 631g, and eventually to settle near the meridian. The gyroscope is now north-seeking. The gyrocompass shown in figure 631h uses this method to seek north.

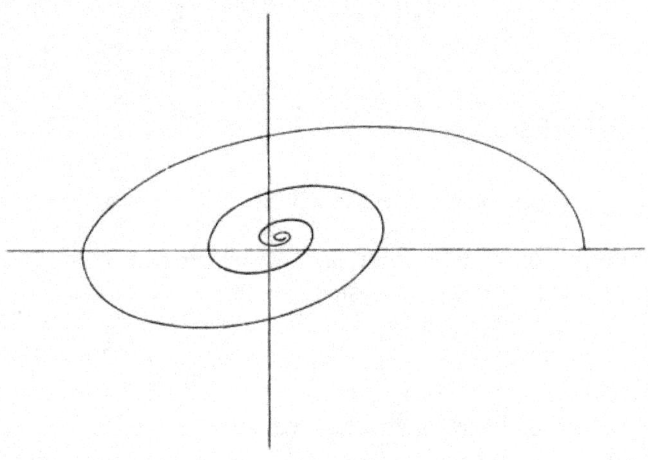

FIGURE 631g.—Spiral path of the axis of spin with damping.

Another method of damping the oscillations caused by the rotation of the earth is to reduce the precessing force of a pendulous gyro as the spin axis approaches the meridian. One way of accomplishing this is to cause oil to flow from one damping tank to another in such a manner as to counteract some of the tendency of an offset pendulous weight to cause precession. Oscillations are completely damped out in approximately one and one-half swings.

632. Gyrocompass errors.—Gyrocompasses are subject to several systematic errors. Some of these can be eliminated or offset in the design of the compass, while others require manual adjustment for their correction.

Courtesy of Sperry Marine Systems.

FIGURE 631h.—The Mark 14 Mod 2 Gyrocompass.

The total combined error (the resultant error) at any time is called **gyro error (GE)**, which is expressed in degrees east or west to indicate the direction in which the axis of the compass is offset from true north. If the gyro error is east, the readings are too low; and if it is west, they are too high. Thus, if GE is 1° W, 1° is subtracted from all readings of the compass, either headings or bearings, to determine the equivalent true directions. One degree is added to all true directions to determine the equivalent gyro directions. The gyro error of modern compasses is generally small. However, significant errors can be introduced in several ways, and it is good practice to compare the gyro heading with the magnetic heading at frequent intervals (as every half hour and after each change of course) and to check the accuracy of the gyrocompass by celestial observation or landmarks from time to time (as every morning and afternoon when means are available).

The errors generally associated with the gyrocompass are speed error, tangent latitude error, ballistic deflection error, ballistic damping error, quadrantal error, and gimballing error. In addition, gyrocompasses are subject to the errors common to directional instruments, such as those introduced by inaccurate graduation of the compass rose or incorrectly located lubber's line. Error may also be introduced, of course, by malfunctioning of the compass.

633. Speed error.—The north-seeking tendency of a gyrocompass depends upon the fact that north is at right angles to the west-to-east direction in which the earth's rotation carries the compass. If the gyrocompass is carried over the earth in some direction other than west to east, it will seek a settling position at right angles to that direction, whatever it may be.

A gyrocompass on the earth's surface is carried from west to east only when it is stationary with respect to the earth's surface, or when it is moving east or west. If the vessel in which the compass is installed is moving in other than an east or west direction the compass is, in effect, being carried in a direction which is a little to the north or south of exactly east. It will then seek a settling position which is at right angles to this direction, and will settle on a line at a small angle off true north.

This error, known as **speed error** and sometimes called **speed-course-latitude error,** is westerly if any component of the vessel's course is north, and easterly if south. Its magnitude depends upon the speed, course, and latitude of the vessel. Refer to figure 633a. If a vessel is at anchor at any point *A,* it is being carried eastward by rotation of the earth at the rate of 902.46 minutes of longitude per hour (with respect to the stars). In terms of knots, this is equal to 902.46 times the cosine of the latitude, approximately. Because of the ellipticity of the earth, the actual value is a little more than this in low latitudes, and a little less in high latitudes. The actual value at any latitude can be found by multiplying the length of a degree of longitude at that latitude (from table 6) by

$$\frac{902.46}{60}=15.041.$$

This eastward motion due to rotation of the earth is shown in figure 633a by the vector *AB.* The north-south axis of the gyrocompass settles in a direction 90° from the direction of motion. Therefore, if the vessel is stationary with respect to the earth, 0° on the compass card coincides with a true meridian, and no error is introduced. This is also true if the vessel is moving due east or due west. In this case the speed of the ship over the surface of the earth is added to or subtracted from the motion due to rotation of the earth, but the direction of motion is unchanged (unless the speed of the vessel is greater than the rotational speed of the earth, and in the opposite direction). The only effect, therefore, is to strengthen or weaken the directive force, usually by a small amount.

If the vessel is on course north or south, as shown by the vector *AC* in figure 633a, the motion in space is tilted toward the north or south of due east. In this case, it is the vector sum (art. 118, vol. II) of the motion due to rotation of the earth and the velocity of the vessel over the surface of the earth, or *AD* in figure 633a. Since *AD* is not due east, the perpendicular to it does not lie in the true meridian, but at some angle δ to it, along AM_v. Since the axis of the gyro lies along AM_v, the "virtual meridian," the angle is the error introduced by the motion of the vessel along its track. Since *AD* is perpendicular to AM_v, and *AB* is perpendicular to *AC,* angle *BAD* is equal to angle δ. Therefore, the angle δ can be found by the formula

$$\tan\,\delta=\frac{AC}{AB}.$$

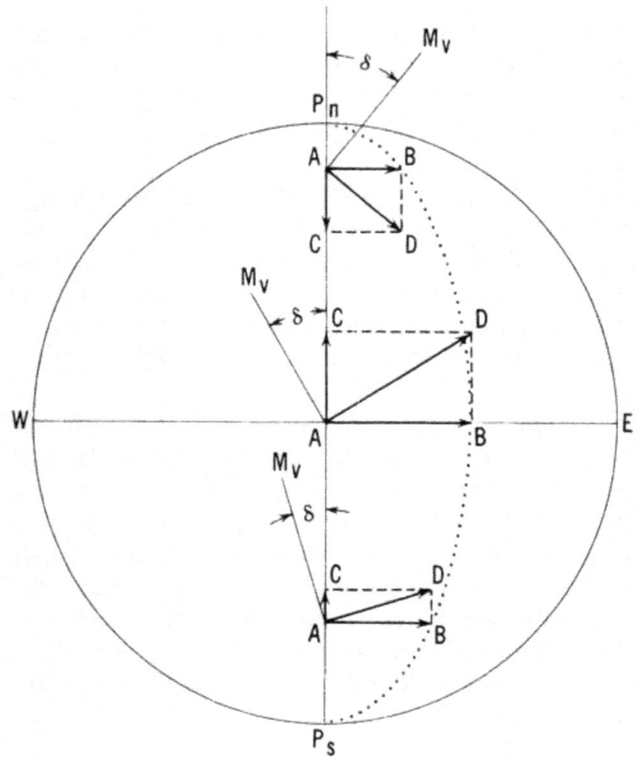

FIGURE 633a.—Speed error.

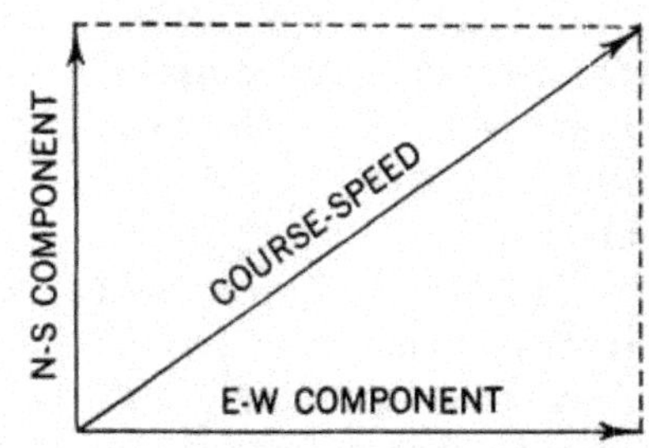

FIGURE 633b.—Components of vessel's motion.

Since *AC* is the speed of the vessel and *AB* is 902.46 cos L, approximately, the formula can be written

$$\tan \delta = \frac{S}{902.46 \cos L}$$

where S is the speed and L the latitude of the vessel.

If the course of the vessel is not a cardinal direction, the resultant is still the vector sum of two speed vectors, and can be found graphically or by computation. One method is to resolve the vessel's speed vector into two components, as shown in figure 633b, obtaining the N–S component along the true meridian, and the E–W component in the direction of rotation of the earth. The N–S component is equal to S cos C, and the E–W component to S sin C, where C is the true course angle. The total N–S motion is then S cos C. The total easterly motion is that due to rotation of the earth plus or minus the E–W component of the ship's speed across the surface of the earth, or 902.46 cos L±S sin C, approximately. The term S sin C is positive (+) for easterly courses and negative (−) for westerly courses. The formula for finding δ now becomes

$$\tan \delta = \frac{S \cos C}{902.46 \cos L \pm S \sin C} \text{ (approximately).}$$

At ship speeds in latitudes less than 70°, the term S sin C is much smaller than 902.46 cos L and has so little effect upon the answer that it can be ignored. The angle δ is small enough that its tangent can be considered the angle itself (expressed in radians). That is, a tangent to a circle can be considered of the same length as an arc of the circle over a short distance from the point of tangency. Therefore, the formula for δ can be written

$$\delta=\frac{57.3\ S\ \cos\ C}{902.46\ \cos\ L}$$

or

$$\delta=0.0635\ S\ \cos C\ \sec L.$$

As shown in this formula, the speed error δ is affected by the three variables, speed, course, and latitude. If the course has a northerly component, the error is westerly; and if it has a southerly component, the error is easterly.

Example.—A ship at latitude 30°N is steaming on true course 045°, at a speed of 20 knots.

Required.—Speed error.

Solution.—

0.0635	log 8.80277
S 20 kn.	log 1.30103
C N45°E	*l* cos 9.84949
L 30°N	*l* sec 10.06247
δ 1°.04W	log 10.01576

Answer.—δ 1°.04W.

In some gyrocompasses this error is corrected mechanically. Speed and latitude are set in by hand, and the cosine of the course is introduced automatically by means of a "cosine cam" running in an eccentric groove on the underside of the azimuth gear. In some compasses these corrections combine to offset the lubber's line by the correct amount. Small changes in speed or latitude have relatively little effect upon the result. Therefore, in normal operations, infrequent changes are sufficient for satisfactory results. If no provision is made for mechanically applying this correction, a table or curves can be used to indicate the correction to be applied mathematically to readings of the compass. These are made up from the formula given above, and are entered with the speed, course, and latitude (art. 639).

634. Tangent latitude error applies only to those gyrocompasses in which damping is accomplished by offsetting the point of application of the force from a mercury ballistic (art. 631). It can be found from the equation

$$\alpha=r\ \tan\ L$$

in which α is the damping error, r is the angle between the vertical through the spin axis of the gyro rotor and a line through this axis and the point of application of the force from the mercury ballistic (1°.7 for Sperry compasses), and L is the latitude. The error is easterly in north latitude and westerly in south latitude.

Example.—A gyrocompass having a value of r of 1°.7 is at latitude 50°N.

Required.—The tangent latitude error.

Solution.—

$$\begin{aligned}\alpha&=r\ \tan\ L\\&=1^\circ.7\times1.1918\\&=2^\circ.03E\end{aligned}$$

Answer.—$\alpha=2^\circ.03$E.

As in the case of speed error, provision is made in most compasses (to which it applies) for correcting this error. An auxiliary latitude-correction scale is provided for this purpose. In some compasses this offsets the lubber's line. In others, it alters the position of a small weight attached to the casing near one end of the axle. The first method is preferable because it is unaffected by changes of gyro speed of rotation.

If this error is not corrected mechanically, it can be combined algebraically with speed error and a single set of tables or graphs made up. This is a method sometimes used in polar regions, beyond the scale of the latitude corrections (arts. 639, 2514).

635. Ballistic deflection error.—When the north-south component of the speed changes, an accelerating force acts upon the compass, causing a surge of mercury from one part of the system to another, or a deflection (along the meridian) of the mass of a pendulous compass. In either case, this is called **ballistic deflection.** It results in a precessing force which introduces a temporary **ballistic deflection error** in the readings of the compass unless it is corrected.

A change of course or speed also results in a change in the speed error, and unless the correcting mechanism responds promptly to this change, a temporary error from this source is also introduced. The sign of this error is opposite that of the ballistic deflection, and so the two tend to cancel each other. If they are of equal magnitude and equal duration, the cancellation is complete and the compass responds immediately and automatically to changes of speed error. This can be accomplished by designing the compass so that

$$\frac{B}{H}=0.0211 \sec L.$$

in which B is the pendulous moment of a pendulous compass and the couple per unit angle applied by a mercury ballistic, H is the angular momentum of the gyro rotor, and L is the latitude.

Gyrocompasses using the fluid ballistic are often designed so that the ratio

$$\frac{B}{H}$$

is correct for some particular latitude (as 41° or 45°) and accept the small residual error that is temporarily present at other latitudes. This is satisfactory for vessels which remain within relatively narrow limits of latitude, or which are seldom subjected to large accelerating forces. However, where these conditions are not met, provision is made for varying the ratio with latitude. In a compass having a mercury ballistic, this is customarily accomplished by moving the mercury reservoirs radially toward or away from the center of the compass, thus altering the value of B. In a pendulous gyro, the value of H is changed by altering the rotational speed of the gyro.

When the ratio

$$\frac{B}{H}$$

is as given in the equation above, the period of oscillation about the vertical axis is given by the equation

$$T=\frac{\pi}{30}\sqrt{\frac{R}{g}}$$

in which T is the period in minutes, R is the radius of the earth in feet (approximately 20,900,000) and g is the acceleration due to gravity (approximately 32.2 feet per second per second). Substituting in the formula,

$$T=0.1047\sqrt{\frac{20{,}900{,}000}{32.2}}$$

$$=84 \text{ minutes (approximately).}$$

This is sometimes stated as the period of a pendulum having a radius equal to the radius of the earth, since the equation for a short pendulum is the same as that given above with *l* (length) being substituted for R. More accurately, it is the period of a pendulum of infinite length with its bottom at the surface of the earth, or the largest period that a simple pendulum can have when acting under the gravitational force of the earth. When a device is adjusted so as to have this period it is said to be "Schuler tuned," after Max Schuler, a German scientist who discovered the relationship. It is because of this tuning of the gyrocompass that one oscillation occurs in about 84 minutes, and that the maximum effect of certain disturbing forces occurs about 21 minutes (one-fourth cycle) after application of the force.

636. Ballistic damping error is a temporary oscillatory error of a gyrocompass introduced during changes of course or speed as a result of the means used to damp the oscillations of the spin axis.

During a change of course or speed the fluid in the ballistic of the nonpendulous compass or in the damping tanks of the pendulous compass (art. 631) is accelerated. As shown in figure 636, during the turn from the westerly to the northerly direction the centrifugal force acting on the mercury causes an excess of mercury to accumulate in the south tanks of the ballistic at *A*. Because of the offset connection of the mercury ballistic, the excess mercury in the south tanks exerts a torque about the vertical axis in addition to the one being exerted about the horizontal axis during the turn. This torque about the vertical axis produces a downward tilt of the north end of the gyro axle at *B* as a result of precession. This tilt of the gyro axle causes an oscillation of the spin axis to start as the centrifugal force diminishes to zero at *C*. This oscillation on a compass with a damped period of about 84 minutes becomes a maximum at *D*, 21 minutes after the change of course is completed, and ends in about 2 hours.

The liquid in the damping tanks of the pendulous compass is subjected to the same centrifugal force on change of course. An excess of liquid collects in one tank. This action causes a torque and consequent movement of the spin axis from the meridian.

The ballistic damping error is eliminated in the nonpendulous compass by automatically moving the point of application of the mercury ballistic from the offset position to the true vertical axis of the gyro whenever rates of change of course or speed exceed certain limits. Moving the point of application of the mercury ballistic to the true vertical axis eliminates the torque about this axis caused by the centrifugal force and prevents the compass from going through a damped oscillation.

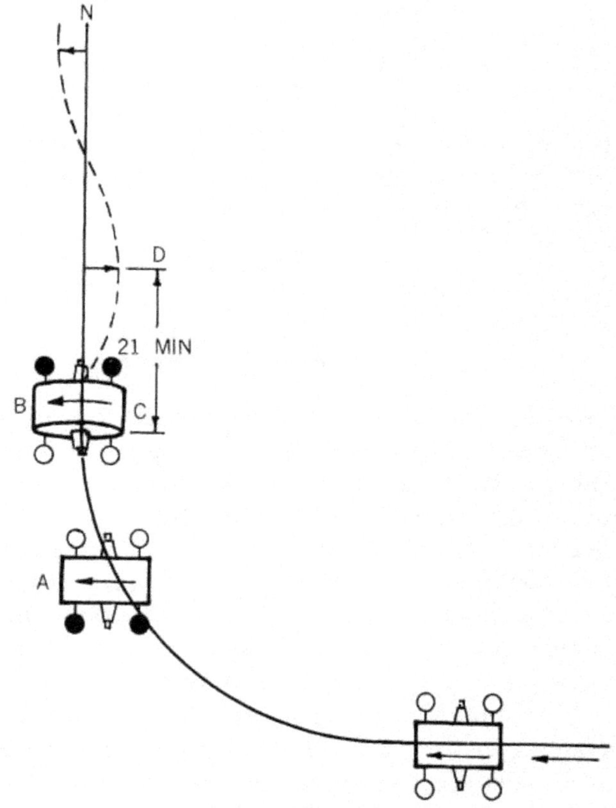

FIGURE 636.—Ballistic damping error.

In the pendulous gyrocompass, the ballistic damping error is eliminated by automatically closing a valve in the pipe line between the damping tanks whenever rates of change of course or speed exceed certain limits.

637. Quadrantal error.—If a body mounted in gimbals is not suitably balanced, a disturbing force causes it to swing from side to side. A swinging body tends to rotate so that its long axis of weight is in the plane of the swing. The rolling of a vessel introduces the force needed to start a gyrocompass swinging. The effect reaches a maximum on intercardinal headings, midway between the two horizontal axes of the compass, and changes direction of error in consecutive quadrants. This is called **quadrantal error,** or sometimes **intercardinal rolling error.** It is corrected by the addition of weights to balance the compass so that the weight is the same in all directions from the center. Without a long axis of weight, there is no tendency to rotate during a swing.

A second cause of quadrantal error is more difficult to eliminate. As a vessel rolls, the apparent vertical is displaced first to one side and then to the other, due to the accelerations involved. The vertical axis of the gyrocompass tends to align itself with the apparent vertical. If the vessel is on a northerly or southerly course, the pivot of the compass is displaced from the vertical, resulting in a precession first to one side, then to the other. The effect is negligible and would be exactly balanced if successive rolls on opposite sides were equal. On an easterly or westerly heading, the pivot remains under the gyro axle, but the dynamic effect of the roll, acting upon the damping mechanism, introduces a precessing force which causes an error. However, the period is short and the error is in opposite directions on opposite rolls, so the effect is negligible. On noncardinal headings, both effects are present, and the relationship is such that the error is in the same direction regardless of the direction of roll. Thus, a persistent error is introduced, which changes direction in successive quadrants. This error is generally eliminated by the use of a second gyroscope. In some compasses, this is in the form of a small gyroscope called a **floating ballistic** which stabilizes the point of application of the mercury ballistic with respect to the true vertical as the vessel rolls. In others, two gyroscopes are used for the directive element and these are so installed that they tend to precess in opposite directions. Thus, they neutralize each other. Another way of eliminating this error is to design the mercury ballistic system so that the surge of liquid due to north-south component of the roll is diminished in amount and delayed so that it is about a quarter of a cycle out of phase with the roll.

638. Gimballing error is that due to tilt of the compass rose. Directions are measured in the horizontal plane. If the compass card is tilted, the projection of its outer rim into the horizontal is an ellipse, and the graduations are not equally spaced with respect to a circle. For normal angles of tilt, this error is small and can be neglected. For accurate results, readings should be made when the card is horizontal. This error applies to the reading of the compass *or its repeaters* (art. 643), rather than to the compass itself. If the compass and its repeaters are installed so that the outer gimbals are in the longitudinal axis of the vessel, this error is minimized.

639. Use of the gyrocompass in polar regions is discussed in article 2514. If means are not available for determining an equivalent setting or correction, a correction graph can be constructed. Ballistic deflection error, quadrantal error, and gimballing error are temporary or corrected in the design of the compass, and so can be ignored. Speed error and tangent latitude error (if it applies to the particular compass involved) can be combined into a single table or curve of corrections, using the formulas of articles 633 and 634. In high latitudes the east-west component of the vessel's speed is significant, and the error may be too large to consider its tangent equal to the angle itself expressed in radians. Therefore, the applicable formulas are:

$$\tan \delta = \frac{S \cos C}{902.46 \cos L \pm S \sin C} \tag{1}$$

$$\alpha = r \tan L. \tag{2}$$

The only approximation remaining is the use of 902.46, which varies slightly with latitude. The error thus introduced is not significant. The U.S. Navy gyrocompass error curves for latitude 80° are shown in figure 639. From the intersection of the appropriate speed curve and the radial line representing the *true* course (interpolating if necessary) a horizontal line is drawn to the vertical line through the origin, where the correction is indicated. To construct the curve for speed 35 knots, proceed as follows:

1. Compute the speed error, δ, for true courses at intervals of perhaps 30°. As an example, the error for course 210° (C S30°W) is:

$$\tan \delta = \frac{35 \times 0.86603}{902.46 \times 0.17365 - 35 \times 0.50000}$$

$$= 0.21773.$$

$$\delta = 12^\circ\!.3\text{E}.$$

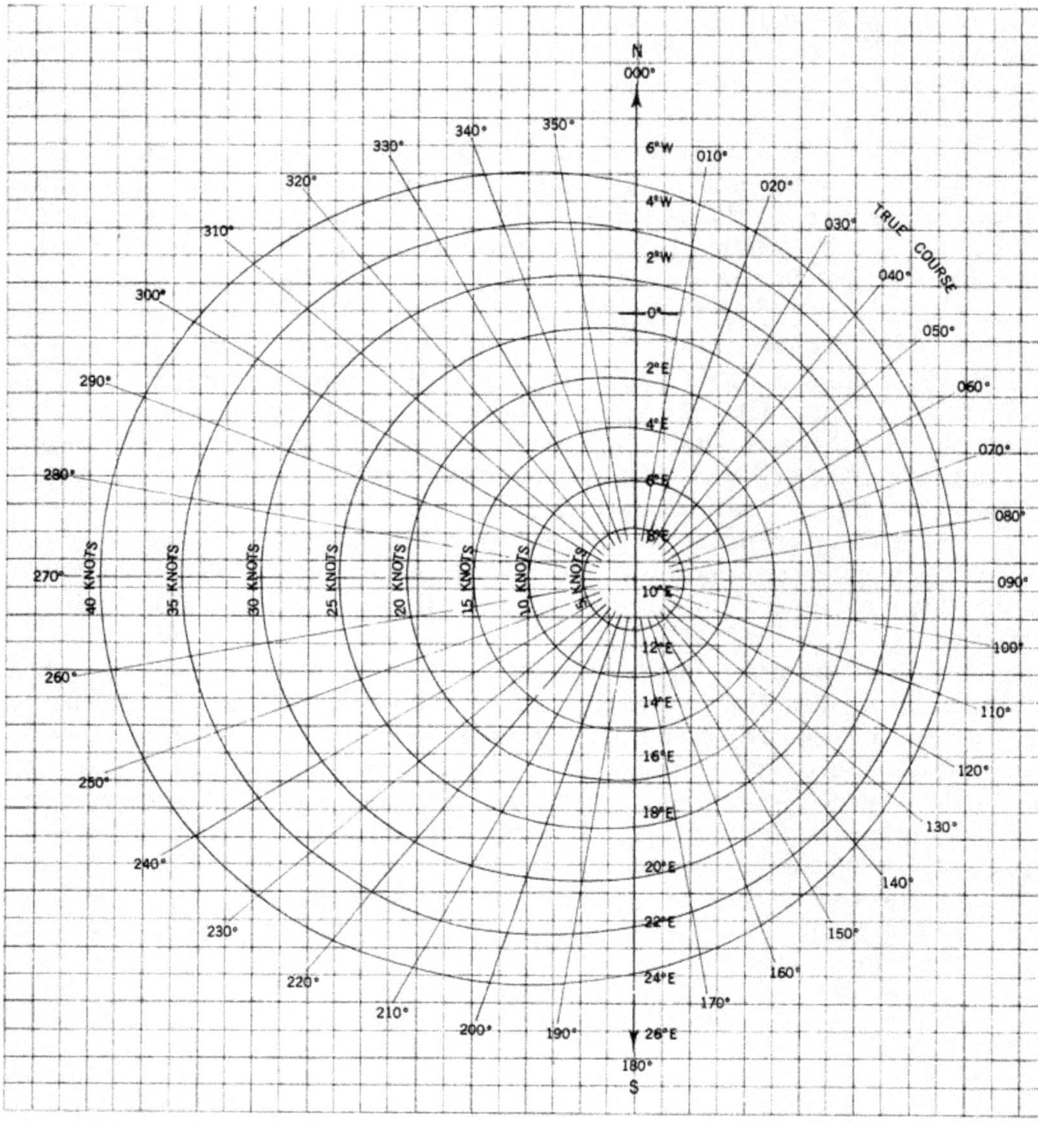

FIGURE 639.—Gyrocompass error curves for latitude 80°.

The error is easterly because the course has a southerly component (art. 633).

2. Compute the tangent latitude error. The curves of figure 639 are for a value of r of 1°.7:

$$\alpha = 1^{\circ}.7 \times 5.6713 = 9^{\circ}.6\text{E}.$$

In northern latitudes tangent latitude error is easterly.

3. Combine δ and α algebraically to obtain gyro error (GE):

TC	δ	α	*GE*
°	°	°	°
000	12.6W	9.6E	3.0W
030	9.9W	9.6E	0.3W
060	5.3W	9.6E	4.3E
090	0.0	9.6E	9.6E

TC	δ	α	GE
°	°	°	°
120	5.3E	9.6E	14.9E
150	9.9E	9.6E	19.5E
180	12.6E	9.6E	22.2E
210	12.3E	9.6E	21.9E
240	7.9E	9.6E	17.5E
270	0.0	9.6E	9.6E
300	7.9W	9.6E	1.7E
330	12.3W	9.6E	2.7W

4. To draw the curve, select a convenient origin and label this with the value of α. Draw a vertical line through the origin and mark off a convenient scale such that all values of δ can be shown both above and below the origin. The zero on this scale is at point α units above the origin (below in the Southern Hemisphere). Label the scale according to GE. Through the origin draw various radial lines at any convenient interval to represent true courses. For each computed course draw a horizontal construction line from the GE on the central scale to the appropriate radial line. The intersection of each pair of lines is one point on the curve. Connect all such points with a smooth curve, and erase the construction lines. If a straightedge or graph paper is used, the construction lines need not be drawn.

It is good practice to draw the curve for the highest speed first, to be sure that succeeding curves will fit on the paper. From such curves the gyro courses corresponding to various true courses can be determined and the radial lines labeled with these values for converting gyro directions to true directions.

The curves described in this article are for use *when all correctors are set on zero,* or if no provision is made for mechanically correcting for speed and damping errors. If the compass does not have a mercury ballistic, the tangent latitude error is omitted from the calculations and curves.

640. Desirable characteristics of the gyrocompass.—Since a gyrocompass is not affected by a magnetic field, it is not subject to magnetic compass errors (ch. VII), nor is it useless near the earth's magnetic poles. If an error is present, it is the same on all headings, and no table of corrections is needed. The directive force is sufficiently strong to permit directional pick-off for use in remote-indicating repeaters, automatic steering, dead reckoning and fire-control equipment, course recorders, etc.

641. Undesirable characteristics of the gyrocompass.—A gyrocompass is dependent upon a source of suitable electric power.

If operation of the compass is interrupted long enough to permit uncertainty in its indications, a considerable period (as much as four hours for some gyrocompasses) may be needed for it to settle on the meridian after it reaches operating speed. This period can be reduced by orienting the compass in the proper direction before it is started. If this is not practicable, the settling period can be hastened by leveling the compass when it reaches the meridian (one-fourth of a cycle or 21 minutes after starting at maximum deflection) or by leveling *and* precessing the gyro to the approximate meridian after its direction and rate of precession are observed for several minutes. Either process may need to be repeated several times and followed by a settling period.

The gyrocompass is subject to certain errors requiring applications of corrections, either manually or automatically (art. 632).

The compass is an intricate mechanism of many parts. Thus, it requires some maintenance. In heavy seas a gyrocompass may become unreliable unless certain features are included in the design—features which are generally omitted from the small, simpler compasses.

The directive force of a gyrocompass decreases with latitude, being maximum at the equator and zero at the geographical poles.

642. Gyrocompass models.—The Mark 19 Gyrocompass System (fig. 642a) consists of four components: the master compass, the control cabinet, the compass failure annunciator, and the solid state power supply.

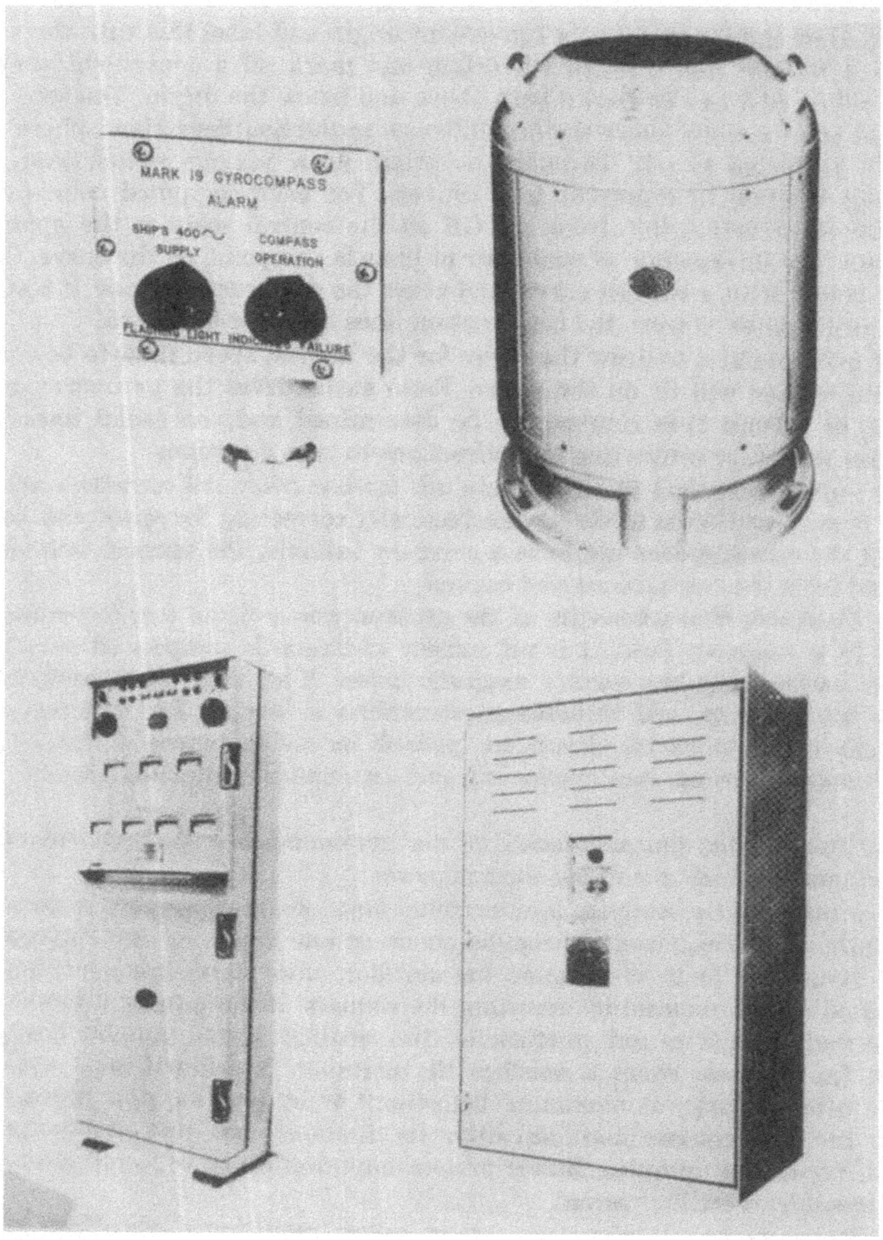

FIGURE 642a.—Mark 19 Gyrocompass System.

The two main elements of the master compass (fig. 642b) are the compass element and the supporting element. The compass element includes the sensitive element (meridian and slave gyros), the phantom or follower element, and the gimbal. The supporting element includes the frame and binnacle which provide a shock-mount support for the compass element.

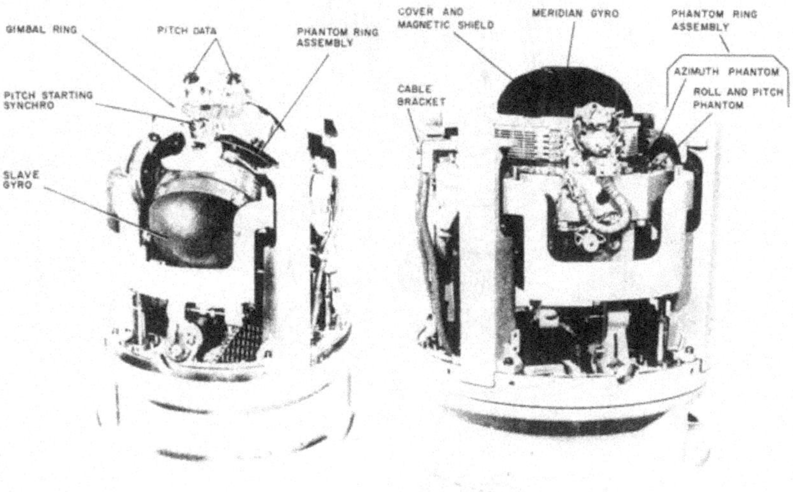

FIGURE 642b.—Mark 19 master compass.

The control cabinet contains all controls and indicators necessary for the operation of the equipment.

The compass failure annunciator contains two indicator lights to indicate a malfunction of the compass system or failure of the vessel's 400 Hertz power supply.

The solid state power supply provides the power necessary for operation of the master compass, control cabinet, compass failure annunciator, and charging batteries used as emergency power. All other gyrocompass system components must be energized from ship power. In the event of loss of ship power, the solid state supply will continue to operate from the batteries until the ship's line is restored or the batteries are discharged.

Fluid suspension of the sensitive element provides high shock tolerance and greatly reduces the effect of accelerations. At running temperature, the specific gravity of the gyrosphere is the same as that of the oil in which it is immersed. Since the gyrosphere is in neutral buoyancy, it exerts no load on the vertical bearings which, therefore, serve only as guides for the sphere.

The Mark 19 Gyrocompass has four modes of operation. The normal mode provides optimum performance up to latitude 75°. The fast settle mode provides accelerated settling of the compass upon starting. The high latitude mode provides optimum performance from latitude 75° to about latitude 86°. The directional gyro mode enables operation of the compass as a free gyro with the spin axis oriented to grid north (art. 2510).

The Mark 19 Gyrocompass consists basically of two gyros (fig. 642c) placed with their spin axes mutually perpendicular in the horizontal plane. The spin axis of one gyro is directed along a north-south line, and the spin axis of the second is slaved to the first along an approximate east-west line. The north-seeking or meridian gyro and the slave gyro are mounted one above the other in a supporting ring. This ring is made to follow the gyros in heading and tilt by azimuth, roll, and pitch servos. These servos also drive the synchro transmitters which serve to supply output data.

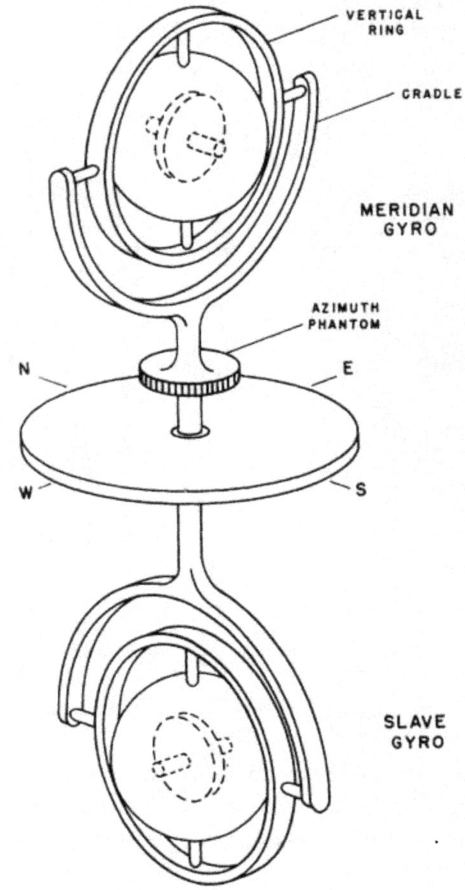

FIGURE 642c.—Simplified diagram of the Mark 19 compass element.

The meridian gyro is essentially a gyrocompass. It furnishes indications of heading as well as tilt about the east-west axis. The slave gyro is essentially a free gyro and furnishes only an indication of tilt about the north-south axis. Thus, the Mark 19 Gyrocompass provides heading, roll, and pitch data.

In the meridian gyro of the Mark 19 Gyrocompass, the tilt is detected by a gravity reference attached to the vertical ring in such a way that it is parallel to the gyro axle. This device (electrolytic level) is a special level which transmits an electrical signal with magnitude and sense according to tilt. Since the gravity reference and axle are parallel and rigidly fixed with respect to one another, the signal emitted by the gravity reference is a measure of the gyro axle tilt about the east-west axis. In the more recent modifications of the gyrocompass, higher accuracy is obtained through the use of linear accelerometers instead of electrolytic levels to sense the direction of gravity.

The tilt signal, after amplification, is applied to the control fields of electrical torquers, as shown in figure 642d, which cause torques about the vertical and horizontal axes. The torquers located about the horizontal axis are known as the azimuth torquers. They apply a torque about the horizontal axis proportional to the amount of tilt of the spin axis and cause the gyro to precess in azimuth. The effect of this torque is the same as making the gyro pendulous by attaching a heavy weight to the bottom of the vertical ring. When one end of the axis is tilted up, the resulting torque about the horizontal axis precesses the gyro in azimuth, i.e., about its vertical axis.

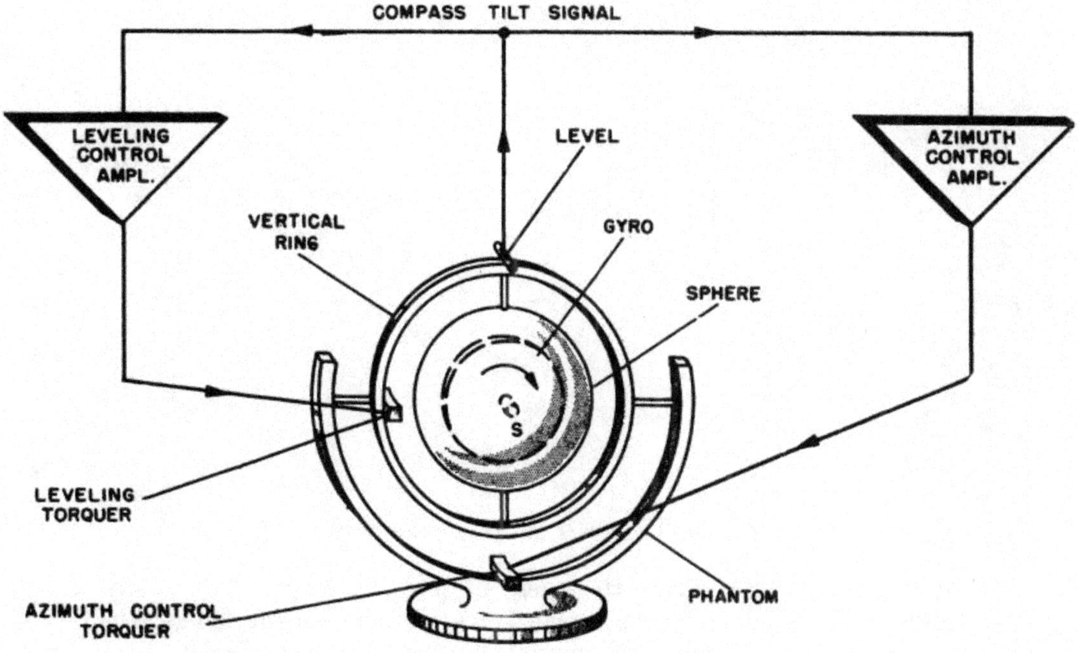

FIGURE 642d.—Simplified diagram of electrical azimuth and leveling controls for meridian gyro.

The leveling torquer, located about the vertical axis of the gyro, applies a torque about the vertical axis proportional to tilt and causes the gyro to precess about the horizontal axis to reduce the tilt to zero. The effect of this torque is the same as attaching a weight to the east side of the sphere (art. 631). When one end of the gyro axle is tilted up, the resulting torque about the vertical axis precesses the high end down. Thus, the effect of these two torques is to continually precess the axis to the meridian and make it level.

The Technical Manual for the Mark 19 Gyrocompass should be referred to for an explanation of the means used to control the compass.

The **Mark 27 Gyrocompass** (fig. 642e) consisting of two major components—the master compass unit and the compass electronics unit—is designed for both military and commercial, small to medium class vessels. The equipment is powered by an internal 400 Hz solid state supply (inverter) operating from a 24 volt DC battery source or from an external 115 volt 60 to 400 Hz converter (rectifier) unit. The master compass unit is an oil-filled sealed unit containing the sensitive element, fluid ballistic, and the supporting gimbals and servo drive. The compass electronics unit contains the compass controls, supporting solid state electronics, and a solid state power supply. A direct-reading heading indication dial on the master compass has red illumination for night viewing. The master compass can be provided with various types of electrical transducers for transmission of the heading data to remote repeaters. The internal static power supply has the capability to power either two step repeaters or two servo synchro repeaters. Additional repeaters can be accommodated with externally supplied power. The master compass unit can be mounted on top of the compass electronics unit, or in a remote location.

FIGURE 642e.—Mark 27 Gyrocompass System.

Fluid suspension of the sensitive element provides high shock tolerance and greatly reduces the effect of accelerations. The sensitive element can be caged when not in use to prevent damage. The manually operated caging element is on top of the master compass.

A fluid ballistic (art. 631) provides the gravitational torques to make the gyro seek north. This ballistic consists of two interconnected brass tanks, partially filled with a 20 centistoke silicone fluid. The small bore of the tubing connecting the tanks retards the free flow of fluid between the tanks. Because of the time it takes for the fluid to flow, the disturbing effects of ship maneuvers and roll and pitch motion are minimized.

To compensate for the effect of changes in vertical earth rate due to change of latitude, a manual latitude dial and a North/South switch is incorporated on the control unit for producing an electrical torque on the gyrosphere. The switch and dial should be properly positioned by the operator.

The **Mark 227 Gyrocompass** (fig. 642f) utilizes the basic Mark 27 design in a configuration designed primarily for large commercial and auxiliary naval vessels. The master compass, which is identical to the Mark 27 master compass except for mounting facilities, is mounted in gimbals atop a deck-mounted console. All of the controls, except for the caging control, power supplies, and repeater switches (for up to 8 repeaters) are contained in the console. Standard units are equipped with a step transmitter and auxiliary equipment to power 8 step repeaters. Synchros, either 60 or 400 Hz, can be added to supply single or single and 36-speed data. Input power to this unit is 115 volt 60 to 400 Hz. Speed correction has been added and is set with a front panel knob. The additional gimbals enable transmission of azimuth data free of error due to deck tilt.

643. Gyrocompass repeaters.—A gyrocompass is customarily located at a favorable position below decks, and its indications transmitted electrically to various positions throughout the vessel. Each repeater consists of a compass rose on a suitable card so mounted that the direction of the ship's head is indicated at a

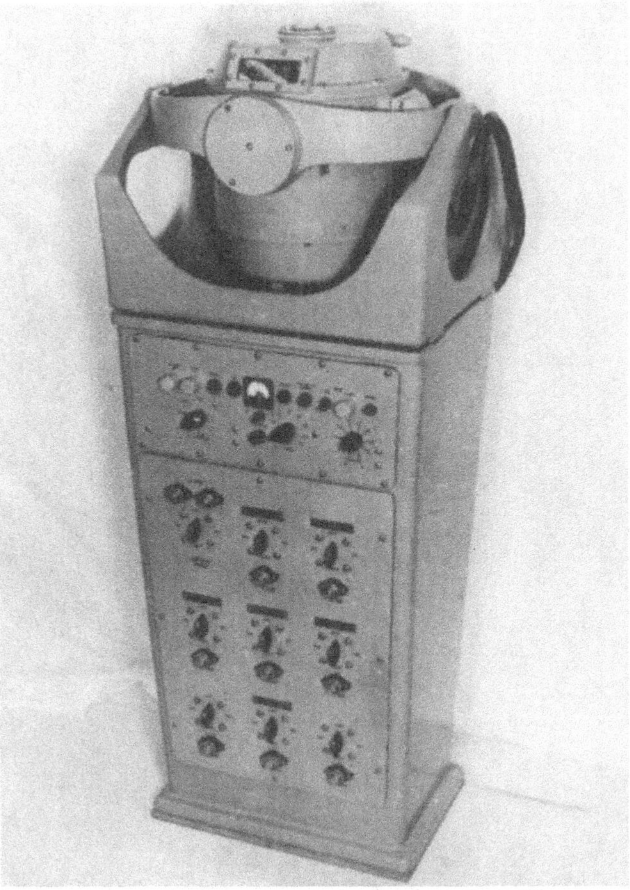

Courtesy of Sperry Marine Systems.

FIGURE 642f.—Mark 227 Gyrocompass System.

lubber's line. Although the repeater may be mounted in any position, including vertically on a bulkhead, it is generally placed in gimbals in a bowl, similar to the mounting of a compass, which it resembles (fig. 643). This is true particularly of repeaters used for obtaining bearings. A gyro repeater used primarily to indicate the gyro heading is sometimes called a **ship's course indicator.**

Gyrocompass indications are also used in automatic steering devices, direction-stabilized radarscopes, wind indicators, fire control equipment, etc.

A compass used to control other equipment, particularly repeaters, is sometimes called a **master compass.** In the case of a gyrocompass, it is usually called a **master gyrocompass.** It is good practice to check all repeaters periodically with the master compass to insure continued synchronization.

644. Gyro repeater accessories.—The bearing circle and azimuth circle (art. 629) are also used with the gyro repeater for bearing and azimuth observations. A **telescopic alidade** (fig. 644) may also be provided for bearing observations from repeaters. The telescopic alidade is basically similar to the bearing circle, except that it is fitted with a telescope instead of sighting vanes. The telescope of the telescopic alidade shown in figure 644 is mounted on a ring that fits on the gyro

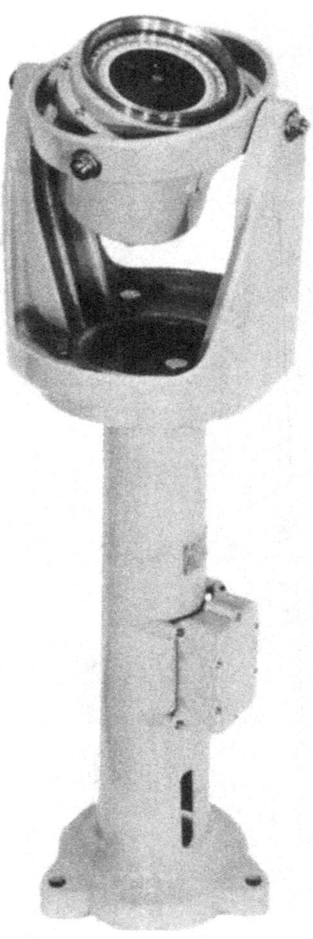

Courtesy of Ahrendt Instrument Co.

FIGURE 643.—A gyro repeater used as a ship's course indicator (Mark 2 Mod 5).

repeater. The erecting telescope is fitted with crosshair, level vial, polarizing light filter, and internal focusing. The optical system projects the image of approximately 25° of the compass card together with a view of the level vial onto the optical axis of the telescope. By this means, both the observed object and its bearing can be viewed at the same time through the eyepiece.

645. Pelorus.—Although it is desirable to have a compass, a compass repeater, or an alidade for obtaining bearings, satisfactory results can be obtained by means of an inexpensive device known as a **pelorus** (fig. 645). In appearance and use this device resembles a compass or compass repeater, with sighting vanes or a sighting telescope attached, but it has no directive properties. That is, it remains at any *relative* direction to which it is set. It is generally used by setting 000° at the lubber's line. Relative bearings are then observed. They can be converted to bearings true, magnetic, grid, etc., by *adding* the appropriate heading. The direct use of relative bearings is sometimes of value. A pelorus is useful, for instance, in determining the moment at which an aid to navigation is broad on the beam. It is also useful in measuring pairs of relative bearings for use with table 7 or for determining distance off and distance abeam without a table.

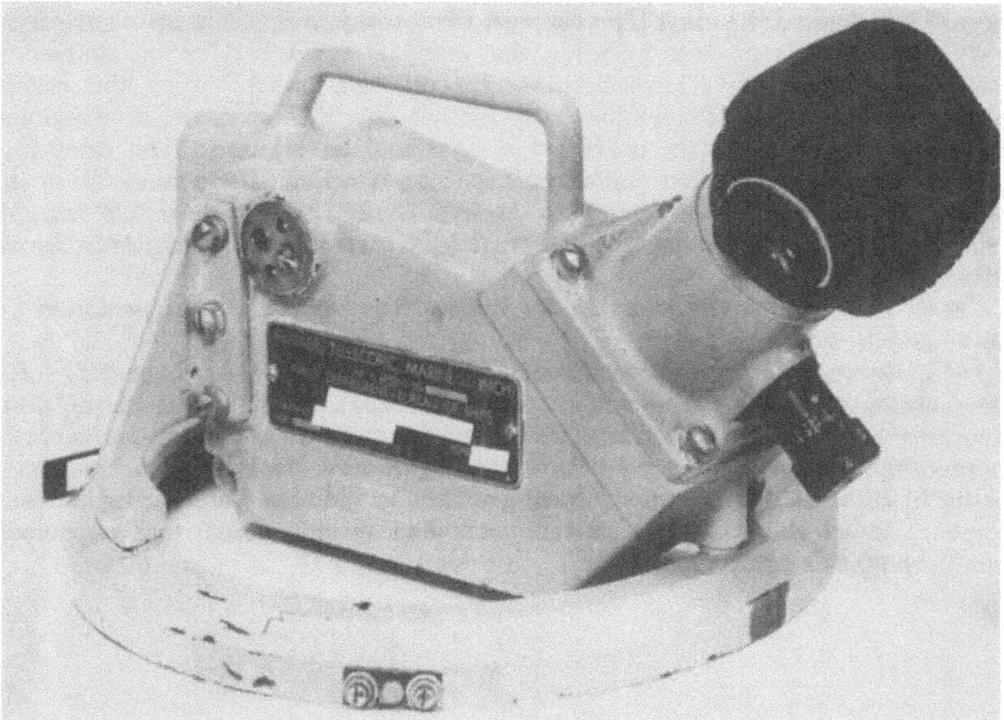

FIGURE 644.—Telescopic alidade.

FIGURE 645. A pelorus.

If the true heading is set at the lubber's line, true bearings are observed directly. Similarly, compass bearings can be observed if the compass heading is set at the lubber's line, etc. However, the vessel must be on the heading to which the pelorus is set if accurate results are to be obtained, or else a correction must be applied to the observed results. Perhaps the easiest way of avoiding error is to have the steersman indicate when the vessel is on course. This is usually done by calling out "mark, mark, mark" as long as the vessel is within a specified fraction of a degree of the desired heading. The observer, who is watching a distant object across the pelorus, selects an instant when the vessel is steady and is on course. An alternative method is to have the observer call out "mark" when the relative bearing is steady, and the steersman note the heading. If the compass is swinging at the moment of observation, the observation should be rejected. The number of degrees between the desired and actual headings is *added* if the vessel is to the *right* of the course, and *subtracted* if to the *left*. Thus, if the course is 060° and the heading is 062° at the moment of observation, a correction of 2° is added to the bearing.

Each observer should determine for himself the technique that produces the most reliable results.

646. Course recorder.—A continuous graphical record of the headings of a vessel can be obtained by means of a **course recorder** (fig. 646). In its usual form, paper with both heading and time graduations is slowly wound from one drum to another, its speed being controlled by a spring-powered clockwork mechanism. A pen is in contact with the paper, tracing a line to indicate the heading at each moment. The pen is attached to an arm controlled by indications from a compass, usually the master gyrocompass.

Courtesy of Sperry Marine Systems.

FIGURE 646.—A course recorder.

647. Dead reckoning equipment.—The primary navigational functions of **dead reckoning equipment (DRE)** are to (1) provide continuous indications of the vessel's present latitude and longitude, and (2) provide a graphical record of the vessel's dead reckoning track. In addition, most types of dead reckoning equipment provide means for tracking one or more other craft, to obtain a graphical record of the other craft's course and speed. This equipment is generally installed only on warships.

Dead reckoning equipment consists in general of four components: (1) an analyzer; (2) latitude and longitude indicator dials; (3) a desk-size unit called a **dead reckoning tracer (DRT)**; and (4) a glass plotting surface over the dead reckoning tracer.

The analyzer receives directional signals from the vessel's gyrocompass, and distance signals from the underwater log. The course and distance data are transformed automatically to electrical signals proportional to the north-south and east-west components of the vessel's movement. These distance signals are transmitted to the latitude and longitude indicators, changing their readings by the correct amount to indicate the new latitude and the new longitude in degrees and minutes. Since the number of miles in the north-south component of distance traveled is nearly equal to the change in latitude expressed in minutes, the latitude indicator is fed directly. Departure (art. 204) is automatically transformed to difference of longitude before being registered on the longitude indicator dials. If the indicator dials are correctly set to latitude and longitude, they continuously show subsequent dead reckoning positions of the vessel.

The north-south and east-west component signals from the analyzer are also transmitted to the DRT (fig. 647), where they control the motion of a pencil which moves across a chart or plotting sheet attached to the DRT base. The pencil draws a line which conforms to the maneuvers of the vessel. The mechanism can be set to plot the track at any scale from ¼ mile per inch (1⁄10 mile on some) to 16 miles per inch. A clock-controlled contact lifts the pencil from the paper for 15 seconds of each minute and for a longer period each 10 minutes, thus providing automatic time measurement. The pencil carriage can be moved manually to any part of the chart for initial setting and the direction of travel can be adjusted so that the chart can be placed with any cardinal direction "up."

The cover of the DRT is a sheet of glass to which a plotting sheet or blank paper can be fastened. An electric lamp on the top of the pencil carriage throws a spot of light through the paper directly over the carriage, thus providing a moving reference scaled to the course and speed of the vessel. If the position of the spot of light is marked periodically on the paper, a second record of the vessel's track is obtained. However, the principal use of this sheet is for plotting successive positions of another craft, using the spot of light as the origin. A polar grid centered on the light may be projected onto the paper to facilitate measurement. The course of the other vessel can be measured directly from the plot, and its speed can be determined by means of the time needed to travel any distance measured on the plot. This process is called **tracking.** If the ranges and bearings are plotted from a fixed point, *relative* movement is determined, a practice commonly followed in connection with radar.

While dead reckoning equipment is a great convenience, particularly when changes of course or speed are numerous, its indications should be checked by graphical plot on the chart or plotting sheet. Reliable dead reckoning is too important to be left entirely to mechanical equipment without an independent check.

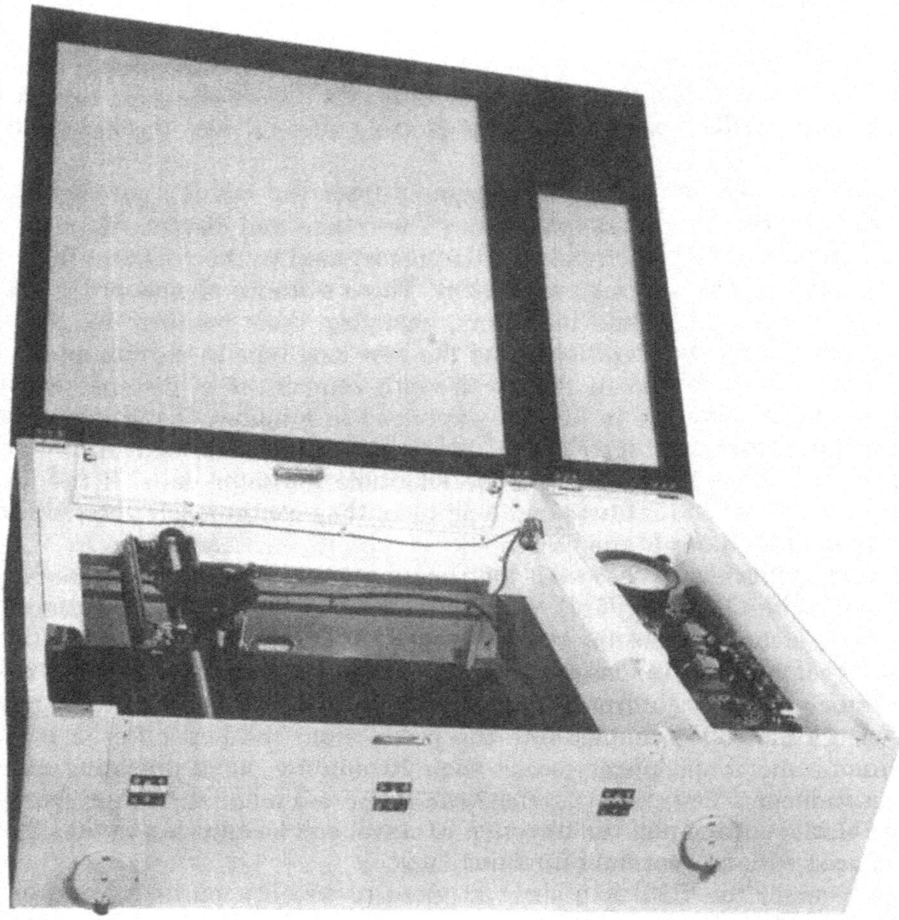

Courtesy of Ahrendt Instrument Co.

FIGURE 647.—A dead reckoning tracer.

CHAPTER VII

COMPASS ERROR

Magnetism

701. Theory of magnetism.—The fact that iron can be magnetized (given the ability to attract other iron) has been known for thousands of years, but the explanation of this phenomenon has awaited the recently acquired knowledge of atomic structure. According to present theory, the magnetic field around a current-carrying wire and the magnetism of a permanent **magnet** are the same phenomenon—fields created by moving electrical charges. This occurs whether the charge is moving along a wire, flowing with the magma of the earth's core, encircling the earth at high altitude as a stream of charged particles, or rotating around the nucleus of an atom.

It has been shown that microscopically small regions, called **domains,** exist in iron and other ferromagnetic substances. In each domain the fields created by electrons spinning around their atomic nuclei are parallel to each other, causing the domain to be magnetized to saturation. In a piece of unmagnetized iron, the directions of the various domains are arranged in a random manner with respect to each other. If the substance is placed in a weak magnetic field, the domains rotate somewhat toward the direction of that field. Those domains which are more nearly parallel to the field increase in size at the expense of the more non parallel ones. If the field is made sufficiently strong, entire domains rotate suddenly by angles of as much as 90° or 180° so as to become parallel to that "crystal axis" which is most nearly parallel to the direction of the field. If the strength of the field is increased to a certain value depending upon individual conditions, all of the domains rotate into parallelism with the field, and the iron itself is said to be magnetically **saturated.** If the field is removed, the domains have a tendency to rotate more or less rapidly to a more natural direction parallel to some crystal axis, and more slowly to random directions under the influence of thermal agitation.

Magnetism which is present only when the material is under the influence of an external field is called **induced magnetism.** That which remains after the magnetizing force is removed is called **residual magnetism.** That which is retained for long periods without appreciable reduction, unless the material is subjected to a demagnetizing force, is called **permanent magnetism.**

Certain substances respond readily to a magnetic field. These **magnetic materials** are principally those composed largely of iron, although nickel and cobalt also exhibit magnetic properties. The best magnets are made of an alloy composed mostly of iron, nickel, and cobalt. Aluminum and some copper may be added. Platinum and silver, properly alloyed with other material, make excellent magnets, but for ordinary purposes the increased expense is not justified by the improvement in performance. Permanent magnets occur in nature in the form of **lodestone,** a form of magnetite (an oxide of iron) possessing magnetic properties. A piece of this material constitutes a **natural magnet.**

702. Hard and soft iron.—In some alloys of iron, the crystals can be so arranged and internally stressed that the domains remain parallel to each other indefinitely, and the metal thus becomes a **permanent magnet.** Such alloys are used for the magnets of a compass. In other kinds of iron, the domains reorient themselves rapidly to conform to the direction of a changing external field, and soon take random directions if the field is removed. A ferromagnetic substance which retains much of its magnetism in the absence of an external field, is said to have high **remanence** or **retentivity.** The strength of a reverse field (one of opposite polarity) required to reduce the magnetism of a magnet to zero is called the **coercivity** or **coercive force** of the magnet. Hence, a compass magnet should have high remanence in order to be strong, and high coercivity so that stray fields will not materially affect it. For convenience, iron is called "hard" if it has high remanence, and "soft" if it has low remanence. **Permeability** (μ) is the ratio of the strength of the magnetic field inside the metal (B) to the strength of the external field (H), or

$$\mu = \frac{B}{H}.$$

703. Lines of force.—The direction of a magnetic field is usually represented by lines, called **lines of force.** Relative intensity in different parts of a magnetic field is indicated by the spacing of the lines of force, a strong field having the lines close together. If a piece of unmagnetized iron is placed in a magnetic field, the lines of force tend to crowd into the iron, following its long axis, and the field is stronger in the vicinity of the iron, somewhat as shown in figure 703a. If the iron becomes permanently magnetized and is removed from this field, the lines of force around the iron follow paths about as shown in figure 703b.

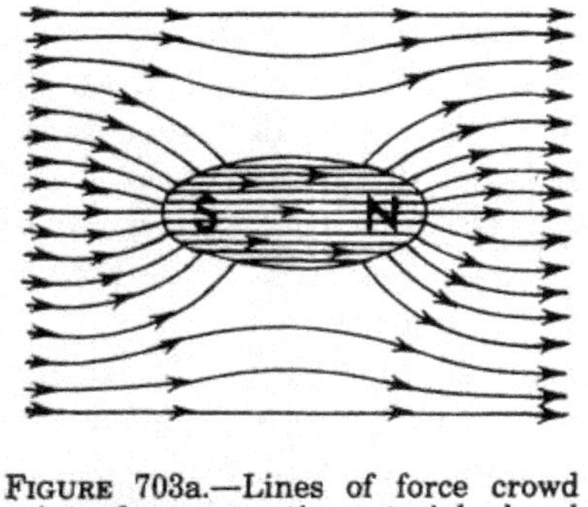

FIGURE 703a.—Lines of force crowd into ferromagnetic material placed in a magnetic field.

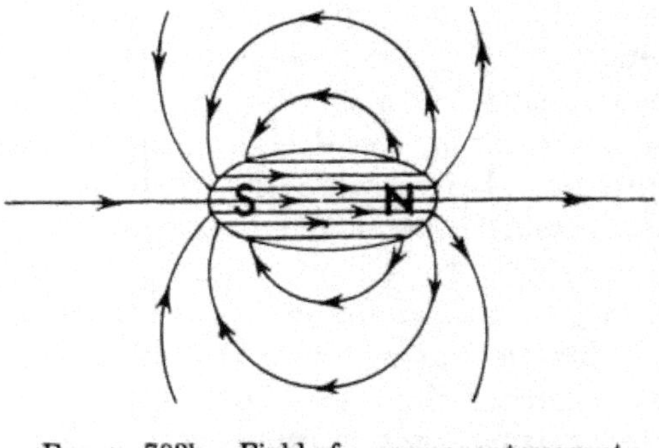

FIGURE 703b.—Field of a permanent magnet.

704. Magnetic poles.—The region in which the lines of force enter the iron is called the **south pole,** and the region in which they leave the iron is called the **north pole.** Thus, the lines of force are directed from south to north within the magnet, and from north to south in the external field. Every magnet has a north pole and a south pole. If a magnet is cut into two pieces, each becomes a magnet with a north pole and south pole. A single pole cannot exist independently. If two magnets are brought close together, *unlike poles attract each other and like poles repel.* Thus, a north pole attracts a south pole but repels another north pole.

The earth itself has a magnetic field (art. 706), with its magnetic poles being some distance from the geographical poles. If a permanent bar magnet is supported so that it can turn freely, both horizontally and vertically, it aligns itself with the magnetic field of the earth, which at most places is in a general north-south direction and inclined to the horizontal. Since the north pole of the magnet points in a northerly direction, the earth's magnetic pole in the Northern Hemisphere has *south* magnetism. Nevertheless, it is called the **north magnetic pole** because of its geographical location. For a similar reason, the pole in the Southern Hemisphere, although it has north magnetism, is called the **south magnetic pole.** To avoid confusion, north magnetism is usually called "red," and south magnetism, "blue." The red (north) pole of a magnet is usually painted red, and in some cases the south (blue) pole is painted blue. The north magnetic pole of the earth is a blue pole, and the south magnetic pole is a red pole.

705. The magnetism of soft iron, in which remanence is low, depends upon the position of the iron with respect to an external field. It is strongest if the long axis is parallel to the lines of force, and decreases to a minimum if the material is rotated so that the long axis is perpendicular to the lines of force. Figure 705 shows three positions of a bar magnet with respect to a magnetic field. At position *X*, the pole at the upper end of the bar is red and relatively strong. As the bar is rotated toward position *Y*, the upper end remains red, but its strength decreases. At position *Y*, no pole is apparent at either end, but a red pole extends along the entire left side of the bar, and a blue pole along the right side. Poles are strongest when concentrated into a small area. Hence, when spread over an entire side, as at position *Y*, they are relatively weak. At position *Z*, the upper end is blue.

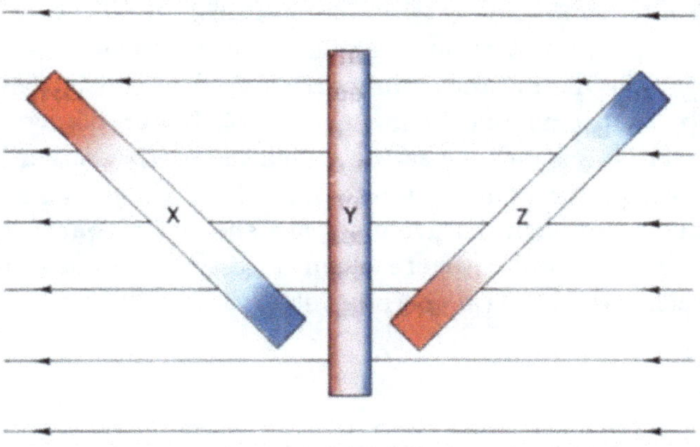

FIGURE 705.—The polarity of a soft iron bar in a magnetic field.

The change in polarity as a bar of soft iron is rotated in a magnetic field can easily be demonstrated. If a bar of soft iron is placed vertical in northern magnetic latitudes (as in any part of the United States), the north (red) end of a compass magnet brought near it will be attracted by the upper end of the bar, and repelled by the lower end. If the bar is inverted, so that its ends are interchanged, the upper end (which as the lower end previously repelled the compass needle) will attract the north end of the needle, and the lower end will repel it. Thus, the polarity of the rod is reversed, *either* end having blue magnetism if it is at the top. This changing polarity of soft iron in the earth's field is a major factor affecting the magnetic compasses of a steel vessel.

706. Terrestrial magnetism.—The earth itself can be considered to be a gigantic magnet. Although man has known for many centuries that the earth has a magnetic field, the origin of the magnetism is not completely understood. Nevertheless, the horizontal component of this field is a valuable reference in navigation, for it provides the directive force for the magnetic compass, which indicates the ship's heading *in relation to the horizontal component of this field.*

The world-wide pattern of the earth's magnetism is roughly like that which would result from a short, powerful, bar magnet near the earth's center, as shown in figure 706. The geographical poles are at the top and bottom, and the magnetic poles are offset somewhat from them. This representation, however, is greatly simplified. The actual field is more complex, and requires measurement of its strength and direction at many places (art. 707) before it can be defined accurately enough to be of practical use to the navigator. Not only are the magnetic poles offset from the geographical poles, but the magnetic poles themselves are not 180° apart and, in general, a magnetic compass aligned with the lines of force does not point toward either magnetic pole. In 1984, the north magnetic pole was located at latitude 78°.9N, longitude 103°.8W, approximately, in the vicinity of the Parry Islands; and the south magnetic pole was at latitude 65°.4S, longitude 139°.5E, approximately, off the coast of the northeastern part of Wilkes Land. However, the magnetic poles are not stationary. The entire magnetic field of the earth, including the magnetic poles, undergoes a small daily or **diurnal change,** and a very slow, progressive **secular change.** In addition, temporary sporadic changes occur from time to time during magnetic storms (art. 2513). During a severe storm, variation may change as much as 5°, or more. However, such disturbances are never so rapid as to cause noticeable deflection of the compass card, and in most navigable waters the change is so little that it is not significant in practical navigation. Even when there is no temporary disturbance, the earth's field is considerably more intricate than indicated by an isomagnetic chart (art. 708). Natural magnetic irregularities occurring over relatively small areas are called **magnetic anomalies** by the magneticians, but the navigator generally refers to these phenomena as **local disturbances.** Notes warning of such disturbances are shown on charts. In addition, artificial disturbances may be quite severe when a vessel is in close proximity to other vessels, piers, machinery, electric currents, etc.

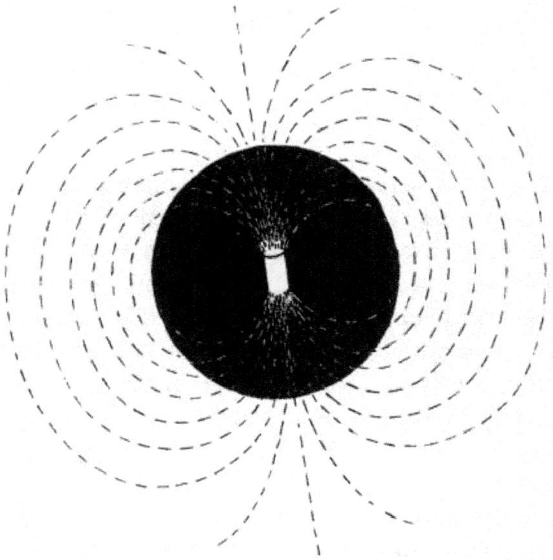

FIGURE 706.—The magnetic field of the earth.

The elements of the earth's field are as follows:

Total intensity (F) is the strength of the field at any point, measured in a direction parallel to the field. Intensity is sometimes measured in **oersteds**, one oersted being equal to a force of one dyne acting on a unit pole. The range of intensity of the earth's field is about 0.25 to 0.70 oersted. For convenience in geomagnetic surveying, a small unit is used, called the **gamma**. One oersted equals 100,000 gammas, so that the range of intensity of the earth's field is about 25,000 to 70,000 gammas.

Horizontal intensity (H) is the horizontal component of the total intensity. At the **magnetic equator**, which corresponds roughly with the geographic equator, the field is parallel to the surface of the earth, and the horizontal intensity is the same as total intensity. At the magnetic poles of the earth, the field is vertical and there is no horizontal component. The direction of the horizontal component at any place defines the **magnetic meridian** at that place. This component provides the desired directive force of a magnetic compass.

North component (X) is the horizontal intensity's component along a geographic (true) meridian.

East component (Y) is the horizontal intensity's component perpendicular to the north component.

Vertical intensity (Z) is the vertical component of the total intensity. It is zero at the magnetic equator. At the magnetic poles it is the same as the total intensity. While the vertical intensity has no direct effect upon the direction indicated by a magnetic compass, it does induce magnetic fields in vertical soft iron, and these may affect the compass.

Variation (V, Var.), called **declination (D)** by magneticians, is the angle between the geographic and magnetic meridians at any place. The expression **magnetic variation** is used when it is necessary to distinguish this from other forms of variation. This element is measured in angular units and named east or west to indicate the side of true north on which the (magnetic) northerly part of the magnetic meridian lies. For computational purposes, easterly variation is sometimes designated positive (+), and westerly variation negative (−). **Grid variation (GV)** or **grivation** is the angle between the grid and magnetic meridians at any place, measured and named in a manner similar to variation.

Magnetic dip (I), called **inclination (I)** by magneticians, is the vertical angle, expressed in angular units, between the horizontal at any point and a line of force through that point. The **magnetic latitude** of a place is the angle having a tangent equal to half that of the magnetic dip of the place.

At a distance of several hundred miles above the earth's surface, the magnetic field surrounding the earth is believed to be uniform, as it appears in figure 706, and centered around two **geomagnetic poles**. These do not coincide with either the magnetic poles (art. 704) or the geographical poles. However, they are 180° apart, the north geomagnetic pole being at latitude 78°.5N, longitude 69°W (near Etah, Greenland) and the south geomagnetic pole being at latitude 78°.5S, longitude 111°E. The great circles through these poles are called **geomagnetic meridians.** That geomagnetic meridian passing through the south geographical pole is the origin for measurement of **geomagnetic longitude**, which is measured eastward through 360°. The complement of the arc of a geomagnetic meridian from the nearer geomagnetic pole to a place is called the **geomagnetic latitude.** When the sun is over the upper branch of the geomagnetic meridian of a place, it is **geomagnetic noon** there, and when it is over the lower branch of the geomagnetic meridian, it is **geomagnetic midnight.** The angle between the lower branch of the geomagnetic meridian of a place and the geomagnetic meridian over which the sun is located is called **geomagnetic time.** The diurnal change is related to geomagnetic time. The auroral zones (art. 4320) are centered on the geomagnetic poles.

707. Measurement of the earth's magnetic field is made continuously at about 70 permanent **magnetic observatories** throughout the world. In addition, large numbers of temporary stations are occupied for short periods to add to man's knowledge of the earth's field. In the past, measurements at sea have been made by means of nonmagnetic ships constructed especially for this purpose. However, this is a slow and expensive method, and quite inadequate to survey properly the 71 percent of the earth's surface covered with water. Since World War II, a satisfactory **airborne magnetometer** has been developed by the U.S. Navy. By means of this instrument, continuous readings can be recorded automatically during long overwater flights.

708. Isomagnetic charts showing lines of equality of some magnetic element are published by the Defense Mapping Agency Hydrographic/Topographic Center. The magnetic data are compiled by the United States Geological Survey with the assistance of the National Oceanic and Atmospheric Administration and in collaboration with the U.S. Naval Oceanographic Office. The three charts of each element consist of one on the Mercator projection (art. 305) covering most of the world, and one on a polar projection (azimuthal equidistant (art. 320) or stereographic) for each of the two polar areas. All charts now included in the series are published at intervals of 10 years, showing the values for the beginning of each year ending in five. Charts showing variation are also published for the years ending in zero (1950, 1960, etc.).

The isomagnetic chart of most concern to a navigator is chart 42, *Magnetic Variation,* a simplified version of which is shown in figure 708a. The lines connecting points of equal magnetic variation are called **isogonic lines.** *These are not magnetic meridians (lines of force).* The line connecting points of zero variation is called the **agonic line.** Variation is also shown on nautical charts. Those of relatively small scale generally show isogonic lines. Those of scale larger than 1:100,000 generally give the information in the form of statements inside compass roses placed at various places on the chart, and sometimes, also, by a **magnetic compass rose** within the true compass rose and offset from it by the amount of the variation. By means of this arrangement, true directions can be plotted without arithmetically applying variation to magnetic directions, or magnetic directions can be read directly from the chart. The magnetic compass rose is generally graduated in both degrees and points. Variation is given to the nearest 15′, and the annual change to the nearest 1′. However, since the rate of change is not constant, a very old chart should not be used, even though it has been corrected for all changes shown in *Notices to Mariners.*

Another isomagnetic chart of value to the mariners is chart 30, *Magnetic Inclination or Dip,* figure 708b. Lines connecting points of equal magnetic dip are called **isoclinal lines.** The line connecting points of zero dip is called the **magnetic equator.**

Other isomagnetic charts are chart 33, showing horizontal intensity in gammas (art. 706); chart 36, showing vertical intensity in gammas; and chart 39, showing total intensity in gammas. Lines connecting points of equal intensity on any of these charts are called **isodynamic lines.**

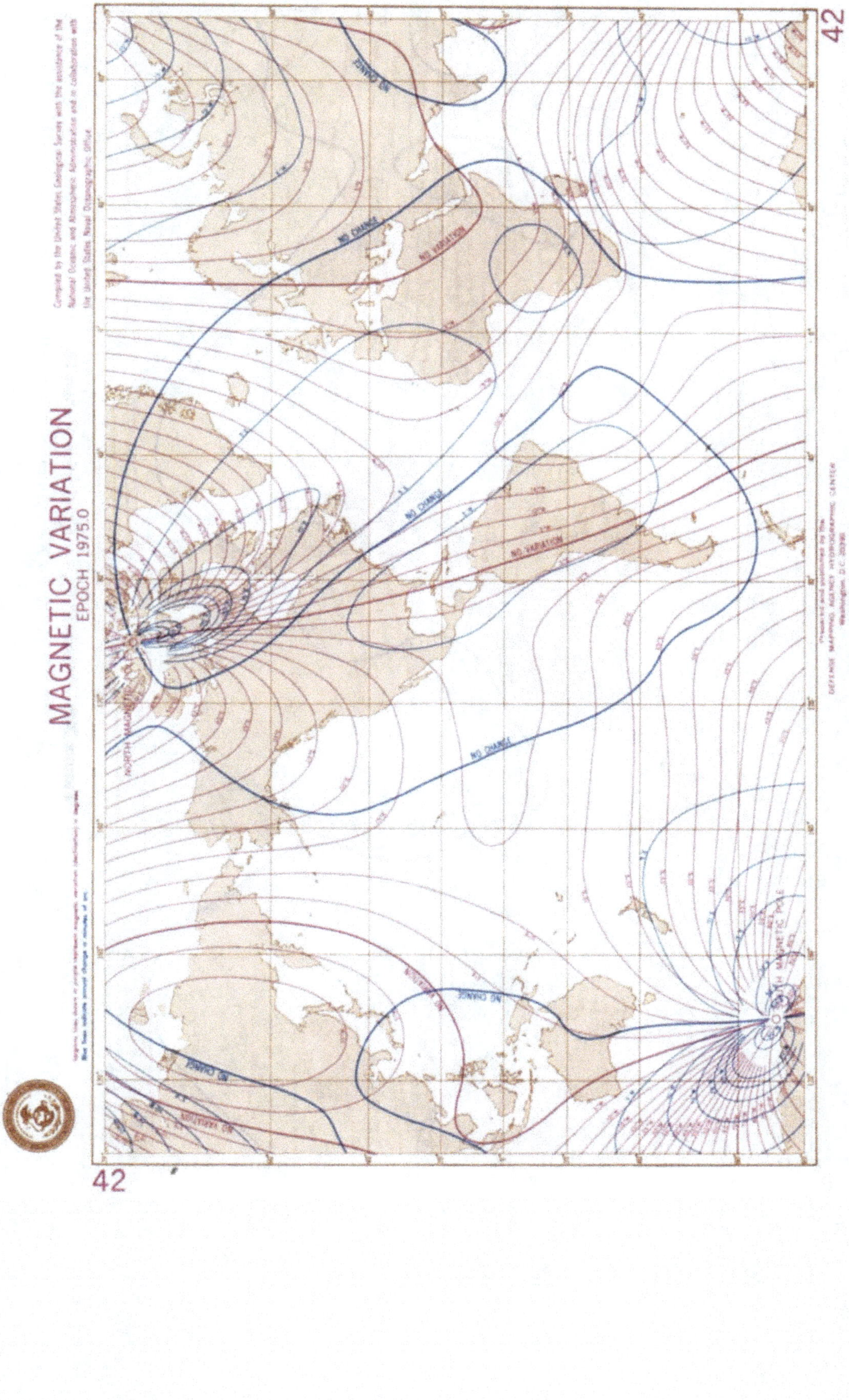

FIGURE 708a.—Variation. A simplification of chart 42.

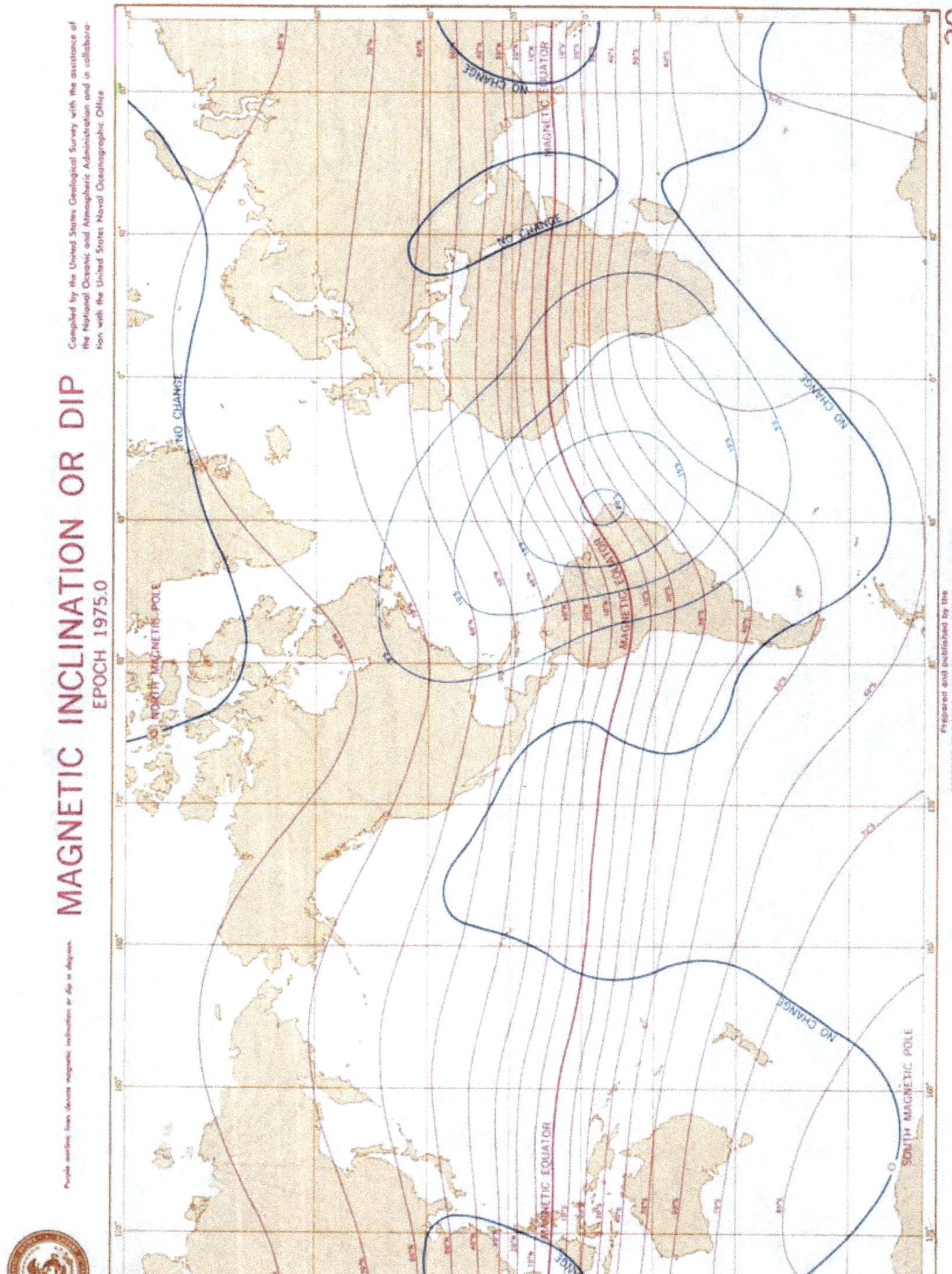

FIGURE 708b.—Magnetic dip. A simplification of chart 30.

Other isomagnetic charts show (1) magnetic inclination in north and south polar areas; (2) horizontal intensity, including horizontal intensity in north and south polar areas; (3) vertical intensity, including vertical intensity in north and south polar areas; (4) total intensity, including total intensity in north and south polar areas; (5) magnetic variation in north and south polar areas; and (6) magnetic grid variation.

All of the isomagnetic charts also show **isopors,** in a distinctive color, connecting points of equal **annual change** of the element at the epoch of the chart.

The charts are as accurate as can be made with available information, except that the lines are smoothed somewhat, rather than depicting every small irregularity. The larger irregularities are reflected in the information shown on nautical charts, but local disturbance is indicated by warning notes at appropriate places. In areas where measurements of the magnetic field have not been made for a long period, the previous information is altered in accordance with the best information available on secular change, with some adjustment to provide continuous smooth curves. When information is thus carried forward for many years, errors may be introduced, particularly in areas where the rate of change is large and variable. Magneticians have not detected a recognizable worldwide pattern in secular change, such as would occur if it were due only to shifting of the positions of the magnetic poles. Rather, these shifts are part of the general complex, little-understood secular change.

The Compass Error

709. Magnetic compass error.—Directions relative to the northerly direction along a geographic meridian are **true.** In this case, true north is the **reference direction.** If a compass card is horizontal and oriented so that a straight line from its center to 000° points to true north, any direction measured by the card is a true direction and has no error (assuming there is no calibration or observational error). If the card remains horizontal but is rotated so that it points in any other direction, the amount of the rotation is the **compass error.** Stated differently, compass error is the angular difference between true north and **compass north** (the direction north as indicated by a magnetic compass). It is named east or west to indicate the side of true north on which compass north lies.

If a magnetic compass is influenced by no other magnetic field than that of the earth, and there is no instrumental error, its magnets are aligned with the magnetic meridian at the compass, and 000° of the compass card coincides with **magnetic north.** All directions indicated by the card are **magnetic.** As stated in article 706, the angle between geographic and magnetic meridians is called **variation (V** or **Var.**). Therefore, if a compass is aligned with the magnetic meridian, compass error and variation are the same.

When a compass is mounted in a vessel, it is generally subjected to various magnetic influences other than that of the earth. These arise largely from induced magnetism in metal decks, bulkheads, masts, stacks, boat davits, guns, etc., and from electromagnetic fields associated with direct current in electrical circuits. Some metal in the vicinity of the compass may have acquired permanent magnetism. The actual magnetic field at the compass is the vector sum, or resultant (art. 118, vol. II), of all individual fields at that point. Since the direction of this resultant field is generally not the same as that of the earth's field alone, the compass magnets do not lie in the magnetic meridian, but in a direction that makes an angle with it. This angle is called **deviation (D** or **Dev.**). Thus, deviation is the angular difference between magnetic north and compass north. It is expressed in angular units and named east or west to indicate the side of magnetic north on which compass north lies. Thus, deviation is the error of the compass in pointing to magnetic north, and all directions measured with compass north as the reference direction are **compass directions.** Since variation and deviation may each be either east or west, the effect of deviation may be to either increase or decrease the error due to variation alone. The algebraic sum of variation and deviation is the total compass error.

For computational purposes (art. 727), deviation and compass error, like variation, may be designated positive (+) if east and negative (−) if west.

Variation changes with location, as indicated in figure 708a. Deviation depends upon the magnetic latitude and also upon the individual vessel, its trim and loading, whether it is pitching or rolling, the heading (orientation of the vessel with respect to the earth's magnetic field), and the location of the compass within the vessel. Therefore, deviation is not published on charts.

710. Deviation table.—In practice aboard ship, the deviation is reduced to a minimum, as explained later in this chapter. The remaining value, called **residual deviation,** is determined on various headings and recorded in some form of **deviation table.** Figure 710 shows both sides of the form used by the United States Navy. This table is entered with the magnetic heading, and the deviation on that heading is determined from the tabulation, separate columns being given for degaussing (now called magnetic silencing) (DG) off and on (art. 740). If the deviation is not more than about 2° on any heading, satisfactory results may be obtained by entering the values at intervals of 45° only.

If the deviation is small, no appreciable error is introduced by entering the table with either magnetic or compass heading. If the deviation on some headings is large, the desirable action is to reduce it, but if this is not practicable, a separate deviation table for compass heading entry may be useful. This may be made by applying the tabulated deviation to each entry value of magnetic heading, to find the corresponding compass heading, and then interpolating between these to find the value of deviation at each 15° compass heading. Another method is to plot the values on cross-section paper and select the desired values graphically.

A nomogram especially designed for interconversion of magnetic and compass headings is called a **Napier diagram,** having been devised by James Robert Napier (1821–79). It consists of a dotted, vertical centerline graduated from 000° to 360° (usually in two parallel parts of 180° each), with two series of crosslines making angles of 60° with the dotted vertical line and with each other. If magnetic headings are used, deviation is measured along a solid crossline; and if compass headings are used, deviation is measured along a dotted crossline. A deviation curve is drawn through the various points. To convert a magnetic heading to a compass heading, one finds the magnetic heading on the vertical centerline, moves parallel to a solid crossline until the curve is reached, and returns to the centerline by moving parallel to a dotted line. The compass heading is the value at the point of return. The reverse process is used for converting a compass heading to a magnetic heading. This nomogram is of particular value where the deviation is large and changing rapidly. It is now possible, however, to reduce deviation to such small values that the Napier diagram has lost much of its appeal and is seldom used.

Another solution is to make a deviation table with one column for magnetic heading, a second column for deviation, and a third for compass heading. Still another solution, most popular among yachtsmen, is to center a compass rose inside

MAGNETIC COMPASS TABLE NAVSEA RPT. 3530-2

NAVSEA 3120/4 (REV. 6-72) (FRONT) *(Formerly NAVSHIPS 1104)*
S/N 0105-601-6527

U.S.S. ______ NO. ______ *(BB, CL, DD, etc.)*

[X] PILOT HOUSE [ ] SECONDARY CONNING STATION [ ] OTHER ______

BINNACLE TYPE [X] NAVY ST'D [ ] OTHER ______

COMPASS 7-1/2" MAKE C.G. Conn SERIAL NO. 8560

TYPE CC COILS "K" DATE 9 September 1975

READ INSTRUCTIONS ON BACK BEFORE STARTING ADJUSTMENT

SHIPS HEAD MAGNETIC	DEVIATIONS DG OFF	DEVIATIONS DG ON	SHIPS HEAD MAGNETIC	DEVIATIONS DG OFF	DEVIATIONS DG ON
0	0.5E	0.5E	180	0.5W	0.0
15	1.0E	1.0E	195	1.0W	0.5W
30	1.5E	1.5E	210	1.0W	1.0W
45	2.0E	1.5E	225	1.5W	1.5W
60	2.0E	2.0E	240	2.0W	2.0W
75	2.5E	2.5E	255	2.0W	2.5W
90	2.5E	3.0E	270	1.5W	2.0W
105	2.0E	2.5E	285	1.0W	1.5W
120	1.5E	2.0E	300	1.0W	1.0W
135	1.5E	1.5E	315	0.5W	0.5W
150	1.0E	1.0E	330	0.5W	0.5W
165	0.0	0.5E	345	0.0	0.0

DEVIATIONS DETERMINED BY: [ ] SUN'S AZIMUTH [X] GYRO [ ] SHORE BEARINGS

B 6 MAGNETS RED [ ] FORE [X] AFT AT 12" FROM COMPASS CARD

C 4 MAGNETS RED [ ] PORT [X] STBD AT 6" FROM COMPASS CARD

D 2-7" [X] SPHERES [ ] CYLS AT 12" [X] ATHWARTSHIP [ ] SLEWED -° [ ] CLOCKWISE [ ] CTR. CLOCKWISE

HEELING MAGNET: [ ] RED UP [X] BLUE UP 6" FROM COMPASS CARD FLINDERS BAR: [X] FORE [ ] AFT 12"

[X] LAT 18°00'N [X] LONG 120°00'E

[ ] H 0.385 [ ] Z 0.151

SIGNED *(Adjuster or Navigator)* APPROVED *(Commanding)*

VERTICAL INDUCTION DATA
(Fill out completely before adjusting)

RECORD DEVIATION ON AT LEAST TWO ADJACENT CARDINAL HEADINGS

BEFORE STATING ADJUSTMENT: N 8 W E 0 S 4 E W 9 E

RECORD BELOW INFORMATION FROM LAST NAVSHIPS 3120/4 DEVIATION TABLE:

DATE 5 December 1974 [ ] LAT 32 53N [ ] LONG 117 18W

[ ] H .260 [ ] Z .420

12" FLINDERS BAR [X] FORWARD [ ] AFT DEVIATIONS N 2.5W E 7E S 6.5E W 5W

RECORD HERE DATA ON RECENT OVERHAULS, GUNFIRE, STRUCTURAL CHANGES, FLASHING, DEPERMING, WITH DATES AND EFFECT ON MAGNETIC COMPASSES:

Shipyard overhaul:
3 Oct - 2 Dec 1974
Depermed at Norfolk, Va.:
3 Dec 1974

PERFORMANCE DATA

COMPASS AT SEA:	[ ] UNSTEADY	[X] STEADY
COMPASS ACTION:	[ ] SLOW	[X] SATISFACTORY
NORMAL DEVIATIONS:	[X] CHANGE	[ ] REMAIN RELIABLE
DEGAUSSED DEVIATIONS:	[X] VARY	[ ] DO NOT VARY

REMARKS:

INSTRUCTIONS

1. This form shall be filled out by the Navigator for each magnetic compass as set forth in Chapter 9240 of NAVAL SHIPS TECHNICAL MANUAL.
2. When a swing for deviations is made, the deviations should be recorded both with degaussing coils off and with degaussing coils energized at the proper currents for heading and magnetic zone.
3. Each time this form is filled out after a swing for deviations, a copy shall be submitted to: Naval Ship Engineering Center Hyattsville, Maryland 20782. A letter of transmittal is not required.
4. When choice of box is given, check applicable box.
5. Before adjusting, fill in section on "Vertical Induction Data" above.

NAVSEA 3120/4 (REV. 6-72) (REVERSE) C-24858

FIGURE 710.—Deviation table.

a larger one so that an open space is between them, and a radial line would connect points of the same graduation on both roses. Each magnetic heading for which deviation has been determined is located on the outer rose, and a straight line is drawn from this point to the corresponding compass heading on the inner rose.

A variation of this method is to draw two parallel lines a short distance apart, and graduate each from 0 to 360 so that a perpendicular between the two lines connects points of the same graduation. Straight lines are drawn from magnetic directions on one line to the corresponding compass directions on the other. If the lines are horizontal and the upper one represents magnetic directions, the slope of the line indicates the direction of the deviation. That is, for westerly deviation the upper part of the connecting line is left (west) of the bottom part, and for easterly deviation it is right.

An important point to remember regarding deviation is that it varies with the *heading*. Therefore, a deviation table is *never* entered with a bearing (art. 1004). If the deviation table converts directly from one type heading to another, deviation is found by taking the difference between the two values. On the compass rose or straight-line type, the deviation can be written alongside the connecting line, and the intermediate values determined by estimate. If one has trouble determining whether to add or subtract deviation when bearings are involved, he has only to note which heading, magnetic or compass, is larger. The same relationship holds between the two values of bearing.

The deviation table should be protected from damage due to handling or weather, and placed in a position where it will always be available when needed. A method commonly used is to mount it on a board, cover it with shellac or varnish, and attach it to the binnacle. Another method is to post it under glass near the compass. It is good practice for the navigator to keep a second copy available at a convenient place for his use.

711. Applying variation and deviation.—As indicated in article 709, a single direction may have any of several numerical values depending upon the reference direction used. One should keep clearly in mind the relationship between the various expressions of a direction. Thus, true and magnetic directions differ by the variation, magnetic and compass directions differ by the deviation, and true and compass directions differ by the compass error. Other relationships are also useful. Thus, grid (art. 2510) and magnetic directions differ by the grid variation or grivation, and true and relative directions differ by the true heading. The use of variation and deviation is considered here. Other relationships are discussed elsewhere in this volume.

If variation or deviation is easterly, the compass card is rotated in a clockwise direction. This brings smaller numbers opposite the lubber's line. Conversely, if either error is westerly, the rotation is counterclockwise and larger numbers are brought opposite the lubber's line. Thus, if the heading is 090° true (fig. 711, A) and variation is 6°E, the magnetic heading is 090°−6°=084° (fig. 711, B). If the deviation on this heading is 2°W, the compass heading is 084°+2°=086° (fig. 711, C). Also, compass error is 6°E−2°W=4°E, and compass heading is 090°−4°=086°. If compass error is easterly, the compass reads too low (in comparison with true directions), and if it is westerly, the reading is too high. Many rules-of-thumb have been devised as an aid to the memory, and any which assist in applying compass errors in the right direction are of value. However, one may forget the rule or its method of application, or may wish to have an independent check. If he understands the explanation given above, he can determine the correct sign without further information. The same rules apply to the use of gyro error. Since variation and deviation are compass errors, the process of removing either from an indication of a direction (converting compass to magnetic or magnetic to true) is often called **correcting.** Conversion in the opposite direction (inserting errors) is then called **uncorrecting.**

Example.—A vessel is on course 215° true in an area where the variation is 7°W. The deviation is as shown in figure 710. Degaussing is off. The gyro error (GE) is 1°E. A lighthouse bears 306°5 by magnetic compass.

Required.—(1) Magnetic heading (MH).

(2) Deviation.

(3) Compass heading (CH).

(4) Compass error.

(5) Gyro heading.

(6) Magnetic bearing of the lighthouse.

(7) True bearing of the lighthouse.

(8) Relative bearing (art. 1004) of the lighthouse.

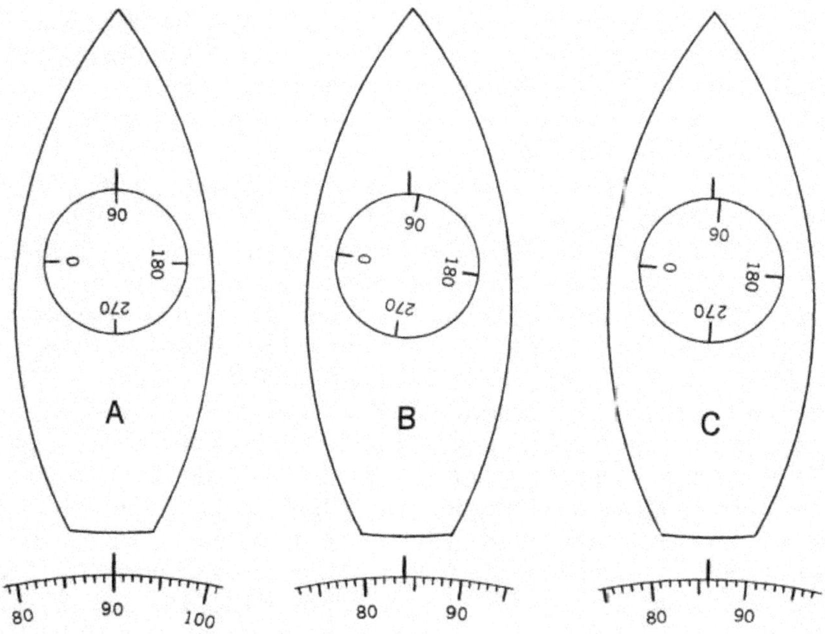

FIGURE 711.—Effect of variation and deviation on the compass card.

Solution.—

TH	215°
V	7°W
(1) MH	222°
(2) D	1°.5W
(3) CH	223°.5

The deviation is taken from the deviation table (fig. 710), to the nearest half degree.

(4) Compass error is 7°W + 1°.5W = 8°.5W.

TH	215°
GE	1°E
(5) H_{pgc}	214°
CB	306°.5W
D	1°.5W
(6) MB	305°
V	7°W
(7) TB	298°

(8) RB = TB − TH = 298° − 215° = 083°.

Answers.—(1) MH 222°, (2) D1°.5W, (3) CH 223°.5, (4) CE 8°.5W, (5) H_{pgc} 214°, (6) MB 305°, (7) TB 298°, (8) RB 083°.

Deviation and its Reduction

712. Magnetism of a steel vessel.—The materials of which a vessel is constructed are not, in general, selected for their magnetic properties. As a result, many degrees of permeability, remanence, and coercivity (art. 702) exist within its structure. Detailed analysis of the complex field existing at a magnetic compass is a specialized study not ordinarily required of the navigator. However, a general knowledge of the basic principles involved is of value to the navigator in helping him understand better the behavior of his magnetic compasses.

For most purposes, a vessel can be considered to be composed of two types of material: "hard iron" and "soft iron."

"Hard iron" is all material having some degree of permanent magnetism. This magnetism is acquired largely during construction of the vessel, when the rearrangement of the domains (art. 701) is facilitated by the bending, riveting, welding, and other violent mechanical processes. Since a vessel remains on a constant magnetic heading while it is on the building ways, a field of permanent magnetism becomes established, the positions of the poles being dependent largely upon the orientation of the hull with respect to the magnetic field of the earth. If a vessel is constructed on a heading of magnetic north, at a place where the magnetic dip is 70°N (the approximate value at the midpoint of the east coast of the United States), its field of permanent magnetism is about as shown at the left of figure 712. The upper and stern portions are magnetically blue, while the lower and forward portions are magnetically red. If the vessel is built on a heading of magnetic east, the starboard and upper portions are blue, and the port and lower portions are red, as shown by the stern view at the right of figure 712. If the heading is magnetic northeast, the upper, starboard, and stern portions are blue, and the lower, port, and forward portions red. The red and blue portions for any given vessel can be visualized by drawing a sketch similar to that of figure 712, with the correct orientation.

The "permanent" magnetism thus acquired during construction is less permanent than that of a permanent magnet such as one of those used in a compass, and is modified somewhat after launching, particularly if the vessel remains on another heading for a considerable time during fitting out. The change is especially rapid during the first few days after launching, when the domains of the softer iron become reoriented. At this stage, deviation due to permanent magnetism may change several degrees. Further changes in the vessel's permanent magnetism may occur during long periods of being moored on a constant heading, or during a run of several days on nearly the same heading. This change is gradual and affects the strength, but usually not the polarity, of the magnetic field. The permanent field may be changed quickly, in polarity as well as in strength, if the vessel grounds, collides with another vessel, is struck by lightning, undergoes magnetic treatment (art. 744), fires its guns, or is struck by shells or bombs, etc.

The effect that the permanent magnetism of hard iron has upon a compass depends upon the position and strength of the poles relative to the compass. When the poles are in line with the north-south axis of the compass card, the only effect is to strengthen or weaken the directive force of the compass. When the compass heading is approximately 90° away, so that the poles are east and west of the compass, the deviating effect is maximum. The direction of the deviation is the same as that of the blue pole with respect to the compass.

"Soft iron" is all that material in which induced magnetism (art. 701) is present. With respect to its effect upon the magnetic compass, it is classed as either vertical or horizontal. Unlike hard iron, its magnetic field changes quickly as its orientation with respect to the earth's field changes. It also changes as the strength of the earth's field changes. For some purposes induced magnetism can be treated as if it were concentrated in two bars of soft iron, one vertical and the other horizontal. The polarity depends upon the position of the vessel relative to the

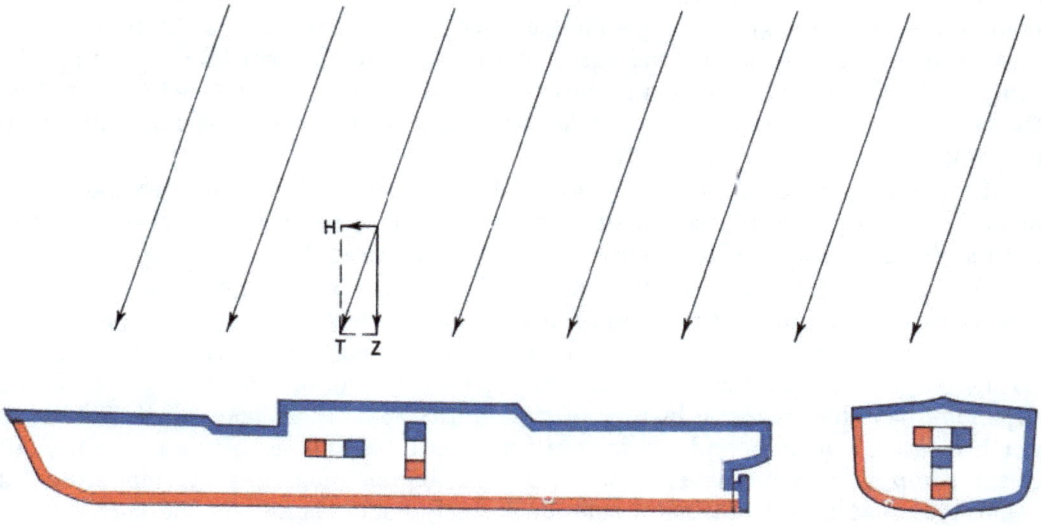

FIGURE 712.—Permanent magnetism of a vessel built on heading magnetic north (left) and magnetic east (right) at a place where the magnetic dip is 70°N.

earth's magnetic field, and the strength depends upon the strength of the vertical and horizontal components of the earth's field. This is illustrated in figure 712. In north magnetic latitude the bottom of the vertical rod has red magnetism and the top has blue magnetism. In south magnetic latitude these are reversed. In both north and south magnetic latitudes the magnetic north end of the horizontal bar has red magnetism, and the magnetic south end has blue magnetism. Thus, whatever the position of the rod, that part in the direction of magnetic north has red magnetism, and that part in the direction of magnetic south has blue magnetism. That is, each end has magnetism opposite to that of the magnetic pole indicated by the direction in which it is pointed.

The effect upon a magnetic compass of the induced magnetism in soft iron depends upon the strength and direction of the field relative to the compass. The cumulative effect of the induced magnetism in vertical soft iron is generally on the centerline of the vessel (if of conventional construction), and for a compass located forward, as on the bridge, is aft of the compass. In magnetic north latitude the effect is generally that of a blue pole at the level of the compass card. In magnetic south latitude the pole is red. On a heading of compass north or south the pole is in line with the magnets of a centerline compass and serves only to strengthen or weaken the directive force. On a heading of compass east or west the pole is perpendicular to the north-south axis of the compass card, and the deviating force is greatest.

For a compass located on the centerline of a vessel of conventional construction, the horizontal soft iron close enough to have appreciable effect upon the compass is arranged in a more-or-less symmetrical manner with respect to the compass. Thus, on any cardinal compass heading, the fore-and-aft and athwartship horizontal soft iron is either in line with the compass magnets or equally and similarly arranged on both sides. No error is introduced by such symmetrical horizontal soft iron because the iron north and south of the compass magnets serves only to strengthen or weaken the directive force, and that east and west of the compass sets up an equal and opposite field on each side. On intercardinal headings, the poles of the induced magnetism are offset and a maximum deviating force occurs. That part of horizontal soft iron which is not symmetrically arranged with respect to the compass—the asymmetrical soft iron—produces deviation which is maximum on the cardinal headings and zero on the intercardinal headings (by compass). This type of deviation is particularly great in a compass not mounted on the centerline of the vessel. It may also produce deviation which is constant on all headings.

In wooden-hulled vessels such as certain yachts and small fishing vessels, one or more of these types of magnetism may be weak or entirely missing, but this does not justify the omission of any part of the correction procedure.

As far as its effect upon the compass is concerned, the magnetic field at a centerline compass located forward on a vessel of conventional construction, and on an even keel, is essentially the same as that which would result from four sources: (1) the earth's magnetism; (2) a single blue pole the location and strength of which depends upon the magnetic history of the vessel; (3) a single pole which is blue in north magnetic latitude and red in south magnetic latitude, is on the centerline aft of the compass, and increases in strength with higher magnetic latitude; and (4) a single blue pole on the starboard side for easterly headings and on the port side for westerly headings, being of zero strength on a heading of north or south and decreasing in strength with increased magnetic latitudes. The single pole concept assumes that the effect of one pole predominates. The locations of the poles depend partly upon the position of the compass to which they apply. The actual field surrounding any magnetic compass may be considerably more complex than indicated.

713. Compass adjustment.—There are at least two possible solutions to the problem of compass error. The error can be permitted to remain, and the various directions interconverted by means of variation and deviation, or compass error, as explained in article 711; or the error can be removed. In practice, a combination of both of these methods is used.

Variation depends upon location of the vessel, and the navigator has no control over it. Provision could be made for offsetting the lubber's line, but this would not be effective in correcting magnetic compass bearings, and this practice is not generally followed. Variation does not affect the operation of the compass itself, and so is not objectionable from this standpoint.

Deviation is undesirable because it is more troublesome to apply, and the magnetic field which causes it partly neutralizes the directive force acting upon the compass, causing it to be unsteady and sluggish. As the vessel rolls and pitches, or as it changes magnetic latitude, the magnetic field changes, producing a corresponding change in the deviation of an unadjusted compass.

Deviation is eliminated, as nearly as practicable, by introducing at the compass a magnetic field that is equal in magnitude and opposite in polarity to that of the vessel. This process is called **compass adjustment,** or sometimes **compass compensation,** although the latter designation is now more generally applied to the process of neutralizing the effect due to degaussing of the vessel (art. 745).

In general, the introduced field is of the same kind of magnetism as well as of the same intensity as those of the field causing deviation. That is, permanent magnets are used to neutralize permanent magnetism, and soft iron to neutralize induced magnetism, so that the adjustment remains effective with changes of heading and magnetic latitude. A relatively small mass of iron near the compass introduces a field equal to that of a much larger mass at a distance.

When a compass is properly adjusted, its remaining or **residual deviation** is small and practically constant at various magnetic latitudes, the directive force is as strong as is obtainable on all headings, and the compass returns quickly from deflections and is comparatively steady as the vessel rolls and pitches.

714. Effect of latitude.—As indicated in article 706, the magnetic field of the earth is horizontal at the magnetic equator, and vertical at the magnetic poles, the change occurring gradually as a vessel proceeds away from the magnetic equator. At any place, the relative strength of the horizontal and vertical components depends upon the magnetic dip. The directive force of a magnetic compass, provided by the horizontal component of the earth's magnetic field, is maximum on or near the magnetic equator and gradually decreases to zero at the magnetic poles. Within a certain area surrounding each magnetic pole the directive force is so weak that the compass is unreliable (art. 2513).

Deviation changes with a change of the relative strength of either the deviating force or the directive force. Thus, with *either* an increase in deviating force or a decrease in directive force, the deviation increases. However, if both the deviating and directive forces change by the same proportion, and with the same sign, there is no change in deviation. Also, if a deviating force is neutralized by an equal and opposite force *of the same kind,* there is no change of deviation with a change of magnetic latitude.

Permanent magnetism is the same at any latitude. If the permanent magnetism of the vessel is neutralized by properly placed permanent magnets of the correct strength, a change of magnetic latitude can be made without introduction of deviation. But if residual deviation due to permanent magnetism is present, it increases with a change to higher latitude. The deviating force remains unchanged while the directive force decreases, resulting in an increase in the relative strength of the deviating force.

As magnetic latitude increases, the vertical component of the earth's magnetic field becomes stronger, increasing the amount of induced magnetism in vertical soft iron. At the same time the directive force of the compass decreases. Both effects result in increased deviation unless the deviating force is neutralized by induced magnetism in vertical soft iron.

As magnetic latitude increases, the induced magnetism in the horizontal soft iron decreases in the same proportion as the decrease in the directive force of the compass, since both are produced by the horizontal component of the earth's magnetic field. Therefore, any deviation due to this cause is the same at any latitude.

715. Parameters.—Compass adjustment might be accomplished by locating the pole of each magnetic field, and establishing another pole of opposite polarity and equal intensity at the same place, or of less intensity and nearer to the compass; or a pole of opposite polarity and suitable intensity might be established at the correct distance on the opposite side of the compass. Thus, a blue pole east of a compass attracts the red northern ends of the compass magnets and repels the blue southern ends. Both effects cause rotation of the compass magnets and the attached compass card in a clockwise direction, producing easterly deviation. Either a red pole east of a compass, or a blue pole west of it, causes westerly deviation. If there are two fields of opposite polarity, one will tend to neutralize the other. If the intensities of the two fields are equal at the compass, one will cancel the other, and no deviation occurs.

Because of the complexities of the magnetic field of a vessel, and the fact that each individual field making up the total is present continuously, the process of isolating individual poles would be a difficult and time-consuming one. Fortunately, this is unnecessary. The vessel's field is resolved into certain specified components.

Each of these components, regardless of its origin or the number of individual fields contributing to it, can be neutralized separately. Each component is called a **parameter,** and the various parameters are designated by letter, as follows:

Permanent magnetism. **Parameter *P*** is the fore-and-aft component. It is positive (+) if it is the equivalent of a blue pole forward of the compass, and negative (−) if red.

Parameter *Q* is the athwartship component. It is positive if it is the equivalent of a blue pole to starboard.

Parameter *R* is the vertical component. It is positive if it is the equivalent of a blue pole below the compass.

Induced magnetism has nine parameters, each the equivalent of that produced by a slender rod of soft iron. Each *end* of a rod is positive if it is forward, to starboard, or below the compass. Each rod is positive if both ends are positive or if both ends are negative, and negative if the two ends are of opposite sign. The rods are as follows:

a, b, c—one end level with the compass and in its fore-and-aft axis, either forward or aft. It is an *a* rod if it extends fore-and-aft, a *b* rod if athwartships, and a *c* rod if vertical.

d, e, f—one end level with the compass and in its athwartships axis, either to starboard or to port. It is a *d* rod if it extends fore-and-aft, an *e* rod if athwartships, and an *f* rod if vertical.

g, h, k—one end in the vertical axis of the compass, either above it or below it. It is a *g* rod if it extends fore-and-aft, an *h* rod if athwartships, and a *k* rod if vertical.

716. Coefficients.—Deviation which is easterly throughout approximately 180° of heading and westerly throughout the remainder is called **semicircular deviation,** indicating that its sign remains unchanged throughout a semicircle. Deviation caused by permanent magnetism and that caused by induced magnetism in vertical soft iron are semicircular. Deviation which changes sign in each quadrant, being easterly in two opposite quadrants and westerly in the other two, is called **quadrantal deviation.** It is caused by induced magnetism in horizontal soft iron. The types of deviation resulting from the various parameters are called coefficients. There are six, as follows:

Coefficient *A* is constant on all headings. If its cause is magnetic, as from an asymmetrical combination of parameters, it is a "true" constant. If its cause is mechanical, as from an incorrectly placed lubber's line, or mathematical, as from an error in computation of magnetic azimuth, it is an "apparent" constant.

Coefficient *B* is semicircular deviation which is proportional to the sine of the compass heading. It is maximum on compass headings east or west, and zero on compass headings north or south. Coefficient *B* is caused by permanent magnetism, and also by induced magnetism in asymmetrical vertical soft iron.

Coefficient *C* is semicircular deviation which is proportional to the cosine of the compass heading. It is maximum on compass headings north or south, and zero on compass headings east or west. Coefficient *C* is caused by permanent magnetism or by induced magnetism in asymmetrical vertical soft iron athwartship of the compass.

Coefficient *D* is quadrantal deviation which is proportional to the sine of twice the compass heading. It is maximum on intercardinal compass headings, and zero on cardinal compass headings. Coefficient *D* is caused by induced magnetism in horizontal soft iron which is symmetrical with respect to the compass.

Coefficient *E* is quadrantal deviation which is proportional to the cosine of twice the compass heading. It is maximum on cardinal compass headings, and zero on intercardinal compass headings. Coefficient *E* is caused by induced magnetism in horizontal soft iron which is asymmetrical with respect to the compass.

Coefficient *J* is the change of deviation for a heel of 1° while the vessel is on compass heading 000°.

The determination and use of the approximate coefficients in the analysis of compass deviation are discussed in article 727. The force components producing these coefficients are called **exact coefficients.** They are designated by the corresponding upper case German letters. The exact coefficients are now little used in practical navigation. They are fully discussed in various books on compass adjustment.

717. Effect of compass location.—The location of a magnetic compass greatly influences the amount and type of deviation, as well as the adjustment. Thus, if a compass is on the centerline, forward, the effective pole of vertical soft iron is aft of it; but if the compass is on the afterpart of the vessel, the effective pole is forward. If the compass is not on the centerline, as the steering compass of an aircraft carrier, the magnetic field of the vessel is not symmetrical with respect to the compass. If a compass is located in a steel pilot house, the surrounding metal acts as a shield and reduces the strength of the magnetic field of the earth. This is of particular significance in high magnetic latitudes, where the directive force is weak.

Many factors influence the selection of a position for the compass. The most important consideration is the use to be made of it. A steering compass is of little use unless it is located so that it can be seen by the steersman. A compass to be used for emergency steering should be at the emergency steering station. A compass to be used for observing bearings or azimuths, or a standard compass to be used for checking other compasses, should be located so as to have a clear view in most directions.

However, some choice is possible. A compass should not be placed off the centerline if it can be placed on the centerline and still serve its purpose. It should not be placed near iron or steel equipment that will frequently be moved, if this can be avoided. Thus, a location near a gun, boat davit, or boat crane is not desirable. The immediate vicinity should be kept free from sources of deviation—particularly those of a changing nature—if this can be done. That is, no source of magnetism, other than the structure of the vessel, should be permitted within a radius of several feet of the magnetic compass. Some sources which might be overlooked are electric wires carrying direct current; magnetic instruments, searchlights, windshield wipers, electronic equipment, or motors; steel control rods, gears, or supports associated with the steering apparatus; fire extinguishers, gas detectors, etc.; and metal coat hangers, flashlights, keys, pocketknives, metal cap devices, or nylon clothing. The effect of some items such as an ammeter or electric windshield wiper varies considerably at different times. If direct current is used to light the compass, the wires should be twisted.

A magnetic compass cannot be expected to give reliable service unless it is properly installed and protected from disturbing magnetic influences.

718. The binnacle.—The compass is housed in a **binnacle.** This may vary from a simple wooden box to an elaborate device of bronze or other nonmagnetic material. Most binnacles provide means for housing or supporting the various objects used for compass adjustment, as well as the equipment for compensating for deviation caused by degaussing. The standard binnacle for the U. S. Navy 7½-inch compass is shown in figure 718. The trays for holding the fore-and-aft and athwartship magnets (art. 719), and the tube for the heeling magnet (art. 724), can be seen through the open door.

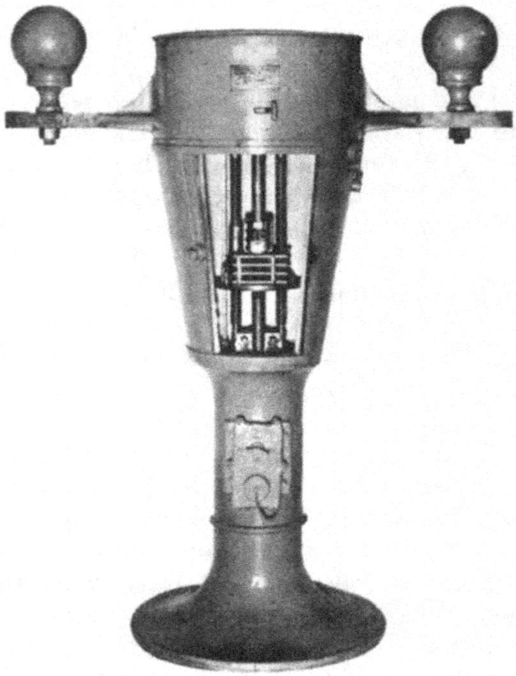

FIGURE 718.—The standard binnacle for a U. S. Navy 7½-inch compass.

719. Adjustment for deviation due to permanent magnetism.—Permanent magnetism can be considered concentrated in a single pole, the position of which depends upon the magnetic heading upon which the vessel was constructed, and the subsequent magnetic history of the vessel. Figure 719a indicates the condition if the permanent magnetism can be considered concentrated in a single blue pole which is directly south of the compass when the vessel is headed magnetic northeast. The only effect on this heading is to weaken the directive force. No deviation is produced because the pole is in line with the compass magnets. On heading magnetic southwest, the pole is also in line with the compass magnets and there is no deviation, but the directive force is strengthened. On any other heading, the pole is not in line with the compass magnets, and deviation occurs, being in the same direction as that of the blue pole from the compass, since the blue pole attracts the red northerly ends of the compass magnets and repels the blue southerly ends. The maximum effect occurs when the *compass* heading is approximately 90° from that of zero deviation. In figure 719a the headings shown on the compass card are the magnetic headings of the vessel. Their offset from the lubber's line shows the direction and relative magnitude of deviation.

If there were no other magnetism in the vessel, the poles might easily be located and neutralized by placing a magnet in such a position that a field of permanent magnetism but opposite polarity would occur at the compass. Although this method of adjustment has been used, it has not proven entirely satisfactory.

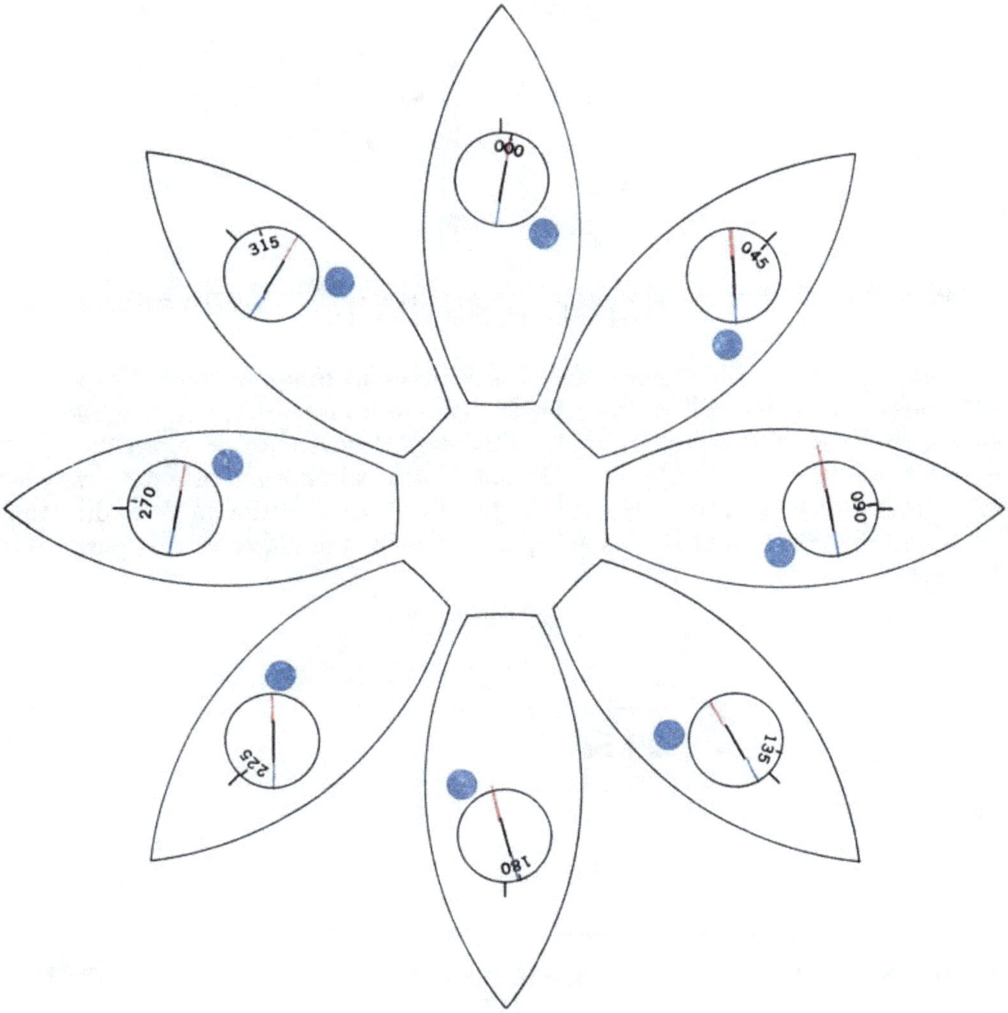

FIGURE 719a.—Deviation due to permanent magnetism if the resultant field is that of a blue pole on the starboard quarter of the vessel.

The usual method is to adjust for the fore-and-aft (parameter *P*) and athwartship (parameter *Q*) components separately. These are shown in figure 719b. The vertical parameter *R* does not produce deviation while the vessel is on an even keel. Its effect when the vessel heels is discussed in article 724. Thus, the effect of a single blue pole at the position shown in figure 719a is the same as that which would be produced by two weaker poles as shown in figure 719b. On heading east or west by the compass, parameter *Q* does not produce deviation directly. However, on easterly headings it does weaken the directive force due to the earth's magnetic field and therefore the deviating force of parameter *P* (causing deviation coefficient *B*) is relatively stronger and has a greater deviating effect. On a westerly heading the directive force would be strengthened, with a corresponding decrease in the *B* coefficient of deviation. By weakening the directive force on easterly headings, parameter *Q* also makes the compass sluggish on these headings. In high latitudes, where the horizontal component of the earth's magnetic field is weak, the compass may lose its directivity at a greater distance from the magnetic pole. Nearer the pole, it might point in the opposite direction.

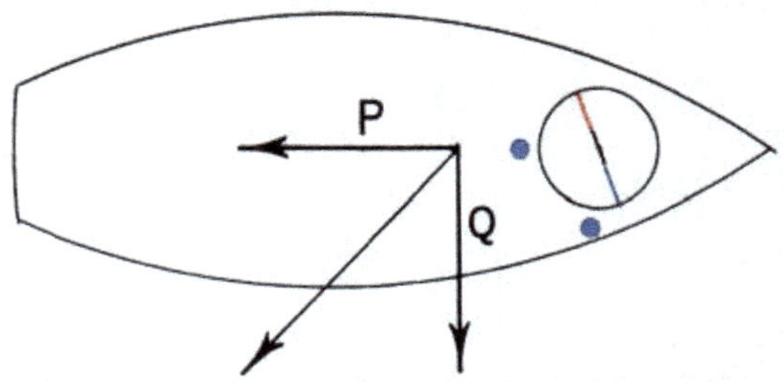

FIGURE 719b.—The horizontal component of the permanent field of figure 719a resolved into its components, parameters *P* and *Q*.

Many binnacles provide a group of several small tubes or "trays" extending in a fore-and-aft direction below the compass. One or more permanent magnets can be inserted in these trays, and the whole assembly moved up or down to vary the effect upon the compass. Figure 719c shows the situation if a single magnet is placed with its red end aft. The field at the compass is in the opposite direction of that of parameter *P*, and if it is of equal strength, the effect of this parameter is eliminated.

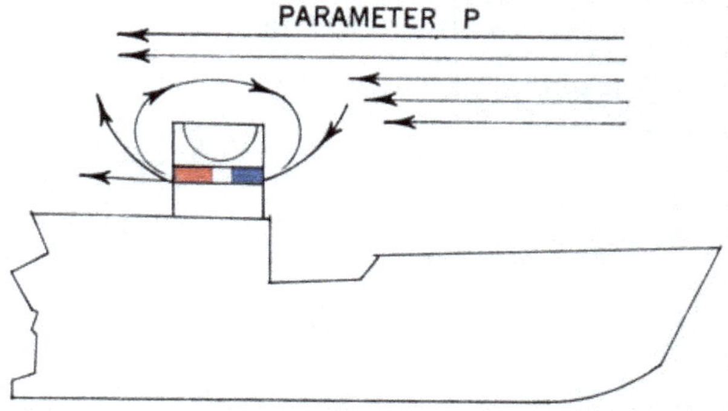

FIGURE 719c.—The field of permanent magnet below the compass and opposing parameter *P* of figure 719b.

If now the vessel is headed north or south by the compass, the only pole remaining is that due to parameter *Q* (causing deviation coefficient *C*), as shown in figure 719d. A set of trays in an athwartship direction below the compass permits insertion of one or more permanent magnets to neutralize the remaining permanent magnetism. The effect of inserting a single magnet with red end to starboard is shown in figure 719e. With both components removed, the field at the compass is completely neutralized.

Both the fore-and-aft (*B*) and athwartship (*C*) trays are in pairs with an equal number of trays on each side of the vertical axis of the compass. In each set of trays it is generally desirable to use an *even* number of magnets equally distributed on each side, to produce a symmetrical field around the compass. However, under some conditions, maximum reduction of deviation occurs with an *odd* number of magnets, particularly when two magnets at maximum distance from the compass overcorrect. If there is a choice, a greater number of magnets at a distance is preferable to a lesser number close to the compass.

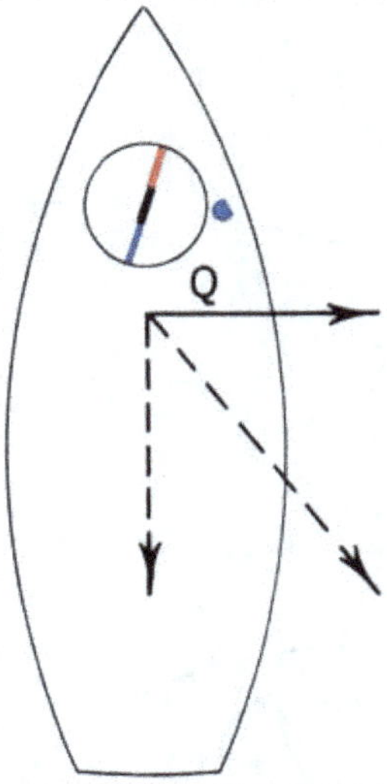

FIGURE 719d.—The permanent field of figure 719a after neutralization of parameter P.

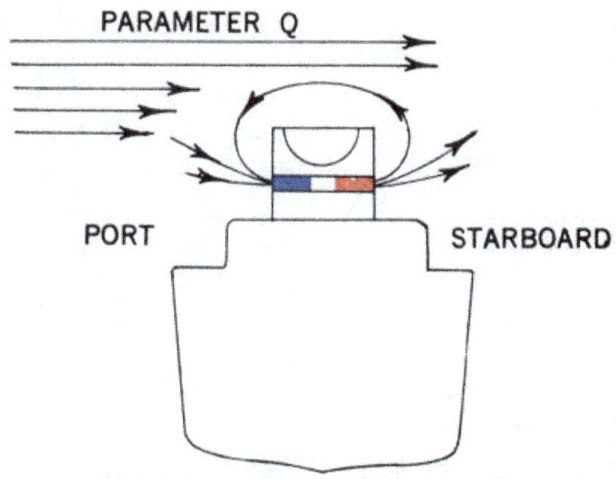

FIGURE 719e.—The field of a permanent magnet below the compass and opposing parameter Q of figure 719b.

With each parameter, the trays to use are those which are approximately perpendicular to the compass magnets. The magnets are placed so that the red ends will be on that side of the compass corresponding to the deviation. Thus, if deviation is easterly, the magnets should be placed so that the red ends will be east of the compass (forward if the heading is east, and to starboard if the heading is north). However, if the wrong end is inserted in the trays, the fact will be immediately apparent because the compass card will rotate in the wrong direction. If the binnacle is not constructed to receive appropriate corrector magnets, these might be secured to some supporting surface near the compass.

During adjustment, the unused magnets should be kept far enough from the compass so that they will not affect it.

720. Adjustment for deviation due to induced magnetism in vertical soft iron.—Figure 720 shows the effect upon the compass of a single blue pole on the centerline of the vessel, aft of the compass. This is a typical situation for induced magnetism in vertical soft iron, for a centerline compass located in the forward part of a vessel in magnetic north latitude. On heading north by compass there is no deviating force, but the directive force is weakened. In high northern latitudes, where this pole becomes strong and the directive force becomes weak, magnetism of this type, if not neutralized, can cause the compass to be unreliable in a much larger area than if the force is neutralized. On a heading of south by compass there is no deviation, but the directive force is strengthened. On headings with an easterly component the deviation is westerly, and on headings with a westerly component the deviation is easterly. In each case the maximum occurs when the vessel is on compass heading approximately east or west. Thus, the deviation due to induced magnetism in vertical soft iron is semicircular, coefficient B. In figure 720 the headings shown on the compass card are the magnetic headings of the vessel. Their offset from the lubber's line shows the direction and relative magnitude of deviation.

The deviating force due to induced magnetism in vertical soft iron is neutralized by placing a bar of soft iron in a vertical position on the opposite side of the

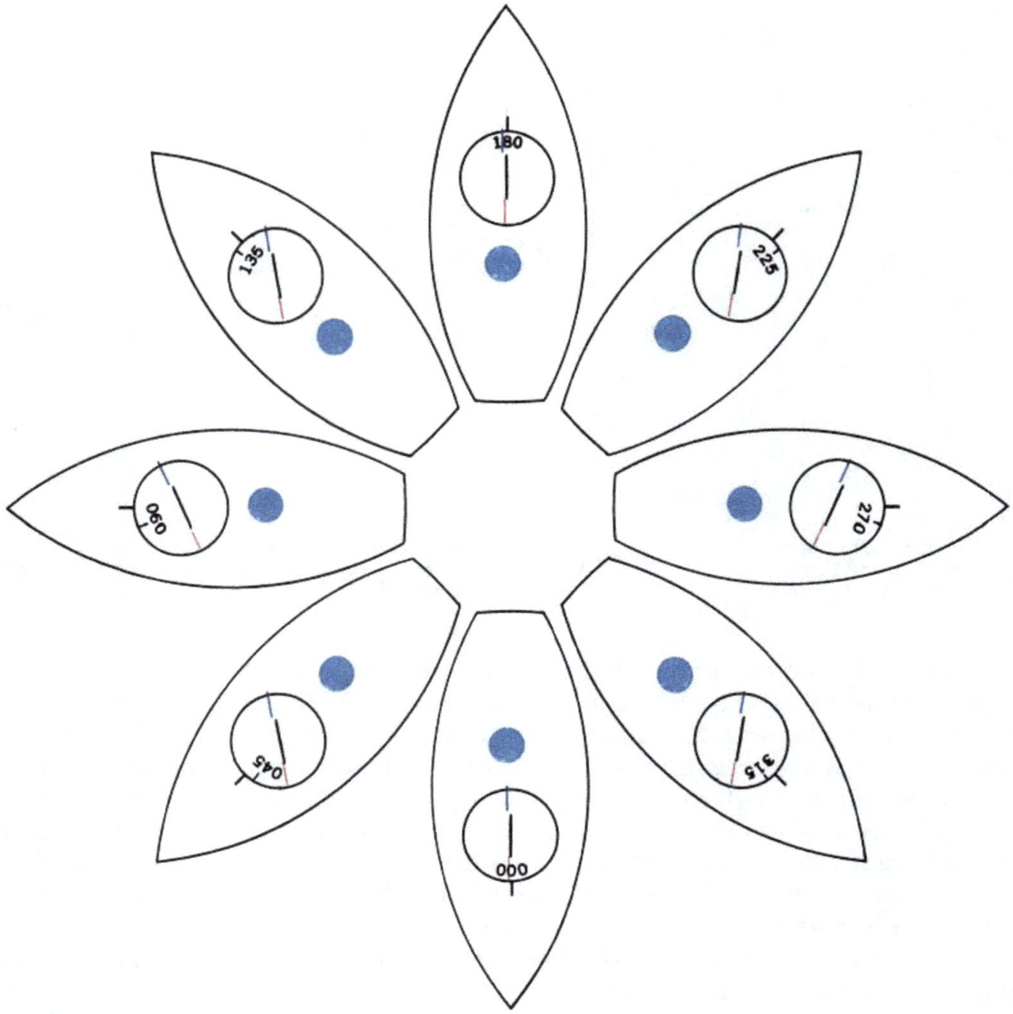

FIGURE 720.—Deviation due to induced magnetism in vertical soft iron if the resultant field is that of a blue pole on the center line aft of the compass.

compass from the effective pole due to the field of the vessel. This piece of metal is called a **Flinders bar**, after Captain Matthew Flinders, RN (1774–1814), an English navigator and explorer who is generally given credit for discovering both the effect and method of adjustment (art. 111). Today, most binnacles for large ships provide a tube for insertion of a Flinders bar. The bar consists of various lengths of soft iron placed end to end; with the remainder of the tube being filled with spacers of nonmagnetic material, usually wood, brass, or aluminum. The standard Flinders bar is two inches in diameter and is divided into six sections, one each of 12, 6, 3, and 1½ inches, and two of ¾ inch. This permits use of any multiple of ¾ inch to 24 inches. All the iron pieces should be above the spacers in the tube, without a gap between pieces, the largest piece being on top. The upper end is then about two inches above the level of the compass card. For short lengths, one or more spacers should be omitted so that about 1/12 of the length of the bar is above the level of the compass card.

The various pieces should be inserted in the tube carefully. If they are dropped, they may acquire some permanent magnetism. This reduces their effectiveness for the purpose intended. Each piece should be tested from time to time to determine whether or not it has acquired permanent magnetism. This can be done by holding it vertical with one end east or west of the compass and very near the compass magnets, noting the reading of the compass, and then inverting the piece so that the ends are interchanged. If the reading differs, permanent magnetism has been acquired by the iron rod. The temporary change of reading while the rod is being inverted should be ignored. In making the test, one should be careful to place the rod in the same position relative to the compass before and after inversion. On an easterly or westerly heading the Flinders bar holder can be used. A small amount of permanent magnetism can be removed by holding the rod approximately parallel to the lines of force of the earth's field, with the blue pole of the rod toward the north, and tapping one end of the rod gently with a hammer. Several alternate tests and treatments may be needed to make the rod magnetically neutral. If this process is not effective in removing the permanent magnetism, the rod should be heated to a dull red and allowed to cool slowly.

An older type Flinders bar, rarely encountered with modern compasses, consists of a number of slender rods of equal length, the *number* of rods being varied rather than the length of a single rod. Another old system consists of using a single rod of fixed length, and varying its distance from the compass.

721. Determination of Flinders bar length.—As indicated in articles 719 and 720, coefficient *B* magnetism may be introduced both by permanent magnetism of the vessel and by induced magnetism in asymmetrical vertical soft iron. A problem thus arises as to what part of the deviation on headings magnetic east and west is due to each cause. If the vessel remains on an even keel at about the same magnetic latitude, adjustment can be made without this knowledge. However, satisfactory performance under all conditions requires separate adjustment for each cause.

There are several possible solutions to this problem. The two sources can be separated by use of the fact that a change of magnetic latitude affects them differently. On the magnetic equator there is no vertical component of the earth's magnetic field, and consequently no induced magnetism in vertical soft iron. Therefore, if the compass is adjusted on the magnetic equator, all coefficient *B* deviation is due to permanent magnetism, and is removed by the fore-and-aft magnets. After a considerable change of magnetic latitude, the deviation on a heading of magnetic east or west is again measured. By means of the curves of figure 721, A, the required amount of standard two-inch Flinders bar is determined. Accurate results will be obtained only if the vessel is magnetically the same at both latitudes. That is, a structural change, an alteration in the number or position of magnets or other devices used in the adjustment, magnetic treatment, etc., invalidates the measurement. After the required amount of Flinders bar has been inserted, some deviation may be present due to mutual induction among the various devices used for adjustment. This should be removed by means of the permanent magnets. Once the correct amount of Flinders bar has been installed, no change should be needed unless there is a substantial change in the amount or location of vertical soft iron, or unless the compass is relocated.

This method is not always practical. If the correct length and location of Flinders bar for another vessel of similar construction and compass location have been determined previously, the same length can be used for the compass being adjusted. If a large change in magnetic latitude can be made without appreciable change of deviation on headings east and west, the amount of Flinders bar is correct. If the deviation changes, readjustment is needed. By studying the structure of the vessel, an experienced compass adjuster may be able to make a reasonably accurate estimate of the length to use.

In the absence of enough reliable information to permit a reasonably accurate determination of the correct length, the Flinders bar may be omitted entirely, and the deviation on east and west headings removed by means of the fore-and-aft permanent magnets. This is common practice for yachts, fishing vessels, and even for some coastal vessels which do not change magnetic latitude more than a few degrees.

The correct length of Flinders bar can be determined by figure 721, B, if reliable data are available on the deviation occurring on magnetic east or west headings at two widely separated magnetic latitudes. The constant *K* is determined by computation, using the formula

$$K=\lambda\frac{H_2 \tan d_2 - H_1 \tan d_1}{Z_2 - Z_1}$$

in which:

K = a constant proportional to the required length of Flinders bar.
λ = shielding factor, or the proportion of the earth's field effective at the compass. Generally, it varies from about 0.7 to 1.0, averaging about 0.9 for compasses in exposed positions, and 0.8 for those surrounded by metal deck houses.
H_1 = horizontal intensity of earth's magnetic field at place of first deviation reading.
H_2 = horizontal intensity of earth's magnetic field at place of second deviation reading.
d_1 = total deviation on heading magnetic east or west at place of first deviation reading.
d_2 = total deviation on heading magnetic east or west at place of second deviation reading.
Z_1 = vertical intensity of earth's magnetic field at place of first deviation reading.
Z_2 = vertical intensity of earth's magnetic field at place of second deviation reading.

The unit of intensity is the oerstead.

The values of horizontal and vertical intensity (H and Z) can be obtained from charts 33 and 36, respectively, by dividing the values in gammas (art. 706) as shown on the charts by 100,000.

The constant *K* represents a mass of vertical soft iron (the *c* rod) causing deviation. From the intersection of the curve of figure 721, B, and a horizontal line through the value of constant *K*, draw a vertical line to the bottom scale, which shows the required length of Flinders bar.

If some length of Flinders bar was in place when the two deviation readings were made, enter the graph of figure 721, B, with this length and determine the corresponding value of *K*. Call this K_2 and that obtained by computation K_1. Algebraically add K_1 and K_2 to determine the value of *K* to use for finding the total length of Flinders bar required. If the Flinders bar is forward of the compass, K_2 is negative (−), and if aft of the compass, K_2 is positive (+). In the computation of K_2, both Z_1 and Z_2 are positive in north magnetic latitude and negative in south magnetic latitude. Also, d_1 and d_2 are positive if deviation is east on magnetic heading east in north latitude or magnetic heading west in south latitude. If *either* the heading or direction of the deviation is reversed, the sign of d_1 or d_2 is negative. If both are reversed, the sign is positive. If the value of *K* is negative, the Flinders bar should be installed forward of the compass, and if positive, it should be installed aft.

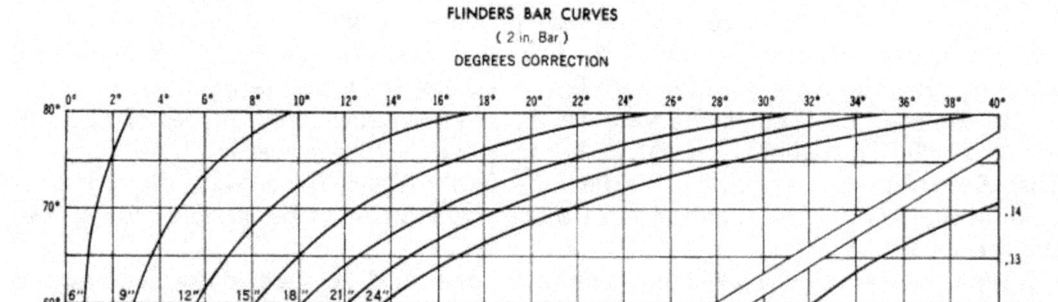

FIGURE 721.—Flinders bar curves: A, if deviation due to induced magnetism in vertical soft iron is known; B, if coefficient K is known.

Example.—The deviation of a magnetic compass of a ship on heading magnetic east is 1°E in an area where H is 0.170 and Z is 0.539. It is 9°E in an area where H is 0.311 and Z is 0.260. The shielding factor is 0.8.

Required.—The correct length of Flinders bar if (1) no Flinders bar is in place during observations, (2) six inches of Flinders bar is in place forward of the compass during observations.

Solution.—

$$(1)\quad K_1 = 0.8\left(\frac{0.311\times 0.15838 - 0.170\times 0.01746}{0.260-0.539}\right)$$

$$= (-)\,0.133$$

$$K_2 = 0$$

$$K = K_1 + K_2 = (-)\,0.133$$

From figure 721, B, the correct amount of Flinders bar is 22 inches. Since the amount used must be a multiple of ¾ inch, the amount to use is 21¾ inches. Since K is negative, the bar should be installed forward of the compass.

(2) From figure 721, B, the value of K_2 corresponding to six inches of Flinders bar is 0.009. The value is negative because the bar is forward of the compass. Therefore, $K_1+K_2=(-)0.133+(-)0.009=(-)0.142$. From figure 721, B, the *total* amount of Flinders bar required is 24 inches, which should be installed forward of the compass.

Answers.—(1) 21¾ inches of Flinders bar installed forward of the compass, (2) 24 inches of Flinders bar installed forward of the compass.

When the length of Flinders bar is determined in this way, accurate results can be expected only if the vessel is magnetically unchanged between deviation readings.

Lord Kelvin suggested the following rule for improving the adjustment for coefficient B if no better method is available:

Remove the deviation observed on magnetic east or west headings by means of fore-and-aft B *magnets when the vessel has arrived at places of weaker vertical magnetic field, and by means of Flinders bar when it has arrived at places of stronger vertical magnetic field, whether in the Northern or Southern Hemisphere.*

After a number of applications of this rule following alternate passage from weaker to stronger fields and then stronger to weaker fields, the amount of Flinders bar should be very nearly correct.

722. Adjustment for deviation due to induced magnetism in symmetrical horizontal soft iron.—That part of horizontal soft iron which is symmetrically arranged with respect to the compass can be considered equivalent to two rods extending through the compass, one in a fore-and-aft direction ($-a$ rod) and the other in an athwartship direction ($-e$ rod). The deviation caused by both of these rods is quadrantal, but of opposite sign. If both rods were equally effective in causing deviation, they would cancel each other and no deviation would result on any heading. In most vessels, however, the athwartships iron dominates, and deviation due to all horizontal soft iron can generally be considered to be that which would result from a single $(-)$ e rod. In figure 722a the deviation resulting from such a rod is shown for various magnetic headings in any latitude. There is no deviation on any cardinal heading, but the directive force is weakened on heading magnetic east or west. The maximum deviation occurs on intercardinal headings by compass, being easterly in the northeast and southwest quadrants, and westerly in the other two quadrants. This is coefficient D deviation. In figure 722a the headings shown on the compass card are the magnetic headings of the vessel. Their offset from the lubber's line shows the direction and relative magnitude of deviation.

The field causing this deviation is neutralized by installing two masses of soft iron abeam of the compass, on opposite sides and equidistant from its center. Such iron is usually in the form of hollow spheres or cylinders, called **quadrantal correctors.** These can be moved in or out in an athwartship direction along brackets on the sides of the binnacle.

Quadrantal correctors act as $(+)$ e parameters which neutralize the $(-)$ e parameter of the athwartships iron. As shown in figure 722b, the portion of the corrector adjacent to the compass is always of opposite polarity to the deflecting force. The amount of the correction can be adjusted by moving the correctors toward or away from the compass card. If the inboard limit of travel is reached without fully removing the deviation, larger correctors are needed. If overcorrection

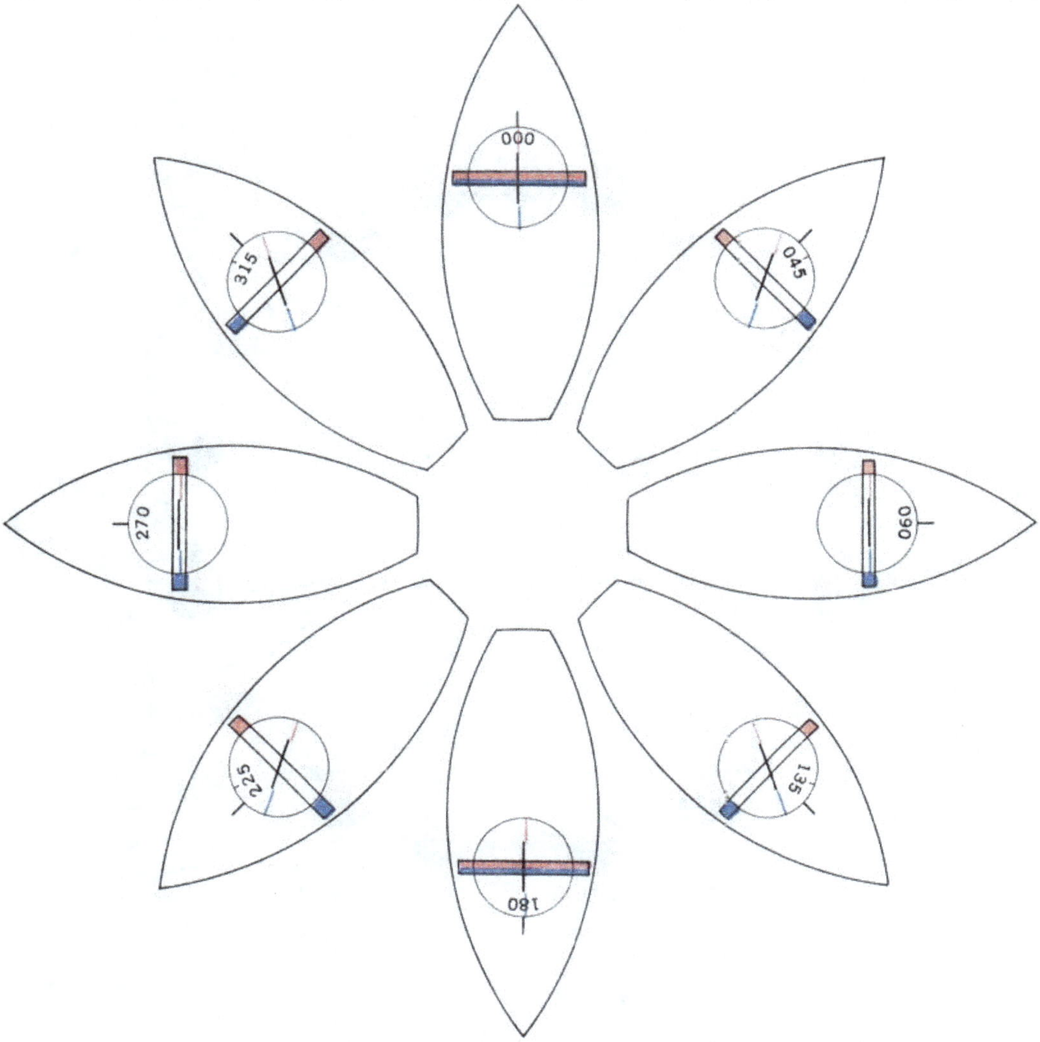

FIGURE 722a.—Deviation caused by induced magnetism in symmetrical horizontal soft iron.

occurs at the outboard limit, smaller correctors are needed. A single corrector can be used, but this produces an unbalanced field which is less desirable than a balanced one. In general, large correctors at a greater distance are preferable to small correctors close up because there is less mutual induction between the correctors if they are widely separated. In the rare case when quadrantal deviation is *westerly* on heading northeast (coefficient D is negative, the fore-and-aft horizontal soft iron predominating), the quadrantal correctors should be mounted fore-and-aft on the binnacle.

Figure 722c shows the approximate amount of deviation correction to be expected from correctors of various sizes, shapes and distance from the center of a standard U. S. Navy 7½-inch compass. The data apply to either the athwartships or fore-and-aft position.

Like the Flinders bar (art. 720), the quadrantal correctors should be handled carefully, and checked from time to time to see if they have acquired permanent magnetism. The test can be made by rotating each corrector through 180° without altering its distance from the center. If the compass heading changes, the correctors

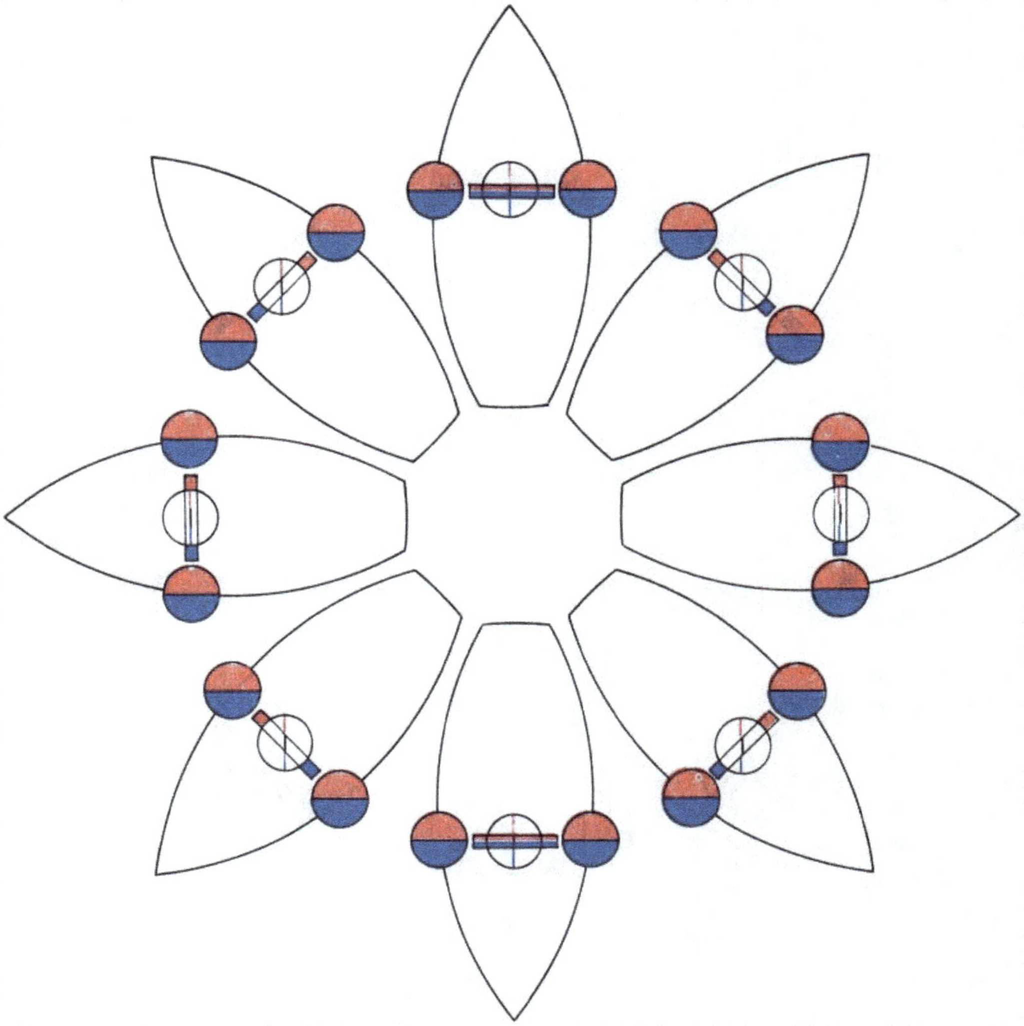

FIGURE 722b.—Adjustment for symmetrical horizontal soft iron.

have acquired permanent magnetism which can be removed by tapping with a hammer when the blue pole is toward the north, or by removing the spheres, heating them to a dull red, and permitting them to cool slowly.

723. Adjustment for deviation due to induced magnetism in asymmetrical horizontal soft iron.—If the horizontal soft iron is not arranged symmetrically with respect to the compass, resulting in an effective pole which is on neither the fore-and-aft nor athwartships axis through the compass, quadrantal deviation with its maximum values on cardinal headings (coefficient *E*) results. Constant deviation (coefficient *A*) may also be used by this arrangement. Either coefficient *E* or *A* is due to a combination of parameters.

For a centerline compass on a ship of conventional construction, any deviation due to induced magnetism in asymmetrical horizontal soft iron is small, and many installations make no provision for neutralizing the effect. However, some binnacles are provided with a pair of ***E*-links**, which are bars that can be attached to the side brackets to permit the quadrantal correctors to be slewed somewhat with respect to

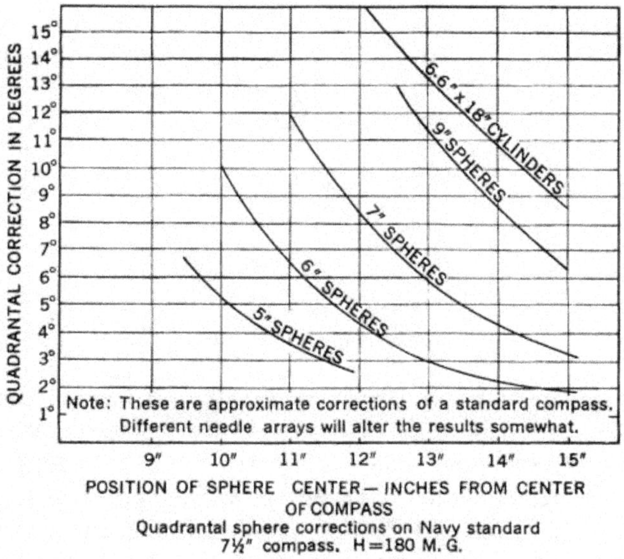

FIGURE 722c.—Effect of various quadrantal correctors.

the compass. When this has been done, the horizontal axis through the correctors and the compass makes an angle with the athwartship axis of the compass.

After a compass has been adjusted, any remaining constant deviation due to magnetic coefficient *A* is likely to be very small. If such deviation exists, its cause is likely to be chiefly mechanical. If a compass is used primarily for determining the heading (as a steering compass), all constant deviation can be removed by realignment of the binnacle so as to rotate the lubber's line by the required amount. However, if a compass is to be used for observing bearings or azimuths, only the mechanical *A*-error should be removed in this manner. This is because such readings are taken on the face of the card itself, and are therefore not affected by misalignment of the lubber's line. The two components of constant deviation can be separated in the following manner: Measure the deviation on various headings by means of bearings or azimuths (art. 1430). This includes only magnetic coefficient *A*. Then measure the deviation on various headings by means of the lubber's line, comparing the heading by compass with the magnetic heading determined by pelorus or gyrocompass. This includes the combined effect of magnetic and mechanical coefficient *A* deviation. The difference between the two values is the mechanical coefficient *A*. For a properly adjusted compass the magnetic coefficient *A* deviation is so small that provision is not made for its removal.

724. Heeling error.—All of the effects discussed previously refer to a vessel on an even keel. When the vessel heels, conditions are altered. Deviation which now appears or the *change* of deviation from that when the vessel was on an even keel, is called **heeling error.** For a constant angle of heel and a steady heading, this error remains essentially unchanged. However, it tends to increase as the heel becomes greater, and to reverse sign as the heel changes from one side to another. Therefore, if a vessel is rolling or pitching, the compass tends to oscillate. This increases the difficulty of reading the compass.

The cause of heeling error is the displacement of the permanent and induced magnetic fields with respect to the compass. Figure 724 shows a vessel heeled to starboard on heading magnetic north or south, in north magnetic latitude. The vessel was constructed in north magnetic latitude. On an even keel the vertical

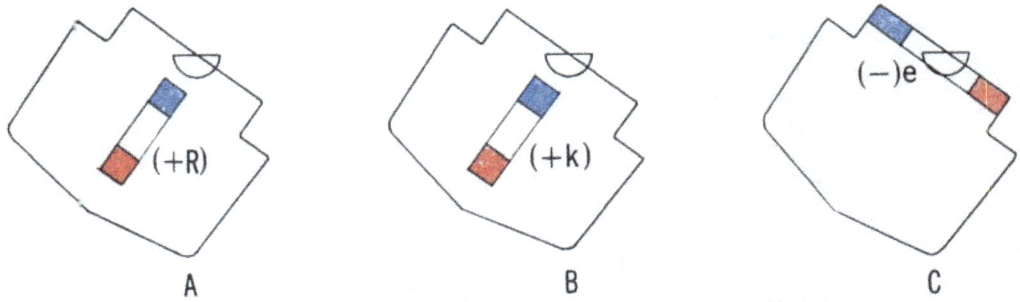

FIGURE 724.—Effect of heel.

parameter R of permanent magnetism for a centrally located compass is directly below the compass, with the blue pole nearer the compass. When the vessel is heeled as shown at A, the blue pole is to port of the compass, causing deviation toward that side. A vertical rod of soft iron below the compass (parameter k) exerts a similar influence, as shown at B. An athwartship horizontal rod through the compass has no deviating effect while the vessel is on an even keel, but when it heels as shown in figure 724, the vertical component of the earth's field causes the port end to acquire a blue pole and the starboard end a red pole (parameter e), as shown at C. Each of the three causes shown in figure 724 results in a blue pole being established on the port or high side of the vessel. This causes the red north ends of the compass magnets to be attracted to this side. If the heading is magnetic north, the deviation is westerly, and if magnetic south, it is easterly. This effect is offset somewhat by the changed magnetic field surrounding the quadrantal correctors. On heading magnetic east or west, these components have no deviating effect, but the directive force of the compass is strengthened or weakened. When the vessel pitches, the effects described for north-south and east-west headings are reversed. On a heading other than a cardinal direction (magnetic) the effect is some combination of the two. The magnetic situation varies not only with the heading, but also with the magnetic latitude and the magnetic history of the vessel.

Although heeling error is due in part to permanent magnetism and in part to induced magnetism, the induced magnetism generally exerts the greater influence. The most effective method of neutralizing this effect would be to attack each parameter separately. This would require the placement of soft iron *above* the compass. Since this would not be a convenient arrangement, the condition is improved by placing a vertical permanent magnet, called a **heeling magnet**, centrally *below* the compass, and adjusting its height until the error is minimized. In north magnetic latitude, the red end is placed uppermost in most installations. As the vessel proceeds to lower magnetic latitudes, parameter R becomes less effective in producing deviation because of the stronger directive force due to the horizontal component of the earth's magnetic field. Parameters k and e become weaker because of decreased intensity of the vertical component of the earth's field, and the strengthening of the horizontal component also reduces their effect. Therefore, the heeling magnet requires readjustment as the magnetic latitude changes. As the vessel approaches the magnetic equator, the heeling magnet should be lowered. After the vessel crosses the magnetic equator, it may be necessary to invert the heeling magnets, so that the opposite end is uppermost. A change in the setting of the heeling magnet may introduce deviation on headings of compass east or west because of altered induction between the heeling magnet and the Flinders bar. This should be removed by means of the fore-and-aft (*B*) magnets in the trays below the compass.

If adjustment for heeling error is made when the vessel is tied up or at anchor, it is best done by listing the vessel on a northerly or southerly heading, and adjusting the heeling magnet until the reading of the compass is restored to what it was before the vessel heeled. If the adjustment is made at sea, the vessel should be placed on a heading of compass north or south. If there is little rolling, the vessel can be listed and the compass reading restored, as at dockside. If the vessel rolls moderately on this heading, the heeling magnet should be placed at that height at which oscillation of the compass card is minimum. If the setting for minimum oscillation is different on north and south headings, the mean position should be used. Any yawing of the vessel should be considered when reading the compass under rolling conditions.

The approximate position of the heeling magnet can be determined by means of an instrument known as a **heeling adjuster** or a **vertical force instrument**, a form of **dip needle.** This consists of a small magnet balanced about a horizontal axis by means of a small adjustable weight. A scale indicates the distance of the weight from the axis. The instrument is taken ashore and balanced at a place where the earth's field is undisturbed, the magnet being in a magnetic north-south direction, approximately. The instrument is then taken aboard ship, the compass removed from its binnacle, and the heeling adjuster installed in its place. The weight is set to a distance equal to the distance determined ashore, multiplied by λ, the shielding factor (art. 721). The heeling magnet is then moved up or down until the magnet of the instrument is level. This should be approximately the correct setting. This method is used principally when the listing of a vessel is difficult or impractical.

725. Soft iron correctors and nearby magnets.—The soft iron correctors used in compass adjustment are near enough to the compass magnets and the magnets used in compass adjustment to be influenced by them.

The Flinders bar acquires a certain amount of induced magnetism from the fields of the heeling magnet and the fore-and-aft (*B*) corrector magnets. The approximate amount of deviation caused by induced magnetism from the heeling magnet of a 7½-inch compass when H=0.165 is shown in figure 725. Because of such induced magnetism, the "drop-in" method of determining the amount of Flinders bar is not accurate. By this method, Flinders bar lengths are added until the compass reading changes by the required amount. Better adjustment is achieved by using the required amount of Flinders bar and removing any remaining deviation on east-west headings by means of the fore-and-aft magnets. The principal reason that it is preferable to use a larger number of magnets at a distance from the compass than a smaller number near it, is that the former arrangement produces less induced magnetism in the Flinders bar and quadrantal correctors. If the Flinders bar length is changed, the deviation on headings of magnetic east and west should be checked, and any needed adjustment made by means of fore-and-aft magnets. When all correctors have been put in place, their positions relative to each other are constant. Therefore, the Flinders bar acts as a permanent magnet, and the resulting deviation is semicircular (coefficient *B*). The Flinders bar may also introduce a small amount of quadrantal deviation (coefficient *D*), its action being somewhat like that of a quadrantal corrector placed in the fore-and-aft axis of the compass.

The quadrantal correctors acquire induced magnetism from the fields of the fore-and-aft (*B*) magnets, the athwartship (*C*) magnets, and the compass magnets.

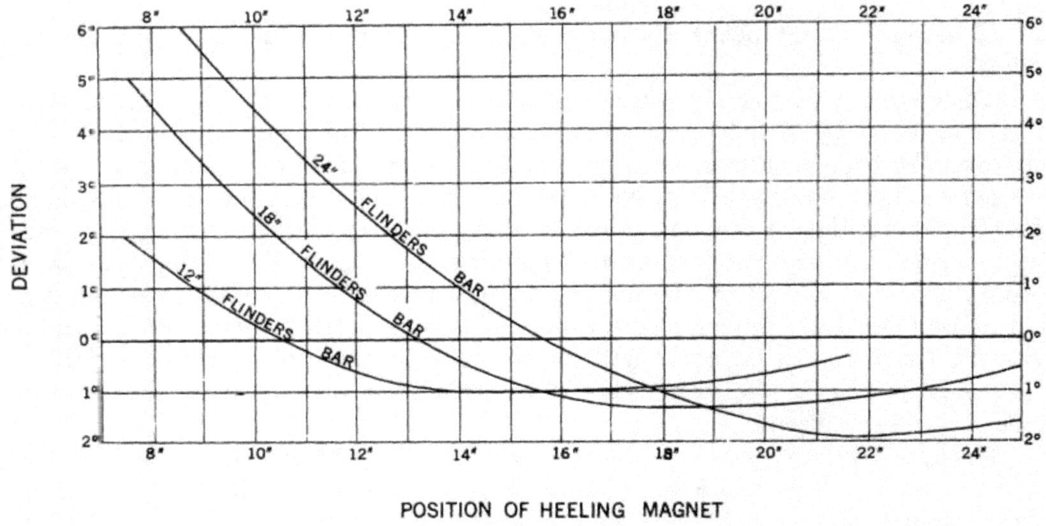

FIGURE 725.—Deviation due to inductive effect of heeling magnet on Flinders bar.

The magnetism acquired from the *B* and *C* magnets is semicircular (coefficient *B* from the *B* magnets, and coefficient *C* from the *C* magnets), and that acquired from the field of the compass magnets is quadrantal (coefficient *D*). The semicircular deviation is minimized by keeping the *B* and *C* magnets as far away from the quadrantal correctors as practicable, and any deviation that does exist is removed by means of these magnets. The quadrantal deviation is removed by means of the quadrantal correctors themselves. The compass magnets of most modern compasses have little effect upon the quadrantal correctors.

Because of the interaction between the various correctors, it is good practice to insert the required amount of Flinders bar, and to install the quadrantal correctors and heeling magnet at their approximate positions before adjusting the compass. If a radical change is subsequently made in any of these adjustments, the settings of the *B* and *C* magnets should be checked and altered if necessary.

Analysis of Deviation

726. Nature and purpose of analysis.—An analysis consists of determining the approximate value of each of the six coefficients, and studying the results. The purpose of the analysis is to give the compass adjuster an understanding of the magnetic properties of the vessel. This provides the basis for the approximate placement of the various correctors, and suggests possibilities for further refinement in the adjustment. Without an analysis, compass adjustment is a more-or-less mechanical process. Fewer mistakes are likely to be made by the person who understands the nature of the magnetic field he seeks to neutralize.

727. The analysis.—The first step in an analysis is to record the deviation on each cardinal and intercardinal heading *by the compass to be analyzed.* For the purpose of analysis, easterly deviation is considered positive (+), and westerly deviation negative (−). Approximate values of the various coefficients are:

Coefficient A—mean of deviation on all headings.

Coefficient B—mean of deviation on headings 090° and 270°, with sign at 270° reversed.

Coefficient C—mean of deviation on headings 000° and 180°, with sign at 180° reversed.

Coefficient D—mean of deviation on intercardinal headings, with signs at headings 135° and 315° reversed.

Coefficient E—mean of deviation on cardinal headings, with signs at 090° and 270° reversed.

Coefficient J—change of deviation for a heel of 1° while the vessel heads 000° by compass. It is considered *positive* if the north end of the compass card is drawn toward the *low* side, and *negative* if toward the *high* side.

Example.—A magnetic compass which has not been adjusted has deviation on cardinal and intercardinal compass headings as follows:

Compass heading	*Deviation*	*Compass heading*	*Deviation*
000°	1°5W	180°	8°0E
045°	34°0E	225°	1°5W
090°	31°0E	270°	29°0W
135°	13°5E	315°	36°0W

On heading compass north the deviation is 13°5W when the vessel heels 10° to starboard.

Required.—The approximate value of each coefficient.

Solution.—

$$A=\frac{-1°5+34°0+31°0+13°5+8°0-1°5-29°0-36°0}{8}=(+)2°3$$

$$B=\frac{31°0+29°0}{2}=(+)30°0$$

$$C=\frac{-1°5-8°0}{2}=(-)4°8$$

$$D=\frac{34°0-13°5-1°5+36°0}{4}=(+)13°8$$

$$E=\frac{-1°5-31°0+8°0+29°0}{4}=(+)1°1$$

$$J=\frac{-13°5+1°5}{10}=(-)1°2$$

Answers.—*A* (+) 2°3, *B* (+) 30°0, *C* (−)4°8, *D* (+) 13°8, *E* (+) 1°1, *J* (−) 1°2.

On any compass heading (CH) the deviation (d) from each coefficient acting alone is:

Coefficient A: $d_A=A$
Coefficient B: $d_B=B \sin CH$
Coefficient C: $d_C=C \cos CH$
Coefficient D: $d_D=D \sin 2CH$
Coefficient E: $d_E=E \cos 2CH$
Coefficient J: $d_J=J \cos CH$.

For a vessel on an even keel, the total deviation on any compass heading is the algebraic sum of the deviation due to each of the first five coefficients:

$$d=d_A+d_B+d_C+d_D+d_E=A+B \sin CH+C \cos CH+D \sin 2CH+E \cos 2CH.$$

For the compass of the example given above, the deviation due to each component, and the total, on various headings is:

CH	A	B	C	D	E	d
°	°	°	°	°	°	°
000	+2.3	0.0	−4.8	0.0	+1.1	−1.4
015	+2.3	+7.8	−4.6	+6.9	+1.0	+13.4
030	+2.3	+15.0	−4.2	+12.0	+0.6	+25.7
045	+2.3	+21.2	−3.4	+13.8	0.0	+33.9
060	+2.3	+26.0	−2.4	+12.0	−0.6	+37.3
075	+2.3	+29.0	−1.2	+6.9	−1.0	+36.0
090	+2.3	+30.0	0.0	0.0	−1.1	+31.2
105	+2.3	+29.0	+1.2	−6.9	−1.0	+24.6
120	+2.3	+26.0	+2.4	−12.0	−0.6	+18.1
135	+2.3	+21.2	+3.4	−13.8	0.0	+13.1
150	+2.3	+15.0	+4.2	−12.0	+0.6	+10.1
165	+2.3	+7.8	+4.6	−6.9	+1.0	+8.8
180	+2.3	0.0	+4.8	0.0	+1.1	+8.2
195	+2.3	−7.8	+4.6	+6.9	+1.0	+7.0
210	+2.3	−15.0	+4.2	+12.0	+0.6	+4.1
225	+2.3	−21.2	+3.4	+13.8	0.0	−1.7
240	+2.3	−26.0	+2.4	+12.0	−0.6	−9.9
255	+2.3	−29.0	+1.2	+6.9	−1.0	−19.6
270	+2.3	−30.0	0.0	0.0	−1.1	−28.8
285	+2.3	−29.0	−1.2	−6.9	−1.0	−35.8
300	+2.3	−26.0	−2.4	−12.0	−0.6	−38.7
315	+2.3	−21.2	−3.4	−13.8	0.0	−36.1
330	+2.3	−15.0	−4.2	−12.0	+0.6	−28.3
345	+2.3	−7.8	−4.6	−6.9	+1.0	−16.0

The various components and the total deviation are shown in graphical form in figure 727. Since the various coefficients are only approximated by the method given above, the curve of total deviation found in this way should not be expected to coincide exactly with a curve drawn from values found by measurement on the various headings.

The *shapes* of the curves of figure 727 are typical of those of an unadjusted compass of a large steel ship. However, an analysis of the results indicates the following:

Coefficient A is normally negligible. The presence of more than 2° of constant error indicates an abnormal condition which should be discovered and corrected. If the vessel has been in service for some time without major structural change, and no misalignment of the lubber's line of the compass or the pelorus or gyrocompass used for measuring deviation has been noted previously, it is probable that a

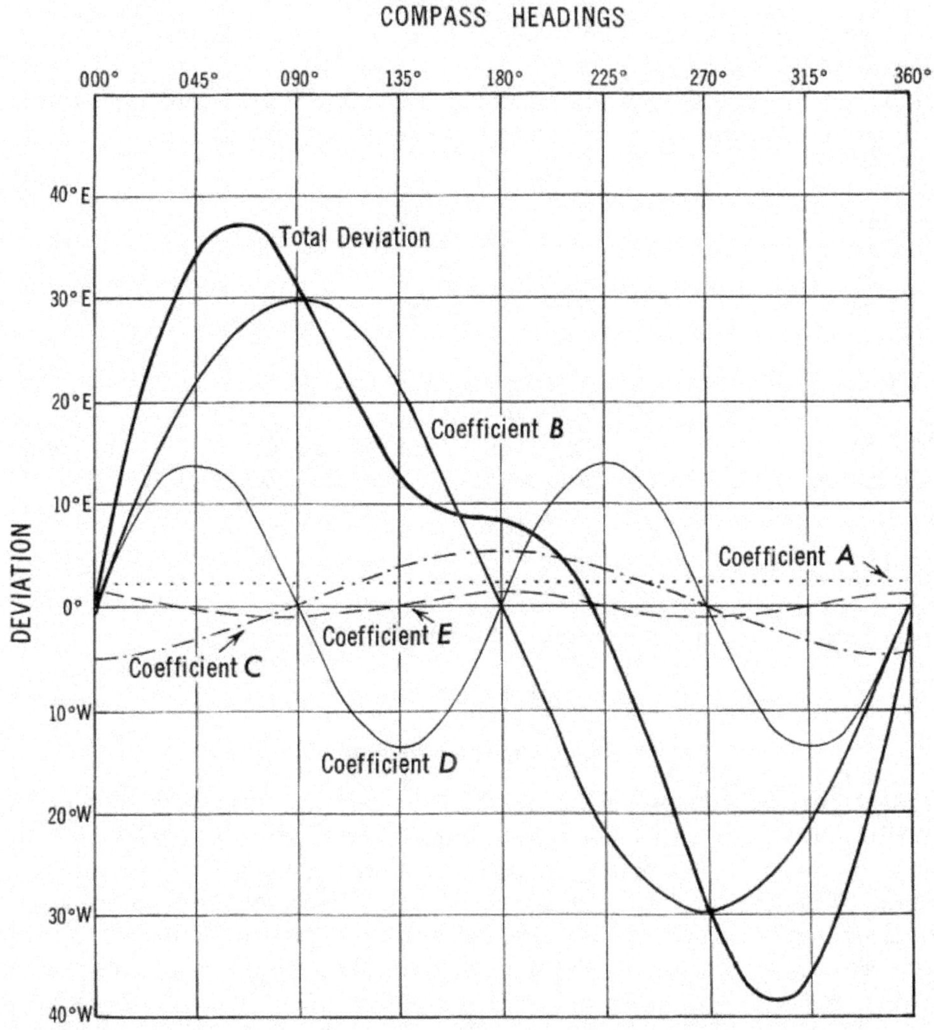

FIGURE 727.—Coefficients and total deviation of an unadjusted magnetic compass.

mistake has been made in determining the azimuth or bearing used for establishing deviation.

Coefficient *E* is normally negligible for a compass located on the centerline of the vessel. This vessel has an excessive amount, which should be corrected by slewing the quadrantal correctors, using an *E*-link.

Since deviation is east on heading 090° and west on 000°, it is probable that the blue pole of the vessel's permanent field is on the port bow.

The compass being unadjusted, no Flinders bar is in place, and the large *B* deviation on heading 090° is a combination of deviation from induced magnetism in vertical soft iron and that due to the permanent magnetism of the vessel. Since the deviation on heading 270° is nearly the same as that on 090°, but of opposite sign, adjustment on one of these headings should result in nearly correct adjustment on the other. Since some *B* and *C* deviation occurs on intercardinal headings, while no *D* deviation occurs on cardinal headings, adjustment for *B* and *C* should be made before that for final *D* adjustment.

A second analysis made after adjustment may reveal possibilities for further refinement in the adjustment.

If heeling error is measured on any heading other than compass north or south, the value of coefficient J can be found by means of the formula:

$$d = J \cos CH$$

converted to

$$J = \frac{d}{\cos CH}$$

or

$$J = d \sec CH.$$

If HE is the total observed change of deviation (heeling error), and i is the angle of heel in degrees (for relatively small angles), the formula becomes

$$J = \frac{HE \sec CH}{i}.$$

If heeling error is sought, the formula becomes

$$HE = Ji \cos CH.$$

Adjustment Procedure

728. Preliminary steps.—Efficient and accurate adjustment is preceded by certain preliminary steps best made while a vessel is moored or at anchor.

The magnetic environment of the compass should be carefully inspected. Stray magnetic influences such as those caused by tools, direct current electric appliances, personal equipment (such as keys, pocketknives, or steel belt buckles), nylon clothing etc., should be eliminated. Permanently installed equipment of magnetic material (such as cargo booms, boat davits, cranes, or guns) should be placed in the positions they normally occupy at sea. The degaussing coils should be secured by the reversing process (art. 743) if this has not already been done.

The compass itself should be checked carefully for bubbles, and to be sure it is centered on the vertical axis of the binnacle. If it is, and the vessel is on an even keel, there is no change of reading as the heeling magnet is raised and lowered in its tube. An adjustment should be made to the gimbal rings if the compass is off center. There should be no play in the position of the compass once it is centered.

The lubber's line, too, should be checked to be sure it is in line with the longitudinal axis of the vessel. This can be done by sighting on the jackstaff if the compass is on the centerline. If it is not, a batten might be erected at a distance from the centerline equal to the distance from the center of the compass to the centerline. Another way is to determine the distance from the compass to the centerline and from this point to the jackstaff. The first distance divided by the second is the natural tangent of the angle at the compass between the line of sight to the jackstaff and the line of sight through the lubber's line. If the compass is in an exposed position where bearings can be taken, and the true heading is known, the observed relative bearing of a distant object can be compared with that obtained by careful measurement on the chart. If the vessel is at anchor or underway, the method explained in article 723 can be used.

If a pelorus or gyrocompass or repeater is to be used in determining deviation of the compass, its lubber's line should be checked in the same manner, or by comparing a relative bearing of a distant object taken by two instruments, the lubber's line of one having previously been checked. If a gyrocompass is to be used, it is checked to see that it is synchronized with a repeater. With accurate synchronization, any error in one will also be present in the other. The speed and latitude adjustments of the gyrocompass should be checked carefully.

All devices to be used in the adjustment should be checked to see that they are on hand and in good condition. The trays for *B* and *C* permanent magnets, the quadrantal correctors, and heeling magnet should be checked for freedom of motion. The Flinders bar and quadrantal correctors should be checked for permanent magnetism. The correct amount of Flinders bar should be placed in its tube. The quadrantal correctors should be placed in their approximate positions, being centered if no better information is available. The heeling magnet is generally placed with the red end uppermost in north magnetic latitude, and the blue end uppermost in south magnetic latitude. If no better information is available, the heeling magnet should be placed near the bottom of the tube.

Plans for the actual adjustment should be made carefully. A suitable time and location should be selected. If landmarks are to be used, suitable ones should be selected to provide the information desired. Areas of heavy traffic should be avoided. If azimuths of the sun are to be used, a time should be selected when the sun will not be too high in the sky for suitable observation. A curve of magnetic azimuths (art. 731) should be made, and just before adjustment begins a comparing watch should be checked and set, if possible, to correct time. Local variation should be checked carefully, and corrected for date, if necessary. Any necessary recording and work forms should be made up. Each person to participate in the adjustment should be instructed regarding the general plan and his specific duties.

729. Underway procedure.—When everything is in order and the vessel has arrived at its adjusting area, final adjustment can begin. Trim should be normal, and the vessel free from list, so that no heeling error is present.

All adjustment headings should be *magnetic*. Compass headings can be used, but this results in a slight turn being required every time an adjustment is altered. Also, the coefficients are not completely separated unless the vessel is on magnetic headings.

Turns to each new heading should be made slowly, swinging slightly beyond the desired heading before steadying on it. If steering is by gyro, the gyro error should be checked on each heading if time and facilities permit. The vessel should remain on each heading for at least two minutes before the deviation is determined or an adjustment made, to permit the compass card to come to rest and the magnetic condition of the vessel to become settled. If observations are made before the vessel's magnetism becomes settled, the reading will be incorrect by an amount called the **Gaussin error.**

Adjustments should be carried out in the correct order, as follows:

1. Steady on magnetic heading 090° (or 270°) and adjust the fore-and-aft permanent magnets until the compass heading coincides with the magnetic heading, thus removing all coefficient *B* on this heading. Use magnets in pairs, from the bottom up, with the trays at the lowest point of travel. When overcorrection occurs, remove the two highest magnets and raise the trays until all deviation has been removed. If two magnets overcorrect, use a single magnet. It is not necessary to determine in

advance which direction the red ends should occupy, for a mistake will be immediately apparent by an *increase* in the deviation.

2. Steady on magnetic heading 180° (or 000°) and adjust the athwartship permanent magnets until the compass heading coincides with the magnetic heading, thus removing all coefficient *C* on this heading. Use the same technique as in step 1.

3. Steady on magnetic heading 270° (090° if 270° was used in step 1) and remove *half* the deviation with the fore-and-aft magnets.

4. Steady on magnetic heading 000° (180° if 000° was used in step 2) and remove *half* the deviation with the athwartship magnets.

5. Steady on any intercardinal magnetic heading and adjust the position of the quadrantal correctors until the compass heading coincides with the magnetic heading, thus removing all coefficient *D* on this heading. Leave the quadrantal correctors at equal distances from the compass.

6. Steady on either intercardinal magnetic heading 90° from that used in step 5 and remove *half* the deviation by adjusting the positions of the quadrantal correctors, leaving them at equal distances from the compass.

7. Secure all correctors in their final positions and record their number, size, positions, and orientation, as appropriate, on the bottom of the deviation table form (if a standard form such as that shown in fig. 710 is used).

8. **Swing ship** for residual deviation. That is, determine the remaining deviation on a number of headings at approximately equal intervals. Every 15° is preferable, but if the maximum deviation is small, every 45° (cardinal and intercardinal headings) may suffice.

9. If the vessel has degaussing, energize the degaussing coils and repeat the swing.

10. Make a deviation table (art. 710) for each condition (degaussing off and on), giving values for headings at 15° intervals if the maximum deviation is large (more than about 2°), or at 45° intervals if the maximum deviation is small. Record values to the nearest half degree.

If preferred, the adjustment may be started on a north or south heading, thus reversing steps 1 and 2 and also 3 and 4.

With patience and skill, the readings can be made at exact headings. However, if some of the headings are off slightly during the swing, this need not invalidate the results. The exact headings should be recorded, and the deviation determined for these values. The results can then be plotted on cross-section paper with the deviation being one coordinate and the heading the other. The deviation at each heading to be recorded can then be read from the curve. This is good practice even when readings are made at exact headings, for if any large errors have been made, the fact will be immediately apparent. Also, such a curve may be of assistance in making an analysis. If a reason cannot be found for any marked irregularity in the curve, readings might be made again at the headings involved.

The deviation of all compasses aboard the vessel can be determined from a single swing if the heading by each compass is recorded at the moment the magnetic direction is noted. If deviation of one compass is determined by means of a magnetic bearing or azimuth (art. 733), the readings of this compass can then be used to establish the magnetic headings for determining the deviation of each other compass (art. 732).

Compass adjustment is best made when the sea is relatively smooth, so that steady headings can be steered, and heeling error is absent. The setting of the heeling magnet can be checked later, preferably at the next time that the vessel is on a north or south heading and rolling moderately.

An analysis of deviation can be made either before or after adjustment. If this reveals an excessive amount of *A* (constant) deviation, the source of the error should be found and corrected (art. 723), if mechanical or mathematical. If an appreciable amount of *E* deviation is present, *E*-links should be used and the spheres slewed. This is particularly to be anticipated for compasses which are not on the centerline.

The procedure outlined above is for initial adjustment aboard a new or radically modified vessel. Deviation on the heading being used for navigation should be checked from time to time and any important differences from the values shown on the deviation table should be investigated. At sea, it is good practice to compare the magnetic and gyrocompasses at intervals not exceeding half an hour. The error of one or both of these compasses should be checked twice a day when means are available. In pilot waters deviation checks should be made as convenient opportunities present themselves.

Whenever there is reason to question the accuracy of the deviation table, the ship should be swung at the first opportunity and a new table made up if there are significant changes in the old one. Suitable occasions for swinging ship would be after a deviation check indicates a significant error or after any event that might result in changes in the magnetic field of the vessel (art. 712). Intervals of swing should not exceed three months even when there is no reason to question the accuracy of the deviation table.

If a swing indicates the presence of large maximum deviation, the compass should be readjusted. Unless there is reason to change it, the Flinders bar length should remain the same. Other adjustments are altered as needed, none of the correctors being removed at the beginning of adjustment. Whenever the vessel crosses the magnetic equator, the opportunity should be used to check the deviation on magnetic headings east and west. Any adjustment needed should be made by means of the fore-and-aft (*B*) magnets. Upon crossing the magnetic equator, the heeling magnet should be inverted.

The Flinders bar and quadrantal correctors should be checked for permanent magnetism at intervals of about a year, or more often if such magnetism is suspected.

Finding the Deviation

730. Placing a vessel on a desired magnetic heading.—As indicated in article 729 compass adjustment is best made with the vessel on *magnetic* headings. The compass being adjusted cannot be used for placing the vessel on a desired magnetic heading because its deviation is unknown, and is subject to change during the process of adjustment. A number of methods are available, including use of (1) another magnetic compass of known deviation, (2) a gyrocompass, (3) bearing of a distant object, and (4) azimuth (art. 1430) of a celestial body.

Magnetic compass. The deviation at the desired magnetic heading is determined from the deviation table for that compass, and applied to the magnetic heading to determine the equivalent compass heading.

Example 1.—It is desired to place a vessel on magnetic heading east, using the standard compass. The deviation table for this compass is shown in figure 710. Degaussing is off.

Required.—Heading per standard compass (psc).

Solution.—From figure 710 the deviation on heading 090° magnetic with degaussing off is found to be 2°.5E. Therefore, the equivalent compass heading is 090° − 2°.5 = 087°.5.

Answer.—H_{psc} 087°.5.

Gyrocompass. The variation is applied to the desired magnetic heading, to determine the equivalent true heading. Any gyro error is then applied to determine the equivalent gyro heading. This is the method commonly used by vessels equipped with a reliable gyrocompass.

Example 2.—It is desired to place a vessel on magnetic heading north, using the gyrocompass. The variation in this area is 6°W, and the gyro error is 1°E.

Required.—Heading per gyrocompass (pgc).

Solution.—The equivalent true heading is 000° − 6° = 354°. The gyro heading is 354° − 1° = 353°.

Answer.—H_{pgc} 353°.

Bearing of distant object. If a vessel remains within a small area during compass adjustment, the bearing of a distant object is essentially constant. The required distance of the object in miles is found by multiplying the cotangent of the maximum tolerable error by the *radius* in miles of the maneuvering circle. Thus, if the maximum error that can be tolerated is 0°.5 (cotangent 114.6), and the vessel can be maneuvered within 200 yards (0.1 mile) of a fixed position such as a buoy, the object selected should be at least 114.6×0.1=11.5 miles away. The 200-yard limit is within radial lines centered at the distant object and tangent to a circle having a radius of 200 yards and its center at the center of the maneuvering area. Thus, a vessel has considerable maneuvering space along the line of sight, but very limited room across this line. However, it is not necessary that the vessel *stay* within the required area, but only that it be there when readings are made. Thus, if the center of the area is marked by a buoy, the vessel might steady on each heading while still some distance away, and note the required readings as the buoy is passed. In this way, a small radius may be practical even for a large vessel.

The object selected should be conspicuous and should have a clearly defined feature of small visible width upon which to observe bearings. The object having been selected, its true bearing from the center of the maneuvering area should be measured on the chart. To this, the variation *at the center of the maneuvering area* should be applied to determine the equivalent magnetic bearing. The desired magnetic heading should be set at the lubber's line of the pelorus, and the far vane set at the magnetic bearing of the distant object. The vessel should then be maneuvered until the object is in line with the vanes.

Example 3.—It is desired to place a vessel on magnetic heading northeast in an area where the variation is 4°E. The true bearing of a distant object is 219°.

Required.—The setting of the pelorus.

Solution.—Set 045° at the lubber's line, and set the far vane at 219° − 4° = 215°.

If preferred, 000° can be set at the lubber's line, and the far vane at the relative bearing, 170° (magnetic bearing minus desired magnetic heading). If a gyro repeater or a magnetic compass is used instead of a pelorus, the true (or magnetic) bearing should be converted to the equivalent gyro (or compass) bearing.

If the distant object selected is not charted, or the position of the vessel is not known accurately, the approximate magnetic bearing of the object can be determined by measuring its *compass* bearing on each cardinal and intercardinal compass heading, and finding the mean of these readings. The value so determined will be incorrect by the amount of any constant deviation (coefficient *A*).

Example 4.—The compass bearings of a distant object are as shown below.

Required.—The magnetic bearing of the object, assuming no constant deviation (coefficient *A*).

Solution.—

CH	*CB*
°	°
000	324.8
045	320.7
090	312.6
135	306.8
180	304.9
225	310.8
270	316.2
315	320.0
sum	2516.8
mean	314.6

Answer.—MB 314°.6.

Azimuth of celestial body. The true azimuth of the celestial body selected should be computed (art. 719, vol. II) for the time of observation. The magnetic variation should then be applied to determine the equivalent magnetic azimuth. The desired magnetic heading should then be set at the lubber's line of the pelorus, and the far vane set at the magnetic azimuth of the celestial body. The vessel should then be maneuvered until the body is in line with the vanes.

Example 5.—It is desired to place a vessel on magnetic heading west in an area where the variation is 17°W, and at a time when the computed true azimuth of the sun is 098°.

Required.—The setting of the pelorus.

Solution.—Set 270° at the lubber's line, and set the far vane at 098°+17°=115°.

If preferred, 000° can be set at the lubber's line, and the far vane at the relative azimuth (magnetic azimuth minus desired magnetic heading). If a gyro repeater or a magnetic compass is used instead of a pelorus, the true (or magnetic) azimuth should be converted to the equivalent gyro (or compass) azimuth.

731. Curve of magnetic azimuths.—During the course of compass adjustment and swinging ship, a magnetic direction is needed many times, either to place the vessel on desired magnetic headings or to determine the deviation of the compass being adjusted. If a celestial body is used to provide the magnetic reference, the azimuth is continually changing as the earth rotates on its axis. Frequent and numerous computations can be avoided by preparing, in advance, a table or **curve of magnetic azimuths.** True azimuths at frequent intervals are computed. The variation at the center of the maneuvering area is then applied to determine the equivalent magnetic azimuths. These are plotted on cross-section paper, with time as the other argument, using any convenient scale. A curve is then faired through the points.

Points at intervals of half an hour (with a minimum of three) are usually sufficient unless the body is near the celestial meridian and relatively high in the sky, when additional points are needed. If the body *crosses* the celestial meridian, the direction of curvature of the line reverses.

Unless extreme accuracy is required, the Greenwich hour angle and declination can be determined for the approximate midtime, the same value of declination used for all computations, and the Greenwich hour angle considered to increase 15° per hour.

An illustration of a curve of magnetic azimuths of the sun is shown in figure 731. This curve is for the period 0700–0900 zone time on May 31, 1975, at latitude 23°09′.5N, longitude 82°24′.1W. The variation in this area is 2°47′E. At the midtime, the meridian angle of the sun is 66°47′.2E, and the declination is 21°52′.3N. Azimuths were computed by Pub. No. 260 (art. 719, vol. II) at half-hour intervals, as follows:

Zone time	*Meridian angle*	*Declination*	*Latitude*	*Magnetic azimuth*
	° ′ (h m)	°	°	° ′
0700	81 47.1E (5 27.1E)	21.9N	23.2N	069 39
0730	74 17.1E (4 57.1E)	21.9N	23.2N	071 57
0800	66 47.2E (4 27.1E)	21.9N	23.2N	074 06
0830	59 17.2E (3 57.1E)	21.9N	23.2N	076 08
0900	51 47.2E (3 27.1E)	21.9N	23.2N	078 07

This curve was constructed on the assumption that the vessel would remain in approximately the same location during the period of adjustment and swing. If the position changes materially, this should be considered in the computation.

732. Deviation by magnetic headings.—If the vessel is placed on a magnetic heading by any of the methods of article 730, compass deviation on that heading is the difference between the magnetic heading and the compass heading. If the compass heading is less than the magnetic heading, deviation is easterly, if the compass heading is greater than the magnetic heading, deviation is westerly.

Example.—A vessel is being maneuvered to determine the deviation of the magnetic steering compass on cardinal and intercardinal headings. The gyrocompass, which has an error of 0°.5W, is used for placing the vessel on each of the magnetic headings. Variation in the area is 27°.5E.

Required.—Deviation on each magnetic heading, using the compass headings given below:

Solution.—

MH	*V*	*TH*	*GE*	*Hpgc*	*CH*	*Dev.*
°	°	°	°	°	°	°
000	27.5E	027.5	0.5W	028	000.3	0.3W
045	27.5E	072.5	0.5W	073	046.1	1.1W
090	27.5E	117.5	0.5W	118	093.6	3.6W
135	27.5E	162.5	0.5W	163	136.7	1.7W
180	27.5E	207.5	0.5W	208	179.6	0.4E
225	27.5E	252.5	0.5W	253	223.8	1.2E
270	27.5E	297.5	0.5W	298	266.5	3.5E
315	27.5E	342.5	0.5W	343	313.2	1.8E

733. Deviation by magnetic bearing or azimuth.—Deviation can be found by comparing a magnetic bearing or azimuth with one measured by compass. The magnetic direction can be obtained as explained in articles 730–731. If the compass direction is less than the magnetic direction, deviation is easterly; if the compass direction is greater than the magnetic direction, deviation is westerly. This method is used for determining deviation on a given *compass* heading. The equivalent

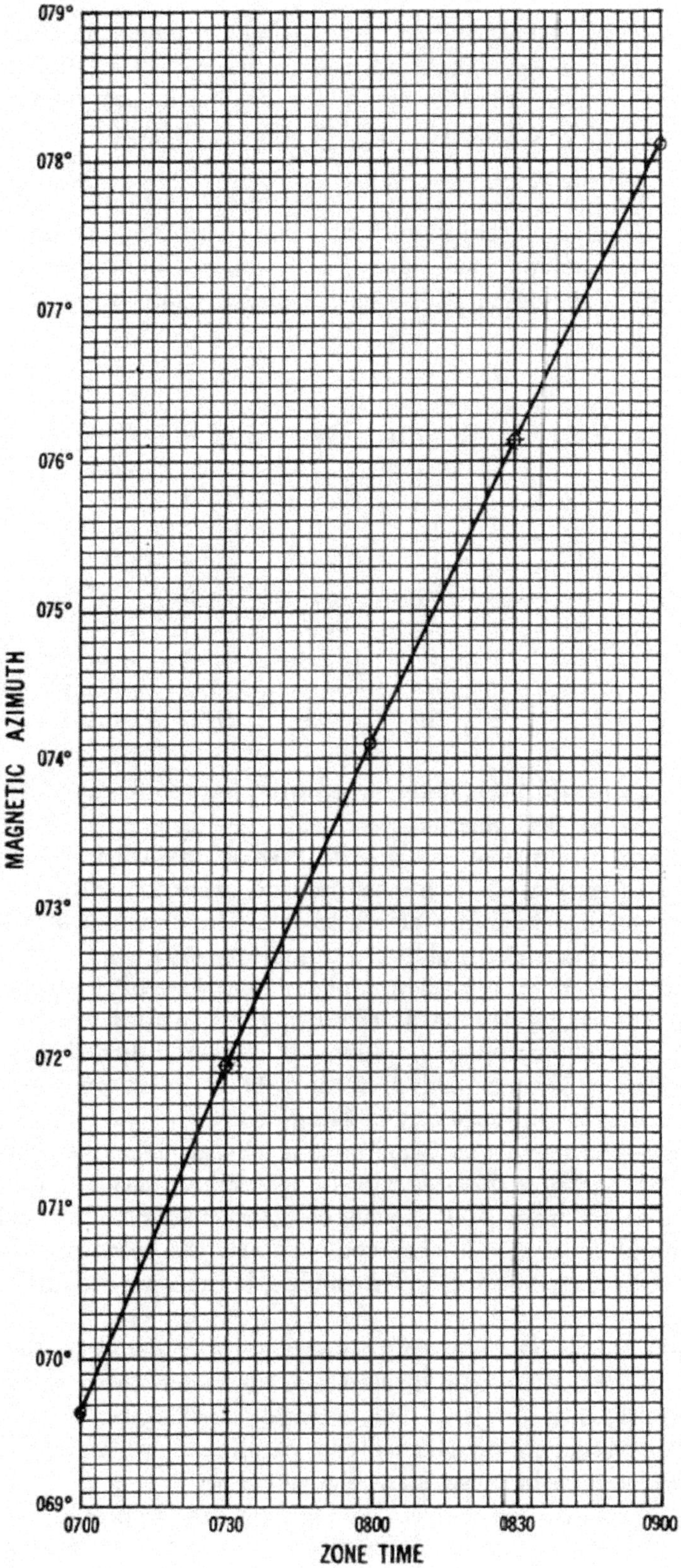

FIGURE 731.—Curve of magnetic azimuths.

magnetic heading can be determined by applying the deviation thus determined. If this method is used for swinging ship, the values can be plotted as explained in article 729. For a well-adjusted compass, the deviation may be so small that the compass headings can be considered magnetic headings, without introducing significant errors.

Example.—The standard compass of a vessel has been adjusted, and the vessel is to be swung for residual deviation during the period and for the place for which the curve of magnetic azimuths of figure 731 has been constructed.

Required.—Find the deviation on each heading given below, at the times indicated.

Solution.—

CH	*Time*	*CZn*	*MZn*	*Deviation*
°	h m s	°	°	°
000	7 35 20	073.2	072.4	0.8W
045	7 41 12	074.0	072.8	1.2W
090	7 50 15	074.2	073.4	0.8W
135	7 57 36	074.0	073.9	0.1W
180	8 04 44	073.7	074.4	0.7E
225	8 10 10	073.5	074.8	1.3E
270	8 16 33	074.3	075.2	0.9E
315	8 24 51	075.8	075.7	0.1W

The magnetic azimuth (MZn) is determined from figure 731, and the deviation from compass azimuth (CZn) and magnetic azimuth.

734. Deviation by a range is a special case of deviation by magnetic bearing. Two objects appearing in line, one behind the other, constitute a **range.** Range markers are established in many places to mark important channels, the extremities of measured miles, etc. In addition, numerous good ranges occur naturally, as when a lighthouse is in line with a tank, or a tower with a chimney. The true direction of such a range can be determined by measurement on the chart, and variation applied to determine the equivalent magnetic direction. In the case of a natural range, the objects should preferably be at least an inch apart as they appear on the chart, to minimize any plotting errors.

A range is superior to the bearing of a single object because it provides a critical indication of when the vessel is in the correct position to take a reading. The vessel crosses the range on various compass headings. At each crossing, the compass bearing of the range is observed, and also the compass heading. It is well to use two ranges nearly 90° apart, if available, because of the difficulty of crossing at small angles.

Example.—A vessel maneuvering to adjust its compass in the Lower Bay of New York Harbor finds the true direction of the range between West Bank Light and Coney Island Light to be 032°. The variation in this area is 11°.2 W. The vessel steams across the range on various compass headings, noting the compass direction of the range at the times of crossing, as shown below.

Required.—The deviation on each compass heading indicated.

Solution.—The magnetic bearing on the range is 032° + 11°.2 = 043°.2.

CH	*MB Range*	*CB Range*	*Deviation*
°	°	°	°
000	043.2	032.9	10.3E
045	043.2	023.7	19.5E
090	043.2	031.9	11.3E
135	043.2	044.2	1.0W
180	043.2	048.5	5.3W
225	043.2	051.0	7.8W
270	043.2	055.6	12.4W
315	043.2	049.8	6.6W

The analysis of these results (art. 727) indicates a constant error of 1.°0E. The mean compass bearing is 042.°2, differing from the correct magnetic bearing by the amount of constant error.

Ranges are widely used to check the deviation on the heading in use as a vessel proceeds through pilot waters. In this manner several checks can be made without advance preparation as a vessel enters or leaves port.

735. Deviation by reciprocal bearings.—Another method of using magnetic bearings is by means of a compass on the beach. This method is particularly useful when no suitable distant object or range is available, or where it may not be practical to remain close to a given bearing line.

A reliable compass is taken ashore to a location which is free from magnetic disturbance. If the location is not marked by a conspicuous object, such as a beacon, flagpole, prominent tree, etc., a temporary marker should be erected. A staff with a flag or bunting should be adequate. The marker should be of sufficient size and nature to be conspicuous at the vessel. At suitable visual or radio signals from the vessel, bearings are observed simultaneously aboard the vessel and ashore. The bearings of the vessel observed by the shore compass are magnetic. The reciprocals of these can be considered magnetic bearings of the shore station from the vessel. The bearings measured aboard the vessel are compass bearings. The difference is deviation. To avoid confusion in the sequence of bearings, the time of each bearing is recorded. Timepieces should be synchronized before the start of observations.

Example.—Simultaneous bearings are observed by a shore compass and the standard compass aboard a vessel, as shown below.

Required.—The deviation of the standard compass on each heading.

Solution.—

CH	*Time*	*MB of vessel*	*MB of shore position*	*CB of shore position*	*Deviation*
°		°	°	°	°
000	1112	307	127	137	10W
045	1120	309	129	131	2W
090	1126	312	132	130	2E
135	1018	296	116	113	3E
180	1029	295	115	109	6E
225	1039	288	108	096	12E
270	1052	288	108	113	5W
315	1104	289	109	115	6W
		mean	118	118	

The analysis of these results indicates no constant deviation. This is further indicated by the fact that the means of the bearings aboard and ashore are equal.

Adjustment by Deflector

736. Principles involved.—As indicated in article 713, the magnetic field of a vessel causes deviation of a magnetic compass, and also alters its directive force, strengthening it on some headings and weakening it on others. *The purpose of compass adjustment is to neutralize the effect of the vessel's magnetic field on the compass. If this is done completely, all deviation is removed, and the directive force is the same on all headings.* The usual procedure, described earlier in this chapter, is to adjust by reducing or eliminating the deviation. By the deflector method, the various correctors are adjusted until the directive force is the same on all cardinal headings. Deviation is then a minimum.

The *relative* directive force on various headings is determined by means of an instrument called a **deflector.** Actual measurement is of the setting of the instrument when the compass card has been rotated or "deflected" through 90° under certain standard conditions. The units are arbitrary "deflector units" which are used only for comparison with readings on other headings.

The deflector method provides a quick adjustment with only four headings being needed, without need for bearings, azimuths, or comparison with other compasses. It is easy to use. However, it is not as thorough as the method described in article 729, and should not be used when the usual method is available. The deflector method makes no provision for determination of coefficient *A* (art. 716), the amount of Flinders bar needed, the setting of the heeling magnet, or the residual deviation. Coefficient *E* can be determined, but is usually ignored. The method has never been popular in the United States. It offers little or no advantage for a vessel equipped with a reliable gyrocompass.

737. Adjustment by deflector.—The preliminary steps of adjustment are the same as indicated in article 728, omitting those relating to peloruses and other compasses. Preparations having been completed, the adjustment should be carried out as follows:

1. Steady on heading 000° (or 180°) *by the compass being adjusted.* Note the heading by another compass and keep the vessel on this heading, steering by means of the second compass. Put the deflector in place over the first compass, and deflect the compass card 90°. Record the reading on the deflector scale, and remove the deflector.

2. Steady on heading 090° (or 270°) by the compass being adjusted, and follow the procedure of step 1.

3. Steady on heading 180° (000° if 180° was used in step 1) by the compass being adjusted, and determine the deflector reading by the procedure of step 1. Leave the deflector in place and set it to the mean of the readings on headings 000° and 180°. Adjust the fore-and-aft permanent magnets until the deflection is 90°. This corrects for coefficient C and the deflector readings on compass headings 090° and 270° should now be the same. Remove the deflector.

4. Steady on heading 270° (090° if 270° was used in step 2) by the compass being adjusted, and determine the deflector reading by the procedure of step 1. Leave the deflector in place and set it to the mean of the readings on headings 090° and 270°. Adjust the athwartship permanent magnets until the deflection is 90°. This corrects for coefficient *C*, and the deflector readings on compass headings 090° and 180° should now be the same.

5. Without changing the heading, set the deflector to the mean of the N-S and E-W means. Adjust the quadrantal correctors until the deflection is 90°. This corrects for coefficient *D*, and the deflector readings on all cardinal headings should be the same. Remove the deflector.

Adjustment is now complete. It can be checked by repeating the five steps, a procedure which is particularly recommended if the difference between deflector readings on opposite headings is more than ten units. If means are available, and time permits, the vessel should be swung for residual deviation. If preferred, a heading of east or west can be used, reversing steps 1 and 2 and also steps 3 and 4.

This method is particularly useful when a quick adjustment is needed following some change that affects the magnetic environment of the compass.

738. The Kelvin deflector was developed in Great Britain by Sir William Thomson (Lord Kelvin). It consists essentially of two permanent magnets hinged like a pair of dividers, with opposite poles at the hinge. The magnets are mounted vertically over the center of the compass, with the hinged end on top. The separation of the lower ends can be varied by means of a screw. The amount of separation, indicated by a scale and vernier drums, is the reading used in the adjustment.

The deflecting force increases as the separation becomes greater. When the deflector is in place over the compass, the blue pole is in line with the north (red) end of the compass magnets, as indicated by a pointer. As the deflecting magnets are rotated around the vertical axis of the instrument, the compass card rotates in the same direction, but at a slower rate. The separation is adjusted until the rotation of the instrument is 170° when the deflection of the compass card is 90°. These are the standard conditions under which readings are made.

The Kelvin type deflector provides adjustment to an accuracy of 2° to 3°.

739. The De Colong deflector was developed in Russia and provides an accuracy of 0°5 to 1°0. Essentially, this instrument consists of two horizontal magnets which are perpendicular to each other. The small magnet is held in a fixed position close to the compass card. The large magnet is mounted in a small tray which can be moved up and down along a vertical spindle mounted over the center of the compass. The red end of this magnet is placed toward the north. When it is positioned so that the directive force is exactly neutralized, the small magnet causes the compass card to be deflected 90°. The height of the large magnet is the deflector reading, the scale being on the vertical spindle, and the index on the movable tray.

Provision is made for mounting the large magnet vertically, to measure the vertical force of the magnetic field at the compass. A separate scale is provided for this purpose. Additional magnets are generally provided for use near the magnetic equator, where the vertical intensity is very small.

In practice, a separate deflector is provided for each compass, and they are not interchangeable. By the addition of an auxiliary scale, the instrument could be made usable for any compass.

Degaussing (Magnetic Silencing) Compensation

740. Degaussing.—As indicated in article 712, a steel vessel has a certain amount of permanent magnetism in its "hard" iron, and induced magnetism in its "soft" iron. Whenever two or more magnetic fields occupy the same space, the total field is the vector sum (art. 118, vol. II) of the individual fields. Thus, within the effective region of the field of a vessel, the total field is the combined total of the earth's field and that due to the vessel. Consequently, the field due to earth's magnetism alone is altered or distorted due to the field of the vessel. This is indicated by a tendency of the lines of force to crowd into the metal of the vessel (art. 703), as shown in figure 741a.

Certain mines and other explosive devices are designed to be triggered by the magnetic influence of a vessel passing near them. It is therefore desirable to reduce to a practical minimum the magnetic field of a vessel. One method of doing this is to neutralize each component by means of an electromagnetic field produced by direct current of electricity in electric cables installed so as to form coils around the vessel. A unit sometimes used for measuring the strength of a magnetic field is the **gauss.** The reduction of the strength of a magnetic field decreases the number of gauss in that field. Hence, the process is one of **degaussing** the vessel.

When a vessel's degaussing coils are energized, the magnetic field of the vessel is completely altered. This introduces large deviation in the magnetic compasses. This is removed, as nearly as practicable, by introducing at each compass an equal and opposite force of the same type—one caused by direct current in a coil—for each component of the field due to the degaussing currents. This is called **compass compensation.** When there is a possibility of confusion with compass adjustment to neutralize the effects of the natural magnetism of the vessel, the expression **degaussing compensation** is used. Since the neutralization may not be perfect, a small amount of deviation due to degaussing may remain on certain headings. This is the reason for swinging ship twice—once with degaussing off and once with it on—and having two separate columns in the deviation table (fig. 710).

741. A vessel's magnetic signature.—A simplified diagram of the distortion of the earth's magnetic field in the vicinity of a steel vessel is shown in figure 741a. The strength of the field is indicated by the spacing of the lines, being stronger as the lines are closer together. If a vessel passes over a device for detecting and recording the strength of the magnetic field, a certain pattern is traced, as shown in figure 741b. Since the magnetic field of each vessel is different, each has a distinctive trace, known as its **magnetic signature.** The simplified signature shown in figure 741b is one that might result from an uncomplicated field such as that shown in figure 741a.

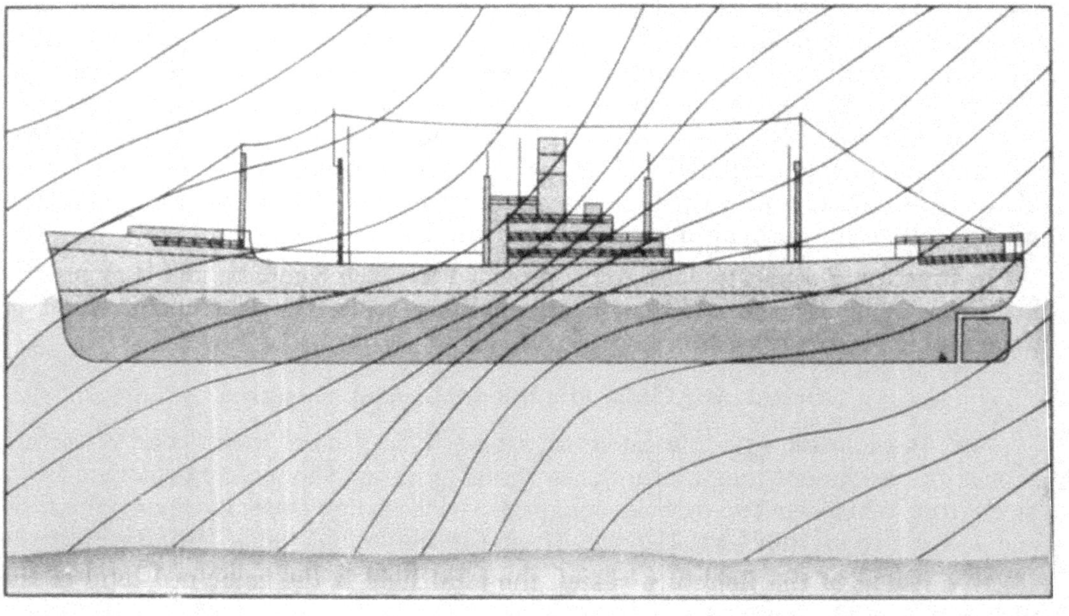

FIGURE 741a.—Simplified diagram of distortion of earth's magnetic field in the vicinity of a steel vessel.

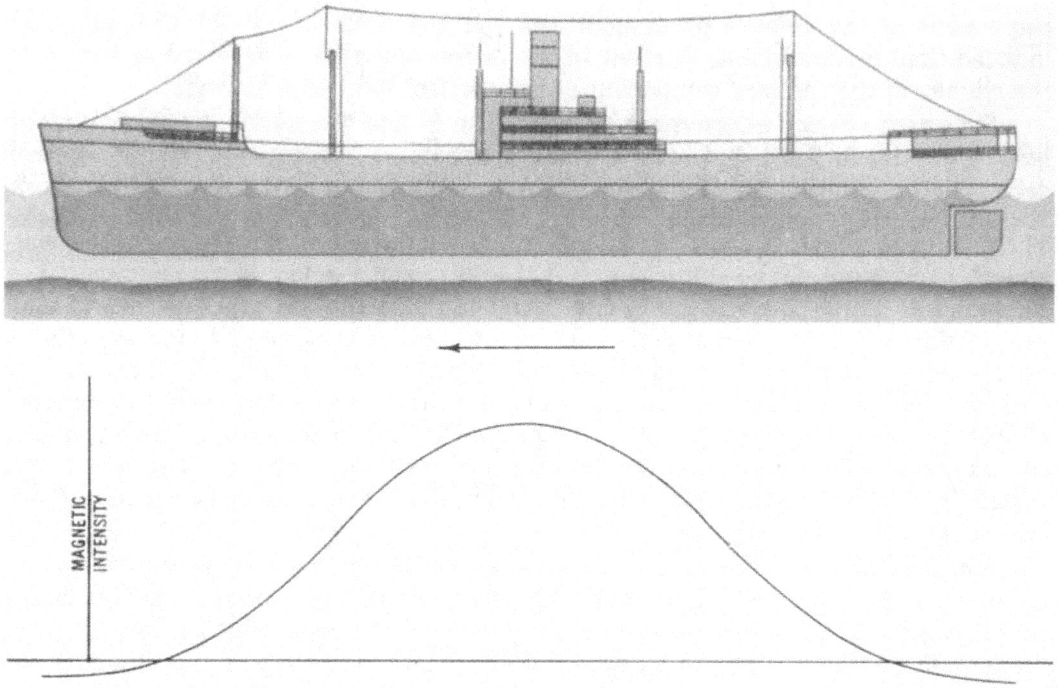

FIGURE 741b.—Simplified signature of vessel of figure 741a.

Several degaussing stations have been established to determine magnetic signatures and recommend the currents needed in the various degaussing coils. Since a vessel's induced magnetism varies with heading and magnetic latitude, the current settings of the coils which neutralize induced magnetism need to be changed to suit the conditions. A "degaussing folder" is provided each vessel to indicate the changes, and to give other pertinent information.

A vessel's permanent magnetism changes somewhat with time and the magnetic history of the vessel. Therefore, the information given in the degaussing folder should be checked from time to time by a return to the magnetic station.

742. Degaussing coils.—For degaussing purposes, the total field of the vessel is divided into three components: (1) vertical, (2) horizontal fore-and-aft, and (3) horizontal athwartships. The positive directions are considered downward, forward, and to port, respectively. These are the normal directions for a vessel headed north or east in north latitude. Each component is opposed by a separate degaussing field just strong enough to neutralize it. Ideally, when this has been done, the earth's field passes through the vessel smoothly and without distortion. The opposing degaussing fields are produced by direct current flowing in coils of wire. Each of the degaussing coils is placed so that the field it produces is directed to oppose one component of the ship's field.

The number of coils installed depends upon the magnetic characteristics of the vessel, and the degree of safety desired. The ship's permanent and induced magnetism may be neutralized separately so that control of induced magnetism can be varied as heading and latitude change, without disturbing the fields opposing the vessel's permanent field. The principal coils employed are the following:

Main (M) coil. The *M*-coil is placed horizontal, and completely encircles the vessel, usually at or near the waterline. Its function is to oppose the vertical component of the vessel's permanent and induced fields combined. Generally the induced field predominates. Current in the *M*-coil is varied or reversed according to the change of the *induced* component of the vertical field with latitude.

Forecastle (F) and quarterdeck (Q) coils. The *F*- and *Q*-coils are placed horizontal just below the forward and after thirds (or quarters), respectively, of the weather deck. The designation "*Q*" for quarterdeck is reminiscent of the days before World War II when the "quarterdeck" of naval vessels was after along the ship's quarter. These coils, in which current can be individually adjusted, remove much of the fore-and-aft component of the ship's permanent and induced fields. More commonly, the combined *F*- and *Q*-coils consist of two parts; one part the *FP*- and *QP*-coils, to take care of the permanent fore-and-aft field, and the other part, the *FI*- and *QI*-coils, to neutralize the induced fore-and-aft field. Generally, the forward and after coils of each type are connected in series, forming a split-coil installation and designated *FP-PQ* coils and *FI-QI* coils. Current in the *FP-QP* coils is generally constant, but in the *FI-QI* coils is varied according to the heading and magnetic latitude of the vessel. In split-coil installations, the coil designations are often contracted to *P*-coil and *I*-coil.

Longitudinal (L) coil. Better control of the fore-and-aft components, but at greater installation expense, is provided by placing a series of vertical, athwartships coils along the length of the ship. It is the *field,* not the coils, which is longitudinal. Current in an *L*-coil is varied as with the *FI-QI* coils. It is maximum on north and south headings, and zero on east and west headings.

Athwartship (A) coil. The *A*-coil is in a vertical fore-and-aft plane, thus producing a horizontal athwartship field which neutralizes the athwartship component of the vessel's field. In most vessels, this component of the permanent field is small and can be ignored. Since the *A*-coil neutralizes the induced field, primarily, the current is changed with magnetic latitude and with heading, being maximum on east or west headings, and zero on north or south headings.

The strength and direction of the current in each coil is indicated and adjusted at a control panel which is normally accessible to the navigator. Current may be controlled directly by rheostats at the control panel or remotely by push buttons which operate rheostats in the engine room.

Since degaussing fields oppose the vessel's fields, the positive directions of the degaussing fields are upward, aft, and to starboard. For positive fields in *M, F, FI, FP, Q, QI,* and *QP* coils, current flows forward on the starboard side of the vessel; and the north end of a small compass placed *above* any of these coils is deflected outboard. For a positive field in the *L*-coil, current flows upward on the starboard side, and the north end of a compass is deflected aft when placed *below* an upper, athwartship portion of the coil. For a positive field in the *A*-coil, current in the upper, fore-and-aft portion flows aft, and the north end of a compass is deflected to starboard when placed *below* this portion of the coil. The *FI-QI* coils are generally connected so that the field in the *FI*-coil is negative when that in the *QI*-coil is positive.

Appropriate values of the current in each coil are determined at a degaussing station, the various currents being adjusted until the vessel's signature is made as flat as possible. Recommended current values and directions for all headings and magnetic latitudes are set forth in the vessel's degaussing folder. This document is normally retained by the navigator, whose responsibility it is to see that the recommended settings are maintained whenever the degaussing system is energized.

743. Securing the degaussing system.—Unless the degaussing system is properly secured, residual magnetism may remain in the metal of the vessel. During degaussing compensation and at other times, as recommended in the degaussing folder, the "reversal" method is used. The steps in the reversal process are as follows:

1. Start with maximum degaussing current used since the system was last energized.
2. Decrease current to zero and increase it in the opposite direction to the same value as in step 1.
3. Decrease the current to zero and increase it to three-fourths maximum value in the original direction.
4. Decrease the current to zero and increase it to one-half maximum value in the opposite direction.
5. Decrease the current to zero and increase it to one-fourth maximum value in the original direction.
6. Decrease the current to zero and increase it to one-eighth maximum value in the opposite direction.
7. Decrease the current to zero and open switch.

744. Magnetic treatment of vessels.—In some instances, the degaussing can be made more effective by changing the magnetic characteristics of the vessel by a process known as **deperming.** Heavy cables are wound around the vessel in an athwartship direction, forming vertical loops around the longitudinal axis of the vessel. The loops are run beneath the keel, up the sides, and over the top of the weather deck at closely spaced equal intervals along the entire length of the vessel. Predetermined values of direct current are then passed through the coils. When the desired magnetic characteristics have been acquired, the cables are removed.

A vessel which does not have degaussing coils, or which has a degaussing system which is inoperative, can be given some temporary protection by a process known as **flashing.** A horizontal coil is placed around the outside of the vessel and energized with large predetermined values of direct current. When the vessel has acquired a vertical field of permanent magnetism of the correct magnitude and polarity to reduce to a minimum the resultant field below the vessel for the particular magnetic latitude involved, the cable is removed. This type protection is not as satisfactory as that provided by degaussing coils because it is not adjustable for various headings and magnetic latitudes, and also because the vessel's magnetism slowly readjusts itself following treatment.

During magnetic treatment it is a wise precaution to remove all magnetic compasses and Flinders bars from the vessel. Permanent adjusting magnets and quadrantal correctors are not materially affected, and need not be removed. If for any reason it is impractical to remove a compass, the cables used for magnetic treatment should be kept as far as practical from it.

745. Degaussing compensation.—The magnetic fields created by the degaussing coils would render the vessel's magnetic compasses useless unless compensated. This is accomplished by subjecting the compass to compensating fields along three mutually perpendicular axes. These fields are provided by small compensating coils adjacent to the compass. In nearly all installations, one of these coils, the heeling coil, is horizontal and on the same plane as the compass card. Current in the heeling coil is adjusted until the vertical component of the total degaussing field is neutralized. The other compensating coils provide horizontal fields perpendicular to each other. Current is varied in these coils until their resultant field is equal and opposite to the horizontal component of the degaussing field. In early installations, these horizontal fields were directed fore-and-aft and athwartships by placing the coils around the Flinders bar and the quadrantal spheres. Compactness and other advantages are gained by placing the coils on perpendicular axes extending 045°–225° and 315°–135° relative to the heading. A frequently used compensating installation, called the type "K," is shown in figure 745. It consists of a heeling coil extending completely around the top of the binnacle, four "intercardinal" coils, and three control boxes. The intercardinal coils are named for their positions relative to the compass when the vessel is on a heading of north, and also for the compass headings on which the current in the coils is adjusted to the correct amount for compensation. The NE–SW coils operate together as one set, and the NW–SE coils operate as another. One control box is provided for each set, and one for the heeling coil.

The compass compensating coils are connected to the power supply of the degaussing coils, and the currents passing through the compensating coils are adjusted by series resistances so that the compensating field is equal to the degaussing field. Thus, a change in the degaussing currents is accompanied by a proportional change in the compensating currents. Each coil has a separate winding for each degaussing circuit it compensates.

Degaussing compensation is carried out while the vessel is moored at the shipyard where the degaussing coils are installed. This is usually done by personnel of the yard, using the following procedure:

1. The compass is removed from its binnacle and a dip needle is installed in its place. The *M*-coil and heeling coil are then energized, and the current in the heeling coil is adjusted until the dip needle indicates the correct value for the magnetic latitude of the vessel. The system is then secured by the reversing process.

2. The compass is restored to its usual position in the binnacle. By means of auxiliary magnets, the compass card is deflected until the compass magnets are parallel to one of the compensating coils or set of coils used to produce a horizontal field. The compass magnets are then perpendicular to the field produced by that coil. One of the degaussing circuits producing a horizontal field, and its compensating winding, are then energized, and the current in the compensating winding is adjusted until the compass reading returns to the value it had before the degaussing circuit was energized. The system is then secured by the reversing process. The process is repeated with each additional circuit used to create a horizontal field. The auxiliary magnets are then removed.

3. The auxiliary magnets are placed so that the compass magnets are parallel to the other compensating coils or set of coils used to produce a horizontal field. The procedure of step 2 is then repeated for each circuit producing a horizontal field.

When the vessel gets under way, it proceeds to a suitable maneuvering area. The vessel is then headed so that the compass magnets are parallel first to one compensating coil or set of coils and then the other, and any needed adjustment is made in the compensating circuits to reduce the error to a minimum. The vessel is then swung for residual deviation, first with degaussing off and then with degaussing on, and the correct current settings for each heading at the magnetic latitude of the vessel. From the values thus obtained, the "DG OFF" and "DG ON" columns of the deviation table (fig. 710) are filled in. If the results indicate satisfactory compensation, a record is made of the degaussing coil settings and the resistances, voltages, and currents in the compensating coil circuits. The control boxes are then secured.

FIGURE 745.—Type "K" degaussing compensation installation.

Under normal operating conditions, the settings need not be changed unless changes are made in the degaussing system, or unless an alteration is made in the amount of Flinders bar or the setting of the quadrantal correctors. However, it is possible for a ground to occur in the coils or control box if the circuits are not adequately protected from sea water or other moisture. If this occurs, it should be reflected by a change in deviation with degaussing on, or by a decreased installation resistance. Under these conditions, compensation should be carried out again. If the compass is to be needed with degaussing on before the ship can be returned to a shipyard where the compensation can be made by experienced personnel, the compensation should be made at sea on the actual headings needed, rather than by deflection of the compass needles by magnets. More complete information related to this process is given in the degaussing folder.

If a vessel has been given magnetic treatment, its magnetic properties have been changed. This necessitates readjustment of each magnetic compass. This is best delayed for several days to permit stabilization of the magnetic characteristics of the vessel. If this cannot be delayed, the vessel should be swung again for residual deviation after a few days. Degaussing compensation should not be made until after compass adjustment has been completed.

Problems

711a. Fill in the blanks in the following:

	TC	*V*	*MC*	*D*	*CC*	*CE*
	°	°	°	°	°	°
(1)	105	15E	—	5W	—	—
(2)	—	—	—	4E	215	14E
(3)	—	12W	—	—	067	7W
(4)	156	—	166	—	160	—
(5)	222	—	216	3W	—	—
(6)	009	—	357	—	—	10E
(7)	—	2W	—	6E	015	—
(8)	—	—	210	—	214	1W

Answers.—(1) MC 090°, CC 095°, CE 10°E; (2) TC 229°, V 10°E, MC 219°; (3) TC 060°, MC 072°, D 5°E; (4) V 10°W, D 6°E, CE 4°W; (5) V 6°E, CC 219°, CE 3°E; (6) V 12°E, D 2°W, CC 359°; (7) TC 019°, MC 021°, CE 4°E; (8) TC 213°, V 3°E, D 4°W.

711b. A vessel is on course 150° by compass in an area where the variation is 19°E. The deviation is as shown in figure 710. Degaussing is on.

Required.—(1) Deviation.

(2) Compass error.

(3) Magnetic heading.

(4) True heading.

Answers.—(1) D 1°E, (2) CE 20°E, (3) MH 151°, (4) TH 170°.

711c. A vessel is on course 055° by gyro and 041° by magnetic compass. The gyro error is 1°W. The variation is 15°E.

Required.—The deviation on this heading.

Answer.—D 2°W.

711d. A vessel is on course 177° by gyro. The gyro error is 0°.5E. A beacon bears 088° by magnetic compass in an area where variation is 11°W. The deviation is as shown in figure 710, degaussing off.

Required.—The true bearing of the beacon.

Answer.—TB 076°.

721a. A magnetic compass is adjusted on the magnetic equator, without any Flinders bar being used. The residual deviation on heading 090° magnetic is 1°E. Some days later, at latitude 37°N, dip 70°, the deviation on heading 090° is 12°W.

Required.—The length and location of Flinders bar required to restore a residual deviation of 1°E (using fig. 721, A) if the magnetic properties of the vessel are unchanged.

Answer.—Fifteen inches of Flinders bar forward of the compass.

721b. The deviation of a magnetic compass of a vessel on heading 270° magnetic is 2°E near Sydney, Australia (south magnetic latitude) and 12°W near Seattle, Wash. (north magnetic latitude). Near Sydney, H=0.258 and Z=0.51. Near Seattle, H=0.188 and Z=0.53. The shielding factor is 0.9.

Required.—The length of Flinders bar to use if (1) no Flinders bar is in place during observations, (2) 12 inches of Flinders bar is in place forward of the compass during observations.

Answers.—(1) 8¼ inches (8.5 inches by computation) of Flinders bar aft of the compass, (2) nine inches (8.8 inches by computation) of Flinders bar forward of the compass.

727. A magnetic compass which has not been adjusted has deviation on cardinal and intercardinal compass headings as follows:

Compass heading	*Deviation*	*Compass heading*	*Deviation*
°	°	°	°
000	2.0E	180	6.0E
045	20.5E	225	5.5W
090	18.5E	270	22.0W
135	8.0E	315	23.5W

On heading compass north the deviation is 6°0W when the vessel heels 7° to starboard.

Required.—(1) The approximate value of each coefficient.

(2) The total deviation to be expected on compass heading 300°, with the vessel on an even keel.

(3) Heeling error on compass heading 060°, with a heel of 10°.

Answers.—(1) *A* (+)0°5, *B* (+)20°2, *C* (−)2°0, *D* (+)7°6, *E* (+)2°9, *J* (−)1°1; (2) d 26°0W; (3) HE 5°5.

730a. It is desired to place a vessel on magnetic heading west, using the magnetic steering compass. The deviation table for this compass is shown in figure 710. Degaussing is on.

Required.—Heading per steering compass (p stg c).

Answer.—$H_{p\ stg\ c}$ 272°.

730b. It is desired to place a vessel on magnetic heading south, using the gyrocompass. The variation in this area is 12°E, and the gyro error is 0°5E.

Required.—Heading per gyrocompass.

Answer.—H_{pgc} 191°5.

730c. It is desired to place a vessel on magnetic heading southeast in an area where the variation is 6°W. The true bearing for a distant object is 047°.

Required.—(1) The magnetic bearing of the object.

(2) The relative bearing of the object when the vessel is on the desired magnetic heading.

Answers.—(1) MB 053°, (2) RB 278°.

730d. The compass bearings of a distant object are as follows:

CH	*CB*	*CH*	*CB*
°	°	°	°
000	358	180	002
045	357	225	006
090	351	270	012
135	353	315	009

Required.—The magnetic bearing of the object, assuming no constant deviation (coefficient *A*).

Answer.—MB 001°.

730e. It is desired to place a vessel on magnetic heading east in an area where the variation is 13°E, and at a time when the computed true azimuth of the sun is 218°.

Required.—(1) The magnetic azimuth of the sun.

(2) The relative azimuth when the vessel is on the desired magnetic heading.

(3) The azimuth by a magnetic compass having deviation as shown in figure 710 (DG on).

(4) The azimuth by a gyrocompass having a gyro error of 1°W.

Answers.—(1) MZn 205°, (2) RZn 115°, (3) CZn 202°, (4) Zn_{pgc} 219°.

732. A vessel is being maneuvered to determine the residual deviation of a magnetic compass. The gyrocompass, which has an error of 1°E, is used for placing the vessel on the magnetic headings indicated below. Variation in the area is 7°.8W. The following readings are obtained:

MH	*CH*	*MH*	*CH*
°	°	°	°
000	000.0	180	180.1
045	044.1	225	225.8
090	088.5	270	271.4
135	134.2	315	315.9

Required.—Gyro heading and deviation on each magnetic heading.
Answers.—

MH	H_{pgc}	*Dev.*	*MH*	H_{pgc}	*Dev.*
°	°	°	°	°	°
000	351.2	0.0	180	171.2	0.1W
045	036.2	0.9E	225	216.2	0.8W
090	081.2	1.5E	270	261.2	1.4W
135	126.2	0.8E	315	306.2	0.9W

733. A vessel is being swung for residual deviation during the period and at the place for which the curve of magnetic azimuths of figure 731 has been constructed. The following readings are obtained:

CH	*Time*	*CZn*	*CH*	*Time*	*CZn*
°	h m s	°	°	h m s	°
000	7 56 13	73.7	180	8 16 36	75.2
045	8 01 22	72.9	225	8 22 19	76.8
090	8 04 55	71.9	270	8 27 12	78.7
135	8 11 01	74.0	315	8 33 27	77.2

Required.—Deviation on each compass heading.
Answers.—

CH	*Deviation*	*CH*	*Deviation*
°	°	°	°
000	0.1E	180	0.0
045	1.3E	225	1.2W
090	2.6E	270	2.8W
135	0.9E	315	0.8W

734. A vessel being swung for residual deviation crosses a range on various compass headings as indicated below, the compass bearing of the range being observed at each crossing. The true direction of the range is 255°. The variation in the vicinity is 24°.5E.

CH	*CB*	*CH*	*CB*
°	°	°	°
000	230.3	180	230.6
045	228.7	225	232.4
090	227.4	270	233.8
135	228.0	315	232.3

Required.—Deviation on each compass heading.

Answers.—

CH	*Deviation*	*CH*	*Deviation*
°	°	°	°
000	0.2E	180	0.1W
045	1.8E	225	1.9W
090	3.1E	270	3.3W
135	2.5E	315	1.8W

735. Bearings of a vessel are taken by means of a compass ashore, and simultaneous bearings of the shore position are taken from the vessel, as follows:

CH	*CB of shore position*	*MB of vessel*	*CH*	*CB of shore position*	*MB of vessel*
°	°	°	°	°	°
000	020	198	180	003	184
045	013	189	225	009	194
090	004	174	270	013	204
135	001	172	315	017	205

Required.—(1) Deviation on each heading.

(2) The value of coefficient *A*.

Answers.—

(1)

CH	*Deviation*	*CH*	*Deviation*
°	°	°	°
000	2W	180	1E
045	4W	225	5E
090	10W	270	11E
135	9W	315	8E

(2) Coefficient *A* is zero.

CHAPTER VIII

DEAD RECKONING

801. Introduction.—Dead reckoning (DR) is the determination of position by advancing a known position for courses and distances. It is reckoning relative to something stationary or "dead" in the water, and hence applies to courses and speeds *through the water.* Because of leeway due to wind, inaccurate allowance for compass error, imperfect steering, or error in measuring speed, the actual motion through the water is seldom determined with complete accuracy. In addition, if the water itself is in motion, the course and speed over the bottom differ from those through the water. It is good practice to use the true course *steered* and the best determination of *measured* speed, which is normally speed *through the water,* for dead reckoning. Hence, geographically, a **dead reckoning position** is an approximate one which is corrected from time to time as the opportunity presents itself. Although of less than the desired accuracy, dead reckoning is the only method by which a position can be determined at *any* time and therefore might be considered *basic* navigation, with all other methods only appendages to provide means for correcting the dead reckoning. The prudent navigator keeps his direction- and speed- or distance-measuring instruments in top condition and accurately calibrated, for his dead reckoning is no more accurate than his measurement of these elements.

If a navigator can accurately assess the disturbing elements introducing geographical errors into his dead reckoning, he can determine a better position than that established by dead reckoning alone. This is properly called an **estimated position (EP).** It may be established either by applying an estimated correction to a dead reckoning position, or by estimating the course and speed being made good over the bottom. The expression "dead reckoning" is sometimes applied loosely to such reckoning, but it is better practice to keep this "estimated reckoning" distinct from dead reckoning, if for no other reason than to provide a basis for evaluating the accuracy of one's estimates. When good information regarding current, wind, etc., is available, it should be used, but the practice of applying corrections based upon information of uncertain accuracy is, at best, questionable, and may introduce an error. Estimates should be based upon judgment and experience. Positional information which is incomplete or of uncertain accuracy may be available to assist in making the estimate. However, before adequate experience is gained, one should be cautious in applying corrections, for the estimates of the inexperienced are often quite inaccurate.

Dead reckoning not only provides means for continuously establishing an approximate position, but also is of assistance in determining times of sunrise and sunset, the celestial bodies available for observation, the predicted availability of electronic aids to navigation, the suitability and interpretation of soundings for checking position, the predicted times of making landfalls or sighting lights, estimates of arrival times, and in evaluating the reliability and accuracy of position-determining information. Because of the importance of accurate dead reckoning, a careful log is kept of all courses and speeds, times of all changes, and compass errors. These may be recorded directly in the log or first in a **navigator's notebook** for later recording in the log, but whatever the form, a careful record is important.

Modern navigators almost invariably keep their dead reckoning by plotting directly on the chart or plotting sheet, drawing lines to represent the direction and distance of travel and indicating dead reckoning and estimated positions from time to time. This method is simple and direct. Large errors are often apparent as inconsistencies in an otherwise regular plot. Before the advent of power vessels, when frequent course and speed changes were common, and when charts were sometimes of questionable accuracy, it was common practice to keep the dead reckoning mathematically by one, or a combination, of the "sailings" (chapter IX). Except for great-circle sailing, and occasionally composite and Mercator sailings, these are of little more than historical interest to modern navigators, other than those of small boats.

In determining distance run in a given time, one may find table 19 useful.

802. Plotting position on the chart.—A position is usually expressed in units of latitude and longitude, generally to the nearest 0ʹ.1, but it may be expressed as bearing and distance from a known position, such as a landmark or aid to navigation.

To plot a position on a Mercator chart, or to determine the coordinates of a point on such a chart, proceed as follows:

To plot a position when its latitude and longitude are known: Mark the given latitude on a convenient latitude scale along a meridian, being careful to note the unit of the smallest division on the scale. Place a straightedge at this point and parallel to a parallel of latitude (perpendicular to a meridian). Holding the straightedge in place, set one point of a pair of dividers at the given longitude on the longitude scale at the top or bottom of the chart (or along any parallel) and the other at a convenient printed meridian. Without changing the spread of the dividers, place one point on the same printed meridian at the edge of the straightedge, and the second point at the edge of the straightedge in the direction of the given longitude. This second point is at the given position. *Lightly* prick the chart. Remove first the straightedge and then the dividers, watching the point to be sure of identifying it. Make a dot at the point, enclose it with a small circle or square as appropriate (art. 805), and label it. If the dividers are set to the correct spread for longitude *before* the latitude is marked, one point of the dividers can be used to locate the latitude and place the straightedge, if one is careful not to disturb the setting of the dividers.

To determine the coordinates of a point on the chart: Place a straightedge at the given point and parallel to a parallel of latitude. Read the latitude where the straightedge crosses a latitude scale. Keeping the straightedge in place, set one leg of a pair of dividers at the given point and the other at the intersection of the straightedge and a convenient printed meridian. Without changing the spread of the dividers, place one end on a longitude scale, at the same printed meridian, and the other point on the scale, in the direction of the given point. Read the longitude at this second point.

Several variations of these procedures may suggest themselves. That method which seems most natural and is least likely to result in error should be used.

803. Measuring direction on the chart.—Since the Mercator chart, commonly used by the marine navigator, is *conformal* (art. 302), directions and angles are correctly represented. It is customary to orient the chart with 000° (north) at the top; other directions are in their correct relations to north and each other.

As an aid in measuring direction, **compass roses** are placed at convenient places on the chart or plotting sheet. A desired direction can be measured by placing a straightedge along the line from the center of a compass rose to the circular graduation representing the desired direction. The straightedge is then in the desired direction, which may be transferred to any other part of the chart by parallel motion, as by parallel rulers or two triangles (art. 603). The direction between two points is determined by transferring that direction to a compass rose. If a drafting machine (art. 606) or some form of plotter (art. 605) or protractor (art. 604) is used, measurement can be made directly at the desired point, without using the compass rose.

Measurement of direction, whether or not by compass rose, can be made at any convenient place on a Mercator chart, since meridians are parallel to each other and a line making a desired angle with any one makes the same angle with all others. Such a line is a **rhumb line,** the kind commonly used for course lines, except in polar regions. For direction on a chart having nonparallel meridians, measurement can be made at the meridian involved if the chart is conformal, or by special technique if it is not conformal. Explanation of the former is given in article 2511. The only nonconformal chart commonly used by navigators is the gnomonic, and instructions for measuring direction on this chart are usually given on the chart itself.

Compass roses for both true and magnetic directions may be given. A drafting machine can be oriented to any reference direction—true, magnetic, compass, or grid. When a plotter or protractor is used for measuring an angle with respect to a meridian, the resulting direction is true unless other than true meridians are used. For most purposes of navigation it is good practice to plot true directions only, and to label them in true coordinates.

804. Measuring distance on the chart.—The length of a line on a chart is usually measured in nautical miles, to the nearest 0.1 mile. For this purpose it is customary to use the latitude scale, considering 1 minute of latitude equal to 1 nautical mile. The error introduced by this assumption is not great over distances normally measured. It is maximum near the equator or geographical poles. Near the equator a ship traveling 180 miles by measurement on the chart would cover only 179 miles over the earth. Near the poles a run of 220 miles by chart measurement would equal 221 miles over the earth.

Since the latitude scale on a Mercator chart expands with increased latitude, measurement should be made at approximately the mid latitude. For a chart covering a relatively small area, such as a harbor chart, this precaution is not important because of the slight difference in scale over the chart. On such charts a separate mile scale may be given, and it may safely be used over the entire chart. However, habit is strong, and mistakes can probably be avoided by *always* using the mid latitude.

For long distances the line should be broken into a number of parts or *legs,* each one being measured at its mid latitude. The length of a line that should be measured in a single step varies with latitude, decreasing in higher latitudes. No realistic numerical value can be given, since there are too many considerations. With experience a navigator determines this for himself. On the larger scale charts this is not a problem because the usual dividers used for this purpose will not span an excessively long distance.

In measuring distance, the navigator spans with his dividers the length of the line to be measured and then, without altering the setting, transfers this length to the latitude scale, carefully noting the graduations so as to avoid an error in reading. This precaution is needed because of the difference from chart to chart. In measuring a desired length along a line, the navigator spans this length on the latitude scale opposite the line and then transfers his dividers to the line, without changing the setting. For a long line the navigator sets his dividers to some convenient distance and steps off the line, counting the number of steps, multiplying this by the length of the step, and adding any remainder. If the line extends over a sufficient spread of latitude to make scale difference a factor, he resets his dividers to the scale for the approximate mid latitude of each leg. The distance so measured is the length of the rhumb line.

For measuring distance on a nearly-constant-scale chart, such as the Lambert conformal, the mid-latitude precaution is usually unnecessary. Such charts generally have a mile scale independent of the latitude scale. On a gnomonic chart a special procedure is needed, and this is usually explained on the chart.

805. Plotting and labeling the course line and positions.—Course is the intended horizontal direction of travel through the water. A **course line** is a line extending in the direction of the course. From a known position of the ship the course line is drawn in the direction indicated by the course. It is good practice to label all lines and points of significance as they are drawn, for an unlabeled line or point can easily be misinterpreted later. Any simple, clear, logical, unambiguous system of labels is suitable. The following is widely used and might well be considered standard.

Label a course line with direction and speed. *Above* the course line place a capital C followed by three figures to indicate the course steered. It is customary to label and steer courses to the nearest whole degree, although they are generally computed to the nearest 0°1. The course label should indicate *true* direction, starting with 000° at true north and increasing clockwise through 360°. *Below* the course line, and under the direction label, place a capital S followed by figures representing the speed in knots. Since the course is *always* given in degrees true and the speed in knots, it is not necessary to indicate the units or the reference direction (fig. 805).

FIGURE 805.—A course line with labels.

A point to be labeled is enclosed by a small circle in the case of a **fix** (an accurate position determined without reference to any former position), a semicircle in the case of a dead reckoning position, and by a small square in the case of an estimated position. It is labeled with the time, usually to the nearest minute; the nature of the position is indicated by the symbol used. Time is usually expressed in four figures without punctuation, on a 24-hour basis (art. 1810). Zone time (art. 1814) is usually used, but Greenwich mean time (art. 1814) may be employed. A course line is a succession of an infinite number of dead reckoning positions. Only selected points are labeled.

The times of fixes and estimated positions are placed horizontally; the times of dead reckoning positions are placed at an angle to the course line.

806. Dead reckoning by plot.—As a vessel clears a harbor and proceeds out to sea, the navigator obtains one last good fix while identifiable landmarks are still available. This is called **taking departure,** and the position determined is called the **departure.** Piloting (ch. X) comes to an end and the course is set for the open sea. The course line is drawn and labeled, and some future position is indicated as a DR position. The number of points selected for labeling depends primarily upon the judgment and individual preference of the navigator. It is good practice to label each point where a change of course or speed occurs. If such changes are frequent, no additional points need be labeled. With infrequent changes, it is good practice to label points at some regular interval, as every hour. From departure, the dead reckoning plot continues unbroken until a new well-established position is obtained, when both DR and fix are shown. The fix serves as the start of a new dead reckoning plot. Although estimated positions are shown, it is generally not good practice to begin a new DR at these points.

A typical dead reckoning plot is shown in figure 806, indicating procedures both when there are numerous changes of course and speed and when there is a long continuous course. It is assumed that no fix is obtained after the initial one at 0800 on September 8. Note that course lines are not extended beyond their limits of usefulness. One should keep a neat plot and leave no doubt as to the meaning of each line and marked point. *A neat, accurate plot is the mark of a good navigator.* The plot of the **intended track** (art. 207) should be kept extended to some future time. A good navigator is always ahead of his ship. In shoal water or when near the shore, aids to navigation, dangers, etc., it is customary to keep the dead reckoning plot on a chart. A chart overprinted with a lattice of a radionavigation system may be used. But on the open sea, with only dead reckoning and celestial navigation available, it is good practice to use a plotting sheet (art. 323).

807. Current.—Water in essentially horizontal motion over the surface of the earth is called **current.** The direction in which the water is moving is called the **set,** and the speed is called the **drift.** In navigation it is customary to use the term "current" to include all factors introducing geographical error in the dead reckoning, whether their immediate effects are on the vessel or the water. When a fix is obtained, one assumes that the current has set *from* the DR position at the same time *to* the fix, and that the drift is equal to the distance in miles between these positions, divided by the number of hours since the last fix. This is true regardless of the number of changes of course or speed since the last fix.

If set and drift since the last fix are known, or can be estimated, a better position can be obtained by applying a correction to that obtained by dead reckoning. This is conveniently done by drawing a straight line in the direction of the set for a distance equal to the drift multiplied by the number of hours since the last fix, as shown in figure 805. The direction of a straight line from the last fix to the EP is the estimated **track made good,** and the length of this line divided by the time is the estimated **speed made good.** The course and speed actually made good over the ground are called the **course over the ground (COG)** and **speed over the ground (SOG),** respectively.

As shown in figure 805, the straight line drawn from the 0900 DR in the direction of the set is constructed as a broken line. The capital S above the line represents the set; the capital D below the line represents the drift.

If a current is setting in the same direction as the course, or its reciprocal, the course over the ground is the same as that through the water. The effect on the speed can be found by simple arithmetic. If the course and set are in the same direction, the speeds are added; if in opposite directions, the smaller is subtracted from the larger. This situation is not unusual when a ship encounters a tidal

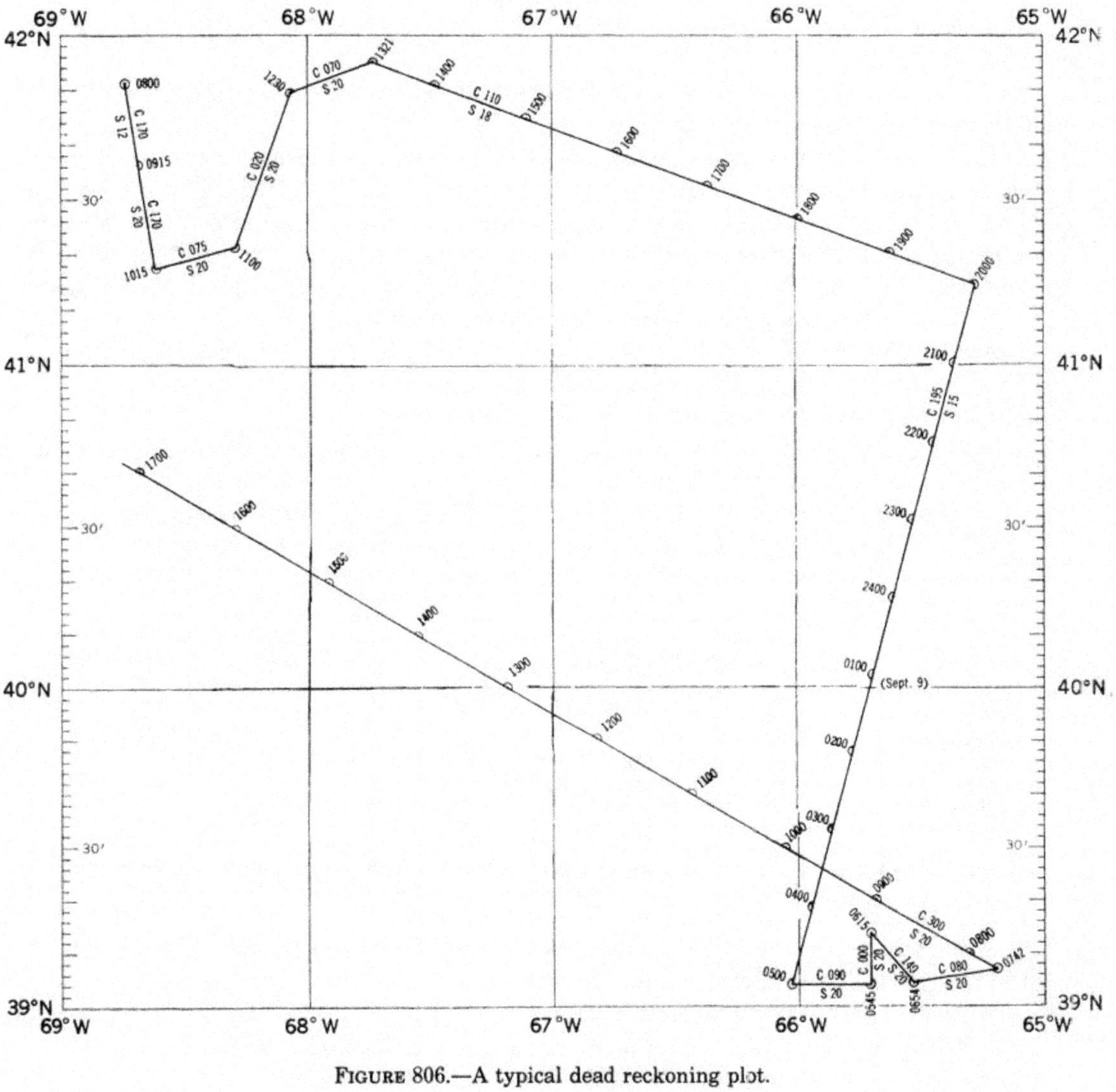

FIGURE 806.—A typical dead reckoning plot.

current while entering or leaving port. If a ship is *crossing* a current, solution can be made graphically by vector diagram (art. 118, vol. II) since velocity over the ground is the vector sum of velocity *through* the water and velocity *of* the water. Although *distances* can be used, it is generally easier to use *speeds.*

Example 1.—A ship on course 080°, speed 10 knots, is steaming through a current having an estimated set of 140° and drift of 2 knots.

Required.—Estimated track and speed made good.

Solution (fig. 807a).—(1) From *A,* any convenient point, draw *AB,* the course and speed of the ship, in direction 080°, for a distance of 10 miles.

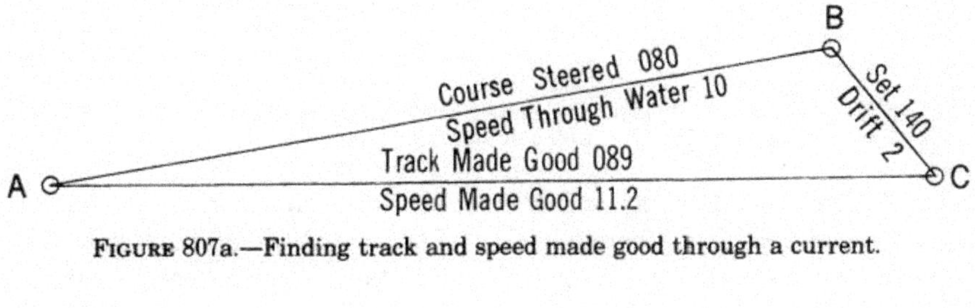

FIGURE 807a.—Finding track and speed made good through a current.

(2) From B draw BC, the set and drift of the current, in direction 140°, for a distance of 2 miles.

(3) The direction and length of AC are the estimated track and speed made good. Determine these by measurement.

Answers.—Estimated track made good 089°, estimated speed made good 11.2 kn.

If it is required to find the course to steer at a given speed to make good a desired course, plot the current vector from the origin, A, instead of from B.

Example 2.—The captain desires to make good a course of 095° through a current having a set of 170° and a drift of 2.5 knots, using a speed of 12 knots.

Required.—The course to steer and the speed made good.

Solution (fig. 807b).—(1) From A, any convenient point, draw line AB extending in the direction of the course to be made good, 095°.

(2) From A draw AC, the set and drift of the current.

(3) Using C as a center, swing an arc of radius CD, the speed through the water (12 knots), intersecting line AB at D.

(4) Measure the direction of line CD, 083°5. This is the course to steer.

(5) Measure the length AD, 12.4 knots. This is the speed made good.

Answers.—Course to steer 083°5, speed made good 12.4 kn.

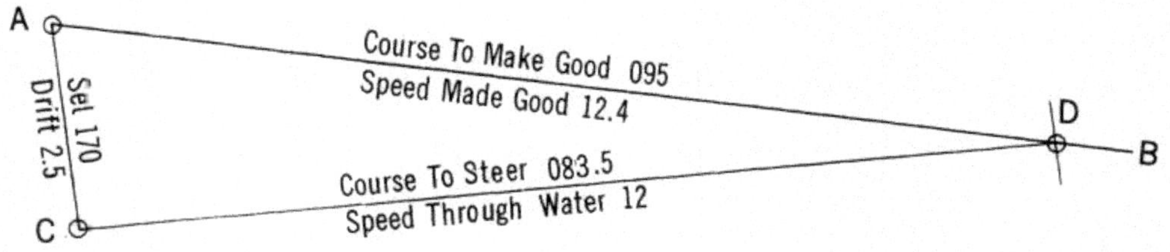

FIGURE 807b.—Finding the course to steer at a given speed to make good a given course through a current.

If it is required to find the course to steer and the speed to use to make good a desired course and speed, proceed as follows:

Example 3.—The captain desires to make good a course of 265° and a speed of 15 knots through a current having a set of 185° and a drift of 3 knots.

Required.—The course to steer and the speed to use.

Solution (fig. 807c).—(1) From A, any convenient point, draw AB in the direction of the course to be made good, 265°, and for length equal to the speed to be made good, 15 knots.

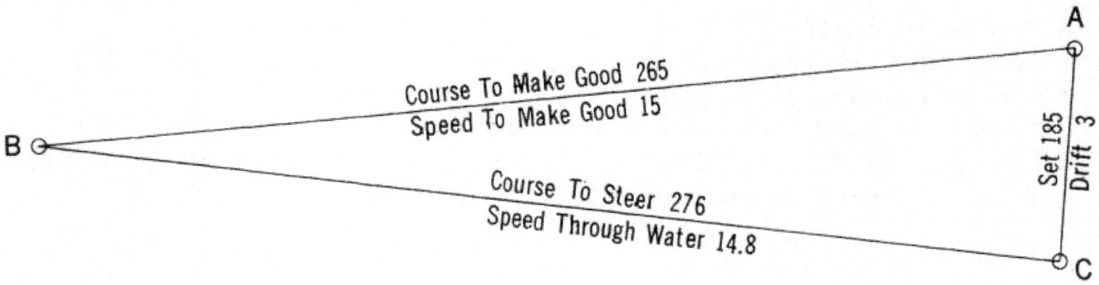

FIGURE 807c.—Finding the course to steer and the speed to use to make good a given course and speed through a current.

(2) From A draw AC, the set and drift of the current.

(3) Draw a straight line from C to B. The direction of this line, 276°, is the required course to steer; and the length, 14.8 knots, is the required speed.

Answers.—Course to steer 276°, speed to use 14.8 kn.

Such vector solutions can be made to any convenient scale and at any convenient place, such as the center of a compass rose, an unused corner of the plotting sheet, a separate sheet, or directly on the plot.

808. Leeway is the leeward motion of a vessel due to wind. It may be expressed as distance, speed, or angular difference between course steered and course through the water. However expressed, its amount varies with the speed and relative direction of the wind, type of vessel, amount of freeboard, trim, speed of the vessel, state of the sea, and depth of water. If information on the amount of leeway to be expected under various conditions is not available for the type vessel involved, it should be determined by observation. When sufficient data have been collected, suitable tables or graphs can be made for quick and convenient estimate. The accuracy of the information should be checked whenever convenient, and corrections made when sufficient evidence indicates the need.

Leeway is most conveniently applied by adding its effect to that of current and other elements introducing geographical error in the dead reckoning. It is customary to consider the combined effect of all such elements as current, and to make allowance for this as explained in article 807. In sailing ship days it was common practice to consider leeway in terms of its effect upon the course only, and to apply it as a correction in the same manner that variation and deviation are applied. While this method has merit even with power vessels, it is generally considered inferior to that of considering leeway as part of current.

809. Automatic dead reckoning.—Several types of devices are in use for performing automatically all or part of the dead reckoning. Perhaps the simplest is the automatic **course recorder**, which provides a graphical record of the various courses steered. In its usual form this device is controlled by the gyrocompass, and so indicates gyro courses.

Dead reckoning equipment receives inputs from the compass, usually the gyrocompass, and a mechanical log or engine revolution counter. It determines *change* in latitude and longitude, the latter by first determining departure and then mechanically multiplying this by the secant of the latitude. The device is provided with counters on which latitude and longitude can be set. As the vessel proceeds, the changes are then mechanically added to or subtracted from these readings to provide a continuous, instantaneous indication of the dead reckoning position. The navigator or an assistant reads these dials at intervals, usually each hour, and records the values in a notebook. Most models of dead reckoning equipment are provided, also, with a tracer for keeping a graphical record of dead reckoning in the form of a plot by moving a pencil or pen across a chart or plotting sheet. This part of the device is called a **dead reckoning tracer.** Whatever the form, dead reckoning equipment is a great convenience, particularly when a ship is maneuvering. However, such mechanical equipment is subject to possible failure. The prudent navigator keeps a hand plot and uses the dead reckoning equipment as a check. In navigation it is never wise to rely upon a single method if a second method is available as a check.

If it were possible to measure, with high accuracy, the direction and distance traveled *with respect to the earth,* an accurate geographical position could be known at all times. The two methods most commonly used are (1) doppler and (2) inertial. By the **doppler** method one or more beams of acoustic energy are directed downward at an angle. The return echo from the bottom is of a slightly different frequency due to the motion of the craft. The amount of the change, or doppler, is proportional to the speed. By proper selection of beams, it is possible to measure speed in a lateral direction as well as in a forward direction. Distance can be determined by mechanical or electronic integration of these measurements, and this can be converted into position. By the **inertial** method, accelerometers measure the acceleration in various directions, and by double integration this is converted to distance, from which position can be determined. Either of these methods can provide considerable accuracy over a period of several hours, but the error increases with time.

Problems

806a. Draw a small area plotting sheet by either method explained in article 324, covering the area between latitude 32°-34°N and longitude 118°-122°W. Plot the following points:

A	L	33°49'.1N	*C*	L	33°38'.0N
	λ	120°52'.0W		λ	118°38'.6W
B	L	32°17'.4N	*D*	L	32°30'.6N
	λ	121°28'.0W		λ	118°36'.2W

Required.—(1) The bearings of *B, C,* and *D* from *A*.

(2) The course and distance of *A, B,* and *C* from *D*.

Answers.—(1) B_{AB} 198°.5, B_{AC} 095°.5, B_{AD} 124°; (2) C_{DA} 304°, D_{DA} 138.8 mi., C_{DB} 264°.5, D_{DB} 145.7 mi., C_{DC} 358°.5, D_{DC} 67.2 mi.

806b. Use the plot of problem 806a. A ship starts from *A* at 1200, and steams as follows:

Time	*Course*	*Speed*
1200	120°	15 kn.
1330	240°	15 kn.
1500	240°	17 kn.
1800	125°	20 kn.
2000	090°	20 kn.
2300	015°	10 kn.
0500		

Plot and label the dead reckoning course line and DR positions.

Required.—(1) The dead reckoning position of the ship at 0500.

(2) The bearing and distance of *D* from the 2300 DR position.

(3) The course and distance from the 0500 DR position to *C*.

(4) Estimated time of arrival (ETA), to the nearest minute, at *C* if the ship proceeds directly from the 0500 DR position at 20 knots.

Answers.—(1) 0500 DR: L 33°35'.1N, λ 119°35'.8W; (2) B 096°, D 66.0 mi.; (3) C 086°, D 48.1 mi.; (4) ETA 0724.

807a. A ship on course 120°, speed 12 knots, is steaming through a current having a set of 350° and a drift of 1.5 knots.

Required.—Track and speed made good.

Answers.—Track made good 114°, speed made good 11.1 kn.

807b. The captain desires to make good a course of 180° through a current having a set of 090° and a drift of 2 knots, using a speed of 11 knots.

Required.—The course to steer and the speed made good.

Answers.—Course to steer 190°.5, speed made good 10.8 kn.

807c. The captain desires to make good a course of 325° and a speed of 20 knots through a current having a set of 270° and a drift of 1 knot.

Required.—The course to steer and the speed to use.

Answers.—Course to steer 327°, speed to use 19.4 kn.

CHAPTER IX

THE SAILINGS

901. Introduction.—Dead reckoning involves the determination of position by means of course and distance from a known position. A closely related problem is that of finding the course and distance from one point to another. Although both of these problems are customarily solved by plotting directly on the chart, it occasionally becomes desirable to solve by computation, frequently by logarithms or traverse table (art. 1002, vol. II). The various methods of solution are collectively called the **sailings.**

The various kinds of sailings are:

1. **Plane sailing** is a method of solving the various problems involving a single course and distance, difference of latitude, and departure, in which the earth, or that part traversed, is regarded as a plane surface. Hence, the method provides solution for latitude of the point of arrival, but not for longitude of this point, one of the spherical sailings being needed for this problem. Because of the basic assumption that the earth is flat, this method should not be used for distances of more than a few hundred miles.

2. **Traverse sailing** combines the plane sailing solutions when there are two or more courses. This sailing is a method of determining the equivalent course and distance made good by a vessel steaming along a series of rhumb lines.

3. **Parallel sailing** is the interconversion of departure and difference of longitude when a vessel is proceeding due east or due west. This was a common occurrence when the sailings were first employed several hundred years ago, but only an incidental situation now.

4. **Middle-** (or **mid-**) **latitude sailing** involves the use of the mid or mean latitude for converting departure to difference of longitude when the course is not due east or due west and it is assumed such course is steered at the mid latitude.

5. **Mercator sailing** provides a mathematical solution of the plot as made on a Mercator chart. It is similar to plane sailing, but uses meridional difference and difference of longitude in place of difference of latitude and departure, respectively.

6. **Great-circle sailing** involves the solution of courses, distances, and points along a great circle between two points, the earth being regarded as a sphere.

7. **Composite sailing** is a modification of great-circle sailing to limit the maximum latitude.

The solutions of the sailings by computations are discussed in more detail in chapter X of volume II.

902. Rhumb lines and great circles.—The principal advantage of a **rhumb line** is that it maintains constant true direction. A ship following the rhumb line between two places does not change true course. A rhumb line makes the same angle with all meridians it crosses and appears as a straight line on a Mercator chart. It is adequate for most purposes of navigation, bearing lines (except long ones, as those obtained by radio) and course lines both being plotted on a Mercator chart as rhumb lines, except in high latitudes. The equator and the meridians are great circles, but may be considered special cases of the rhumb line. For any other case, the difference between the rhumb line and the great circle connecting two points increases (1) as the latitude increases, (2) as the difference of latitude between the two points decreases, and (3) as the difference of longitude increases. It becomes very great for two places widely separated on the same parallel of latitude far from the equator.

A **great circle** is the intersection of the surface of a sphere and a plane through the center of the sphere. It is the largest circle that can be drawn on the surface of the sphere, and is the shortest distance, along the surface, between any two points on the sphere. Any two points are connected by only one great circle unless the points are antipodal (180° apart on the earth), and then an infinite number of great circles passes through them. Thus, two points on the same meridian are not joined by any great circle other than the meridian, unless the two points are antipodal. If they are the poles, *all* meridians pass through them. Every great circle bisects every other great circle. Thus, except for the equator, every great circle lies half in the Northern Hemisphere and half in the Southern Hemisphere. Any two points 180° apart on a great circle have the same latitude numerically, but contrary names, and are 180° apart in longitude. The point of greatest latitude is called the **vertex.** For each great circle there is one of these in each hemisphere, 180° apart. At these points the great circle is tangent to a parallel of latitude, and hence its direction is due east-west. On each side of these vertices the direction changes progressively until the intersection with the equator is reached, 90° away, where the great circle crosses the equator at an angle equal to the latitude of the vertex. As the great circle crosses the equator, its change in direction reverses, again approaching east-west, which it reaches at the next vertex.

On a Mercator chart a great circle appears as a sine curve extending equal distances each side of the equator. The rhumb line connecting any two points of the great circle on the same side of the equator is a chord of the curve, being a straight line nearer the equator than the great circle. Along any intersecting meridian the great circle crosses at a higher latitude than the rhumb line. If the two points are on opposite sides of the equator, the direction of curvature of the great circle relative to the rhumb line changes at the equator. The rhumb line and great circle may intersect each other, and if the points are equal distances on each side of the equator, the intersection takes place at the equator.

903. Great-circle sailing is used when it is desired to take advantage of the shorter distance along the great circle between two points, rather than to follow the longer rhumb line. The arc of the great circle between the points is called the **great-circle track.** If it could be followed exactly, the destination would be dead ahead throughout the voyage (assuming course and heading were the same). The rhumb line *appears* the more direct route on a Mercator chart because of chart distortion. The great circle crosses meridians at higher latitudes, where the distance between them is less.

The decision as to whether or not to use great-circle sailing depends upon the conditions. The saving in distance should be worth the additional effort, and of course the great circle should not cross land, or carry the vessel into dangerous waters or excessively high latitudes. A slight departure from the great circle or a modification called composite sailing (art. 901) may effect a considerable saving over the rhumb line track without leading the vessel into danger. If a fix indicates the vessel is a considerable distance to one side of the great circle, the more desirable practice often is to determine a new great-circle track, rather than to return to the original one.

Since a great circle is continuously changing direction as one proceeds along it, no attempt is customarily made to follow it exactly, except in polar regions (ch. XXV). Rather, a number of points are selected along the great circle, and rhumb lines are followed from point to point, taking advantage of the fact that for short distances a great circle and a rhumb line almost coincide.

The number of points to use is a matter of personal preference, a large number of points providing closer approximation to the great circle but requiring more frequent change of course. As a general rule, each 5° of longitude is a convenient length. Legs of equal length are not provided in this way, but this is not objectionable under normal conditions.

If a magnetic compass is used, the variation for the middle of the leg is usually used for the entire leg.

The problems of great-circle sailing can be solved by (1) chart (art. 904), (2) computation (art. 1016, vol. II), (3) table (art. 905), (4) graphically, or (5) mechanically. Of these, (4) and (5) are but graphical or mechanical solutions of (2). They usually provide solution only for initial course and the distance, and are not in common use.

904. Great-circle sailing by chart.—Problems of great-circle sailing, like those of rhumb line sailing, are most easily solved by plotting directly on a chart. For this purpose the Defense Mapping Agency Hydrographic/Topographic Center publishes a number of charts on the gnomonic projection (art. 317), covering the principal navigable waters of the world. On this projection any straight line is a great circle, but since the chart is not conformal (art. 302), directions and distances cannot be measured directly, as on a Mercator chart. An indirect method is explained on each chart.

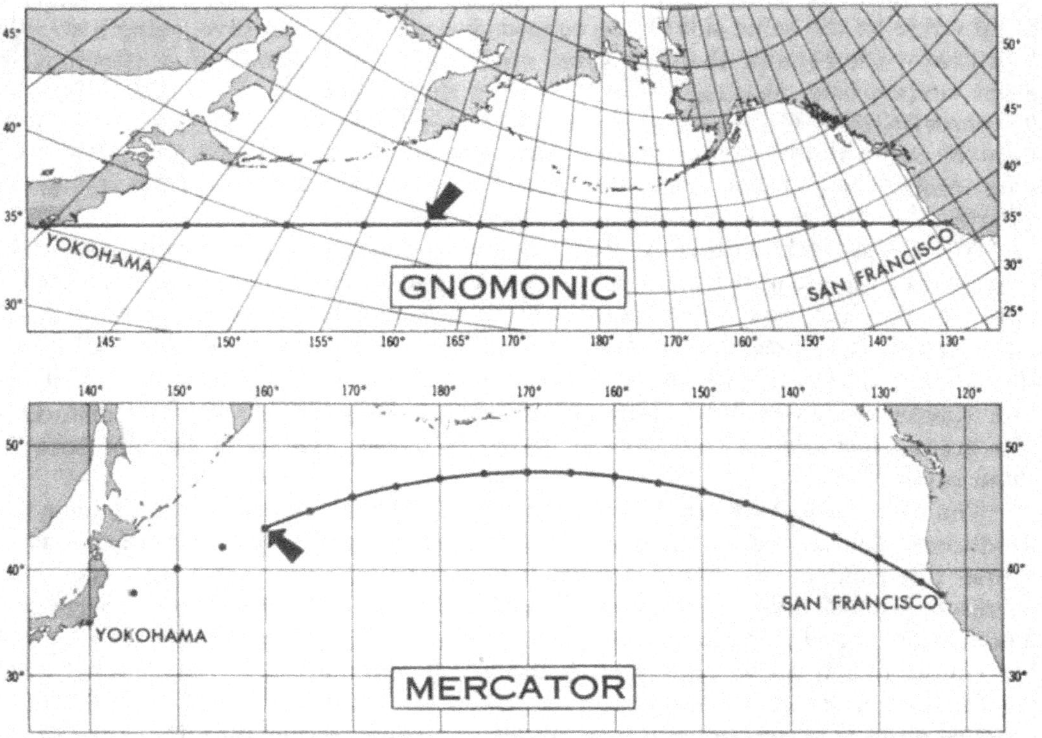

FIGURE 904.—Transferring great-circle points from a gnomonic chart to a Mercator chart.

The usual method of using a gnomonic chart is to plot the great circle and, if it provides a satisfactory track, to determine a number of points along the track, using the latitude and longitude scales in the immediate vicinity of each point. These points are then transferred to a Mercator chart or plotting sheet and used as a succession of destinations to be reached by rhumb lines. The course and distance for each leg is determined by measurement on the Mercator chart or plotting sheet. This method is illustrated in figure 904, which shows a great circle plotted as a straight line on a gnomonic chart and a series of points transferred to a Mercator chart. The arrows represent corresponding points on the two charts. The points can be plotted directly on plotting sheets without the use of a small-scale chart, but the use of the chart provides a visual check to avoid large errors, and a visual indication of the suitability of the track.

Since gnomonic charts are normally used only because of their great-circle properties, they are often popularly called **great-circle charts.**

A projection on which a straight line is *approximately* a great circle can be used in place of a gnomonic chart with negligible error. If such a projection is conformal, as in the case of the Lambert conformal (art. 314), measurement of course and distance of each leg can be made directly on the chart, as explained in article 2511.

Some great circles are shown on pilot charts and certain other charts, together with the great-circle distances. Where tracks are recommended on charts or in sailing directions, it is good practice to follow such recommendations.

905. Great-circle sailing by table.—Any method of solving the astronomical triangle of celestial navigation can be used for solving great-circle sailing problems. When such an adaptation is made, the point of departure replaces the assumed position of the observer, the destination replaces the geographical position of the body, difference of longitude replaces meridian angle or local hour angle, initial course angle replaces azimuth angle, and great-circle distance replaces zenith distance (90°—altitude), as shown in figure 905. Therefore, any table of azimuths (if the entering values are meridian angle, declination, and latitude) can be used for determining initial great-circle course. Pubs. Nos. 214, 229, 249, 260, and 261 are examples of tables that can be used for this purpose. Tables which provide solution for altitude, such as Pubs. Nos. 214, 229, and 249, can be used for determining great-circle distance. The required distance is $90° - $altitude ($90° + $negative altitudes).

In inspection tables such as Pubs. Nos. 214, 229, 249, 260, and 261, the given combination of L_1, L_2, and DLo may not be tabulated. In this case reverse the name of L_2 and use $180° - $DLo for entering the table. The required course angle is then 180° minus the tabulated azimuth, and distance is 90° plus the altitude. If neither combination can be found, solution cannot be made by that method. By interchanging L_1 and L_2, one can find the supplement of the final course angle.

Solution by table often provides a rapid approximate check, but accurate results usually require triple interpolation (art. 204, vol. II). Except for Pub. No. 229, inspection tables do not provide a solution for points along the great circle. Pub. No. 229 provides solutions for these points only if interpolation is not required.

906. Great-circle sailing by Pub. No. 229.—By entering Pub. No. 229 with the latitude of the point of departure as latitude, latitude of destination as declination, and difference of longitude as LHA, the tabular altitude and azimuth angle may be extracted and converted to great-circle distance and course. As in sight reduction, the tables are entered in accordance to whether the name of the latitude of the point of departure is the same as or contrary to the name of the latitude of the destination (declination). If after so entering the tables, the respondent values

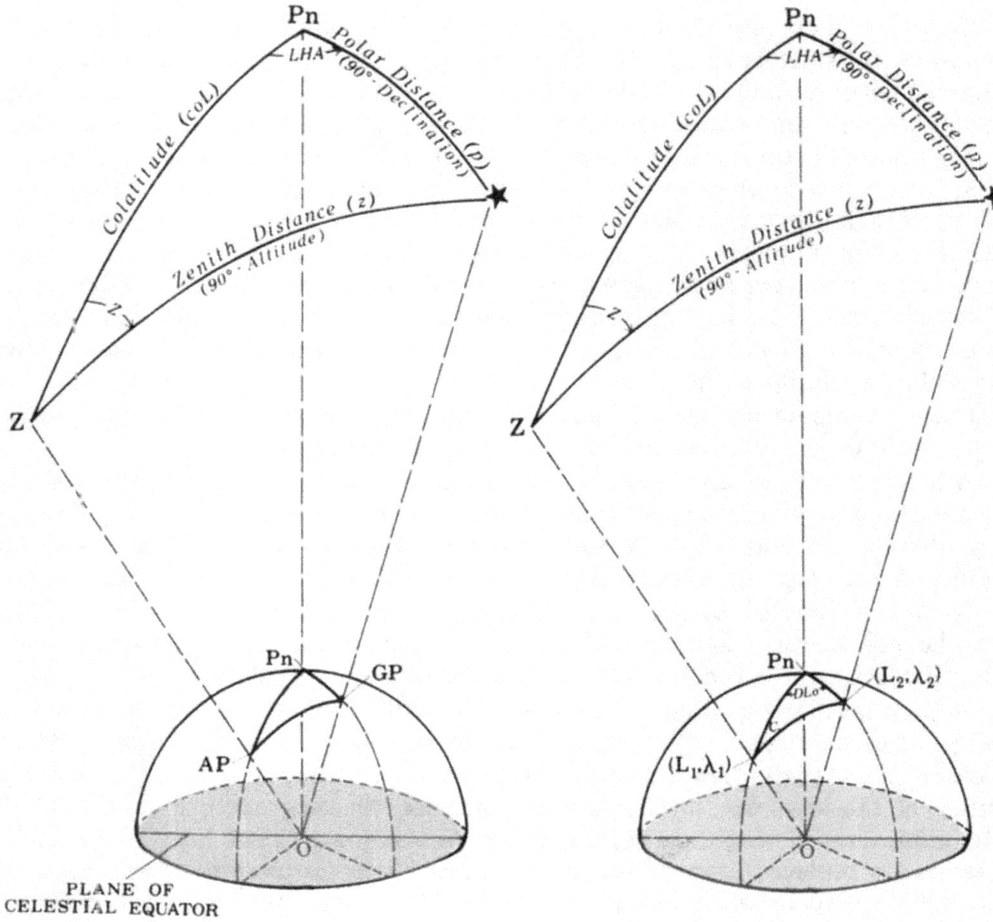

FIGURE 905.—Adapting the astronomical triangle to the navigational triangle of great-circle sailing.

correspond to those of a celestial body *above* the celestial horizon, 90° minus the arc of the tabular altitude becomes the distance; the tabular azimuth angle becomes the initial great-circle course angle. If the respondents correspond to those of a celestial body *below* the celestial horizon, the arc of the tabular altitude plus 90° becomes the distance; the supplement of the tabular azimuth angle becomes the initial great-circle course angle.

When the C–S Line is crossed in either direction, the altitude becomes negative; the body lies below the celestial horizon. For example: If the tables are entered with the LHA (DLo) at the bottom of a right-hand page and declination (L_2) such that the respondents lie above the C–S Line, the C–S Line has been crossed. Then the distance is 90° plus the tabular altitude; the initial course angle is the supplement of the tabular azimuth angle. Similarly, if the tables are entered with the LHA (DLo) at the top of a right-hand page and the respondents are found below the C–S Line, the distance is 90° plus the tabular altitude; the initial course angle is the supplement of the tabular azimuth angle. If the tables are entered with the LHA (DLo) at the bottom of a right-hand page and the name of L_2 is contrary to L_1, the respondents are found in the column for L_1 on the facing page. In which case, the C–S Line has been crossed; the distance is 90° plus the tabular altitude; the initial course angle is the supplement of the tabular azimuth angle.

Inspection of figures 2007a and 2007b reveals that the data in a latitude column are continuous with the data in the column for the same latitude on the facing page.

The tabular azimuth angle, or its supplement, is prefixed N or S for the latitude of the point of departure and suffixed E or W depending upon the destination being east or west of the point of departure.

If all entering arguments are integral degrees, the distance and course angle are obtained directly from the tables without interpolation. If the latitude of the destination is nonintegral, interpolation for the additional minutes of latitude is done as in correcting altitude for any declination increment; if the latitude of departure or difference of longitude is nonintegral, the additional interpolation is done graphically.

Since the latitude of destination becomes the declination entry, and all declinations appear on every page, the great-circle solution can always be extracted from the volume which covers the latitude of the point of departure.

Example 1.—By Pub. No. 229 find the distance and initial great-circle course from lat. 32°S, long. 116°E to lat. 30°S, long. 31°E.

Solution.—(1) Refer to figure 905. The point of departure (lat. 32°S, long. 116°E) replaces the AP of the observer; the destination (lat. 30°S, long. 31°E) replaces the GP of the celestial body; the difference of longitude (DLo 85°) replaces local hour angle (LHA) of the body.

(2) The solution by Pub. No. 229 is effected by entering volume 3 with lat. 32° (Same Name), LHA 85°, and declination 30°. The respondents as so found correspond to those of a celestial body *above* the celestial horizon. Therefore, 90° minus the tabular altitude (90° − 19°12′.4 = 70°47′.6) becomes the distance; the tabular azimuth angle (S66°.0W) becomes the initial great-circle course angle, prefixed S for the latitude of the point of departure and suffixed W due to the destination being west of the point of departure.

Answers.—(1) D 4248 nautical miles
C S66°.0W
(2) Cn 246°.0.

Example 2.—By Pub. No. 229 find the distance and initial great-circle course from lat. 38°N, long. 122°W to lat. 24°S, long. 151°E.

Solution.—(1) Refer to figure 905. The point of departure (lat. 38°N, long. 122°W) replaces the AP of the observer; the destination (lat. 24°S, long. 151°E) replaces the GP of the celestial body; the difference of longitude (DLo 87°) replaces local hour angle (LHA) of the body.

(2) The solution by Pub. No. 229 is effected by entering volume 3 with lat. 38° (Contrary Name), LHA 87°, and declination 24°. The respondents as so found correspond to those of a celestial body *below* the celestial horizon. Therefore, the tabular altitude plus 90° (12°17°.0 + 90° = 102°17′.0) becomes the distance; the *supplement* of tabular azimuth angle (180° − 69°.0 = 111°.0) becomes the initial great-circle course angle, prefixed N for the latitude of the point of departure and suffixed W due to the destination being west of the point of departure.

That the tabular data corresponds to a celestial body below the celestial horizon is indicated by the fact that the data is extracted from those tabulations across the C–S Line from the entering argument (LHA 87°).

Answers.—(1) D 6137 nautical miles
C N111°.0W
(2) Cn 249°.

Example 3.—By Pub. No. 229 find the distance and initial great-circle course from Fremantle (32°03′S, 115°45′E) to Durban (29°52′S, 31°04′E).

Solution.—(1) Refer to figure 905. Since the latitude of the point of departure, the latitude of the destination, and the difference of longitude (DLo) between the point of departure and destination are not integral degrees, the solution is effected from an adjusted point of departure or assumed position of departure chosen as follows: the latitude of the assumed position (AP) is the integral degrees of latitude nearest to the point of departure; the longitude of the AP is chosen to provide integral degrees of DLo. This AP, which should be within 30′ of the longitude of the point of departure, is at latitude 32°S, longitude 116°04′E. The DLo is 85°.

(2) Enter the tables with 32° as the latitude argument (Same Name), 85° as the LHA argument, and 29° as the declination argument.

(3) From the tables extract the tabular altitude, altitude difference, and azimuth angle; interpolate altitude and azimuth angle for declination increment. The Dec. Inc. is the minutes that the latitude of the destination is in excess of the integral degrees used as the declination argument.

	ht (Tab. Hc)	d	Z
LHA 85°, Lat. 32° (Same), Dec. 29°	18°45′.4	(+)27′.0	66°.9
Dec. Inc. 52′, d(+)27′0	Tens (+)17′.3		
	Units (+) 6′.1		
Interpolated for Dec. Inc.	19°08′.8		C S66°.1W
Initial great-circle course from AP			Cn 246°.1
Great-circle distance from AP(90° − 19°08′.8)			4251.2 n.mi.

(4) Using the graphical method for interpolating altitude for latitude and LHA increments, the course line is drawn from the AP in the direction of the initial great-circle course from the AP (246°.1). As shown in figure 906a, a line is drawn from the point of departure perpendicular to the initial great-circle course line or its extension.

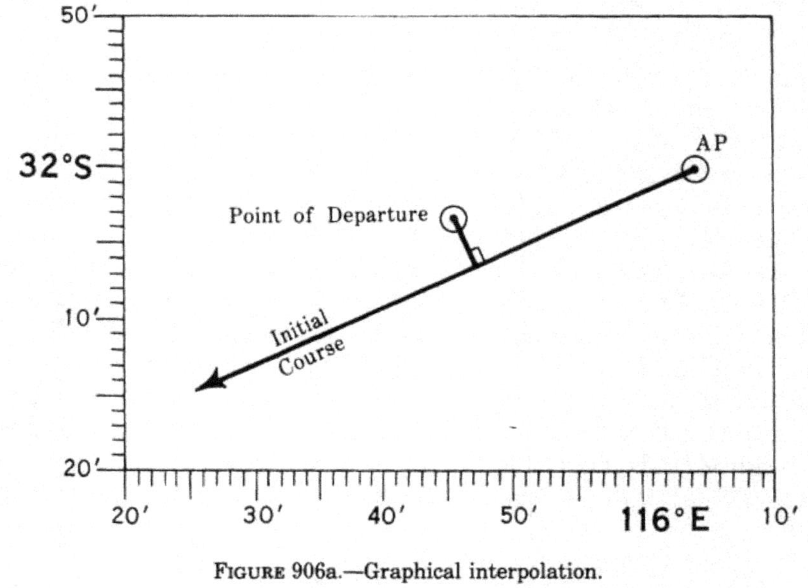

FIGURE 906a.—Graphical interpolation.

(5) The required correction, in units of minutes of latitude, for the latitude and DLo increments is the length along the course line between the foot of the perpen-

dicular and the AP. The correction as applied to the distance from the AP is −15′.8; the great-circle distance is 4235 nautical miles.

(6) The azimuth angle interpolated for declination, LHA, and latitude increments is S66°.3W; the initial great-circle course from the point of departure is 246°.3.

Example 4.—By Pub. No. 229 find the distance and initial great-circle course from San Francisco (37°49′N, 122°25′W) to Gladstone (23°51′S, 151°15′E).

Solution.—(1) Refer to figure 905. Since the latitude of the point of departure, the latitude of the destination, and the difference of longitude (DLo) between the point of departure and destination are not integral degrees, the solution is effected from an adjusted point of departure or assumed position of departure chosen as follows: the latitude of the assumed position (AP) is the integral degrees of latitude nearest to the point of departure; the longitude of the AP is chosen to provide integral degrees of DLo. This AP, which should be within 30′ of the longitude of the point of departure, is at latitude 38°N, longitude 122°45′W. The DLo is 86°.

(2) Enter the tables with 38° as the latitude argument (Contrary Name), 86° as the LHA argument, and 23° as the declination argument.

(3) From the tables extract the tabular altitude, altitude difference, and azimuth angle; interpolate altitude for Dec. Inc. as if the altitude were positive, adhering strictly to the sign given d. After interpolation regard the results as negative. Subtract tabular azimuth angle from 180°; interpolate for Dec. Inc.

	ht(Tab. Hc)	d	Z
LHA 86°, Lat. 38° (Contrary), Dec. 23°	10°57′.0	(+)35′.9	69°.3
Dec. Inc. 51′, d(+)35′.9	Tens (+)25′.5		180° − Z = 110°.7
	Units (+) 5′.1		
Interpolated for Dec. Inc.	(−)11°27′.6		C N111°.4W
Initial great-circle course from AP			Cn 248°.6
Great-circle distance from AP (90° + 11°27′.6)			6087.6 n.mi.

(4) Using the graphical method for interpolating altitude for latitude and LHA increments, the course line is drawn from the AP in the direction of the initial great-circle course from the AP (248°.6). As shown in figure 906b, a line is drawn from the point of departure perpendicular to the course line or its extension.

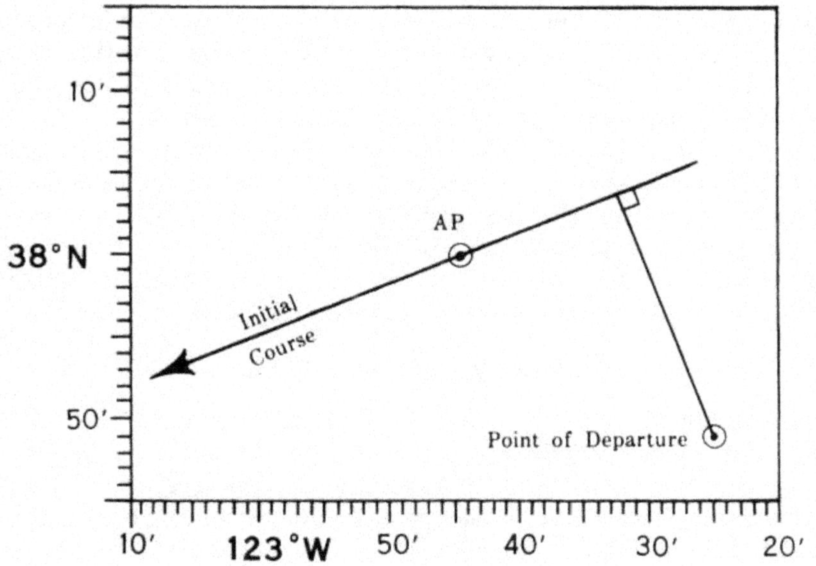

FIGURE 906b.—Graphical interpolation.

(5) The required additional correction, in units of minutes of latitude, for the latitude and DLo increments is the length along the course line between the foot of the perpendicular and the AP. The correction as applied to the distance from the AP is +10′.7; the great-circle distance is 6098 nautical miles.

(6) The azimuth angle interpolated for declination, LHA, and latitude increments is 111°.2; the initial great-circle course from the point of departure is 248°.8.

Example 5.—By Pub. No. 229 find the distance and initial great-circle course from Cabo Pilar (52°43′S, 74°41′W) to Wake Island (19°17′N, 166°39′E).

Solution.—(1) Refer to figure 905. Since the latitude of the point of departure, the latitude of the destination, and the difference of longitude (DLo) between the point of departure and destination are not integral degrees, the solution is effected from an adjusted point of departure or assumed position of departure chosen as follows: the latitude of the assumed position (AP) is the integral degrees of latitude nearest to the point of departure; the longitude of the AP is chosen to provide integral degrees of DLo. This AP, which should be within 30′ of the longitude of the point of departure, is at latitude 53°S, longitude 74°21′W; the DLo is 119°.

(2) Enter the tables with 53° as the latitude argument (Contrary Name), 119° as the LHA argument, and 19° as the declination argument. Since the tables are entered with the LHA (DLo) at the bottom of a right-hand page and the name of L_2 is contrary to the name L_1, the respondents are found in the column for L_1 on the facing page. In which case the C–S Line has been crossed, and the respondents correspond to those of a celestial body below the celestial horizon.

(3) From the table, extract the tabular latitude, altitude difference, and azimuth angle; interpolate altitude for Dec. Inc. as if the altitude were positive, adhering strictly to the sign given d. After interpolation regard the results as negative. Subtract tabular azimuth angle from 180°; interpolate for Dec. Inc.

	ht (Tab. Hc)	d	Z
LHA 119°, Lat. 53° (Contrary), Dec. 19°	32°24′.2	(+)46′.8	101°.6
Dec. Inc. 17′, d(+)47′.1	Tens (+)11′.3		180° − Z = S78°.4W
	Units (+) 2′.0		
Interpolated for Dec. Inc.	(−)32°37′.5		C S78°.4W
Initial great-circle course from AP			Cn 258°.4
Great-circle distance from AP (90° + 32°37′.5)			7357.5 n.mi.

(4) Using the graphical method for interpolating altitude for latitude and LHA increments, the course line is drawn from the AP in the direction of the initial great-circle course from the AP (258°.6). As shown in figure 906c a line is drawn from the point of departure perpendicular to the course line or its extension.

(5) The required additional correction, in units of minutes of latitude, for the latitude and DLo increments is the length along the course line between the foot of the perpendicular and the AP. The correction as applied to the distance from the AP is −8′.5; the great-circle distance is 7349 nautical miles.

(6) The azimuth angle interpolated for declination, LHA, and latitude increments is 79°.1; the initial great-circle course from the point of departure is 259°.1.

Points Along Great Circle

If, as in examples 1 and 2, the latitude of the point of departure and the initial great-circle course angle are integral degrees, points along the great circle are found by entering the tables with the latitude of departure as the latitude argument (always Same Name), the initial great-circle course angle as the LHA argument, and 90° minus distance to a point on the great circle as the declination

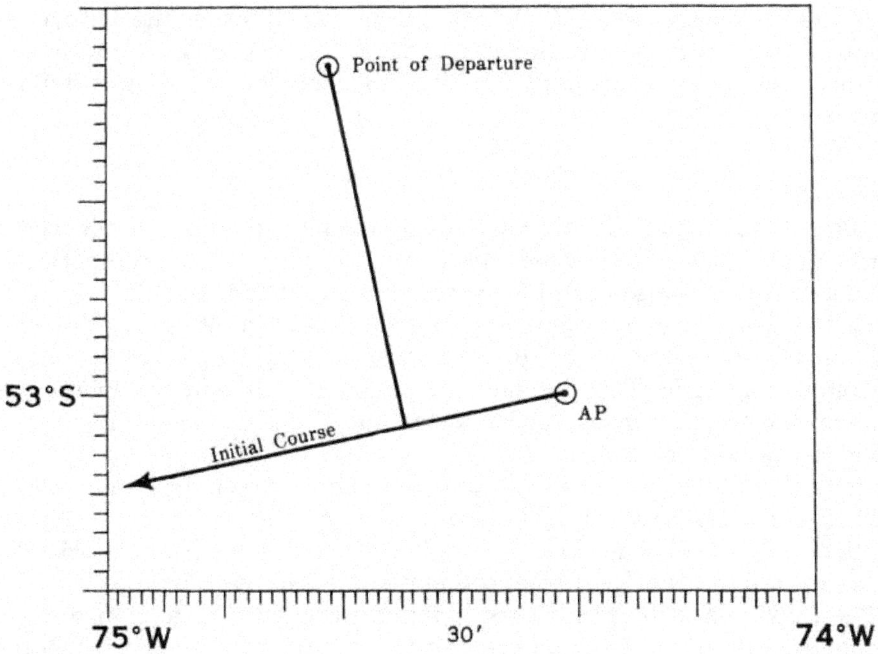

FIGURE 906c.—Graphical interpolation.

argument. The latitude of the point on the great circle and the difference of longitude between that point and the point of departure are the tabular altitude and azimuth angle respondents, respectively. If, however, the respondents are extracted from across the C–S Line, the tabular altitude corresponds to a latitude on the side of the equator opposite from that of the point of departure; the tabular azimuth angle is the supplement of the difference of longitude.

Example 6.—Find a number of points along the great-circle from latitude 38°N, longitude 125°W when the initial great-circle course angle is N111°W.

Solution.—(1) Entering the tables with latitude 38° (Same Name), LHA 111°, and with successive declinations of 85°, 80°, 75° . . . the latitudes and differences in longitude, from 125°W, are found as tabular altitudes and azimuth angles respectively:

Distance n.mi. (arc)	*300(5°)*	*600(10°)*	*900(15°)*	*3600(60°)*	*4800(70°)*
Latitude	36°.1N	33°.9N	31°.4N	3°.6N	3°.1S
DLo	5°.8	11°.3	16°.5	54°.1	61°.5
Longitude	130°.8W	136°.3W	141°.5W	179°.1W	173°.5E

Example 7.—Find a number of points along the great-circle from latitude 38°N, long. 125°W when the initial great-circle course angle is N69°W.

Solution.—Entering the tables with latitude 38° (Same Name), LHA 69°, and with successive declinations of 85°, 80°, 75° ... the latitudes and differences of longitude, from 125°W, are found as tabular altitudes and azimuth angles, respectively:

Distance n.mi. (arc)	*300(5°)*	*600(10°)*	*900(15°)*	*6600(110°)*	*7200(120°)*
Latitude	39°.6N	40°.9N	41°.9N	3°.1N	3°.6S
DLo	6°.1	12°.4	18°.9	118°.5	125°.9
Longitude	131°.1W	137°.4W	143°.9W	116°.5E	109°.1E

The latitude and difference of longitude of the point 6600 miles from the point of departure are found among the data for the latitude of departure continued on the facing page. Since the respondents for the point 7200 miles from the point of departure are found across the C–S Line on the facing page, the tabular altitude corresponds to a latitude on the side of the equator opposite from that of the point of departure; the tabular azimuth angle is the supplement of the difference of longitude.

Finding The Vertex

The use of Pub. No. 229 to find the approximate position of the vertex of a great-circle track provides a rapid check on the solution by computation. This approximate solution is also useful for voyage planning purposes.

Using the procedures for finding points along the great circle, the column of data for the latitude of the point of departure is inspected to find the maximum value of tabular altitude. This maximum tabular altitude and the tabular azimuth angle correspond to the latitude of the vertex and the difference of longitude of the vertex and the point of departure.

Example 8.—Find the vertex of the great-circle track from lat. 38°N, long. 125°W when the initial great-circle course angle is N69°W.

Solution.—(1) Enter Pub. No. 229 with lat. 38° (Same Name), LHA 69°, and inspect the column for lat. 38° to find the maximum tabular altitude.

(2) The maximum tabular altitude is found to be 42°38.′1 at a distance of 1500 nautical miles (90° − 65° = 25°) from the point of departure. The corresponding tabular azimuth angle is 32.°4. Therefore, the difference of longitude of vertex and point of departure is 32.°4.

Answers.—(1) Latitude of vertex 42°38.′1N.
(2) Longitude of vertex 157.°4W.

907. Altering a great-circle track to avoid obstructions.—Great-circle sailing cannot be used unless the great-circle track is free from obstructions. It does not start until one clears the harbor and takes his departure (art. 806), and often ends near the entrance to the destination. However, islands, points of land, or other obstructions may prevent the use of great-circle sailing over the entire distance. One of the principal advantages of solution by great-circle chart is that the presence of any obstructions is immediately apparent.

Often a relatively short run by rhumb line is sufficient to reach a point from which the great-circle track can be followed. Where a choice is possible, the rhumb line selected should conform as nearly as practicable to the direct great circle.

If the great circle crosses a small island, one or more legs may be altered slightly, or perhaps the drift of the vessel will be sufficient to make any planned alteration unnecessary. The possible use of the island in obtaining an en route fix should not be overlooked. If a larger obstruction is encountered, as in the case of the Aleutian Islands on a great circle from Seattle to Yokohama, some judgment may be needed in selecting the track. It may be satisfactory to follow a great circle to the vicinity of the obstruction, one or more rhumb lines along the edge of the obstruction, and another great circle to the destination. Another possible solution is the use of composite sailing (art. 908), and still another the use of two great circles, one from the point of departure to a point near the maximum latitude of unobstructed water, and the second from this point to the destination.

It is sometimes desirable to alter a great-circle track to avoid unfavorable winds or currents. The shortest route is not always the quickest.

Whatever the problem, a great-circle chart can be helpful in its solution.

908. Composite sailing.—When the great circle would carry a vessel to a higher latitude than desired, a modification of great-circle sailing, called **composite sailing,**

may be used to good advantage. The composite track consists of a great circle from the point of departure and tangent to the limiting parallel, a course line along the parallel, and a great circle tangent to the limiting parallel and through the destination.

Solution of composite sailing problems is most easily made by means of a great-circle chart. Lines from the point of departure and the destination are drawn tangent to the limiting parallel. The coordinates of various selected points along the composite track are then measured and transferred to a Mercator chart, as in great-circle sailing (art. 904).

Composite sailing problems can also be solved by computation (art. 1017, vol. II).

CHAPTER X

PILOTING

General

1001. Introduction.—On the high seas, where there is no immediate danger of grounding, navigation is a comparatively leisurely process. Courses and speeds are maintained over relatively long periods, and fixes are obtained at convenient intervals. Under favorable conditions a vessel might continue for several days with no positions other than those obtained by dead reckoning, or by estimate, and with no anxiety on the part of the captain or navigator. Errors in position can usually be detected and corrected before danger threatens.

In the vicinity of shoal water the situation is different. Frequent or continuous positional information is usually essential to the safety of the vessel. An error, which on the high seas may be considered small, may in what are called **pilot waters** be intolerably large. Frequent changes of course and speed are common. The proximity of other vessels increases the possibility of collisions and restricts movements.

In some waters the services of a specially qualified navigator having local knowledge may be necessary to insure safe navigation. **Local knowledge** extends beyond that publicly available in charts and publications, being more detailed, intimate, and current. The pilot's knowledge of his waters is gained not only through his own experience and familiarity, but by his availing himself of all local information resources, public and private, recent and longstanding, particularly concerning underwater hazards and obstructions, uncharted above-water landmarks and topographical configurations, local tides and currents, recent shoaling, temporary changes or deficiencies in aids to navigation, and similar matters of local concern. This local knowledge should enable a pilot to traverse his waters safely without reliance on man-made aids to navigation and to detect any unusual conditions or departure from a safe course. This service does not substitute for the ship's own safe navigation, but complements it. Prudence may also dictate the use of this specially qualified navigator to better insure safe navigation in situations where local knowledge is not essential. This navigator specially qualified for specific waters is called a **pilot;** his services are referred to as **pilotage** or **piloting.**

In its more general sense, the term **piloting** is used to mean the art of safely conducting a vessel on waters the hazards of which make necessary frequent or continuous positioning with respect to *charted* features and close attention to the vessel's draft with respect to the depth of water. Except for special circumstances, such as proceeding along a **range** (art. 1004), this positioning normally must be effected by constructing a plot on the chart based upon accurate navigational observations of charted features.

No other form of navigation requires the continuous alertness needed in piloting. At no other time is navigational experience and judgment so valuable. The ability to work rapidly and to correctly interpret all available information, always keeping "ahead of the vessel," may mean the difference between safety and disaster.

1002. Preparation for piloting.—Because the time element is often of vital importance in piloting, adequate preparation is important. Long-range preparation includes the organization and training of those who will assist in any way. This includes the steersman, who will be granted less tolerance in straying from the prescribed course than when farther offshore.

The more immediate preparation includes a study of the charts and publications of the area to familiarize oneself with the channels, shoals, tides, currents, aids to navigation, etc. One seldom has time to seek such information once he is proceeding in pilot waters. This preparation also includes the development of a definite plan for transiting the hazardous waters. Since the services provided by pilots having local knowledge are usually *advisory*, prudence dictates that the regularly assigned navigational personnel be advised of the pilot's plan. Otherwise, their ability to counteract any imprudent action on the part of the pilot may be severely limited. Also, knowledge of the pilot's plan enables the regularly assigned navigational personnel to act more effectively in verifying that the pilot is making a safe passage. The more detailed preparation required for leaving or entering port is given in chapter XXIII.

Position

1003. Lines of position.—As in celestial and radionavigation, piloting makes extensive use of **lines of position.** Such a line is one on some point of which the vessel may be presumed to be located, as a result of observation or measurement. It may be highly reliable, or of questionable accuracy. Lines of position are of great value, but one should always keep in mind that *they can be in error* because of imperfections in instruments used for obtaining them and human limitations in those who use the instruments and utilize the results. The extent to which one can have confidence in various lines of position is a matter of judgment acquired from experience.

A line of position might be a straight line (actually a part of a great circle), an arc of a circle, or part of some other curve such as a hyperbola (art. 134, vol. II). An appropriate label should be placed on the plot of a line of position *at the time it is drawn,* to avoid possible error or confusion. A label should include all information essential for identification, but no extraneous information. *The labels shown in this volume are recommended.*

1004. Bearings.—A **bearing** is the horizontal direction of one terrestrial point from another. It is usually expressed as the angular difference between a reference direction and the given direction. In navigation, north is generally used as the reference direction, and angles are measured clockwise through 360°. It is customary to express all bearings in three digits, using preliminary zeros where needed. Thus, north is 000° or 360°, a direction 7° to the right of north is 007°, east is 090°, southwest is 225°, etc.

For plotting, *true* north is used as the reference direction. A bearing measured from this reference is called a **true bearing.** A **magnetic, compass,** or **grid bearing** results from using magnetic, compass, or grid north, respectively, as the reference direction. This is similar to the designation of courses. In the case of bearings, however, one additional reference direction is often convenient. This is the heading of the ship. A bearing expressed as angular distance from the heading is called a **relative bearing.** It is usually measured clockwise through 360°. A relative bearing may be expressed in still another way, as indicated in figure 1004. Except for dead ahead and points at 45° intervals from it, this method is used principally for indicating directions obtained visually, without precise measurement. An even more general indication of relative bearing may be given by such directions as "ahead," "on the starboard bow," "on the port quarter," "astern." The term *abeam* may be used as the equivalent of either the general "on the beam" or, sometimes, the more precise "broad on the beam." Degrees are sometimes used instead of points to express relative bearings by the system illustrated in figure 1004. However, if degrees are used, a better practice is to use the 360° system. Thus, a relative bearing of "20° forward of the port beam" is better expressed as "290°."

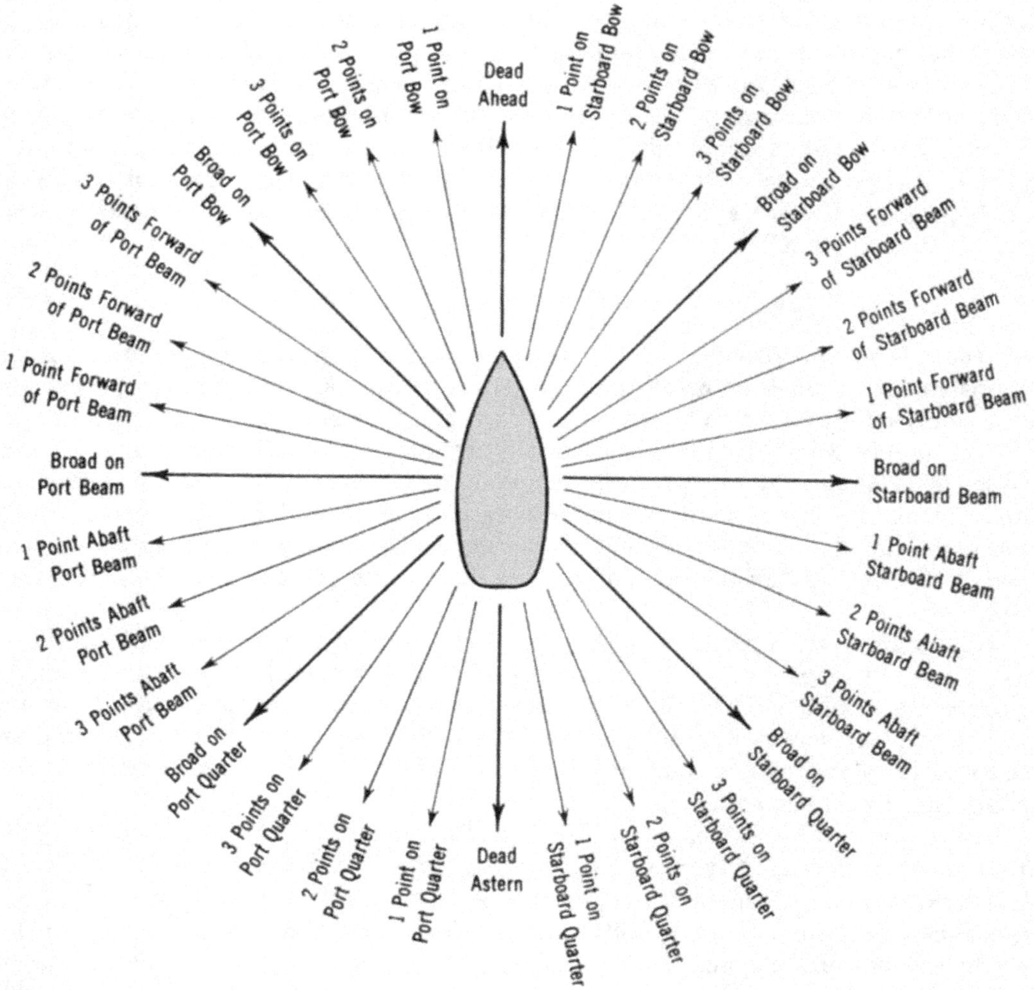

FIGURE 1004.—One method of expressing relative bearings.

True, magnetic, and compass bearings are interconverted by the use of variation and deviation, or compass error, in the same manner as courses. Interconversion of relative and other bearings is accomplished by means of the heading. If true heading is added to relative bearing, true bearing results. If magnetic, compass, or grid heading is added to relative bearing, the corresponding magnetic, compass, or grid bearing is obtained.

A **bearing line** extending in the direction of an observed bearing of a charted object is one of the most widely used lines of position. If one knows that an identified landmark has a certain bearing from his vessel, the vessel can only be on the line at which such a bearing might be observed, for at any other point the bearing would be different. This line extends outward from the landmark, along the *reciprocal* of the observed bearing. Thus, if a lighthouse is *east* of a ship, that ship is *west* of the lighthouse. If a beacon bears 156°, the observer must be on a line extending 156°+180°=336° from the beacon. Since observed bearing lines are great circles, this relationship is not strictly accurate, but the error is significant only where the great circle departs materially from the rhumb line, as in high latitudes (ch. XXV).

Bearings are obtained by compass, gyro repeater, pelorus, alidade, radar, etc. One type of bearing can be obtained by eye without measurement. When two objects appear directly in line, one behind the other, they are said to be "in range," and together they constitute a **range.** For accurately charted objects, a range may provide the most accurate line of position obtainable, and one of the easiest to observe. Tanks, steeples, towers, cupolas, etc., sometimes form **natural ranges.** A navigator should be familiar with prominent ranges in his operating area, particularly those which can be used to mark turning points, indicate limits of shoals, or define an approach heading or let-go point of the anchorage of a naval vessel. So useful is the range in marking a course that artificial ranges, usually in the form of two lighted beacons, have been installed in line with channels in many ports. A vessel proceeding along the channel has only to keep the beacons in range to remain in the center of the channel. If the *farther* beacon (customarily the higher one) appears to "open out" (move) to the right of the forward (lower) beacon, one knows that he is to the right of his desired track. Similarly, if it opens out to the left, the vessel is off track to the left.

The line defined by the range is called a **range line** or **leading line.** Range daybeacons (art. 412) and other charted objects forming a range are often called **leading marks.** Range lights (art. 402) are often called **leading lights.**

It is good practice to plot only a short part of a line of position in the vicinity of the vessel, to avoid unnecessary confusion and to reduce the chart wear by erasure. Particularly, one should avoid the drawing of lines through the chart symbol indicating the landmark used. In the case of a range, a straightedge is placed along the two objects, and the desired portion of the line is plotted. One need not know the numerical value of the bearing represented by the line. However, if there is any doubt as to the identification of the objects observed, the measurement of the bearing should prove useful.

A single bearing line is labeled with the time above the line.

1005. Distance.—If a vessel is known to be a certain distance from an identified point on the chart, it must be somewhere on a circle with that point as the center and the distance as the radius. A single distance (range) arc is labeled with the time above the line.

Distances are obtained by radar, range finder, stadimeter, synchronized sound and radio signals, synchronized air and water sounds, vertical angles (table 9), etc. If vertical angles are used, measurement should be made from the top of the object to the visible sea horizon, if it is available. If measurement is made to a water line not vertically below the top of the object, a problem may be encountered because distance from table 9 is to the point vertically below the top of the object, while the distance used for entering table 22 to determine dip short of the horizon is to the water line. Generally, any differences in these two distances can be determined from the chart. This problem may, in some cases, be avoided by decreasing the height of eye sufficiently to bring the horizon between the observer and the object.

1006. The fix.—A line of position, however obtained, represents a series of possible positions, but not a single position. However, if *two simultaneous, nonparallel* lines of position are available, the only position that satisfies the requirements of being on both lines at the same time is the intersection of the two lines. This point is one form of **fix.** Examples of several types of fixes are given in the illustrations. In figure 1006a a fix is obtained from two bearing lines. The fix of figure 1006b is obtained by two distance circles. Figure 1006c illustrates a fix from a range and a distance. In figure 1006d a bearing and distance of a single object are used. A small circle is used to indicate the fix at the intersection of the lines of position. The time of the fix is the time at which the lines of position were established.

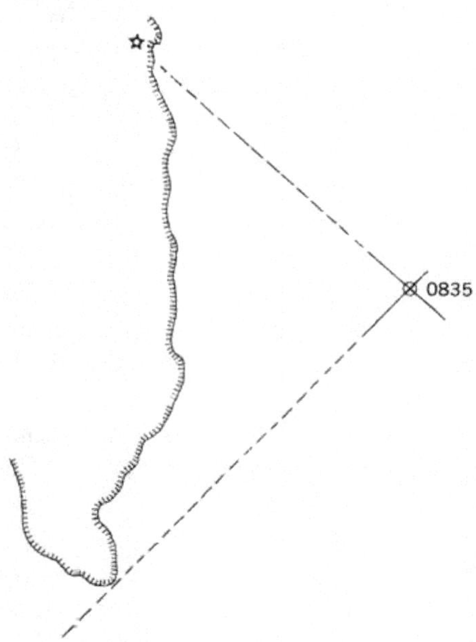

FIGURE 1006a.—A fix by two bearing lines.

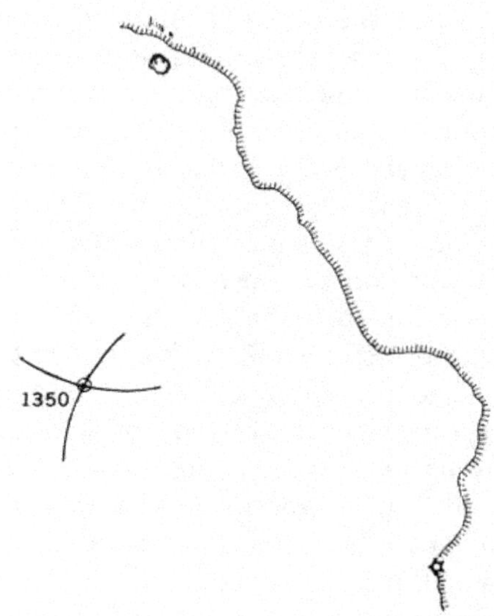

FIGURE 1006b.—A fix by two distances.

Some consideration should be given to the selection of objects to provide a fix. It is essential, for instance, that the objects be identified. The angle between lines of position is important. The ideal is 90°. If the angle is small, a slight error in measuring or plotting either line results in a relatively large error in the indicated position. In the case of a bearing line, nearby objects are preferable to those at a considerable distance, because the linear (distance) error resulting from an angular error increases with distance.

Another consideration is the type of object. Lighthouses, spires, flagpoles, etc., are good objects because the point of observation is well defined. A large building, most nearby mountains, a point of land, etc., may leave some reasonable doubt as to the exact point used for observation. If a tangent is used (fig. 1006a), there is a possibility that a low spit may extend seaward from the part observed. A number of towers, chimneys, etc., close together require careful identification. A buoy or a lightship may drag anchor and be out of position. Most buoys are secured by a

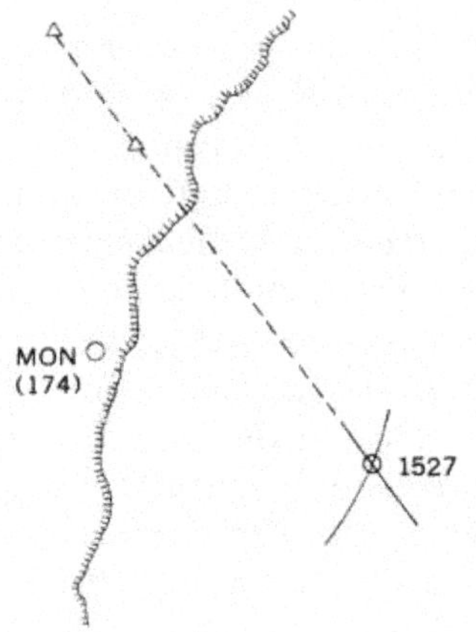

FIGURE 1006c.—A fix by a range and distance.

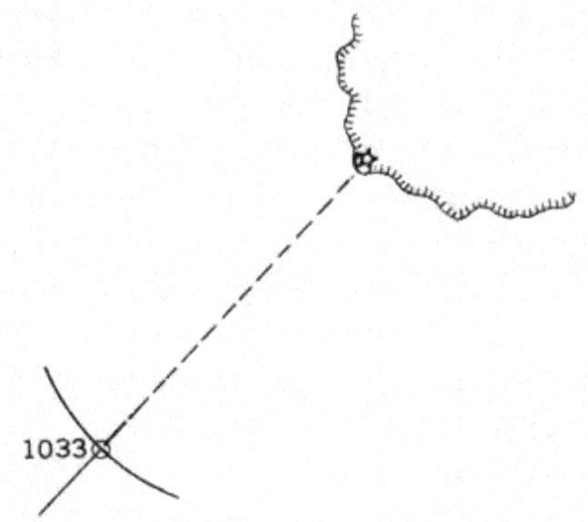

FIGURE 1006d.—A fix by distance and bearing of single object.

single anchor and so have a certain radius of swing as the tide, current, and wind change.

Although two accurate nonparallel lines of position completely define a position, if they are taken at the same time, an element of doubt always exists as to the accuracy of the lines. Additional lines of position can serve as a check on those already obtained, and, usually, to reduce any existing error. If three lines of position cross at a common point, or form a small triangle, it is usually a reasonable assumption that the position is reliable, and defined by the center of the figure. However, this is not *necessarily* so, and one should be aware of the possibility of an erroneously indicated position.

A single bearing line of an accurately charted object will be offset from the observer's actual position by an amount dependent upon the net angular error of the observation and plot, and the distance of the charted object from the observer. The amount of offset is expressed approximately in the **Rule of Sixty**, which may be stated as follows: *The offset of the plotted bearing line from the observer's actual position is 1/60th of the distance to the object observed for each degree of error.* In the derivation of the Rule of Sixty, the assumption is made that the angular error is small, i.e., not more than the small errors normally associated with compass observations and plotting. Using this assumption, the sine function of the angular error is taken as equal to the same number of radians as the error. As shown in figure 1006e, the offset is equal to 1/60th of the distance to the charted object observed times the sine of the angular error of the bearing line as plotted. Thus, an error of 1° represents an error of about 100 feet if the object is 1 mile distant, 1,000 feet if the object is 10 miles away, and 1 mile if the object is 60 miles from the observer.

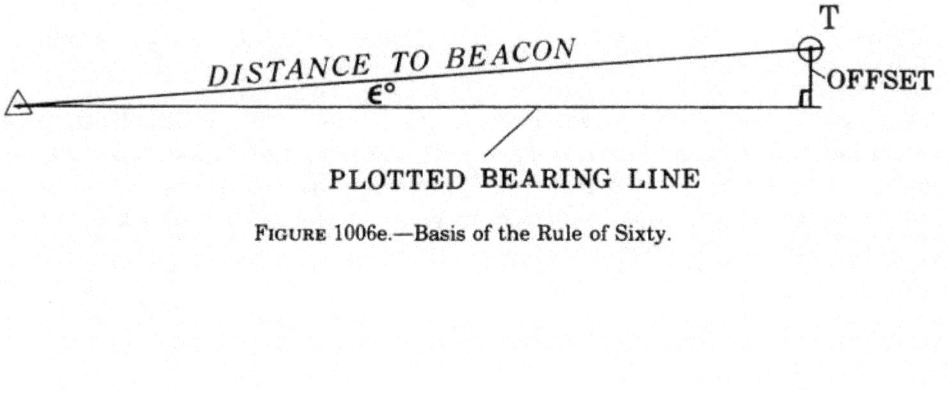

FIGURE 1006e.—Basis of the Rule of Sixty.

1007. Two-bearing plot.—If as shown in figure 1007a, the observer is located at point T and the bearings of a beacon and cupola are observed and plotted without error, the intersection of the bearing lines lies on the circumference of a circle passing through the beacon, cupola, and the observer. With *constant error*, i.e., an error of fixed magnitude and sign (or direction) for a given set of observations, the angular difference of the bearings of the beacon and the cupola is not affected. Thus, the angle formed at point F by the bearing lines plotted with constant error is equal to the angle formed at point T by the bearing lines plotted without error. From geometry it is known that angles having their apexes on the circumference of a circle and that are subtended by the same chord are equal. Since the angles at points T and F are equal and the angles are subtended by the same chord, the intersection at point F lies on the circumference of a circle passing through the beacon, cupola, and the observer.

Assuming only constant error in the plot, the direction of displacement of the two-bearing fix from the position of the observer is in accordance with the sign (or direction) of the constant error. However, a third bearing is required to determine the direction of the constant error.

Assuming only constant error in the plot, the two-bearing fix lies on the circumference of the circle passing through the two charted objects observed and the observer. The fix error, i.e., the length of the chord FT in figure 1007b, is dependent upon the magnitude of the constant error ϵ, the distance between the charted objects, and the cosecant of the angle of cut, angle θ. In figure 1007b,

$$\text{fix error} = FT = \frac{\epsilon\ BC \csc\theta}{2} \quad \text{(where } \epsilon \text{ is a small angle).}$$

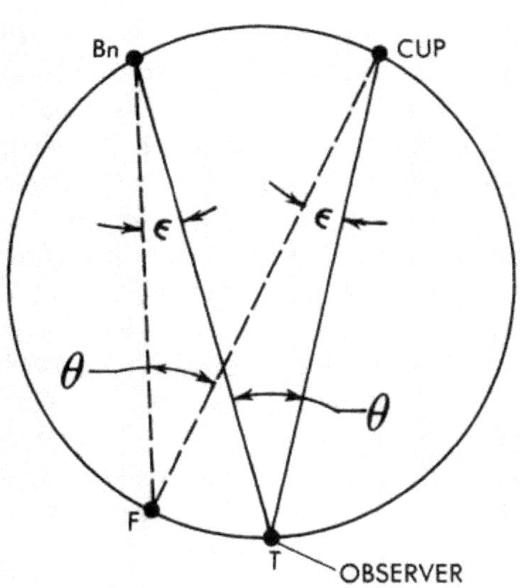

FIGURE 1007a.—Two-bearing plot.

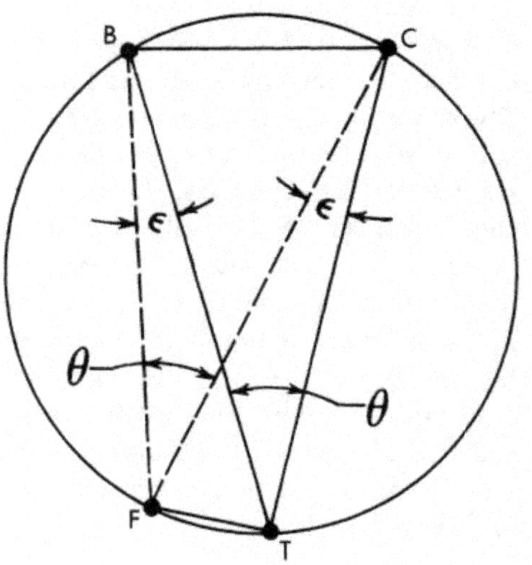

FIGURE 1007b.—Two-bearing plot with constant error.

Thus, the fix error is least when the angle of cut is 90°. As illustrated in figure 1007c, the error increases in accordance with the cosecant function as the angle of cut decreases. The increase in the error becomes quite rapid after the angle of cut has decreased to below about 30°. With an angle of cut of 30°, the fix error is about twice that at 90°.

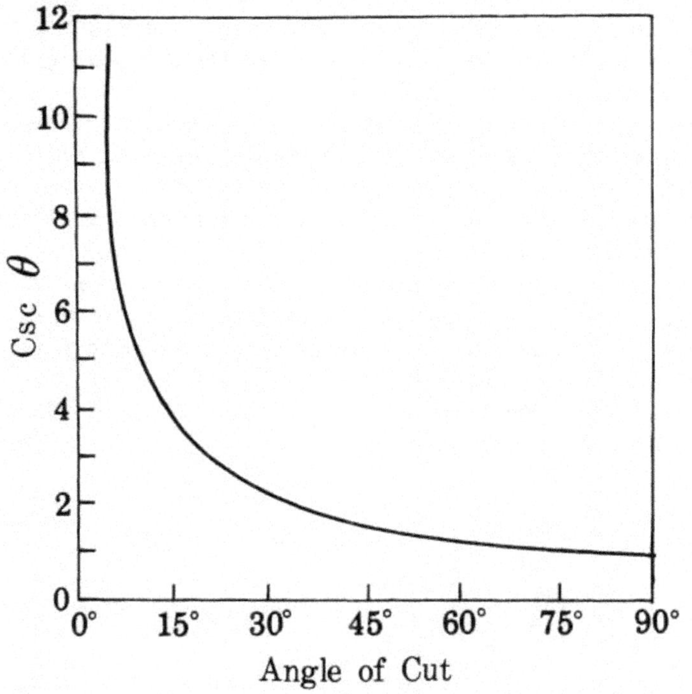

FIGURE 1007c.—Error of two-bearing plot.

1008. Three-bearing plot.—When three nonsimultaneous bearings are taken from a moving ship, it is necessry to advance (art. 1010) all but the last line of position by amounts equal to the course and distance made good from the last observation. The point where the two advanced lines of position cut the third one is the position of the ship at the time of obtaining the last line of position.

If there is a constant error in the plot, the three lines of position form a **triangle of error** called a **cocked hat**. The causes of the cocked hat may be any of the following:

(a) error in identifying the object,
(b) error in plotting the line of bearing,
(c) inaccuracy of observation resulting from the limitations of the compass,
(d) compass error or gyro error unknown or incorrectly applied,
(e) excessive time interval between observations,
(f) inaccuracy of the survey or the chart.

If the cocked hat is large, the work should be revised to ensure that errors (a) and (b) are eliminated.

Error (c) should never be greater than ¼° and may generally be neglected.

Error (d) should be eliminated by taking every opportunity of checking the deviation or gyro error.

1009. Adjusting a fix for constant error.—If several fixes obtained by bearings on three objects produce triangles of error of about the same size, one might reasonably suspect a constant error in the observation of the bearings, particularly if the same instrument is used for all observations, or in the plotting of the lines. If the application of a constant error to all bearings results in a point, or near-point fix, the navigator is usually justified in applying such a correction. This situation is illustrated in figure 1009a, where the solid lines indicate the original plot, and the broken lines indicate each line of position moved 3° in a clockwise direction. If different instruments are used for observation, one of them might be consistently in error. This might be detected by altering all bearings observed by that instrument by a fixed amount and producing good fixes.

When there is indication of constant error in a cross-bearing plot, the **Franklin Piloting Technique** can be used for rapid determination of the direction and magnitude of the error. The use of the technique avoids the delays associated with the trial and error method (fig. 1009a). Being a rapid method for determining with sufficient accuracy the normally small errors of the compass, the technique provides a means for maintaining the cross-bearing plot even when the compass error is variable from one round of bearings to the next. To apply the technique when there is indication of constant error in the plot, the navigator selects three objects for his next round of bearings in accordance with the following rules:

1. Two of the three objects should be nearby so that the displacements of their plotted bearing lines from the observer's actual position, as a result of a small constant error, will be small.

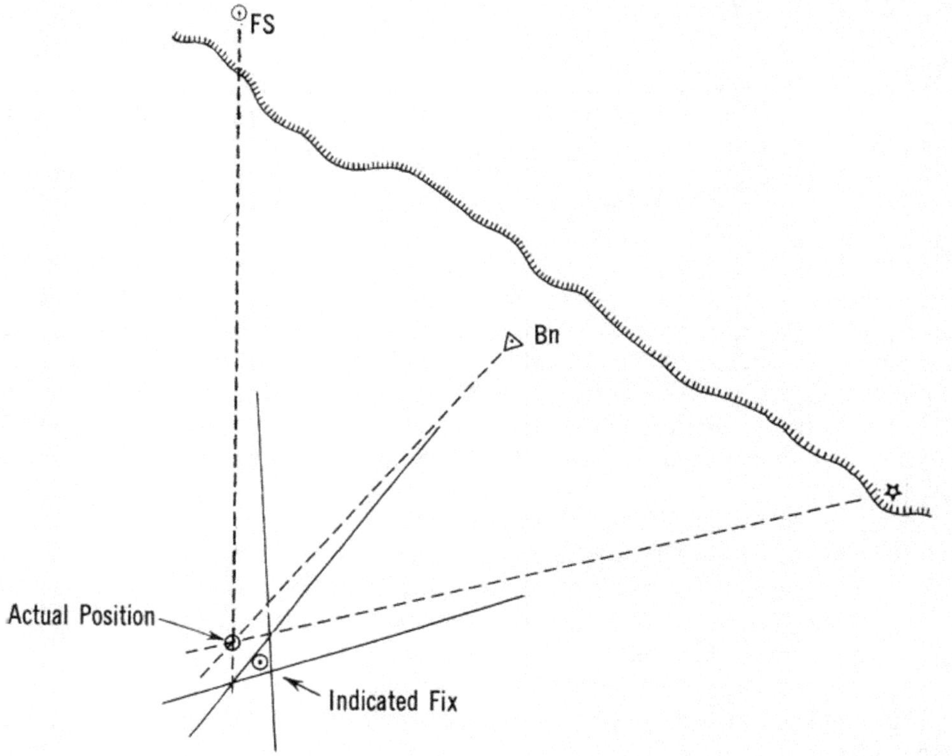

FIGURE 1009a.—Adjusting a fix for constant error.

2. The third object should be at least two and one-half times the distance of the farther of the two nearby objects so that the displacement of its plotted bearing line from the observer's actual position, as a result of a small constant error, will be relatively large.

3. Preferably, the third object should lie in a direction from the observer approximately parallel to a line between the two nearby objects.

With selection as in (3) above, the line drawn from the third object to the intersection of the bearing lines of the two nearby objects is nearly tangent to the circle through these nearby objects and the observer's actual position (fig. 1009b).

Being nearly tangent to this circle, the line drawn from the most distant object passes close to the observer's actual position, which is close to the point of tangency for small constant errors.

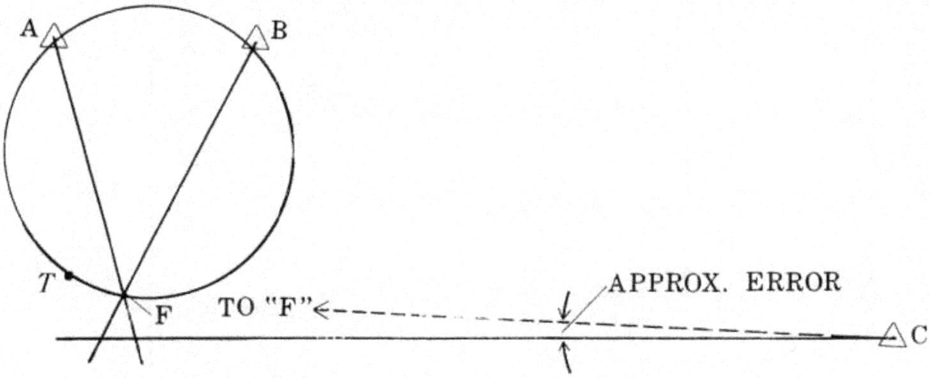

FIGURE 1009b.—Selection of charted objects for observation.

As shown in figure 1009b, the bearing lines through nearby objects A and B intersect at F on the circle through A, B, and the observer's actual position at point T. The acute angle between the plotted bearing line through distant object C and a line from C to T is exactly equal to the error, assuming that all error is due to a constant compass error. It follows that the acute angle between the plotted bearing line through distant object C and the line from C to F is approximately equal to the error.

Because the objects observed in figure 1009b lie within a bearing spread of 180°, the most probable position of the fix is outside the triangle of error. The fix is determined from the set of adjacent exterior angles (and opposite interior angle) which are immediately adjacent to the area formed by the three lines extending from the triangle of error toward the observed objects.

From geometry, the fix cannot lie within the shaded area (fig. 1009c); the fix cannot lie within the triangle of error; the fix is determined by the set of adjacent exterior angles of the triangle of error which are immediately adjacent to the shaded area. The adjacent set establishes the fix at T.

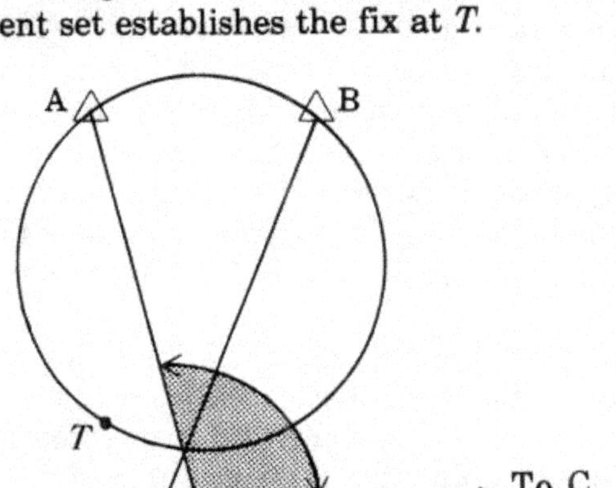

FIGURE 1009c.—Determining the most probable position of the fix.

Although the foregoing use of the geometry of the cross-bearing plot could be used to find the direction of the constant error, the Franklin technique provides a simpler means. However, a reference direction must first be established. This reference direction is used to determine whether the plotted bearing line of one of the three objects lies to the left or right of the intersection of the bearing lines of the other two. For example, in the left-hand plot of figure 1009d the intersection at *F* is taken as being to the right of the plotted bearing line of object *C*.

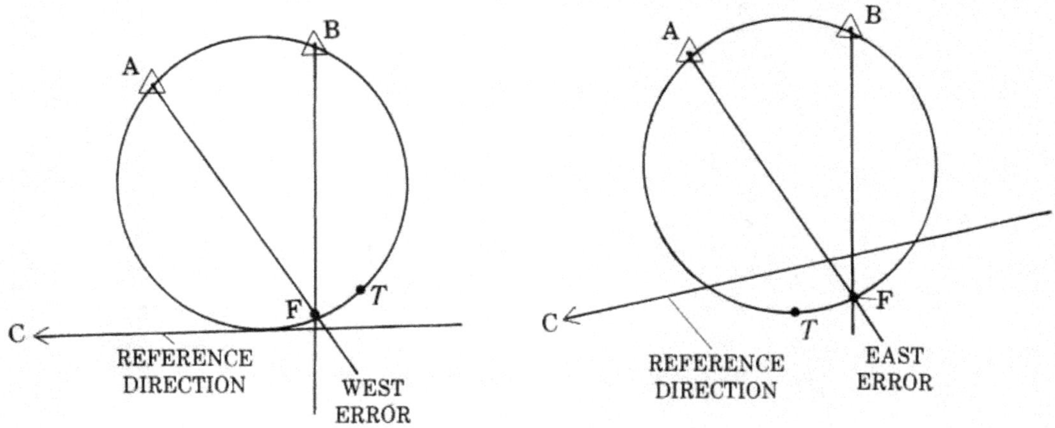

FIGURE 1009d.—Determining constant error when bearing spread is less than 180°.

When the bearing spread is less than 180°, the direction of either the extreme left-hand or right-hand object from the observer's approximate position is used as the reference direction. When the bearing spread is greater than 180°, the direction of any one of the three objects from the observer's position can be used as the reference direction.

If the bearing spread is less than 180° and the plotted bearing line extended through the extreme left-hand or right-hand object lies to the right of the intersection of the other two plotted bearing lines (fig. 1009d), the error is east; otherwise the error is west.

If the bearing spread is greater than 180° and the plotted bearing line through one object lies to the right of the intersection of the other two plotted bearing lines, the error is east; otherwise the error is west.

1010. Nonsimultaneous observations.—For fully accurate results, observations made to fix the position of a moving vessel should be made simultaneously, or nearly so. On a slow-moving vessel, relatively little error is introduced by making several observations in quick succession. A wise precaution is to observe the objects more nearly ahead or astern first, since these are least affected by the motion of the observer. However, when it is desired to obtain a good estimate of the speed being made good, it may be desirable to observe the most rapidly changing bearing first, assuming that such observation can be better coordinated with the time "mark."

Sometimes it is not possible or desirable to make simultaneous or nearly simultaneous observations. Such a situation may arise, for instance, when a single object is available for observation, or when all available objects are on nearly the same or reciprocal bearings, and there is no means of determining distance. Under such conditions, a period of several minutes or more may be permitted to elapse between observations to provide lines of position crossing at suitable angles. When this occurs, the lines can be adjusted to a common time to obtain a **running fix.** Refer to figure 1010a. A ship is proceeding along a coast on course 020°, speed 15 knots. At 1505 lighthouse *L* bears 310°. If the line of position is accurate, the ship is somewhere on it at the time of observation. Ten minutes later the ship will have traveled 2.5 miles in direction 020°. If the ship was at A at 1505, it will be at A' at 1515. However, if the position at 1505 was B, the position at 1515 will be B'. A similar relationship exists between C and C', D and D', E and E', etc. Thus, if *any* point on the original line of position is moved a distance equal to the distance run, and in the direction of the motion, a line through this point, parallel to the original line of position, represents all possible positions of the ship at the later time. This process is called **advancing** a line of position. The moving of a line *back* to an *earlier* time is called **retiring** a line of position.

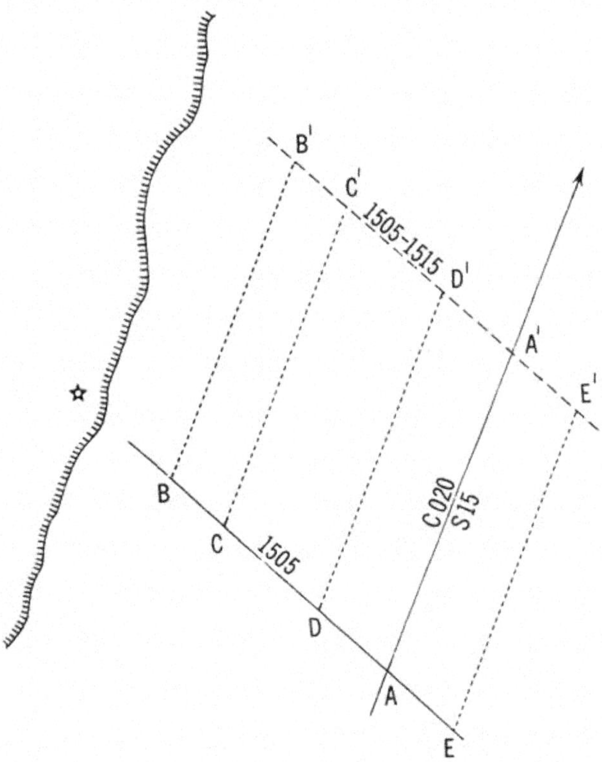

FIGURE 1010a.—Advancing a line of position.

The accuracy of an adjusted line of position depends not only upon the accuracy of the original line, but also upon the reliability of the information used in moving the line. A small error in the track made good has little effect upon the accuracy of a bearing line of an object near the beam, but maximum effect upon the bearing line of an object nearly ahead or astern. Conversely, the effect of an error in speed is maximum upon the bearing line of an object abeam. The opposite is true of circles of position. The best estimate of track and speed made good should be used in advancing or retiring a line of position.

If there are any changes of course or speed, these should be considered, for the motion of the line of position should reflect as accurately as possible the motion of the observer between the time of observation and the time to which the line is adjusted. Perhaps the easiest way to do this is to measure the direction and distance between dead reckoning or estimated positions at the two times, and use these to adjust some point on the line of position. This method is shown in figure 1010b. In this illustration allowance is made for the estimated combined effect of wind and current, this effect being plotted as an additional course and distance. If courses and speeds made good over the ground are used, the separate plotting of the wind and current effect is not used. In the illustration, point *A* is the DR position at the time of observation, and point *B* is the estimated position (the DR position adjusted for wind and current) at the time to which the line of position is adjusted. Line *A'B'* is of the same length and in the same direction as line *AB*.

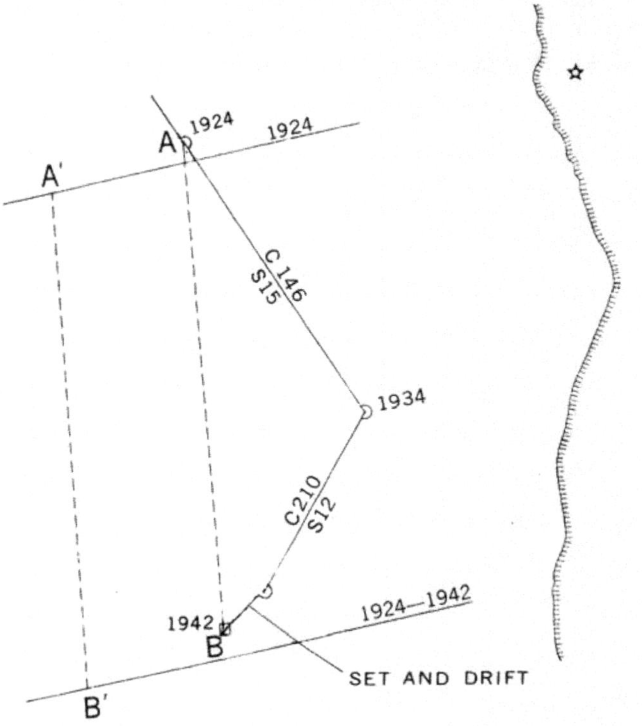

FIGURE 1010b.—Advancing a line of position with a change in course and speed, and allowing for current.

Other techniques may be used. The position of the object observed may be advanced or retired, and the line of position drawn in relation to the adjusted position. This is the most satisfactory method for a circle of position, as shown in figure 1010c. When the position of the landmark is adjusted, the advanced line of position can be laid down without plotting the original line, which need be shown only if it serves a useful purpose. This not only eliminates part of the work, but reduces the number of lines on the chart, and thereby decreases the possibility of error. Another method is to draw any line, such as a perpendicular, from the dead reckoning position at the time of observation to the line of position. A line of the same length and in the same direction, drawn from the DR position or EP at the time to which the line is adjusted, locates a point on the adjusted line, as shown in figure 1010d. If a single course and speed is involved, common practice is to measure from the intersection of the line of position and the course line. If the dividers are set to the distance run between bearings and placed on the chart so that one point is on the first bearing line and the other point is on the second bearing line, and the line connecting the points is parallel to the course line, the points will indicate the positions of the vessel at the times of the bearings.

An adjusted line of position is labeled the same as an unadjusted one, except that both the time of observation and the time to which the line is adjusted are

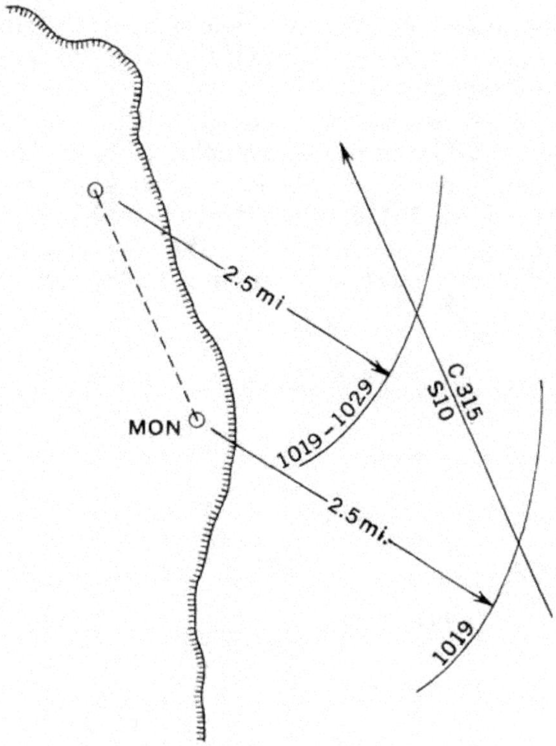

FIGURE 1010c.—Advancing a circle of position.

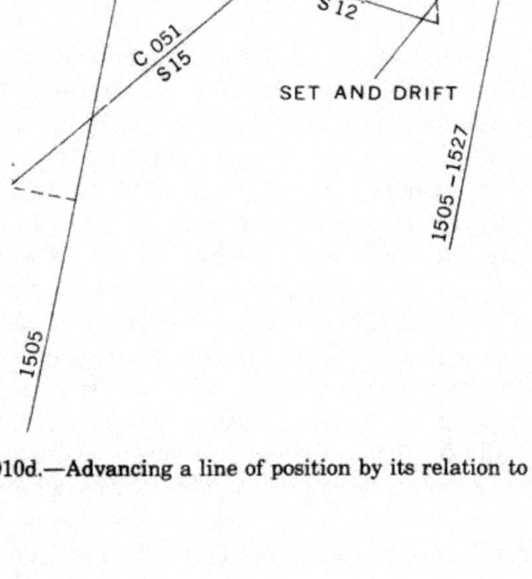

FIGURE 1010d.—Advancing a line of position by its relation to the dead reckoning.

shown, as in the illustrations of this article and article 1011. Because of additional sources of error in adjusted lines of position, they are not used when satisfactory simultaneous lines can be obtained.

1011. The running fix.—As stated in article 1010, a fix obtained by means of lines of position taken at different times and adjusted to a common time is called a **running fix.** In piloting, common practice is to *advance* earlier lines to the time of the last observation. Figure 1011a illustrates a running fix obtained from two bearings of the same object. In figure 1011b the ship changes course and speed between observations of two objects. A running fix by two circles of position is shown in figure 1011c.

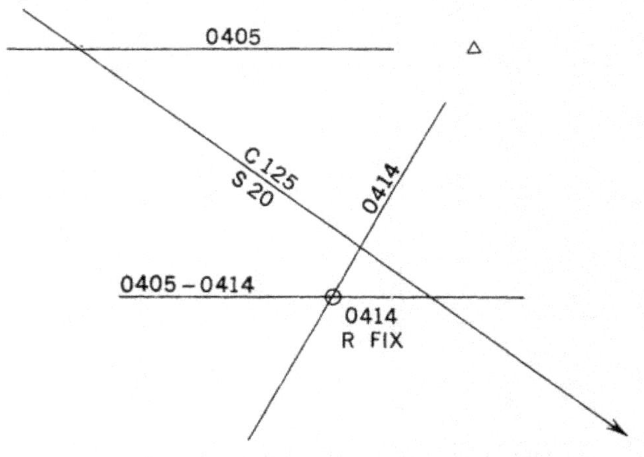

FIGURE 1011a.—A running fix by two bearings on the same object.

When simultaneous observations are not available, a running fix may provide the most reliable position obtainable. The time between observations should be no longer than about 30 minutes, for the uncertainty of course and distance made good increases with time.

The errors applicable to a running fix are those resulting from errors of the individual lines of position. However, a given error may have quite a different effect upon the fix than upon the line of position. Consider, for example, the situation of an unknown head current. In figure 1011d a ship is proceeding along a coast, on course 250°, speed 12 knots. At 0920 light *A* bears 190°, and at 0930 it bears 143°. If the earlier bearing line is advanced a distance of 2 miles (10 minutes at 12 knots) in the direction of the course, the running fix is as shown by the solid lines. However, if there is a head current of 2 knots, the ship is making good a speed of only 10 knots, and in 10 minutes will travel a distance of only 1⅔ miles. If the first bearing line is advanced this distance, as shown by the broken line, the actual position of the ship is at *B*. *This is nearer the beach than the running fix,* and therefore a dangerous situation. A following current gives an indication of position too far from the object. Therefore, if a current *parallel* to the course (either head or following) is suspected, a *minimum* estimate of speed made good will result in a possible margin of safety. If the second bearing is of a different object, a *maximum* estimate of speed should be made if the second object is on the same side and farther forward, or on the opposite side and farther aft, than the first object was when observed. All of these situations assume that danger is on the same side as the object observed first. If there is either a head or following current, a series of running fixes based upon a number of bearings of the same object will plot in a straight line parallel to the

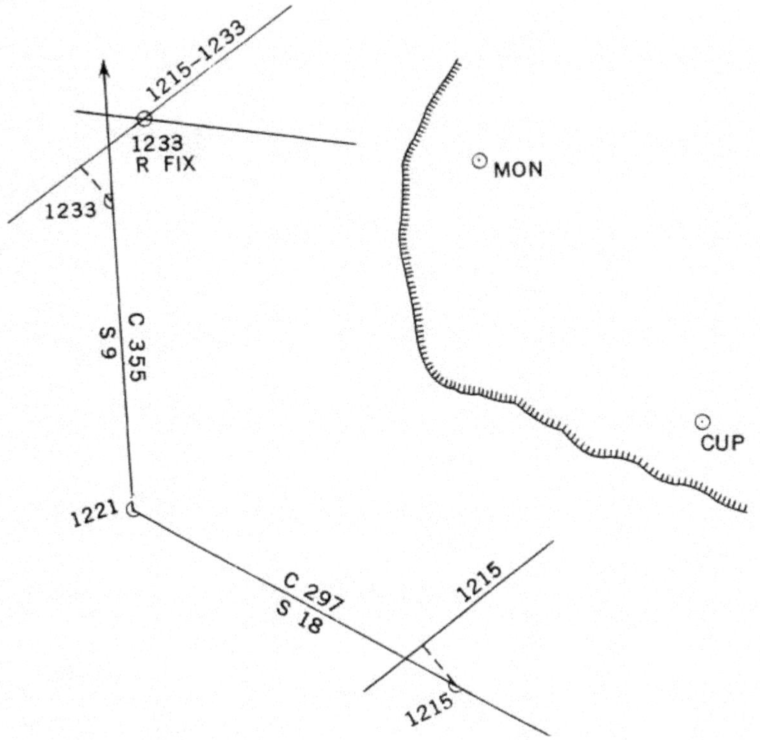

FIGURE 1011b.—A running fix with a change of course and speed between observations on separate landmarks.

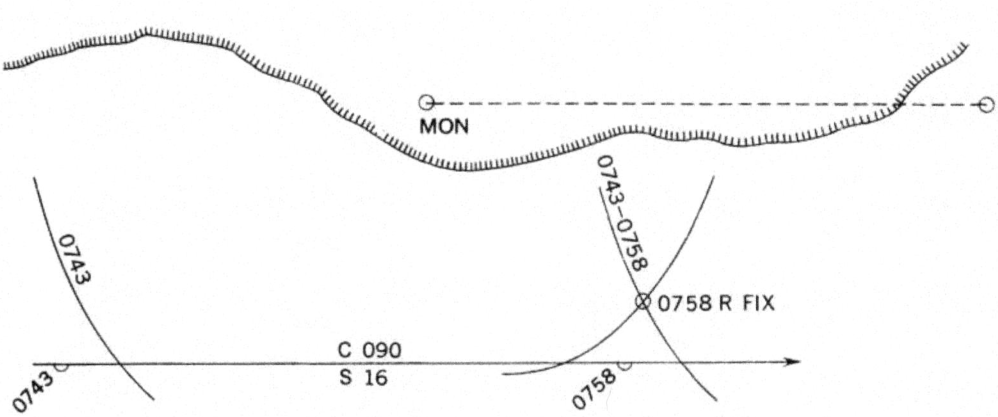

FIGURE 1011c.—A running fix by two circles of position.

course line, as shown in figure 1011e. The plotted line will be too close to the object observed if there is a following current, and too far out if there is a head current. The existence of the current will not be apparent unless the actual speed over the ground is known. The position of the plotted line relative to the dead reckoning course line is not a reliable guide.

A current oblique to the course will result in an incorrect position, but the direction of the error is indeterminate. In general, the effect of a current with a

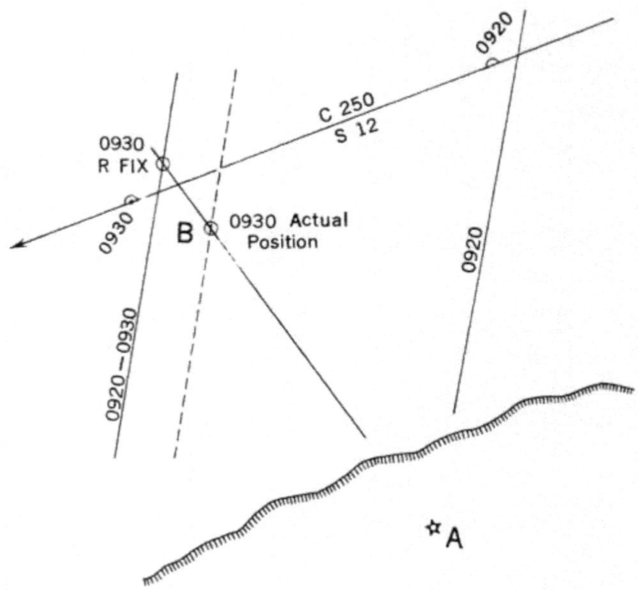

FIGURE 1011d.—Effect of a head current on a running fix.

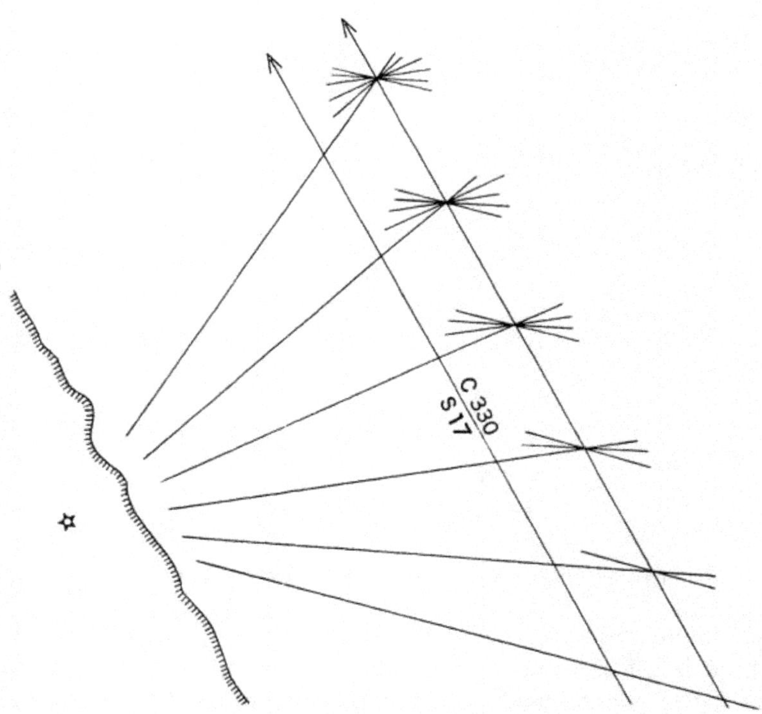

FIGURE 1011e.—A number of running fixes with a following current.

strong head or following component is similar to that of a head or following current, respectively. The existence of an oblique current, but not its amount, can be detected by observing and plotting several bearings of the same object. The running fix obtained by advancing one bearing line to the time of the next one will not agree with the running fix obtained by advancing an earlier line. Thus, if bearings *A, B,* and *C* are observed at five-minute intervals, the running fix obtained by advancing *B* to the time of *C* will not be the same as that obtained by advancing *A* to the time of *C,* as shown in figure 1011f.

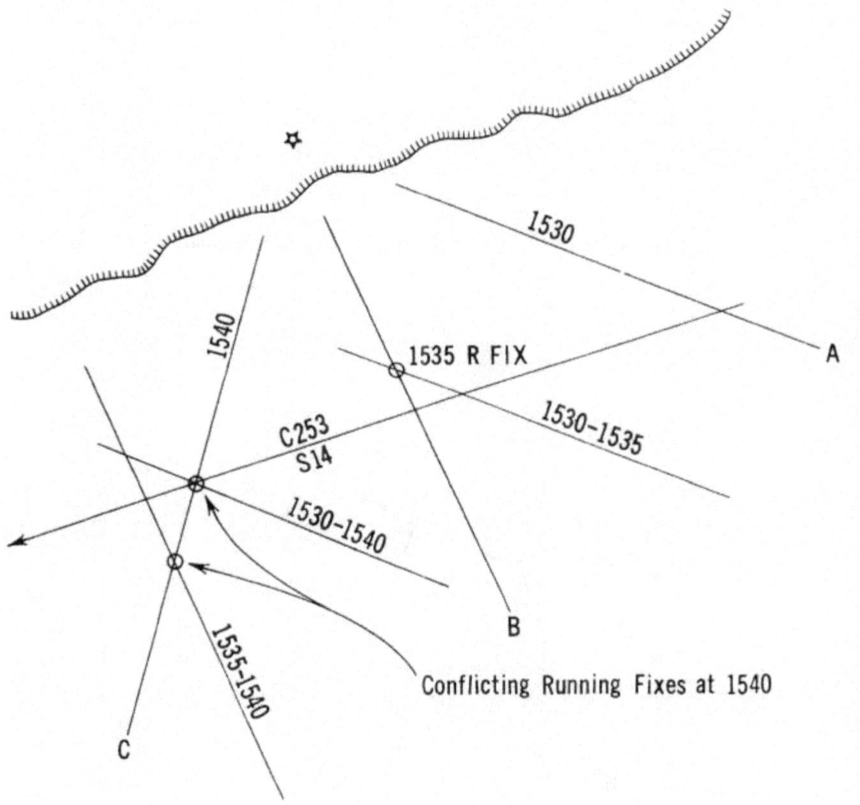

FIGURE 1011f.—Detecting the existence of an oblique current, by a series of running fixes.

Whatever the current, the *direction* of the track made good (assuming constant current and constant course and speed) can be determined. Three bearings of a charted object *O* are observed and plotted (fig. 1011g). Through *O* draw *XY* in any direction. Using a convenient scale, determine points *A* and *B* so that *OA* and *OB* are proportional to the time intervals between the first and second bearings and the second and third bearings, respectively. From *A* and *B* draw lines parallel to the second bearing line, intersecting the first and third bearing lines at *C* and *D,* respectively. The direction of the line from *C* and *D* is the track being made good.

The principle of the method shown in figure 1011g is based on the property of similar triangles. A frequently desirable variation of the method is to use the first bearing line as the side of the triangle that is divided in proportion to the time intervals between bearings (fig. 1011h). This method of solution of the **three-bearing problem** is presented in *The Complete Coastal Navigator* by Charles H. Cotter.

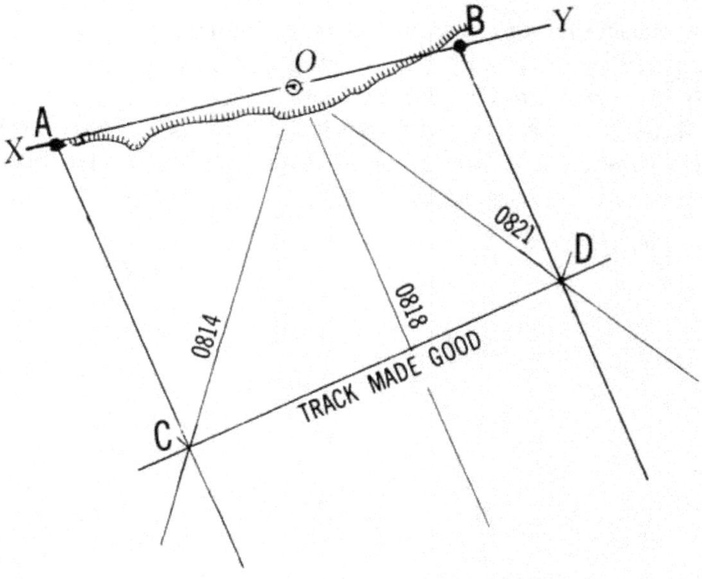

FIGURE 1011g.—Determining the track made good.

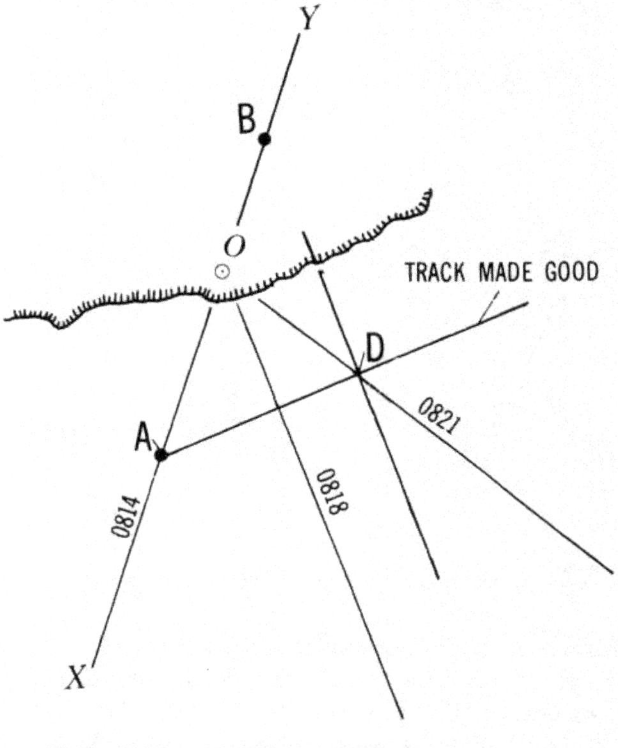

FIGURE 1011h.—Determining the track made good.

The distance of the line *CD* in figure 1011g from the track is in error by an amount proportional to the ratio of the speed being made good to the speed assumed for the solution. If a good fix (not a running fix) is obtained at some time before the first bearing for the running fix, and the current has not changed, the track can be determined by drawing a line from the fix, in the direction of the track made good. The intersection of the track with any of the bearing lines is an actual position.

The current can be determined whenever a dead reckoning position and fix are available for the same time. The direction *from* the dead reckoning position *to* the fix is the set of the current. The distance between these two positions, divided by the time (expressed in hours and tenths) since the last fix, is the drift of the current in knots. For accurate results, the dead reckoning position must be run up from the previous fix without any allowance for current. Any error in either the dead reckoning position (such as poor steering, unknown compass error, inaccurate log, wind, etc.) or the fix will be reflected in the determination of current. *When the dead reckoning position and fix are close together, a relatively small error in either may introduce a large error in the apparent set of the current.*

1012. Distance of an object by two bearings.—A running fix can be obtained by utilizing the mathematical relationships involved. A ship steams past landmark *D* (fig. 1012). At any point *A* a bearing of *D* is observed and expressed as degrees right or left of the course (*a relative bearing if the ship is on course).* At some later time, at *B,* a second bearing of *D* is observed and expressed as before. At *C* the landmark is broad on the beam. The angles at *A, B,* and *C* are known, and also the distance run between points. The various triangles could be solved by trigonometry to find the distance from *D* at any bearing. Distance and bearing provide a fix. Table 7 provides a quick and easy solution.

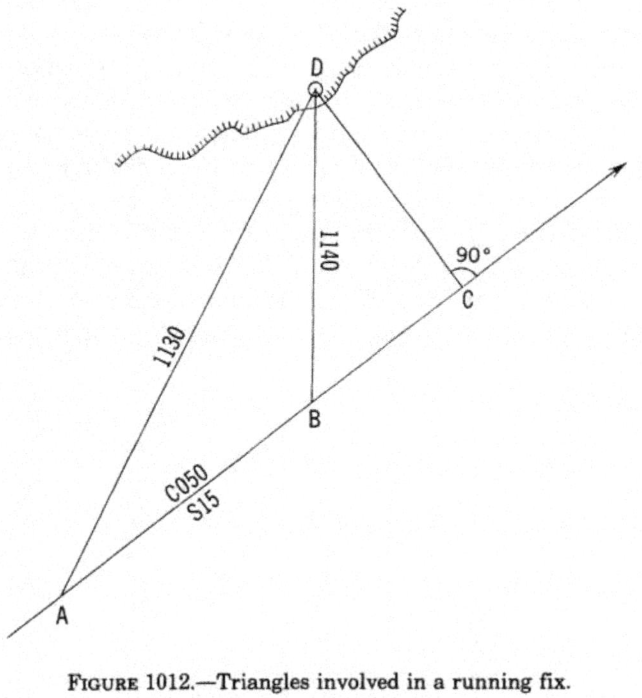

FIGURE 1012.—Triangles involved in a running fix.

Solution by table 7 or one of the special cases is accurate only *if* a *steady course has been steered,* the *vessel is unaffected by current, and the speed used is the speed over ground.*

There are certain *special cases* arising under the method of obtaining a running fix from two bearings and the intervening run which do not require the use of tables. Two of these cases arise when the multiplier is equal to unity, and the distance run is therefore equal to the distance from the object.

If the second difference (angle *CBD* of figure 1012) is double the first difference (angle *BAD*), triangle *BAD* is isosceles (art. 128, vol. II) with equal angles at *A* and *D*. Therefore, side *AB* (the run) is equal to side *BD* (the distance off at the time of the second bearing). This is called **doubling the angle on the bow.** If the first angle is 45° and the second 90°, the distance run equals the distance when broad on the beam. These are called **bow and beam bearings.** Solutions by special cases are discussed in more detail in article 505 of volume II.

1013. Safe piloting without a fix.—A fix or running fix is not always necessary to insure safety of the vessel. If a ship is proceeding up a dredged channel, for instance, the only knowledge needed to prevent grounding is that the ship is within the limits of the dredged area. This information might be provided by a range in line with the channel. A fix is not needed except to mark the point at which the range can no longer be followed with safety.

Under favorable conditions a **danger bearing** might be used to insure safe passage past a shoal or other danger. Refer to figure 1013. A vessel is proceeding along a coast, on intended track *AB*. A shoal is to be avoided. A line *HX* is drawn from lighthouse *H*, tangent to the outer edge of the danger. As long as the bearing of light *H* is *less* than *XH*, the danger bearing, the vessel is in safe water. An example is *YH*, no part of the bearing line passing through the danger area. Any bearing *greater* than *XH*, such as *ZH*, indicates a *possible* dangerous situation. If the object is passed on the port side, the safe bearing is *less* than the danger bearing, as shown in figure 1013. If the object is passed on the starboard side, the danger bearing represents the minimum bearing, safe ones being *greater*. To be effective, a danger bearing should not differ greatly from the course, and the object of which bearings are to be taken should be easily identifiable and visible over the entire area of usefulness of the danger bearing. A margin of safety might be provided by drawing line *HX* through a point a short distance off the danger. In figure 1013, the danger bearing is labeled NMT 074 to indicate that the bearing of the light should not be more than 074°. The hazardous side of the bearing line is hatched. If a natural or artifical range is available as a danger bearing, it should be used.

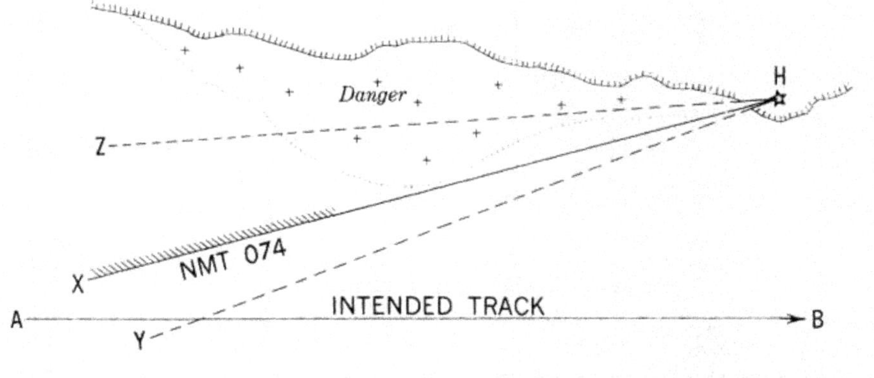

FIGURE 1013.—A danger bearing.

A vessel proceeding along a coast may be in safe water as long as it remains a minimum distance off the beach. This information may be provided by any means available. One method useful in avoiding particular dangers is the use of a **danger angle** (art. 1109).

A vessel may sometimes be kept in safe water by means of a **danger sounding.** The value selected depends upon the draft of the vessel and the slope of the bottom. It should be sufficiently deep to provide adequate maneuvering room for the vessel to reach deeper water before grounding, once the minimum depth is obtained. In an area where the shoaling is gradual, a smaller margin of depth can be considered than in an area of rapid shoaling. Where the shoaling is very abrupt, as off Point Conception, California, no danger sounding is practical. It is good practice to prominently mark the danger sounding line on the chart. A colored pencil is useful for this purpose.

If it is desired to round a point marked by a prominent landmark, without approaching closer than a given minimum distance, this can be done by steaming until the minimum distance is reached and then immediately changing course so as to bring the landmark broad on the beam. Frequent small changes of course are then used to keep the landmark near, *but not forward of,* the beam. This method is not reliable if the vessel is being moved laterally by wind or current.

An approximation of the distance off can be found by noting the rate at which the bearing changes. If the landmark is kept abeam, the change is indicated by a change of heading. During a change of 57°5, the distance off is about the same as the distance run. For a change of 28°5, the distance is about twice the run; for 19° it is about three times the run; for 14°5 it is about four times the run; and for 11°5 it is about five times the run. Another variation is to measure the number of seconds required for a change of 16°. The distance off is equal to this interval multiplied by the speed in knots and divided by 1,000. That is,

$$D = \frac{St}{1{,}000},$$

where D is the distance in nautical miles, S is the speed in knots, and t is the time interval in seconds. This method can also be used for straight courses (with bearings 8° forward and abaft the beam), but with somewhat reduced accuracy.

1014. Soundings.—The most important use of soundings is to determine whether the depth is sufficient to provide a reasonable margin of safety for the vessel. For this reason, soundings should be taken continuously in pilot waters. A study of the chart and the establishment of a danger sounding (art. 1013) should indicate the degree of safety of the vessel at any time.

Under favorable conditions, soundings can be a valuable aid in establishing the position of the vessel. Their value in this regard depends upon the configuration of the bottom, the amount and accuracy of information given on the chart, the type and accuracy of the sounding equipment available aboard ship, and the knowledge and skill of the navigator. In an area having a flat bottom devoid of distinctive features, or in an area where detailed information is not given on the chart, little positional information can be gained from soundings. However, in an area where depth curves run roughly parallel to the shore, a sounding might indicate distance from the beach. In any area where a given depth curve is sharply defined and relatively straight, it serves as a line of position which can be used with other lines, such as those obtained by bearings of landmarks, to obtain a fix. The 100-fathom curve at the outer edge of the continental shelf might be crossed with a line of position from celestial observation or Loran. The crossing of a sharply defined trench, ridge, shoal, or flat-topped seamount (a **guyot**) might provide valuable positional information.

In any such use, identification of the feature observed is important. In an area of rugged underwater terrain, identification might be difficult unless an almost continuous determination of position is maintained, for it is not unusual for a number of features within a normal radius of uncertainty to be similar. If the echo sounder produces a continuous recording of the depth, called a **bottom profile,** this can be matched to the chart in the vicinity of the course line. If no profile is available, a rough approximation of one can be constructed as follows: Record a series of soundings at short intervals, the length being dictated by the scale of the chart and the existing situation. For most purposes the interval might be each minute, or perhaps each half-mile or mile. Draw a straight line on transparent material and, at the scale of the chart, place marks along the line at the distance intervals at which soundings were made. For this purpose the line might be superimposed over the latitude scale or a distance scale of the chart. At each mark record the corresponding sounding. Then place the transparency over the chart and, by trial and error, match the recorded soundings to those indicated on the chart. Keep the line on the transparency parallel or nearly parallel to the course line plotted on the chart. A current may cause some difference between the plotted course line and the track made good. Also, speed over the bottom might be somewhat different from that used for the plot. This should be reflected in the match. This method should be used with caution, because it may be possible to fit the **line of soundings** to several places on the chart.

Exact agreement with the charted bottom should not be expected at all times. Inaccuracies in the soundings, tide, or incomplete data on the chart may affect the match, but general agreement should be sought. Any marked discrepancy should be investigated, particularly if it indicates less depth than anticipated. If such a discrepancy cannot be reconciled, the wisest decision might well be to haul off into deeper water or anchor and wait for more favorable conditions or additional information.

1015. Most probable position (MPP).—Since information sufficient to establish an *exact* position is seldom available, the navigator is frequently faced with the problem of establishing the most probable position of the vessel. If three reliable bearing lines cross at a point, there is usually little doubt as to the position, and little or no judgment is needed. But when conflicting information or information of questionable reliability is received, a decision is required to establish the MPP. At such a time the experience of the navigator can be of great value. Judgment can be improved if the navigator will continually try to account for all apparent discrepancies, even under favorable conditions. If a navigator habitually analyzes the situation whenever positional information is received, he will develop judgment as to the reliability of various types of information, and will learn something of the conditions under which certain types should be treated with caution.

When complete positional information is lacking, or when the available information is considered of questionable reliability, the most probable position might well be considered an **estimated position (EP)**. Such a position might be determined from a single line of position, from a line of soundings, from lines of position which are somewhat inconsistent, from a dead reckoning position with a correction for current or wind, etc.

Whether the most probable position is a fix, running fix, estimated position, or dead reckoning position, it should be kept continually in mind, together with some estimate of its reliability. The practice of continuing a dead reckoning plot from one good fix to another is advisable, whether or not information is available to indicate a most probable position differing from the dead reckoning position, for the DR plot provides an indication of current and leeway. A series of estimated positions may not be consistent because of the continual revision of the estimate as additional information is received. However, it is good practice to plot all MPP's, and sometimes to maintain a separate EP plot based upon the best estimate of track and speed being made good over the ground, for this should furnish valuable information to indicate whether the present course is a safe one.

1016. Allowing for turning characteristics of vessel.—When precise piloting is necessary (as in an area where maneuvering space is limited, when a specified anchorage is approached, or when steaming in formation with other ships), the turning characteristics of the vessel should be considered. That is, a ship does not complete a turn instantaneously, but follows a curve the characteristics of which depend upon the vessel's length, beam, underwater contour, draft, trim, rudder angle, speed, effects of wind and sea, etc. At the moment the rudder is put over, the vessel begins to follow a spiral path (fig. 1016a). This path becomes circular when the vessel has turned about 90°. The distance the vessel moves in the direction of the original course until the new course is reached is called **advance.** The distance the vessel moves perpendicular to the original course during the turn is called **transfer.** The **tactical diameter** is the distance gained to the right or left of the original course after a turn of 180° with a constant rudder angle. The **final diameter** is the diameter of a circle traversed by a vessel after turning through 360° and maintaining the same speed and rudder angle. This diameter is always less than the tactical diameter. It is measured perpendicular to the original course and between the tangents at the points where 180° and 360° of the turn have been completed. The vessel turns with its bow inside and its stern outside the tangent to the path of its center of gravity. The angle between the tangent to this path, the **turning circle,** and the centerline of the vessel is the **drift angle.** After the vessel has assumed its drift angle in a turn, the point on the centerline between the bow and the center of gravity at which the resultant of the velocities of rotation and translation is directed along the centerline is the **pivot point.** To an observer on board, the vessel appears to rotate about this point, which is normally at one-third to one-sixth of the distance from the bow to the center of gravity.

The amount of advance and transfer for a given vessel depends primarily upon the amount of rudder used and the angle through which the ship is to be turned. The speed of the vessel has little effect. Figure 1016b is a simplified illustration of advance and transfer for a turn of less than 90°. This figure does not include the initial drift away from the center of turning due to a lateral force caused by rudder action.

Allowance for advance and transfer is illustrated in the following example.

Example (fig. 1016c).—A ship proceeding on course 100° is to turn 60° to the left to come on a range which will guide it up a channel. For a 60° turn and the amount of rudder used, the advance is 920 yards and the transfer is 350 yards.

Required.—The bearing of flagpole "FP." when the rudder is put over.

Solution.—(1) Extend the original course line, *AB.*

(2) At a perpendicular distance of 350 yards, the transfer, draw a line *A'B'* parallel to the original course line *AB.* The point of intersection, *C,* of *A'B'* with the new course line (located by the range) is the place at which the turn is to be completed.

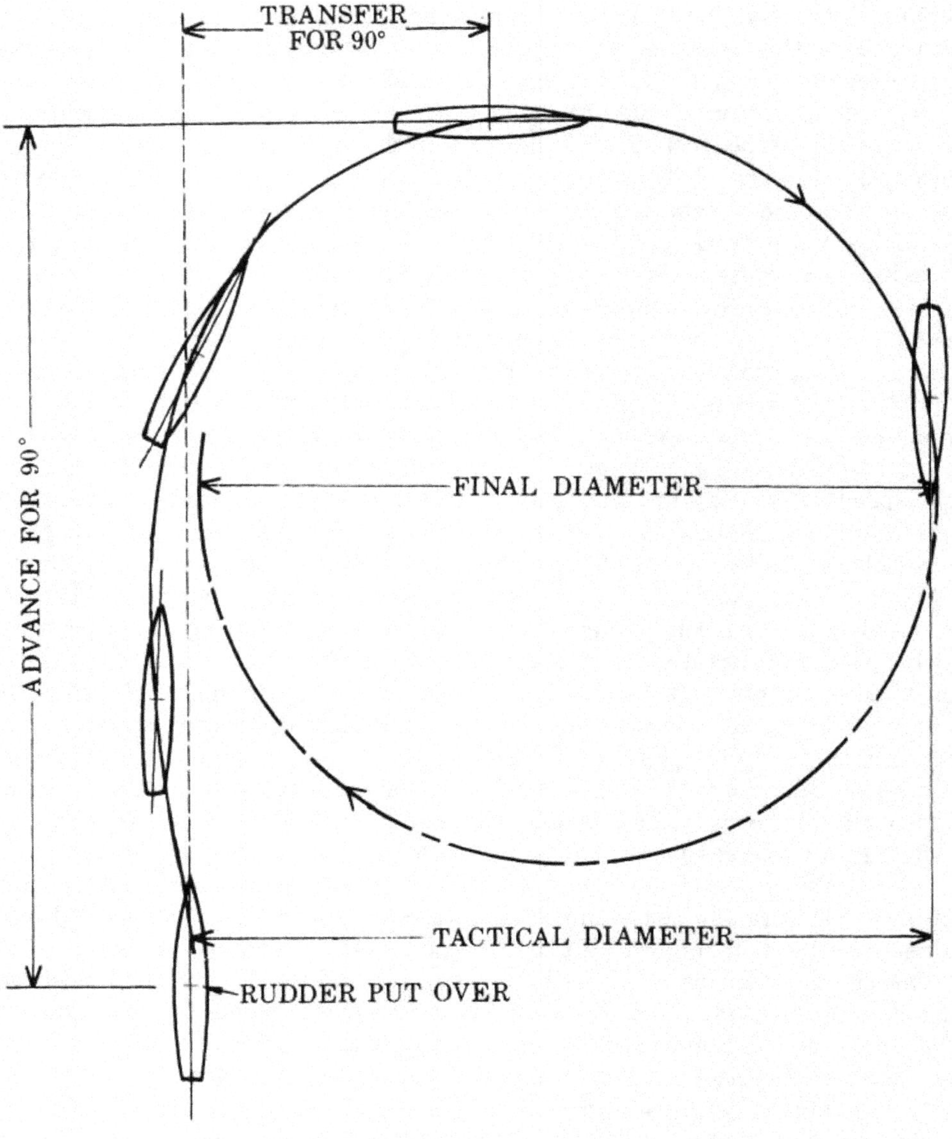

FIGURE 1016a.—Turning circle.

(3) From *C* draw a perpendicular, *CD,* to the original course line, intersecting it at *D.*

(4) From *D* measure the advance, 920 yards, *back* along the original course line. This locates *E,* the point at which the turn should be started.

(5) The direction of "FP." from *E,* 058°, is the bearing when the turn should be started.

Answer.—B 058°.

A frequently useful alternative procedure is the **distance to new course method.** From the vessel's tactical characteristics, a table is constructed to indicate, for various course alterations, speeds, and rudder angles, the distance from the point where the rudder is put over to the intersection (fig. 1016d) of the original course line and the extension of the intended new course line.

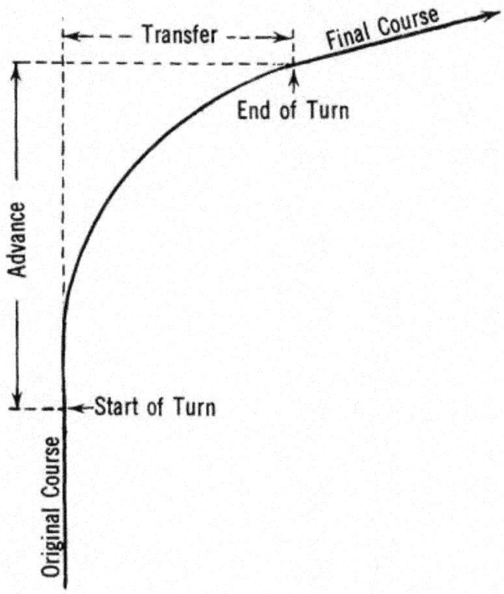

FIGURE 1016b.—Advance and transfer.

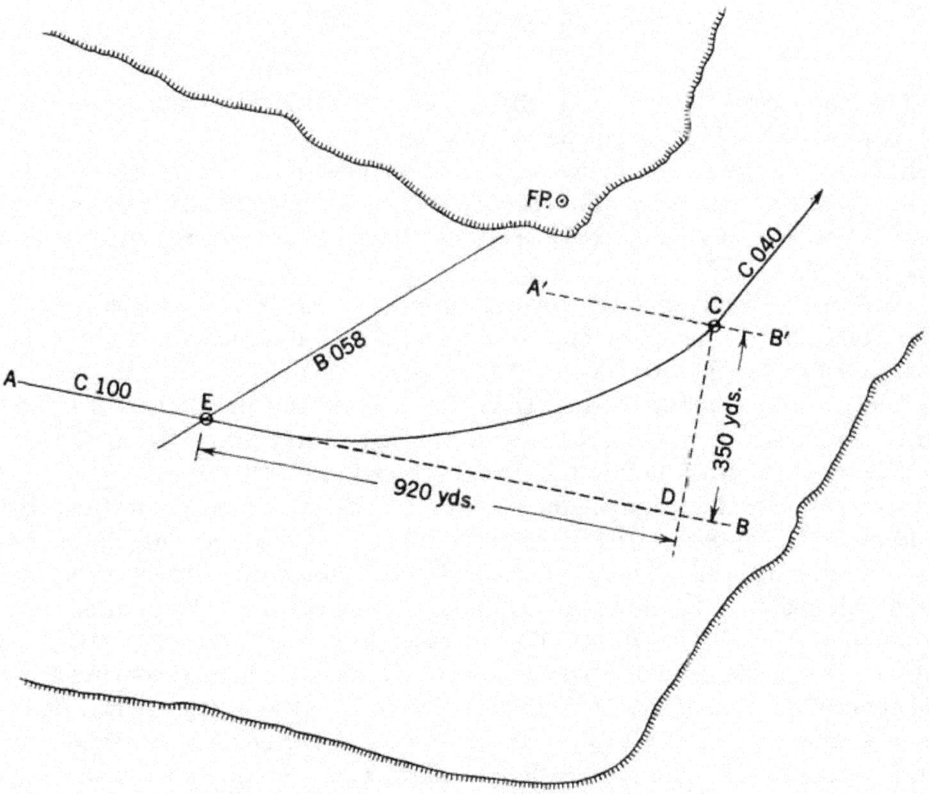

FIGURE 1016c.—Allowing for advance and transfer.

In application, the extension of the intended new course line is drawn to intersect the original course line. The table is referred to to determine the distance

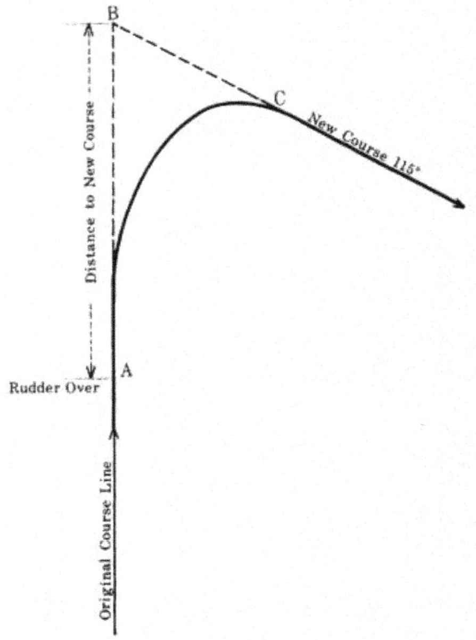

FIGURE 1016d.—Distance to new course method.

from this intersection to the point where the rudder should be put over. When the vessel reaches this point, the rudder is put over.

Although the distance to new course method does not indicate where the vessel will be first on the new course line, it is simpler and faster to use than advance and transfer. However, practical considerations limit its use with course alterations greater than about 120°.

1017. Turning bearing.—The turning bearing is the predetermined bearing to a charted object from that point on the original track at which the rudder must be put over in order to effect the desired turn.

If the turning bearing is such that it is not nearly parallel to the predetermined course to which the vessel is turning, a large error may occur if the vessel is not on the intended track before the turn as shown in figure 1017a.

In order to assist the navigator in revising the turn bearing so that the ship will end up on the approach course shown in figure 1017a, a technique known as the **slide bar** can be utilized. The slide bar consists of a short line representing the next course paralleled back to the predetermined turning point on the original course as shown in figure 1017b and allows the navigator to revise the turn bearing without having to recompute advance and transfer. By measuring the bearing from the object where the actual track intercepts the slide bar, the navigator can determine the revised turning bearing to keep the ship in the center of a narrow channel or on the intended track as shown in figure 1017b. This technique can also be revised for those situations when the navigator desires to remain left or right of the intended track.

1018. Anchoring.—If a vessel is to anchor at a predetermined point, as in an assigned berth, an established procedure should be followed to insure accuracy of placing the anchor. Several procedures have been devised. The following is repre-

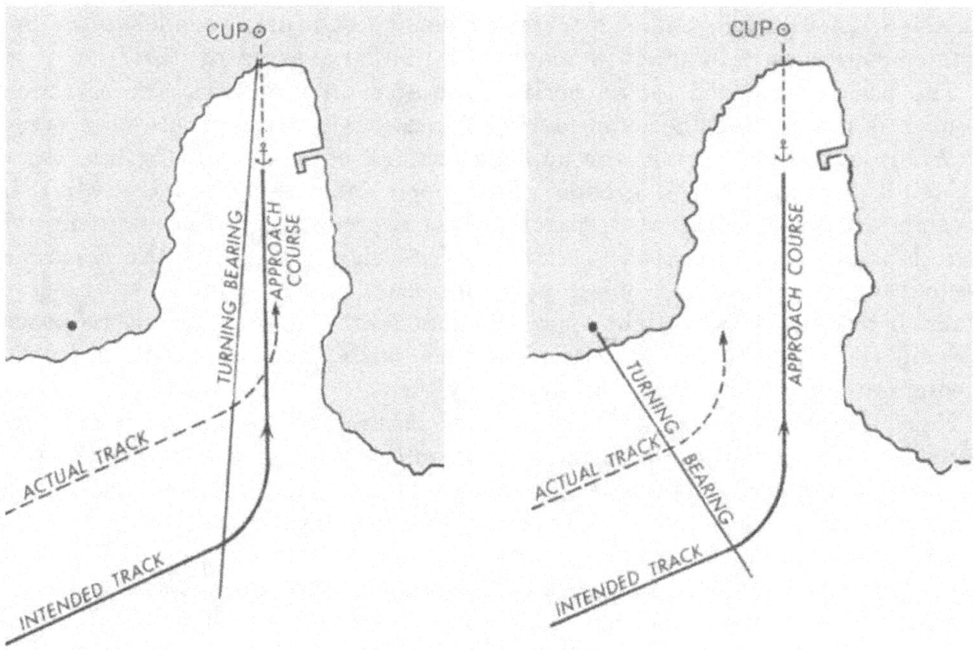

FIGURE 1017a.—Selection of turning bearing to avoid large error when turning to predetermined course.

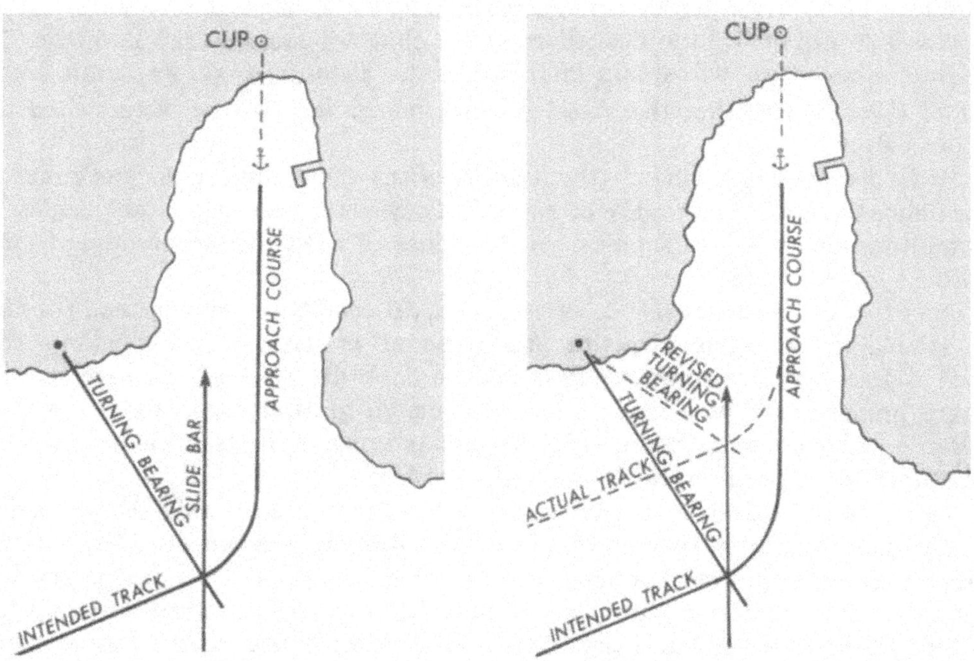

FIGURE 1017b.—Slide bar technique.

sentative (fig. 1018). The use of a turning bearing not nearly parallel to the predetermined course, as in figure 1018, may result in a large error (art. 1017).

The position selected for anchoring is located on the chart. The direction of approach is then determined, considering limitations of land, shoals, other vessels, etc. Where conditions permit, the approach should be made heading into the current or, if the wind has a greater effect upon the vessel, into the wind. It is desirable to approach from such direction that a prominent object, or preferably a range, is available dead ahead to serve as a steering guide. It is also desirable to have a range or prominent object near the beam at the point of letting go the anchor. If practicable a straight approach of at least 500 yards should be provided to permit the vessel to steady on the required course. The track is then drawn in, allowing for advance and transfer during any turns.

Next, a circle is drawn with the selected position of the anchor as the center, and with a radius equal to the distance between the hawsepipe and pelorus, alidade, etc., used for measuring bearings. The intersection of this circle and the approach track, point *A*, is the position of the vessel (bearing-measuring instrument) at the moment of letting go. A number of arcs of circles are then drawn and labeled as shown in figure 1018. The desired position of the anchor is the common center of these arcs. The selected radii may be chosen at will. Those shown in figure 1018 have been found to be generally suitable. In each case the distance indicated is from the small circle. Turning bearings may also be indicated.

During the approach to the anchorage, fixes are plotted at frequent intervals, the measurement and plotting of bearings going on continuously, usually to the nearest half or quarter degree. The navigator advises the captain of any tendency of the vessel to drift from the desired track, so that adjustments can be made. The navigator also keeps the captain informed of the distance to go, to permit adjustment of the speed so that the vessel will be nearly dead in the water when the anchor is let go.

At the moment of letting go, the position of the vessel should be determined as accurately as possible, preferably by two simultaneous horizontal sextant angles, or by simultaneous or nearly simultaneous bearings of a number of prominent landmarks.

The exact procedure to use depends upon local conditions, number and training of available personnel, equipment, and personal preference of individuals concerned. Whatever the procedure, it should be carefully planned, and any needed advance preparations should be made early enough to avoid haste and the attendant danger of making a mistake. Teamwork is important. Each person involved should understand precisely what is expected of him.

1019. Piloting and electronics.—Many of the familiar electronic aids to navigation are used primarily in piloting. The radio direction finder provides bearings through fog and at greater distance from the aids. The sonic depth finder provides frequent or continuous soundings. Radar provides bearings, distances, and information on the location and identity of various targets. Some of the longer range

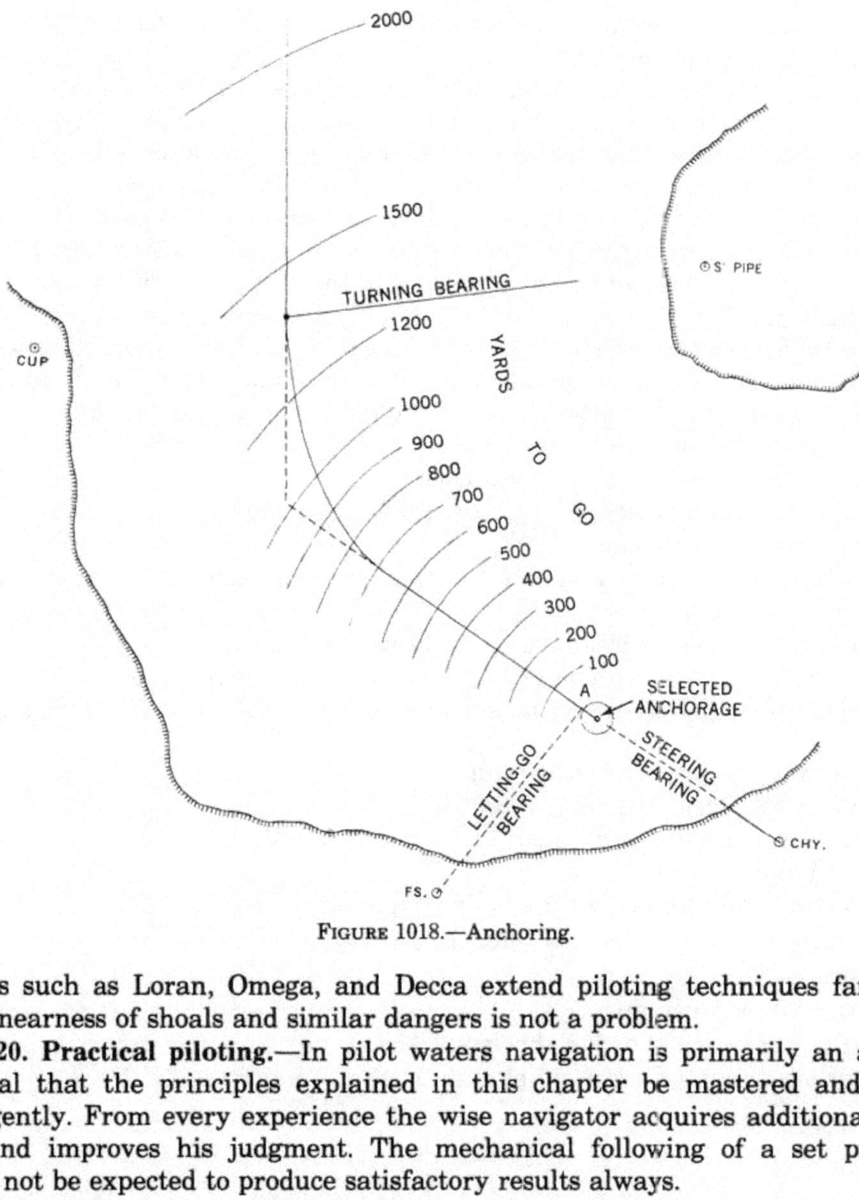

FIGURE 1018.—Anchoring.

systems such as Loran, Omega, and Decca extend piloting techniques far to sea, where nearness of shoals and similar dangers is not a problem.

1020. Practical piloting.—In pilot waters navigation is primarily an art. It is essential that the principles explained in this chapter be mastered and applied intelligently. From every experience the wise navigator acquires additional knowledge and improves his judgment. The mechanical following of a set procedure should not be expected to produce satisfactory results always.

While piloting, the successful navigator is somewhat of an opportunist, fitting his technique to the situation at hand. If a vessel is steaming in a large area having relatively weak currents and moderate traffic, like Chesapeake Bay, fixes may be obtained at relatively long intervals, with a dead reckoning plot between. In a narrow channel with swift currents and heavy traffic, an almost continuous fix is needed. In such an area the navigator may draw the desired track on the chart and obtain fixes every few minutes, or less, directing the vessel back on the track as it begins to drift to one side.

If the navigator is to traverse unfamiliar waters, he studies the chart, sailing directions or coast pilot, tide and tidal current tables, and light lists to familiarize himself with local conditions. The experienced navigator learns to interpret the signs around him. The ripple of water around buoys and other obstructions, the direction and angle of tilt of buoys, the direction at which vessels ride at anchor, provide meaningful information regarding currents. The wise navigator learns to interpret such signs when the position of his vessel is not in doubt. When visibility is poor, or available information is inconsistent, the ability developed at favorable times can be of great value.

With experience, a navigator learns when a danger angle or danger bearing is useful, and what ranges are reliable and how they should be used. However familiar one is with an area, he should not permit himself to become careless in the matter of timing lights for identification, plotting his progress on a chart, or keeping a good recent position. Fog sometimes creeps in unnoticed, obscuring landmarks before one realizes its presence. A series of frequent fixes obtained while various aids are visible provides valuable information on position and current.

Practical piloting requires a thorough familiarity with principles involved and local conditions, constant alertness, and judgment. A study of avoidable groundings reveals that in most cases the problem is not lack of knowledge, but failure to use or interpret available information. Among the more common errors are:

1. Failure to obtain or evaluate soundings.
2. Failure to identify aids to navigation.
3. Failure to use available navigational aids effectively.
4. Failure to correct charts.
5. Failure to adjust a magnetic compass or maintain an accurate table of corrections.
6. Failure to apply deviation, or error in its application.
7. Failure to apply variation, or to allow for change in variation.
8. Failure to check gyro and magnetic compass readings at frequent and regular intervals.
9. Failure to keep a dead reckoning plot.
10. Failure to plot information received.
11. Failure to properly evaluate information received.
12. Poor judgment.
13. Failure to do own navigating (following another vessel).
14. Failure to obtain and use information available on charts and in various publications.
15. Poor ship organization.
16. Failure to "keep ahead of the vessel."

Further discussion on practical piloting is given in chapter XXIII.

CHAPTER XI

USE OF SEXTANT IN PILOTING

1101. Introduction.—The marine sextant provides the most accurate means generally available to the mariner for fixing his position in confined waters. But following the widespread use of the highly reliable gyrocompass, the mariner's use of the sextant for piloting has declined to such an extent that it is seldom if ever used by many. This is unfortunate because the sextant can be used to advantage in situations where other methods or tools, including the gyrocompass, are inadequate.

The applications of the sextant during daylight in coastal waters, harbor approaches, and more confined waters may be summarized as follows:

1. fixing to make a safe transit of hazardous waters;
2. fixing to take a specific geographic position;
3. fixing to establish accurately the position of the anchor on anchoring;
4. fixing to determine whether or not the ship is dragging anchor;
5. using horizontal and vertical danger angles;
6. using vertical angles to determine distance off;
7. fixing to determine the positions of uncharted objects, or to verify the positions of charted features;
8. using the sextant to evaluate the accuracy of navigation by other means.

Because the use of the sextant has declined, many navigators, unfortunately, do not have the proficiency necessary to use it to advantage in those situations where other methods are inadequate. Proficiency in the use of the sextant can be invaluable in situations where even a small error in either observing or plotting cross bearings can result in a grounding.

1102. Three-point problem.—Normally, three charted objects are selected for measuring horizontal sextant angles to determine the observer's position, one of the objects being common to each angular measurement. With simultaneous or nearly simultaneous measurements of the horizontal angles between each pair of charted objects, the observer establishes two circles of position. For each pair of objects, there is only one circle which passes through the two objects and the observer's position. Thus, there are two circles, intersecting at two points as shown in figure 1102a, which pass through the observer's position at *T.*

Since the observer knows that he is not at the intersection at *B,* he must be at *T.*

The solution of what is known as the **three-point problem** is effected by placing the hairlines of the arms of a plastic three-arm protractor over the three observed objects on the chart as shown in figure 1102b. With the arms so placed, the center of the protractor disk is over the observer's position on the chart at the time of the measurements.

1103. Solution without three-arm protractor.—Although the conventional solution of the three-point problem is obtained by placing the arms of a three-arm protractor over the three observed objects on the chart, the use of the protractor is not necessary. The use of the protractor may not be practicable because of limited room and facilities for plotting, as in a small open boat. Where a common charted

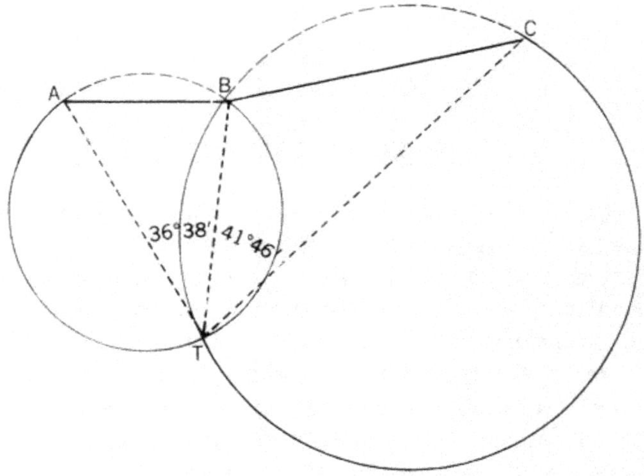

FIGURE 1102a.—Solving the three-point problem.

object cannot be used in the horizontal angle observations, a means other than the three-arm protractor must be employed to determine the position of the observer. Also, point fixes as obtained from the three-arm protractor can be misleading if the navigator has limited skill in evaluating the strengths of the three-point solutions.

In plotting the three-point fix without a three-arm protractor, the procedure is to find the center of each circle of position, sometimes called **circle of equal angle** (fig. 1103a), and then, about such center, to strike an arc of radius equal to the distance on the chart from the circle center to one of the two objects through which the circle passes. The same procedure is applied to the other pair of objects to establish the fix at the intersection of the two arcs.

Some of the methods for finding the center of a circle of equal angle are described in the following text.

The center of the circle of equal angle lies on the perpendicular bisector of the baseline of the pair of objects. With the bisector properly graduated (fig. 1103b), one need only to place one point of the compasses at the appropriate graduation, the other point at one of the observed objects, and then to strike the circle of equal angle or an arc of it in the vicinity of the DR.

The bisector can be graduated through calculation or by means of either the simple protractor or the three-arm protractor.

As shown in figure 1103a when the observed angle is 90°, the center of the circle of equal angle lies at the center of the baseline or at the foot of the perpendicular bisector of the baseline. When the observed angle is less than 90°, for example 40°, the center of the circle lies on the perpendicular bisector on the same side of the baseline as the observer. When the observed angle is 26°34′, the center of the circle lies on the bisector at a distance from its foot equal to the distance between the two objects. When the observed angle is greater than 90°, the center of the circle lies on the perpendicular bisector on the side of the baseline opposite from the observer. The center for 100° is the same distance from the baseline as the center for 80°; the center for 110° is the same distance as the center for 70°, etc. These facts can be used to construct a nomogram for finding the distances of circles of equal angle from the foot of the perpendicular for various angles.

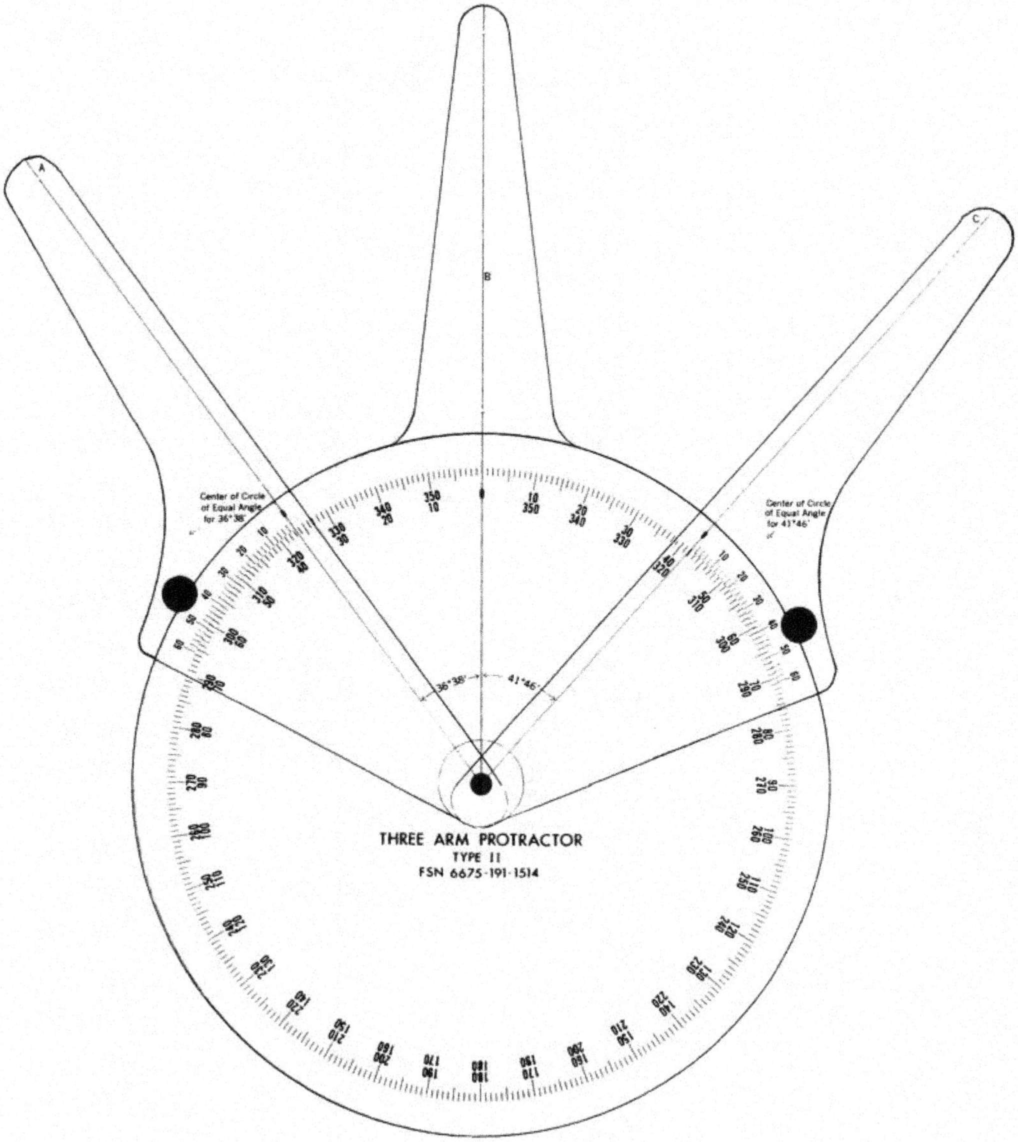

FIGURE 1102b.—Use of three-arm protractor.

From geometry the central angle subtended by a chord is twice the angle with its vertex on the circle and subtended by the same chord. Therefore, when the observed horizontal angle is 30°, the central angle subtended by the baseline is 60°. Or, the angle at the center of the circle between the perpendicular bisector and the line in the direction of one of the observed objects is equal to the observed angle, or 30° as shown in figure 1103c. The angle at the object between the baseline and the center of the circle on the bisector is 90° minus observed angle, or 60°.

The 30° graduation can be located quickly using a suitable protractor as shown in figure 1103d. The placement of the protractor as shown in the upper part of the figure requires moving its center along the bisector until the straightedge passes through the object. The method shown in the lower part enables more rapid

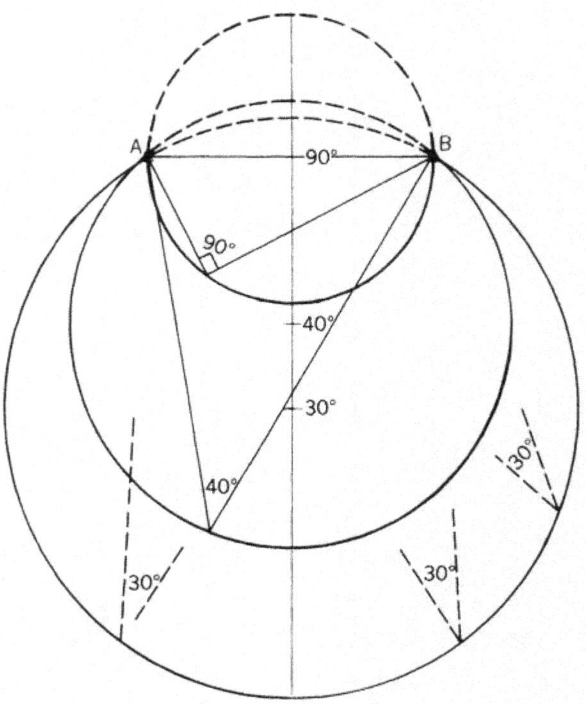

FIGURE 1103a.—Circles of equal angle.

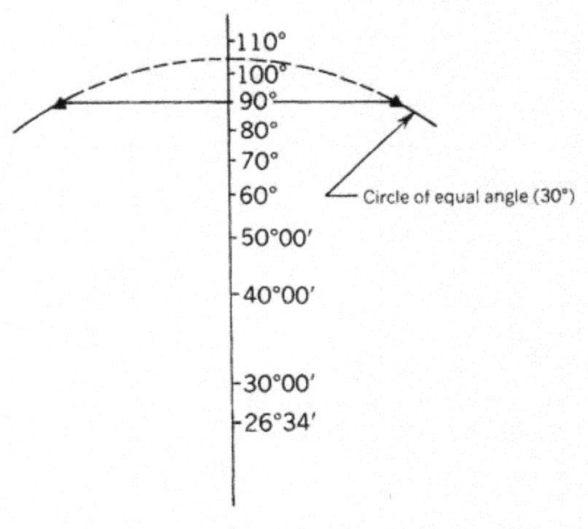

FIGURE 1103b.—Graduated perpendicular bisector.

location of the 30° graduation because there is no need to slide the protractor after its center is placed over the object and the angle is set.

1104. Split fix.—Occasions when a common charted object cannot be used in horizontal angle observations are not infrequent. On these occasions the mariner must obtain what is called a **split fix** through observation of two pairs of charted objects, with no object being common. As with the three-point fix, the mariner will obtain two circles of equal angle, intersecting at two points. As shown in figure 1104, one of these two intersections will fix the observer's position.

1105. Conning aid.—Preconstructed circles of equal angle can be helpful in conning the vessel to a specific geographic position when fixing by horizontal

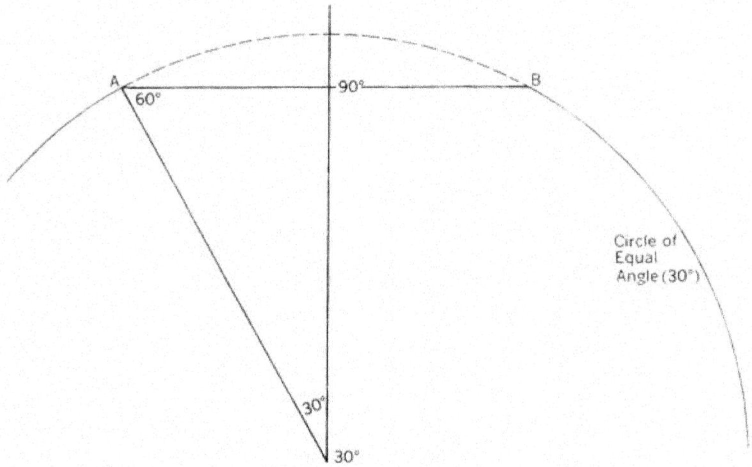

FIGURE 1103c.—Circle of equal angle (30°).

angles. In one application, the vessel is conned to keep one angle constant, or nearly constant, in order to follow the circumference of the associated circle of equal angle to the desired position; the other angle is changing rapidly and is approaching the value for the second circle of equal angle passing through the desired position.

1106. Strength of three-point fix.—Although an experienced hydrographer can readily estimate the strength of a three-point fix, and is able to select the objects providing the strongest fix available quickly, others often have difficulty in visualizing the problem and may select a weak fix when strong ones are available. The following generally useful but not infallible rules apply to selection of charted objects to be observed:

1. The strongest fix is obtained when the observer is inside the triangle formed by the three objects. And in such case the fix is strongest where the three objects form an equilateral triangle (fig. 1106, view A), the observer is at the center, and the objects are close to the observer.

2. The fix is strong when the sum of the two angles is equal to or greater than 180° and neither angle is less than 30°. The nearer the angles are equal to each other, the stronger is the fix (view B).

3. The fix is strong when the three objects lie in a straight line and the center object is nearest the observer (view C).

4. The fix is strong when the center object lies between the observer and a line joining the other two, and the center object is nearest the observer (view D).

5. The fix is strong when two objects a considerable distance apart are in range and the angle to the third object is not less than 45° (view E).

6. Small angles should be avoided as they result in weak fixes in most cases and are difficult to plot. However, a strong fix is obtained when two objects are nearly in range and the nearest one is used as the common object. The small angle must be measured very accurately, and the position of the two objects in range must be very accurately plotted. Otherwise, large errors in position will result. Such fixes are strong only when the common object is nearest the observer. The fix will become very weak when the observer moves to a position where the distant object is the common object (view F).

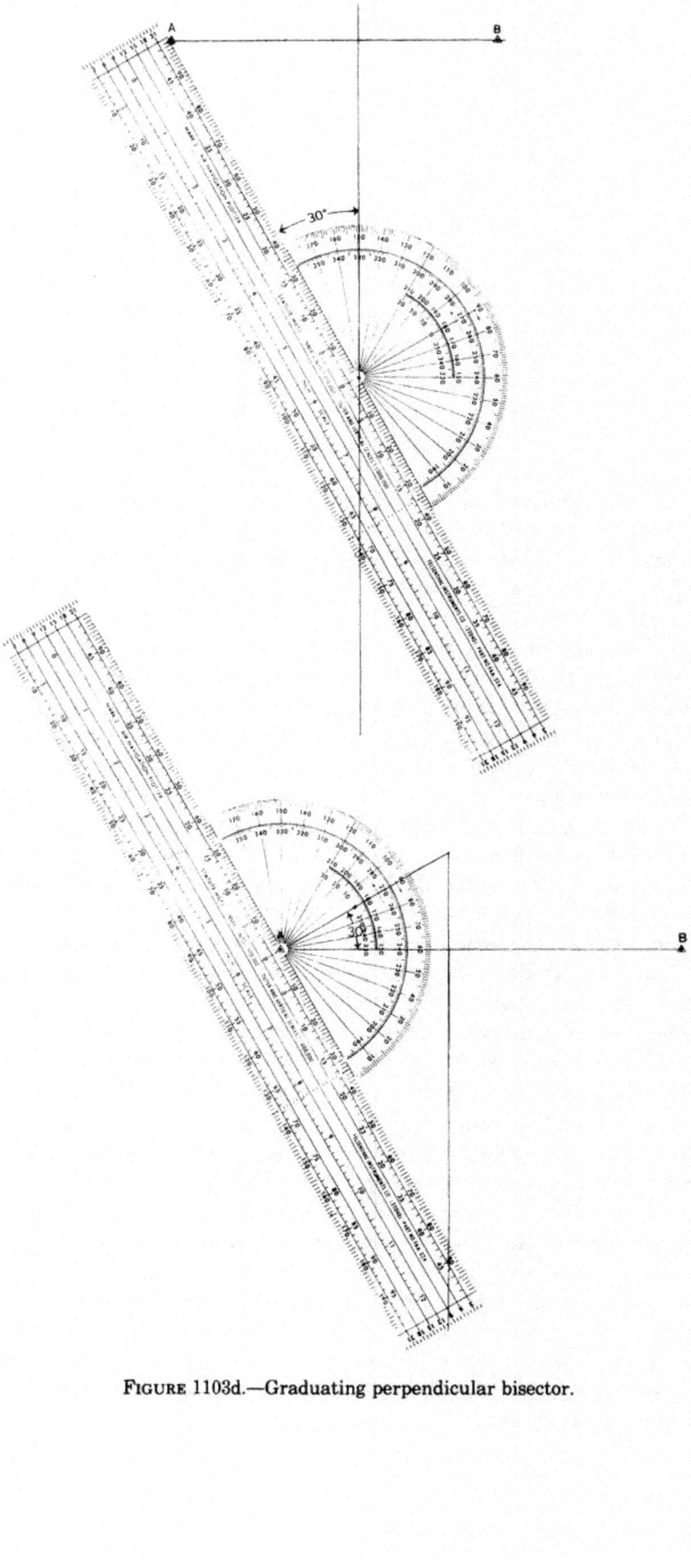

FIGURE 1103d.—Graduating perpendicular bisector.

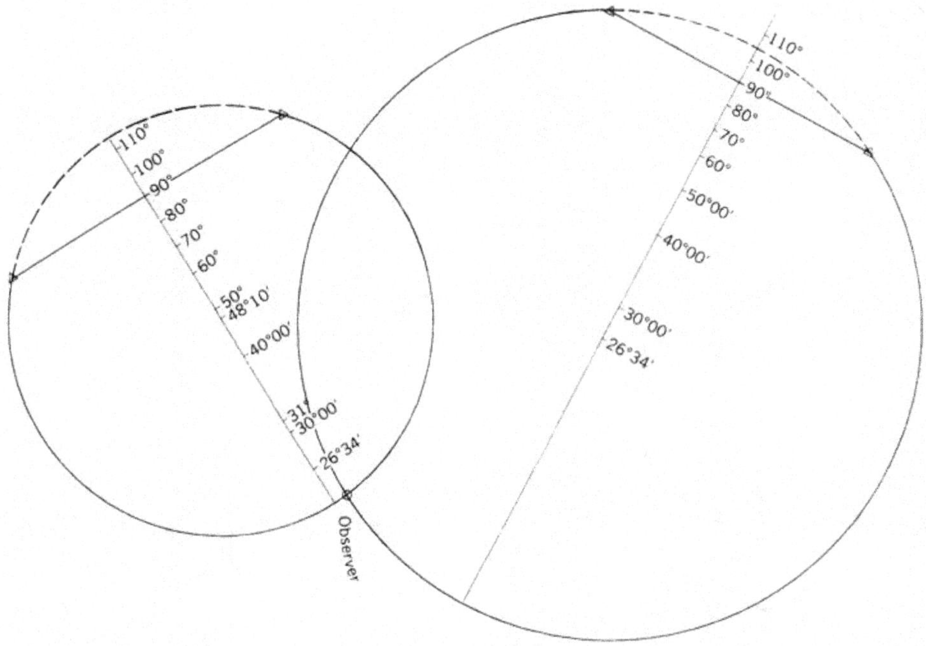

FIGURE 1104.—Split fix.

7. A fix is strong when at least one of the angles changes rapidly as the vessel moves from one location to another.

8. The sum of the two angles should not be less than 50°; better results are obtained when neither angle is less than 30°.

9. Do not observe an angle between objects of considerably different elevation. Indefinite objects such as tangents, hill-tops, and other poorly defined or located points should not be used. Take care to select prominent objects such as major lights, church spires, towers or buildings which are charted and are readily distinguished from surrounding objects.

Beginners should demonstrate the validity of the above rules by plotting examples of each and their opposites. It should be noted that a fix is strong if, in plotting, a slight movement of the center of the protractor moves the arms away from one or more of the stations, and is weak if such movement does not appreciably change the relation of the arms to the three points. An appreciation of the accuracy required in measuring angles can be obtained by changing one angle about five minutes in arc in each example and noting the resulting shift in the plotted positions.

The **error of the three-point fix** will be due to:

1. error in measurement of the horizontal angles;

2. error resulting from observer and observed objects not lying in a horizontal plane;

3. instrument error; and

4. plotting error.

The magnitude of the error varies directly as the error in measurement, the distance of the common object from the observer, (D) and inversely as the sine function of the angle of cut (θ). The magnitude of the error also depends upon the following ratios:

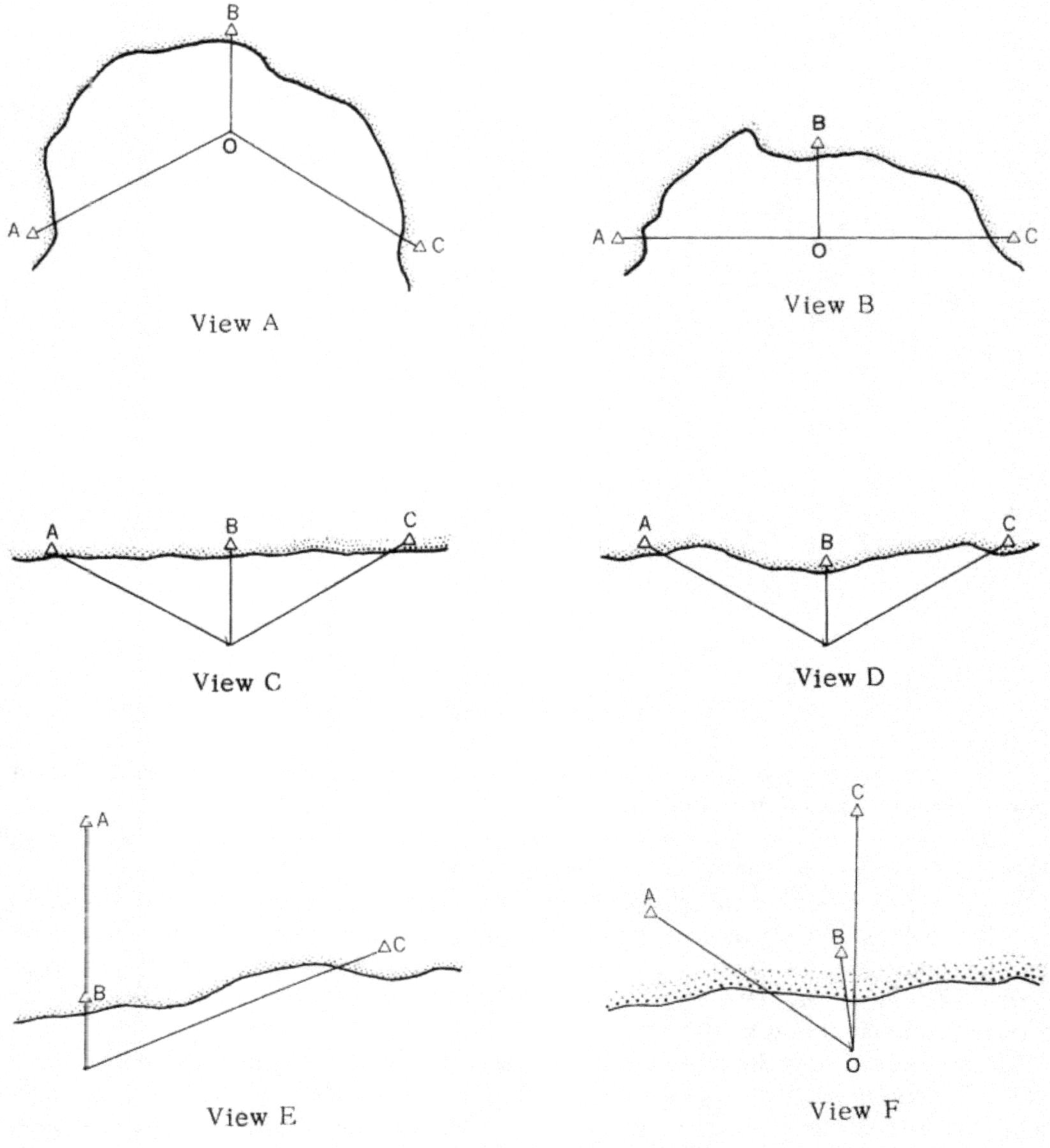

FIGURE 1106.—Strengths of three-point fixes.

1. The distance to the object to the left of the observer divided by the distance from this object to the center object (r_1).

2. The distance to the object to the right of the observer divided by the distance from this object to the center object (r_2).

Assuming that each horizontal angle has the same error (α), the magnitude of the error (E) is expressed in the formula

$$E = \frac{\alpha D}{\sin \theta} \sqrt{r_1^2 + r_2^2 + 2 r_1 r_2 \cos \theta},$$

where error in measurement (α) is expressed in radians.

The magnitude of the error (E) is expressed in the formula

$$E=\frac{.00029a\ D}{\sin\ \theta}\quad\sqrt{r_1^2+r_2^2+2r_1r_2\ \cos\ \theta},$$

where error in measurement (a) is expressed in minutes of arc.

To avoid mistakes in the identification of charted objects observed, either a check bearing or a check angle should be used to insure that the objects used in observation and plotting are the same.

1107. Avoiding the swinger.—Avoid a selection of objects which will result in a "**revolver**" or "**swinger**"; that is, when the three objects observed on shore and the ship are all on, or near, the circumference of a circle (fig. 1107). In such a case the ship's position is indeterminate by three-point fix.

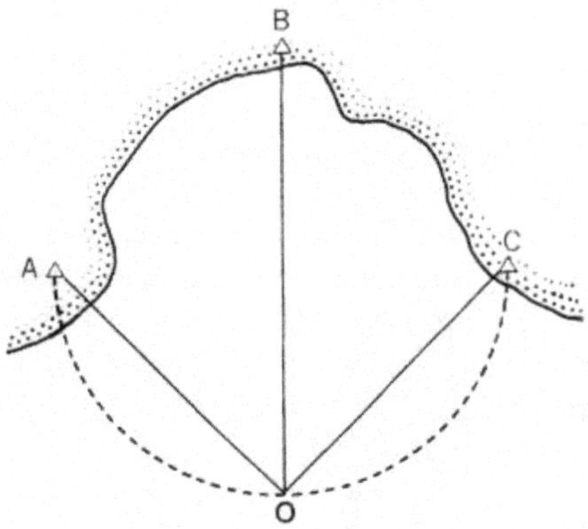

FIGURE 1107.—Revolver or swinger.

If *bearings* as plotted are affected by unknown and uncorrected compass error, the bearing lines may intersect at a point when the objects observed ashore and the ship are all on, or near, the circumference of a circle.

1108. Cutting in uncharted objects.—To cut in or locate on the chart uncharted objects, such as newly discovered offshore wrecks or objects ashore which may be useful for future observations, proceed as follows:

1. Fix successive positions of each ship or ship's boat by three-point fixes, i.e., by horizontal sextant angles. At each fix, simultaneously measure the sextant angle between one of the objects used in the fix and the object to be charted (fig. 1108a). For more accurate results, the craft from which the observations are made should be either lying to or proceeding slowly.

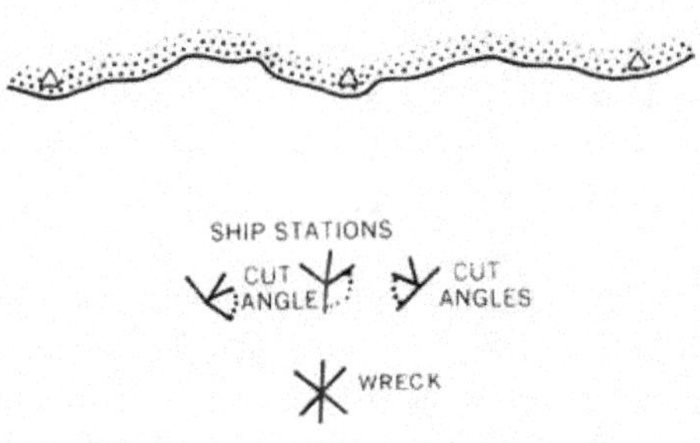

FIGURE 1108a.—Cutting in uncharted objects.

2. For best results, the angles should be measured simultaneously. If verification is undertaken, the angles observed should be interchanged among observers.

3. The fix positions should be selected carefully to give strong fixes, and so that the cuts to the object will provide a good intersection at the next station taken for observations. A minimum of three cuts should be taken.

An alternative procedure is to select observing positions so that the object to be charted will be in range with one of the charted objects used to obtain the three-point fix (fig. 1108b). The charted objects should be selected to provide the best possible intersections at the position of the uncharted object.

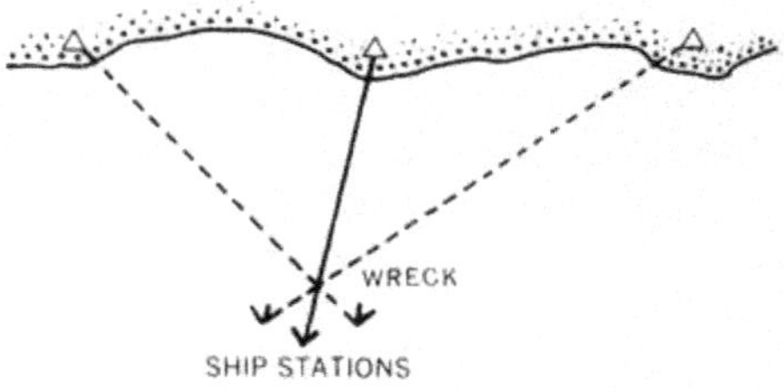

FIGURE 1108b.—On range method.

1109. Horizontal and vertical danger angles.—A vessel proceeding along a coast may be in safe water as long as it remains a minimum distance off the beach. This information may be provided by any means available. One method useful in avoiding particular dangers is the use of a **danger angle.** Refer to figure 1109. A ship is proceeding along a coast on course line *AB*, and the captain wishes to remain outside a danger *D*. Prominent landmarks are located at *M* and *N*. A circle is drawn through *M* and *N* and tangent to the outer edge of the danger. If *X* is a point on this circle, angle *MXN* is the same as at any other point on the circle (except that part between *M* and *N*). Anywhere within the circle the angle is *larger* and anywhere outside the circle it is *smaller*. Therefore, any angle smaller than *MXN* indicates a safe position and any angle larger than *MXN* indicates possible danger. Angle *MXN* is therefore a maximum **horizontal danger angle.** A minimum horizontal danger angle is used when a vessel is to pass *inside* an offlying danger, as at *D′* in figure 1109. In this case the circle is drawn through *M* and *N* and tangent to the *inner* edge of the danger area. The angle is kept larger than *MYN*. If a vessel is to pass *between* two danger areas, as in figure 1109, the horizontal angle should be kept smaller than *MXN* but larger than *MYN*. The minimum danger angle is effective only while the vessel is inside the larger circle through *M* and *N*. Bearings on either landmark might be used to indicate the entering and leaving of the larger circle. A margin of safety can be provided by drawing the circles through points a short distance off the dangers. Any method of measuring the angles, or difference of bearing of *M* and *N*, can be used. Perhaps the most accurate is by horizontal sextant angle. If a single landmark of known height is available, similar procedure can be used with a **vertical danger angle** between top and bottom of the object. In this case the charted position of the object is used as the center of the circles.

1110. Distance by vertical angle.—Table 9 provides means for determining the distance of an object of known height above sea level when the object lies beyond the sea horizon. The vertical angle between the top of the object and the visible (sea) horizon (the sextant altitude) is measured and corrected for index error and dip only. The table is entered with the difference in the height of the object and the height of the eye of the observer and the corrected vertical angle. The distance in nautical miles is taken directly from the table.

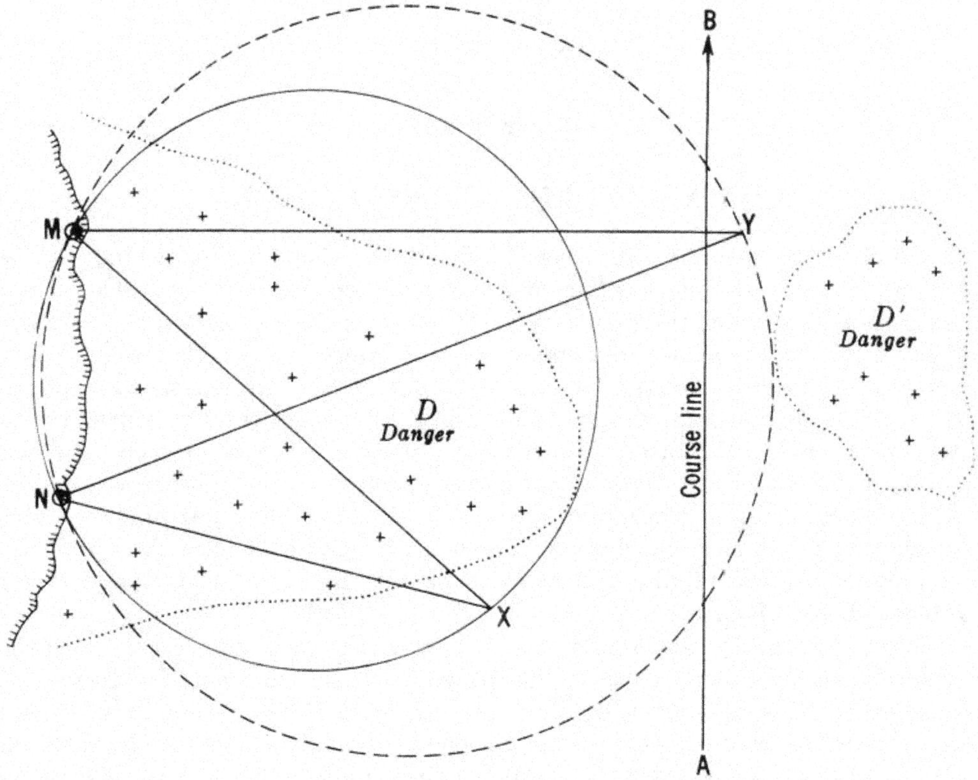

FIGURE 1109.—Horizontal danger angles.

Table 39 provides means for determining the distance to an object lying within or short of the horizon when the height of the eye of the observer is known. The vertical angle between the waterline at the object and the visible (sea) horizon beyond is measured and corrected for index error. The table is entered with the corrected vertical angle and the height of the eye of the observer in feet; the distance in yards is taken directly from the table.

Table 41 provides the solution of a plane right triangle having its right angle at the base of the observed object and its altitude coincident with the vertical dimension of the observed object. The solutions are based upon the following simplifying assumptions: (1) the eye of the observer is at sea level, (2) the sea surface between the observer and the object is flat, (3) atmospheric refraction is negligible, and (4) the waterline at the object is vertically below the peak of the object. The error due to the height of the eye of the observer does not exceed 3 percent of the distance-off for sextant angles less than 20° and the height of eye less than one-third of the object height. The error due to the waterline not being below the peak of the object does not exceed 3 percent of the distance-off when the height of eye is less than one-third of the object height and the offset to the waterline from the base of the object is less than one-tenth of the distance off. Errors due to the earth's curvature and atmospheric refraction are negligible for cases of practical interest. Use of table 41 is explained in article 503 of volume II.

1111. Evaluation.—As time and conditions permit, it behooves the navigator to use the sextant to evaluate the accuracy of navigation by other means in pilot waters. Such accuracy comparisons tend to provide navigators with better appreciation of the limitations of fixing by various methods in a given piloting situation.

CHAPTER XII

TIDE AND CURRENT PREDICTIONS

1201. Tidal effects.—The daily rise and fall of the **tide,** with its attendant flood and ebb of **tidal current,** is familiar to every mariner. He is aware, also, that at **high water** and **low water** the depth of water is momentarily constant, a condition called **stand.** Similarly, there is a moment of **slack water** as a tidal current reverses direction. As a general rule, the *change* in height or the current speed is at first very slow, increasing to a maximum about midway between the two extremes, and then decreasing again. If plotted against time, the height of tide or speed of a tidal current takes the general form of a sine curve. Sample curves, and more complete information about causes, types, and features of tides and tidal currents, are given in chapter XXXI. The present chapter is concerned primarily with the application of tides and currents to piloting, and predicting the tidal conditions that might be encountered at any given time.

Although tides and tidal currents are caused by the same phenomena, the time relationship between them varies considerably from place to place. For instance, if an estuary has a wide entrance and does not extend far inland, the time of maximum speed of current occurs at about the mid time between high water and low water. However, if an extensive tidal basin is connected to the sea by a small opening, the maximum current may occur at about the time of high water or low water outside the basin, when the difference in height is maximum.

The *height of tide* should not be confused with *depth of water.* For reckoning tides a reference level is selected. Soundings shown on the largest scale charts are the vertical distances from this level to the bottom. At any time the actual depth is this charted depth *plus* the height of tide. In most places the reference level is some form of low water. But all low waters at a place are not the same height, and the selected reference level is seldom the *lowest* tide that occurs at the place. When lower tides occur, these are indicated by a negative sign. Thus, at a spot where the charted depth is 15 feet, the actual depth is 15 feet plus height of tide. When the tide is three feet, the depth is $15+3=18$ feet. When it is (−) 1 foot, the depth is $15-1=14$ feet. It is well to remember that *the actual depth can be less than the charted depth.* In an area where there is a considerable **range of tide** (the difference between high water and low water), the height of tide might be an important consideration in using soundings to assist in determining position, or whether the vessel is in safe water.

One should remember that heights given in the tide tables are *predictions,* and that when conditions vary considerably from those used in making the predictions, the heights shown may be considerably in error. Heights lower than predicted are particularly to be anticipated when the atmospheric pressure is higher than normal, or when there is a persistent strong offshore wind. Along coasts where there is a large inequality between the two high or two low tides during a tidal day the height predictions are less reliable than elsewhere.

The current encountered in pilot waters is due primarily to tidal action, but other causes are sometimes present. The tidal current tables give the best prediction of total current, regardless of cause. The predictions for a river may be considerably in error following heavy rains or a drought. The effect of current is to alter the course and speed made good over the bottom. Due to the configuration of land (or shoal areas) and water, the set and drift may vary considerably over different parts of a harbor. Since this is generally an area in which small errors in position of a vessel are of considerable importance to its safety, a knowledge of predicted currents can be critical, particularly if the visibility is reduced by fog, snow, etc. If the vessel is proceeding at reduced speed, the effect of current with respect to distance traveled is greater than normal. Strong currents are particularly to be anticipated in narrow passages connecting larger bodies of water. Currents of more than five knots are encountered from time to time in the Golden Gate at San Francisco. Currents of more than 13 knots sometimes occur at Seymour Narrows, British Columbia.

In straight portions of rivers and channels the strongest currents usually occur in the middle, but in curved portions the swiftest currents (and deepest water) usually occur near the outer edge of the curve. Countercurrents and eddies may occur on either side of the main current of a river or narrow passage, especially near obstructions and in bights.

In general, the range of tide and the speed of tidal current are at a minimum upon the open ocean or along straight coasts. The greatest tidal effects are usually encountered in rivers, bays, harbors, inlets, bights, etc. A vessel proceeding along a coast can be expected to encounter stronger sets toward or away from the shore while passing an indentation than when the coast is straight.

1202. Predictions of tides and currents to be expected at various places are published annually by the National Ocean Service. These are supplemented by eleven sets of tidal current charts (art. 1211), each set consisting of charts for each hour of the tidal cycle. On these charts the set of the current at various places in the area is shown by arrows, and the drift by numbers. Since these are *average* conditions, they indicate in a general way the tidal conditions on any day and during *any* year. They are designed to be used with tidal current diagrams (art. 1211) or the tidal current tables (except those for New York Harbor, and Narragansett Bay, which are used with the tide tables). These charts are available for Boston Harbor, Narragansett Bay to Nantucket Sound, Narragansett Bay, Long Island Sound and Block Island Sound, New York Harbor, Delaware Bay and River, upper Chesapeake Bay, Charleston Harbor, San Francisco Bay, Puget Sound (northern part), and Puget Sound (southern part). Current arrows are sometimes shown on nautical charts. These represent average conditions and should not be considered reliable predictions of the conditions to be encountered at any given time. When a strong current sets over an irregular bottom, or meets an opposing current, ripples may occur on the surface. These are called **tide rips.** Areas where they occur frequently are shown on charts.

Usually, the mariner obtains tidal information from tide and tidal current tables. However, if these are not available, or if they do not include information at a desired place, the mariner may be able to obtain locally the **mean high water lunitidal interval** or the **high water full and change.** The approximate *time* of high water can be found by adding either interval to the time of transit (either upper or lower) of the moon (art. 1902). Low water occurs approximately ¼ tidal day (about 6^h12^m) before and after the time of high water. The actual interval varies somewhat from day to day, but approximate results can be obtained in this manner. Similar information for tidal currents (**lunicurrent interval**) is seldom available.

1203. Tide tables for various parts of the world are published in four volumes by the National Ocean Service. Each volume is arranged as follows:

Table 1 contains a complete list of the predicted times and heights of the tide for each day of the year at a number of places designated as **reference stations.**

Table 2 gives differences and ratios which can be used to modify the tidal information for the reference stations to make it applicable to a relatively large number of **subordinate stations.**

Table 3 provides information for use in finding the approximate height of the tide at any time between high water and low water.

Table 4 is a sunrise-sunset table at five-day intervals for various latitudes from 76°N to 60°S (40°S in one volume).

Table 5 provides an adjustment to convert the local mean time of table 4 to zone or standard time.

Table 6 (two volumes only) gives the zone time of moonrise and moonset for each day of the year at certain selected places.

Certain astronomical data are contained on the inside back cover of each volume.

Extracts from tables 1, 2, and 3 for the East Coast of North and South America are given in appendix L.

1204. Tide predictions for reference stations.—The first page of appendix L is the table 1 daily predictions for New York (The Battery) for the first quarter of 1975. As indicated at the bottom of the page, times are for Eastern Standard Time (+5 zone, time meridian 75°W). Daylight saving time is not used. Times are given on the 24-hour basis. The tidal reference level for this station is mean low water.

For each day, the date and day of week are given, and the time and height of each high and low water are given in chronological order. Although high and low waters are not labeled as such, they can be distinguished by the relative heights given immediately to the right of the times. Since *two* high tides and *two* low tides occur each tidal day, the type of tide at this place is *semidiurnal.* The *tidal* day being longer than the *civil* day (because of the revolution of the moon eastward around the earth), any given tide occurs *later* from day to day. Thus, on Saturday, March 29, 1975, the first tide that occurs is the lower low water (−1.2 feet at 0334). The following high water (lower high water) is 4.9 feet above the reference level (a 6.1 foot rise from the preceding low water), and occurs at 0942. This is followed by the higher low water (−0.9 feet) at 1547, and then the higher high water of 5.5 feet at 2206. The cycle is repeated on the following day with variations in height, and later times.

Because of later times of corresponding tides from day to day, certain days have only one high water or only one low water. Thus, on January 17 high tides occur at 1120 and 2357. The next following high tides are at 1154 on January 18 and 0029 on January 19. Thus, only one high tide occurs on January 18, the previous one being shortly before midnight on the seventeenth, and the next one occurring early in the morning of the nineteenth, as shown.

1205. Tide predictions for subordinate stations.—The second page of appendix L is a page of table 2 of the tide tables. For each subordinate station listed, the following information is given:

Number. The stations are listed in geographical order and given consecutive numbers. At the end of each volume an alphabetical listing is given, and for each entry the consecutive number is shown, to assist in finding the entry in table 2.

Place. The list of places includes both subordinate and reference stations, the latter being given in bold type.

Position. The approximate latitude and longitude are given to assist in locating the station. The latitude is north or south, and the longitude east or west, depending upon the letters (N, S, E, W) next *above* the entry. These may not be the same as those at the *top* of the column.

Differences. The differences are to be applied to the predictions for the reference station shown in bold capitals next *above* the entry on the page. Time and height differences are given separately for high and low waters. Where differences are omitted, they are either unreliable or unknown.

The time difference is the number of hours and minutes to be applied to the time at the reference station to find the time of the corresponding tide at the subordinate station. This interval is added if preceded by a plus sign (+), and subtracted if preceded by a minus sign (−). The results obtained by the application of the time differences will be in the zone time of the time meridian shown directly above the difference for the subordinate station. Special conditions occurring at a few stations are indicated by footnotes on the applicable pages. In some instances, the corresponding tide falls on a different date at reference and subordinate stations.

Height differences are shown in a variety of ways. For most entries separate height differences in feet are given for high water and low water. These are applied to the height given for the reference station. In many cases a *ratio* is given for either high water or low water, or both. The height at the reference station is multiplied by this ratio to find the height at the subordinate station. For a few stations, *both* a ratio and difference are given. In this case the height at the reference station is first multiplied by the ratio, and the difference is then applied. An example is given in each volume of tide tables. Special conditions are indicated in the table or by footnote. Thus, a footnote on the second page of appendix L indicates that "Values for the Hudson River above George Washington Bridge are based upon averages for the six months May to October, when the fresh-water discharge is a minimum."

Ranges. Various ranges are given, as indicated in the tables. In each case this is the difference in height between high water and low water for the tides indicated.

Example.—List chronologically the times and heights of all tides at Yonkers. (No. 1531) on January 2, 1975.

Solution.—

Date	January 2, 1975
Subordinate station	Yonkers
Reference station	New York
High water time difference	(+) 1^h09^m
Low water time difference	(+) 1^h10^m
High water height difference	(−) 0.8 ft.
Low water height difference	0.0 ft.

	New York		*Yonkers*	
HW	2321 (1st)	4.6 ft.	0030	3.8 ft.
LW	0516	(−) 0.6 ft.	0626	(−) 0.6 ft.
HW	1138	4.9 ft.	1247	4.1 ft.
LW	1749	(−) 0.9 ft.	1859	(−) 0.9 ft.

1206. Finding height of tide at any time.—Table 3 of the tide tables provides means for determining the approximate height of tide at any time. It is based upon the assumption that a plot of height versus time is a sine curve (art. 140, vol. II). Instructions for use of the table are given in a footnote below the table, which is reproduced in appendix L.

Example 1.—Find the height of tide at Yonkers (No. 1531) at 1000 on January 2, 1975.

Solution.—The given time is between the low water at 0626 and the high water at 1247 (example of art. 1205). Therefore, the tide is rising. The duration of rise is $1247-0626=6^h21^m$. The range of tide is $4.1-(-0.6)=4.7$ feet. The given time is 2^h47^m *before* high water, the nearest tide. Enter the upper part of the table with duration of rise 6^h20^m (the nearest tabulated value to 6^h21^m), and follow the line horizontally to 2^h45^m (the nearest tabulated value to 2^h47^m). Follow this column vertically downward to the entry 1.8 feet in the line for a range of tide of 4.5 feet (the nearest tabulated value to 4.7 feet). This is the correction to be applied to the nearest tide. Since the nearest tide is high water, subtract 1.8 from 4.1 feet. The answer, 2.3 feet, is the height of tide at the given time.

Answer.—Ht. of tide at 1000, 2.3 ft. A suitable form (fig. 1206a) is used to facilitate the solution.

Interpolation in this table is not considered justified.

It may be desired to know at what time a given depth of water will occur. In this case, the problem is solved in reverse.

Example 2.—The captain of a vessel drawing 22 feet wishes to pass over a temporary obstruction near Days Point, Weehawken (No. 1521), having a charted depth of 21 feet, passage to be made during the morning of January 31, 1975. Refer to figure 1206b.

Required.—The earliest time after 0800 that this passage can be made, allowing a safety margin of 2 feet.

Solution.—The least acceptable depth of water is 24 feet, which is 3 feet more than the charted depth. Therefore, the height of tide must be 3 feet or more. At the New York reference station a low tide of (−)0.9 foot occurs at 0459, followed by a high tide of 4.9 feet at 1120. At Days Point the corresponding low tide is (−)0.9 foot at 0522, and the high tide is 4.6 feet at 1144. The duration of rise is 6^h22^m, and the range of tide is 5.5 feet. The least acceptable tide is 3.0 feet, or 1.6 feet less than high tide. Enter the *lower* part of table 3 with range 5.5 feet and follow the horizontal line until 1.6 feet is reached. Follow this column vertically *upward* until the value of 2^h19^m is reached on the line for a duration of 6^h20^m (the nearest tabulated value to 6^h22^m). The minimum depth will occur about 2^h19^m *before* high water or at about 0925.

Answer.—A depth of 24 feet occurs at 0925.

If the range of tide is more than 20 feet, *half* the range (*one third* if the range is greater than 40 feet) is used to enter table 3, and the correction to height is *doubled* (*tripled* if one third is used).

A diagram for a graphical solution is given in figure 1206c. Eye interpolation can be used if desired. The steps in this solution are as follows:

1. Enter the upper graph with the duration of rise or fall. This is represented by a horizontal line.

2. Find the intersection of this line and the curve representing the interval from the nearest *low* water (point *A*).

3. From *A*, follow a vertical line to the sine curve of the lower diagram (point *B*).

TIDE AND CURRENT TABLES
SRNC-USNA-NC&M-3161/31(1-71)

NAVIGATION DEPARTMENT DIVISION OF NAVAL COMMAND AND MANAGEMENT

COMPLETE TIDE TABLE

Date: Jan. 2, 1975

Substation	Yonkers
Reference Station	New York
HW Time Difference	(+) 1h 09m
LW Time Difference	(+) 1h 10m
Difference in height of HW	(−) 0.8ft.
Difference in height of LW	0.0ft.

Reference Station			Substation	
HW	2321	4.6ft.	0030	3.8ft.
LW	0516	(−) 0.6ft.	0626	(−) 0.6ft.
HW	1138	4.9ft.	1247	4.1ft.
LW	1749	(−) 0.9ft.	1859	(−) 0.9ft.
HW				
LW				

HEIGHT OF TIDE AT ANY TIME

Locality: Yonkers Time: 1000 Date: Jan. 2, 1975

Duration of Rise or Fall:	6h 21m
Time from Nearest Tide:	2h 47m
Range of Tide:	4.7ft.
Height of Nearest Tide:	4.1ft.
Corr. from Table 3:	1.8ft.
Height of Tide at: 1000	2.3ft.

FIGURE 1206a.—U.S. Naval Academy tide form.

4. From *B*, follow horizontally to the vertical line representing the range of tide (point *C*).

5. Using *C*, read the correction from the series of curves.

6. Add (algebraically) the correction of step 5 to the *low* water height, to find the height at the given time.

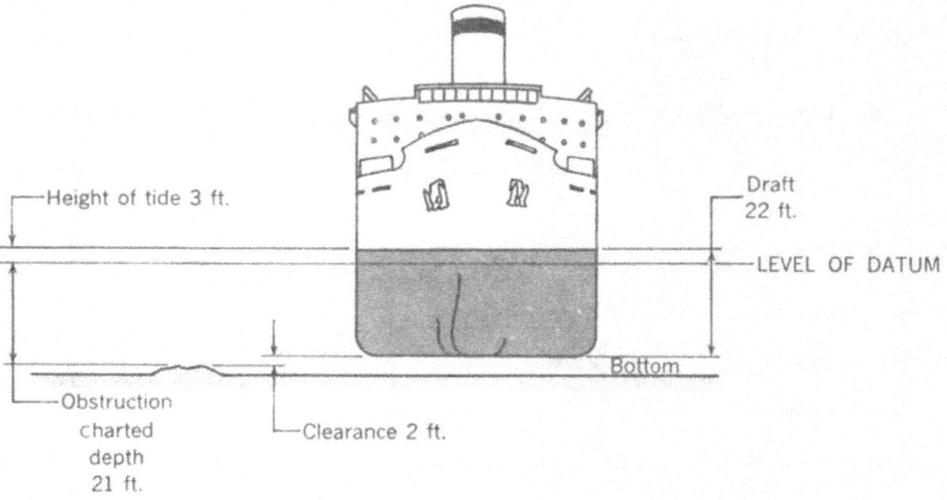

FIGURE 1206b.—Height of tide required to pass clear of charted obstruction.

The problem illustrated in figure 1206c is similar to that of example 1 given above. The duration of rise is 6^h25^m, and the interval from *low* water is 5^h23^m. The range of tide is 6.1 feet. The correction (by interpolation) is 5.7 feet. If the height of the preceding low tide is $(-)0.2$ foot, the height of tide at the given time is $(-)0.2+5.7=5.5$ feet. To solve example 2 by the graph, enter the lower graph and find the intersection of the vertical line representing 5.5 feet and the curve representing 3.9 feet (the minimum acceptable height above low water). From this point follow horizontally to the sine curve, and then vertically to the horizontal line in the upper figure representing the duration of rise of 6^h22^m. From the curve, determine the interval 4^h00^m. The earliest time is about 4^h00^m *after* low water, or at about 0922.

1207. Tidal current tables are somewhat similar to tide tables, but the coverage is less extensive, being given in two volumes. Each volume is arranged as follows:

Table 1 contains a complete list of predicted times of maximum currents and slack, with the velocity (speed) of the maximum currents, for a number of reference stations.

Table 2 gives differences, ratios, and other information related to a relatively large number of subordinate stations.

Table 3 provides information for use in finding the speed of the current at any time between tabulated entries in tables 1 and 2.

Table 4 gives the number of minutes the current does not exceed stated amounts, for various maximum speeds.

Table 5 (Atlantic Coast of North America only) gives information on rotary tidal currents.

Each volume contains additional useful information related to currents. Extracts from the tables for the Atlantic Coast of North America are given in appendix M.

1208. Tidal current predictions for reference stations.—The extracts of appendix M are for The Narrows, New York Harbor. Times are given on the 24-hour basis, for meridian 75°W. *Daylight saving time is not used.*

For each day, the date and day of week are given, with complete current information. Since the cycle is repeated twice each tidal day, currents at this place

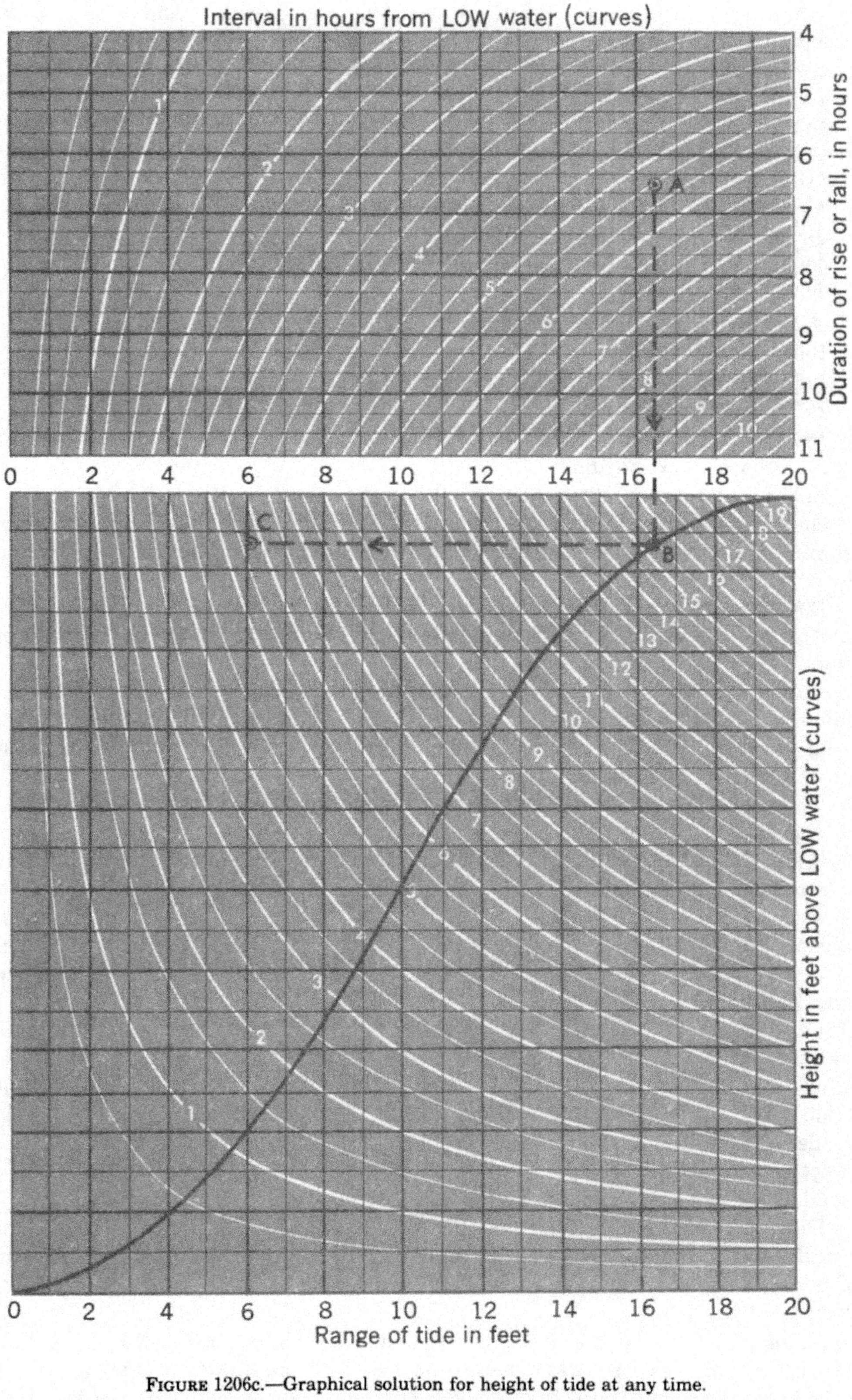

FIGURE 1206c.—Graphical solution for height of tide at any time.

are semidiurnal. On most days there are four slack waters and four maximum currents, two of them floods (F) and two of them ebbs (E). However, since the tidal day is longer than the civil day, the corresponding condition occurs later from day to day, and on certain days there are only three slack waters or three maximum currents. At some places, the current on some days runs maximum flood twice, but ebb only once, a minimum flood occurring in place of the second ebb. The tables show this information.

As indicated by appendix M, the sequence of currents at The Narrows on Monday, February 3, 1975, is as follows:

0000 Flood current, 5^m after maximum velocity (speed).
0305 Slack, ebb begins.
0621 Maximum ebb of 2.0 knots, setting 160°.
1005 Slack, flood begins.
1222 Maximum flood of 1.5 knots, setting 340°.
1516 Slack, ebb begins.
1839 Maximum ebb of 1.9 knots, setting 160°.
2216 Slack, flood begins.
2400 Flood current, 56^m before maximum velocity (speed).

Only one maximum flood occurs on this day, the previous one having occurred 5 minutes before the day began, and the following one predicted for 56 minutes after the day ends.

1209. Tidal current predictions for subordinate stations.—For each subordinate station listed in table 2 of the tidal current tables, the following information is given:

Number. The stations are listed in geographical order and given consecutive numbers, as in the tide tables (art. 1205). At the end of each volume an alphabetical listing is given, and for each entry the consecutive number is shown, to assist in finding the entry in table 2.

Place. The list of places includes both subordinate and reference stations, the latter being given in bold type.

Position. The approximate latitude and longitude are given to assist in locating the station. The latitude is north or south and the longitude east or west as indicated by the letters (N, S, E, W) next *above* the entry. The current given is for the center of the channel unless another location is indicated by the station name.

Time difference. Two time differences are tabulated. One is the number of hours and minutes to be applied to the tabulated times of slack water at the reference station to find the times of slack waters at the subordinate station. The other time difference is applied to the times of maximum current at the reference station to find the times of the corresponding maximum current at the subordinate station. The intervals, which are added or subtracted in accordance with their signs, include any difference in time between the two stations, so that the answer is correct for the standard time of the subordinate station. Limited application and special conditions are indicated by footnotes.

Velocity (speed) ratios. Speed of the current at the subordinate station is found by multiplying the speed at the reference station by the tabulated ratio. Separate ratios may be given for flood and ebb currents. Special conditions are indicated by footnotes.

As indicated in appendix M, the currents at The Battery (No. 2375) can be found by *adding* 1^h30^m for slack water and 1^h35^m for maximum current to the times for The Narrows, and multiplying flood currents by 0.9 and ebb currents by 1.2.

Applying these to the values for Monday, February 3, 1975, the sequence is as follows:

0000 Flood current, 1^h30^m before maximum velocity (speed).
0130 Maximum flood of 1.8 knots, setting 015°.
0435 Slack, ebb begins.
0756 Maximum ebb of 2.4 knots, setting 195°.
1135 Slack, flood begins.
1357 Maximum flood of 1.4 knots, setting 015°.
1646 Slack, ebb begins.
2014 Maximum ebb of 2.3 knots setting 195°.
2346 Slack, flood begins.
2400 Flood current, 14^m after slack.

1210. Finding speed of tidal current at any time.—Table 3 of the tidal current table provides means for determining the approximate velocity (speed) at any time. Instructions for its use are given below the table, which is reproduced in appendix M.

Example 1.—Find the speed of the current at The Battery at 1500 on February 3, 1975.

Solution.—The given time is between the maximum flood of 1.4 knots at 1357 and the slack at 1646 (art. 1209). The interval between slack and maximum current (1646−1357) is 2^h49^m. The interval between slack and the desired time (1646−1500) is 1^h46^m. Enter the table (A) with 2^h40^m at the top, and 1^h40^m at the left side (the nearest tabulated values to 2^h49^m and 1^h46^m, respectively), and find the factor 0.8 in the body of the table. The approximate speed at 1500 is $0.8 \times 1.4 = 1.1$ knots, and it is flooding.

Answer.—Speed 1.1 kn. A suitable form (fig. 1210) is used to facilitate the solution.

It may be desired to determine the period during which the current is less (or greater) than a given amount. Table 4 of the tidal current tables can be used to determine the period during which the speed does not exceed 0.5 knot. For greater speeds, and for more accurate results under some conditions, table 3 of the tidal current tables can be used, solving by reversing the process used in example 1.

Example 2.—During what period on the evening of February 3, 1975, does the ebb current equal or exceed 1.0 knot at The Battery?

Solution.—The maximum ebb of 2.3 knots occurs at 2014. This is preceded by a slack at 1646, and followed by the next slack at 2346. The interval between the earlier slack and the maximum ebb is 3^h28^m, and the interval between the ebb and following slack is 3^h32^m. The desired factor is:

$$\frac{1.0}{2.3} = 0.4.$$

Enter table A with 3^h20^m (the nearest tabulated value to 3^h28^m) at the top, and follow down the column to 0.4 (midway between 0.3 and 0.5). At the left margin the interval between slack and the desired time is found to be 0^h50^m (midway between 0^h40^m and 1^h00^m). Therefore, the current becomes 1.0 knot at $1646 + 0^h50^m = 1736$. Next, enter table A with 3^h40^m (the nearest tabulated value to 3^h32^m) at the top, and follow down the column to 0.4. Follow this line to the left margin, where the interval between slack and desired time is found to be 1^h00^m. Therefore, the current

NAVIGATION DEPARTMENT DIVISION OF NAVAL COMMAND AND MANAGEMENT

COMPLETE CURRENT TABLE

Locality: The Battery Date: Feb. 3, 1975

Reference Station: The Narrows

Time Difference: Slack Water: (+) $1^h 30^m$
Maximum Current: (+) $1^h 35^m$

Velocity Ratio: Maximum Flood: 0.9
Maximum Ebb: 1.2

Flood Direction: 015°
Ebb Direction: 195°

Reference Station: The Narrows		Locality: The Battery	
		0000	F
2355	2.0F	0130	1.8F
0305	0	0435	0
0621	2.0E	0756	2.4E
1005	0	1135	0
1222	1.5F	1357	1.4F
1516	0	1646	0
1839	1.9E	2014	2.3E
2216	0	2346	0
2400	F	2400	F

VELOCITY OF CURRENT AT ANY TIME

Int. between slack and desired time: $1^h 46^m$
Int. between slack and maximum current: $2^h 49^m$ (Ebb) (Flood)
Maximum current: 1.4kn
Factor, Table 3: 0.8
Velocity: 1.1kn
Direction: 015°

DURATION OF SLACK

Times of maximum current:	0756	1357
Maximum current:	2.4 kn	1.4 kn
Desired maximum:	0.3	0.3
Period – Table 4:	35^m	46^m
Sum of periods:		81^m
Average period:		40^m
Time of slack:		1135

Duration of slack: From: 1115 To: 1155

FIGURE 1210.—U.S. Naval Academy tidal current form.

is 1.0 knot or greater until $2346-1^h00^m=2246$. If the two intervals between maximum current and slack were nearest the same 20^m interval, table A would have to be entered only once.

Answer.—The speed equals or exceeds 1.0 knot between 1736 and 2246.

The predicted times of slack water given in the tidal current tables indicate the instant of zero velocity. There is a period each side of slack water, however, during which the current is so weak that for practical purposes it may be considered as negligible. Table 4 of the tidal current tables gives, for various maximum currents, the approximate period of time during which weak currents not exceeding 0.1 to 0.5 knot will be encountered. This duration includes the last of the flood or ebb and the beginning of the following flood or ebb, that is, half of the duration will be before and half after the time of slack water.

When there is a difference between the velocities of the maximum flood and ebb preceding and following the slack for which the duration is desired, it will be sufficiently accurate for practical purposes to find a separate duration for each maximum velocity and take the average of the two as the duration of the weak current.

Of the two subtables of table 4, table A should be used for all places *except* those listed for table B; table B should be used for all places listed, and all stations in table 2 which are referred to them.

Example 3.—Find the period from just before until just after the slack at The Battery at 1135 on February 3, 1975, that the current does not exceed 0.3 kn.

Solution.—Refer to table 4. Table A of table 4 of the tidal current tables is entered with the maximum current before the slack to find the period during which the current does not exceed 0.3 kn. Since there is a difference between the velocities of the maximum ebb and flood preceding and following the slack for which the duration is desired, table A is re-entered with the maximum current after the slack to find the period during which the current does not exceed 0.3 kn. The average of the two values so found is taken as the duration of the weak current. The form shown in figure 1210 is used to facilitate the solution.

Answer.—Duration 40 min. (from 1115 to 1155).

1211. Tidal current charts present a comprehensive view of the hourly speed and direction of the current in 11 bodies of water (art. 1202). They also provide a means for determining the speed and direction of the current at various localities throughout these bodies of water. The arrows show the direction of the current; the figures give the speed in knots at the time of spring tides, that is, during the time of new or full moon when the currents are stronger than average. When the current is given as weak, the speed is less than 0.1 knot. The decimal point locates the position of the station.

The charts depict the flow of the tidal current under normal weather conditions. Strong winds and freshets, however, bring about nontidal currents which may modify considerably the speed and direction shown on the charts.

The speed of the tidal current varies from day to day principally in accordance with the phase, distance, and declination of the moon. Therefore, to obtain the speed for any particular day and hour, the *spring speeds* shown on the charts must be modified by correction factors. A correction table given in the charts can be used for this purpose.

The **tidal current diagrams** are a series of 12 monthly diagrams to be used with the tidal current charts. There is one diagram for each month of the year. A new set of diagrams must be used each year. The diagrams are computer constructed lines that locate each chart throughout all hours of every month. The diagrams indicate directly the chart and the speed correction factor to use at any desired time.

1212. Current diagrams.—A current diagram is a graph showing the speed of the current along a channel at different stages of the tidal current cycle. The current tables include such diagrams for Vineyard and Nantucket Sounds (one diagram); East River, New York; New York Harbor; Delaware Bay and River (one diagram); and Chesapeake Bay. The diagram for New York Harbor is reproduced in appendix M.

On this diagram each vertical line represents a given instant identified in terms of the number of hours before or after slack at The Narrows. Each horizontal line represents a distance from Ambrose Channel Entrance, measured along the usually traveled route. The names along the left margin are placed at the correct distances from Ambrose Channel Entrance. The current is for the center of the channel opposite these points. The intersection of any vertical line with any horizontal line represents a given moment in the current cycle at a given place in the channel. If this intersection is in a shaded area, the current is flooding; if in an unshaded area, it is ebbing. The speed in knots can be found by interpolation (if necessary) between the numbers given in the body of the diagram. The given values are *averages*. To find the value at any given time, multiply the speed found from the diagram by the ratio of *maximum speed of the current involved* to the *maximum shown on the diagram*, both values being taken for The Narrows. If the diurnal inequality is large, the accuracy can be improved by altering the width of the shaded area to fit conditions. The diagram covers 1½ current cycles, so that the right-hand third is a duplication of the left-hand third.

If the current for a single station is desired, table 1 or 2 should be used. The current diagrams are intended for use in either of two ways: First, to determine a favorable time for passage through the channel. Second, to find the average current to be expected during any passage through the channel. For both of these uses a number of "speed lines" are provided. When the appropriate line is transferred to the correct part of the diagram, the current to be encountered during passage is indicated along the line.

Example.—During the morning of January 3, 1975, a ship is to leave Pier 83 at W. 42nd St., and proceed down the bay at 10 knots.

Required.—(1) Time to get underway to take maximum advantage of a favorable current, allowing 15 minutes to reach mid channel.

(2) Average speed over the bottom during passage down the bay.

Solution.—(1) Transfer the line (slope) for 10 knots southbound to the diagram, locating it so that it is centered on the unshaded ebb current section between W. 42nd St. and Ambrose Channel Entrance. This line crosses a horizontal line through W. 42nd St. about one-half of the distance between the vertical lines representing 3 and 2 hours, respectively, after ebb begins at The Narrows. The setting is not critical. Any time within about half an hour of the correct time will result in about the same current. Between the points involved, the entire speed line is in the ebb current area.

(2) Table 1 indicates that on the morning of January 3 ebb begins at The Narrows at 0132. Two hours twenty-eight minutes after ebb begins, the time is 0400. Therefore, the ship should reach mid channel at 0400. It should get underway 15 minutes earlier, at 0345.

(3) To find the average current, determine the current at intervals (as every 2 miles), add, and divide by the number of entries.

Distance	*Current*
18	1.2
16	1.4
14	1.9
12	1.5
10	2.0
8	1.9
6	1.3
4	1.2
2	1.4
0	1.2
sum	15.0

The sum of 15.0 is for 10 entries. The average is therefore 15.0÷10=1.5 knots.

(4) This value of current is correct only if the ebb current is an average one. From table 1 the maximum ebb involved is 2.2 knots. From the diagram the maximum value at The Narrows is 2.0 knots. Therefore, the average current found in step (3) should be increased by the ratio 2.2÷2.0=1.1. The average for the run is therefore 1.5×1.1=1.6 knots. Speed over the botton is 10+1.6=11.6 knots.

Answers.—(1) T 0345, (2) S 11.6 kn.

In the example, an ebb current is carried throughout the run. If the transferred speed line had been partly in a flood current area, all ebb currents (those increasing the ship's speed) should be given a positive sign (+), and all flood currents a negative sign (−). A separate ratio should be determined for each current (flood or ebb), and applied to the entries for that current. In Chesapeake Bay it is not unusual for an outbound vessel to encounter three or even four separate currents during passage down the bay. Under the latter condition, it is good practice to multiply *each* current taken from the diagram by the ratio for the current involved.

If the time of starting the passage is fixed, and the current during passage is desired, the starting time is identified in terms of the reference tidal cycle. The speed line is then drawn through the intersection of this vertical time line and the horizontal line through the place. The average current is then determined in the same manner as when the speed line is located as described above.

Problems

1202. The mean high water lunitidal interval at a certain port is 2^h17^m.

Required.—The approximate times of each high and low water on a day when the moon transits the local meridian at 1146.

Answers.—HW at 0139 and 1403, LW at 0751 and 2015.

1204. List chronologically the times and heights of all tides at New York (The Battery) on February 11, 1975.

Answer.—

Time	*Tide*	*Height*
0222	LW	(−) 0.4 ft.
0829	HW	4.6 ft.
1449	LW	(−) 0.6 ft.
2053	HW	4.2 ft.

1205. List chronologically the times and heights of all tides at Castle Point, Hoboken, N.J. (No. 1519) on March 18, 1975.

Answer.—

Time	*Tide*	*Height*
0533	LW	0.2 ft.
1141	HW	3.5 ft.
1724	LW	0.3 ft.
0003	HW	4.1 ft.

1206a. Find the height of tide at Union Stock Yards, New York (No. 1523) at 0600 on February 6, 1975.

Answer.—Ht. of tide at 0600, 3.8 ft.

1206b. The captain of a vessel drawing 24 feet wishes to pass over a temporary obstruction near Bayonne, N.J. (No. 1505) having a charted depth of 23 feet, passage to be made during the afternoon of March 5, 1975.

Required.—The earliest and latest times that the passage can be made, allowing a safety margin of 2 feet.

Answers.—Earliest time 1316, latest time 1531.

1208. Determine the sequence of currents at The Narrows on January 15, 1975.

Answer.—

0000 Ebb current, 42^m after slack.
0231 Maximum ebb of 1.9 knots.
0557 Slack, flood begins.
0822 Maximum flood of 1.7 knots.
1137 Slack, ebb begins.
1455 Maximum ebb of 2.1 knots.
1836 Slack, flood begins.
2051 Maximum flood of 1.5 knots.
2400 Flood current, 2^m before slack.

1209. Determine the sequence of currents at Ambrose Channel Entrance (No. 2310) on January 12, 1975.

Answer.—

0000 Ebb current, 42^m after maximum velocity (speed).
0241 Slack, flood begins.
0533 Maximum flood of 2.0 knots, setting 310°.
0828 Slack, ebb begins.
1155 Maximum ebb of 2.5 knots.
1527 Slack, flood begins.
1801 Maximum flood of 1.5 knots, setting 310°.
2040 Slack, ebb begins.
2400 Ebb current, 3^m before maximum velocity (speed).

1210a. Find the speed of the current at Bear Mountain Bridge (No. 2445) at 0900 on February 19, 1975.

Answer.—Speed 0.8 kn.

1210b. At about what time during the afternoon of February 3, 1975, does the flood current northwest of The Battery (No. 2375) reach a speed of 1.0 knot?

Answer.—T 1245.

1212. A vessel arrives at Ambrose Channel Entrance 2 hours after flood begins at The Narrows on the morning of February 16, 1975.

Required.—(1) The speed through the water required to take fullest advantage of the flood tide in steaming to Chelsea Docks.

(2) The average current to be expected.

(3) Estimated time of arrival off Chelsea Docks.

Answers.—(1) S 9 kn., (2) S 1.4 kn., (3) ETA 1035.

CHAPTER XIII

SAILING DIRECTIONS AND LIGHT LISTS

1301. Introduction.—Sailing directions (pilots) and light lists provide the information that cannot be shown graphically on the nautical chart and that is not readily available elsewhere. In pilot waters, the prudent navigator makes effective use of all three tools: the nautical chart, sailing directions, and light lists. He does not use one to the exclusion of the others.

Sailing Directions

1302. Format.—The format of the 70 volumes of the sailing directions produced by the Defense Mapping Agency Hydrographic/Topographic Center prior to 1971 was such that each volume, as it related to specific foreign areas, provided detailed descriptions of coasts, channels, dangers, aids, winds, currents, tides, port facilities, signal systems, pilotage, instructions for approaching and entering harbors, as well as a variety of other material required by mariners. This format differed little from sailing directions of centuries past. There were the same geographic divisions and lengthy descriptions of approaches or harbors even though improved charts had obviated the necessity for such detail.

In the earlier format the limited geographic coverage of a given volume precluded inclusion of important information pertaining to transoceanic passages.

Using a new format, the Defense Mapping Agency Hydrographic/Topographic Center is replacing the previous 70 volumes with 47 volumes: 37 *Sailing Directions (Enroute)* and 10 *Sailing Directions (Planning Guide)*. Port facilities data is contained in Pub. No. 150, *World Port Index*.

The old sailing directions described and located features by bearing and distance from previously described landmarks and formed a maze of descriptive hydrography covering the coasts of the world. Sometimes, the description amounted to a mass of verbosity, especially when it pertained to an archipelago. The new Index-Gazetteer, listing each feature by its coordinates, eliminates the need for lengthy, unwieldy descriptive text. Another innovation is the fact that the features described are referred to in the text by page numbers rather than by the chapter-paragraph method used in the old sailing directions.

1303. *Sailing Directions (Planning Guide)*.—Each of the *Sailing Directions (Planning Guide)* contains five chapters, the titles of which are shown in figure 1303a.

The Planning Guides are relatively permanent because of the nature of the material they contain. The *Sailing Directions (Enroute)* must be updated by relatively frequent new editions, and so must the *World Port Index*.

The new sailing directions are designed to assist the navigator in planning a voyage of any extent, particularly if it involves an ocean passage. Each of the *Sailing Directions (Planning Guide)* covers one of the world's great land-sea areas based on an arbitrary division of the world's seaways into eight "ocean basins" as shown in figure 1303b.

Chapter 1 of the Planning Guide, COUNTRIES, contains useful information about all of the countries adjacent to the particular ocean basin being covered by one of the 10 publications. This is the chapter concerned with pratique, pilotage, signals, and pertinent regulations for shipping. A treatment of Search and Rescue includes graphics showing all lifesaving stations and radio stations open to public correspondence.

CHAPTER	
1. COUNTRIES	Governments Regulations Search & Rescue Communications Signals
2. OCEAN BASIN ENVIRONMENT	Oceanography Magnetic Disturbances Climatology
3. WARNING AREAS	Operating Areas, Firing Areas Reference Guide to Warnings and Cautions
4. OCEAN ROUTES	Route Chart & Text Traffic Separation Schemes
5. NAVAID SYSTEMS	Electronic Navigation Systems Systems of Lights & Buoyage

FIGURE 1303a.—Table of contents of *Sailing Directions (Planning Guide).*

Chapter 2 of the Planning Guide, OCEAN BASIN ENVIRONMENT, contains important information relative to the physical environment of an ocean basin. It consists of Ocean Summaries and local coastal phenomena not found in referenced atlases, and provides the mariner with general, concise information concerning the physical forces he must consider in planning a route.

Chapter 3 of the Planning Guide, WARNING AREAS, includes firing danger areas published in foreign sailing directions and not already shown on nautical charts or in other Defense Mapping Agency Hydrographic/Topographic Center publications. A graphic key identifies Submarine Operating Areas. References are made to publications and periodicals which list danger areas, for example *Notice to Mariners No. 1* which gives an annual listing of Atlantic and Pacific danger areas. General cautions pertinent to navigation are given.

In Chapter 4 of the Planning Guide, ROUTES, the recommended steamship routes are described and shown graphically. To facilitate planning, the new publication shows entire routes as they originate from all major U.S. ports and naval bases and terminate at foreign ports in the Planning Guide area. The new concept is in sharp contrast to the localized method used in the old sailing directions. Chapter 4 also includes all applicable Traffic Separation Schemes.

The Planning Guide concludes with Chapter 5, NAVAID SYSTEMS. In keeping with the principles of the new concept, all radionavigation systems pertaining to

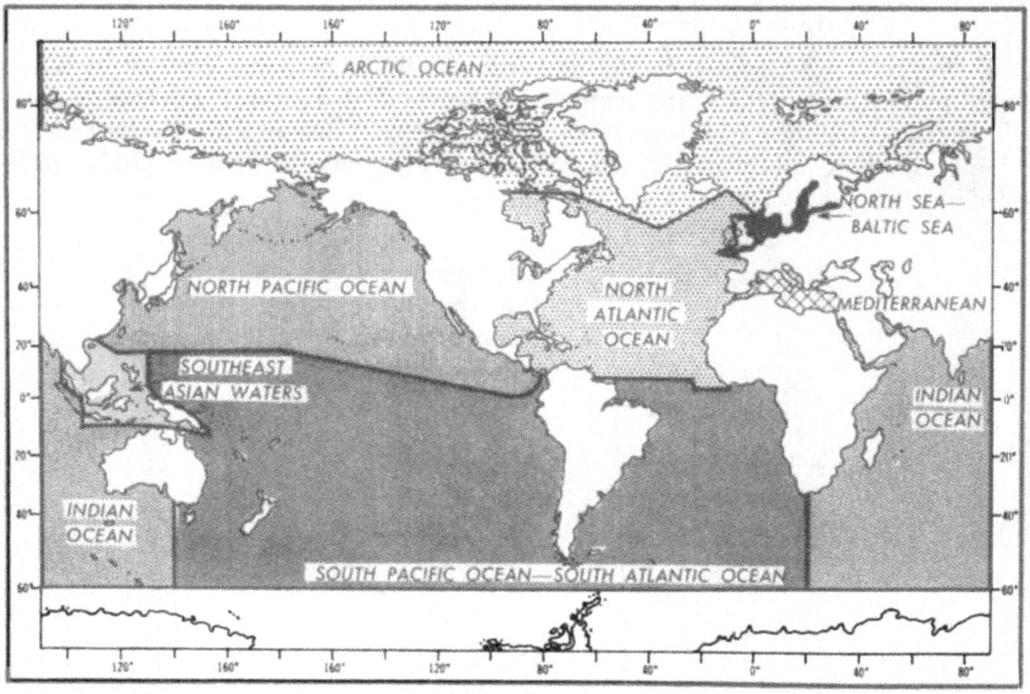

FIGURE 1303b.—Division of world's great land-sea areas into eight ocean basins.

the ocean area are described. The national and international systems of lights, beaconage, and buoyage are also described and illustrated.

1304. *Sailing Directions (Enroute).*—Each volume of the *Sailing Directions (Enroute)* is divided into numbered sectors. Figure 1304a shows a portion of the sectors covered by one of the two *Sailing Directions (Enroute)* covering the English Channel. Figure 1304b illustrates a typical table of contents for a sector.

Chart Information, the first subtitle in the sample table of contents, refers to a graphic key to charts pertaining to a sector. Figure 1304c is an actual sample of the graphic key for Sector 1, "English Channel—Scilly Isles to Start Point." The graduation of the border scale of the chartlet in five-minute increments enables navigators to quickly identify the largest scale chart for a location and to find a feature listed in the publications Index-Gazetteer.

The Index-Gazetteer is simply an alphabetical listing, including both described features and charted features. Each feature is listed with its geographic coordinates and sector number for use with the graphic key. Only features mentioned in the text are given a page number.

Coastal Winds and Currents, the second subtitle in the table of contents, refers to a graphic depicting coastal winds, weather, tides, and currents. Figure 1304d is a reproduction of the graphic for a sector covering the coast of Libya and which appears in one of the new *Sailing Directions (Enroute).* In the new format all of the information previously scattered throughout a chapter of the old sailing directions is given on a single graphic to facilitate use by navigators.

Outer Dangers, the third subtitle in the table of contents, refers to that part of Sector 1 which describes dangers to navigation in the outer portion of a harbor, bay, river, etc. In the old format all dangers, both inner and outer, were described at length. In the new format the outer dangers are fully described, but inner dangers which are well-charted are for the most part omitted. The greatest offshore distance

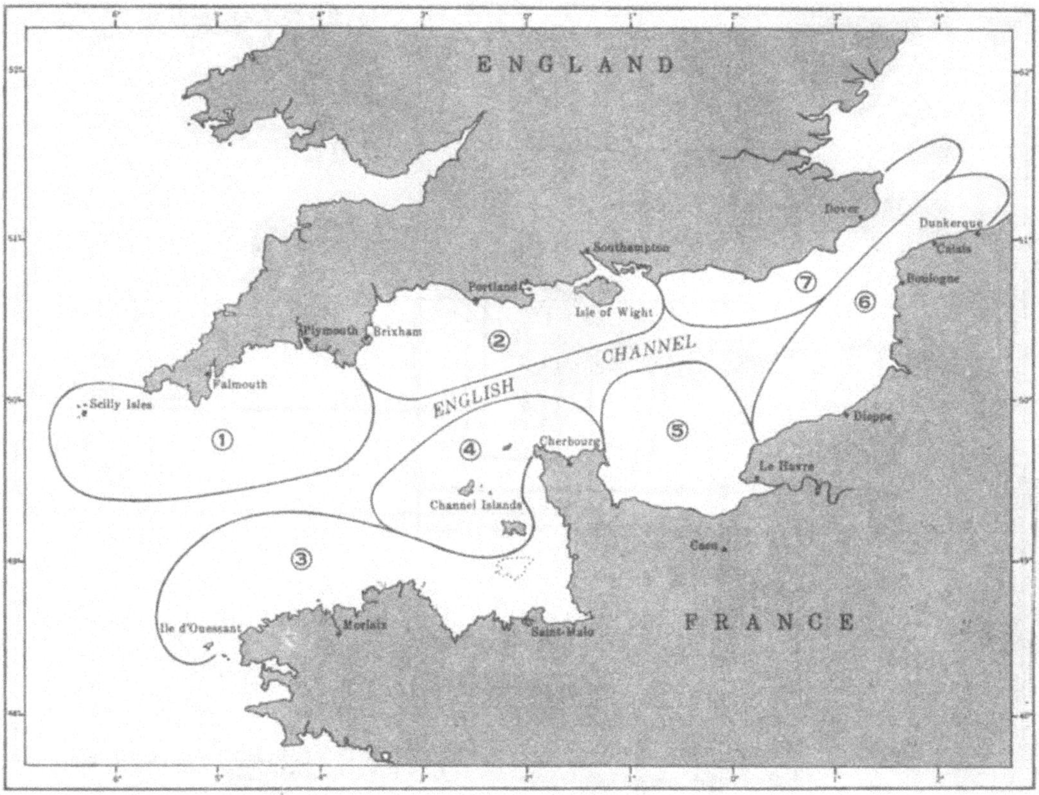

FIGURE 1304a.—Typical diagram of sector limits.

SECTOR 1
- CHART INFORMATION GRAPHIC
- COASTAL WINDS & CURRENTS GRAPHIC
- OUTER DANGERS
- COASTAL FEATURES
- ANCHORAGES (COASTAL)
- MAJOR PORTS
  - —Directions; Landmarks; Navaids; Depths; Limitations; Restrictions; Pilotage; Regulations; Winds; Tides; Currents; Anchorages

FIGURE 1304b.—Typical table of contents of *Sailing Directions (Enroute).*

of the 20-meter (10-fathom) curve, or other appropriate curve, is stated. Numerous offshore dangers, grouped together, are mentioned only in general terms. Dangers adjacent to a coastal passage or fairway are described along with supplementary information to ensure safe passage.

Coastal Features, the fourth subtitle in the table of contents, consists of both text and graphics and includes information in geographical sequence that supplements the charted landmarks, aids to navigation, salient points, fringing reefs, shoals, river mouths, coastal islets, inlets, and bays. In compiling this section it is assumed that the majority of ships have radar and, hence, annotated radarscope photographs have been included whenever possible. Aerial and surface views of harbors and approaches are included to aid mariners in identifying features. Where no photographs or sketches are available, features are described in the text.

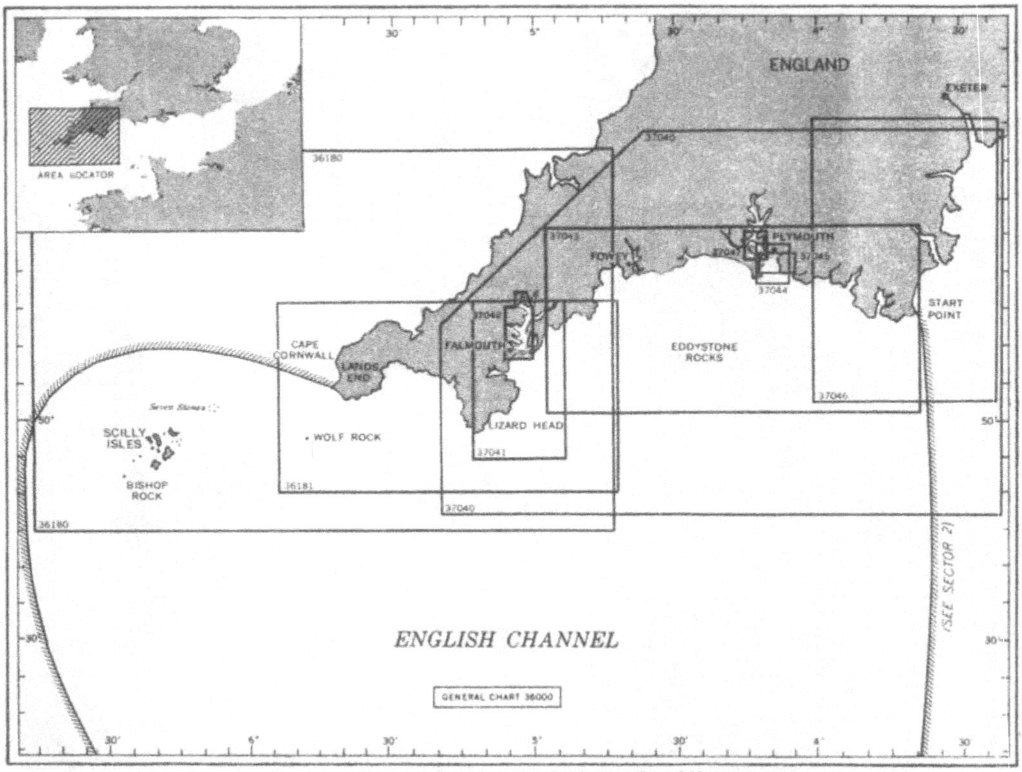

Additional chart coverage may be found in CATP2 Catalog of Nautical Charts.

FIGURE 1304c.—Graphic key to charts within a sector.

Anchorages, the fifth subtitle in the table of contents, describes in geographical sequence all coastal anchoring information pertaining to a sector. A tabulated listing of these anchorages is included in an appendix at the back of the book.

Major Ports, the sixth and final subtitle in the table of contents, gives specific information for the major seaports within a sector. In keeping with the precepts of the new format every effort is made to limit such information to essential facts, thus permitting a significant reduction in the textual material presented. An example is the use of graphic directions for entering a particular port. These graphic directions consist of an annotated chartlet with line drawings of aids to navigation and prominent landmarks. Orientation photos may be included. Port facilities are given in the *World Port Index.*

1305. ***United States Coast Pilots*** published by the National Ocean Service supplement the navigational information shown on nautical charts. These *sailing directions* for United States coastal and intracoastal waters provide information that cannot be shown graphically on nautical charts and that is not readily available elsewhere. *Coast Pilot* subjects include navigation regulations, outstanding landmarks, channel and anchorage peculiarities, dangers, weather, ice, freshets, routes, pilotage, and port facilities.

Each of the eight *Coast Pilots* is corrected through the dates of *Notice to Mariners* shown on the title page, and should not be used without reference to the *Notices to Mariners* issued subsequent to those dates.

The ***Great Lakes Pilot,*** also published by the National Ocean Service, provides similar information for the Great Lakes. Distances given in this publication are expressed in statute miles.

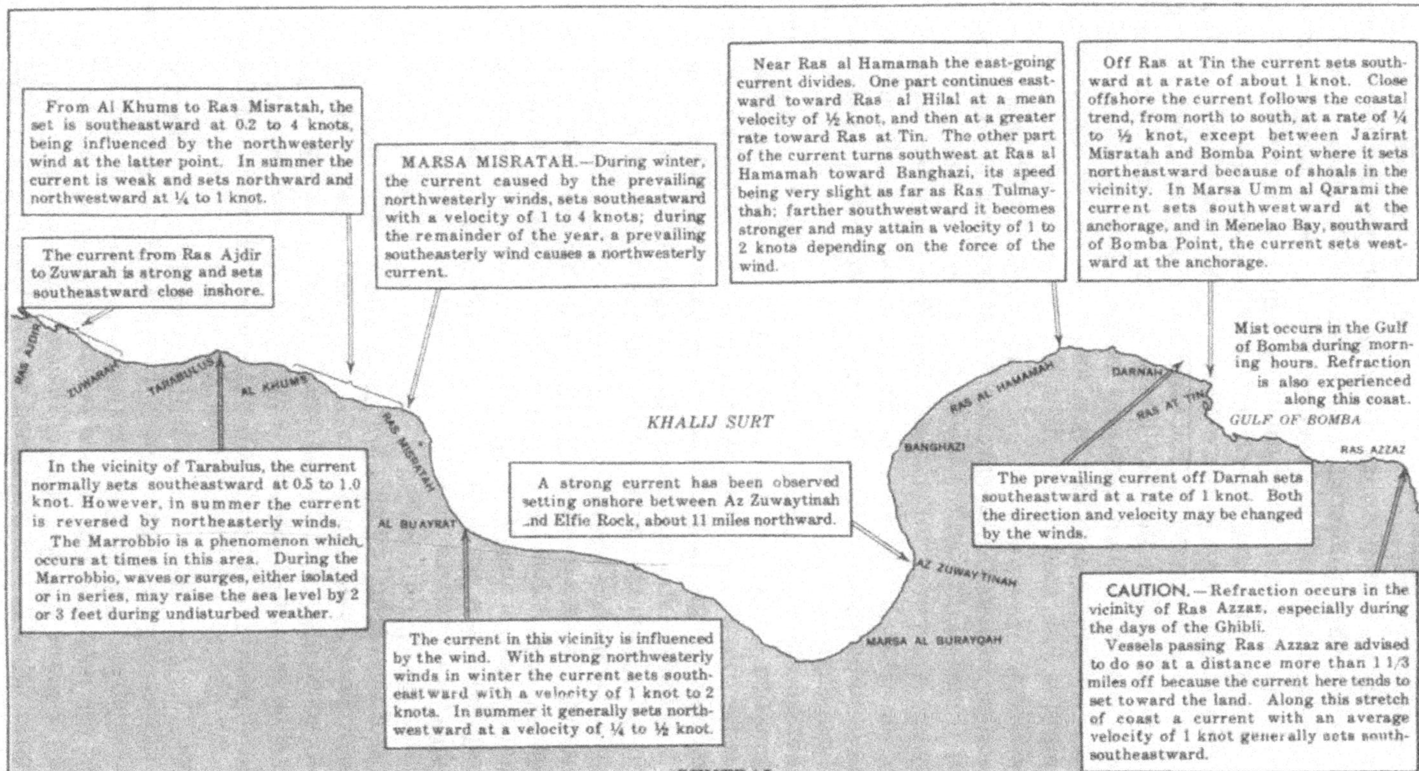

GENERAL

WINDS—WEATHER

During summer, winds are generally light or moderate, with the frequency of stronger winds increasing in late September and October. From November to April, gales are not uncommon.

The summers are hot and dry with hardly any clouds, and the winters are mild. Rain, not too abundant along the African Coast, occurs during the winter season.

During the winter season, vessels working cargo in roadsteads during daylight hours, are advised to anchor farther out at night because of strong northeasterly winds that sometimes reach gale force.

TIDES—CURRENTS

The current generally sets eastward along the Libyan Coast, the flow being augmented by water coming from the Ionian Sea. This easterly setting current is strongly influenced by winds. Reversal of the current is most common in autumn and spring when for short periods southeasterly winds cause the currents to set westward and northwestward along the greater part of the coast.

Because of small tides on the Libyan Coast, tidal currents are generally weak to negligible. The flood current apparently sets eastward along the coast, and the ebb current westward.

FIGURE 1304d.—Graphic depicting coastal winds, weather, tides and currents.

Light Lists

1306. Light lists furnish more complete information concerning aids to navigation than can be conveniently shown on charts. They are not intended to be used for navigation in place of charts and sailing directions (pilots), and should not be so used. The charts should be consulted for the location of all aids to navigation. It may be dangerous to use aids to navigation without reference to charts. Likewise, the charts should not be used without reference to the more detailed information given in the light list, even during daylight. For example: Only the light list may reveal that certain channel buoys are actually located some 50 yards beyond the charted channel limits. Or, only the light list may reveal that a certain charted lighted buoy with a radar reflector is replaced by a nun buoy when endangered by ice. Since this replacement is indicated in the light list, it would not normally be included in *Notice to Mariners*.

Light lists give detailed information regarding navigational lights and sound signals. The U. S. Coast Guard *Light List* for the United States and its possessions, including the Intracoastal Waterway, the Great Lakes (both United States and certain aids on Canadian shores), and the Mississippi River and its navigable tributaries also gives information on unlighted buoys, radiobeacons, radio direction finder calibration stations, daybeacons, racons, and Loran stations. The *Light List* does not include aeronautical lights.

In addition to information on lighted aids to navigation and sound signals, the DMAHTC *List of Lights* for coasts other than the United States and its possessions provides information on storm signals, signal stations, racons, radiobeacons, and radio direction finder calibration stations located at or near lights. However, for detailed information on radio aids the navigator should refer to Pubs. Nos. 117A and 117B, *Radio Navigational Aids*. The *List of Lights* does not include information on lighted buoys in harbors. Those aeronautical lights situated near the coast are listed in the *List of Lights* in order that the marine navigator may be able to obtain more complete information concerning their description. However, it should be borne in mind that these lights are not designed or maintained for marine navigation, and they are subject to changes of which the marine navigator may not receive prompt notification.

Within each volume of the *Light List* aids to navigation are listed in geographic order from north to south along the Atlantic coast, from east to west along the gulf coast, and from south to north along the Pacific coast. Seacoast aids are listed first, followed by entrance and harbor aids listed from seaward to the head of navigation. In volumes I and II, Intracoastal Waterway aids are listed last and in geographic order from north to south along the Atlantic coast and south to north and east to west along the gulf coast.

The introductions to the light lists contain useful information pertaining to the contents which should be carefully studied by the user. In addition to the notes in the remarks columns of the lists, *the user should be sure to refer to all other notes, such as those which may be given near the head of the location column.*

The U.S. Coast Guard *Light List* is published in five volumes; the DMAHTC *List of Lights* is published in seven volumes. The data in both lists are corrected through the *Notice to Mariners* specified in the preface of each volume. For example, the 1984 edition of Pub. 116, *List of Lights and Fog Signals for the Baltic Sea* cancels the previous edition of Pub. 116. The new edition contains information available to DMAHTC up to January 28, 1984, including *Notice to Mariners* No. 4 of 1984. Corrections which have accumulated since the latter date are included in section II weekly. All of these corrections should be applied in the appropriate places and their insertion noted in the "Record of Corrections."

1307. Visual range of lights.—Usually a navigator wants to know not only the identity of a light, but also the area in which he might reasonably expect to observe it. His track is planned to take him within range of lights which can prove useful during periods of darkness. If lights are not sighted within a reasonable time after prediction, a dangerous situation may exist, requiring resolution or action to insure safety of the vessel.

The area in which a light can be observed is normally a circle with the light as the center, and the visual range as the radius. However, on some bearings the range may be reduced by obstructions. In this case the obstructed arc might differ with height of eye and distance. Also, lights of different colors may be seen at different distances. This fact should be considered not only in predicting the distance at which a light can be seen, but also in identifying it. The condition of the atmosphere has a considerable effect upon the distance at which lights can be seen. Sometimes lights are obscured by fog, haze, dust, smoke, or precipitation which may be present at the light, or between it and the observer, but not at the observer, and possibly unknown to him. There is always the possibility of a light being extinguished. In the case of unwatched lights, this condition might not be detected and corrected at once. During periods of armed conflict, certain lights might be deliberately extinguished if they are considered of greater value to the enemy than to one's own vessels.

On a dark, clear night the visual range is limited primarily by one of two ways: (1) luminous intensity and (2) curvature of the earth. A weak light cannot normally be expected to be seen beyond a certain range, regardless of the height of eye. This distance is called luminous range. Light travels in almost straight lines, so that an observer below the visible horizon of the light should not expect to see the light, although the loom extending upward from the light can sometimes be seen at greater distances. Table 8 gives the distance to the horizon at various heights. A condensed version of table 8 is given in the light lists. The tabulated distances assume normal refraction. Abnormal conditions might extend this range somewhat (or in some cases reduce it). Hence, the geographic range, as the luminous range, is not subject to exact prediction at any given time.

The **luminous range** is the maximum distance at which a light can be seen under existing visibility conditions. This luminous range takes no account of the elevation of the light, the observer's height of eye, the curvature of the earth, or interference from background lighting. The luminous range is determined from the known nominal luminous range, called the nominal range, and the existing visibility conditions. The **nominal range** is the maximum distance at which a light can be seen in clear weather as defined by the International Visibility Code (meteorological visibility of 10 nautical miles). The **geographic range** is the maximum distance at which the curvature of the earth permits a light to be seen from a particular height of eye *without* regard to the luminous intensity of the light. The geographic range *sometimes* printed on charts or tabulated in light lists is the maximum distance at which the curvature of the earth permits a light to be seen from a height of eye of 15 feet above the water when the elevation of the light is taken above the height datum of the largest scale chart of the locality.

The geographic range depends upon the height of both the light and the observer, as shown in figure 1307a. In this illustration a light 150 feet above the water is shown. At this height, the distance to the horizon, by table 8, is 14.3 miles.

Within this range the light, *if powerful enough and atmospheric conditions permit,* is visible regardless of the height of eye of the observer (if there is no obstruction). Beyond this range, the visual range depends upon the height of eye. Thus, by table 8 an observer with height of eye of 5 feet can see the light on his horizon if he is 2.6 miles beyond the horizon of the light, or a total of 16.9 miles. For a height of 30 feet the distance is 14.3+6.4=20.7 miles. If the height of eye is 70 feet, the geographic range is 14.3+9.8=24.1 miles.

Except for range and some directional lights, the nominal range is listed in the U. S. Coast Guard *Light List.* The **Luminous Range Diagram** shown in the *Light List* and figure 1307b is used to convert the nominal range to the luminous range. When using this diagram, it must be remembered that the ranges obtained are approximate, the transmissivity of the atmosphere may vary between the observer and the light, and *glare from background lighting will reduce considerably the range at which lights are sighted.* After estimating the meteorological visibility with the aid of the Meteorological Optical Range Table shown in table 1307, the Luminous Range Diagram is entered with the nominal range on the horizontal nominal range scale; a vertical line is followed until it intersects the curve or reaches the region on the diagram representing the meteorological visibility; from this point or region a horizontal line is followed until it intersects the vertical luminous range scale.

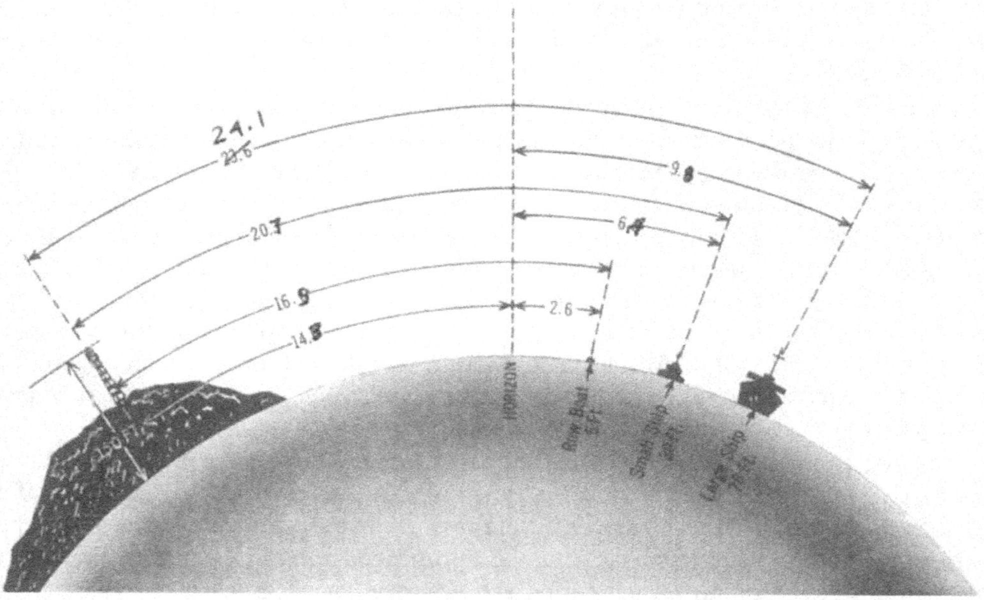

FIGURE 1307a.—Geographic range of a light.

Example 1.—The nominal range of a light as extracted from the *Light List* is 15 nautical miles.

Required.—The luminous range when the meteorological visibility is (1) 11 nautical miles and (2) 1 nautical mile.

Solution.—To find the luminous range when the meteorological visibility is 11 nautical miles, the Luminous Range Diagram is entered with nominal range 15 nautical miles on the horizontal nominal range scale; a vertical line is followed until it intersects the curve on the diagram representing a meteorological visibility of 11 nautical miles; from this point a horizontal line is followed until it intersects the vertical luminous range scale at 16 nautical miles. A similar procedure is followed to find the luminous range when the meteorological visibility is 1 nautical mile.

Answers.—(1) 16 nautical miles; (2) 3 nautical miles.

In predicting the range at which a light can be seen, one should first determine the geographic range to compare this range with the luminous range, if known. If the geographic range is less than the luminous range, the geographic range must be taken as the limiting range. If the luminous range is less than the geographic range, the luminous range must be taken as the limiting range.

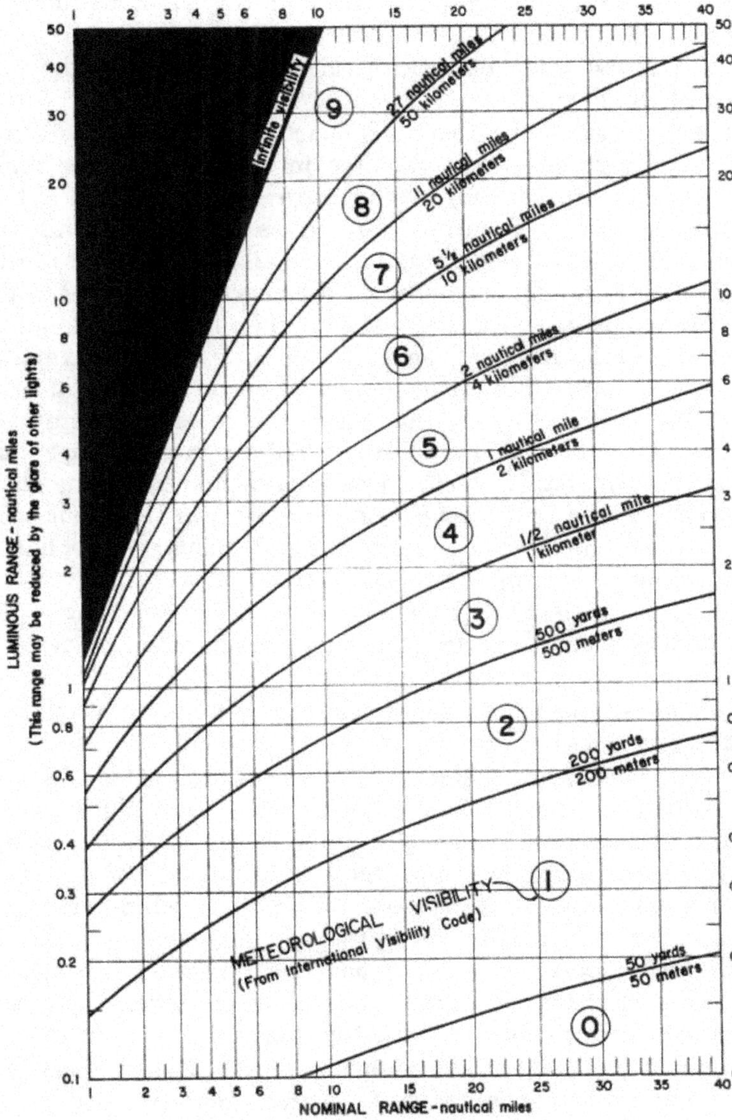

FIGURE 1307b.—Luminous Range Diagram.

These predictions are simple when using the U. S. Coast Guard *Light List* because only nominal ranges are tabulated. Also the current practice of the National Ocean Service is to follow the *Light List* when printing the range of a light on a chart.

Example 2.—The nominal range of a navigational light 120 feet above the chart datum is 20 nautical miles. The meteorological visibility is 27 nautical miles.

Required.—The distance at which an observer at a height of eye of 50 feet can expect to see the light.

Solution.—The maximum range at which the light may be seen is the lesser of the luminous and geographic ranges.

At 120 feet the distance to the horizon, by table 8, is 12.8 miles. Adding 8.3 miles, the distance to the horizon at a height of eye of 50 feet, the geographic range (12.8 mi.+8.3 mi.=21.1 mi.) is found to be less than the luminous range, which is 40 nautical miles.

Answer.—21 nautical miles. Because of various uncertainties, the range is given only to the nearest whole mile.

If the range of a light as printed on a chart, particularly a foreign chart or a reproduction of a foreign chart, or tabulated in a light list other than the U. S. Coast Guard *Light List,* approximates the geographic range for a 15-foot height of eye of the observer, one is generally safe in assuming that this range is the geographic range. With lesser certainty, one may also assume that the lesser of the geographic and nominal ranges is printed on the chart or tabulated in the light list. Using these assumptions, the predicted range is then found by adding the distance to the horizon for both the light and the observer, or approximately, by the *difference* between 4.5 miles (the distance to the horizon at a height of 15 feet) and the distance for the height of eye of the observer (a constant for any given height) and *adding* this value to the tabulated or charted geographic range (subtracting if the height of eye is less than 15 feet). In making a prediction, one should keep in mind the possibility of the luminous range being *between* the tabulated or charted geographic range and the predicted range. The intensity of the light, if known, should be of assistance in identifying this condition.

Example 3.—The range of a light as printed on a foreign chart is 17 miles. The light is 120 feet above chart datum. The meteorological visibility is 10 nautical miles.

Required.—The distance at which an observer at a height of eye of 50 feet can expect to see the light.

Solution.—At 120 feet the distance to the horizon, by table 8, is 12.8 miles. Adding 4.5 miles (the distance to the horizon at a height of 15 feet), the geographic range is found to approximate the range printed on the chart. Then assuming that the latter range is the geographic range for a 15-foot height of eye of the observer and that the nominal range is the greater value, the predicted range is found by adding the distance to the horizon for both the light and the observer (predicted range=12.8 mi.+8.3 mi.=21.1 mi.). The additional distance, i.e., the distance in excess of the assumed charted geographic range, is dependent upon the luminous intensity of the light and the meteorological visibility.

If one is approaching a light, and wishes to predict the *time* at which it should be sighted, he first predicts the range. It is then good practice to draw an arc indicating the visual range. The point at which the course line crosses the arc of visual range is the predicted position of the vessel at the time of sighting the light. The predicted time of arrival at this point is the predicted time of sighting the light. The direction of the light from this point is the predicted bearing at which the light should be sighted. Conversion of the true bearing to a relative bearing is usually helpful in sighting the light. The accuracy of the predictions depends upon the accuracy of the predicted range, and the accuracy of the predicted time and

Code No.	Weather	Yards
0	Dense fog	Less than 50
1	Thick fog	50–200
2	Moderate fog	200–500
3	Light fog	500–1000
		Nautical Miles
4	Thin fog	½–1
5	Haze	1–2
6	Light haze	2–5½
7	Clear	5½–11
8	Very clear	11.0–27.0
9	Exceptionally clear	Over 27.0

From the International Visibility Code

TABLE 1307.—Meteorological Optical Range Table.

place of crossing the visual range arc. If the course line crosses the visual range arc at a small angle, a small lateral error in track may result in a large error of prediction, both of bearing and time. This is particularly apparent if the vessel is *farther* from the light than predicted, in which case the light might be passed without being sighted. Thus, if a light is not sighted at the predicted time, the error *may* be on the side of safety. However, such an interpretation should not be given unless confirmed by other information, for there is always the possibility of reduced meteorological visibility, or of the light being extinguished.

When a light is first sighted, one might determine whether it is on the horizon by immediately reducing the height of eye by several feet, as by squatting or changing position to a lower height. If the light disappears, and reappears when the original height is resumed, it is on the horizon. This process is called **bobbing a light.** If a vessel has considerable vertical motion due to the condition of the sea, a light sighted on the horizon may alternately appear and disappear. This may lead the unwary to assign faulty characteristics and hence to err in its identification. The true characteristics should be observed after the distance has decreased, or by increasing the height of eye of the observer.

PART THREE
CELESTIAL NAVIGATION

CHAPTER XIV

NAVIGATIONAL ASTRONOMY

Preliminary Considerations

1401. Introduction.—Astronomy is that science which deals with the size, constitution, motions, relative positions, etc., of celestial bodies. **Navigational astronomy** is that part of astronomy of direct use to a navigator, comprising principally celestial coordinates, time, and the apparent motions of celestial bodies with respect to the earth. Sometimes it is called **nautical astronomy.**

1402. Apparent and absolute motions.—All celestial bodies of which man has knowledge are in motion. Since the earth itself is one of these moving bodies, the motion of other bodies, as seen by an observer on the earth, is **apparent motion.** If the earth were stationary in space, any change in the position of another body, relative to the earth, would be due only to the motion of that body. This would be **absolute motion,** or motion relative to a fixed point. But since it has been impossible to identify a fixed point in space, all motion of which man is aware is apparent, made up of a combination of the movement of the other body and the motions of the observer. A person without suitable instruments is not aware of motion in the line of sight, and therefore only motions across the line of sight are observed.

Since all motion is relative, one should be cognizant of the position of the observer when motions are discussed. When one speaks of planets following their orbits around the sun, he is placing the observer at some distant point in space, usually one of the poles of the ecliptic (art. 1419). When he speaks of a body rising or setting, the observer is on the earth. If he refers to a particular rising or setting, he must locate the observer at a particular point on the earth, since the setting sun for one observer may be the rising sun for another. At the same time it may be crossing the meridian of a third observer.

1403. The celestial sphere.—As one looks at the sky on a dark night, he is not aware of the differences in the distances to the various celestial bodies. They might easily be imagined as being equally distant from the earth, all located on the inner surface of a vast hollow sphere of infinite radius, with the earth at its center. This is the **celestial sphere** (fig. 1403). For most purposes of navigation it can be considered an actuality. Since the navigator is concerned primarily with apparent motion for an observer on the earth, this geocentric universe of Ptolemy (art. 121) is a useful concept. While the motions of various bodies relative to each other are important to the astronomer who predicts future positions of celestial bodies, and perhaps to the navigational scientist who designs navigation tables, the navigator speaks of bodies rising, crossing the celestial meridian, and setting, as though these were absolute motions.

1404. Units of astronomical distance.—The distances between celestial bodies, even those within a single family such as the solar system, are so great that terrestrial units are unsatisfactory to express them. The units commonly used for astronomical distances are:

Astronomical unit (**AU**), the mean distance between the earth and the sun, approximately 92,960,000 statute miles. The astronomical unit is often used as a unit of measurement for distances within the solar system.

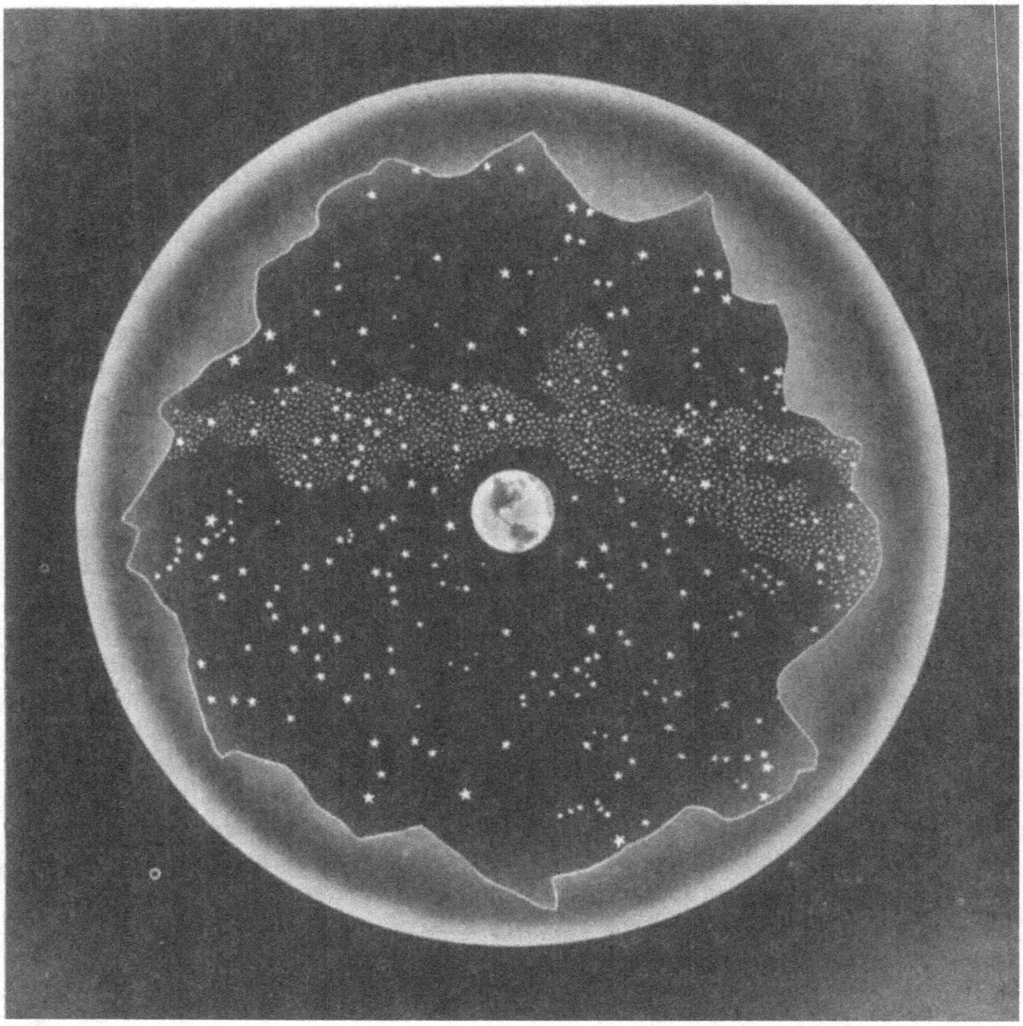

FIGURE 1403.—The celestial sphere.

Light-year, the distance light travels in one year. Since the speed of light is about 186,000 statute miles per *second* and there are about 31,600,000 seconds per year, the length of one light-year is about 5,880,000,000,000 (5.88×10^{12}) statute miles, or 63,280 astronomical units. The light-year is commonly used for expressing distances to the stars and galaxies. Alpha Centauri and its neighbor Proxima, generally considered the nearest stars, are 4.3 light-years away. Relatively few stars are less than 100 light-years away, and the most distant galaxies are in excess of one billion light-years away. However, most navigational stars are relatively close. Light travels from the sun to the earth in about 8⅓ *minutes,* and from the moon to the earth in about 1¼ *seconds.*

Parsec, the distance at which the **heliocentric parallax** (difference in apparent position as viewed from the earth and the sun) is 1″. At this distance a star would appear to change its position 2″ among the distant stars, if observed from points 180° apart on the earth's orbit. The name is derived from the first letters of the words **par**allax and **sec**ond. One parsec is equal to about 3.26 light-years. Hence, even the nearest star is more than one parsec away. This unit is used to express distances to stars and galaxies.

The difficulty of illustrating astronomical distances and sizes is indicated by the fact that if the earth were represented by a circle one inch in diameter, the moon would be a circle one-fourth inch in diameter at a distance of 30 inches, the sun would be a circle nine feet in diameter at a distance of nearly a fifth of a mile, and Pluto would be a circle half an inch in diameter at a distance of about seven miles. The nearest star would be one-fifth the actual distance to the moon.

1405. Magnitude.—The relative brightness of celestial bodies is indicated by a scale of stellar magnitudes. In the *Almagest* (art. 121) Ptolemy divided the stars into six groups according to brightness, the 20 brightest being classified as of the first magnitude, and the dimmest being of the sixth magnitude. In modern times, when it became desirable to define more precisely the limits of magnitude, a first magnitude star was considered 100 times brighter than one of the sixth magnitude, the approximate value of Ptolemy's ratio. Since the fifth root (art. 109, vol. II) of 100 is 2.512, this number is considered the **magnitude ratio.** A first magnitude star is 2.512 times as bright as a second magnitude star, which is 2.512 times as bright as a third magnitude star, etc. A second magnitude is $2.512 \times 2.512 = 6.310$ times as bright as a fourth magnitude star. A first magnitude star is $2.512^{20} \times 100^{4} =$ 100,000,000 times as bright as a star of the 21st magnitude, the dimmest that can be seen through the 200-inch telescope.

Brightness is normally tabulated to the nearest 0.1 magnitude, about the smallest change that can be detected by the unaided eye of a trained observer. All stars of magnitude 1.50 or brighter are popularly called "first magnitude" stars. Those between 1.51 and 2.50 are called "second magnitude" stars, those between 2.51 and 3.50 are called "third magnitude" stars, etc. Sirius, the brightest star, has a magnitude of (−)1.6. The only other star with a negative magnitude is Canopus, (−)0.9. At greatest brilliance Venus has a magnitude of about (−)4.4. Mars, Jupiter, and Saturn are sometimes of negative magnitude. The full moon has a magnitude of about (−)12.6, but varies somewhat. The magnitude of the sun is about (−)26.7.

The Universe

1406. The solar system.—The **sun,** the most conspicuous celestial object in the sky, is the central body of the solar system. Associated with it are at least nine principal **planets,** of which the earth is one; a number of **satellites** accompanying some of the planets; thousands of **minor planets** or **asteroids;** multitudes of **comets;** and vast numbers of **meteors.**

1407. Motions of bodies of the solar system.—Astronomers distinguish between the two principal motions of celestial bodies, as follows: **rotation** is a spinning motion about an axis within the body, while **revolution** is the motion of a body in its elliptical **orbit** around another body, called its **primary.** For the satellites, the primary is a planet. For the planets and other bodies of the solar system, the primary is the sun. The entire solar system is held together by the gravitational force of the sun. The whole system revolves around the center of its galaxy (art. 1415) as a unit, and the galaxy is probably in motion relative to its neighboring galaxies. The motion of bodies of the solar system relative to surrounding stars is called **space motion.**

Rotation and revolution may be further classified as **synodic** or **sidereal.** During one synodic *rotation* the body makes one complete turn relative to the sun. On the earth it is called an **apparent solar day.** During one sidereal rotation the body makes one complete turn relative to the stars. Because of motion of the body in its orbit, a sidereal rotation is either longer or shorter, by a small amount, than a synodic rotation. If both rotation and revolution are in the same direction (in the solar system they are both *east* for most bodies, that is, counter clockwise as seen from above the North Pole) the sidereal rotation is shorter. During a synodic *revolution* a celestial body makes one trip around the sun, *as viewed from the earth.* Hence, the earth cannot have a synodic revolution. During a sidereal revolution, a celestial body makes one trip around its orbit with respect to the stars; to an observer on the celestial body, the sun would appear to make one trip around the celestial sphere, with respect to the stars. On the earth this is one year.

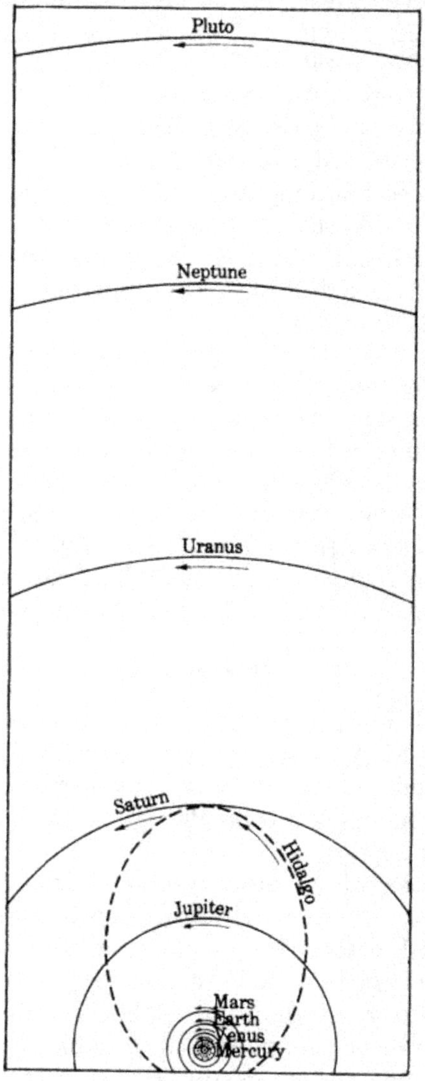

FIGURE 1407a.—Relative size of planetary orbits.

All of the planets revolve around the sun in nearly circular orbits. The flattening or **eccentricity** of the earth's orbit is only 0.017 (zero would be a circle). Some of the minor planets have orbits more eccentric than that of any principal planet (note the orbit of Hidalgo in fig. 1407a). The orbits of comets are highly eccentric. The orbits of all known planets except Pluto are in nearly the same plane, that of the ecliptic (art. 1419). The orbit of Pluto is inclined more than 17° to the ecliptic.

The laws governing the motions of planets in their orbits were discovered by Johannes Kepler, and are now known as **Kepler's laws:**

1. *The orbits of the planets are ellipses, with the sun at a common focus.*
2. *The straight line joining the sun and a planet (the* **radius vector***) sweeps over equal areas in equal intervals of time.*
3. *The squares of the sidereal periods of any two planets are proportional to the cubes of their mean distances from the sun.*

In 1687 Isaac Newton stated three "laws of motion," which he believed were applicable to the planets. **Newton's laws of motion** are:

1. *Every body continues in a state of rest or of uniform motion in a straight line unless acted upon by an external force.*
2. *When a body is acted upon by an external force, its acceleration is directly proportional to that force, and inversely proportional to the mass of the body, and acceleration takes place in the direction in which the force acts.*
3. *To every action there is an equal and opposite reaction.*

Newton also stated a single **universal law of gravitation,** which he believed applied to all bodies, although it was based upon observation within the solar system only:

Every particle of matter attracts every other particle with a force that varies directly as the product of their masses and inversely as the square of the distance between them.

From these fundamental laws of motion and gravitation, Newton derived Kepler's empirical laws. He proved rigorously that the gravitational interaction between any two bodies results in an orbital motion of each body about the barycenter of the two masses that is some form of conic section, that is a circle, ellipse, parabola, or hyperbola.

Circular and parabolic orbits are unlikely to occur in nature because of the precise speeds required. Hyperbolic orbits are open, that is one body, due to its speed, recedes into space. Therefore, a planet's orbit must be elliptical as found by Kepler.

Both the sun and each body revolve about their common center of mass. Because of the preponderance of the mass of the sun over that of the individual planets, the common center of the sun and each planet except Jupiter lies within the sun. The common center of the combined mass of the solar system moves in and out of the sun.

The various laws governing the orbits of planets apply equally well to the orbit of any body with respect to its primary.

In each planet's orbit that point nearest the sun is called the **perihelion.** That point farthest from the sun is called the **aphelion** (ă·fē′lē·ŏn). The line joining periphelion and aphelion is called the **line of apsides** (ăp′sĭ·dēz). In the orbit of the moon, that point nearest the earth is called the **perigee,** and that point farthest from the earth is called the **apogee.** Figure 1407b shows the orbit of the earth (with exaggerated eccentricity), and the orbit of the moon around the earth.

1408. The sun is the dominant member of the solar system because its mass is nearly a thousand times that of all other bodies of the solar system combined. It supplies heat and light to the entire system.

The diameter of the sun is about 866,000 miles. At the distance of the earth, varying between 91,300,000 and 94,500,000 miles, the visible diameter is about 32′. At the closest approach early in January the sun appears largest, being 32′.6 in diameter. Six months later the apparent diameter is 31′.5, the minimum.

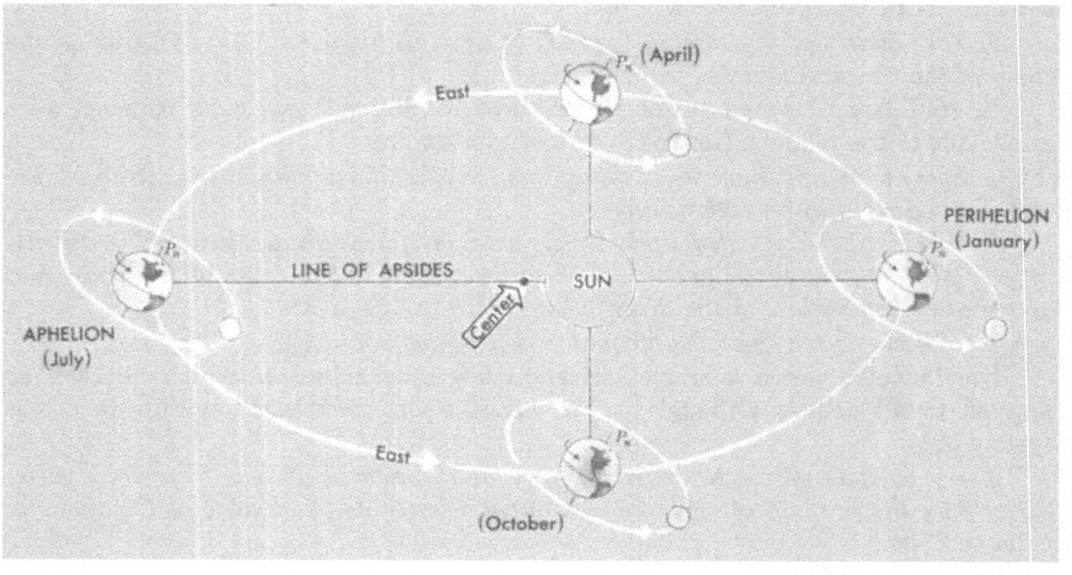

FIGURE 1407b.—Orbits of the earth and moon.

Of the various physical features of the sun, one of particular interest is the appearance from time to time of **sun spots** on the surface (fig. 1408). These spots are apparently areas of cooler gas which have risen to the surface and appear dark in contrast to the hotter gases around them. In size they vary from perhaps 50,000 miles in diameter to the smallest spots that can be detected (a few hundred miles in diameter), and perhaps smaller. They generally appear in groups. At the start of each cycle of about 11 years the spots appear at a maximum distance of about 40° on each side of the solar equator. Succeeding spots of the cycle appear progressively closer to the solar equator, until a minimum solar latitude of 5° may be reached. The maximum number of sun spots occurs about midway in the cycle, when the spots are about 16° from the solar equator. Large sun spots can be seen without a telescope if the eyes are protected, as by the shade glasses of a sextant. The shade glasses of inexpensive "training" sextants may not have the optical quality required for protection. Sun spots have magnetic properties. For one cycle all spots north of the solar equator are of positive polarity, and all those to the south are of negative polarity. During the next cycle, which may begin before the last spots of the old cycle have disappeared, the polarity is reversed. Sun spots are related to magnetic storms which adversely affect radio, including radio aids to navigation, on the

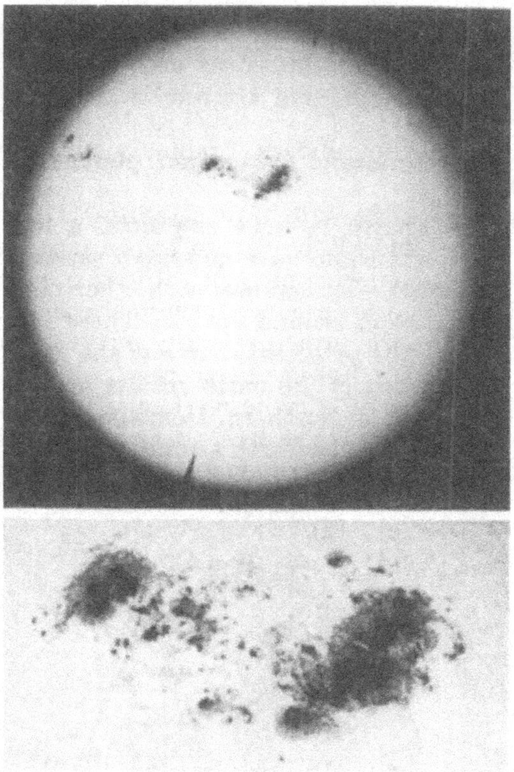

Courtesy of Mt. Wilson and Palomar Observatories.

FIGURE 1408.—Whole solar disk and an enlargement of the great spot group of April 7, 1947.

earth. At such times the **auroras** (art. 3823) are particularly brilliant and widespread.

The sun rotates on its axis, the period of rotation varying from about 25 days at the solar equator to 34 days at the poles, but this fact has little or no navigational significance beyond its effect upon the changing positions of sun spots relative to the earth. The sun is moving approximately toward Vega at about 12 miles per second, or about two-thirds as fast as the earth moves in its orbit around the sun. The path of the sun toward Vega is called the **sun's way.** This is in addition to the motion of the sun around the center of its galaxy (art. 1415).

1409. Planets.—The principal bodies having nearly circular orbits around the sun are called **planets,** from a Greek word meaning "wandering." They were so named because they were observed to change position or "wander" among the "fixed stars" which remained in about the same positions relative to each other. Because the sun and moon had a similar wandering motion, the ancients considered them planets, also.

Nine principal planets are known. In order of increasing distance from the sun, these are Mercury, Venus, Earth, Mars, Jupiter, Saturn, Uranus, Neptune, and Pluto. Of these, only four are commonly used for celestial navigation. These are Venus, Mars, Jupiter, and Saturn, sometimes called the **navigational planets.** The two planets with orbits smaller than that of the earth are called **inferior planets,** and those with orbits larger than that of the earth are called **superior planets.** The four planets nearest the sun are sometimes called the **inner planets,** and the others the **outer planets.** Jupiter, Saturn, Uranus, and Neptune are so much larger than the others that they are sometimes classed as **major planets.** Neptune and Pluto are not visible to the unaided eye, and Uranus is barely so, being of the sixth magnitude.

The orbits of the many thousand tiny **minor planets** lie chiefly between the orbits of Mars and Jupiter.

Six of the planets are known to have satellites, a total of 33 having been discovered. Mercury, Venus, and Pluto have no known satellites.

1410. The earth as a planet.—In common with other planets, the earth rotates on its axis and revolves in its orbit around the sun. These actual motions (discussed in articles 1416 and 1417) are the principal source of the apparent motions of other celestial bodies. Also, the rotation of the earth results in a deflection of water and air currents to the right in the Northern Hemisphere and to the left in the Southern Hemisphere. Because of the earth's rotation, the high tides on the open sea lag behind the meridian transit of the moon.

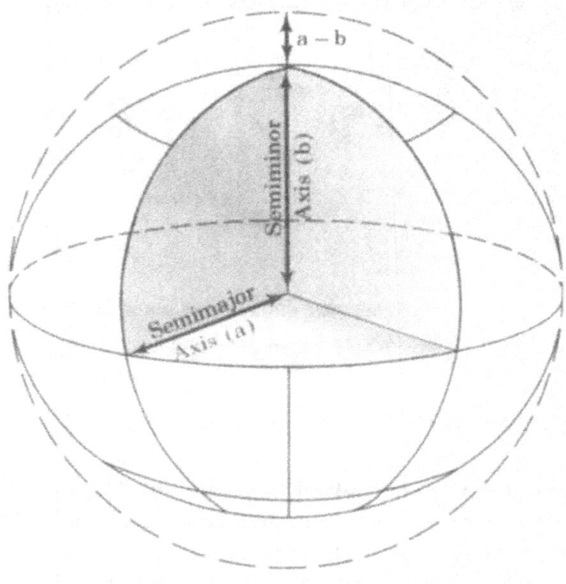

FIGURE 1410.—Oblate spheroid or ellipsoid of revolution.

For most navigational purposes, the earth can be considered a sphere, but, like the other planets, the earth is approximately an **oblate spheroid,** or **ellipsoid of revolution,** being flattened at the poles and bulged at the equator. Therefore, the polar diameter is less than the equatorial diameter, and the meridians are slightly elliptical, rather than circular. The dimensions of the earth are recomputed from time to time, as additional and more precise measurements become available. Since the earth is not *exactly* an ellipsoid, results differ slightly when equally precise and extensive measurements are made on different parts of the surface. Hence, different "spheroids" are used for mapping various parts of the earth. That used for charts of North America was computed by the English geodesist A. R. Clarke in 1866. However, since Clarke did not clearly define his units, the U.S. Coast and Geodetic Survey in 1880 considered it desirable to adopt standard values which probably added about 170 feet to the diameter computed by Clarke. In 1880, also, Clarke himself made a new estimate of the size and shape of the earth but this has not been adopted by the United States. Although the Clarke spheroid of 1866 is still used for charting North America, the World Geodetic System (WGS) constants provides a slightly better approximation of the size and shape of the earth. According to these constants, the dimensions of the earth are (fig. 1410):

Equatorial radius *(a)*	=3,963.189 statute miles=3,443.917 nautical miles
Polar radius *(b)*	=3,949.902 statute miles=3,432.371 nautical miles
Mean radius	=3,958.760 statute miles=3,440.068 nautical miles

$$\text{Flattening or ellipticity } \left(f=1-\frac{b}{a}\right)=\frac{1}{298.26}$$

The mean radius is the average of the polar radius and two equatorial radii perpendicular to each other (the three dimensions of the solid), or

$$\frac{(2a+b)}{3}.$$

1411. Other planets and the minor planets.—Mercury in some ways resembles the moon more than it does other planets. Its diameter is only about 50 percent larger than that of the moon and about the same as those of Jupiter's two largest satellites. Like the moon it has little or no atmosphere, and is believed to keep the same side turned toward its primary. Mercury's mass is only about five percent that of the earth, and its orbit is so small that the planet is never seen more than about 28° from the sun. It is for this reason that Mercury is not commonly used for navigation. Near greatest elongation (art. 1422) it appears near the western horizon after sunset or the eastern horizon before sunrise. At these times it resembles a first magnitude star, and is sometimes reported as a new or strange object in the sky. As seen from the earth, Mercury goes through all the phases similar to those of the moon, and occasionally **transits** (crosses) the face of the sun, appearing as a tiny, dark, inconspicuous dot on the surface. Mercury has no known satellite. Mercury rotates with a period of about 59 days.

Venus, like Mercury, has no known satellite, goes through the various phases similar to those of the moon, and may transit the sun. In size of orbit, sidereal period of revolution, diameter, volume, mass, density, and surface gravity it resembles the earth more than any other planet. Its orbit is more nearly circular than that of any other planet (eccentricity 0.007). At maximum brilliance, about five weeks before and after inferior conjunction (art. 1422), it has a magnitude of about (−)4.4 and is brighter than any other object in the sky except the sun and moon. At these times it can be seen during the day, and is sometimes observed for a celestial line of position.

Mars (fig. 1411) has a diameter only a little more than half that of the earth, and a mass of 11 percent as much, although its density is nearly 72 percent that of the earth. It has a thin atmosphere, but few clouds. Its day is only slightly longer than that on the earth, but its year is nearly twice as long. Being a superior planet (art. 1409), it is seen only in the full or gibbous phase (art. 1423). When nearest the earth, its apparent diameter is about eight times that at conjunction (art. 1422). Mars has two satellites. **Phobos** is about 10 miles in diameter and has an orbit the diameter of which is only about three times the diameter of Mars. To an observer on Mars it would appear to have a diameter about one-third the diameter of the moon as seen from the earth, and would appear to rise in the west and set in the east, going through three-fourths of its cycle of phases while above the horizon. It would do this three times each day, since its sidereal period of revolution is only about one-third the period of rotation of Mars. No other natural satellite is known to *revolve* faster than its primary *rotates.* **Deimos** is only about 5 miles in diameter, and at greatest brilliance would appear as a very bright star. About 66 hours would elapse between rising and setting, during which it would go through the various phases twice.

Jupiter (fig. 1411), largest of the known planets, has more than twice the mass of all other known planets combined. Its density is low and its rotation fast (9^h50^m), resulting in a pronounced equatorial bulge. It is believed to have a dense, solid core, surrounded by lighter material, and a deep atmosphere of ammonia, methane, helium, and hydrogen. Two of Jupiter's 13 known satellites are about the same size as Mercury, and may have atmospheres. The four outermost satellites revolve from east to west, opposite to the general direction of revolution within the solar system.

Saturn (fig. 1411) is the only planet having a density less than that of water, yet it has a mass of nearly one-third that of Jupiter, and nearly three times that of all other known planets combined. Its composition is believed to be similar to that of Jupiter. It is more oblate than any other known planet. Perhaps the most interesting feature of this planet is its rings, composed of a great number of small solid particles spread out in three thin, flat rings more than 170,000 miles in diameter. The particles nearest the planet revolve more rapidly than those farther out, the innermost ones completing a revolution in less time than the planet completes a rotation. During half the 29.5-year sidereal period of revolution of the planet one side of the rings is visible to observers on the earth, and during the second half of the period the opposite side is visible. Saturn has ten known satellites, the outermost one of which revolves from east to west.

Uranus is barely visible to the unaided eye, being of the sixth magnitude. It is a comparatively large planet, and probably is similar in composition to Jupiter and Saturn. The inclination of the equator of Uranus to the plane of the ecliptic is 98°, or 82° if the revolution is considered from east to west. Its five known satellites, all small, revolve in the equatorial plane, in the same direction as that of rotation of the planet.

Neptune is slightly smaller than Uranus, but has greater mass, and a longer period of rotation. Relatively little is known of this remote planet of the eighth magnitude. However, it is known to have two satellites, the larger (probably bigger than the moon) revolving from east to west.

Pluto (fig. 1411) was identified in 1930. It is of the 15th magnitude, and cannot be seen in small telescopes. In all but the 200-inch telescope it appears as a point of light. Its diameter is less than half that of the earth. Its orbit is the most eccentric and has the greatest inclination to the ecliptic of any of the known planets. At perihelion it is closer to the sun than Neptune, and there is some evidence to support the view that it was at one time a satellite of the larger planet.

Minor planets. About 1,800 of these tiny planets have been discovered, but it is estimated that there may be as many as 50,000 bright enough to be seen by the largest telescopes, when they are nearest the earth. The largest, **Ceres,** has a diameter of about 480 miles. All but a few are less than 100 miles in diameter. Since there is no known lower limit, there may be no distinction between minor planets and meteors. The combined mass of all minor planets probably does not exceed 0.1 percent that of the earth. The orbits, of various degrees of eccentricity and inclination to the ecliptic, lie mostly between those of Mars and Jupiter. However, at perihelion some of the minor planets are inside the earth's orbit. The orbit of **Hidalgo** is shown in figure 1407a.

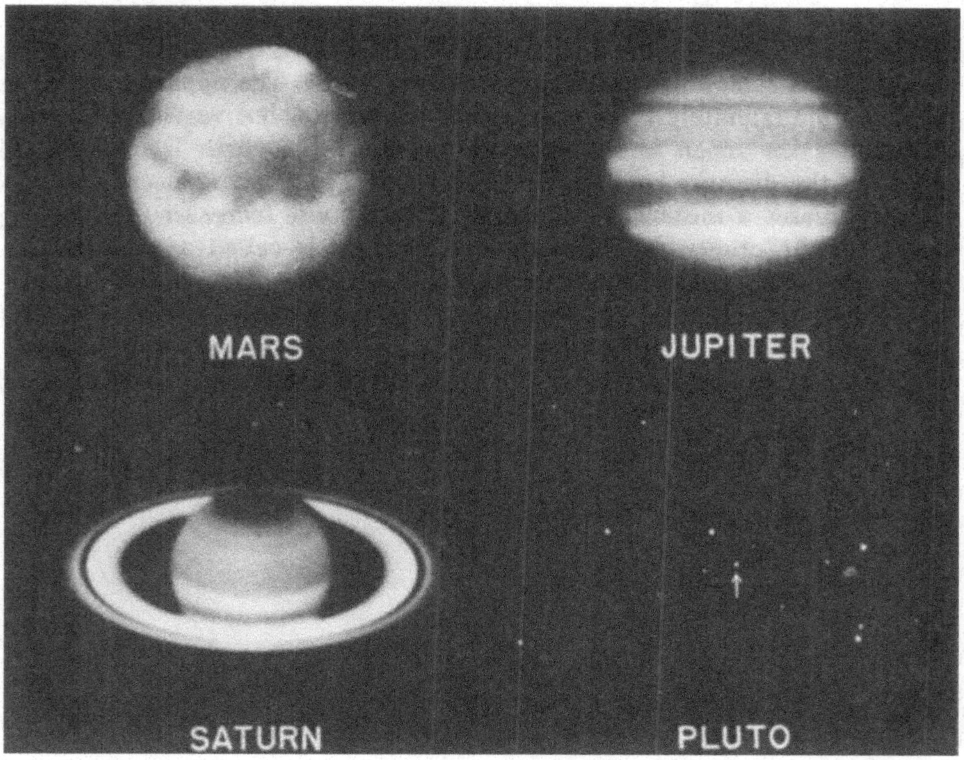

Courtesy of Mt. Wilson and Palomar Observatories.

FIGURE 1411.—Mars, Jupiter, Saturn, and Pluto. First three photographed with 100-inch telescope, Pluto with 200-inch telescope.

1412. The moon is the only satellite of direct navigational interest, although the satellites of Jupiter were at one time used to determine Greenwich mean time, so that longitude could be found (art. 126). The rotation and revolution of the moon are both west to east, and both are of the same duration, $27^d07^h43^m11^s.5$ with respect to the stars (the **sidereal month**) and $29^d12^h44^m02^s.8$ with respect to the sun (the **synodical month**). Because there is no difference in the periods of rotation and revolution, the same side of the moon is always turned toward the earth. However, about 59 percent of the moon's surface has been seen from earth, due to **libration. Libration in latitude** occurs because the axis of rotation is tilted about $6^\circ.5$ with respect to the axis of revolution. **Libration in longitude** occurs because the speed of revolution varies in accordance with Kepler's second law (art. 1407), while the rotational speed is essentially constant. **Diurnal libration** occurs because of the changing position of the observer relative to the moon, due to rotation of the earth. **Physical libration** is a small pendulum-like rotational oscillation of the moon with respect to its radius vector.

At **perigee** the moon is about 221,000 statute miles from the earth's center, and at **apogee** it is about 253,000 miles distant. The average distance is about 238,862 miles. Because of the relative nearness of the moon, its **geocentric parallax** (difference in position relative to the background of stars, as observed from the surface and center of the earth) is comparatively large. It is at maximum when the moon is on the horizon, when it is called **horizontal parallax.** The **equatorial horizontal parallax** for an observer at the equator, where the maximum radius of the earth is involved, is tabulated in the *Nautical Almanac* and the *Astronomical Almanac,* and used in sextant altitude corrections given in the nautical and air almanacs. The **parallax** varies from a maximum at the horizon to zero at the zenith. The parallax at any altitude is sometimes called **parallax in altitude.** The apparent diameter of the moon is approximately the same as that of the sun, but varies through wider limits. Because the moon is so near, the radius of the earth is an appreciable percentage of the distance between earth and moon, and the apparent diameter of the moon increases a measurable amount as its altitude increases (decreasing the distance from the observer). This apparent increase is called **augmentation** (fig. 1412). A similar effect for the sun is very small.

As with the planets and sun, the moon and earth both revolve around their common center of mass, which is about 2,900 miles from the center of the earth. It is this center of mass that describes the orbit of the earth (and moon) around the sun.

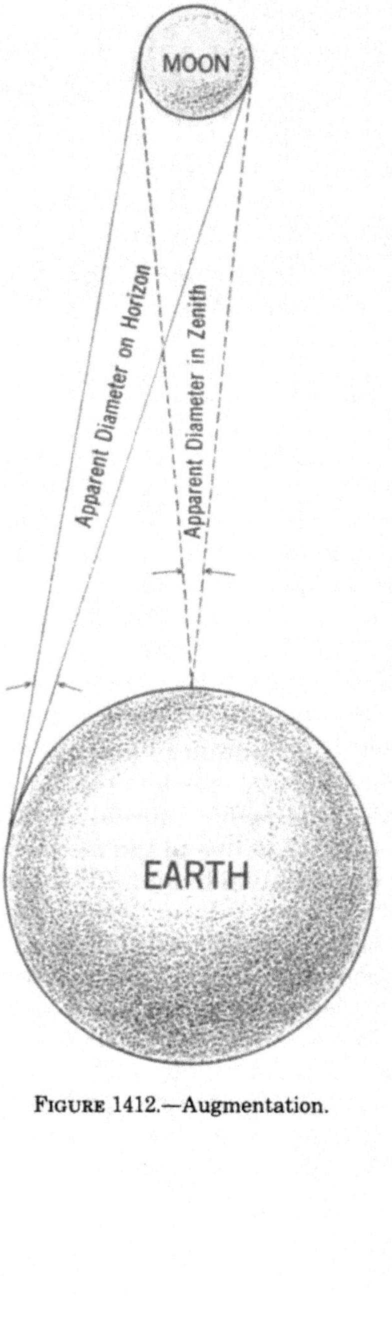

FIGURE 1412.—Augmentation.

Because of its relative nearness and size, the moon is the principal source of the gravitational attraction that causes tides, although the sun has an appreciable effect, also. The action of these bodies in causing tides is described in article 3103. Because of the frictional action of tides, the rotation of the earth is slowing, the length of the day increasing about $0\overset{s}{.}001$ per century.

On the moon, the day is equal in length to the synodical month (about 29½ days). The earth would remain almost stationary in the sky for an observer on the 41 percent of the moon's surface always visible from the earth, would rise and set at about the same point on the horizon for one on the 18 percent which is sometimes visible, and would never appear for one on the 41 percent not seen from the earth.

1413. Comets and meteors.—Comets are swarms of relatively small, widely separated, solid bodies held together by mutual attraction. Around this **nucleus,** a more spectacular, gaseous head or **coma** and **tail** may form as the comet approaches the sun. The tail is directed *away* from the sun, so that it follows the head while the comet is approaching the sun, and precedes the head while the comet is receding. The total mass of a comet is very small, and the tail is so thin that stars can easily be seen through it. In 1910 the earth passed through the tail of Halley's comet (fig. 1413) without noticeable effect.

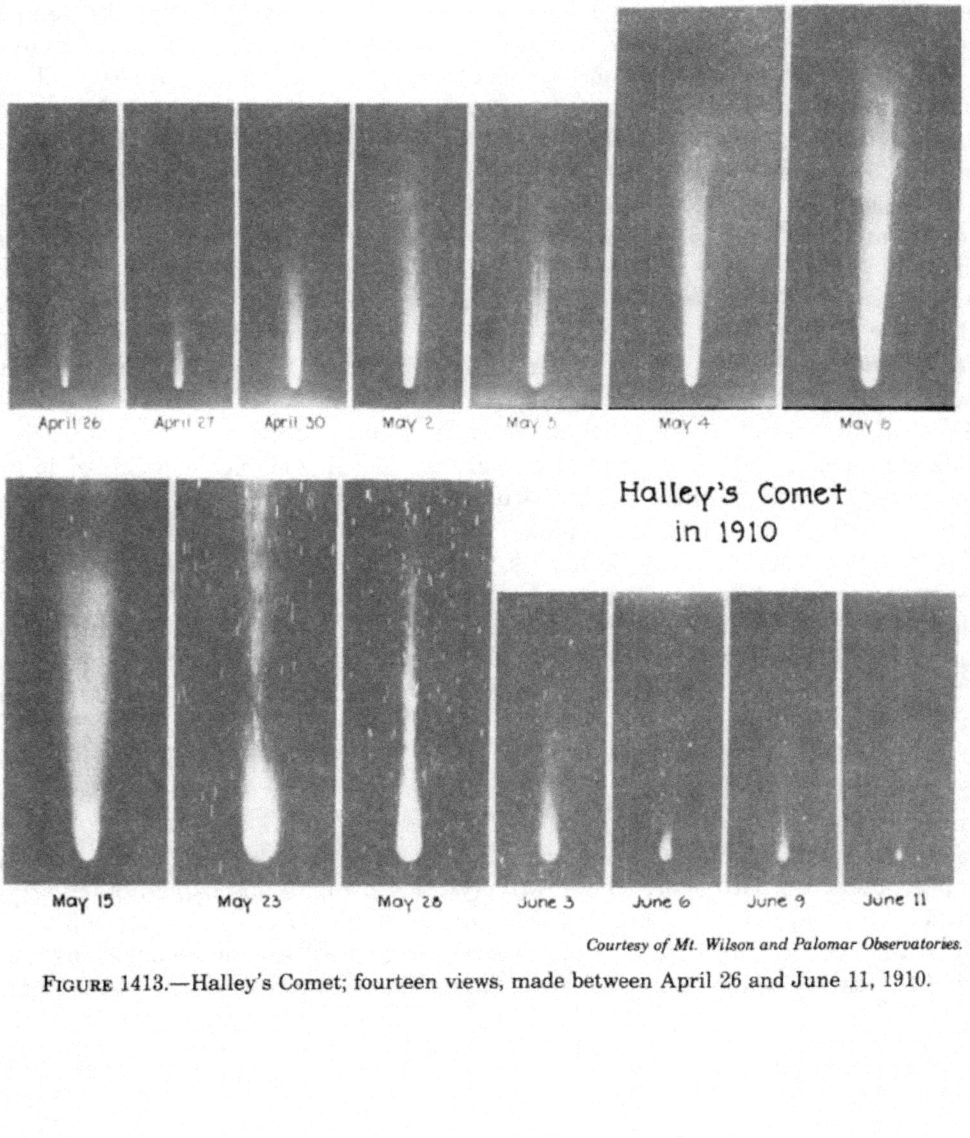

Courtesy of Mt. Wilson and Palomar Observatories.

FIGURE 1413.—Halley's Comet; fourteen views, made between April 26 and June 11, 1910.

Comets are erratic and inconsistent. Some travel east to west and some west to east, in highly eccentric orbits inclined at any angle to the ecliptic. The shortest period of revolution is about 3.3 years. Some periods are so long that astronomers speculate as to whether some comets may not come in from outside the solar system for a single trip around the sun, and then leave the solar system, never to return. In such a case the orbit would be approximately a **parabola** (art. 134, vol. II).

Without their tails, which exist only when near the sun, comets are not spectacular. Because of the small size of their nuclei, which shine by reflected light from the sun, comets are visible for only a small part of their period of revolution, and this is the part of most rapid motion, in accordance with Kepler's second law (art. 1407). An average of about five comets is observed each year, and about two-thirds of these are identified as previously observed comets. Very few comets are ever visible without a telescope. The spectacular Halley's comet reached aphelion in 1948 and started back toward the sun. It is expected to reach perihelion about February, 1986.

Meteors, popularly called **shooting stars,** are tiny, solid bodies too small to be seen until heated to incandescence by air friction while passing through the earth's atmosphere. A particularly bright meteor is called a **fireball.** One that explodes is called a **bolide.** A meteor that is not consumed during its fall through the atmosphere, but lands as a solid particle, is called a **meteorite.** These are composed principally of iron, with some nickel, and smaller quantities of other material.

Vast numbers of meteors exist. It has been estimated that an average of about 1,000,000 bright enough to be seen enter the earth's atmosphere each hour, and many times this number undoubtedly enter, but are too small to attract attention. A faint glow sometimes observed extending upward approximately along the ecliptic before sunrise and after sunset has been attributed to the reflection of sunlight from quantities of such material. This glow is called **zodiacal light.** A faint glow at that point of the ecliptic 180° from the sun is called the **gegenschein** or **counterglow.** Comets may be an assemblage of a large number of meteors traveling together, and minor planets (art. 1411) may be larger meteors. **Meteor showers** occur at certain times of the year when the earth is believed to be passing through **meteor swarms,** the scattered remains of comets that have broken up. At these times the number of meteors observed is many times the usual number.

Since such large amounts of this material are in existence, much of it in an orbit near the ecliptic, and since the orbits of most minor planets lie between those of Mars and Jupiter, where astronomers compute the orbit of another planet should be located, it is possible that another planet may have existed there at one time and been disrupted, perhaps by an atomic explosion of hydrogen or other material. The estimated total mass of all meteors, comets, and minor planets would make a small planet, but if the material which has fallen on other planets and satellites, and perhaps some or all of the satellites themselves, are added, a sizeable planet might be accounted for.

1414. Stars are distant suns, in many ways resembling the body which provides the earth with most of its light and heat. Even the nearest star is too distant to be seen as more than a point of light in the largest telescope. If planets, satellites, comets, etc., accompany those distant suns, as they do the one nearby, they have not been detected. However, comparatively dark companions of planetary size are known to accompany some stars. Nonluminous stars may exist, since most of the **radio stars** (points from which radio energy emanates) are not marked by a body visible to observers on the earth. The distance of the stars is so great that none is known to have a **heliocentric parallax** (difference in apparent position as observed from the earth and the sun) of as much as 1″.

Stars differ in size from gaseous giants having diameters greater than that of the *orbit* of the earth, to dense dwarfs which may be no larger than the major planets. Although the size and density cover wide ranges, the mass does not differ greatly. Relatively few stars have more than five times or less than one-fifth the mass of the sun, which is also about average in size, density, and temperature. The color varies with the temperature. A very hot star, having a surface temperature of perhaps 20,000° K (Celsius absolute) or more, is bluish-white; while a cooler star, having a temperature of perhaps 2,000° K, is faintly reddish. In *Orion,* blue Rigel and red Betelgeuse, located on opposite sides of the belt, constitute a noticeable contrast.

Under ideal viewing conditions, the dimmest star that can be seen with the unaided eye is of the sixth magnitude. In the entire sky there are about 6,000 stars of this magnitude or brighter. Half of these are below the horizon at any time. Because of the greater absorption of light near the horizon, where the path of a ray travels for a greater distance through the atmosphere, not more than perhaps 2,500 stars are visible to the unaided eye at any time. The 200-inch telescope on Palomar Mountain permits stars as dim as the twenty-first magnitude to be seen. It has been estimated that there are about 1,000,000,000 of this magnitude or brighter. A long-term photographic exposure with the 200-inch telescope permits observation of about twice this number. There is no indication that this is more than a tiny fraction of the total number. However, the average navigator seldom uses more than perhaps 20 or 30 of the brighter stars. Stars which exhibit a noticeable change of magnitude are called **variable stars.** A star which suddenly becomes several magnitudes brighter and then gradually fades is called a **nova.** A particularly bright one is called a **supernova.**

Two stars which appear to be very close together are called a **double star.** If more than two stars are included in the group, it is called a **multiple star;** and if a large number appear in approximately spherical shape, it is called a **globular cluster.** A group of stars moving through space together, but not exhibiting the intimate relationship of a globular cluster, is called an **open cluster.** The Pleiades and some stars of the Big Dipper (with certain other stars) are examples of open clusters. A group of stars which *appear* close together, regardless of actual distances, is popularly called a **constellation,** particularly if the group forms a striking configuration. Among astronomers a constellation is now considered a region of the sky having precise boundaries so arranged that all of the sky is covered, without overlap. The ancient Greeks recognized 48 constellations covering only certain groups of stars. Modern astronomers recognize 88 constellations. The constellation names and meanings are given in appendix K.

A cloudy patch of matter in the heavens is called a **nebula** (plural *nebulae*). If it is within the galaxy of which the sun is a part, it is called a **galactic nebula;** if outside, it is called an **extragalactic nebula.**

Stars rotate on their axes, and revolve around the center of their galaxy, in addition to influencing and being influenced by surrounding stars. Motion of a star through space, like that of any celestial body, is called **space motion.** That component in the line of sight is called **radial motion;** while that component across the line of sight, causing a star to change its apparent position relative to the background of more distant stars, is called **proper motion.**

1415. Galaxies.—A great number of the nebulae have been identified as extragalactic, and as telescopes became more powerful, it was discovered that these small cloudy patches are groups of stars, in many ways resembling the group of stars of which the sun is a part. Each such vast assemblage of stars constitutes an island universe as widely separated from others, comparatively, as individual stars in one group. Such a group is called a **galaxy.** It was not until well within the twentieth century that the sun was recognized as a part of such a galaxy, the **Milky Way.** In a galaxy the stars tend to congregate in groups called **star clouds** arranged in long spiral arms. The spiral nature is believed due to revolution of the stars about the center of the galaxy, the inner stars revolving more rapidly than the outer ones (fig. 1415). At the position of the sun, about two-thirds of the way out from the center, and nearly midway between "top" and "bottom," the period of revolution is about 200,000,000 years at the present speed of about 175 miles per second. This is nearly ten times the speed of the earth in its orbit. An average estimate of the size of a galaxy is that it is about 100,000 light years in diameter, 15,000 light years thick at the center, and 5,000 light years thick near the outer edge, and that it contains perhaps 100,000,000,000 stars. This is about 100 times the number of stars that can be seen through the 200-inch telescope. Within the radius of 1,600,000,000 light years that man is able to penetrate there are perhaps 100,000,000 galaxies, although only a small fraction of this number has been actually observed.

Courtesy of Mt. Wilson and Palomar Observatories.

FIGURE 1415.—Spiral nebula Messier 51, in *Canes Venetici.*
Satellite nebula is NGC 5195.

The galaxies which have been discovered are observed to congregate in groups, somewhat similar to stars in a galaxy. Whether the part seen is but a small portion of a larger unit too vast to be seen with present instruments has not been established. Through his progress in astronomy man hopes to see much more of what surrounds him in space, and perhaps to answer some of the questions which confront him.

Apparent Motion

1416. Apparent motion due to rotation of the earth is much greater than any other observed motion of celestial bodies. It is this motion that causes celestial bodies to appear to rise somewhere along the eastern half of the horizon, climb to maximum altitude as they cross the meridian, and set along the western horizon, at about the same point relative to due west as the rising point was to due east. This apparent motion along the daily path, or **diurnal circle,** of the body is approximately parallel to the plane of the equator. It would be exactly so if rotation of the earth were the only motion, and the axis of rotation of the earth were stationary in space (arts. 1417 and 1419).

The apparent effect due to rotation of the earth varies with the latitude of the observer. At the equator, where the equatorial plane is vertical (since the axis of rotation of the earth is parallel to the plane of the horizon), bodies appear to rise and set vertically. Every celestial body is above the horizon approximately half the time. The celestial sphere as seen by an observer at the equator is called the **right sphere,** shown in figure 1416a. Several unique relationships of the right sphere are discussed in article 1433.

For an observer at one of the poles, bodies having constant declination neither rise nor set (neglecting precession of the equinoxes and changes in refraction), but circle the sky, always at the same altitude, making one complete trip around the horizon each day. At the North Pole the motion is clockwise, and at the South Pole it is counterclockwise. Approximately half the stars are always above the horizon and the other half never are. This is modified somewhat by actual conditions, a description of which is given in chapter XXV. The **parallel sphere** at the poles is illustrated in figure 1416b.

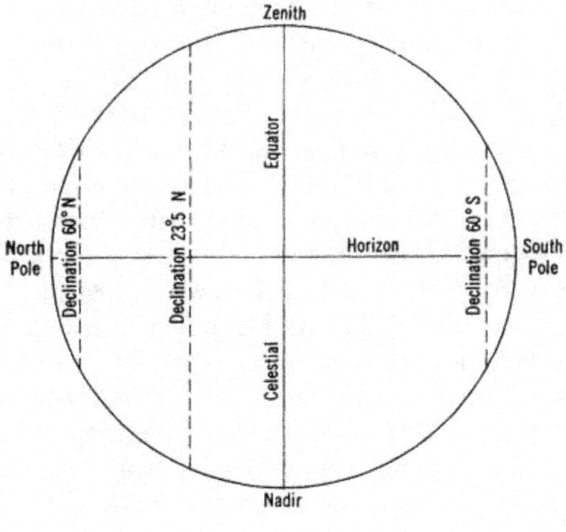

FIGURE 1416a.—The right sphere.

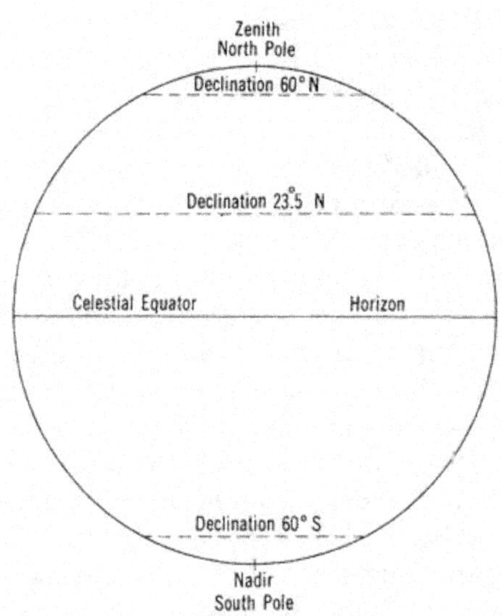

FIGURE 1416b.—The parallel sphere.

Between these two extremes, the apparent motion is a combination of the two. On this **oblique sphere,** illustrated in figure 1416c, **circumpolar** celestial bodies remain above the horizon during the entire 24 hours, circling the elevated celestial pole (art. 1426) each day. The stars of the Big Dipper and *Cassiopeia* are circumpolar for many observers in the United States. An approximately equal part of the celestial sphere remains below the horizon during the entire day. The Southern Cross is not visible to most observers in the United States. Other bodies rise obliquely along the eastern horizon, climb to maximum altitude at the celestial meridian, and set along the western horizon. The length of time above the horizon, and the altitude at meridian transit, vary with both the latitude of the observer and the declination of the body. Several useful relationships of the oblique sphere are indicated in article 1432. At the polar circles (art. 1419) of the earth and beyond, even the sun becomes circumpolar. This is the land of the **midnight sun**, where the sun does not set during part of the summer, and does not rise during part of the winter.

The increased obliquity at higher latitudes explains why days and nights are always about the same length in the tropics, and the change of length of the day becomes greater as the latitude increases. It also explains why twilight lasts longer in higher latitudes. **Twilight** is that period of incomplete darkness following sunset and preceding sunrise. **Evening twilight** starts at sunset, and **morning twilight** ends at sunrise. The darker limit of twilight occurs when the center of the sun is a stated number of degrees below the celestial horizon. Three kinds of twilight are defined, depending upon the darker limit. These are:

Twilight	*Lighter limit*	*Darker limit*	*At darker limit*
civil	−0°50′	−6°	Horizon clear and bright stars visible
nautical	−0°50′	−12°	Horizon not visible
astronomical	−0°50′	−18°	Full night

The conditions at the darker limit are relative and vary considerably under different atmospheric conditions.

In figure 1416d the twilight band is shown, with the darker limits of the various kinds indicated. The nearly vertical celestial equator line is for an observer at latitude 20°N. The nearly horizontal celestial equator line is for an observer at latitude 60°N. The broken line in each case is the diurnal circle of the sun when its declination is 15°N. The relative duration of any kind of twilight at the two latitudes is indicated by that portion of the diurnal circle between the horizon and the darker limit, although it is not directly proportional to the relative length of line shown, since the projection is orthographic (art. 319). The duration of twilight at the higher latitude is longer, proportionally, than shown. Note that complete darkness does not occur at latitude 60°N when the declination of the sun is 15°N.

1417. Apparent motion due to revolution of the earth.—If it were possible to stop the *rotation* of the earth so that the celestial sphere would appear stationary, the effects of the *revolution* of the earth would become more noticeable. In one year the sun would appear to make one complete trip around the earth, from west to east. Hence, it would seem to move eastward a little less than 1° per day. This motion can be observed by watching the changing position of the sun among the stars. But since both sun and stars generally are not visible at the same time, a better way is to observe the constellations at the same time each night. On any night a star rises nearly four minutes earlier than on the previous night. Thus, the celestial sphere appears to shift westward nearly 1° each night, so that different constellations are associated with different seasons of the year.

Apparent motions of planets and the moon are due to a combination of their motions and those of the earth. If the rotation of the earth were stopped, the combined apparent motion due to the revolutions of the earth and other bodies would be similar to that occurring if both rotation and revolution of the earth were

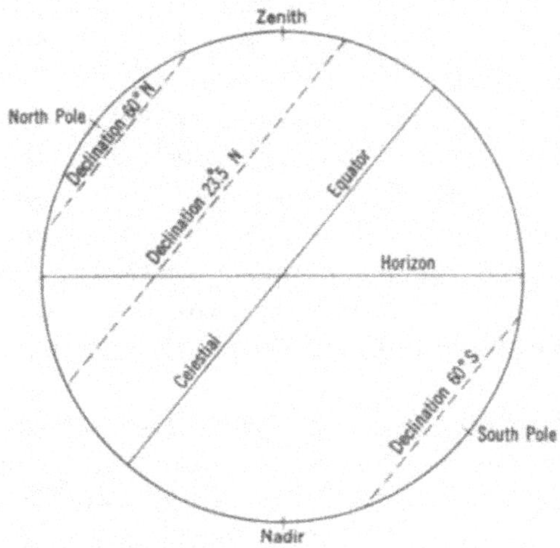

FIGURE 1416c.—The oblique sphere at lat. 40°N.

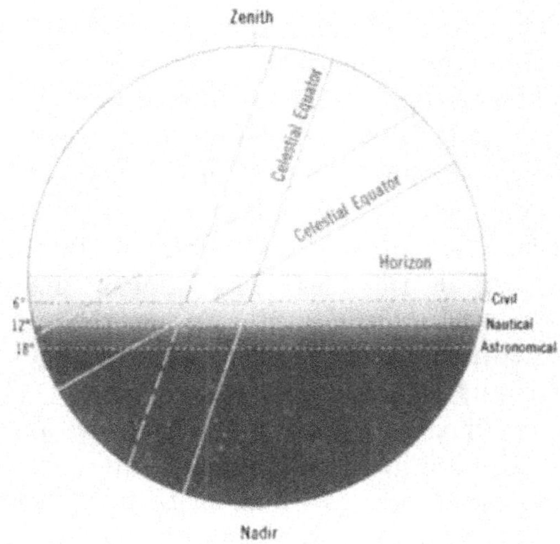

FIGURE 1416d.—The various twilights at lat. 20°N and lat. 60°N.

stopped, as discussed in article 1418, but with different timing. Stars would appear nearly stationary in the sky, but would undergo a small annual cycle of change due to **aberration.** The motion of the earth in its orbit is sufficiently fast to cause the light from stars to appear to shift slightly in the direction of the earth's motion. This is similar to the illusion one has when walking in rain that is falling vertically, but appearing to come from ahead due to his own motion. The apparent direction of the light ray from the star is the vector difference (art. 118, vol. II) of the motion of light and the motion of the earth, similar to that of apparent wind on a moving vessel. This effect is most apparent for a body perpendicular to the line of travel of the earth in its orbit, for which it reaches a maximum value of 20″.5. The effect of aberration can be noted by comparing the coordinates (declination and sidereal hour angle) of various stars throughout the year. A change is observed in some bodies as the year progresses, but at the end of the year the values have returned almost to what they were at the beginning. That they do not return exactly is due to proper motion (art. 1418), precession of the equinoxes (art. 1403), and **nutation,** which is an irregularity in the motion of the earth due to the disturbing effect of other celestial bodies, principally the moon. **Eulerian motion** is a slight wobbling of the earth about its axis of rotation, often called **polar motion,** and sometimes **wandering of the poles.** This motion, which does not exceed 40 feet from the mean position, produces slight **variation of latitude and longitude** of places on the earth.

By the **calendar,** one year is of 365 days duration for **common years** and 366 days for **leap years.** A leap year is any year divisible by four, unless it is a century year, which must be divisible by 400 to be a leap year. Thus, 1900 was not a leap year, but 2000 will be. This calendar, now in general use, is called the **Gregorian calendar.** Astronomically, the year is not divisible into a whole number of days, and the present system will introduce an error of three days in about 10,000 years. The length of the year with respect to the vernal equinox (art. 1419) is about 365 days, 5 hours, 48 minutes, 46 seconds. This is the **tropical, astronomical, equinoctial, natural,** or **solar year.** Since the vernal equinox is in motion on the celestial sphere (art. 1419), this does not quite agree with the **sidereal year** of about 365 days, 6 hours, 9 minutes, 10 seconds, with respect to the stars. The period of revolution from perihelion to perihelion, about 365 days, 6 hours, 13 minutes, 53 seconds, is called the **anomalistic year.** These values vary slightly from year to year, and progressively over the years, as shown in appendix D.

1418. Apparent motion due to movement of other celestial bodies.—Even if it were possible to stop both the rotation and revolution of the earth, celestial bodies would not appear stationary on the celestial sphere. The moon would make one revolution about the earth each sidereal month (art. 1412), rising in the west and setting in the east. The inferior planets would appear to move eastward and westward relative to the sun, as explained in article 1422, staying within the zodiac. Superior planets would appear to make one revolution around the earth, from west to east, each sidereal period.

Since the sun (and the earth with it) and all other stars, as far as is known, are in motion relative to each other, slow apparent motions would result in slight changes in the positions of the stars relative to each other. This **space motion** (art. 1414) is, in fact, observed by telescope. That component of such motion across the line of sight, called **proper motion,** produces a change in the apparent position of the star. The maximum which has been observed is that of "Barnard's Star," which is moving at the rate of 10″.3 per year. This is a tenth-magnitude star, and hence not visible to the unaided eye. Of the 57 stars listed on the daily pages of the almanacs, Rigil Kentaurus has the greatest proper motion, about 3″.7. Arcturus, with 2″.3, has the greatest proper motion of the navigational stars in the Northern Hemisphere. In a few thousand years proper motion will be sufficient to materially alter some familiar configurations of stars, notably the Big Dipper.

1419. The ecliptic is the path the sun appears to take among the stars due to the annual revolution of the earth in its orbit. It is considered a great circle of the celestial sphere, inclined at an angle of about 23°27′ to the celestial equator, but undergoing a continuous slight change. This angle is called the **obliquity of the ecliptic.** This inclination is due to the fact that the axis of rotation of the earth is not perpendicular to its orbit. It is this inclination which causes the sun to appear to move north and south during the year, giving the earth its **seasons,** and changing lengths of periods of daylight. This seasonal variation is one of the factors making the earth a desirable place on which to live.

Refer to figure 1407b. The earth is at perihelion early in January and at aphelion six months later. On or about June 21, about ten or eleven days before reaching aphelion, the northern part of the earth's axis is tilted toward the sun. The north polar regions are having continuous sunlight; the Northern Hemisphere is having its **summer** with long, warm days and short nights; the Southern Hemisphere is having winter with short days and long, cold nights; and the south polar region is in continuous darkness. This is the **summer solstice.** Three months later, about September 23, the earth has moved a quarter of the way around the sun, but its axis of rotation still points in about *the same direction in space.* The sun shines equally on both hemispheres, and days and nights are the same length over the entire world. The sun is setting at the North Pole, and rising at the South Pole. The Northern Hemisphere is having its **autumn,** and the Southern Hemisphere its spring. This is the **autumnal equinox.** In another three months, on or about December 22, the Southern Hemisphere is tilted toward the sun and conditions are the reverse of those six months earlier, the Northern Hemisphere having its **winter,** and the Southern Hemisphere its summer. This is the **winter solstice.** Three months later, when both hemispheres again receive equal amounts of sunshine, the

Northern Hemisphere is having **spring** and the Southern Hemisphere autumn, the reverse of conditions six months before. This is the **vernal equinox.**

The word "equinox," meaning "equal nights," is applied because it occurs at the time when days and nights are of approximately equal length all over the earth. The word "solstice," meaning "sun stands still," is applied because the sun stops its apparent northward or southward motion and momentarily "stands still" before it starts in the opposite direction. This action, somewhat analogous to the "stand" of the tide (art. 3104), refers to the motion in a north-south direction only, and not to the daily apparent revolution around the earth. Note that it does not occur when the earth is at perihelion and aphelion (fig. 1407b). Refer to figure 1419a. At the time of the vernal equinox, the sun is directly over the equator, crossing from the Southern Hemisphere to the Northern Hemisphere. It rises due east and sets due west, remaining above the horizon about 12 hours. It is not exactly 12 hours because of refraction, semidiameter, and the height of the eye of the observer. These cause it to be above the horizon a little longer than below the horizon. Following the vernal equinox, the northerly declination increases, and the sun climbs higher in the sky each day (at the latitudes of the United States), until the summer solstice, when a declination of about 23°27′ north of the celestial equator is reached. The sun then gradually retreats southward until it is again over the equator at the autumnal equinox, at about 23°27′ south of the celestial equator at the winter solstice, and back over the celestial equator again at the next vernal equinox.

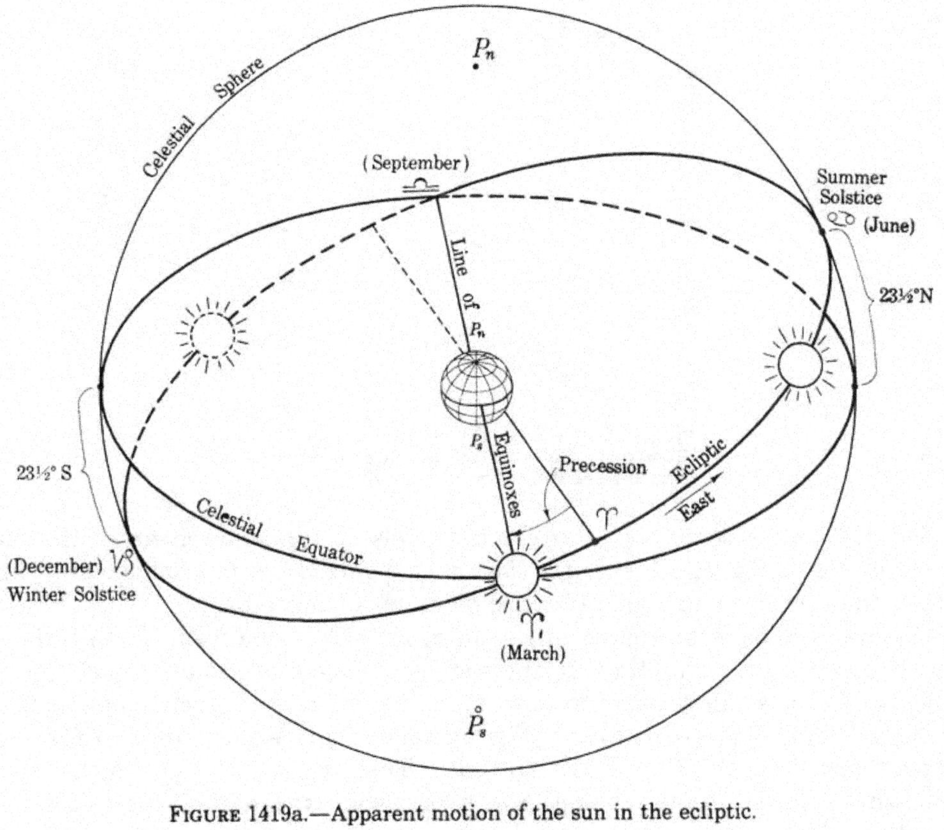

FIGURE 1419a.—Apparent motion of the sun in the ecliptic.

The sun is nearest the earth during the northern hemisphere winter. Hence, it is not the distance that is responsible for the difference in temperature during the different seasons. The reason is to be found in the altitude of the sun in the sky and the length of time it remains above the horizon. During the summer the rays are more nearly vertical, and hence more concentrated, as shown in figure 1419b. Since the sun is above the horizon more than half the time, heat is being added by absorption during a longer period than it is being lost by radiation. This explains the lag of the seasons. Following the longest day, the earth continues to receive more heat than it dissipates, but at a decreasing proportion. Gradually the proportion decreases until a balance is reached, after which the earth cools, losing more heat than it gains. This is analogous to the day, when the highest temperatures normally occur several hours after the sun reaches maximum altitude at meridian transit, and for the same reason. A similar lag occurs at other seasons of the year. *Astronomically,* the seasons *begin* at the equinoxes and solstices. *Meteorologically,* they differ from place to place.

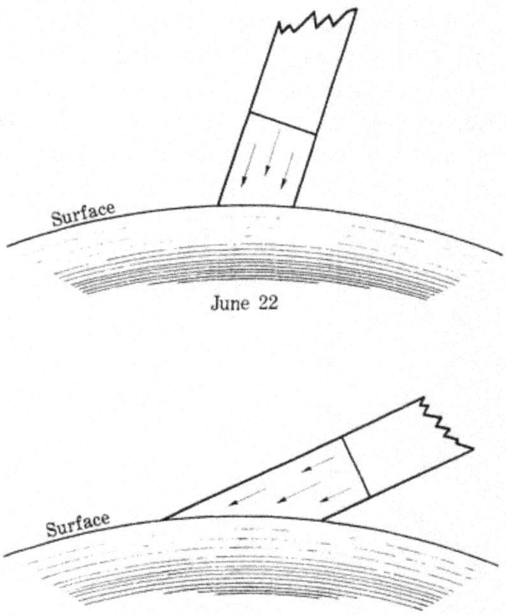

FIGURE 1419b.—Sunlight in summer and winter. Compare the surface covered by the same amount of sunlight on the two dates.

By Kepler's second law, the earth travels faster when nearest the sun, as shown in figure 1419c. Hence, the northern hemisphere (astronomical) winter is shorter than its summer, the difference being about seven days.

Everywhere between the parallels of about 23°27′N and about 23°27′S the sun is directly overhead at some time during the year. Except at the extremes, this occurs twice, once as the sun appears to move northward, and the second time as it moves southward. This is the **torrid zone.** The northern limit is the **Tropic of Cancer,** and the southern limit the **Tropic of Capricorn.** These names come from the constellations which the sun entered at the solstices when the names were first applied, more than 2,000 years ago. Today, the sun is in the next constellation toward the west, because of precession of the equinoxes, described below. The parallels about

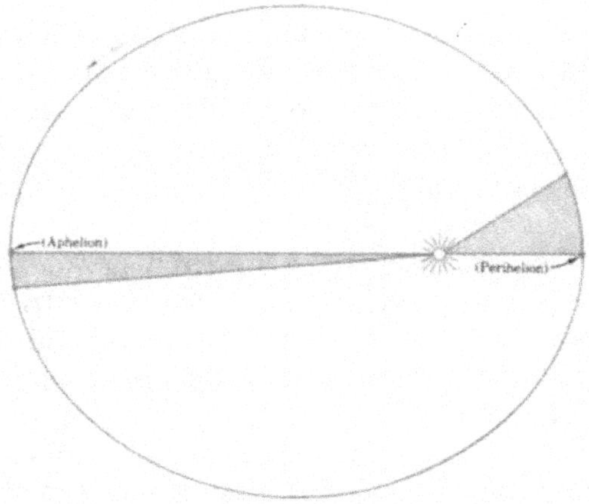

FIGURE 1419c.—Kepler's second law. Since the shaded areas are equal, speed at perihelion is greater than at aphelion.

23°27′ from the poles, marking the approximate limits of the circumpolar sun, are called **polar circles,** the one in the Northern Hemisphere being the **Arctic Circle** and the one in the Southern Hemisphere the **Antarctic Circle.** The areas inside the polar circles are the north and south **frigid zones.** The regions between the frigid zones and the torrid zones are the north and south **temperate zones.**

The expression "vernal equinox," and associated expressions, are applied both to the *times* and *points* of occurrence of the various phenomena. Navigationally, the vernal equinox is sometimes called the **first point of Aries,** because, when the name was given, the sun entered the constellation *Aries,* the ram (♈), at this time. This point is of interest to navigators because it is the origin of measurement of sidereal hour angle (art. 1426). The expressions **March equinox, June solstice, September equinox,** and **December solstice** are occasionally applied as appropriate, because the more common names are associated with the seasons in the Northern Hemisphere, and are six months out of step for the Southern Hemisphere.

The axis of the earth is undergoing a precessional motion similar to that of a top spinning with its axis tilted. In about 25,800 years the axis completes a cycle and returns to the position from which it started. Since the celestial equator is 90° from the celestial poles, it too is moving. The result is a slow westward movement of the equinoxes and solstices, which has already carried them about 30°, or one constellation, along the ecliptic from the positions they occupied when named more than 2,000 years ago. Since sidereal hour angle (art. 1426) is measured from the vernal equinox, and declination (art. 1426) from the celestial equator, the coordinates of celestial bodies would be changing even if the bodies themselves were stationary. This westward motion of the equinoxes along the ecliptic is called **precession of the equinoxes** (fig. 1419a). The total amount, called **general precession,** is about 50″27 per year (in 1975). It may be considered divided into two components, **precession in right ascension** (about 46″10 per year) measured along the celestial equator, and **precession in declination** (about 20″04 per year) measured perpendicular to the celestial equator. The annual change in the coordinates of any given star, due to precession alone, depends upon its position on the celestial sphere, since these coordinates are measured relative to the polar axis while the precessional motion is relative to the ecliptic axis.

Due to precession of the equinoxes, the celestial poles are describing circles in the sky. The north celestial pole is moving closer to Polaris, which it will pass at a distance of approximately 28′ about the year 2102. Following this, the polar distance will increase, and eventually other stars, in their turn, will become the Pole Star. Similarly, the south celestial pole will some day be marked by stars of the false Southern Cross.

The precession of the earth's axis (fig. 1419d) is the result of gravitational forces exerted principally by the sun and moon on the earth's equatorial bulge. The spinning earth responds to these forces in the manner of a gyroscope (art. 630). Regression of the nodes (art. 1423) introduces certain irregularities known as nutation in the precessional motion.

1420. The zodiac is a circular band of the sky extending 8° on each side of the ecliptic. The navigational planets and the moon are within these limits. The zodiac is divided into 12 sections of 30° each, each section being given the name and symbol ("sign") of the constellation within it. These are shown in figure 1420. The complete list of signs and names is given in appendix C.

The sun remains in each part for approximately one month. When the names were assigned, more than 2,000 years ago, the sun entered *Aries* (♈) at the vernal equinox, *Cancer* (♋) at the summer solstice, *Libra* (♎) at the autumnal equinox, and *Capricornus* (♑) at the winter solstice. Even though this is no longer true because of precession of the equinoxes, these constellations are still used for the position of the sun at the times of the equinoxes and solstices. The pseudo science of **astrology** assigns additional significance, not recognized by scientists, to the positions of the sun and planets among the signs of the zodiac.

1421. Time.—Traditionally, astronomy has furnished the basis for measurement of time, a subject of primary importance to the navigator. The **year** is associated with the revolution of the earth in its orbit. The **day** is one rotation of the earth about its axis.

The *duration* of one rotation of the earth depends upon the external reference point used. One rotation relative to the sun is called a **solar day.** However, rotation relative to the **apparent sun** (the actual sun that *appears* in the sky) does not provide time of uniform rate, because of variations in the rate of revolution and rotation of the earth. The error due to lack of uniform rate of revolution is removed by using a fictitious **mean sun.** Thus, **mean solar time** is nearly equal to the *average* **apparent solar time.** Because the accumulated difference between these times, called **equation of time,** is continually changing, the period of daylight is shifting slightly, in addition to its increase or decrease in length due to changing declination. Apparent and mean suns seldom cross the celestial meridian at the same time. The earliest sunset (in latitudes of the United States) occurs about two weeks *before* the winter solstice, and the latest sunrise about two weeks *after* winter solstice. A similar but smaller apparent discrepancy occurs at the summer solstice.

Universal Time is a particular case of the measure known in general as mean solar time. Universal Time is the mean solar time on the Greenwich meridian, reckoned in days of 24 mean solar hours beginning with 0^h at midnight. Universal Time and sidereal time are rigorously related by a formula so that if one is known the other can be found.

Universal Time, in principle, is determined by the *average* rate of the apparent daily motion of the sun relative to the meridian of Greenwich; but in practice the numerical measure of Universal Time at any instant is computed from sidereal time.

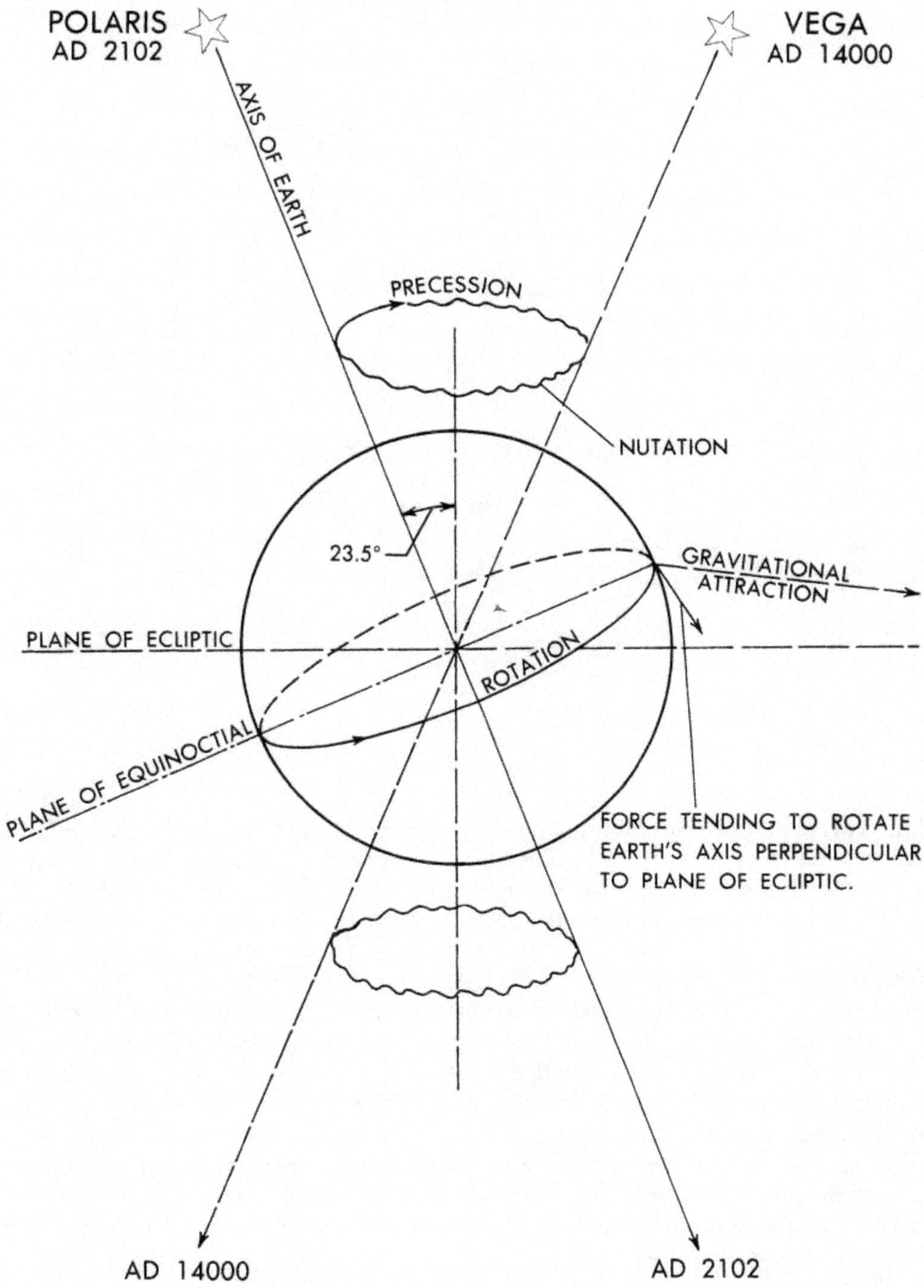

FIGURE 1419d.—Precession and nutation.

Universal Time is the standard in the application of astronomy to navigation. Observations of Universal Times are made by observing the times of transit of stars.

If the vernal equinox is used as the reference, a **sidereal day** is obtained, and from it, **sidereal time.** This indicates the approximate positions of the stars, and for this reason is the basis of star charts (art. 2204) and star finders (art. 2210). Because

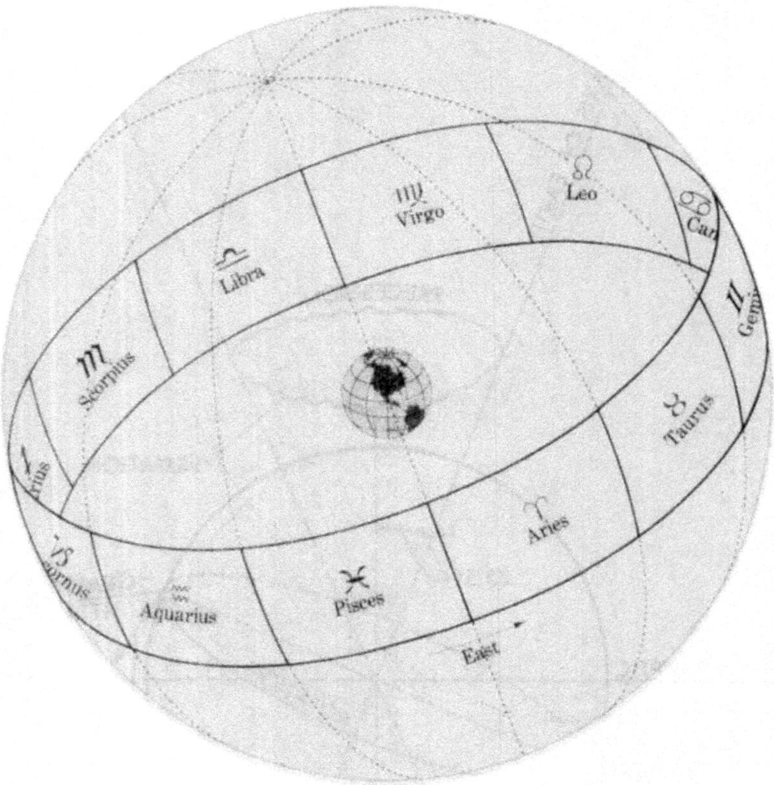

FIGURE 1420.—The zodiac.

of the revolution of the earth around the sun, a sidereal day is about 3^m56^s *shorter* than a solar day, and there is one more sidereal than solar days in a year. One mean solar day equals 1.00273791 mean sidereal days. Because of precession of the equinoxes, one rotation of the earth with respect to the stars is not quite the same as one rotation with respect to the vernal equinox. One mean solar day averages 1.0027378118868 rotations of the earth with respect to the stars.

In tide analysis, the moon is sometimes used as the reference, producing a **lunar day** averaging 24^h50^m (mean solar units) in length, and **lunar time.**

Since each kind of day is divided arbitrarily into 24 hours, each hour having 60 minutes of 60 seconds, the length of each of these units differs somewhat in the various kinds of time.

Time is also classified according to the terrestrial meridian used as a reference. **Local time** results if one's own meridian is used, **zone time** if a nearby reference meridian is used over a spread of longitudes, and **Greenwich** or **Universal Time** if the Greenwich meridian is used.

The subject of time is discussed in more detail in XVIII.

1422. Planetary configurations.—Since the orbit of an inferior planet lies within that of the earth, the planet and sun are nearly in line twice each synodic period of revolution of the inferior planet. When the sun is between the earth and the other planet, that planet is at **superior conjunction.** When the planet is between the earth and sun, it is at **inferior conjunction.** If the orbit of the planet had no inclination to the ecliptic, the planet would cross or **transit** the face of the sun at inferior conjunction and be eclipsed or **occulted** by the sun at superior conjunction. Occasionally this does occur.

Refer to figure 1422, showing orbits of the earth, Venus (an inferior planet,) and Mars (a superior planet). As shown, the relative sizes of the orbits are correct, and the relative sizes of the planets are correct, but the planets are too large for their orbits and the sun, and the sun is too large for the orbits of the planets. The earth is considered stationary in its orbit. The positions of Venus are shown at superior and inferior conjunctions. In moving eastward from one to the other. Venus appears to move to the left of the sun. As observed from the earth, the angle between lines to the sun and a planet, particularly an inferior planet, is called the planet's **elongation,** which may be designated east or west to indicate the apparent position of the planet relative to the sun. As Venus continues along its orbit, its elongation increases slowly until the planet arrives at the point where a straight line from the earth is tangent to its orbit, when the elongation becomes maximum. Here it is called **greatest elongation east.** As Venus continues along its orbit, its elongation decreases rapidly, becoming zero at inferior conjunction. Through the second half of its synodic period its elongation increases rapidly to **greatest elongation west,** and then decreases slowly to zero at the next superior conjunction. The greatest elongation of Venus is about 46°, but varies because its orbit and that of the earth are elliptical, and the phenomenon occurs at different points on the orbits.

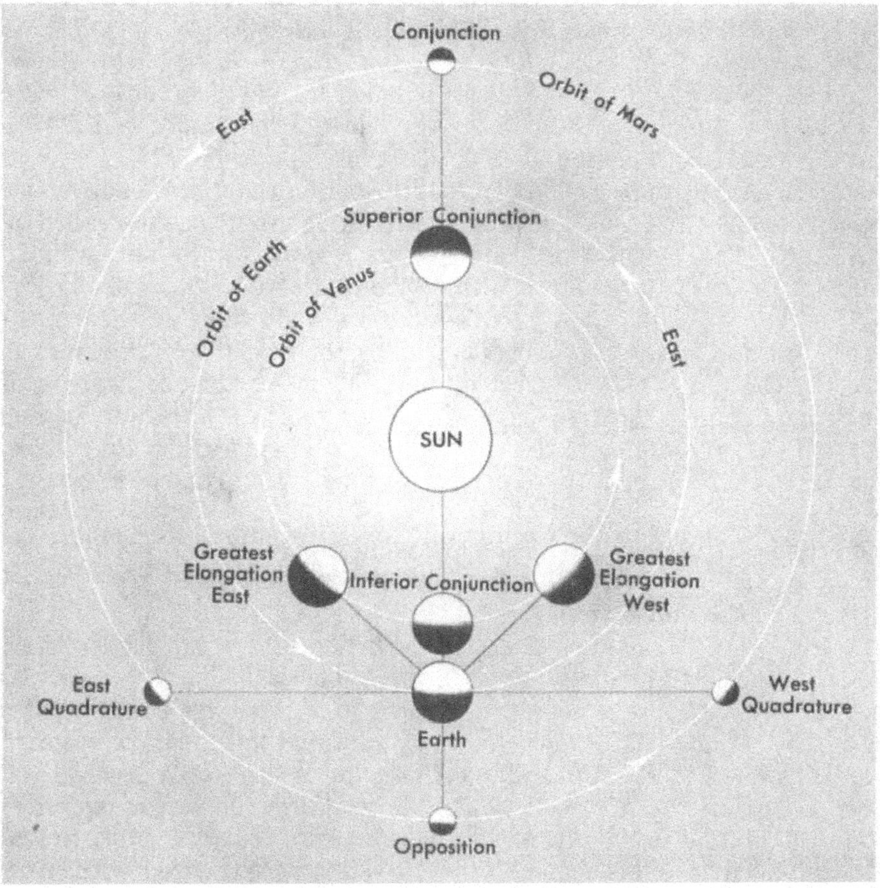

FIGURE 1422.—Planetary configurations.

The orbit of the planet Mercury lies inside that of Venus, and hence the greatest elongation is not as great, being about 28°. It is because the apparent position of Mercury is never far from the sun that this planet is not considered navigationally important. Since its synodic period of revolution is only 115.9 days, it is at conjunction a little oftener than once every two months. By comparison, Venus is at conjunction a little oftener than once every ten months, having a synodic period of revolution of 583.9 days.

As shown in figure 1422, an inferior planet goes through all phases similar to those of the moon (art. 1423), being "full" at superior conjunction, "new" at inferior conjunction, and at "quarter" when it reaches greatest elongation. A telescope is needed to see the phases.

For a superior planet the situation is different. Refer again to figure 1422. When the sun is between the earth and the planet, that planet (Mars in the illustration) is at **conjunction** (☌). The adjective "superior" is not needed because a superior planet, when on the opposite side is *away* from the sun, or at **opposition** (☍) and can never be at inferior conjunction. When its elongation is 90°, a superior planet is at east or west **quadrature** (□), depending upon its apparent position relative to the sun. Since a superior planet has a longer period of revolution than the earth, it *appears* to move westward around the sun, being at conjunction, east quadrature, opposition, west quadrature, and back to conjunction. It is at "full" phase at conjunction and opposition, and gibbous between.

Unless a planet is in the ecliptic, it is not directly in line with the earth and sun at conjunction and opposition. These points are defined as those at which either the sidereal hour angles (art. 1426) or the celestial longitudes (art. 1430) are the same (in the case of conjunction) or 180° apart (at opposition).

The apparent positions of the planets in relation to other members of the solar system, particularly the relationships shown in figure 1422, are called **planetary configurations.** The motions of planets with respect to the sun would be true, generally with respect to the stars, also, if the earth were stationary in its orbit, as shown. However, because of the earth's motion around the sun, the sun appears to move eastward among the stars. This is usually the direction of apparent motion of the planets, too, and is called **direct motion.** When a planet is near opposition or inferior conjunction, its apparent *westerly* motion relative to the sun is greater than the apparent *easterly* motion of the sun relative to the stars, and the planet appears to move in a *westerly* direction relative to the stars. This is called **retrograde motion.**

The brightest planet in the western sky following sunset is popularly called the **evening star,** and the brightest planet in the eastern sky preceding sunrise is popularly called the **morning star.**

1423. Phases of the moon.—Relative to the sun, the moon makes one complete trip around the celestial sphere each synodical month (about 29½ days). As it does so, it goes through a cycle of aspects or **phases** to an observer on the earth, because the moon, like the planets, shines chiefly by reflected light from the sun. The orbit of the moon is inclined about 5° to the ecliptic, and undergoes a precessional motion called **regression of the nodes.** It is similar to precession of the equinoxes of the earth (art. 1419), and is chiefly responsible for nutation (art. 1417). However, the cycle is completed in a little more than 18 years, as compared with about 25,800 years for the earth.

Because of the small inclination of its orbit, the moon is never far from the ecliptic. At conjunction, when the moon passes nearly between the earth and sun, its illuminated portion is *away* from the earth (*toward* the sun), as shown in figure 1423. (In this illustration, the outer figures show various positions of the moon relative to the earth and sunlight. The inner circle of moons shows the appearance from the earth.) It is then a **new moon,** and may be barely visible because of **earthshine,** which is sunlight reflected from the illuminated side of the earth. To an observer on the moon, the "full earth" would be visible at this time, three and one-half times as great in diameter and nearly 40 times as bright as the full moon appears to an observer on the earth. Since it is at conjunction, the new moon rises, transits the celestial meridian, and sets at approximately the same time as the sun.

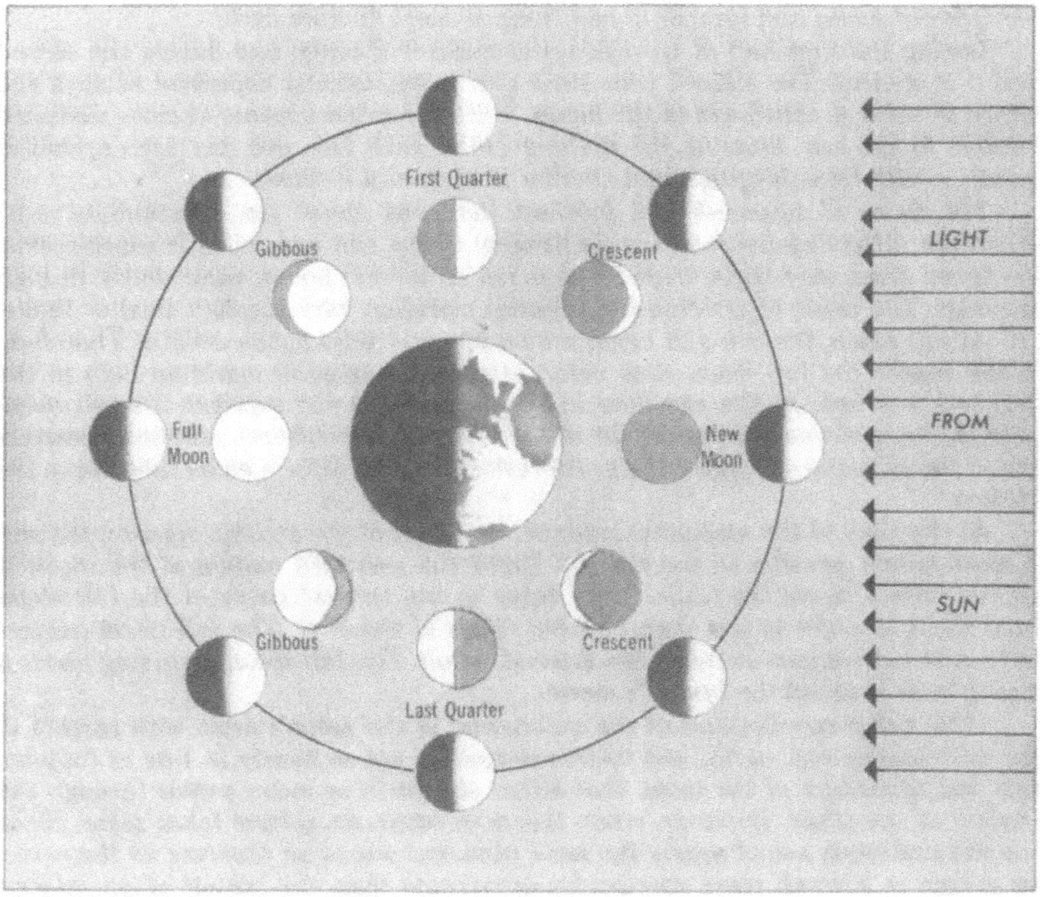

FIGURE 1423.—Phases of the moon. The inner figures of the moon represent its appearance from the earth.

A day later the moon has moved about 12°2 eastward of the sun and a thin **crescent** appears on the side toward the sun, with the horns or **cusps** pointing away from the sun. The moon is low in the western sky after sunset. Because of glow from this illuminated portion, and the fact that the side of the earth toward the moon is not quite "full," that part of the moon illuminated by earthshine is not quite as bright. Each day the moon moves approximately 12°2 east, relative to the sun. As it does so, the crescent grows fatter, and the earthshine less conspicuous.

When the moon reaches quadrature, about a week after new moon, it is at **first quarter.** That half of the moon toward the sun is illuminated. The moon is now about 90° or six hours *behind* the sun. It rises about noon, is on the celestial meridian about 6 PM, and sets about midnight.

As the moon continues eastward on successive days, the line separating the illuminated and dark portions, called the **terminator,** moves on across the moon. The moon is now in the **gibbous** phase, which continues until the moon is at opposition, or **full moon.** It now rises about the time of sunset, reaches the celestial meridian about midnight, and sets about the time of sunrise.

On succeeding days the moon again becomes gibbous, and at quadrature it is at **last quarter,** rising about midnight, crossing the celestial meridian about 6 AM, and setting about noon. During the remainder of its cycle the moon again goes through the crescent phase and returns to new moon to start another cycle.

During the first half of the cycle, the moon is **waxing,** and during the second half it is **waning.** The elapsed time since new moon, usually expressed as days and tenths of a day is called **age of the moon.** Since the moon appears to move *eastward* relative to the sun, crossing the meridian *later* each day, one day each synodical month is without a moonrise, and another is without a moonset.

The times of moonrise and moonset indicated above are approximate only. When the difference between the declination of the sun and moon is considerable, the times given may be in error by as much as several hours, particularly in high latitudes. The times of crossing the celestial meridian vary through smaller limits.

At full moon, the sun and moon are on opposite sides of the ecliptic. Therefore, in the winter the full moon rises early, crosses the celestial meridian high in the sky, and sets late; as the sun does in the summer. In the summer the full moon rises in the southeastern part of the sky (Northern Hemisphere), remains relatively low in the sky, and sets along the southwestern horizon after a short time above the horizon.

At the time of the autumnal equinox, that part of the ecliptic opposite the sun is most nearly parallel to the horizon. Since the eastward motion of the moon is approximately along the ecliptic, the delay in the time of rising of the full moon from night to night is less than at other times of the year. The full moon nearest the autumnal equinox is called the **harvest moon.** The full moon occurring about a month later is called the **hunter's moon.**

1424. Eclipses.—Because of the inclination of the moon's orbit with respect to the ecliptic, the sun, earth, and moon are usually not so nearly in line at conjunction and opposition of the moon that either the earth or moon passes through the shadow of the other. However, when this does occur, an **eclipse** takes place. Since the sun and moon are of nearly the same apparent size to an observer on the earth, an eclipse is a much more spectacular occurrence than the transit of an inferior planet across the face of the sun, or the occultation of a star or planet by the sun or moon (art. 1422).

When conditions are suitable, the moon passes between the sun and earth, as shown in figure 1424a. If the moon's apparent diameter is larger than that of the sun, the moon being near perigee, its shadow reaches the earth as a nearly round dot only a few miles in diameter. The dot moves rapidly across the earth, from west to east, as the moon continues in its orbit. Within the dot, the sun is completely hidden from view, and a **total eclipse** of the sun occurs. For a considerable distance around the shadow, part of the surface of the sun is obscured, and a **partial eclipse** occurs. In the line of travel of the shadow a partial eclipse occurs as the round disk of the moon appears to move slowly across the surface of the sun, hiding an ever-increasing part of it, until the total eclipse occurs. Because of the uneven edge of the mountainous moon, the light is not cut off evenly, but several last illuminated portions appear through the valleys or passes between the mountain peaks. These are called **Baily's Beads.** A total eclipse is a spectacular phenomenon. As the last light from the sun is cut off, the solar **corona,** or envelope of thin, illuminated gas around the sun, becomes visible. Wisps of more dense gas may appear as **solar prominences** (fig. 1424b). The only light reaching the observer is that diffused by the atmosphere surrounding the shadow. As the moon appears to continue on across the face of the sun, the sun finally emerges from the other side, first as Baily's Beads, and then as an ever widening crescent until no part of its surface is obscured by the moon.

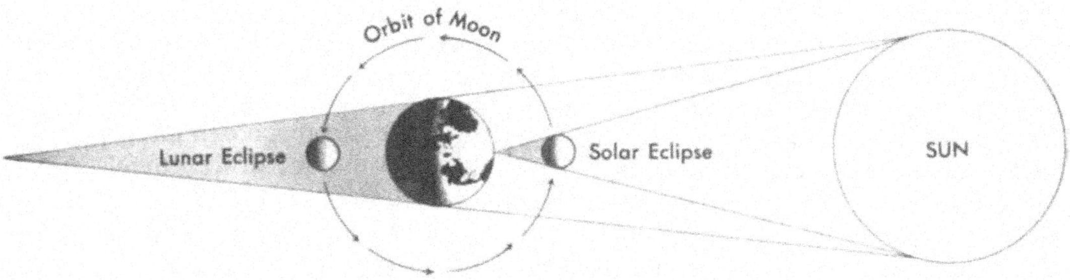

FIGURE 1424a.—Eclipses of the sun and moon.

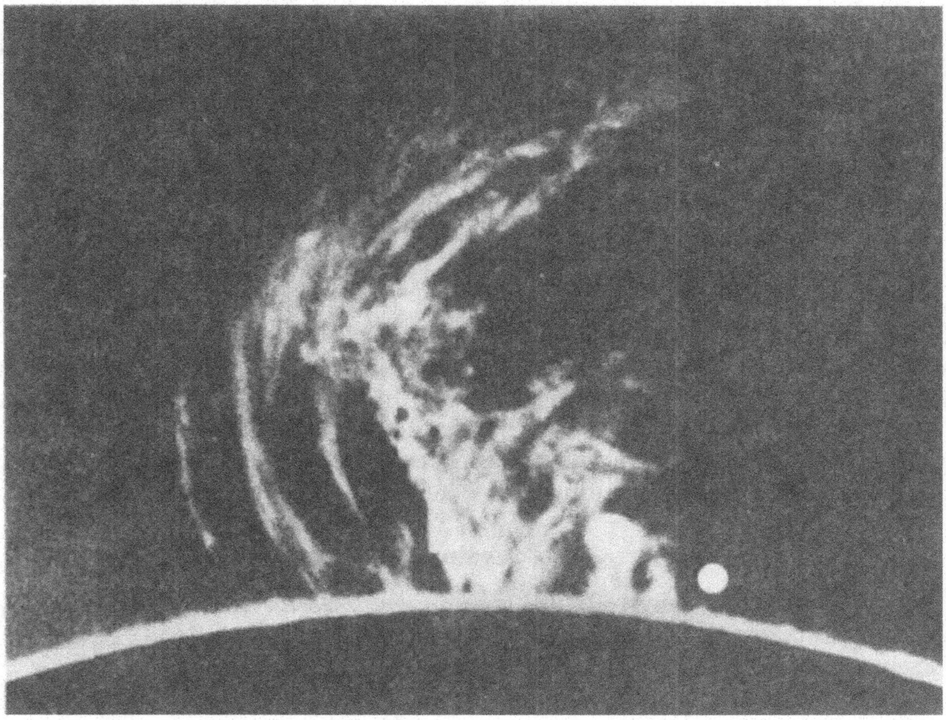

Courtesy of Mt. Wilson and Palomar Observatories.

FIGURE 1424b.—Solar prominence, 140,000 miles high, photographed in light of calcium. July 9, 1917. Small white disk shows relative size of earth.

The duration of a total eclipse depends upon how nearly the moon crosses the center of the sun, the location of the shadow on the earth, the relative orbital speeds of the moon and earth, and (principally) the relative apparent diameters of the sun and moon. The maximum length that can occur is a little more than seven minutes.

If the apparent diameter of the moon is less than that of the sun, its shadow does not quite reach the earth. Over a small area of the earth directly in line with the moon and sun, the moon appears as a black disk almost covering the surface of the sun, but with a thin ring of the sun around its edge. This is an **annular eclipse,** and occurs a little more often than a total eclipse.

If the shadow of the moon passes close to the earth, but not directly in line with it, a partial eclipse may occur without a total or annular eclipse.

An eclipse of the moon occurs when the moon passes through the shadow of the earth, as shown in figure 1424a. Since the diameter of the earth is about three and one-half times that of the moon, the earth's shadow at the distance of the moon is much larger than that of the moon. A total eclipse of the moon can last nearly one and three-quarters hours, and some part of the moon may be in the earth's shadow for almost four hours. During a total **solar eclipse** no part of the sun is visible because a body (the moon) intervenes in the line of sight. During a **lunar eclipse** some light does reach the moon because of diffraction by the atmosphere of the earth, and hence the eclipsed full moon is visible as a faint reddish disk. A lunar eclipse is visible over the entire hemisphere of the earth facing the moon. Anyone who can see the moon can see the eclipse.

During any one year there may be as many as five eclipses of the sun, and always there are at least two. There may be as many as three eclipses of the moon, or none. The total number of eclipses during a single year does not exceed seven, and can be as few as two. There are more solar than lunar eclipses, but the latter are more numerous at any one place because of the restricted areas over which solar eclipses are visible.

The two points of intersection of the moon's orbit and the ecliptic are called **nodes,** and the line connecting them, the **line of nodes.** Eclipses occur when the sun, earth, and moon are nearly on this line, twice each **eclipse year** of 346.6 days. This is less than a calendar year because of regression of the nodes (art. 1423). In a little more than 18 years the line of nodes returns to approximately the same position with respect to the sun, earth, and moon. During an almost equal period, called the **saros,** a cycle of eclipses occurs. During the following saros the cycle is repeated with only minor differences.

Eclipses have considerable value in establishing additional facts about the sun and moon, and in determining distances between two widely separated points on the earth, at which accurate timing of the eclipse is made.

Coordinates

1425. Latitude and longitude are coordinates used for locating positions on the earth. Several types, differing slightly from each other, are defined. Three of these are discussed here.

Astronomic latitude is the angle (*ABQ*, fig. 1425) between a line in the direction of gravity (*AB*) at a station and the plane of the equator (*QQ'*). **Astronomic longitude** is the angle between the plane of the celestial meridian at a station and the plane of the celestial meridian at Greenwich. These coordinates are customarily found by means of celestial observations. If the earth were perfectly homogeneous and level, these positions would be consistent and satisfactory. However, because of **deflection of the vertical** (app. X) due to uneven distribution of the mass of the earth, lines of equal astronomic latitude and longitude are not circles, although the irregularities are small. In the United States the prime-vertical component (affecting longitude) may be a little more than 18″, and the meridional component (affecting latitude) as much as 25″.

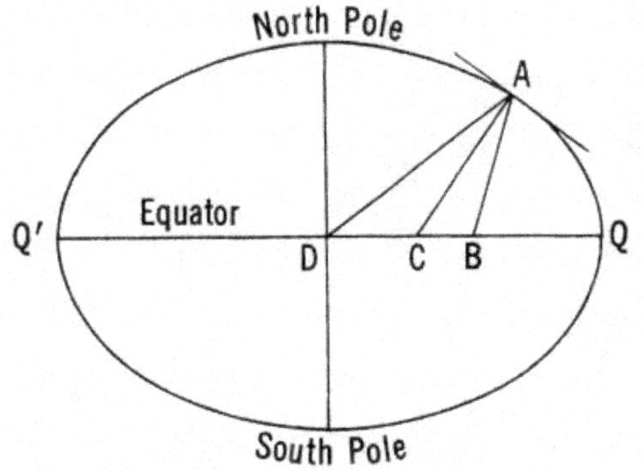

FIGURE 1425.—Three kinds of latitude at point *A*.

Geodetic latitude is the angle (*ACQ*, fig. 1425) between a normal to the spheroid (*AC*) at a station and the plane of the geodetic equator (*QQ′*). **Geodetic longitude** is the angle between the plane defined by the normal to the spheroid and the axis of the earth, and the plane of the geodetic meridian at Greenwich. These values are obtained when astronomical latitude and longitude are corrected for deflection of the vertical. These coordinates are the ones used for charting, and are frequently referred to as **geographic latitude** and **geographic longitude**, although these expressions are sometimes used to refer to astronomical latitude

Geocentric latitude is the angle (*ADQ*, fig. 1425) at the center of the ellipsoid between the plane of its equator (*QQ′*) and a straight line (*AD*) to a point on the surface of the earth. This differs from geodetic latitude because the earth is a spheroid, rather than a sphere, and the meridians are ellipses. Since the parallels of latitude are considered to be circles, geodetic longitude is geocentric, and a separate expression is not used. The difference between geocentric and geodetic latitudes is a maximum of about 11′.6 at latitude 45°.

Because of the oblate shape of the ellipsoid, the length of a degree of geodetic latitude is not everywhere the same, increasing from about 59.7 nautical miles at the equator to about 60.3 nautical miles at the poles, as shown by table 6. The value of 60 nautical miles customarily used everywhere by the navigator is correct at about latitude 45°.

Measurement on the Celestial Sphere

1426. Elements of the celestial sphere.—The celestial sphere (art. 1403) is an imaginary sphere of infinite radius with the earth at its center (fig. 1426a). The north and south **celestial poles** of this sphere are located by extension of the earth's axis. The **celestial equator** (sometimes called **equinoctial**) is formed by projecting the plane of the earth's equator to the celestial sphere. A **celestial meridian** is formed by the intersection of the plane of a terrestrial meridian, extended, and the celestial sphere. It is the arc of a great circle through the poles of the celestial sphere.

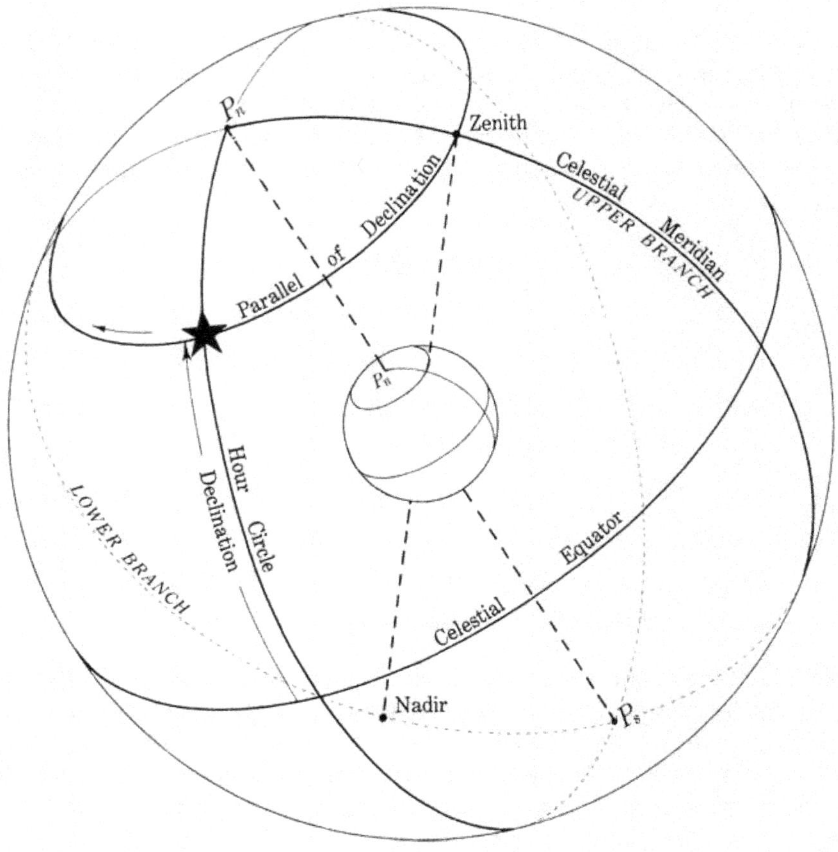

FIGURE 1426a.—Elements of the celestial sphere. The celestial equator is the primary great circle.

The point on the celestial sphere vertically overhead of an observer is the **zenith** and the point on the opposite side of the sphere, vertically below him, is the **nadir.** The zenith and nadir are the extremities of a diameter of the celestial sphere through the observer and the common center of the earth and the celestial sphere. The arc of a celestial meridian between the poles is called the **upper branch** if it contains the zenith and the **lower branch** if it contains the nadir. The upper branch is frequently used in navigation and references to a celestial meridian are understood to mean only its upper branch unless otherwise stated. Celestial meridians take the names, as 65° west, of their terrestrial counterparts.

An **hour circle** is a great circle through the celestial poles and a point or body on the celestial sphere. It is similar to a celestial meridian, but moves with the celestial sphere as it rotates about the earth, while a celestial meridian remains fixed with respect to the earth.

The location of a body along its hour circle is defined by the body's angular distance from the celestial equator. This distance, called **declination,** is measured north or south of the celestial equator in degrees, from 0° through 90°, similar to latitude on the earth.

A circle parallel to the celestial equator is called a **parallel of declination,** since it connects all points of equal declination. It is similar to a parallel of latitude on the earth. The path of a celestial body during its daily apparent revolution around the earth is called its **diurnal circle.** It is not actually a circle if a body changes its declination. Since the declination of all navigational bodies is continually changing, the bodies are describing flat, spherical spirals as they circle the earth. However, since the change is relatively slow, a diurnal circle and a parallel of declination are usually considered identical.

A point on the celestial sphere may be identified at the intersection of its parallel of declination and its hour circle. The parallel of declination is identified by the declination.

Two basic methods of locating the hour circle are in use. Its angular distance west of a reference hour circle through a point on the celestial sphere called the **vernal equinox** or **first point of Aries** is called **sidereal hour angle (SHA)** (fig. 1426b). This angle measured eastward from the vernal equinox is called **right ascension,** and is usually expressed in time units.

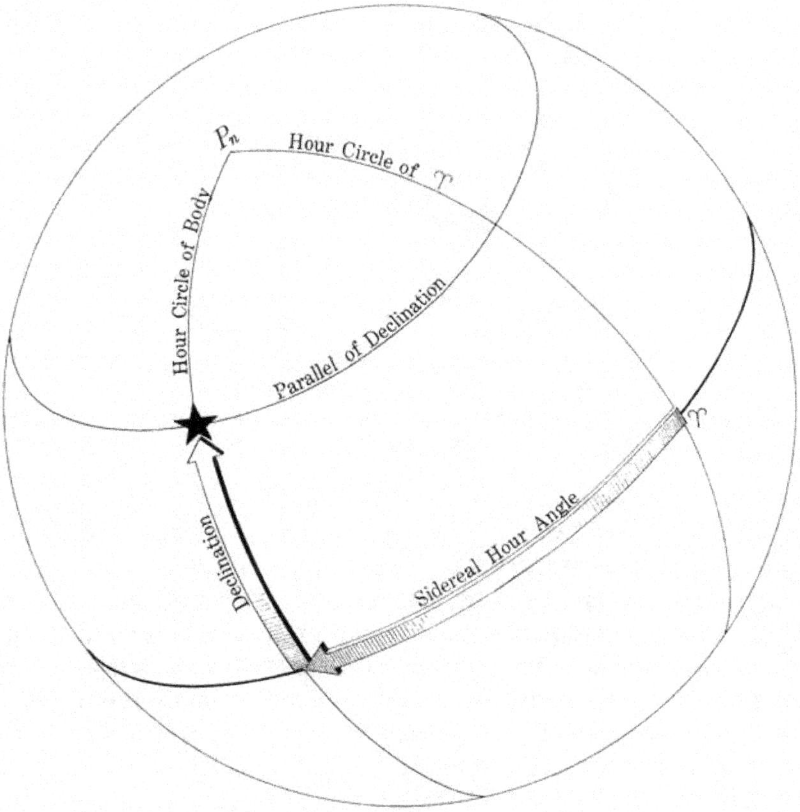

FIGURE 1426b.—A point on the celestial sphere can be located by its declination and sidereal hour angle.

The second method of locating the hour circle is to indicate its angular distance *west* of a celestial meridian (fig. 1426c). If the Greenwich celestial meridian is used as the reference, the angular distance is called **Greenwich hour angle (GHA)**, and if the meridian of the observer, it is called **local hour angle (LHA)**. It is sometimes more convenient to measure hour angle either eastward or westward, as longitude is measured on the earth, in which case it is called **meridian angle** (t). These coordinates are discussed further in article 1428.

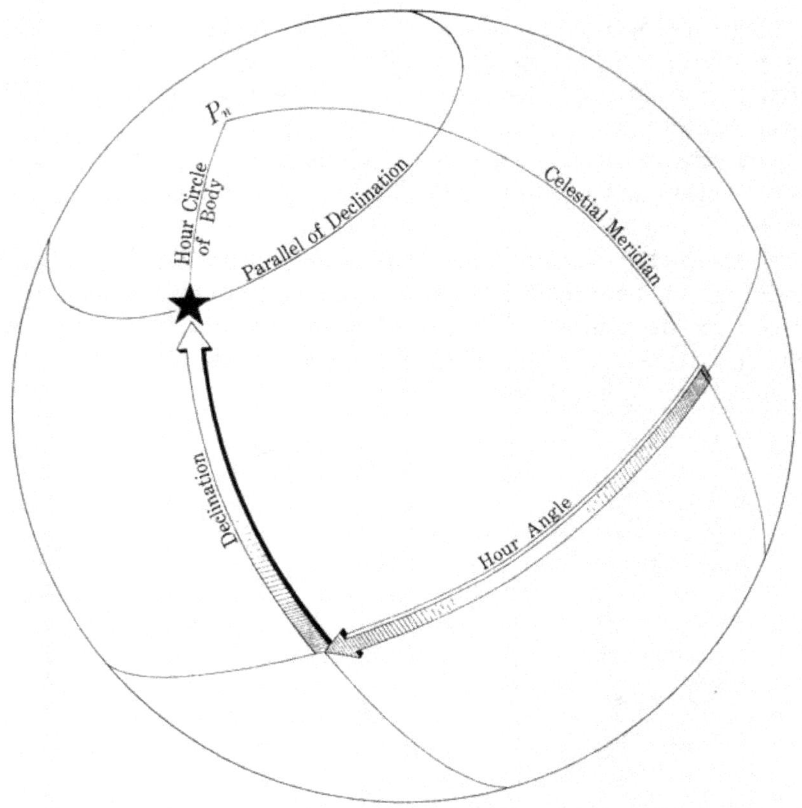

FIGURE 1426c.—A point on the celestial sphere can be located by its declination and hour angle.

A point on the celestial sphere may also be located by means of **altitude** and **azimuth**, coordinates based upon the horizon as the primary great circle, instead of the celestial equator. This system is discussed in article 1430.

Two additional systems used by astronomers are based upon the ecliptic (art. 1419) and the galactic equator (the approximate mid great circle of the galaxy). The coordinates of the ecliptic system are **celestial latitude** and **celestial longitude** and those of the galactic system are **galactic latitude** and **galactic longitude.**

1427. Coordinate systems.—Various systems of coordinates on the celestial sphere, all of them similar to the familiar latitude and longitude on the earth, were discussed briefly in article 1426. Of these, the navigator is rarely concerned with any but the celestial equator system and the horizon system. The former is but an extension to the celestial sphere of the geographical system of the earth. The latter is a similar system in which the horizon replaces the celestial equator as the primary great circle, and the zenith and nadir are the poles. These two systems are the almost constant companions of the celestial navigator.

1428. The celestial equator system of coordinates.—If the familiar graticule of latitude and longitude lines is expanded until it reaches the celestial sphere of infinite radius, it forms the basis of the celestial equator system of coordinates, as explained in article 1426. On the celestial sphere the familiar *latitude* becomes **declination** (**Dec.** or **d**), and *longitude,* measured always toward the west, through 360°, becomes **sidereal hour angle** (**SHA**) if measured from the vernal equinox.

Declination (**Dec.** or **d**) is angular distance north or south of the celestial equator (d in fig. 1428a). It is measured along an hour circle, from 0° at the celestial equator through 90° at the celestial poles, and is labeled N or S to indicate the direction of measurement. All points having the same declination lie along a **parallel of declination.**

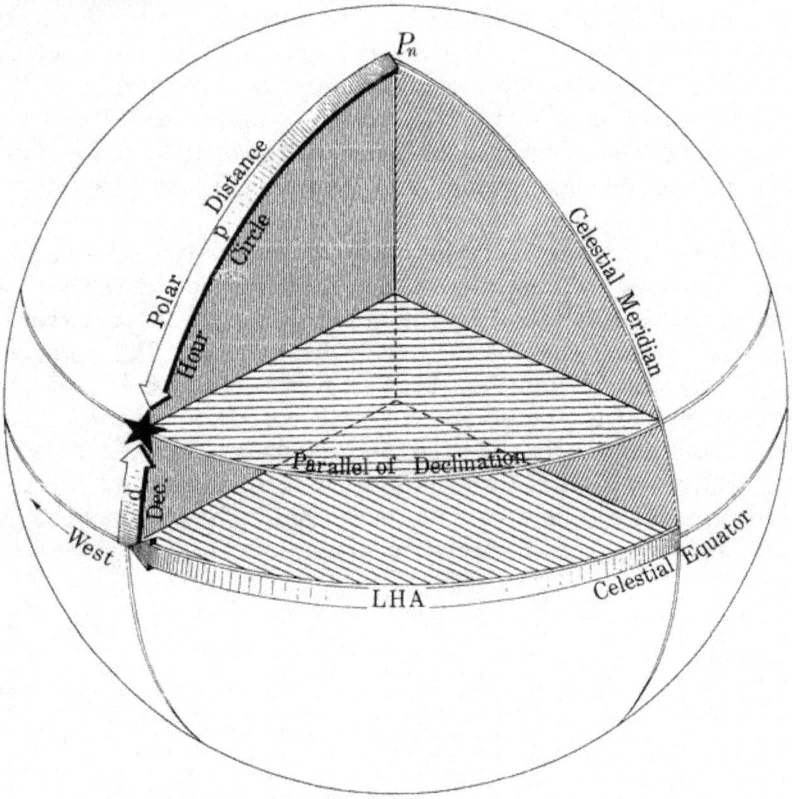

FIGURE 1428a.—The celestial equator system of coordinates, showing measurements of declination, polar distance, and local hour angle.

Polar distance (**p**) is angular distance from a celestial pole, or the arc of an hour circle between the celestial pole and a point on the celestial sphere. It is measured along an hour circle and may vary from 0° to 180°, since either pole may be used as the origin of measurement. It is usually considered the complement of declination, though it may be either 90°−d or 90°+d, depending upon the pole used.

Local hour angle (**LHA**) is angular distance west of the local celestial meridian, or the arc of the celestial equator between the upper branch of the local celestial meridian and the hour circle through a point on the celestial sphere, measured westward from the local celestial meridian, through 360°. It is also the similar arc of the parallel of declination and the angle at the celestial pole, similarly measured. If the Greenwich (0°) meridian is used as the reference, instead of the local meridian, the expression **Greenwich hour angle** (**GHA**) is applied. It is sometimes convenient to measure the arc or angle in *either* an easterly or westerly direction from the local meridian, through 180°, when it is called **meridian angle** (**t**) and labeled E or

W to indicate the direction of measurement. All bodies or other points having the same hour angle lie along the same **hour circle.**

Because of the apparent daily rotation of the celestial sphere, hour angle continually increases, but meridian angle increases from 0° at the celestial meridian to 180°W, which is also 180°E, and then decreases to 0° again. The rate of change for the mean sun (art. 1421) is 15° per hour. The rate of all other bodies except the moon is within 3′ of this value. The average rate of the moon is about 14°.5.

As the celestial sphere rotates, each body crosses each branch of the celestial meridian approximately once a day. This crossing is called **meridian transit** (sometimes called **culmination**). It may be called **upper transit** to indicate crossing of the upper branch of the celestial meridian, and **lower transit** to indicate crossing of the lower branch.

The time diagram shown in figure 1428b illustrates the relationship between the various hour angles and meridian angle. The circle is the celestial equator as seen from above the *South Pole,* with the upper branch of the observer's meridian (P_SM) at the top. The radius P_SG is the Greenwich meridian, P_S♈ the hour circle of the vernal equinox, and P_SS and P_SS' the hour circles of celestial bodies to the *west* and *east,* respectively, of the observer's celestial meridian. Note that when LHA is less than 180°, t is numerically the same and is labeled W, but that when LHA is greater than 180°, t=360°−LHA and is labeled E. In figure 1428b arc *GM* is the longitude, which in this case is west. The relationships shown apply equally to other arrangements of radii, except for relative magnitudes of the quantities involved.

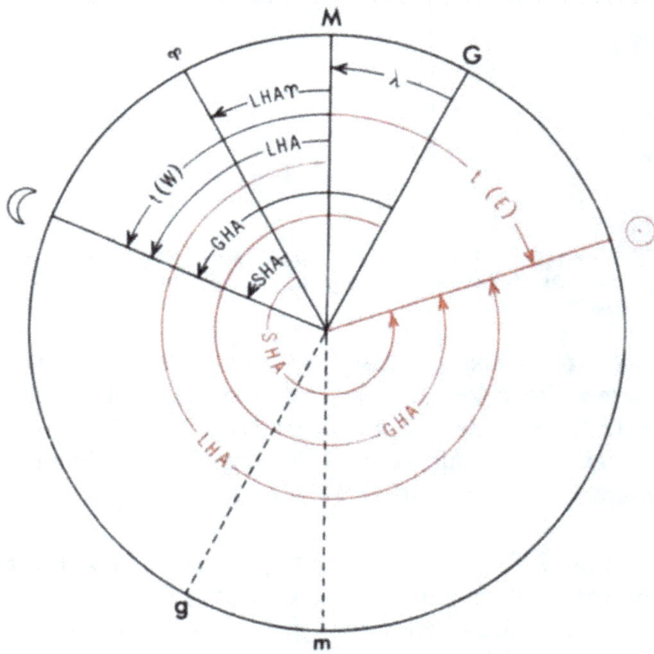

FIGURE 1428b.—Time diagram. Local hour angle, Greenwich hour angle, and sidereal hour angle are measured westward through 360°. Meridian angle is measured eastward or westward through 180° and labled E or W to indicate the direction of measurement.

1429. The horizons.—The second set of celestial coordinates with which the navigator is directly concerned is based upon the **horizon** as the primary great circle. However, since several different horizons are defined, these should be thoroughly understood before proceeding with a consideration of the horizon system of coordinates.

The line where earth and sky appear to meet is called the **visible** or **apparent horizon.** On land this is usually an irregular line unless the terrain is level. At sea the visible horizon appears very regular and often very sharp. However, its *position relative to the celestial sphere* depends primarily upon (1) the refractive index of the air, and (2) the height of the observer's eye above the surface.

Figure 1429 shows a cross section of the earth and celestial sphere through the position of an observer at *A* above the surface of the earth. A straight line through *A* and the center of the earth *O* is the vertical of the observer, and contains his zenith (Z) and nadir (Na). A plane perpendicular to the true vertical is a horizontal plane, and its intersection with the celestial sphere is a horizon. It is the **celestial horizon** if the plane passes through the center of the earth, the **geoidal horizon** if it is tangent to the earth, and the **sensible horizon** if it passes through the eye of the observer at *A*. Since the radius of the earth is considered negligible with respect to that of the celestial sphere, these horizons become superimposed, and most measurements are referred only to the celestial horizon. This is sometimes called the **rational horizon** from the latin word "ratio," reckoning.

If the eye of the observer is at the surface of the earth, his visible horizon coincides with the plane of the geoidal horizon; but when elevated above the surface, as at *A*, his eye becomes the vertex of a cone which, neglecting refraction, is tangent to the earth at the small circle *BB,* and which intersects the celestial sphere in *B'B'*, the **geometrical horizon.** This expression is sometimes—but less appropriately—applied to the celestial horizon.

Because of refraction (art. 1606), the visible horizon *C'C'* appears above but is actually slightly below the geometrical horizon as shown in figure 1429.

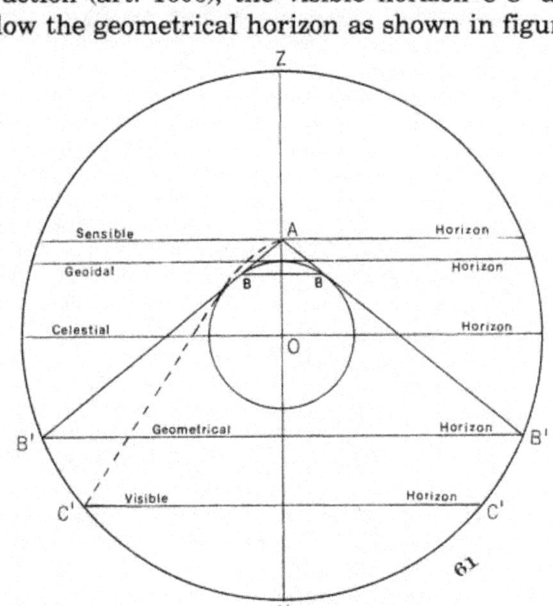

FIGURE 1429.—The horizons used in navigation.

For any elevation above the surface, the celestial horizon is usually *above* the geometrical and visible horizons, the difference increasing as elevation increases. It is thus possible to observe a body which is above the visible horizon but below the celestial horizon. That is, the body's altitude is negative and its zenith distance is greater than 90° (art. 1430).

1430. The horizon system of coordinates is based upon the celestial horizon as the primary great circle, and a series of secondary **vertical circles,** which are great circles through the zenith and nadir of the observer and hence perpendicular to his horizon (fig. 1430a). Thus, the celestial horizon is similar to the equator, and the vertical circles are similar to meridians, but with one important difference. The celestial horizon and vertical circles are dependent upon the position of the observer and hence move with him as he changes position, while the primary and secondary great circles of both the geographical and celestial equator systems are independent of the observer. The horizon and celestial equator systems coincide for an observer at the geographical pole of the earth, and are mutually perpendicular for an observer on the equator. At all other places the two are oblique.

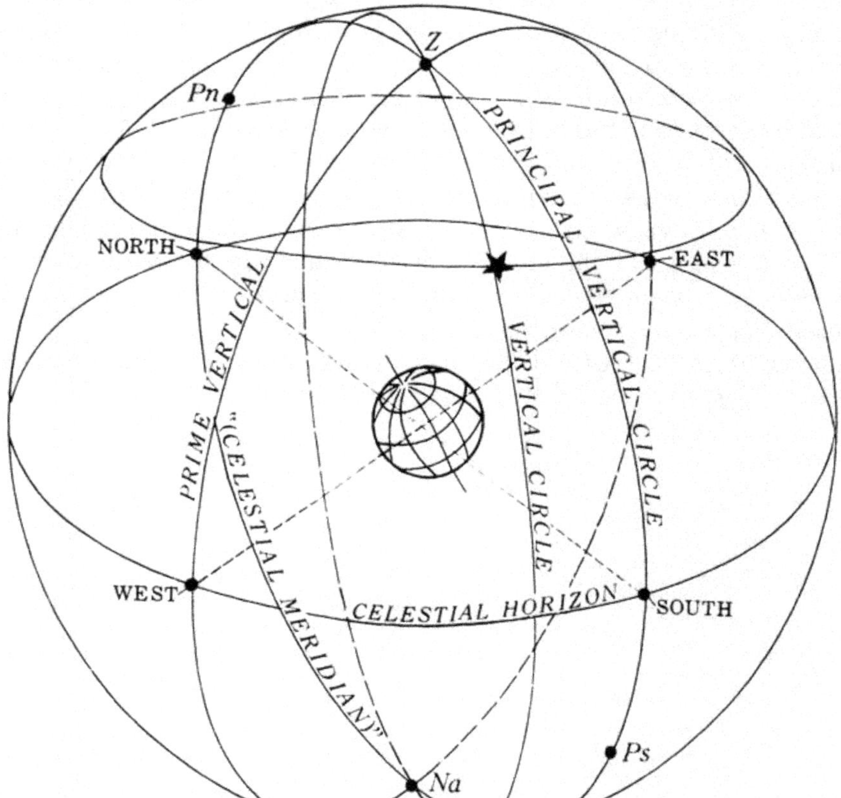

FIGURE 1430a.—Elements of the celestial sphere. The celestial horizon is the primary great circle.

The vertical circle through the north and south points of the horizon passes through the poles of the celestial equator system of coordinates. One of these poles (having the same name as the latitude) is above the horizon and is called the **elevated pole.** The other, called the **depressed pole,** is below the horizon. Since this vertical circle is a great circle through the celestial poles, and includes the zenith of the observer, it is also a celestial meridian. In the horizon system it is called the **principal vertical circle.** The vertical circle through the east and west points of the horizon, and hence perpendicular to the principal vertical circle, is called the **prime vertical circle,** or simply the **prime vertical.**

As shown in figure 1430b, **altitude** is angular distance above the horizon. It is measured along a vertical circle, from 0° at the horizon through 90° at the zenith. Altitude measured from the visible horizon may exceed 90° because of the dip of the horizon, as shown in figure 1429. Angular distance below the horizon, called **negative altitude,** is provided for by including certain negative altitudes in some tables for use in celestial navigation, such as Pub. No. 249. All points having the same altitude lie along a **parallel of altitude** or **almucantar.**

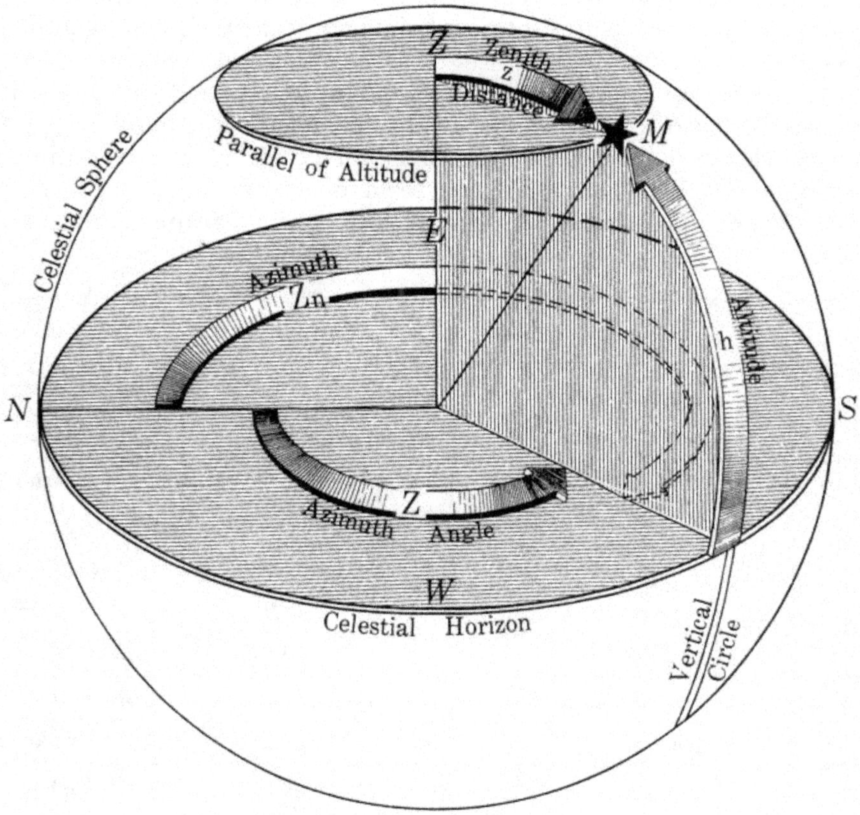

FIGURE 1430b.—The horizon system of coordinates, showing measurement of altitude, zenith distance, azimuth, and azimuth angle.

Zenith distance (z) is angular distance from the zenith, or the arc of a vertical circle between the zenith and a point on the celestial sphere. It is measured along a vertical circle from 0° through 180°. It is usually considered the complement of altitude. For a body above the celestial horizon it is equal to 90° − h and for a body below the celestial horizon it is equal to 90° − (− h) or 90° + h; or 90° + a negative altitude.

The horizontal direction of a point on the celestial sphere, or the bearing of the geographical position is called **azimuth** or **azimuth angle** depending upon the method of measurement. In both methods it is an arc of the horizon (or parallel of altitude) or an angle at the zenith. It is **azimuth (Zn)** if measured clockwise through 360°, starting at the north point on the horizon; and **azimuth angle (Z)** if measured *either* clockwise or counterclockwise through 180°, starting at the north point of the horizon in north latitude and the south point of the horizon in south latitude.

The **ecliptic system** is based upon the **ecliptic** as the primary great circle, analogous to the equator. The points 90° from the ecliptic are the north and south **ecliptic poles.** The series of great circles through these poles, analogous to meridians, are **circles of latitude.** The circles parallel to the plane of the ecliptic, analogous to parallels on the earth, are **parallels of latitude** or **circles of longitude.** Angular distance north or south of the ecliptic, analogous to latitude, is **celestial latitude. Celestial longitude** is measured *eastward* along the ecliptic through 360°, starting at the vernal equinox. This system of coordinates is of interest chiefly to astronomers. Another system of interest primarily to astronomers is known as the **galactic system.**

1431. Summary of coordinate systems.—The four systems of celestial coordinates are analogous to each other and to the terrestrial system, although each has distinctions such as differences in directions, units, and limits of measurement. The following table indicates the analogous term or terms under each system. For differences, see the description of each system, given earlier in the chapter, or appendix E.

Earth	Celestial Equator	Horizon	Ecliptic
equator	celestial equator	horizon	ecliptic
poles	celestial poles	zenith, nadir	ecliptic poles
meridians	hour circles, celestial meridians	vertical circles	circles of latitude
prime meridian	hour circle ♈, Greenwich celestial meridian, local celestial meridian	principal vertical circle, prime vertical circle	circle of latitude through ♈
parallels	parallels of declination	parallels of altitude	parallels of latitude
latitude	declination	altitude	celestial latitude
colatitude	polar distance	zenith distance	celestial colatitude
longitude	SHA, RA, GHA, LHA, t	azimuth, azimuth angle, amplitude	celestial longitude

1432. Diagram on the plane of the celestial meridian.—From a point outside the celestial sphere (if this were possible) and over the celestial equator, at such a distance that the view would be orthographic, the great circle appearing as the outer limit would be a celestial meridian. Other celestial meridians would appear as ellipses. The celestial equator would appear as a diameter 90° from the poles, and parallels of declination as straight lines parallel to the equator. The view would be similar to the orthographic view of the earth, as shown in figure 319b.

A number of useful relationships can be demonstrated by drawing a **diagram on the plane of the celestial meridian** showing this orthographic view. Arcs of circles can be substituted for the ellipses without destroying the basic relationships. Refer to figure 1432a. In the lower diagram the circle represents the celestial meridian, *QQ'* the celestial equator, *Pn* and *Ps* the north and south celestial poles, respectively. If a star has a declination of 30° N, an angle of 30° can be measured from the celestial equator, as shown. It could be measured either to the right or left, and would have been toward the south pole if the declination had been south. The parallel of declination is a line through this point and parallel to the celestial equator. The star is somewhere on this line (actually a circle viewed on edge).

To locate the hour circle, draw the upper diagram so that *Pn* is directly above *Pn* of the lower figure (in line with the polar axis *Pn Ps*), and the circle is of the same diameter as that of the lower figure. This is the plan view, looking down on the celestial sphere from the top. The circle is the celestial equator. Since the view is from above the *north* celestial pole, west is clockwise. The diameter *QQ'* is the celestial meridian shown as a circle in the lower diagram. If the *right* half is considered the upper branch, local hour angle is measured clockwise from this line to the hour circle, as shown. In this case the LHA is 80°. The intersection of the hour circle and celestial equator, point *A*, can be projected down to the lower diagram (point *A'*) by a straight line parallel to the polar axis. The elliptical hour circle can be represented approximately by an arc of a circle through *A'*, *Pn, Ps*. The center of this circle is somewhere along the celestial equator line *QQ'*, extended if necessary. It is usually found by trial and error. The intersection of the hour circle and parallel of declination locates the star.

Since the upper diagram serves only to locate point *A'* in the lower diagram, the two can be combined. That is, the LHA arc can be drawn in the lower diagram, as shown, and point *A* projected *upward* to *A'*. In practice, the upper diagram is not drawn, being shown here for illustrative purposes only.

In this example the star is on that half of the sphere toward the observer, or the *western* part. If LHA had been greater than 180°, the body would have been on the *eastern* or "back" side.

From the east or west point over the celestial horizon, the orthographic view of the horizon system of coordinates would be similar to that of the celestial equator system from a point over the celestial equator (fig. 1432a), since the celestial meridian is also the principal vertical circle. The horizon would appear as a diameter, parallels of altitude as straight lines parallel to the horizon, the zenith and nadir as poles 90° from the horizon, and vertical circles as ellipses through the zenith and nadir, except for the principal vertical circle, which would appear as a circle, and the prime vertical, which would appear as a diameter perpendicular to the horizon.

A celestial body can be located by altitude and azimuth in a manner similar to that used with the celestial equator system. If the altitude is 25°, this angle is measured from the horizon toward the zenith and the parallel of altitude is drawn as a straight line parallel to the horizon, as shown at *hh'* in the lower diagram of figure 1432b. The plan view from above the zenith is shown in the upper diagram. If north is taken at the left, as shown, azimuths are measured clockwise from this point. In the figure the azimuth is 290° and the azimuth angle is N70°W. The vertical circle is located by measuring either arc. Point *A* thus located can be projected vertically downward to *A'* on the horizon of the lower diagram, and the vertical circle represented approximately by the arc of a circle through *A'* and the zenith and nadir. The center of this circle is on *NS*, extended if necessary. The body is at the intersection of the parallel of altitude and the vertical circle. Since the upper diagram serves only to locate *A'* on the lower diagram, the two can be combined, point *A* located on the lower diagram and projected upward to *A'*, as shown. Since the body of the example has an azimuth greater than 180°, it is on the western or "front" side of the diagram.

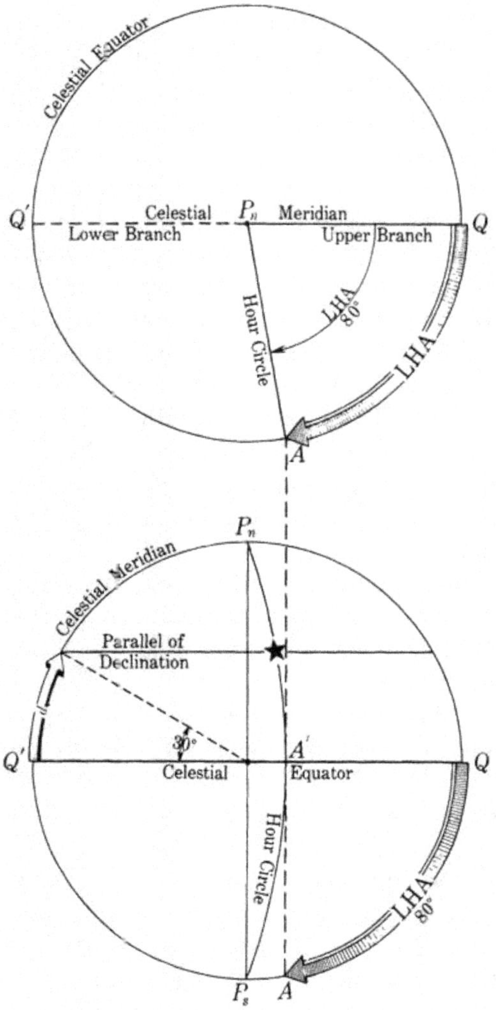

FIGURE 1432a.—Measurement of celestial equator system of coordinates.

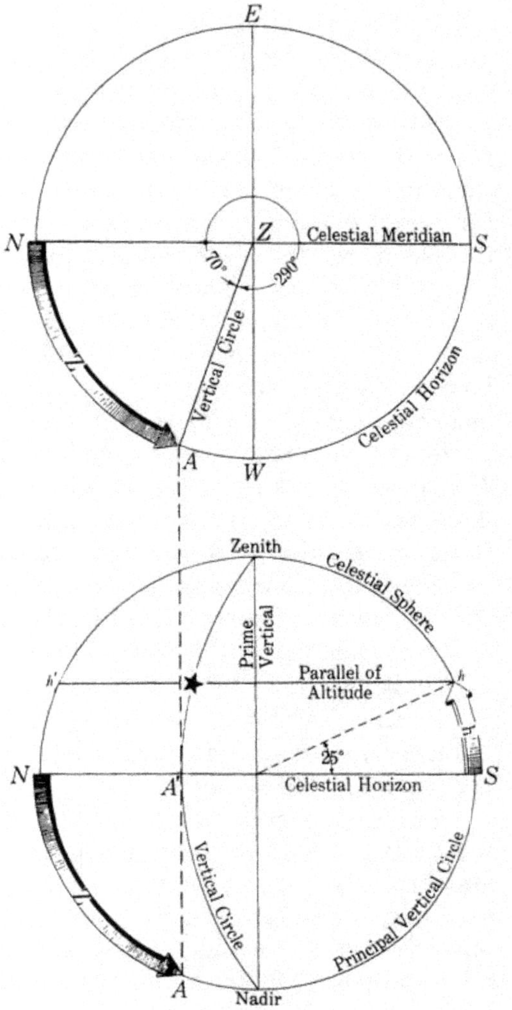

FIGURE 1432b.—Measurement of horizon system of coordinates.

Since the celestial meridian appears the same in both the celestial equator and horizon systems, the two diagrams can be combined and, if properly oriented, a body can be located by one set of coordinates, and the coordinates of the other system can be determined by measurement.

Refer to figure 1432c, in which the black lines represent the celestial equator system, and the red lines the horizon system. By convention, the zenith is shown at the top and the north point of the horizon at the left. The west point on the horizon is at the center, and the east point directly behind it. In the figure the latitude is 37°N. Therefore, the zenith is 37° north of the celestial equator. Since the zenith is established at the top of the diagram, the equator can be found by measuring an arc of 37° toward the south, along the celestial meridian. If the declination is 30°N and the LHA is 80°, the body can be located as shown by the black lines, and described above.

The altitude and azimuth can be determined by the reverse process to that described above. Draw a line *hh'* through the body and parallel to the horizon, *NS*. The altitude, 25°, is found by measurement, as shown. Draw the arc of a circle

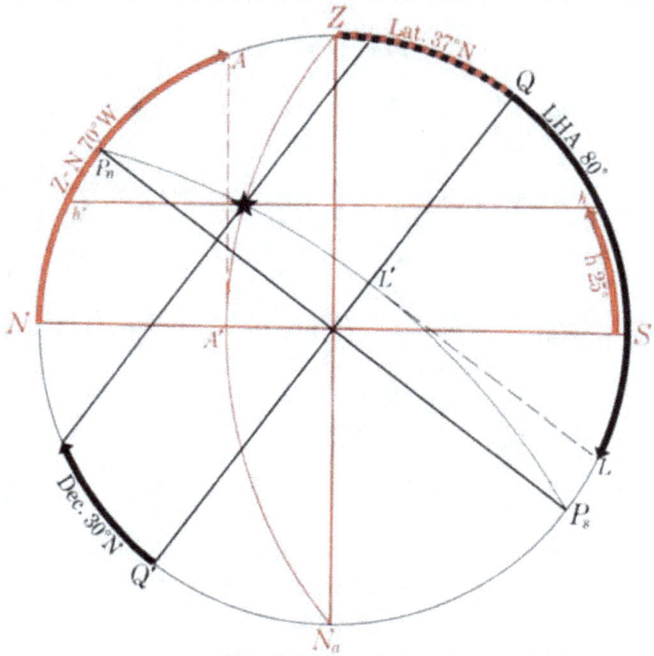

FIGURE 1432c.—Diagram on the plane of the celestial meridian.

through the body and the zenith and nadir. From *A'*, the intersection of this arc with the horizon, draw a vertical line intersecting the circle at *A*. The azimuth, N70°W, is found by measurement, as shown. The prefix N is applied to agree with the latitude. The body is left *(north)* of *ZNa*, the prime vertical circle. The suffix W applies because the LHA, 80°, shows that the body is west of the meridian.

If altitude and azimuth are given, the body is located by means of the red lines. The parallel of declination is then drawn parallel to *QQ'*, the celestial equator, and the declination determined by measurement. Point *L'* is located by drawing the arc of a circle through *Pn*, the star, and *Ps*. From *L'* a line is drawn perpendicular to *QQ'*, locating *L*. The meridian angle is then found by measurement. The declination is known to be north because the body is between the celestial equator and the north celestial pole. The meridian angle is west to agree with the azimuth, and hence LHA is numerically the same.

Since *QQ'* and *PnPs* are perpendicular, and *ZNa* and *NS* are also perpendicular, arc *NPn* is equal to arc *ZQ*. That is, *the altitude of the elevated pole is equal to the declination of the zenith, which is equal to the latitude.* This relationship is the basis of the method of determining latitude by an observation of Polaris (art. 2105).

The diagram on the plane of the celestial meridian is useful in approximating a number of relationships. Consider figure 1432d. The latitude of the observer (*NPn* or *ZQ*) is 45°N. The declination of the sun (*Q4*) is 20°N. Neglecting the change in declination for one day, note the following: At sunrise, position 1, the sun is on the horizon (*NS*), at the "back" of the diagram. Its altitude, h, is 0°. Its azimuth angle, Z, is the arc *NA*, N63°E. This is prefixed N to agree with the latitude and suffixed E to agree with the meridian angle of the sun at sunrise. Hence, Zn = 0° + 63° = 063°. The amplitude, A, is the arc *ZA*, E27°N. The meridian angle, t, is the arc *QL*, 110°E. The suffix E is applied because the sun is east of the meridian at rising. The LHA is 360° − 110° = 250°.

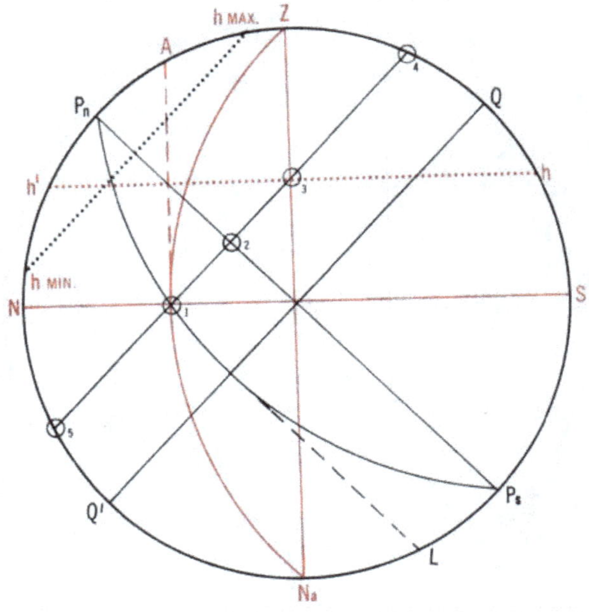

FIGURE 1432d.—A diagram on the plane of the celestial meridian for lat. 45° N.

As the sun moves upward along its parallel of declination, its altitude increases. It reaches position 2 at about 0600, when t=90°E. At position 3 it is on the prime vertical, *ZNa*. Its azimuth angle, Z, is N90°E, and Zn=090°. The altitude is *Nh'* or *Sh*, 27°.

Moving on up its parallel of declination, it arrives at position 4 on the celestial meridian about noon—when t and LHA are both 0°, by definition. On the celestial meridian a body's azimuth is 000° or 180°. In this case it is 180° because the body is south of the zenith. The maximum altitude occurs at meridian transit, in this case the arc *S*4, 65°. The zenith distance, z, is the arc *Z*4, 25°. A body is not in the zenith at meridian transit unless its declination is numerically, and by name, the same as the latitude.

Continuing on, the sun moves downward along the "front" or western side of the diagram. At position 3 it is again on the prime vertical. The altitude is the same as when previously on the prime vertical, and the azimuth angle is numerically the same, but now measured toward the west. The azimuth is 270°. The sun reaches position 2, six hours after meridian transit, and sets at position 1, when the azimuth angle is numerically the same as at sunrise, but westerly, and Zn=360°−63°=297°. The amplitude is W27°N.

After sunset the sun continues on downward along its parallel of declination until it reaches position 5, on the lower branch of the celestial meridian, about midnight. Its negative altitude, arc *N*5, is now greatest, 25°, and its azimuth is 000°. At this point it starts back up along the "back" of the diagram, arriving at position 1 at the next sunrise, to start another cycle.

Half the cycle is from the crossing of the 90° hour circle (the *PnPs* line, position 2) to the upper branch of the celestial meridian (position 4) and back to the *PnPs* line (position 2). When the declination and latitude have the **same name** (both north or both south), more than half the parallel of declination (position 1 to 4 to 1) is above the horizon, and the body is above the horizon more than half the time, crossing the 90° hour circle above the horizon. It rises and sets on the same side of the prime vertical as the elevated pole. If the declination is of the same name but numerically smaller than the latitude, the body crosses the prime vertical above the horizon. If the declination and latitude have the same name and are numerically equal, the body is in the zenith at upper transit. If the declination is of the same name but numerically *greater* than the latitude, the body crosses the upper branch of the celestial meridian between the zenith and elevated pole, and does not cross the prime vertical. If the declination is of the same name as the latitude and complementary to it ($d+L=90°$), the body is on the horizon at lower transit, and does not set. If the declination is of the same name as the latitude and numerically *greater* than the colatitude, the body is above the horizon during its entire daily cycle, and has maximum and minimum altitudes, as shown by the black dotted line in figure 1432d.

If the declination is 0° at any latitude, the body is above the horizon half the time, following the celestial equator *QQ'*, and rising and setting on the prime vertical. If the declination is of **contrary name** (one north and the other south), the body is above the horizon less than half the time, and crosses the 90° hour circle below the horizon. It rises and sets on the opposite side of the prime vertical from the elevated pole. If the declination is of contrary name and numerically smaller than the latitude, the body crosses the prime vertical below the horizon. This is the situation with the sun in winter, when days are short. If the declination is of contrary name and numerically equal to the latitude, the body is in the nadir at lower transit. If the declination is of contrary name and complementary to the latitude, the body is on the horizon at upper transit. If the declination is of contrary name and numerically greater than the colatitude, the body does not rise.

All of these relationships, and those that follow, can be derived by means of a diagram on the plane of the celestial meridian. They are modified slightly by atmospheric refraction, height of eye, semidiameter, parallax, changes in declination, and apparent speed of the body along its diurnal circle.

It is customary to keep the same orientation in south latitude, as shown in figure 1432e. In this illustration the latitude is 45°S, and the declination of the body is 15°N. Since *Ps* is the elevated pole, it is shown above the southern horizon, with both *SPs* and *ZQ* equal to the latitude, 45°. The body rises at position 1, on the opposite side of the prime vertical from the elevated pole; moves upward along its parallel of declination to position 2, on the upper branch of the celestial meridian, bearing north; and then downward along the "front" of the diagram to position 1, where it sets; remaining above the horizon for less than half the time because declination and latitude are of contrary name. The azimuth at rising is arc *NA,* the amplitude *ZA,* and the azimuth angle *SA*. The altitude circle at meridian transit is shown at *hh'*.

A diagram on the plane of the celestial meridian can be used to demonstrate the effect of a change in latitude. As the latitude increases, the celestial equator becomes more nearly parallel to the horizon. The colatitude becomes smaller, increasing the number of circumpolar bodies and those which neither rise nor set, and also increasing the difference in the length of the days between summer and winter. At the poles (fig. 1416b), celestial bodies circle the sky, parallel to the horizon. At the equator (fig. 1416a) the 90° hour circle coincides with the horizon. Bodies rise and set vertically; and are above the horizon half the time. At rising and setting the amplitude is equal to the declination. At meridian transit the altitude is equal to the codeclination. As the latitude changes name, the same-

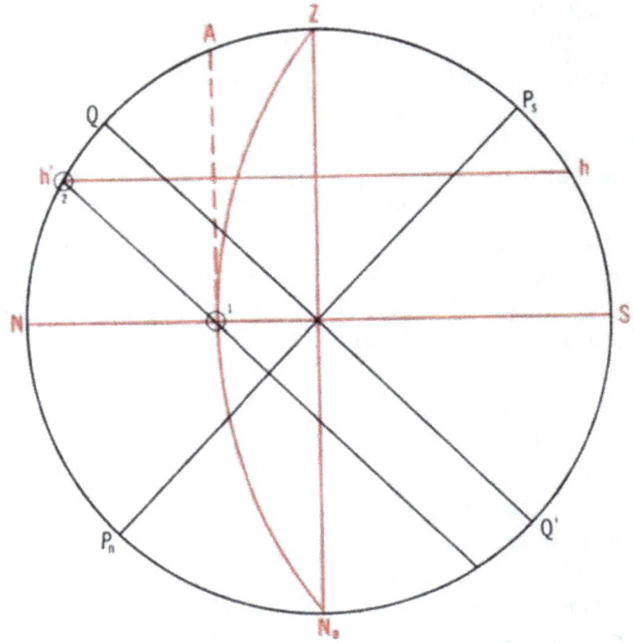

FIGURE 1432e.—A diagram on the plane of the celestial meridian for lat. 45° S.

contrary name relationship with declination reverses. This accounts for the fact that one hemisphere has winter while the other is having summer.

The error arising from showing the hour circles and vertical circles as arcs of circles instead of ellipses increases with increased declination or altitude. More accurate results can be obtained by measurement of azimuth on the parallel of altitude instead of the horizon, and of hour angle on the parallel of declination instead of the celestial equator. Refer to figure 1432f. The vertical circle shown is for a body having an azimuth angle of S60°W. The arc of a circle is shown in black, and the ellipse in red. The black arc is obtained by measurement around the horizon, locating *A′* by means of *A,* as previously described. The intersection of this arc with the altitude circle at 60° places the body at *M.* If a semicircle is drawn with the altitude circle as a diameter, and the azimuth angle measured around this, to *B,* a perpendicular to the hour circle locates the body at *M′,* on the ellipse. By this method the altitude circle, rather than the horizon, is, in effect, rotated through 90° for the measurement. This refinement is seldom used because actual values are usually found mathematically, the diagram on the plane of the meridian being used primarily to indicate relationships.

With experience, one may mentally visualize the diagram on the plane of the celestial meridian without making an actual drawing. Devices with two sets of spherical coordinates, on either the orthographic (art. 319) or stereographic (art. 318) projection, pivoted at the center, have been produced commercially to provide a mechanical diagram on the plane of the celestial meridian. However, since the diagram's principal use is to illustrate certain relationships, such a device is not a necessary part of the navigator's equipment.

1433. The navigational triangle.—A triangle formed by arcs of great circles of a sphere is called a **spherical triangle.** A spherical triangle on the celestial sphere is called a **celestial triangle.** The spherical triangle of particular significance to navi-

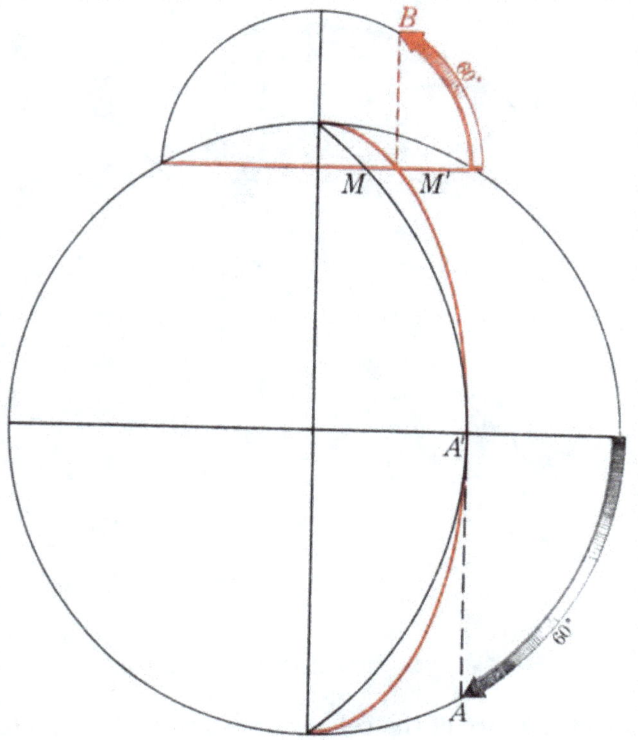

FIGURE 1432f.—Locating a point on an ellipse of a diagram on the plane of the celestial meridian.

gators is called the **navigational triangle.** It is formed by arcs of a celestial meridian, an hour circle, and a vertical circle. Its vertices are the elevated pole, the zenith, and a point on the celestial sphere (usually a celestial body). The terrestrial counterpart is also called a navigational triangle, being formed by arcs of two meridians and the great circle connecting two places on the earth, one on each meridian. The vertices are the two places and a pole. In great-circle sailing these places are the point of departure and the destination. In celestial navigation they are the **assumed position (AP)** of the observer and the **geographical position (GP)** of the body (the place having the body in its zenith). The GP of the sun is sometimes called the **subsolar point,** that of the moon the **sublunar point,** that of a satellite (either natural or artificial) the **subsatellite point,** and that of a star its **substellar** or **subastral point.** When used to solve a celestial observation, either the celestial or terrestrial triangle may be called the **astronomical triangle.**

The navigational triangle is shown in figure 1433a on a diagram on the plane of the celestial meridian, labeled as in article 1432, but with the hour circle and vertical circle properly shown as ellipses. The earth is at the center, *O*. The star is at *M*, *dd'* is its parallel of declination, and *hh'* its altitude circle.

In the figure, arc *QZ* of the celestial meridian is the latitude of the observer, and *PnZ*, one side of the triangle, is the **colatitude.** Arc *AM* of the vertical circle is the altitude of the body, and side *ZM* of the triangle is the zenith distance, or **coaltitude.** Arc *LM* of the hour circle is the declination of the body, and side *PnM* of the triangle is the polar distance, or **codeclination.**

The angle at the elevated pole, *ZPnM*, having the hour circle and the celestial meridian as sides, is the meridian angle, t. The angle at the zenith, *PnZM*, having

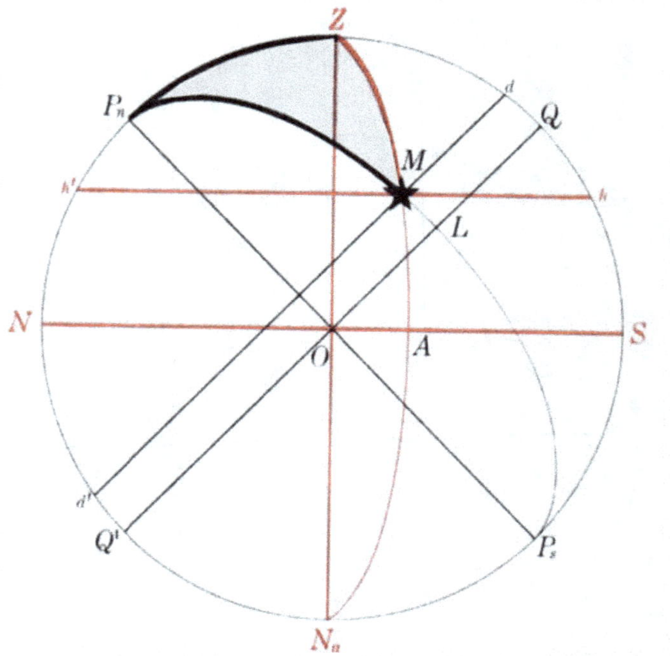

FIGURE 1433a.—The navigational triangle.

the vertical circle and that arc of the celestial meridian which includes the elevated pole as sides, is the azimuth angle. The angle at the celestial body, *ZMPn,* having the hour circle and the vertical circle as sides, is the **parallactic angle** (X) (sometimes called the **position angle**), which is not generally used by the navigator.

A number of problems involving the navigational triangle are encountered by the navigator, either directly or indirectly. Of these, the most common are:

1. Given latitude, declination, and meridian angle, to find altitude and azimuth angle. This is used in the reduction of a celestial observation, to establish a line of position (ch. XX).

2. Given latitude, altitude, and azimuth angle, to find declination and meridian angle. This is used to identify an unknown celestial body (ch. XXII).

3. Given meridian angle, declination, and altitude, to find azimuth angle. This may be used to find azimuth when the altitude is known (ch. XX).

4. Given the latitude of two places on the earth and the difference of longitude between them, to find the initial great-circle course and the great-circle distance (ch. IX). This involves the same parts of the triangle as in 1, above, but in the terrestrial triangle, and hence defined differently.

Both celestial and terrestrial navigational triangles are shown in perspective in figure 1433b.

Problems

1428. *Given.*—An observer is at longitude 77°E. The sun is 60° east of the meridian. GHA ♈ is 37°.

Required.—(1) LHA of the sun.

(2) GHA of the sun.

(3) SHA of the sun.

(4) Approximate time at the local meridian.

Answers.—(1) LHA 300°, (2) 223°, (3) SHA 186°, (4) T 0800.

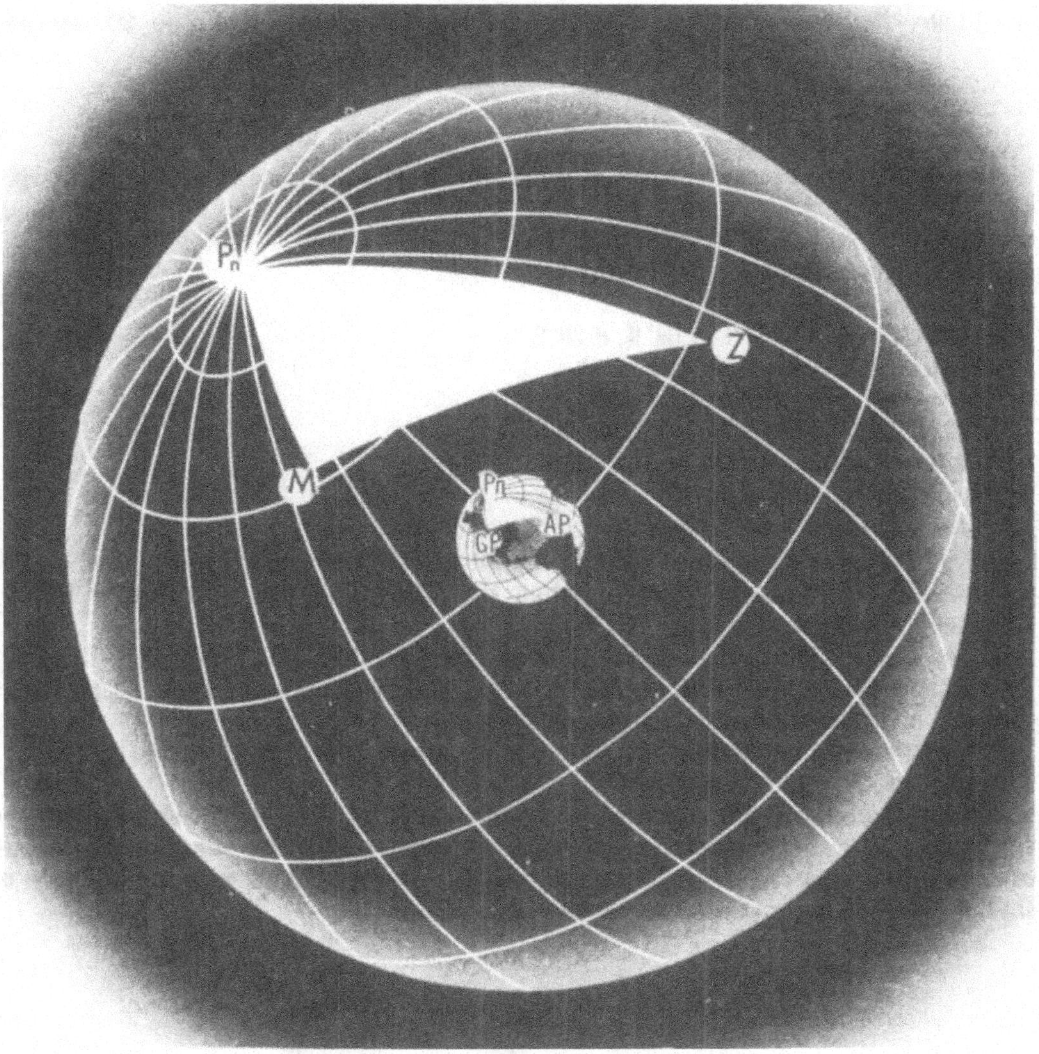

FIGURE 1433b.—The navigational triangle in perspective.

1430a. *Required.*—Convert Z to Zn in the following:

(1) N174°E	(4) S39°E
(2) S1°E	(5) N106°W
(3) S90°W	(6) N90°W

Answers.—(1) Zn 174°, (2) Zn 179°, (3) Zn 270°, (4) Zn 141°, (5) Zn 254°, (6) Zn 270°.

1430b. *Required.*—Convert Zn to Z in the following, using the 180° system:

Zn	*Lat.*	*Zn*	*Lat.*
(1) 214°	N	(4) 333°	S
(2) 163°	S	(5) 206°	N
(3) 007°	N	(6) 206°	S

Answers.—(1) Z N146°W, (2) Z S17°E, (3) Z N7°E, (4) Z S153°W, (5) Z N154°W, (6) Z S26°W.

1430c. *Required.*—Convert Zn to Z in the following, using the 90° system:

(1) 051°	(3) 251°
(2) 151°	(4) 351°

Answers.—(1) Z N51°E, (2) Z S29°E, (3) Z S71°W, (4) Z N9°W.

Solve the following problems by diagrams on the plane of the celestial meridian:

1432a. *Given.*—L 32°N, t 71°W, d 27°N.

Required.—Altitude and azimuth.

Answers.—h 28°, Zn 288°.

1432b. *Given.*—L 17°S, t 64°E, d 28°S.

Required.—Altitude and azimuth.

Answers.—h 28°, Zn 115°.

1432c. *Given.*—L 59°N, h 27°, Zn 052°.

Required.—Declination and meridian angle.

Answers.—d 41°N, t 111°E.

1432d. *Given.*—L 31°N, declination of sun 18°S.

Required.—(1) Azimuth at sunrise, (2) maximum altitude, (3) altitude when the azimuth is 234°, (4) azimuth angle when the altitude in the afternoon is 10°, (5) amplitude at sunset.

Answers.—(1) Zn 111°, (2) h 41°, (3) h 18°, (4) Z N118°W, (5) A W21°S.

1432e. *Given.*—The declination of the star Dubhe is approximately 62°N. When observed at lower transit, its altitude is 43°.

Required.—(1) Latitude of the observer, (2) azimuth at upper transit.

Answers.—(1) L 71°N, (2) Zn 180°.

1432f. *Required.*—For an observer at latitude 39°N, find for the sun at summer and winter solstices, respectively: (1) LHA at sunrise, (2) LHA when on the prime vertical during the morning, (3) maximum altitude, (4) LHA at sunset, (5) length of daylight if the sun moves 15° per hour.

Answers.—

	Summer	*Winter*	
(1) LHA	248°	292°	
(2) LHA	304°	236°	(below horizon)
(3) h	74°	28°	
(4) LHA	112°	68°	
(5) T	14^h56^m	9^h04^m	

1432g. *Given.*—L 83°N, sun's declination 4°S.

Required.—(1) LHA at sunrise, (2) maximum altitude, (3) LHA at sunset, (4) length of daylight (sun moving 15° per hour).

Answers.—(1) LHA 305°, (2) max h 3°, (3) LHA 55°, (4) T 7^h20^m.

References

ELEMENTARY

Abell, G. O. *Exploration of the Universe.* 3rd ed. New York, Holt, Rinehart and Winston, 1975.

Alter, D., and C. H. Cleminshaw. *Pictorial Astronomy.* New York, Crowell, 1952.

Baker, R. H. *An Introduction to Astronomy.* 6th ed. Princeton, Van Nostrand, 1961.

Hood, P. *Observing the Heavens.* New York, Oxford, 1951.

Mayall, R. N., and M. W. Mayall. *A Beginner's Guide to the Skies.* New York, Putnam, 1960.

Struve, O., B. Lynds, and H. Pillans. *Elementary Astronomy.* New York, Oxford, 1959.

Wright, H. *Palomar.* New York, Macmillan, 1952.

ADVANCED

Baker, R. H. *Astronomy.* 8th ed. Princeton, Van Nostrand, 1964.

Brouwer, D., and G. M. Clemence. *Methods of Celestial Mechanics.* New York, Academic, 1961.

Jones, Sir Harold Spencer. *General Astronomy.* 4th ed. New York, Longmans, 1961.

Krogdahl, W. S. *The Astronomical Universe.* New York, Macmillan, 1952.

Rudaux, L., and G. De Vaucouleurs. *Larousse Encyclopedia of Astronomy.* New York, Prometheus. 1959.

Russell, H. N., R. S. Dugan, and J. Q. Stewart. *Astronomy.* Vol. I, *The Solar System.* Rev. ed. 1945. Vol. II, *Astrophysics and Stellar Astronomy.* Boston, Ginn, 1938.

CHAPTER XV

INSTRUMENTS FOR CELESTIAL NAVIGATION

1501. The marine sextant is a hand-held instrument for measuring the angle between the lines of sight to two points by bringing into coincidence at the eye of the observer the direct ray from one point, and a double-reflected ray from the other, the measured angle being twice the angle between the reflecting surfaces. Its principal use is to measure the altitudes of celestial bodies above the visible sea horizon. Sometimes it is turned on its side and used for measuring the *difference* in bearing of two terrestrial objects. Because of its great value for determining position at sea, the sextant has been a symbol of navigation for more than 200 years. The quality of his instrument, the care he gives it, and the skill with which he makes observations are to the navigator matters of professional pride.

The name "sextant" is from the Latin *sextans,* "the sixth part." The arc of early marine sextants is approximately the sixth part of a circle, but because of the optical principle involved (art. 1502), the instrument measures angles of 120°. Most modern instruments measure something more than this.

1502. Optical principles.—When a ray of light is reflected from a plane surface, the **angle of reflection** is equal to the **angle of incidence** (fig. 1502a). From optics the angle between the first and final directions of a ray of light that has undergone double reflection in the same plane is twice the angle that the two reflecting surfaces make with each other (fig. 1502b).

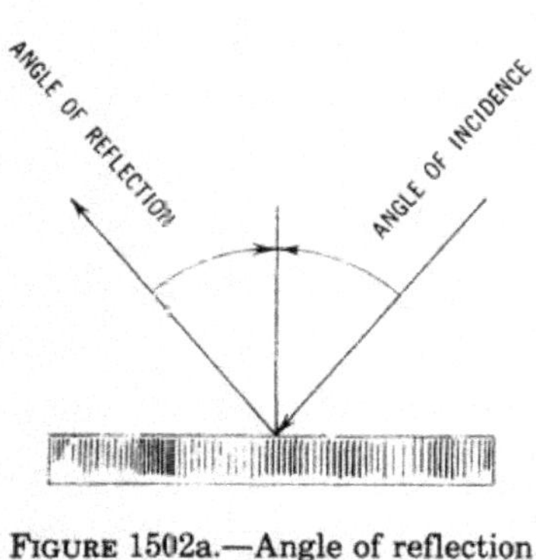

FIGURE 1502a.—Angle of reflection equals angle of incidence.

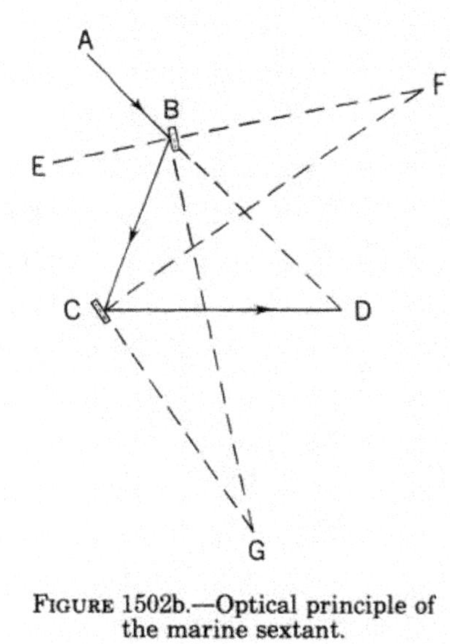

FIGURE 1502b.—Optical principle of the marine sextant.

In figure 1502b, *AB* is a ray of light from a celestial body. The index mirror of the sextant is at *B,* the horizon glass at *C,* and the eye of the observer at *D.* Construction lines *EF* and *CF* are perpendicular to the index mirror and horizon glass, respectively, and lines *BG* and *CG* are parallel to these mirrors. Therefore, angles *BFC* and *BGC* are equal because their sides are mutually perpendicular (art.

127, vol. II). Angle *BGC* is the inclination of the two reflecting surfaces. The ray of light *AB* is reflected at mirror *B*, proceeds to mirror *C*, where it is again reflected, and then continues on to the eye of the observer at *D*. Since the angle of reflection is equal to the angle of incidence,

$$ABE=EBC, \text{ and } ABC=2EBC$$
$$BCF=FCD, \text{ and } BCD=2BCF.$$

Since an exterior angle of a triangle equals the sum of the two nonadjacent interior angles (art. 128, vol. II),

$$ABC=BDC+BCD, \text{ and } EBC=BFC+BCF.$$

Transposing,

$$BDC=ABC-BCD, \text{ and } BFC=EBC-BCF.$$

Substituting 2 *EBC* for *ABC*, and 2 *BCF* for *BCD* in the first of these equations,

$$BDC=2\ EBC-\ 2\ BCF, \text{ or } BDC=\ 2\ (EBC-BCF).$$

Since

$$BFC=EBC-BCF, \text{ and } BFC=BGC,$$

therefore

$$BDC=\ 2\ BFC=\ 2\ BGC.$$

That is, *BDC*, the angle between the first and last directions of the ray of light, is equal to 2*BGC*, twice the angle of inclination of the reflecting surfaces. Angle *BDC* is the altitude of the celestial body.

If the two mirrors are parallel, the incident ray from any observed body must be parallel to the observer's line of sight through the horizon glass; i.e., the altitude of the body is zero. Accordingly, the 0° graduation on the arc coincides with that position of the index arm when the index mirror is parallel to the horizon glass. Since the angle that these two reflecting surfaces make with each other is one-half the angle actually observed, the arc is so graduated that 10° of arc on the limb is labelled 20°, 20° of arc is labelled 40°, etc.

1503. Micrometer drum sextant.—A modern marine sextant, called a **micrometer drum sextant,** is shown in figure 1503a. In most marine sextants, the **frame,** *A*, is made of brass or aluminum. There are several variations of the design of the frame, nearly all conforming generally to that shown. The **limb,** *B*, is cut on its outer edge with teeth, each representing one degree of altitude. The altitude graduations, *C*, along the limb, are called the **arc.** Some sextants have an arc marked in a strip of brass, silver, or platinum inlaid in the limb.

The **index arm,** *D*, is a movable bar of the same material as the frame. It is pivoted about the center of curvature of the limb. The **tangent screw,** *E*, is mounted perpendicularly on the end of the index arm, where it engages the teeth of the limb. Because the index arm can be moved through the length of the arc by rotating the tangent screw, this is sometimes called an "endless tangent screw," in contrast with the limited-range device on older instruments. The **release,** *F*, is a spring-actuated clamp which keeps the tangent screw engaged with the teeth of the limb. By applying pressure on the legs of the release, one can disengage the tangent screw. The index arm can then be moved rapidly along the limb. Mounted on the end of the tangent screw is a **micrometer drum,** *G*, which is graduated in minutes of altitude. One complete turn of the drum moves the index arm one degree of altitude along the arc. Adjacent to the micrometer drum and fixed on the index arm is a **vernier,** *H*, which aids in reading fractions of a minute. The vernier shown

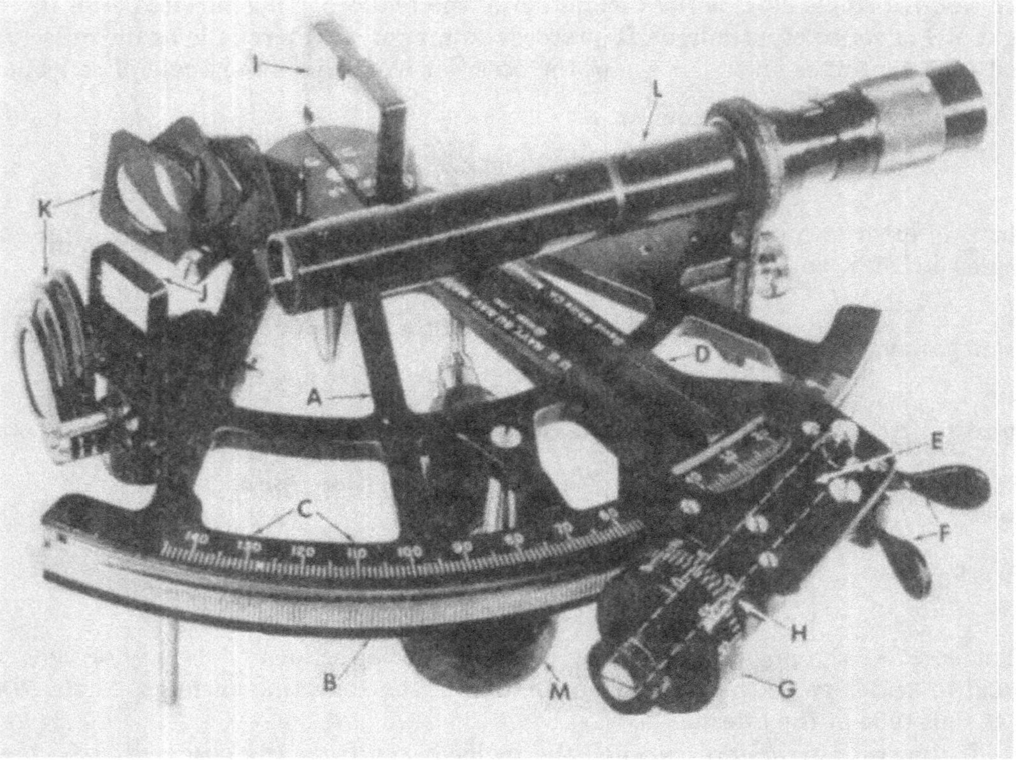

FIGURE 1503a.—U. S. Navy Mark 2 micrometer drum sextant.

is graduated into ten parts, permitting readings to six seconds. Other sextants (generally of European manufacture) have verniers graduated into only five parts, permitting readings to 12 seconds.

The **index mirror,** *I,* is a piece of silvered plate glass mounted on the index arm, perpendicular to the plane of the instrument, with the center of the reflecting surface directly over the pivot of the index arm. The **horizon glass,** *J,* is a piece of optical glass silvered on its half nearer the frame. It is mounted on the frame, perpendicular to the plane of the sextant. The index mirror and horizon glass are mounted so that their surfaces are parallel when the micrometer drum is set at 0°, if the instrument is in perfect adjustment. **Shade glasses,** *K,* of varying or variable darkness, are mounted on the frame of the sextant in front of the index mirror and horizon glass. They can be moved into the line of sight at will, to reduce the intensity of light reaching the eye of the observer. Some sextants have two sets of shade glasses, as shown in figure 1504.

The **telescope,** *L,* screws into an adjustable collar in line with the horizon glass, and should then be parallel to the plane of the instrument. Most modern sextants are provided with only one telescope, but some are equipped with two or more. When only one telescope is provided, it is of the "erect image type," either such as shown or one with a wider "object glass" (far end of telescope), which generally is shorter in length and gives a greater field of view. The second telescope, if provided, may be the "inverting type." The inverting telescope, having one lens less than the erect type, absorbs less light, but at the expense of producing an inverted image. A small colored glass cap is sometimes provided, to be placed over the "eyepiece" (near end of telescope) to reduce the glare. With this in place, shade glasses are generally not needed. A "peep sight" may be provided. It is a clear tube which serves to direct the line of sight of the observer when no telescope is used.

The telescope shown in figure 1503a is fitted with a "spiral focusing mechanism." Other sextants substitute a "draw" for this mechanism. The draw is fitted inside the telescope tube without threads and is slid in or out as necessary to focus the instrument. The spiral focusing mechanism is easily adjusted each time the sextant is used, but on the draw type, the navigator should mark the draw to indicate the correct extension for his eyes.

The **handle,** *M,* of most sextants is made of wood or plastic. Sextants are designed to be held in the right hand. Some are equipped with a small light on the index arm to assist in reading altitudes. The batteries for this light are fitted inside a recess in the sextant handle.

Figure 1503b shows the U. S. Navy Mark 3 micrometer drum sextant from the handle side. This figure shows parts not clearly shown in figure 1503a, such as the legs and the tangent screw. This sextant, which is generally similar to sextants used in the merchant marine, is fitted with a four power telescope. A diopter scale is provided on the focusing mechanism of the telescope for adjustment to the individual user's eye. The "anatomical" handle is designed for decreased arm fatigue and hand tremor. A light switch and batteries are conveniently located in the handle. Large coated optics are provided for more light transmission and improved field of view, resolution, and magnification.

There are two basic designs commonly used for mounting and adjusting mirrors on marine sextants. On the U. S. Navy Mark 3 and other sextants, the mirror is mounted so that it can be moved against retaining or mounting springs within its frame. Only one perpendicular adjustment screw is required. On the U. S. Navy Mark 2 and other sextants the mirror is fixed within its frame. Two perpendicular adjustment screws are required. *One screw must be loosened before the other screw bearing on the same surface is tightened.*

Figure 1503c shows a sextant with a silver arc inserted in the limb, a micrometer drum graduated oppositely to the one in figure 1503a, a vernier graduated into six parts, a shorter telescope with a wider object glass than that in figure 1503a, a telescope draw substituted for a spiral focusing mechanism, and a light fitted on the index arm.

1504. Vernier sextant.—Nearly all marine sextants of recent manufacture are of the type described in article 1503. At least two older-type sextants are still in use. These differ from the micrometer drum sextant principally in the manner in which the final reading is made. They are called **vernier sextants.**

The **clamp screw vernier sextant** is the older of the two. In place of the modern "release," a **clamp screw** is fitted on the underside of the index arm. To move the index arm, one loosens the clamp screw, releasing the arm. When the arm is placed at the appropriate altitude of the body being observed, the clamp screw is tightened. Fixed to the clamp screw and engaged with the index arm is a long tangent screw. When this screw is turned, the index arm moves slowly, permitting accurate setting. Movement of the index arm (by the tangent screw) is limited to the length of the screw (several degrees of arc). Before an altitude is measured, this screw should be set to the approximate mid-point of its range. The final reading is made on a vernier set in the index arm below the arc. A small microscope or magnifying glass fitted to the index arm is used in making the final reading. Figure 1504 shows a clamp screw vernier sextant.

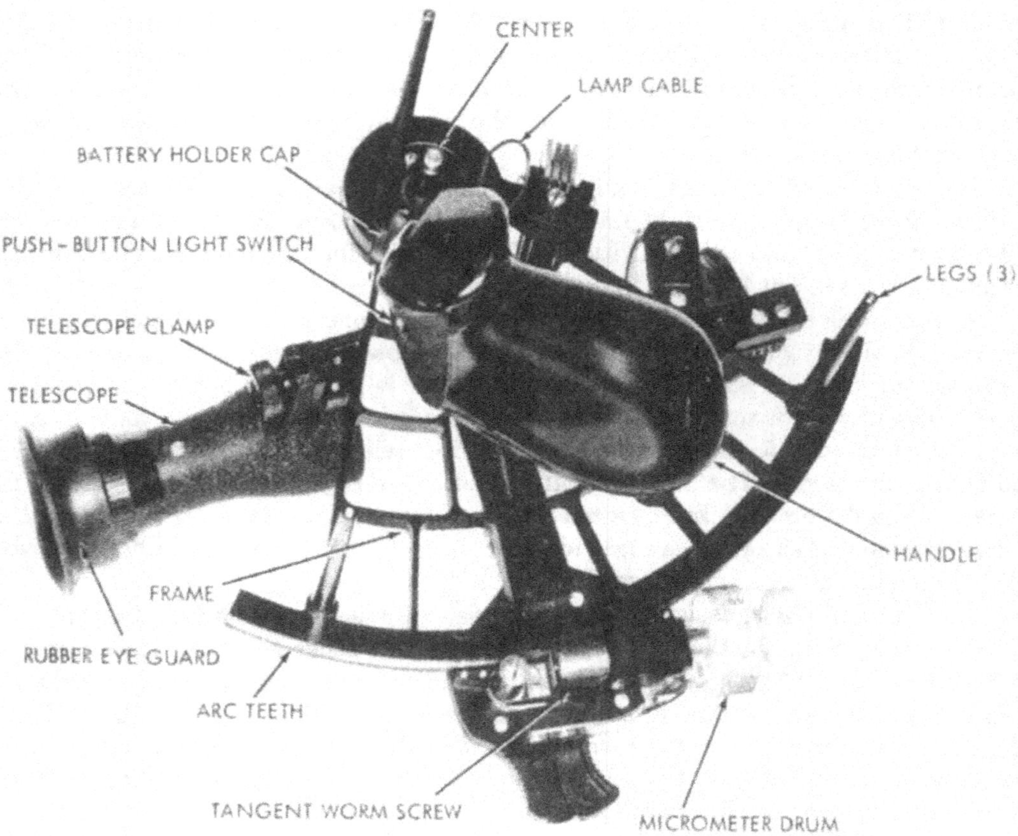

FIGURE 1503b.—U. S. Navy Mark 3 micrometer drum sextant.

The **endless tangent screw vernier sextant** is identical to the micrometer drum sextant, except that it has no drum, and the fine reading is made by a vernier along the arc, as with the clamp screw vernier sextant. The release is the same as on the micrometer drum sextant and teeth are cut into the underside of the limb which engage with the endless tangent screw. The vernier itself is explained in article 1506.

1505. Use of the sextant.—When the *sun* is observed, the sextant is held vertically in the right hand, and the line of sight is directed at the point on the horizon directly below the body. Suitable shade glasses are moved into the line of sight, and the index arm is moved outward from near the 0° point until the reflected image of the sun appears in the horizon glass, near the direct view of the horizon. The sextant is then tilted slightly to the right and left to check its perpendicularity. As the sextant is tilted, the image of the sun appears to move in an arc, and the observer may have to change slightly the direction in which he is facing, to prevent the image from moving out of the horizon glass. When the sun appears at the *bottom* of its apparent arc resulting from this **swinging the arc**, or **rocking the sextant**, the sextant is vertical, and in the correct position for making the observation. If the sextant is tilted, too *great* an angle will be measured. When

FIGURE 1503c.—A micrometer drum sextant used in the merchant marine.

FIGURE 1504.—A clamp screw vernier sextant.

the sextant is vertical, and the observer is facing directly toward the sun, its reflected image appears at the center of the horizon glass, half on the silvered part, and half on the clear part. The index arm is then moved slowly until the sun appears to be resting exactly on the horizon, which is tangent to the **lower limb.** Occasionally, the sun image is brought *below* the horizon, and the **upper limb** observed. It is good practice to make several observations, moving the limb away from the horizon, alternately above and below it, between readings. Practice is needed to determine the appearance at tangency, which occurs at only one point, to avoid the common error of beginners of bringing the image down too far (too little for an upper-limb observation). Some navigators get more accurate observations by letting the body contact the horizon by its own apparent motion, bringing it slightly below the horizon if rising, and above if setting. At the instant the horizon is tangent to the disk, the time is noted. The **sextant altitude** is the uncorrected reading of the sextant. Figure 1505a illustrates the major steps in making an observation of the sun. At the left, the index arm has been moved a short distance from 0°. In the center, it has been clamped with the sun in the approximate position for a reading, and the sextant is being rocked. At the right, the sun is in the correct position for a reading.

FIGURE 1505a.—*Left,* view through telescope with index arm set near zero. *Center,* "swinging the arc" after the sun has been brought close to the horizon. *Right,* sun at the instant of tangency.

When the *moon* is observed, the procedure is the same as for the sun, except that shade glasses are usually not required. The upper limb of the moon is observed more often than that of the sun, because of the phases of the moon. When the terminator (art. 1423) is nearly vertical, care should be exercised in selecting the limb that is illuminated, if an inaccurate reading is to be avoided. Sights of the moon are best made during daylight hours, or during that part of twilight in which the moon is least luminous. During the night, false horizons nearly always appear below the moon, due to illumination of the water by moonlight.

When a *star* or *planet* is observed, three methods of making the initial approximation of the altitude are in common use. In a common method, the index arm and micrometer drum are set on zero and the line of sight is directed at the body to be observed. Then, while keeping the reflected image of the body in the mirrored half of the horizon glass, the index arm is slowly swung *out* and the frame of the sextant is rotated *down.* The reflected image of the body is kept in the mirror until the horizon appears in the clear part of the horizon glass.

When there is little contrast between brightness of the sky and the body, this procedure is difficult, for if the body is "lost" while it is being brought down, it may not be recovered without starting again at the beginning of the procedure. An alternative method frequently used consists of holding the sextant upside down in the left hand, directing the line of sight at the body, and slowly moving the index arm out until the horizon appears in the horizon glass. This is illustrated in figure 1505b. After contact is made, the sextant is inverted and the sight taken in the usual manner.

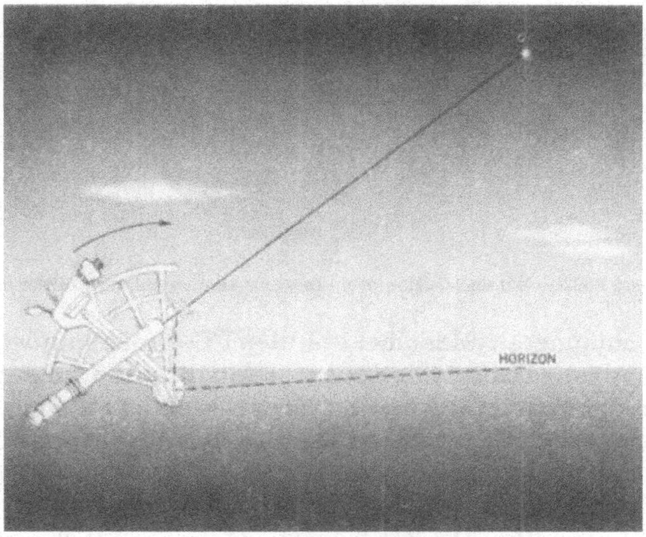

FIGURE 1505b.—Method of bringing horizon "up" to body.

A third method consists of determining in advance the approximate altitude and azimuth of the body by a **star finder** such as No. 2102-D (art. 2210). The sextant is set at the indicated altitude, and the observer faces in the direction indicated by the azimuth. After a short search, during which the index arm is moved backward and forward a few degrees, and the azimuth in which the observer faces is changed a little to each side, the image of the body should appear in the horizon glass. The best method to use for any observation is that which produces the desired result with the least effort. It is largely a matter of personal preference.

Measurement of the altitude of a star or planet differs from that of the sun or moon in that the *center* of a star or planet, rather than a limb, is brought into coincidence with the horizon. Figure 1505c shows the reflected image of a star as it should appear at the time of observation. Because of this difference, and the limited time usually available for observation during twilight, the method of letting a star or planet intersect the horizon by its own motion is little used. As with the sun and moon, however, the navigator should not forget to swing the arc to establish perpendicularity of the sextant.

Occasionally, fog, haze, or other ships may obscure the horizon directly below a body which the navigator wishes to observe. If the arc of the sextant is sufficiently long, a **back sight** might be obtained, using the opposite point of the horizon as the reference. The observer faces *away* from the body and observes the *supplement* of the altitude. If the sun or moon is observed in this manner, what appears in the horizon glass to be the lower limb is in fact the upper limb. In the case of the sun, it is usually preferable to observe what appears to be the upper limb. The arc that appears when rocking the sextant for a back sight is inverted; that is, the *highest* point indicates the position of perpendicularity.

If more than one telescope is furnished with the sextant, the erecting telescope is used to observe the sun. Generally, the inverting telescope will produce the best results for daylight observations, although some navigators prefer not to use any

FIGURE 1505c.—Correct position of planet or star at moment of observation.

telescope, thus obtaining a wider field of view. The collar into which the sextant telescope fits may be adjusted in or out in relation to the frame. When moved in, more of the mirrored half of the horizon glass is visible to the navigator, and a star or planet is more easily observed when the sky is relatively bright. Near the darker limit of twilight, the telescope can be moved out, giving a broader view of the clear half of the glass, and making the less distinct horizon more easily discernible. If both eyes are kept open until the last moments of an observation, eye strain will be lessened. But in making the final measurement, the nonsighting eye should be closed to permit full ocular concentration. Practice will permit observations to be made quickly, reducing inaccuracy due to eye fatigue. If several observations are made in succession, with a short rest between them, the best results should be obtained. With experience, the observer should be able to "call his shots," identifying the better ones.

When an altitude is being measured, it is desirable to have an assistant note the time, so that simultaneous values of time and altitude will be available. He should be given a warning "stand-by" when the measurement is nearly completed, and a "mark" at the moment a reading is made. He should be instructed to read the three hands in order of their rapidity of motion; the second hand first, then the minute hand, and finally the hour hand. If it is sufficiently dark that a light is needed to make the reading, the assistant should read both the time, and then the altitude, *behind* the observer and facing away from him, to avoid impairment of the observer's eye adaption to sky and horizon lighting conditions.

If an assistant is not available to time the observations, the observer holds the watch in the palm of his left hand, leaving his fingers free to manipulate the tangent screw of the sextant. After making the observation, he quickly shifts his view to the watch, and notes the positions of the second, minute, and hour hands, respectively. The delay between completing the altitude observation and noting the time should not be more than one or two seconds. The average time should be determined by having someone measure it for several observations, or by counting the half seconds (learning to count with the half-second beats of a chronometer). This interval can then be subtracted from the observed time of each sight.

1506. Reading the sextant.—The reading of a micrometer drum sextant is made in three steps. The degrees are read by noting the position of the arrow on the index arm in relation to the arc. The minutes are read by noting the position of the zero on the vernier with relation to the graduations on the micrometer drum. The fraction of a minute is read by noting which mark on the vernier most nearly coincides with one of the graduations on the micrometer drum. This is similar to reading the time by means of the hour, minute, and second hands of a watch. In both, the relationship of one part of the reading to the others should be kept in mind. Thus, if the hour hand of a watch were *about* on "4," one would know that the time was about four o'clock. But if the minute hand were on "58," one would know that the time was 0358 (or 1558), not 0458 (or 1658). Similarly, if the arc indicated a reading of about 40°, and 58′ on the micrometer drum were opposite zero on the vernier, one would know that the reading was 39°58′, not 40°58′. Similarly, any doubt as to the correct minute can be removed by noting the fraction of a minute from the position of the vernier. In figure 1506a the reading is 29°42′.5. The arrow on the index mark is between 29° and 30°, the zero on the vernier is between 42′ and 43′, and the "0′.5" graduation on the vernier coincides with one of the graduations on the micrometer drum.

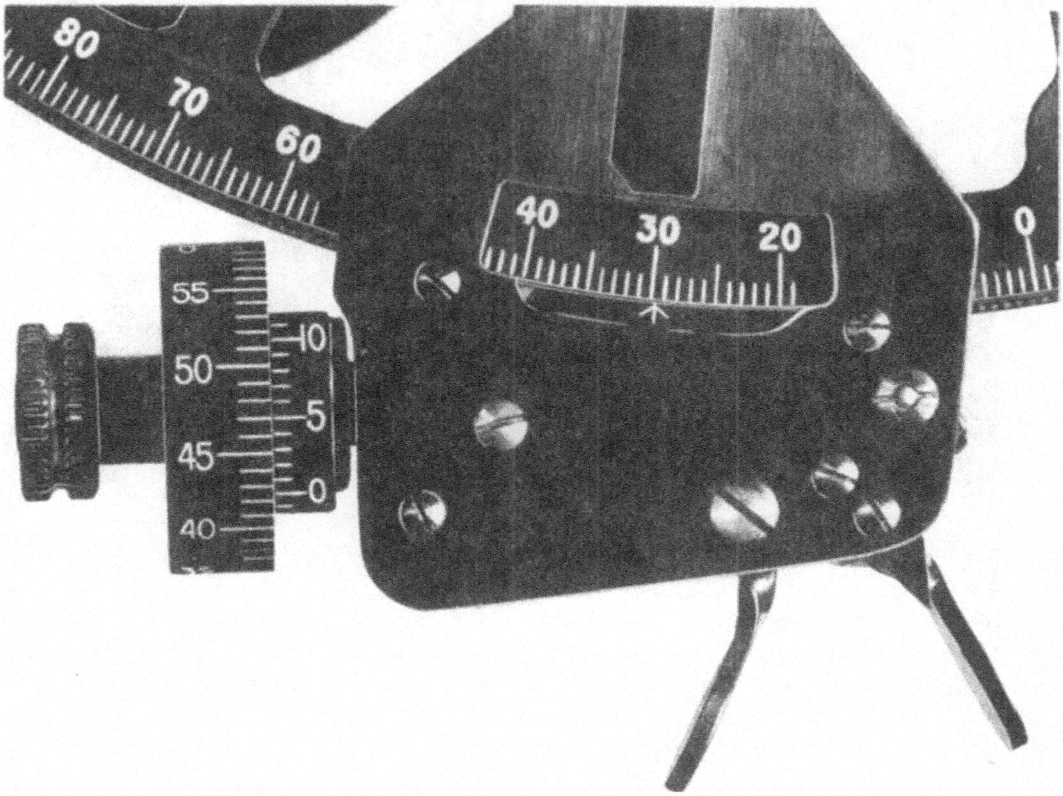

FIGURE 1506a.—Micrometer drum sextant set at 29°42′.5.

The principle of reading a vernier type sextant is the same, but the reading is made in two steps. Figure 1506b shows a typical altitude setting on this type sextant. Each degree on the arc of this sextant is graduated into three parts, permitting an initial reading by the reference mark on the index arm to the nearest full 20 minutes of arc. In this illustration the reference mark lies between 29°40′ and 30°00′, indicating a reading between these values. The reading for the fraction of 20′ is made by means of the vernier, which is engraved on the index arm and has the small reference mark as its zero graduation. On this vernier, 40 graduations coincide with 39 graduations on the arc. Each graduation on the vernier is equivalent to 1/40 of one graduation (20′) on the arc, or 0′.5 (30″). In the illustration, the vernier graduation representing 2½ minutes (2′30″) most nearly coincides with one of the graduations on the arc. Therefore, the reading is 29°42′30″, or 29°42.′5, as before. When a vernier of this type is used, any doubt as to which mark on the vernier coincides with a graduation on the arc can usually be resolved by noting the position of the vernier mark on each side of the one that seems to be in coincidence.

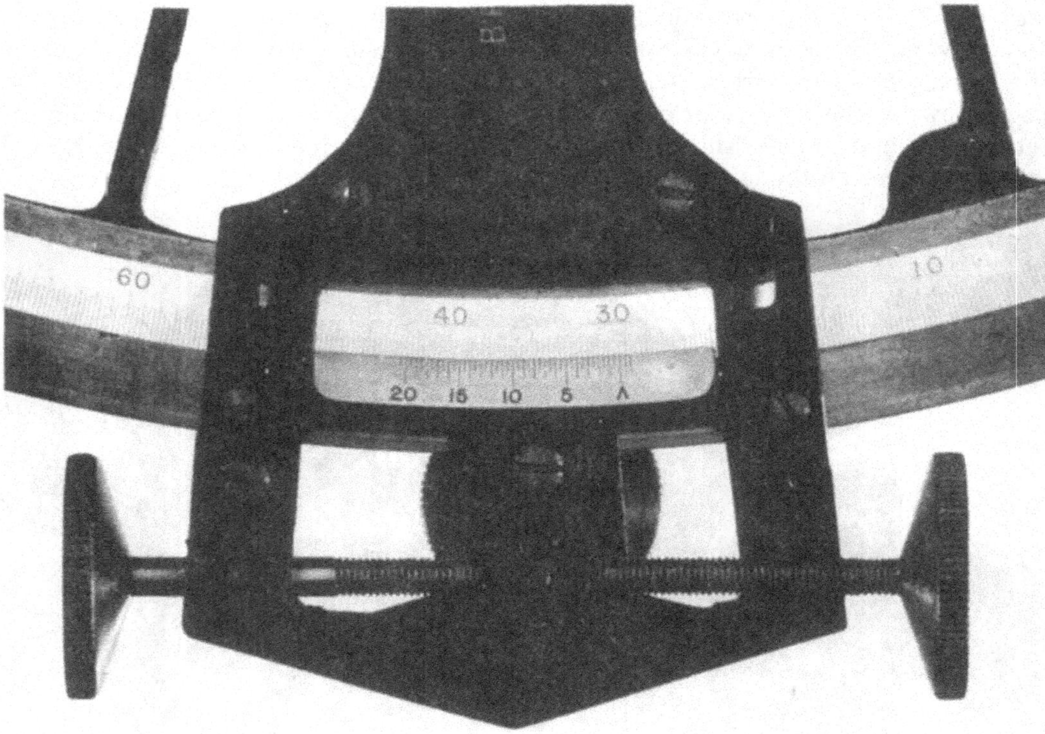

FIGURE 1506b.—Vernier sextant set at 29°42′30″.

Negative readings (as in determining index correction, art. 1603) are made in the same manner as positive readings, the various parts being added algebraically (art. 106, vol. II). Thus, if the three parts of a micrometer drum reading are (−)1°, 56′ and 0.′3, the total reading is (−)1°+56′+0.′3=(−)3.′7.

1507. Developing observational skill.—A well-constructed marine sextant is capable of measuring angles with an instrument error not exceeding 0.′1. Lines of position from altitudes of this accuracy would not be in error by more than about 200 yards. However, there are various sources of error, other than instrumental, in altitudes measured by sextant. One of the principal sources is the observer himself. There is probably no single part of his work that the navigator regards with the same degree of professional pride as his ability to make good celestial observations. Probably none of his other tasks requires the same degree of skill.

The first fix a student navigator obtains by his observation of celestial bodies is likely to be disappointing. Most navigators require a great amount of practice to develop the skill needed to make good observations. But practice alone is not sufficient, for if a mistake is repeated many times, it will be difficult to eradicate. Early in his career a navigator would do well to establish good observational technique—and continue to develop it during the remainder of his days as navigator. Many good pointers can be obtained from experienced navigators, but it should be remembered that each develops his own technique, and a practice that proves highly successful for one observer may not help another. Also, an experienced navigator is not necessarily a good observer, although he may consider himself such. Navigators have a natural tendency to judge the accuracy of their observations by the size of the figure formed when the lines of position are plotted. Although this is some indication, it is an imperfect one, because it does not indicate the errors of individual observations, and may not reflect constant errors. Also, it is a compound of a number of errors, some of which are not subject to control by the navigator.

When a student first begins to use the sextant, he can eliminate gross errors of principle in its use, and gain some ability in making observations, by accepting the coaching of an experienced navigator. By watching the novice make observations, the experienced navigator can observe a tendency to hold the instrument incorrectly, swing the arc improperly, or make other mistakes. When a celestial body is near the celestial meridian, the experienced navigator might make an observation and quickly transfer the sextant to the inexperienced one, who can see how the sight should appear. The two might make simultaneous observations and compare results. At first it is well to select bodies of low altitude, if they are available.

This procedure is helpful in detecting gross mistakes, but since the observations of the experienced navigator are not without error, this method is not suitable for final polishing of technique. For this purpose, observations should be compared with a more exact standard. Lines of position from celestial observations can be compared with good positions obtained by electronics or by piloting, if near a shore. Although this is good practice and provides a means of checking one's skill from time to time, it does not provide the large number of comparisons in a short time needed if technique is to be perfected.

This can sometimes be accomplished when a vessel is at anchor, or at a pier, if a stretch of open horizon is available. In advance, the altitude of a celestial body which will be over the open horizon at a time favorable for observation is computed at intervals of perhaps eight minutes (change in hour angle of 2°). If the body will be near the meridian, a smaller interval should be used. The altitude is determined for the position of the vessel, and all sextant altitude corrections (ch. XVI) are applied with reversed sign. These altitudes are then plotted versus time on cross-section paper, to a large scale, and a curve drawn through the points. At the selected time, a large number of observations are made at short intervals, allowing only enough time between observations for resting the eyes and arms. These observations are then plotted on the cross-section paper and compared with the curve.

An analysis of the results should be instructive. Erratic results indicate poor observational conditions or the need for practice and more care in making observations. If the measured altitudes are consistently too great, the sextant may not be rocked properly, the condition of tangency of the lower limb of the sun or moon may not be judged accurately, a false horizon in the water may have been used, subnormal refraction (dip) might be present, the eye might be higher above water than estimated, time might be in error, the index correction may have been determined incorrectly, the sextant might be out of adjustment, an error may have been made in the computation, the horizontal (vertical) may be tilted slightly by nearby mountains, etc. If the measured altitudes are consistently too low, the condition of tangency of the upper limb of the sun or moon may not be judged accurately, a low cloud may have been used as the horizon, abnormal refraction (dip) might be present, height of eye might be lower than estimated, time might be in error, the index error may have been determined incorrectly, the sextant might be out of adjustment, an error may have been made in the computation, the waves or swell at the horizon might be higher than at the ship, the horizontal (vertical) may be tilted slightly, a planet or bright star may have been placed "tangent" to the horizon rather than centered on the horizon, etc.

A single test of this type, while instructive, may not be conclusive. Several tests should be made with different celestial bodies, at various altitudes, under various conditions of weather and sea, and at different places. Generally, it is possible and desirable to correct any errors being made in the technique of observation, but occasionally a **personal error** (sometimes called **personal equation**) will persist. This might be different for the sun and moon than for planets and stars, and might vary with degree of fatigue of the observer, and other factors. For this reason, a personal error should be applied with caution. However, if a relatively constant personal error persists, and experience indicates that observations are improved by applying a correction to remove its effect, better results might be obtained by this procedure than by attempting to eliminate it from one's observations.

When lines of position of great reliability are desired, even an experienced navigator can usually improve his results by averaging to reduce random error. A number of observations, preferably not less than ten, are made in quick succession. These can then be plotted versus time, on cross-section paper, and a curve faired through the points. Unless the body is near the celestial meridian, this curve should be very nearly a straight line. *Any* point on the curve can be used as the observation, using the time and altitude indicated by the point. It is best to use a point near the middle of the line, to avoid possible errors in its slope.

The slope can be determined by means of Pub. No. 214, using Δt, which is the change of altitude relative to change in meridian angle (time). Meridian angle changes at the rate of 1′ in 4^s. Therefore, the change in altitude, in minutes of arc per second of time, is equal to Δt (expressed as minutes of arc) divided by 4^s, or $\frac{\Delta t'}{4^s}$. Thus, if Δt is 0.66, the altitude changes $\frac{0'.66}{4^s}=0'.165$ per second, or $15'\times 0'.66=9'.9$ per minute of time, increasing if the body is rising, and decreasing if it is setting. This rate may be altered by motion of the ship, the amount being the distance traveled in one minute, multiplied by the natural cosine of the relative azimuth of the body. Thus, if the speed is 15 knots, the ship moves 0.25 mile per minute. If the body is 30° on the bow, the altitude changes $0.25\times 0.86603=0'.2$ per minute due to motion of the ship, in addition to its own apparent motion due to rotation of the earth. If the body is forward of the beam, the effect of the ship's motion is to *increase* the altitude; if abaft the beam, to decrease it. The total effect is the algebraic sum of the separate effects due to rotation of the earth and motion of the vessel, since rate *at the vessel* is desired. Rapid change of Δt indicates a curved rate line. If a large number of observations is made, the slope of the line should be apparent from the plotted points.

A somewhat simpler variation is generally available if observations are made at equal intervals, unless the body is near the meridian. It is based upon the assumption that the change in altitude should be equal for equal intervals of time. A number of observations might be made by having an assistant give a warning "stand-by" and then a "mark" at equal intervals of time, as every ten or 20 seconds. Perhaps a better procedure is to make the observations at equal altitude increments. After the first observation, the altitude is changed by a set amount according to its rate of change, as 5′. The setting is *increased* if the body is rising, and *decreased* if it is setting. The body is then permitted to cross the horizon by its own motion, and at the instant of doing so, the time is noted. If time intervals are constant, the *mid time* and the *average altitude* are used as the observation. If altitude increments are constant, the *average time* and *mid altitude* are used. An uneven number of observations simplifies the finding of the mid value, but with ten observations the finding of the average value is easier.

If only a small number of observations is available, as three, it is usually preferable to solve all observations and plot the resulting lines of position, adjusting them to a common time. The *average* position of the line might be used, but it is generally better practice to use the middle line (or a line midway between the two middle ones if there are an even number).

In this discussion of averaging, it has been assumed that all observations are considered of nearly equal value. Any observation considered unreliable, either in the judgment of the observer or as a result of a plot, should be rejected in finding an average.

1508. Care of the sextant.—The modern marine sextant is a well-built, precision instrument capable of rendering many years of reliable service, with minimum attention. However, its usefulness can easily be impaired by careless handling or neglect. If it is ever dropped, it may never again provide reliable information. If this occurs, the instrument should be taken to an expert for careful testing and inspection.

When not in use, a sextant should invariably be kept in its case and properly stowed. The sextant case should be a well-constructed hardwood box fitted on its exterior with a lock, a handle, and two hooks, preferably the type having safety catches. The interior of the case should be fitted with blocks in which the handle or legs, or both, are placed when the sextant is stowed. Some sextant cases are fitted with catches which clamp over the handle when the sextant is stowed, and some are fitted with felt-lined blocks on the inside of the cover, to clamp down on the extreme ends of the arc when the case is closed. The case should be so constructed that it can be closed with the shade glasses and index arm in nearly any normal position, and preferably with the telescope in place. The last is particularly valuable to the navigator on an overcast day when only one opportunity to observe the sun may present itself, and the sight may have to be taken quickly. A case such as the plastic case (fig. 1508) for the U. S. Navy Mark 3 sextant is an adequate alternative. In this case, the sextant is stowed snugly within a polyurethane cushion. A keeper prevents movement of the sextant. However, the index arm must be placed at 20° before the sextant can be placed in the recesses of the cushion.

The case itself should be securely stowed in a convenient place away from excessive heat, dampness, and vibration. A shelf with built-up sides into which the case fits snugly is a good stowage place. The practice of leaving the sextant in its case on a chart room settee is a bad one, and the instrument should *never* be left unattended on the chart table.

To remove the sextant from its case, grasp the frame firmly with the left hand, making sure that no pressure is applied to the index arm, and lift the instrument from the box. Then take the sextant in the right hand, by its handle, leaving the left hand free to make any adjustments necessary before taking a sight. The instrument should never be held by its limb, index arm, or telescope.

Next to careless handling, the greatest enemy of the sextant is moisture. The mirrors, especially, and the arc should be wiped dry after each use. A new sheet of

FIGURE 1508.—Case for Mark 3 sextant showing sextant stowed in polyurethane cushion.

plain lens paper is best to use for this purpose, and linen second best. Over a period of time, however, linen collects dust, which may contain abrasives that will scratch the surface of the mirrors. For this reason, linen, if it is used, should be kept in a small bag to protect it from dust in the air. Chamois leather and silk are particularly likely to collect abrasive dusts from the air and they should not be used to clean the mirrors or telescope lenses. Should the mirrors become particularly dirty, they can be cleaned with a small amount of alcohol, applied with a clean piece of lens paper. The arc can be cleaned, when necessary, with ammonia, but never with a polishing compound. In cleaning or drying the mirrors and arc, care should be taken that excessive pressure is not applied to any part of the instrument.

A small bag of silica gel kept in the sextant case will help in keeping the air in the case free from moisture, and will help to preserve the mirrors. Occasionally, the silica gel should be heated in an oven to remove the absorbed moisture.

It may be necessary to wash the sextant with fresh water if it is subjected to sea spray. After washing, the sextant should be wiped gently, using a soft cotton cloth. Then, the optics should be gently polished using lens paper.

Glass optics do not transmit the whole of the light received. This is due to the tendency of air-to-glass surfaces to reflect a portion of the light received with resultant decrease in the light transmitted. This loss of light reduces the brightness of the image of the object viewed through such a glass optic. If the object is viewed through several glass optics, as is the case with the sextant telescope assembly, then the brightness of the image seen will be seriously affected and the image will be indistinct. The reflection also causes a glare which obscures the image of the object being viewed.

To reduce to a minimum the effect of this reflection from air-to-glass surfaces, and thus, to improve light transmission, the glass optics are treated with an anti-reflection coating. The coatings are extremely thin and can be easily damaged. Therefore, only light pressure should be applied when polishing the coated optics with lens paper. It is good practice to blow loose dust off the lens before attempting to clean. This will insure that there will be no grit under the lens paper.

The tangent screw and the teeth on the side of the limb should be kept clean and lightly oiled, using the oil provided with the sextant. It is good practice to set occasionally the index arm of an endless tangent screw at one extremity of the limb and then to rotate the tangent screw over the length of the arc. This will clean the teeth and spread the oil through them. At any time that the sextant is to be stowed for a long period, the arc should be protected with a thin coat of petroleum jelly.

If the mirrors need resilvering, they are best taken to an instrument shop where a professional job can be done. However, on rare occasions, it may be necessary to resilver the mirrors of a sextant at sea. In anticipation of this possibility, the navigator should obtain the necessary materials in advance, as makeshift substitutes cannot be relied upon to do the job adequately. The required materials are xylene (available in most pharmacies), dilute nitric acid (optional), alcohol, cotton, tin foil about 0.005 inch thick, a small amount of mercury, a clean blotter, and some tissue paper. Do not substitute aluminum foil commonly used in packaging candy and cigarettes.

First, remove the protective coating with alcohol (or better, acetone) from the back of the mirror to be resilvered, and clean the glass with xylene or acid. If the old silvering is difficult to remove, soak it in water. Place the blotter on a flat surface and turn up and seal the edges to form a tray. This will serve to contain the mercury if the vessel should roll during the operation. Using cotton, clean and smooth out both sides of a piece of tin foil slightly larger than the glass to be silvered, first with alcohol and then with xylene (do not use acid). Make certain that no lint adheres to the foil, and place it on the blotter. Clean the mercury by squeezing it through cheese cloth, and apply a drop to the foil. Carefully spread it over the surface with a finger, making sure that none of the mercury gets under the foil. Add a few more drops of mercury until the entire surface of the foil is covered and tacky. The mercury combines with some of the tin to form an amalgam. Place the chemically cleaned glass on a piece of clean tissue paper with the side to be silvered face down. Then place the glass and the paper on the amalgam. Apply slight pressure to the glass and withdraw the tissue paper. Following this, grasp the edge of the tin foil and lift it and the mirror from the blotter. Invert the glass and the tin foil and place in an inclined position, silvered side up. Any mercury remaining on the blotter is no longer pure and should be disposed of. Five or six hours later any loose foil may be scraped from the sides of the mirror, and the following day a coat of commercial varnish or lacquer should be applied to the silvered surface. Should the mirrored half of the horizon glass require silvering, the clear half may be protected by a strip of cellulose or adhesive tape.

1509. Sextant adjustments.—There are at least seven sources of error in the marine sextant, three nonadjustable by the navigator, and four adjustable.

The **nonadjustable errors** are: "prismatic error," "graduation error," and "centering error."

The **prismatic error** is present if the two faces of the shade glasses and mirrors are not parallel. Error due to lack of parallelism in the shade glasses may be called **shade error.** Shade error in the shade glasses near the index mirror can be determined by comparison of an angle measured when a shade glass is in the line of sight with the same angle measured when the glass is not in the line of sight. In this manner, the error for each shade glass can be determined and recorded. If shade glasses are used in combination, their combined error should be determined separately. If additional shading is needed for the observations, use the colored telescope eyepiece cover. This does not introduce an error because direct and reflected rays are traveling together when they reach it, and are therefore affected equally by any lack of parallelism of its two sides.

Lack of parallelism of the two faces of the index mirror can be detected by carefully measuring a series of angles; then removing the index mirror, inverting it, and replacing it; and then measuring the same angles again. Half the difference is the prismatic error. After the index mirror has been inverted, it should be checked carefully for perpendicularity to the frame of the sextant, as explained below.

Lack of parallelism of the two faces of the horizon glass will appear as part of the index error, and so need not have separate attention. The same is true of prismatic error in the shade glasses located near the horizon glass, but unless index error is determined with the shade glasses in place, the measured index error will not be the correct value for the combined error.

Graduation errors occur in the arc, micrometer drum, and vernier of a sextant which is improperly cut or incorrectly calibrated. Normally, the navigator cannot determine whether the arc of a sextant is improperly cut, but the principle of the vernier makes it possible to determine the existence of graduation errors in the micrometer drum or vernier and is a useful guide in detecting a poorly made instrument. The first and last markings on any vernier should align perfectly with one less graduation on the adjacent micrometer drum. In figure 1503a, the vernier is graduated in ten units. When the zero point is aligned with any graduation on the micrometer drum, the "ten" graduation should be in perfect alignment with a micrometer graduation nine units greater than the one in line with zero on the vernier. In figure 1503c, the vernier is graduated in six units and should align perfectly with any two graduations five units apart on the micrometer.

Centering error results if the index arm is not pivoted at the exact center of curvature of the arc. It can be determined by measuring known angles, after the adjustable errors have been removed. Horizontal angles can be used by determining the accurate value by careful measurement with a theodolite. Several readings by both theodolite and sextant should minimize errors. An alternative method is to measure angles between the lines of sight to stars, comparing the measured angles with computed values. To minimize refraction errors, one should select stars at about the same altitude, and avoid stars near the horizon.

The same shade glasses, if any, used for determining or eliminating index error should be used for measuring centering error. The errors determined in this manner include any error due to faulty graduation, and prismatic error of the index mirror, unless corrections are applied for these errors. However, since all vary with the angle measured, they need not be separated. Usually, it is preferable to make a single correction table for all three errors, called **instrument error.** Customarily, such a table is determined by the manufacturer and attached to the inside cover of the sextant case. The sign of the error is reversed, so that the values given are for **instrument correction** (**I**).

The **adjustable errors** in the sextant are those related to *perpendicularity* of (1) the frame and the index mirror, and (2) the frame and the horizon glass, and *parallelism* of (3) the index mirror and horizon glass to each other at zero setting, and of (4) the telescope to the frame. Each of these errors, if it exists, can be removed from the sextant by careful adjustment. In making these adjustments, *never tighten one adjusting screw without first loosening the other screw which bears on the same surface.* The adjustments should be made in the order indicated.

The first adjustment is for *perpendicularity of the index mirror* to the frame of the sextant. To test for perpendicularity, place the index arm at about 35° on the arc, and hold the sextant on its side, with the index mirror "up" and toward the eye. Observe the direct and reflected views of the sextant arc, as illustrated in figure 1509a. If the two views do not appear to be joined in a straight line, the index mirror is not perpendicular. If the reflected image is above the direct view, the mirror is inclined forward. If the reflected image is below the direct view, the mirror is inclined backward. An alternative and sometimes more satisfactory method of determining perpendicularity involves the use of two small vanes, or similar objects, of exactly the same height. Figure 1509b illustrates this method. Again the index arm is set at about 35°. The vanes are placed upright on the extremities of the limb, in such a way that the observer can, by placing his eye near the index mirror, see the direct view of one vane and the reflected image of the other. The tops of the objects are then observed for alignment. The use of vanes permits observation in the plane of adjustment, rather than at an angle. Adjustment is made by means of two screws at the back of the index mirror.

The second adjustment is for *perpendicularity of the horizon glass* to the frame of the sextant. An error resulting from the horizon glass not being perpendicular is called **side error.** To test for perpendicularity, set the index arm at zero and direct the line of sight at a star. Then rotate the tangent screw back and forth so that the reflected image passes alternately above and below the direct view. If, in changing from one position to the other, the reflected image passes directly over the star as seen without reflection, no side error exists, but if it passes to one side, the horizon glass is not perpendicular to the frame of the sextant. Figure 1509c illustrates observations without side error (left) and with side error (right). Whether the sextant reads zero when the true and reflected images are in coincidence is immaterial in this test. An alternative method is to observe a vertical line, such as one edge of the mast of another vessel (or the sextant can be held on its side and the horizon used). If the direct and reflected portions do not form a continuous line, the horizon glass is not perpendicular to the frame of the sextant. A third method is to hold the sextant vertical, as in observing the altitude of a celestial body, and bring the reflected image of the horizon into coincidence with the direct view, so that it

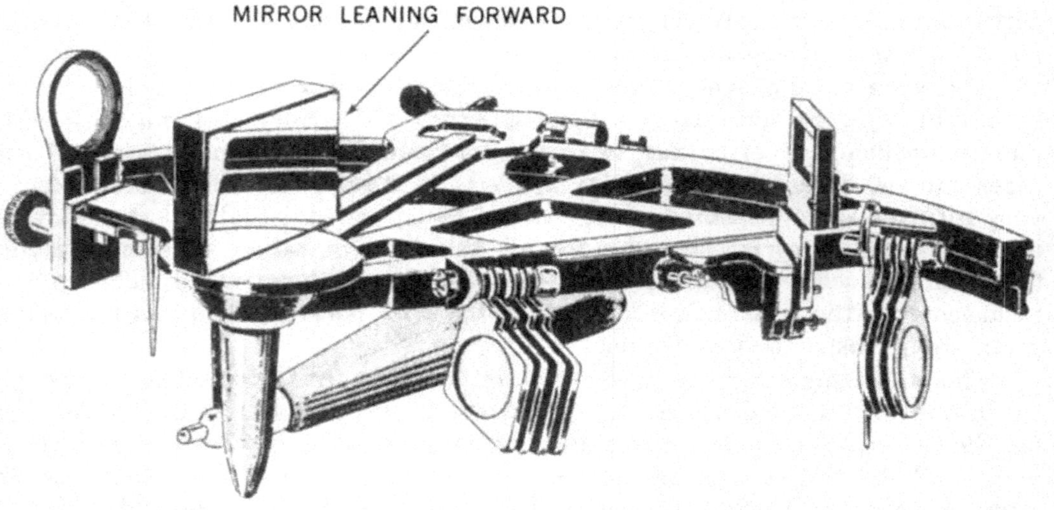

FIGURE 1509a.—Testing the perpendicularity of the index mirror. Here the mirror is not perpendicular.

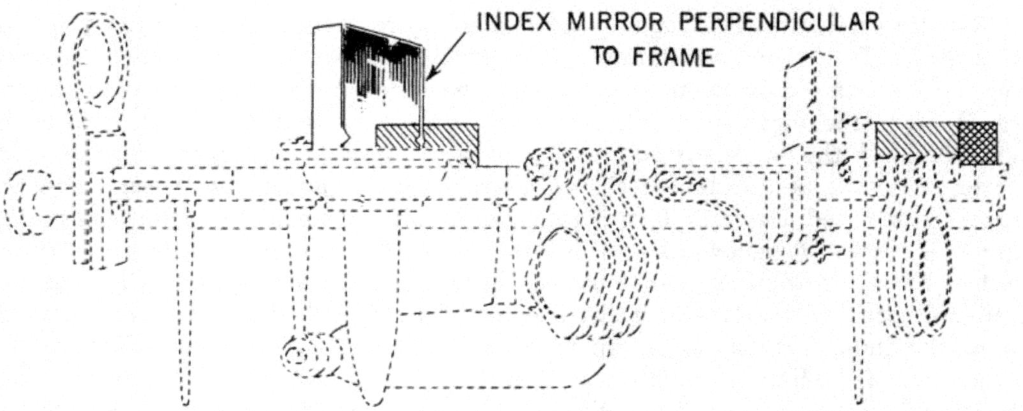

FIGURE 1509b.—Alternative method of testing the perpendicularity of the index mirror. Here the mirror is perpendicular.

appears as a continuous line across the horizon glass. Then tilt the sextant right or left. If the horizon still appears continuous, the horizon glass is perpendicular to the frame, but if the reflected portion appears above or below that part seen direct, the glass is not perpendicular. Adjustment is made by means of two screws at the back of the horizon glass.

The third adjustment is to make the *index mirror and horizon glass parallel* when the index arm is set exactly at zero. The error which results when the two are not parallel is the principal cause of **index error,** the total error remaining after the four adjustments have been made. Index error should be determined each time the sextant is used and need not be removed if its value is known accurately. To make the test for parallelism of the mirrors, set the instrument at zero, and direct the line of sight at the horizon or a star. Side error having been eliminated, the direct view and reflected image of the horizon appear as a continuous line, or the star as a single point, if the two mirrors are parallel. If the mirrors are not parallel, the horizon appears broken at the edge of the mirrored part of the horizon glass, one part being higher than the other. The reflected image of a star appears

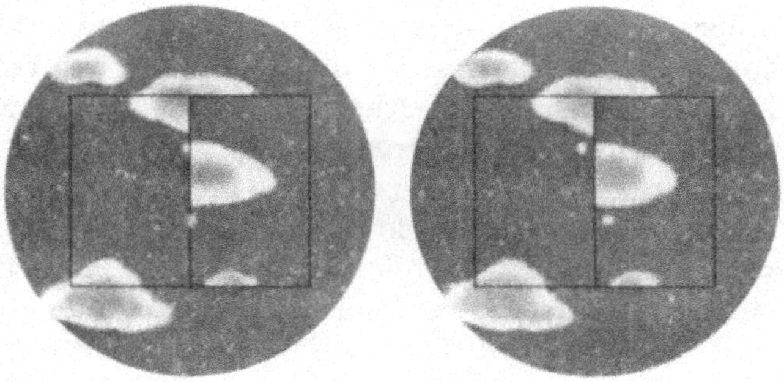

FIGURE 1509c.—Testing the perpendicularity of the horizon glass. *Left,* side error does not exist. *Right,* side error does exist.

above or below the star seen without reflection. If the star appears as a single point, move the tangent screw a small amount to be sure both direct view and reflected image are in the range of vision. The sun can be used by noting the reading when the reflected image is tangent to the sun as seen direct, first above it and then below. These should be numerically equal but of opposite sign (one positive and the other negative). To avoid variations in refraction, do not use low altitudes; or turn the sextant on its side and use the two sides of the sun. Adjustment is made by two screws at the back of the horizon glass. If the error is not to be removed, turn the tangent screw until direct view and reflected image of the horizon or a star are in coincidence. The reading of the sextant is the index error. It is positive if the reading is "on the arc" (positive angle), and negative if "off the arc" (negative angle). In the case of the sun it is *half* the numerical difference (algebraic sum) of the readings, positive or negative to agree with the larger reading. **Index correction** (**IC**) is numerically the same as index error, but of opposite sign. Since both the second and third adjustments involve the position of the horizon glass, it is good practice to recheck for side error after index error has been eliminated. Index error should always be checked after adjustment for side error.

The fourth adjustment is to make the *telescope parallel* to the frame of the sextant. If the line of sight through the telescope is not parallel to the plane of the instrument, an **error of collimation** will result, and altitudes will be measured as greater than their actual values. To check for parallelism of the telescope, insert it in its collar, and observe two stars 90° or more apart, bringing the reflected image of one into coincidence with the direct view of the other, near either the right or left edge of the field of view (the upper or lower edge if the sextant is horizontal). Then tilt the sextant so that the stars appear near the opposite edge. If they remain in coincidence, the telescope is parallel to the frame, but if they separate, it is not. An alternative method is to place the telescope in its collar and then lay the sextant on a flat table. Sight along the frame of the sextant and have an assistant place a mark on the opposite bulkhead, in line with the frame. Place another mark above the first at a distance equal to the distance from the center of the telescope to the frame. This second line should be in the center of the field of view of the telescope if the telescope is parallel to the frame. Adjustment for nonparallelism is made to the collar, by means of the two screws provided for this purpose.

Determination of any of the errors should be based upon a series of observations, rather than a single one. This is particularly true in the case of index error, which should be determined by approaching coincidence from opposite directions (up and down) on alternate readings. If adjustments are made carefully, and the sextant is given proper handling, it should remain in adjustment over a long period of time. Unless the navigator has reason to question the accuracy of the adjustments, they need not be checked at intervals of less than several months, except in the case of index error, which has the greatest effect on accuracy of readings, and should be checked each time the sextant is used. If the horizon is used for determining index error, this check should be made *before* evening twilight observations, and *after* morning twilight observations, while the horizon is sharp and distinct. If a star is used, the index error should be determined *after* evening observations and *before* morning sights are taken. During the day, it should be checked both before and after observations.

Frequent manipulation of the adjusting screws should be avoided, as it may cause excessive wear. Except in the case of index error, slight lack of adjustment has little effect on the results, and should be ignored. If adjustments are needed at frequent intervals, the sextant is not receiving proper care, or has worn parts which should be replaced at a navigation instrument shop. If index error is not constant, it should not be removed, but index correction should be determined before or after every observation and applied to the readings, until the sextant can be repaired. A small variable error might well be accepted, but should be watched to see that it does not become unduly large.

1510. Selection of a sextant.—For satisfactory results a sextant should be selected carefully. For accurate work the radius of the arc should be about 6½ inches or more. The instrument should be light, but strongly built. The various moving parts should fit snugly, but move freely without binding or gritting. If the index arm is either too loose or too tight at either end of the arc, the pivot may not be perpendicular to the frame of the sextant. The telescope should be easy to insert or remove from its holder, and to focus.

The use to be made of a sextant should be considered in its selection. For ordinary use in measuring altitudes of celestial bodies, an arc of 90° or slightly more is sufficient. A longer arc is desirable if back sights are to be made, or if horizontal angles are to be measured. If use of the sextant is to be limited to horizontal angles, less accuracy is required. The arc can be of smaller radius, and small nonadjustable errors are unimportant.

If practicable, a sextant should be examined by an expert, and tested for nonadjustable errors before acceptance.

1511. Octants, quintants, and quadrants.—Originally, the term "sextant" was applied to the navigator's double-reflecting, altitude-measuring instrument only if its arc was 60° in length—a sixth of a circle—permitting measurement of angles from 0° to 120°. In modern usage the term is applied to all navigational altitude-measuring instruments, regardless of angular range or principles of operation, although some are octants (angular range 90°), some quintants (144°), some quadrants (180°), and many have an intermediate range.

1512. The artificial horizon.—Measurement of altitude requires a horizontal reference. In the case of the marine sextant this is commonly provided by the visible sea horizon. If this is not clearly visible, reliable altitudes cannot be measured unless a different horizontal reference is available. Such a reference is commonly called an **artificial horizon.** If it is attached to, or part of, the sextant, altitudes can be measured at sea, on land, or in the air, whenever celestial bodies are available for observations. On land, where the visible horizon is not a reliable indication of the horizontal, an external artificial horizon can be devised.

Any horizontal reflecting surface will serve the purpose. A pan of mercury, heavy oil, molasses, or other viscous liquid sheltered from the wind is perhaps simplest. A piece of plate glass fitting snugly across the top of the container is usually the best shelter. If there is any reasonable doubt as to the parallelism of the two sides of the glass, two readings should be made with the glass turned 180° in azimuth between readings, and the average value taken. The pan and liquid should be clean, as foreign material on the surface of the liquid is likely to distort the image and introduce an error in the reading.

To use an external artificial horizon, the observer stands or sits in such a position that the celestial body to be observed is reflected in the liquid, and is also visible by direct view. By means of the sextant, the double-reflected image is brought into coincidence with the image appearing in the liquid. In the case of the sun or moon the *bottom* of the double-reflected image is brought into coincidence with the *top* of the image in the liquid, if a lower-limb observation is desired. For an upper-limb observation, the opposite sides are brought into coincidence. If one image is made to cover the other, the observation is of the *center* of the body.

When the observation has been made, apply the index correction and any other instrumental correction, as well as any correction for personal error. Then take *half* the remaining angle and apply all other corrections except dip (height of eye) correction, since this is not applicable. If the *center* of the sun or moon is observed, omit, also, the correction for semidiameter. Chapter XVI explains the various corrections and their applications.

A commercial artificial horizon consisting of a metal tray, mercury, cover of two sloping glass slides held in a metal frame, metal bottle to hold the mercury when not in use, and a funnel for pouring, was at one time a familiar part of a navigator's equipment, but the modern navigator might experience difficulty in locating such a device.

1513. Artificial-horizon sextants.—Shortly after the marine sextant was invented (art. 124), an attempt was made to extend its use to periods of darkness. This was done by providing a spirit level attachment. The observer brought the double-reflected image of the celestial body being observed into coincidence with the bubble of the spirit level. Such devices have been made available from time to time, and are still being manufactured. However, they have never come into general use, and are of questionable value.

Charles A. Lindbergh's historic solo flight across the North Atlantic in 1927 demonstrated the practicability of long over-water flights. The development of a suitable instrument for observing altitudes of celestial bodies during darkness and when the horizon was obscured by clouds or haze became a virtual requirement. Various forms of artificial horizon have been used, including a bubble, gyroscope, and pendulum. Of these, the bubble has been most widely used. Figure 1513a illustrates a modern periscopic sextant permitting observation with only a small tube protruding through the top of the aircraft. Figure 1513b shows the optical principle of a different type aircraft sextant.

With an artificial horizon of the bubble or pendulum type, considerable skill is needed to make an observation. The image of the horizontal reference (a circle or horizontal line) and the celestial body both appear in the field of view, and both may seem unsteady. An observation is made by matching the two near the center of the field of view. The appearance at coincidence depends upon the instrument. Some bubbles appear dark and are placed on a level with the body. Others have a clear center and are placed over the body. One pendulum type has a horizontal line

FIGURE 1513a.—Aircraft periscopic sextant.

that is customarily placed directly across the body, although a limb observation can be made if desired. Bubbles can be regulated in size, and the instructions provided with the instrument should be followed. In general, the bubble diameter should be about one-sixth to one-fourth the size of the field of view. This is about three to four times the size of the sun or full moon as seen through the eyepiece. A very small bubble should be avoided because it tends to lag sextant movements so much that it is unreliable as a horizontal reference.

A considerable amount of practice is needed to develop skill in making reliable observations with an artificial-horizon sextant, even on land or other steady platform. At sea or in the air the motions of the craft greatly increase the difficulty of observation. In addition to compounding the difficulty of making coincidence, the craft motion introduces a sometimes large and rapidly varying **acceleration error.** That is, motions of the craft produce an acceleration on the pendulum or the liquid of the bubble chamber, causing a false indication of the horizontal. In smooth air the accelerations tend to follow a cycle of about one to two minutes in length. They are largely eliminated by use of an averaging device. In making an observation, the observer attempts to maintain coincidence continuously over a period, usually two minutes. The *average* altitude, generally indicated on a dial or drum, is used with the *mid* time of observation. Thus, perhaps 60 individual observations, or a continuously integrated altitude, are available to smooth out errors of individual observations.

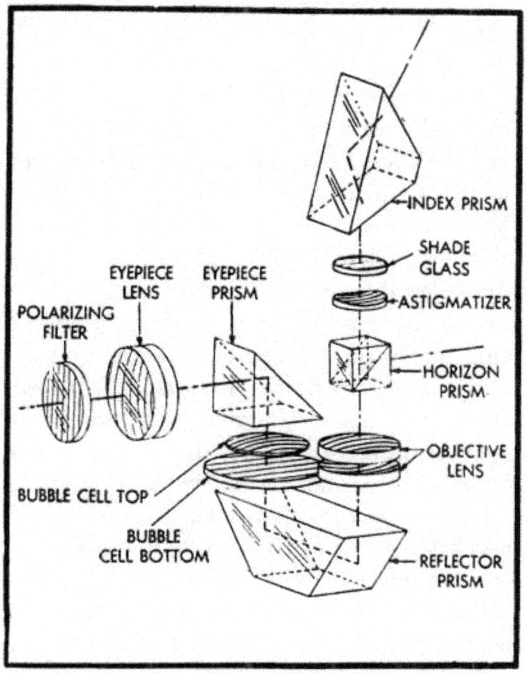

FIGURE 1513b.—Optical principle of a typical bubble sextant.

On land or other steady platform a skillful observer using a 2-minute averaging bubble or pendulum sextant can measure altitudes to an accuracy of perhaps 2′ (2 miles). This, of course, refers to the accuracy of measurement only, and does not include additional errors such as abnormal refraction, deflection of the vertical, computational and plotting errors, etc. In steady flight through smooth air the error of a 2-minute observation is increased to perhaps 5 to 10 miles. At sea, conditions are different. In a glassy sea with virtually no roll or pitch, results should approach those on land. However, with even a slight, gentle roll the accelerations to which a vessel is subjected are quite complex, as indicated by the difficulty one not accustomed to the sea has in getting his "sea legs" during the early part of a voyage. If the vessel is yawing, a large Coriolis error (art. 815, vol. II) may be introduced. Under these conditions observational errors of 10-15 miles are not unreasonable. With a moderate sea, errors of 30 miles or more are common. In a heavy sea, any useful observations are virtually impossible to obtain. Single altitude observations in a moderate sea can be in error by a matter of *degrees.*

Because of the difficulty of observing, and the large acceleration errors encountered aboard a vessel, bubble and pendulum type sextants have very limited use at sea. A submarine on war patrol, surfacing only during darkness, may have use for such an instrument. A large number of observations on a reasonably calm night can produce results of some value. However, even under these conditions some navigators report better results with a marine sextant and dark-adapted eyes. In pack ice a ship generally provides a reasonably steady platform. When the horizon is obscured by ice or haze, polar navigators can sometimes obtain better results with an artificial-horizon sextant than with a marine sextant. Some artificial-horizon sextants have provision for making observations with the natural horizon as a reference, but since this is a secondary usage, results are not generally as satisfactory as by marine sextant. Because of their more complicated optical systems, and the need for providing a horizontal reference, artificial-horizon sextants are generally much more costly to manufacture than marine sextants. Designed for use in the air, they serve a useful purpose there, but for ordinary use aboard ship they have little to recommend them.

Altitudes observed by artificial-horizon sextant are subject to the same errors as those observed by marine sextant, except that dip (height of eye) correction does not apply. Also, when the center of the sun or moon is observed, no correction for semidiameter should be made. Chapter XVI explains the various sextant altitude corrections and their applications.

Adjustment of an artificial-horizon sextant should not be attempted by other than an instrument man qualified to handle the particular type instrument involved. An exception is the adjustment of the size of the bubble. Also, with some instruments an easily movable index permits elimination or reduction of index error. This error can best be determined in an instrument shop equipped with a collimator. If one is not available, the error can be determined by comparing the average of a number of observations made at a known point on land with the computed values. A precomputed curve of altitude versus time is useful for this purpose. Altitude corrections equal to the errors but with reversed sign should be applied to computed altitudes. With normal usage, the index error should not change. In most artificial-horizon sextants there is no index error.

The care and operation of various types of instruments vary considerably. The instruction booklet provided with each instrument should serve as the guide.

1514. The spring-driven marine chronometer is a timepiece having a nearly constant rate. It is used aboard ship to provide accurate time, primarily for timing celestial observations for lines of position, and secondarily for setting the ship's other timepieces. It differs from a watch principally in that it contains a variable lever device to maintain even pressure on the mainspring, and a special balance designed to compensate for temperature variations. A ship in which celestial navigation is used carries one or more chronometers.

A spring-driven chronometer is set approximately to Greenwich mean time (GMT) and is not reset until the instrument is overhauled and cleaned, usually at three-year intervals. Resetting might disturb the rate. Instead, the difference between GMT and **chronometer time (C)** is carefully determined, and applied as a correction to all chronometer readings. This difference, called **chronometer error (CE)**, is "fast" (F) if chronometer time is later than GMT, and "slow" (S) if earlier. The amount by which chronometer error changes in 1 day is called **chronometer rate,** or sometimes **daily rate,** considered "gaining" or "losing" as the chronometer is running faster or slower than the correct rate. An erratic rate indicates a defective instrument, or need for overhaul. The methods of determining and applying chronometer error and chronometer rate are explained in chapter XVIII.

A spring-driven chronometer is mounted in gimbals in a box, which should be carefully stowed to protect the instrument from damage due to heavy rolling and pitching, vibration, temperature variations, and electrical and magnetic influences. Usually this is done by fitting the box snugly into a heavily padded case suitably located in the chart room of merchant ships, and below decks, near the center of motion, in U.S. Navy ships.

The principal maintenance requirement aboard ship is regular winding at about the same time each day. Aboard United States naval vessels this is customarily done at about 1130 each morning, and reported to the commanding officer at 1200. Aboard merchant ships it is usually wound at about 0800. Although a spring-driven chronometer is designed to run for more than 2 days, daily winding helps insure a uniform rate, and constitutes a daily routine that decreases the possibility of letting the instrument run down. On the face of each chronometer is a small dial that indicates the number of hours elapsed since the instrument was wound. To wind the chronometer, gently turn the instrument on its side, and slide back the guard covering the keyhole. Insert the key and carefully wind in a counterclockwise direction. Seven half-turns should suffice. If a chronometer should run down, wait until GMT is nearly the same as the time indicated before winding. If the chronometer does not start after winding, move the case back and forth gently. Check the error and rate carefully.

At maximum intervals of about three years, a spring-driven chronometer should be sent to a good chronometer repair shop for cleaning and overhaul. When transported by hand, a chronometer should be clamped in its gimbals and stowed in the large case provided. When shipped, it should be allowed to run down, and the balance secured by a cork before the chronometer is stored in the large case.

1515. Quartz crystal marine chronometers are being used as replacements for the spring-driven chronometers aboard ships of the U.S. Navy. The accuracy of these instruments is such that the time can be read without resort to chronometer rate. Should the second hand be in error by a readable amount, it can be reset electrically.

The quartz crystal chronometer designed for U.S. Navy use (fig. 1515a) displays time using a 24-hour dial. It indicates the day of the week. The chronometer is furnished in a case suitable for direct installation in shipboard chronometer lockers or modular type chart tables without gimbaling or other special mounting or restraining features.

The basic element for time generation is a quartz crystal oscillator. The quartz crystal is temperature compensated and is hermetically sealed in an evacuated envelope. A calibrated adjustment capability is provided to adjust for the aging of the crystal. The performance requirements for this chronometer are shown in figure 1515b.

The chronometer is designed to operate a minimum of 1 year on a single set of batteries. A built-in battery test meter is operated by means of a push button. The meter indicates the relative strength of the battery. The dial is marked to indicate the point at which the battery should be replaced. The chronometer continues to operate and keep the correct time for at least 5 minutes while the batteries are being changed. The chronometer is designed to accommodate the gradual voltage drop during the life of the batteries by which it is powered while maintaining accuracy requirements.

A two-position setting mechanism is provided to set the three hands. One position of the mechanism permits the minutes and hours to be set without any movement of the second hand. Return of the mechanism to the other (normal) position results in resumption of the measurement of time.

Two electrical push buttons are provided for making small corrections in time by electrical means. Depression of one button stops the chronometer movement and depression of the other speeds up the movement. Operation of the push buttons does not upset the synchronization of the minute and second hands.

1516. Watches.—In the interest of accuracy, a spring driven chronometer is not disturbed more than necessary. Celestial observations are timed and ship's clocks set by means of a **comparing watch.** This is a high-grade pocket watch which is set by comparison with a chronometer, and then carried to the place where accurate

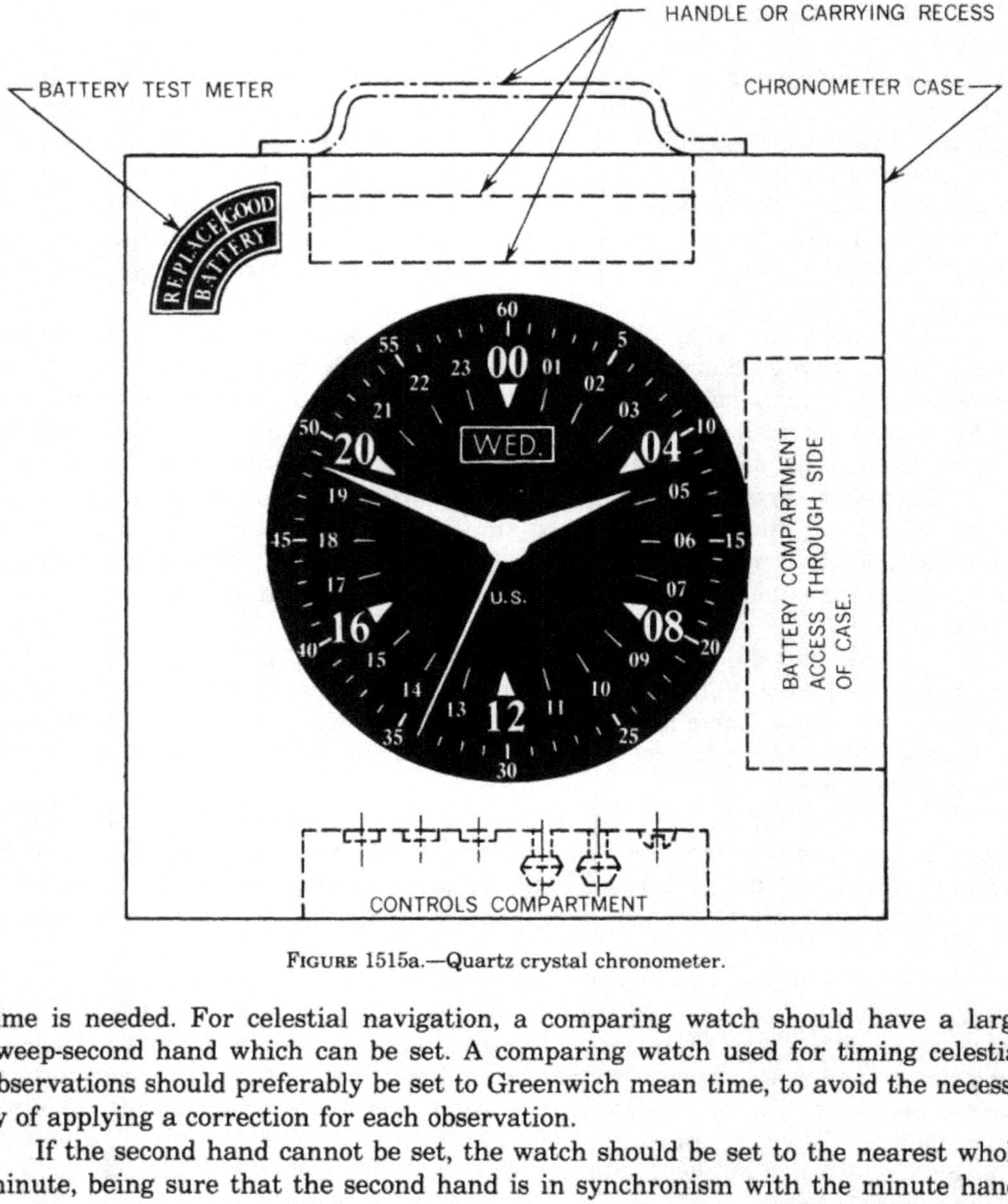

FIGURE 1515a.—Quartz crystal chronometer.

time is needed. For celestial navigation, a comparing watch should have a large sweep-second hand which can be set. A comparing watch used for timing celestial observations should preferably be set to Greenwich mean time, to avoid the necessity of applying a correction for each observation.

If the second hand cannot be set, the watch should be set to the nearest whole minute, being sure that the second hand is in synchronism with the minute hand, and the **watch error (WE)** determined. If a watch is to be used for other purposes than timing of celestial observations, it might preferably be set to zone time. A comparing watch should be set, or watch error determined, immediately before or after celestial observations are made, to avoid the necessity for determining and applying a correction for **watch rate,** and to eliminate a possible error due to an inaccurate or variable rate. If a watch set to GMT is used for timing celestial observations, care should be taken to avoid a possible error of 12 hours or 24 hours. The mental application of zone description of ship's time indicates the approximate GMT and the Greenwich date. The subject of time is discussed more fully in chapter XVIII. A stop watch may also be used for celestial observations.

1517. Other instruments.—The sextant, chronometer, and comparing watch (or stop watch) are the principal instruments of celestial navigation. The azimuth circle for observing azimuths of celestial bodies is discussed in article 629. Plotting equip-

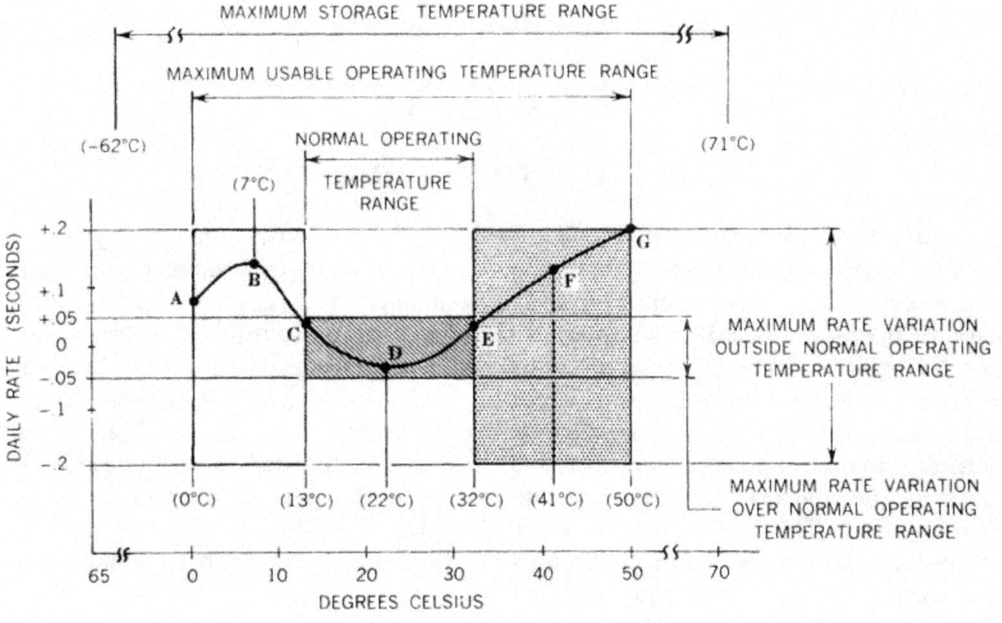

FIGURE 1515b.—Performance requirements.

ment is the same as that for dead reckoning (arts. 602–606). A flashlight might be needed for reading the sextant and the comparing watch. A pocket notebook is desirable for recording predicted positions of celestial bodies if a star finder is used, and for recording the observations. A workbook is desirable for solving celestial observations so that a permanent record is available. Work forms are desirable, but should form part of the work book, and not be kept separately. These might be provided by rubber stamp, or by printing. In the latter case a looseleaf work book may be desirable to permit arrangement of the various papers in chronological order.

CHAPTER XVI

SEXTANT ALTITUDE CORRECTIONS

1601. Need for correction.—Altitudes of celestial bodies, obtained aboard ship for the purpose of establishing lines of position, are normally measured by a hand-held **sextant**, described in chapter XV. The uncorrected reading of a sextant after such an operation is called **sextant altitude (hs)**. If the sextant is in proper adjustment, certain sources of error are eliminated, as explained in article 1509. There remains, however, a number of sources of error over which the observer has little or no control. For each of these he applies a correction. When all of these **sextant altitude corrections** have been applied, the value obtained is the altitude of the center of the celestial body above the celestial horizon, for an observer at the center of the earth. This value, called **observed altitude (Ho)**, is compared with the **computed altitude (Hc)** to find the **altitude intercept (*a*)** used in establishing a line of position as explained in chapter XVII.

Articles 1602–1614 describe the various corrections. For highly accurate results, all of these are needed to the greatest accuracy obtainable. The needs of ordinary practical navigation, however, make no such exacting requirements, and in the course of his usual day's work at sea, the navigator has relatively few corrections to apply, from conveniently-arranged tables readily accessible to him. The more detailed information in chapter VIII of volume II is given to (1) provide the basis for a better understanding of the problem, (2) furnish the information needed for evaluation of results, and (3) provide a source of reference material beyond that given in the usual navigation text.

1602. Instrument correction (I) is the combined correction for nonadjustable errors (prismatic error, graduation error, and centering error) of the sextant, as explained in article 1509. Usually, this correction is determined by the manufacturer, and recorded on a card attached to the inside of the top of the sextant box. It varies with the angle, may be either positive or negative, and is applied to all angles measured by that instrument. For a well-made instrument, the maximum value is so small that this correction can be ignored for all except the most accurate work. Normally, instrument error of artificial-horizon sextants is so small, considering the precision to which angles can be measured by such instruments, that no correction is provided.

1603. Index correction (IC), due primarily to lack of parallelism of the horizon glass and index mirror at zero reading, is discussed in article 1509. Until the adjustment is disturbed, the index correction remains constant for all angles, and is applicable to all angles measured by the instrument. It may be either positive or negative. Normally, artificial-horizon sextants do not have index corrections.

1604. Personal correction (PC) is numerically the same as personal error (art. 1507), but of opposite sign, either positive or negative. If experience indicates the need for such a correction, it should be made to altitudes of the bodies to which it applies. However, the observer should be sensitive to changes in its value. Unless the observer has sufficient evidence to be sure of the existence and relative constancy of a personal error, no correction should be applied.

1605. Dip (D) of the horizon is the angle by which the visible horizon (art. 1429) differs from the horizontal at the eye of the observer (the sensible horizon, art. 1429). Thus, it applies only when the visible horizon is used as a reference, and not when an artificial horizon, either internal or external to the sextant, is used. It applies to all celestial bodies. If the eye of the observer were at the surface of the earth, visible and sensible horizons would coincide, and there would be no dip. This is never the situation aboard ship, however, and at any height *above* the surface, the visible horizon is normally *below* the sensible horizon, as shown in figure 1605a. Normally, then, an altitude measured from the visible horizon is too *great,* and the correction is *negative.* It increases with greater height of the observer's eye. Because of this, it is sometimes called **height of eye correction.**

If there were no atmospheric refraction (art. 1606), dip would be the angle between the horizontal at the eye of the observer, and a straight line from this point tangent to the surface of the earth. With refraction, dip is the angle (∠ HAC′ of figure 1605b) between the horizontal at the eye of the observer and a straight line tangent at the eye of the observer to the curved ray of light from the visible horizon.

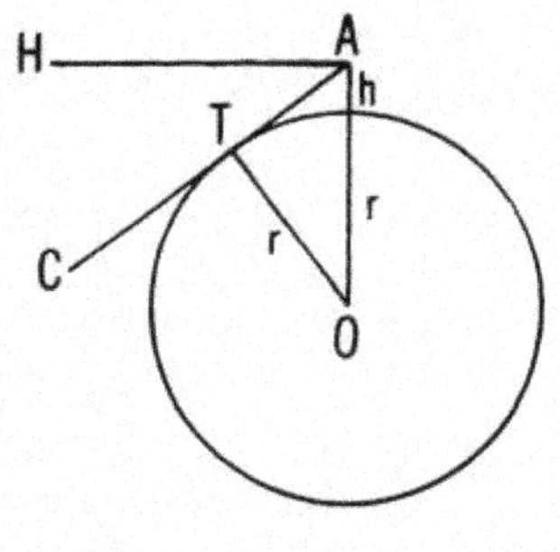

FIGURE 1605a.—Dip without refraction.

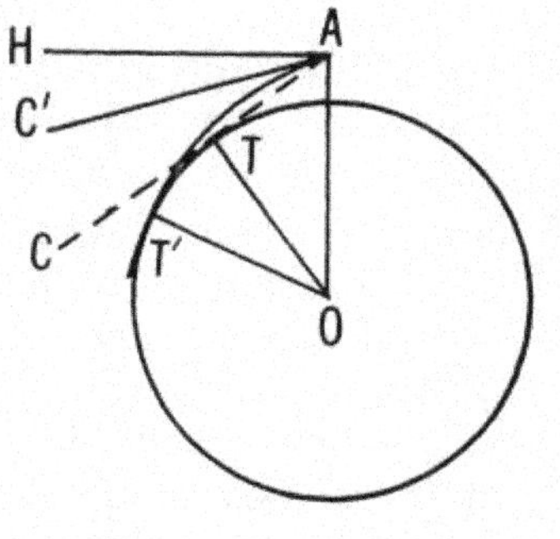

FIGURE 1605b.—Dip with refraction.

The amount by which refraction alters dip varies with changing atmospheric conditions. Even the *average* value has not been established with certainty, and several methods of computing dip have been proposed. The values given in the critical table on the inside front cover of the *Nautical Almanac* were computed by the equation

$$D=0.97\sqrt{h}$$

where D is the dip, in minutes of arc; and h is the height of eye of the observer, in feet. Part of this table is repeated on the page facing the inside back cover. The *Air Almanac* table was computed independently by a different method, to a precision of whole minutes. The minor discrepancies thus introduced are not important in practical navigation.

The values given in the table are satisfactory for practical navigation under most conditions. An investigation by the Carnegie Institution of Washington showed that of 5,000 measurements of dip at sea, no value differed from the tabulated value by more than 2′.5, except for one difference of 10′.6. Extreme values of more than 30′ have been reported and even values of several *degrees* have been encountered in polar regions. Greatest variations from tabulated values can be expected in calm weather, with large differences between sea and air temperatures, particularly if mirage effects are present. Irregularities in the shape of the rising or setting sun may indicate abnormal conditions. Large variations may also be present shortly after passage of a squall line, when errors of as much as 15′ have been reported. When a temperature inversion is known to exist, the tabulated dip may be too small, numerically. The effect of sea-air temperature difference is discussed in greater detail in chapter VIII of volume II.

In the determination of height of eye, position on the ship should be considered, and also the condition of loading and trim. If an observation is made from a position differing from the usual place, the altered height of eye should not be overlooked. Momentary changes due to rolling and pitching can be neutralized, to a large extent, by making observations from a point on the centerline of the vessel, at the axis of pitch.

Since variations from normal dip may be one of the principal sources of error in celestial observations, the observer should be alert to conditions affecting terrestrial refraction. Any observation taken within half an hour after passage of a squall line should be regarded as unreliable.

If dip cannot be measured, the effects of abnormal conditions can be minimized by observing three bodies differing in azimuth by about 120° (or four bodies by 90°, five bodies by 72°, etc.). If the error is constant in all directions, its effect is to increase (or possibly to decrease) the size of the closed figure formed by the lines of position without altering the position of its center. Hence, the size of the figure is not necessarily an indication of the accuracy of the fix.

1606. Refraction (R).—Light, or other radiant energy, is assumed to travel in a straight line at uniform speed, if the medium in which it is traveling has uniform properties. But if light enters a medium of different properties, particularly if the density is different, the speed of light changes somewhat. Light from a single point source travels outward in all directions, in an expanding sphere. At great distances, a small part of the surface of this sphere can be considered flat, and light continuing to emanate from the source can be considered similar to a series of waves, in some respects resembling the ocean waves encountered at sea. If these light "waves" enter a more dense medium, as when they pass from air into water, the speed decreases. If the light is traveling in a direction perpendicular to the surface separating the two media (in this case vertically downward), all parts of each wave front enter the new medium at the same time, and so all parts change speed together, as shown in figure 1606a. But if the light enters the more dense medium at an oblique angle, as shown in figure 1606b, the change in speed occurs progressively along the wave front as the different parts enter the more dense medium. This results in a change in the direction of travel, as shown. This change in direction of motion is called **refraction.** If light enters a more dense medium, it is refracted *toward* the normal (*NN′*), as in figure 1606b. If it enters a less dense medium, it is refracted *away* from the normal, as light traveling in the opposite direction to that shown in figure 1606b.

The *amount* of the change in direction is directly proportional to the angle between the direction of travel and the normal (angle *ABN* in figure 1606b). The ratio of this angle to the similar angle after refraction takes place (angle *CBN′* in figure 1606b) is constant, so that as one increases, the other increases at the same *rate.* Hence, the *difference* between them (the change in direction) also increases at the same rate. Therefore, if the *incident* ray (*AB*) is nearly parallel to the surface at which refraction takes place, relatively large amounts of refraction occur.

The amount of refraction is also directly proportional to the relative speed of travel in the two media. Various substances are compared by means of a number

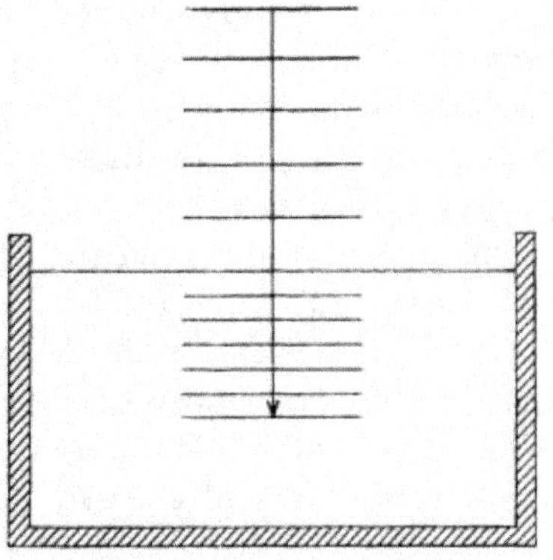

FIGURE 1606a.—No refraction occurs when light enters denser medium normal to the surface.

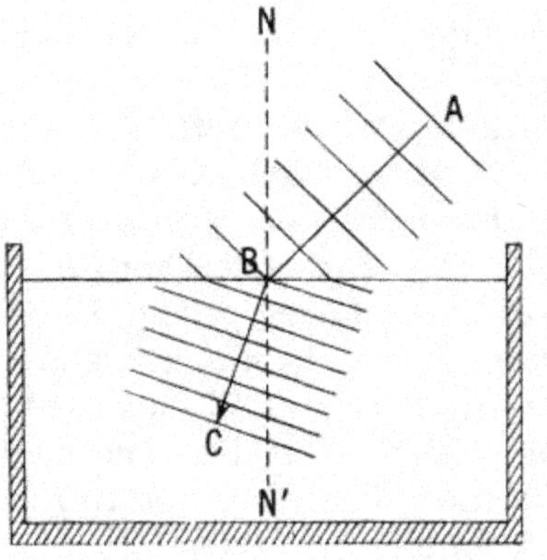

FIGURE 1606b.—A ray entering a denser medium at an oblique angle is bent toward the normal.

called the **index of refraction (μ)**, which depends primarily upon the density of the substance. In figure 1606b, angle *ABN* is called the **angle of incidence (ϕ)** and angle *CBN'* the **angle of refraction (θ)**. These are related by Snell's law, which states that *the sines of the angle of incidence and angle of refraction are inversely proportional to the indices of refraction of the substances in which they occur.* Thus, if μ_1 is the index of refraction of the substance in which ϕ occurs, and μ_2 is the index of refraction of the substance in which θ occurs

$$\frac{\sin \phi}{\sin \theta} = \frac{\mu_2}{\mu_1}.$$

If the index of refraction changes suddenly, as along the surface separating water and air (as shown in fig. 1606b), the change in direction is equally sudden. However, if a ray of light travels through a medium of gradually changing index of refraction, its path is curved, undergoing increased refraction as the index of refraction continues to change. This is the situation in the earth's atmosphere, which generally decreases in density with increased height. The gradual change of direction occurring there is called **atmospheric refraction.** The bending of a ray of light traveling from a point on or near the surface of the earth, to the eye of the observer, is called **terrestrial refraction.** This affects dip of the horizon, as discussed in article 1605. A ray of light entering the atmosphere from outside, as from a star, undergoes a similar bending called **astronomical refraction.**

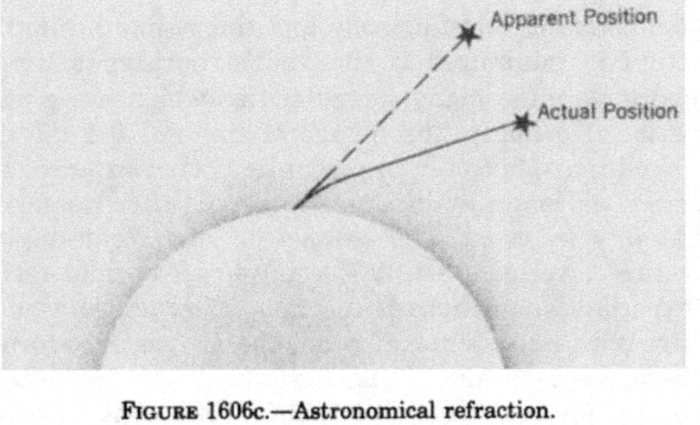

FIGURE 1606c.—Astronomical refraction.

The effect of astronomical refraction is to make a celestial body appear *higher* in the sky than it otherwise would, as shown in figure 1606c. If a body is in the zenith, its light is not refracted, except for a very slight amount when the various layers of the atmosphere are not exactly horizontal. As the zenith distance increases, the refraction becomes greater. At an altitude of 20° it is about 2'.6; at 10°, 5'.3; at 5°, 9'.9; and at the horizon, 34'.5. A table of refraction is given on the inside front cover and facing page of the *Nautical Almanac,* in the columns headed "Stars and Planets." As height above the surface of the earth increases, light from an outside source travels through less of the atmosphere, and refraction decreases. At shipboard heights the difference is negligible, but at aircraft heights the change is a consideration. Therefore, the refraction table given near the back of the *Air Almanac* is a double-entry table.

The values given in the tables are for average conditions. This is called **mean refraction.** A considerable amount of research has been conducted to determine the mean values, the conditions under which values differ from the mean, and the amount of such differences. A number of different mean refraction tables have been produced. Values in the various tables differ slightly because of different assumptions, different methods of observation, and different observed results under apparently similar conditions. This last source of difference is due primarily to the fact that conditions could be determined at the position of the observer, but not at various points along the line traveled by the ray of light in passing through the atmosphere. Nevertheless, the various tables agree very well down to a minimum altitude of 2°. Below this, the refraction is erratic, and differences between values in the various tables are not as important as differences between mean and instantaneous values. The values given in the almanac tables are in excellent agreement with those actually measured.

Because of their variability, refraction and dip (also affected by refraction) are the principal uncertainties in the accuracy of celestial observations of a careful observer. As a result of this uncertainty, navigators formerly avoided all observations below some arbitrary altitude, usually 15°. While this is still good practice if higher bodies are available, the growing knowledge of atmospheric refraction has increased the confidence with which navigators can use low-altitude sights. There is little reason for lack of confidence in sights as low as 5°. Below this, other available corrections should be applied (art. 1626). If altitudes below 2° are used, larger probable errors should be anticipated, even with the use of additional corrections. Generally, the error in tabulated refraction should not exceed two or three minutes, even at the horizon. However, a knowledge of conditions affecting refraction is helpful in determining the confidence to be placed in such observations. Since refraction elevates both the celestial body and the visible horizon, the error due to abnormal refraction is minimized if the visible horizon is used as a reference.

The atmosphere contains many irregularities which are erratic in their influence upon refraction. Normally, the navigator has not the information needed to correct for such conditions, but only to recognize their existence. He must recognize that those observations made within half an hour after passage of a squall line might be considerably in error. The passage of any front might have a similar effect. A temperature inversion (art. 3815) may upset normal refraction. Abnormal values may be expected when there is a large difference between the temperature of the sea and air. With an absence of wind, the air tends to form in layers. When this condition becomes extreme, mirage effects occur. Sometimes the rising or setting sun or moon appears distorted. Multiple horizons may appear, and other ships or islands may seem to float a short distance above the water. Under any such conditions large errors in refraction might be encountered.

Conditions causing abnormal refraction can be expected to occur with considerable frequency in the vicinity of the Grand Banks, along the west coast of Africa from Mogador to Cap Blanc and from the Congo to the Cape of Good Hope, in the Red Sea and the Persian Gulf, and over ice-free water in polar regions. Abnormal refraction may be encountered when offshore winds blow from high, snow-covered mountains to nearby tropical seas, as along the west coast of South America; where cold water from large rivers such as the Mississippi flows into warm sea water; when a strong current flows past a bay or coast, causing colder water to be drawn to the surface, as in the Bay of Rio de Janeiro and Santos, and along the Atlantic coast of Africa between Cape Palmas and Cape Three Points during the time of the southwest monsoon; and along the east coast of Africa in the vicinity of Capo Guardafui during the summer. In the temperate zones abnormal refraction is most common during the spring and summer.

Of the more systematic errors which affect refraction, two can be evaluated, and corrections applied. These are for air temperature (art. 1607) and atmospheric pressure (art. 1608). However, these corrections are based upon assumed standard *gradients* (changes) with height.

Since refraction causes celestial bodies to appear elevated in the sky, they are above the horizon longer than they otherwise would be. The mean diameter of the sun and moon are each about 32′, and horizontal refraction is 34!5. Therefore, the entire sun or moon is actually below the visible horizon when the lower limb appears tangent to the horizon. The effect of dip is to further increase the time above the horizon. Near the horizon the sun and moon appear flattened because of the rapid change of refraction with altitude, the lower limb being raised by refraction to a greater extent than the upper limb.

As a correction to sextant altitudes, refraction is negative because it causes the measured altitude to be too *great.* It *decreases* with increased altitude, and applies to all celestial bodies, regardless of sextant or horizon used.

1607. Air temperature correction (T).—The *Nautical Almanac* refraction table is based upon an air temperature of 50°F (10°C) at the surface of the earth. At other temperatures the refraction differs somewhat, becoming greater at lower temperatures, and less at higher temperatures. Table 23 provides the correction to be applied to the altitude to correct for this condition. If preferred, this correction can be applied with *reversed sign* to the refraction from the almanac, and a single refraction applied to the altitude. A combined correction for nonstandard air temperature and nonstandard atmospheric pressure (art. 1608) is given on page A4 of the *Nautical Almanac.* The correction for air temperature varies with the temperature of the air and the altitude of the celestial body, and applies to all celestial bodies, regardless of the method of observation. However, except for extreme temperatures or low altitudes, this correction is not usually applied unless results of unusual accuracy are desired.

1608. Atmospheric pressure correction (B).—The *Nautical Almanac* refraction table is based upon an atmospheric pressure of 29.83 inches of mercury (1010 millibars) at sea level. At other pressures the refraction differs, becoming greater as pressure increases, and smaller as it decreases. Table 24 provides the correction to be applied to the altitude for this condition. A combined correction for nonstandard air temperature (art. 1607) and nonstandard atmospheric pressure is given on page A4 of the *Nautical Almanac.* If the correction is to be applied to the refraction, reverse the sign. This correction varies with atmospheric pressure and altitude of the celestial body, and is applicable to all celestial bodies, regardless of the method of observation. However, except for extreme pressures or low altitudes, this correction is not usually applied unless results of unusual accuracy are desired.

1609. Irradiation correction (J).—When a bright surface is observed adjacent to a darker one, a physiological effect in the eye causes the brighter area to appear to be larger than is actually the case; conversely, the darker area appears smaller. This is called **irradiation.** Thus, since the sun is considerably brighter than the sky background, the sun appears larger than it really is; and when the sky is considerably brighter than the water, the horizon appears slightly depressed. The effects on the horizon and lower limb of the sun are in the same direction and tend to cancel each other while the effect on the upper limb of the sun is in the opposite direction to that on the horizon and tends to magnify the effect.

From 1958–1970 a correction of 1'.2 was included in the *Nautical Almanac* data for the upper limb of the sun as an average correction for the effect of irradiation. Recent investigations have not supported that average value and have revealed that the magnitude of the effect depends on the individual observer, the size of the ocular, the altitude of the sun, and other variables. In summary, the accuracy of observations of the limb of the sun at low altitudes may be affected systematically by irradiation, but the size of the correction is so dependent upon the variables enumerated above that it is not feasible to include an average correction in the tables.

1610. Semidiameter (SD) of a celestial body is half the angle, at the observer's eye, subtended by the visible disk of the body. The position of the lower or upper limb of the sun or moon with respect to the visible horizon can be judged with greater precision than that of the center of the body. For this reason it is customary, when using a marine sextant and the visible horizon, to observe one of the limbs of these two bodies, and apply a correction for semidiameter. Normally, the lower limb is used if it is visible. In the case of a gibbous or crescent moon, however, only the upper limb may be available. Semidiameter is shown in figure 1612.

The semidiameter of the sun varies from a little less than 15'.8 early in July, when the earth is at its greatest distance from the sun, to nearly 16'.3 early in January, when the earth is nearest the sun. In the *Nautical Almanac* the semidiameter of the sun at GMT 12^h on the middle day of each page opening of the daily page section is given to the nearest 0'.1 at the bottom of the sun's GHA column. The altitude correction tables of the sun, given on the inside front cover and facing page, are divided into two parts, to be used during different periods of the year. The mean semidiameter of each period is included in the tables of both upper and lower limb corrections. The semidiameter each day is listed to the nearest 0".01 in the *Astronomical Almanac.* In the *Air Almanac* the semidiameter to the nearest 0'.1 is given near the lower right-hand corner of each daily page.

The moon undergoes a similar change in semidiameter as its distance from the earth varies. However, because of the greater eccentricity of the moon's orbit than that of earth, the variation in semidiameter is also greater, varying between about 14'.7 and 16'.8. The variation is more rapid, partly because of the greater spread of values, but principally because the moon completes its revolution in approximately one month, while the earth makes one revolution per year. In the *Nautical Almanac,* semidiameter of the moon at 12^h each day is given to the nearest 0'.1 at the bottom of the moon data columns. The correction for semidiameter of the moon is included in the corrections given on the inside back cover and facing page. In the *Air Almanac,* semidiameter is given to the nearest whole minute, being shown on the daily pages, immediately below the value for the sun. The semidiameter at intervals of a day is given to the nearest 0.″01 in the *Astronomical Almanac.*

The navigational planets have small semidiameters. For Venus it varies between about 5″ and 32″; for Mars, 2.″7 to 12.″6; for Jupiter, 16″ to 25″; and for Saturn, 7″ to 10″. The value for any date is given in the *Astronomical Almanac,* but not in the *Nautical Almanac* or *Air Almanac* because the apparent *centers* of these bodies are customarily observed.

Stars have no measurable semidiameter.

The computed altitude of a body refers to the *center* of that body, since the coordinates listed in the almanacs are for the center. If the *lower* limb is observed, the sextant altitude is *less* than the altitude of the center of the body, and hence the correction is *positive.* If the *upper* limb is observed, the correction is *negative.* The correction does not apply when the center of the body is observed, which is usually the case when an artificial-horizon sextant is used. With a marine sextant and either the natural or an artificial horizon, semidiameter is customarily applied to observations of the sun and moon, but not other celestial bodies.

1611. Phase correction (F).—Because of phase (fig. 1612), the actual centers of planets and the moon may differ somewhat from the apparent centers. Average corrections for this difference are included in the additional corrections for Venus and Mars given on the inside front cover of the *Nautical Almanac.* They should be applied only when these bodies are observed during twilight. At other times the magnitude and even the sign of the correction might differ from those tabulated, because of a different relationship between the body and the horizon. The phase correction for navigational planets other than Venus and Mars is too small to be significant.

A phase correction may apply to observations of the moon if the apparent center of the body is observed, as with an artificial-horizon sextant. However, no provision is made for a correction in this case; the need for it can be avoided by observing one of the limbs of the body.

Phase correction does not apply to observations of the sun or stars.

1612. Augmentation (A).—As indicated in article 1610, semidiameter changes with distance of the celestial body from the observer, becoming greater as the distance decreases. The semidiameter given in the ephemeris and used in the almanacs is for a fictitious observer at the center of the earth. If the celestial body is on the actual observer's horizon, its distance is approximately the same as from the center of the earth; but if the body is in the zenith, its distance is less by about the radius of the earth (fig. 1612). Therefore, the semidiameter *increases* as the altitude becomes greater. This increase is called **augmentation.** For the moon, the augmentation from horizon to zenith is about 0.′3 at the mean distance of the moon. At perigee it is about 2″ greater, and at apogee about 2″ less. Augmentation of the sun from horizon to zenith is about 1/24 of 1 second of arc. For planets it is correspondingly small, varying with the positions of the planets and the earth in their orbits. At any altitude the augmentation is equal to the sine of the altitude times the value at the zenith.

Augmentation increases the size of the semidiameter correction, whether positive or negative. It is included in the moon correction tables on the inside back cover and facing page of the *Nautical Almanac.* It is not included in the correction tables of other bodies or in the *Air Almanac* tables.

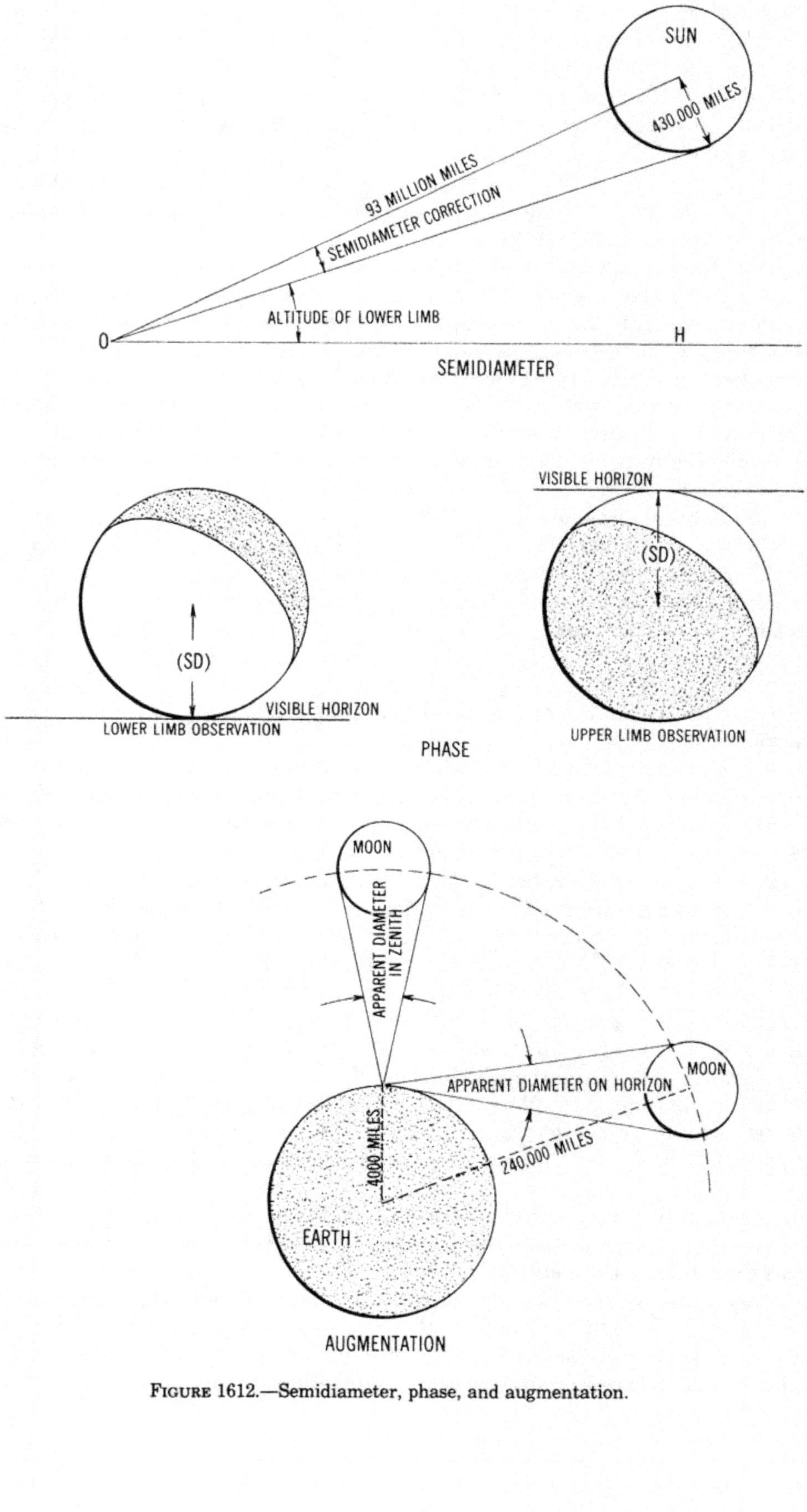

FIGURE 1612.—Semidiameter, phase, and augmentation.

1613. Parallax (P) is the difference in apparent position of a point as viewed from two different places. If a finger is held upright at arm's length and the right and left eyes closed alternately, the finger appears to move right and left a short distance. Similarly, if one of the nearer stars were observed from the earth and from the sun, it would appear to change slightly with respect to the background of more distant stars. This is called **heliocentric parallax** or **stellar parallax.** The nearest star has a parallax of less than 1″. Even if the value were greater, no correction to sextant altitudes would be needed, for the difference would be reflected in the tabulated position of the body.

However, positions of celestial bodies are given relative to the *center* of the earth, while observations are made from its surface. The difference in apparent position from these two points is called **geocentric parallax.** If a body is in the zenith, at *Z* in figure 1613, there is virtually no parallax, for the line from the body to the center of the earth passes approximately through the observer at *A*. Suppose, however, the moon is at *M*. From *A* it appears to be along the line *AM*, while at the center of the earth it would appear to be along *OM*. The altitude at *A* would be the angle *SAM*, and that at *O* the angle *COM*. Angle *COM* is equal to angle *SBM*, which is exterior to the triangle *ABM*, and hence equal to the sum of angles *SAM* and *AMO*.

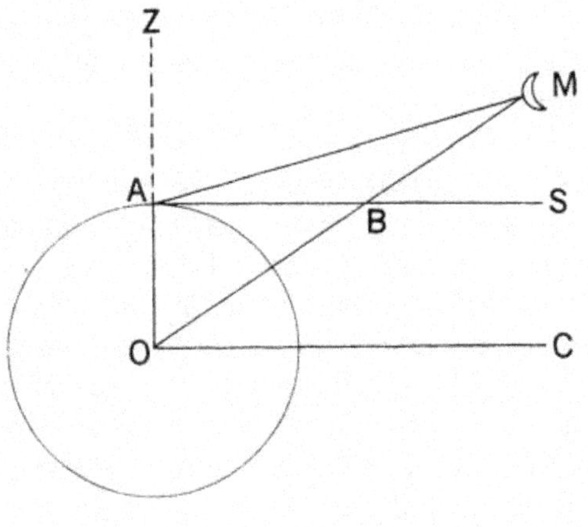

FIGURE 1613.—Geocentric parallax.

Since

$$\angle COM = \angle SBM = \angle SAM + \angle AMO,$$

then

$$\angle AMO = \angle COM - \angle SAM.$$

That is, the angle at the body between lines to the observer and the center of the earth is equal to the difference in altitude at the two places. Angle *AMO* is the geocentric parallax. Since it varies with altitude, it is sometimes called **parallax in altitude (P in A).** The maximum value for a visible body occurs when that body is on the horizon, at *S*. At this position the value is called **horizontal parallax (HP).**

The sine of horizontal parallax is equal to

$$\frac{r}{D},$$

where r is the radius of the earth, and D the distance of the body from the center of the earth. Thus, the sine of the horizontal parallax is directly proportional to the radius of the earth, and inversely proportional to the distance of the body. Since the earth is an oblate spheroid, and not a sphere, the parallax varies slightly over different parts of the earth. The value at the equator, called **equatorial horizontal parallax,** is greatest, and the value at the poles, called **polar horizontal parallax,** is least. The difference is not enough to be of practical navigational significance. The parallax in altitude is equal almost exactly to the horizontal parallax times the cosine of the altitude (h). That is,

$$\text{P in A} = \text{HP} \cos h.$$

The moon, being nearest the earth, has the greatest parallax of any celestial body used for navigation. The equatorial horizontal parallax at mean distance is 57′02″.70. As the distance of the moon varies, so does the parallax, becoming greater as the moon approaches closer to the earth, and less as it recedes, horizontal parallax varying several minutes each side of the value at mean distance. For the sun, mean equatorial horizontal parallax, called **solar parallax,** is 8″.794. Differences in position on the earth, and distance from the sun, have small effect, the maximum variation due to the latter being about 0″.15. Horizontal parallax of the planets varies considerably because of the large differences in their distances from the earth. For Venus the value varies between 5″ and 32″; for Mars, 3″ and 24″; for Jupiter, 1″ and 2″; and for Saturn, 0″.8 and 1″.0. The geocentric parallax of stars is too small to be measured, even by the most precise telescopes, since the value for the nearest star is only 0″.00003.

Daily values of horizontal parallax for the sun, moon, and planets are given in the *Astronomical Almanac,* to a precision of 0″.01. In the *Nautical Almanac,* mean values for the sun are included in the two sun correction tables given on the inside front cover and facing page. Horizontal parallax of the moon is tabulated at intervals of one hour on the daily pages. This value is used to enter the lower part of the moon correction tables on the inside back cover and facing page. The additional corrections for Venus and Mars given on the inside front cover are partly for parallax. No correction is given for parallax of Jupiter and Saturn. The *Air Almanac* gives parallax corrections only for the moon. These values are given in the "Moon's P in A" column on each daily page.

Because of the geocentric parallax, a body appears too *low* in the sky. Therefore, the correction is always positive. It applies regardless of the method of observation.

1614. Summary of corrections.—The essential information regarding the application of the various corrections may be tabulated as shown below. In the "Bodies" column, the symbols are: ☉, sun; ☾, moon; P, planets; ☆, stars. In the "Sextants" column, M refers to a marine sextant with visible horizon, A refers to a marine sextant with artificial horizon, and B refers to an artificial-horizon sextant. The tabulation assumes that completely accurate results are desired and that corrections are to be made in the usual manner, where they are available. Some of the entries need qualification or explanation which may be found in the preceding articles or chapter VIII of volume II.

These corrections can be considered to fall into five groups:

1. *Corrections for inaccuracies in reading.* Instrument correction, index correction*, and personal correction.

2. *Corrections for inaccuracies in reference level.* Dip* and sea-air temperature difference.

Correction	*Symbol*	*Sign*	*Increases with*	*Bodies*	*Sextants*	*Source*
Instrument	I	±	changing altitude	⊙, ☾, P, ☆	M, A, B	sextant box
Index	IC	±	constant	⊙, ☾, P, ☆	M, A, B	measurement
Personal	PC	±	constant	⊙, ☾, P, ☆	M, A, B	measurement
Dip	D	−	higher height of eye	⊙, ☾, P, ☆	M	almanacs
Sea-air temp. diff.	S	±	greater temp. diff.	⊙, ☾, P, ☆	M	computation
Refraction	R	−	lower altitude	⊙, ☾, P, ☆	M, A, B	almanacs
Air temp.	T	±	greater diff. from 50° F	⊙, ☾, P, ☆	M, A, B	almanacs, table 23
Atmospheric pressure	B	±	greater diff. from 29.83 inches of mercury	⊙, ☾, P, ☆	M, A, B	*Nautical Almanac,* table 24
Irradiation	J	−		⊙	M, A	
Semidiameter	SD	±	lesser dist. from earth	⊙, ☾	M, A	almanacs
Phase	F	±	phase	P	M, A, B	*Nautical Almanac*
Augmentation	A	±	higher altitude	☾	M, A	*Nautical Almanac*
Parallax	P	+	lower altitude	⊙, ☾, P	M, A, B	almanacs

3. *Corrections for bending of ray of light from body.* Refraction*, air temperature, atmospheric pressure.

4. *Adjustment to equivalent reading at center of body.* Irradiation, semidiameter*, phase, augmentation.

5. *Adjustment to equivalent reading at center of earth.* Parallax*.

In the ordinary practice of seamen, extreme accuracy is not required, and only the principal correction of each group is applied (except that augmentation is applied for the moon). These principal corrections are indicated by asterisks. For low altitudes, additional corrections are applied, as indicated in chapter VIII of volume II.

1615. Order of applying corrections.—For purposes of ordinary navigation, sextant altitudes can be applied in any order desired, using sextant altitude for the entering argument whenever altitude is required. This practice is not strictly accurate, but for altitudes usually observed, the error thus introduced is too small to be of practical significance. When extreme accuracy is desired, however, or at low altitudes, where small changes in altitude result in significant changes in correction, the order of applying corrections is important. Corrections from the first two groups of article 1614 are applied to sextant altitude (hs) to obtain **apparent (rectified) altitude (ha),** which is then used as an entering argument for obtaining corrections of the third group. For strictest accuracy, all corrections of the first three groups and, in addition, irradiation and semidiameter, should be applied before augmentation, and all other corrections before parallax.

1616. Marine sextant corrections.—Under normal conditions and when the highest accuracy is not required, it is necessary to apply only a few corrections. Several of these corrections may be combined within a single altitude correction table. In addition to corrections for index error, dip, and mean refraction, the normal altitude corrections when using the *Nautical Almanac* are: phase and parallax for Venus and Mars; semidiameter and parallax for the sun; and semidiameter, augmentation, and parallax for the moon.

1617. Artificial-horizon corrections.—When an artificial horizon is used, index correction (and any others of the first group of article 1614) is first applied. The result is then divided by two. Other corrections are then applied to the result, as applicable, in the same manner as for observations using the visible horizon. The sun and full moon are normally observed by bringing the lower limb of one image tangent to the upper limb of the other image. The lower limb is observed if the image seen in the horizon mirror is *above* the image seen in the artificial horizon, unless an inverting telescope is used, when the opposite relationship holds. With a gibbous or crescent moon, judgment may be needed to establish the positions of the limbs. In some cases better results may be obtained by superimposing one image over the other, as with a planet or star. When this is done, the *center* of the body has been observed, and no correction is applied for semidiameter (or irradiation, phase, or augmentation). There is no correction for dip (or sea-air temperature) when an artificial horizon is used.

1618. Artificial-horizon sextant corrections are the same as those for observations made by the use of the visible horizon, with two notable exceptions. First, there is no correction for dip (or sea-air temperature difference or wave height), none for semidiameter (or irradiation, phase, or augmentation), and usually none for index correction (or instrument correction). Second, because of the lower accuracy normally obtainable by artificial-horizon sextant, corrections are normally made only to the nearest whole minute of arc. As a result of these differences, refraction is the only correction normally applied, except in the case of the moon, where parallax is also applied.

1619. Corrections by *Nautical Almanac*.—In the *Nautical Almanac,* certain corrections or parts of corrections are combined. Index correction, of course, is not included because this depends upon adjustment of the sextant. The various correction tables are as follows:

"*Sun,*" on the inside front cover and facing page, gives mean refraction, mean semidiameter for each of two periods during the year, and mean solar parallax. The table on the inside front cover, and repeated on the loose bookmark, is of the critical type, with altitude as the entering value. Thus, a tabulated correction applies to any value of altitude between that given half a line above it and that half a line below it. If an exact tabulated altitude is used to enter the table, the correction half a line *above* it should be used. In ordinary navigation, index correction, dip, and the correction from this table are needed for correcting marine sextant observations of the sun. For low altitudes or extremes of temperature or atmospheric pressure, a correction from the table on almanac page A4 (or tables 23 and 24 of volume II) should be applied.

"*Stars and planets,*" on the inside front cover and repeated on the loose bookmark, gives mean refraction only, for the main tabulation. This is a critical type table, with altitude as the entering argument. The correction is always negative. In ordinary navigation, index correction, dip, and the correction from this table are the only ones needed for stars and the planets Jupiter and Saturn. For Venus and Mars, an additional correction for parallax and phase is given to the right of the main tabulation. The entering altitudes are limited to those occurring during twilight. If observations are made at other times, this additional correction should not be applied even though the altitude may fall within the tabulated range.

"*Dip,*" on the inside front cover and repeated on the loose bookmark, is for dip of the horizon. An abbreviated dip table is also given on the page facing the inside back cover. The tables are of the critical type, and the entering argument is the height of the observer's eye, in feet and meters, above the surface of the sea. The correction, always negative, applies to all observations made with the visible sea horizon as a reference.

"*Additional Correction Tables*" for *nonstandard conditions,* given on almanac page A4, provides an additional correction for nonstandard temperature and atmospheric pressure. The sign of each correction is indicated. Equivalent information is given, with increased range of entering values, in tables 23 and 24 of volume II.

"*Altitude Correction Tables—Moon,*" on the inside back cover and facing page, gives mean refraction, semidiameter, augmentation, and parallax. The entering argument is altitude for the upper portion of the table, and altitude and horizontal parallax for the lower portion. The combined correction is always positive, but 30′ is to be subtracted from the altitude of the upper limb. In ordinary navigation, index correction, dip, and the correction from this table are needed in correcting marine sextant observations of the moon.

The various separate corrections available from the *Nautical Almanac* can be found as follows:

Dip. Dip table on inside front cover and repeated on loose bookmark, and on the page facing the inside back cover.

Refraction. Mean refraction from "Stars and Planets" table on inside front cover and repeated on loose bookmark, and on the facing page.

Semidiameter. For the sun, the semidiameter for the middle day of each page opening of this daily page section is given at the bottom of the sun GHA column. For the moon, semidiameter for each day is given at the bottom of the moon data columns. The values given are for GMT 1200 on the dates indicated.

Parallax. For the sun, parallax in altitude can be considered 0′.1 for altitudes 0° to 70°07′, and 0′.0 for higher altitudes, with negligible error. This is based upon the mean value of 8″.794. For the moon, horizontal parallax each hour is tabulated on the daily pages. Parallax in altitude is this value multiplied by the cosine of the altitude.

If artificial-horizon sextant altitudes of the sun or moon are corrected by *Nautical Almanac,* the upper and lower limb corrections can be found and the average computed.

1620. Corrections by *Air Almanac.*—In the *Air Almanac,* various corrections as applicable to hand-held marine sextant observations are given separately in critical type tables, to the nearest whole minute (nearest two or five minutes of refraction for low altitudes), as follows:

Dip. Inside back cover.

Refraction. Near the back. Aboard ship use the values for zero height.

Air temperature. Near the back. This is shown, not as a separate correction, but as an adjustment to mean refraction. Instructions for use of the table are given within the table.

Semidiameter. For the sun and moon, on the A.M. and P.M. pages, below the moon's P in A. Values given are for GMT 1200.

Parallax. For the moon, in the P in A table on the A.M. and P.M. pages. Horizontal parallax is the value for 0° altitude.

1621. Correcting altitudes of the sun.—In the normal practice of navigation, sun observations obtained by marine sextant with the visible horizon as reference are corrected as shown in the following examples:

*Example 1.—*On June 2, 1975, the lower limb of the sun is observed with a marine sextant having an IC of (−)2′.0, from a height of eye of 38 feet. The hs is 51°28′.4.

*Required.—*Ho using (1) *Nautical Almanac,* and (2) *Air Almanac.*

Solution.—

(1)

	+	$\underline{\odot}$	−
IC			2′.0
D			6′.0
$\underline{\odot}$	15′.2		
sum	15′.2		8′.0
corr.		(+)7′.2	
hs		51°28′.4	
Ho		51°35′.6	

(2)

	+	$\underline{\odot}$	−
IC			2′
D			6′
R			1′
SD	16′		
sum	16′		9′
corr.		(+)7′	
hs		51°28′	
Ho		51°35′	

Example 2.—On June 2, 1975, the upper limb of the sun is observed with a marine sextant having an IC of (+)1′.0, from a height of eye of 45 feet. The hs is 32°47′.9.

Required.—Ho using (1) *Nautical Almanac,* and (2) *Air Almanac.*

Solution.—

(1)

	+	$\overline{\odot}$	−
IC	1′.0		
D			6′.5
$\overline{\odot}$			17′.3
sum	1′.0		23′.8
corr.		(−)22′.8	
hs		32°47′.9	
Ho		32°25′.1	

(2)

	+	$\overline{\odot}$	−
IC	1′		
D			7′
R			2′
SD			16′
sum	1′		25′
corr.		(−)24′	
hs		32°48′	
Ho		32°24′	

A convenient work form is helpful in the solution. Once the form is prepared, the corrections can be entered in any order desired. The symbols $\underline{\odot}$ and $\overline{\odot}$ are used for the corrections from the sun table on the inside front cover of the *Nautical Almanac.* If additional corrections are used, they are included in the same manner as those shown. Observations by artificial horizon and by artificial-horizon sextant, and low-altitude observations and back sights, are discussed elsewhere in this chapter.

1622. Correcting altitudes of the moon.—Moon observations by marine sextant with the visible horizon as reference are normally corrected as shown in the following examples:

Example 1.—At about GMT 1100 on June 2, 1975, the lower limb of the moon is observed with a marine sextant having an IC of (+) 3′.2, from a height of eye of 32 feet. The hs is 18°04′.6.

Required.—Ho using (1) *Nautical Almanac,* and (2) *Air Almanac.*

Solution.—

(1)

	+	☾	−
IC	3′.2		
D			5′.5
☾	62′.5		
L	0′.8		
sum	66′.5		5′.5
corr.		(+)1°01′.0	
hs		18°04′.6	
Ho		19°05′.6	

(2)

	+	☾	−
IC	3′		
D			6′
R			3′
SD	15′		
P	51′		
sum	69′		9′
corr.		(+)60′	
hs		18°05′	
Ho		19°05′	

Example 2.—At about GMT 0900 on June 2, 1975, the upper limb of the moon is observed with a marine sextant having an IC of (—)1′.6, from a height of eye of 70 feet. The hs is 66°47′.3.

Required.—Ho using (1) *Nautical Almanac,* and (2) *Air Almanac.*

Solution.—

(1)	+	☾̄	−
IC			1′.6
D			8′.1
☾	33′.1		
U	3′.2		
add'l			30′.0
sum	36′.3		39′.7
corr.			(−)3′.4
hs			66°47′.3
Ho			66°43′.9

(2)	+	☾̄	−
IC			2′
D			8′
R			–
SD			15′
P	21′		
sum	21′		25′
corr.			(−)4′
hs			66°47′
Ho			66°43′

The typical work forms shown are useful in problems of this type. The symbol ☾ is used for the correction from the upper part of the moon correction table on the inside back cover, and facing page, of the *Nautical Almanac.* The letters L and U are used for the corrections from the lower part of this table. Observations by artificial horizon, and by artificial-horizon sextant, and low-altitude observations and back sights, are discussed elsewhere in this chapter, as are additional corrections for use when unusual accuracy is desired.

1623. Correcting altitudes of planets.—When Venus and Mars are observed by marine sextant using the visible horizon as reference, sextant altitudes are normally corrected as shown in the following example:

Example.—On June 19, 1975, Venus is observed with a marine sextant having no IC, from a height of eye of 28 feet. The hs is 44°21′.3.

Required.—Ho using (1) *Nautical Almanac,* and (2) *Air Almanac.*

Solution.—

(1)	+	V	−
IC	–		–
D			5′.1
☆-P			1′.0
add'l	0′.3		
sum	0′.3		6′.1
corr.			(−)5′.8
hs			44°21′.3
Ho			44°15′.5

(2)	+	V	−
IC	–		–
D			5′
R			1′
sum	–		6′
corr.			(−)6′
hs			44°21′
Ho			44°15′

For Jupiter and Saturn, no additional correction is given. Correction of observations of these bodies is the same as corrections of star observations (art. 1624). Work forms are useful. The symbol ☆-P is used for the correction taken from the "Star-Planet" table on the inside front cover of the *Nautical Almanac.* If additional corrections are to be used, for results of unusual accuracy or low altitudes, they are included in the form in the same manner as those shown. Observations by artificial horizon and by artificial-horizon sextant, and low-altitude observations and back sights are discussed elsewhere in this chapter.

1624. Correcting altitudes of stars.—Star observations by marine sextant, using the visible horizon as reference, are normally corrected as shown in the following example:

Example.—Miaplacidus is observed with a marine sextant having an IC of (+)1′.0, from a height of eye of 50 feet. The hs is 27°54′.0.

Required.—Ho using (1) *Nautical Almanac,* and (2) *Air Almanac.*

Solution.—

(1)

	+	☆	–
IC	1′.0		
D			6′.9
☆-P			1′.8
sum	1′.0		8′.7
corr.			(–)7′.7
hs			27°54′.0
Ho			27°46′.3

(2)

	+	☆	–
IC	1′		
D			7′
R			2′
sum	1′		9′
corr.			(–)8′
hs			27°54′
Ho			27°46′

Work forms for such problems are helpful. Additional corrections, used when unusual accuracy is desired, are included in the same manner as those shown. Observations by artificial horizon and by artificial-horizon sextant, and low-altitude observations and back sights, are discussed elsewhere in this chapter.

1625. Low altitudes are normally avoided because of large and variable refraction. But sometimes these are the only observations available. This is particularly true in polar regions, where the sun may be the only celestial body available, and may not reach an altitude of more than a few degrees over a considerable period. In lower latitudes the sun may appear briefly just before sunset or just after sunrise. Low-altitude observations can supply useful information if additional corrections are applied. Reliable lines of position can generally be obtained from low-altitude observations, but when conditions are abnormal, the errors introduced are generally larger than for higher altitudes, and the precautions of article 806 of volume II should be particularly observed.

In correcting low-altitude observations, which for normal conditions can be defined as those less than 5°, first apply corrections from the first two groups of article 1614 to obtain apparent altitude (ha). Normally, this includes only index correction and dip. Then apply the remaining corrections, using apparent altitude when an altitude is needed for entering correction tables. The corrections normally applied are mean refraction, air temperature, atmospheric pressure, semidiameter (as applicable), and parallax (for the sun and moon).

In practice, sextant altitudes are corrected in the usual manner, except that additional corrections are applied, and the process is divided into two parts. The use of apparent altitude for finding parallax introduces an error but this is too small (less than 0′.1) for practical consideration. If the *Nautical Almanac* is used, corrections for altitudes between the horizon and 10° are given in a noncritical type table on almanac page A3. The correction for a negative altitude can be obtained by extrapolation without introducing a significant error for values obtained at ship heights of eye. A combined temperature-atmospheric pressure correction can be obtained from the table on almanac page A4. This table is intended for use without interpolation between columns. Separate corrections can be obtained from tables 23 and 24 of volume II, which provide interpolated values for greater accuracy. They also provide greater range of temperature and atmospheric pressure.

To correct a low altitude of the sun, then, apply index correction and dip to sextant altitude to find apparent altitude. Using this altitude as an entering value, find the following corrections and apply them to apparent altitude:

sun correction (⊙ or ⊙̄), from page A3 of the *Nautical Almanac;*

combined temperature-atmospheric pressure correction (TB), from page A4 of the *Nautical Almanac* (separate corrections for temperature (T) and atmospheric pressure (B) from tables 23 and 24, respectively, can be used *in place of* the combined correction).

If the *Air Almanac* is used, the mean refraction and air temperature corrections can be combined by using the factor in the refraction table. A semidiameter correction of 16′ is added if the lower limb is observed, and subtracted if the upper limb is observed. Since corrections are to whole minutes only, parallax is not used for the sun. In summary, apply index correction and dip to sextant altitude to find apparent altitude. Using this altitude as an entering value, where needed, apply the following corrections to apparent altitude:

refraction (adjusted for air temperature) (R), from table near back of *Air Almanac;*

atmospheric pressure (B), from table 24;

semidiameter (SD), 16′ (add if lower limb, and subtract if upper limb).

Example 1.—On June 2, 1975, the lower limb of the sun is observed with a marine sextant having an IC of (+)1′.8 from a height of eye of 45 feet. The hs is 1°24′.4, air temperature 88°F, and atmospheric pressure 29.78 inches.

Required.—Ho using (1) *Nautical Almanac,* (2) tables 23 and 24, and (3) *Air Almanac.*

Solution.—

(1)

	+ ⊙	−
IC	1′.8	
D		6′.5
sum	1′.8	6′.5
corr.		(−)4′.7
hs		1°24′.4
ha		1°19′.7
⊙		6′.0
TB	2′.5	
sum	2′.5	6′.0
corr.		(−)3′.5
ha		1°19′.7
Ho		1°16′.2

(2)

	+ ⊙	−
IC	1′.8	
D		6′.5
sum	1′.8	6′.5
corr.		(−)4′.7
hs		1°24′.4
ha		1°19′.7
⊙		6′.0
T	1′.5	
B	–	
sum	1′.5	6′.0
corr.		(−)4′.5
ha		1°19′.7
Ho		1°15′.2

(3)

	+ ⊙	−
IC	2′	
D		7′
sum	2′	7′
corr.		(−)5′
hs		1°24′
ha		1°19′
R		18′
B	–	
SD	16′	
sum	16′	18′
corr.		(−)2′
ha		1°19′
Ho		1°17′

The larger intervals given in the *Air Almanac* refraction table may introduce additional error. In this example, the temperature is changed to Celsius (centigrade), giving a value of 31°. The factor at a height of 0 feet corresponding to this temperature is 0.9. With this and the apparent altitude, the combined refraction and air temperature correction is found to be as shown. Approximately the same result would have been obtained by correcting for mean refraction (without the factor) and temperature (from table 23) separately.

If the moment at which either limb is tangent to the horizon is noted, an observation of 0° altitude has been made without a sextant.

Example 2.—On June 2, 1975, the sun is observed at sunset as the upper limb drops below the horizon, from a height of eye of 38 feet. The air temperature is (−)10°F, and atmospheric pressure 30.06 inches. Double extrapolation would be needed to solve this problem by the *Nautical Almanac*. A better solution is provided by means of tables 23 and 24.

Required.—Ho using (1) tables 23 and 24, and (2) *Air Almanac*.

Solution.—

(1)

	+	$\overline{\odot}$	−
IC	–		–
D			6′.0
sum	–		6′.0
corr.			(−)6′.0
hs			0°00′.0
ha			(−)0°06′.0
$\overline{\odot}$			51′.5
T			4′.8
B			0′.3
sum	–		56′.6
corr.			(−)56′.6
ha			(−)0°06′.0
Ho			(−)1°02′.6

(2)

	+	$\overline{\odot}$	−
IC	–		–
D			6′
sum			6′
corr.			(−)6′
hs			0°00′
ha			(−)0°06′
R			42′
B			–
SD			16′
sum	–		58′
corr.			(−)58′
ha			(−)0°06′
Ho			(−)1°04′

Corrections are applied algebraically. Therefore, for negative altitudes a negative correction is *numerically* added, and a positive correction is *numerically* subtracted.

To correct low altitudes of the moon, apply index correction and dip to sextant altitude to find apparent altitude. Using this altitude as an entering value, find the following corrections and apply them to apparent altitudes:

moon correction (☾), from inside back cover, and facing page, of *Nautical Almanac;*

lower or upper limb correction (L or U), from inside back cover, and facing page, of *Nautical Almanac;*

additional correction (add'l, (−)30′, for upper limb observation only);

combined temperature-atmospheric pressure correction (TB), from page A4 of the *Nautical Almanac* (separate corrections for temperature (T) and atmospheric pressure (B) from tables 23 and 24, respectively, can be used in place of the combined correction).

If the *Air Almanac* is used, correct the apparent altitude by applying the following corrections:

refraction (adjusted for air temperature) (R), from table near back of *Air Almanac;*

atmospheric pressure (B), from table 24;

semidiameter, from daily page;

parallax, from daily page.

Example 3.—At GMT 17ʰ14ᵐ27ˢ on June 2, 1975, the upper limb of the moon is observed with a marine sextant having no IC, from a height of eye of 33 feet. The hs is 2°35′.4, air temperature 63° F, and atmospheric pressure 29.81 inches.

Required.—Ho using (1) *Nautical Almanac,* (2) tables 23 and 24, and (3) *Air Almanac.*

Solution.—

(1)	+	$\overline{☾}$ −
IC	–	–
D		5′.6
sum	–	5′.6
corr.		(–)5′.6
hs		2°35′.4
ha		2°29′.8
☾	52′.1	
U	1′.1	
add'l		30′.0
TB	0′.4	
sum	53′.6	30′.0
corr.		(+)23′.6
ha		2°29′.8
Ho		2°53′.4

(2)	+	$\overline{☾}$ −
IC	–	–
D		5′.6
sum	–	5′.6
corr.		(–)5′.6
hs		2°35′.4
ha		2°29′.8
☾	52′.1	
U	1′.1	
add'l		30′.0
T	0′.4	
B	–	
sum	53′.6	30′.0
corr.		(+)23′.6
ha		2°29′.8
Ho		2°53′.4

(3)	+	$\overline{☾}$ −
IC	–	–
D		6′
sum	–	6′
corr.		(–)6′
hs		2°35′
ha		2°29′
R		16′
B	–	
SD		15′
P	54′	
sum	54′	31′
corr.		(+)23′
ha		2°29′
Ho		2°52′

A lower limb solution would be the same, except that an L correction would have been used from the *Nautical Almanac* and there would be no "add'l" correction, and in the *Air Almanac* solution the sign of the semidiameter correction would be reversed. The moon correction table on the inside back cover, and facing page, of the *Nautical Almanac* extends to a minimum altitude of 0°. The corrections for negative altitudes can be found by extrapolation.

To correct low altitudes of the planets Venus and Mars, apply index correction and dip to sextant altitude to find apparent altitude. Using this altitude as an entering value, find the following corrections and apply them to apparent altitude:

star-planet correction (☆-P), from page A3 of the *Nautical Almanac;*

additional correction (add'l), from page A2 of the *Nautical Almanac;*

combined temperature-atmospheric pressure correction (TB), from page A4 of the *Nautical Almanac* (separate corrections for temperature (T) and atmospheric pressure (B) from tables 23 and 24, respectively, can be used *in place of* the combined correction).

If the *Air Almanac* is used, correct the apparent altitude by applying the following corrections:

refraction (adjusted for air temperature) (R), from table near back of *Air Almanac;*

atmospheric pressure (B), from table 24.

Example 4.—On November 28, 1975, Mars is observed with a marine sextant having an IC of (+)3′.5, from a height of eye of 17 feet. The hs is 4°02′.6, air temperature 2°F, and atmospheric pressure 29.67 inches.

Required.—Ho using (1) *Nautical Almanac,* (2) tables 23 and 24, and (3) *Air Almanac.*

Solution.—

(1)	+	M	−
IC	3′.5		
D			4′.0
sum	3′.5		4′.0
corr.			(−)0′.5
hs			4°02′.6
ha			4°02′.1
☆-P			11′.7
add'l	0′.3		
TB			1′.5
sum	0′.3		13′.2
corr.			(−)12′.9
ha			4°02′.1
Ho			3°49′.2

(2)	+	M	−
IC	3′.5		
D			4′.0
sum	3′.5		4′.0
corr.			(−)0′.5
hs			4°02′.6
ha			4°02′.1
☆-P			11′.7
add'l	0′.3		
T			1′.2
B	0′.1		
sum	0′.4		12′.9
corr.			(−)12′.5
ha			4°02′.1
Ho			3°49′.9

(3)	+	M	−
IC	4′		
D			4′
sum	4′		4′
corr.			−
hs			4°03′
ha			4°03′
R			14′
B	−		
sum	−		14′
corr.			(−)14′
ha			4°03′
Ho			3°49′

The solution for Jupiter and Saturn, and for stars, is identical with that of example 4, except that the additional correction (phase and parallax) is omitted.

1626. Back sights.—An altitude measured by facing *away* from the celestial body being observed is called a **back sight.** It may be used when an obstruction, such as another vessel, obscures the horizon under the body; when that horizon is indistinct; or when observations are made in both directions, either to determine dip or to avoid error due to suspected abnormal dip. Such an observation is possible only when the arc of the sextant is sufficiently long to permit measurement of the angle, which is the supplement of the altitude. For such an observation of the sun or moon, the lower limb is observed when the image is brought below the horizon, appearing as a normal upper limb observation, and vice versa. To correct such an altitude, subtract it from 180° and reverse the sign of corrections of the first two groups of article 1614 (normally only index correction and dip).

Example.—On June 2, 1975, a back sight is taken of the lower limb of the sun, with a marine sextant having an IC of (−)2′.0, from a height of eye of 24 feet. The measured sextant altitude is 118°41′.4.

Required.—Ho using (1) *Nautical Almanac,* and (2) *Air Almanac.*

Solution.—

(1)	+	☉	−
IC	2′.0		
D	4′.8		
☉	15′.4		
sum	22′.2		−
corr.			(+)22′.2
180° − hs			61°18′.6
Ho			61°40′.8

(2)	+	☉	−
IC	2′		
D	5′		
R			1′
SD	16′		
sum	23′		1′
corr.			(+)22′
180° − hs			61°19′
Ho			61°41′

1627. Correcting horizontal angles.—When a marine sextant is used to measure the horizontal angle between two objects, the result is not usually desired to a precision that makes correction necessary, unless the sextant has an unusually large index error. However, if precise results are desired, corrections of the first group only of article 1614 are applied. If a personal error exists, it is not likely to be the same as for altitudes. For measuring angles between two objects differing widely in altitude, as between two stars, it is not likely that results will be required to such precision that additional correction for the third, fourth, and fifth groups of article 1614 will be needed. If they are, the method of application can be determined from the principles of spherical trigonometry. In this case, the altitudes of both bodies will also be needed. Corrections for the second group of article 1614 are not applicable.

Problems

1617a. At about GMT 0800 on June 2, 1975, the following bodies are observed with marine sextants having an IC of (+)2′.2, using an artificial horizon: sun (lower limb) hs 134°33′.9, moon (upper limb) hs 77°23′.4, Venus hs 98°04′.6, Schedar hs 43°24′.4.

Required.—Ho of each observation using (1) *Nautical Almanac,* and (2) *Air Almanac.*

Answers.—(1) Sun Ho 67°33′.6, moon Ho 39°09′.1, Venus Ho 49°02′.6, Schedar Ho 21°40′.9; (2) sun Ho 67°34′, moon Ho 39°09′, Venus Ho 49°03′, Schedar Ho 21°41′.

1617b. At about GMT 0300 on June 2, 1975, the following bodies are observed with bubble sextants having no IC: sun hs 23°51′, moon hs 52°20′, Jupiter hs 63°18′, Eltanin hs 24°45′.

Required.—Ho of each observation using (1) *Nautical Almanac,* and (2) *Air Almanac.*

Answers.—(1) and (2) Sun Ho 23°49′, moon Ho 52°52′, Jupiter Ho 63°18′, Eltanin Ho 24°43′.

1621a. On June 2, 1975, the lower limb of the sun is observed with a marine sextant having an IC of (+)1′.8, from a height of eye of 34 feet. The hs is 41°34′.8.

Required.—Ho using (1) *Nautical Almanac,* and (2) *Air Almanac.*

Answers.—(1) Ho 41°45′.8; (2) Ho 41°46′.

1621b. On June 2, 1975, the upper limb of the sun is observed with a marine sextant having no IC, from a height of eye of 30 feet. The hs is 15°21′.7.

Required.—Ho using (1) *Nautical Almanac,* and (2) *Air Almanac.*

Answers.—(1) Ho 14°57′.1; (2) Ho 14°57′.

1621c. On June 2, 1975, the lower limb of the sun is observed with a marine sextant having an IC of (−)1′.3, from a height of eye of 43 feet. Another ship is between the observer and the horizon, at a distance of 1.4 miles from the observer. The water line of this ship is used as the horizontal reference. The hs is 25°18′.2.

Required.—Ho using table 22 and (1) *Nautical Almanac,* and (2) *Air Almanac.*

Answers.—(1) Ho 25°12′.9; (2) Ho 25°13′.

1622a. At about GMT 2100 on June 2, 1975, the lower limb of the moon is observed with a marine sextant having an IC of (−)2′.5, from a height of eye of 55 feet. The hs is 47°35′.5.

Required.—Ho using (1) *Nautical Almanac,* and (2) *Air Almanac.*

Answers.—(1) Ho 48°16′.3; (2) Ho 48°16′.

1622b. At about GMT 2300 on June 2, 1975, the upper limb of the moon is observed with a marine sextant having an IC of (+)4′.0, from a height of eye of 12 feet. The hs is 22°58′.3.

Required.—Ho using (1) *Nautical Almanac,* and (2) *Air Almanac.*

Answers.—(1) Ho 23°31′.8; (2) Ho 23°32′.

1623a. On June 18, 1975, Mars is observed with a marine sextant having an IC of (+)2′.2, from a height of eye of 60 feet. The hs is 34°11′.7.

Required.—Ho using (1) *Nautical Almanac,* and (2) *Air Almanac.*

Answers.—(1) Ho 34°05′.1; (2) Ho 34°05′.

1623b. Jupiter is observed with a marine sextant having an IC of (−)1′.0, from a height of eye of 27 feet. The hs is 11°23′.9.

Required.—Ho using (1) *Nautical Almanac,* and (2) *Air Almanac.*

Answers.—(1) Ho 11°13′.2; (2) Ho 11°13′.

1624. Alpheratz is observed with a marine sextant having no IC, from a height of eye of 42 feet. The hs is 38°20′.3.

Required.—Ho using (1) *Nautical Almanac,* and (2) *Air Almanac.*

Answers.—(1) Ho 38°12′.8; (2) Ho 38°13′.

1625a. On June 2, 1975, the lower limb of the sun is observed with a marine sextant having an IC of (−)2′.3, from a height of eye of 24 feet. The hs is 2°04′.6, air temperature 65°F, and atmospheric pressure 30.81 inches.

Required.—Ho using (1) *Nautical Almanac,* (2) tables 23 and 24, and (3) *Air Almanac.*

Answers.—(1) Ho 1°55′.1; (2) Ho 1°55′.1; (3) Ho 1°55′.

1625b. On July 2, 1975, the sun is observed as the upper limb drops below the horizon at sunset, from a height of eye of 19 feet. The air temperature is 16°F, and atmospheric pressure 29.90 inches.

Required.—Ho using (1) *Nautical Almanac,* (2) tables 23 and 24, and (3) *Air Almanac.*

Answers.—(1) Ho (−)0°59′.9; (2) Ho (−)0°57′.8; (3) Ho (−)0°58′.

1625c. At GMT $6^h03^m29^s$ on June 2, 1975, the upper limb of the moon is observed with a marine sextant having an IC of (+)2′.6, from a height of eye of 35 feet. The hs is 1°12′.6, air temperature (−)23°F, and atmospheric pressure 29.04 inches.

Required.—Ho using (1) tables 23 and 24, and (2) *Air Almanac.*

Answers.—(1) Ho 1°22′.3; (2) Ho 1°20′.

1625d. At GMT $12^h44^m01^s$ on June 2, 1975, the lower limb of the moon is observed with a marine sextant having an IC of (+)3′.2, from a height of eye of 22 feet. The hs is 0°24′.4, air temperature 40°F, and atmospheric pressure 29.94 inches.

Required.—Ho using (1) *Nautical Almanac,* (2) tables 23 and 24, and (3) *Air Almanac.*

Answers.—(1) Ho 1°01′.2; (2) Ho 1°01′.5; (3) Ho 0°58′.

1625e. On January 19, 1975, Venus is observed with a marine sextant having an IC of (−)0′.5, from a height of eye of 31 feet. The hs is 3°29′.8, air temperature 55°F, and atmospheric pressure 30.15 inches.

Required.—Ho using (1) *Nautical Almanac,* (2) tables 23 and 24, and (3) *Air Almanac.*

Answers.—(1) Ho 3°10′.7; (2) Ho 3°10′.7; (3) Ho 3°10′.

1625f. Saturn is observed with a marine sextant having an IC of (−)2′.3, from a height of eye of 37 feet. The hs is 4°39′.2, air temperature 76°F, and atmospheric pressure 28.89 inches.

Required.—Ho using (1) *Nautical Almanac,* (2) tables 23 and 24, and (3) *Air Almanac.*

Answers.—(1) Ho 4°21.′1; (2) Ho 4°21.′1; (3) Ho 4°21′.

1625g. Gienah is observed with a marine sextant having no IC, from a height of eye of 44 feet. The hs is 2°46.′1, air temperature 35°F, and atmospheric pressure 29.92 inches.

Required.—Ho using (1) *Nautical Almanac,* (2) tables 23 and 24, and (3) *Air Almanac.*

Answers.—(1) Ho 2°23.′4; (2) Ho 2°36.′6; (3) Ho 2°21′.

1626. On June 2, 1975, a back sight is taken of the lower limb of the sun, with a marine sextant having an IC of (+)1.′7, from a height of eye of 49 feet. The measured sextant altitude is 141°04.′9.

Required.—Ho using (1) *Nautical Almanac,* and (2) *Air Almanac.*

Answers.—(1) Ho 39°15.′0; (2) Ho 39°15′.

1627. The horizontal angle between two objects is measured with a marine sextant having an IC of (+)4.′0. The measured angle is 85°14.′6.

Required.—Corrected angle.

Answer.—Corrected angle 85°18.′6.

CHAPTER XVII

LINES OF POSITION FROM CELESTIAL OBSERVATIONS

1701. Circles of equal altitude.—For every point on the earth there is a zenith (art. 1426) vertically overhead on the celestial sphere (art. 1403). Likewise, every point on the celestial sphere is vertically over *some* terrestrial point, called its **geographical position (GP).** However, since the earth rotates on its axis, causing *apparent* rotation of the celestial sphere, the GP of any point on the celestial sphere is continually moving to the westward, at the rate of about 15° per hour. If a celestial body is changing its apparent position on the celestial sphere, this motion is added to that caused by rotation, so that the rates of motion of the GP's of various bodies differ slightly. Further, this motion may not be exactly westward, having a small northerly or southerly component as the body changes declination, due either to its own proper motion or precession of the equinoxes (art. 1419), or a combination of the two.

At any moment the declination of a celestial body is equal to the latitude of its GP. The Greenwich hour angle (GHA) of the body, if not greater than 180°, is equal to the longitude (west) of the GP. If the GHA is greater than 180°, its explement (360° − GHA) is equal to the longitude (east). Thus, if it is established that a body of known coordinates is in the zenith of an observer, the position of the observer is known. However, for the celestial bodies used in navigation, this condition rarely occurs for any individual observer, and is difficult to determine when it does occur.

More commonly, the altitude (art. 1430) is measured, and from this the zenith distance (art. 1430) can be determined. This value defines a circle on the earth, as shown in figures 1701a and 1701b. Thus, if the observer is one mile from the GP, in *any* direction, he is 1′ from it, and his zenith is 1′ from the celestial body. Anywhere on a circle of one mile (1′) radius, with the GP as the center, the zenith distance is 1′. Similarly, if the zenith distance is 10°, the observer may be anywhere on a circle (assuming a spherical earth) of radius $10 \times 60 = 600$ miles, with the GP as the center. If the zenith distance is 30°, the radius is 1,800 miles; if 60°, the radius is 3,600 miles; and if 90° (body on the celestial horizon), the radius is 5,400 miles. This is a great circle dividing the earth into two hemispheres. Anywhere within that hemisphere having the GP as its center the celestial body is above the celestial horizon. Anywhere within the opposite hemisphere the body is below the celestial horizon.

These **circles of equal altitude** are **circles of position,** or circular **lines of position.** Two such circles for different celestial bodies, or for the same body at different times, may intersect at two points, as shown in figure 1701c. If these circles have radii equal to the zenith distances at the observer, the position of the observer is established at one of the two intersections. Normally, these intersections are separated by such great distances that no question arises as to which represents the position of the observer. However, uncertainty can be removed if additional altitude circles can be established by observation of other celestial bodies. It would be a rare coincidence for a third such circle to pass through both intersections of the first two. The third observation also serves as a check on the accuracy of the first two. The ambiguity might also be resolved by noting the azimuth of either or

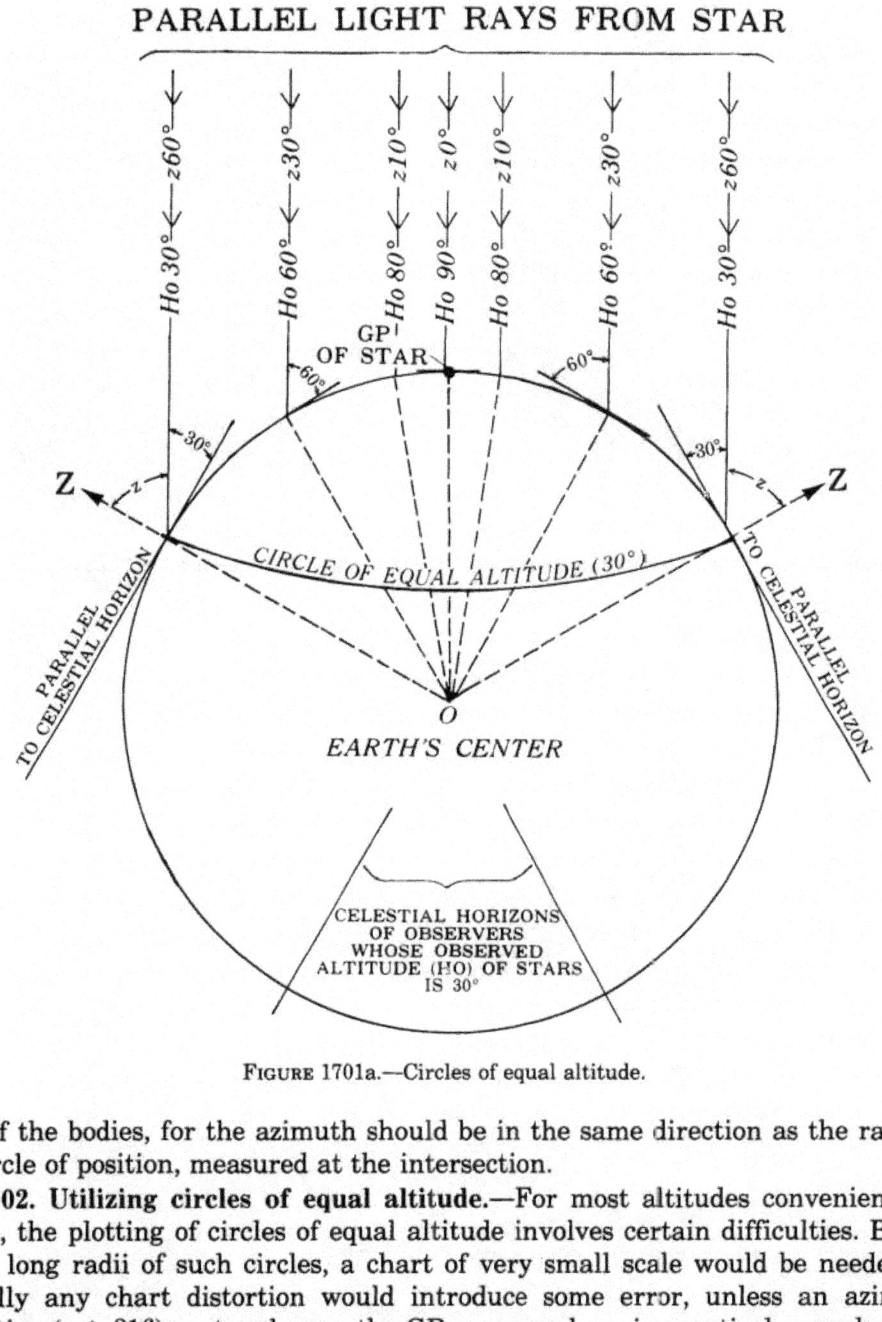

FIGURE 1701a.—Circles of equal altitude.

both of the bodies, for the azimuth should be in the same direction as the radius of the circle of position, measured at the intersection.

1702. Utilizing circles of equal altitude.—For most altitudes conveniently observed, the plotting of circles of equal altitude involves certain difficulties. Because of the long radii of such circles, a chart of very small scale would be needed, and virtually any chart distortion would introduce some error, unless an azimuthal projection (art. 316) centered upon the GP were used, an impractical procedure with a moving GP for each body. The appearance of two circles of equal altitude plotted on a Mercator chart is shown in figure 1702.

It has been suggested that the second difficulty, that of distortion, might be overcome by plotting directly on a sphere, using equipment designed for this purpose. While theoretically sound, this procedure does not overcome the first difficulty, that of scale, and has not proved practical. A variation of this has been the use of movable arcs, by which a small-scale model of one or more navigational triangles (art. 1433) is mechanically produced. The coordinates are carefully measured by means of sliding indices controlled by verniers or micrometers. Another variation

FIGURE 1701b.—Circles of equal altitude.

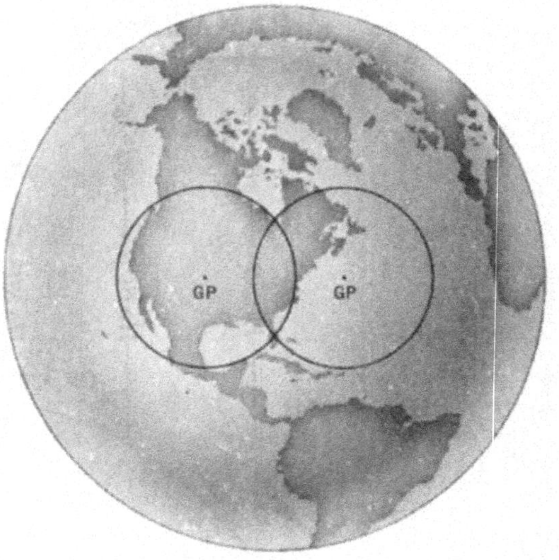

FIGURE 1701c.—Intersections of two circles of equal altitude.

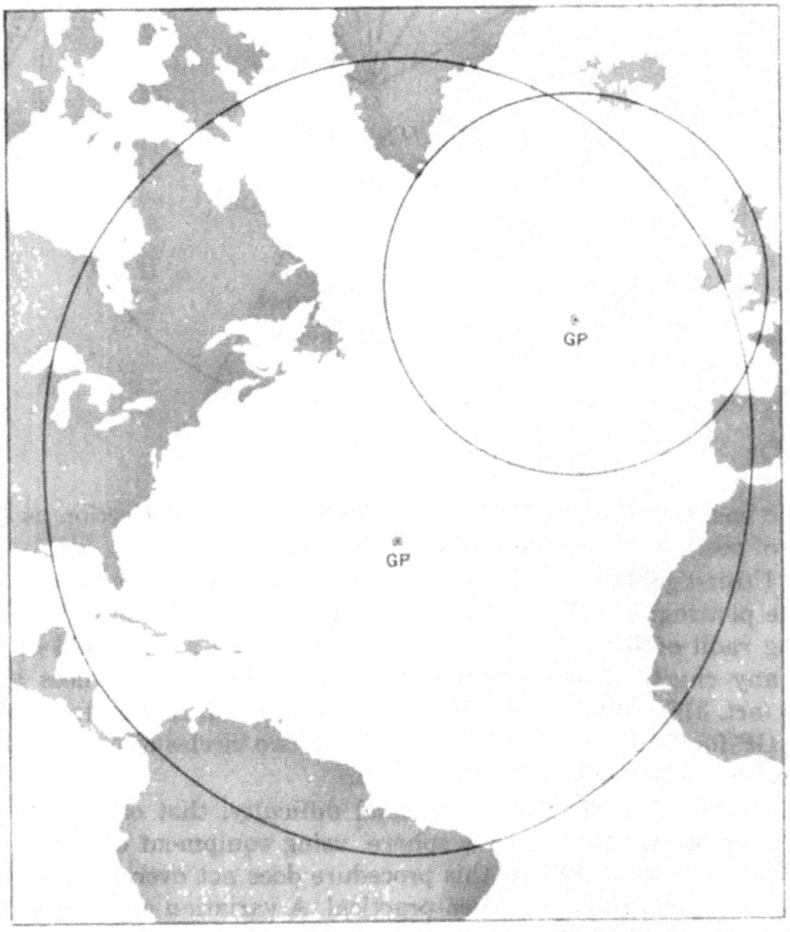

FIGURE 1702.—Circles of equal altitude on a Mercator chart.

has been a graphical solution based upon the drawing of a diagram according to any of various principles. Although a number of mechanical and graphical solutions have been devised, and some have proved practical (ch. XXI), none has been generally accepted as superior to the commonly used tabular methods of solution.

However, as the altitude of a body increases, reducing the zenith distance, both distortion and scale difficulties decrease. Also, on a Mercator chart, they decrease as the GP approaches the equator. The observation of a celestial body near the zenith is difficult, but in the case of the sun no alternative may be available near noon in the Tropics. Such a situation does provide an easy solution and may permit obtaining of a fix from two observations of the same body, with only a few minutes between observations. This solution is discussed further in article 2011.

1703. The line of position.—For zenith distances too great to plot conveniently, a line of position can be laid down in another manner.

The altitude of a celestial body may be measured. After appropriate corrections are applied, this is called **observed altitude (Ho).** For the instant of observation, the altitude and azimuth at some convenient **assumed position (AP)** near the actual position of the observer are determined by calculation or equivalent process. The difference between this **computed altitude (Hc)** and Ho is the **altitude intercept** ***(a)***, sometimes called **altitude difference.** Since a is the difference in altitude at the assumed and actual positions, it is also the difference in zenith distance, and therefore the difference in radii of the circles of equal altitude at the two places. The position having the greater altitude is on the circle of smaller radius, and hence is closer to the GP of the body. In figure 1703a the AP is shown on the inner circle. Hence, Hc is greater than Ho.

The line of position can be plotted by using the altitude intercept portion of the information of figure 1703a, as shown in figure 1703b. First, the AP is plotted. The circle of equal altitude through this position is not needed, and is not plotted. From the AP the azimuth line is measured toward or away from the GP as appropriate, and the altitude intercept is measured along this line. At the point thus located, a line is drawn perpendicular to the azimuth line. For several miles on each side of the azimuth line, this perpendicular can be considered part of the circle of position through the observer, as shown in figure 1703a. This perpendicular is the line of position. It is labeled with the time of observation above the line, and the name of the celestial body below the line, as shown in figure 1703b.

For neatness of plot the azimuth line should not be extended beyond the line of position for the AP, unless it is extended a short distance in the direction of the body, and the symbol of the body observed is shown to indicate whether a "toward" or "away" observation. This method is used in the examples of Pub. No. 229. Some navigators may omit the azimuth line, showing only the AP and line of position, and using a straightedge as a guide for the dividers in measuring the altitude difference. This is good practice, for it reduces the number of lines on the plotting sheet, and therefore minimizes the possibility of making an error. However, until one gains confidence in plotting lines of position, it is desirable to show the azimuth line.

For plotting a line of position from a celestial observation, then, only the assumed position, altitude intercept (with an indication of which altitude is greater), and azimuth are needed.

The assumed position is chosen somewhat arbitrarily. It may be the dead reckoning position, an estimated position, or any arbitrarily chosen position nearby. Most commonly, however, the **assumed latitude (αL)** is taken as the *nearest* whole

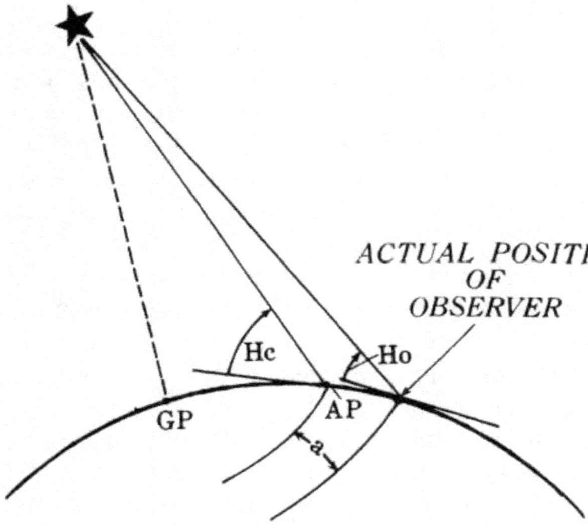

FIGURE 1703a.—The basis for the line of position from a celestial observation.

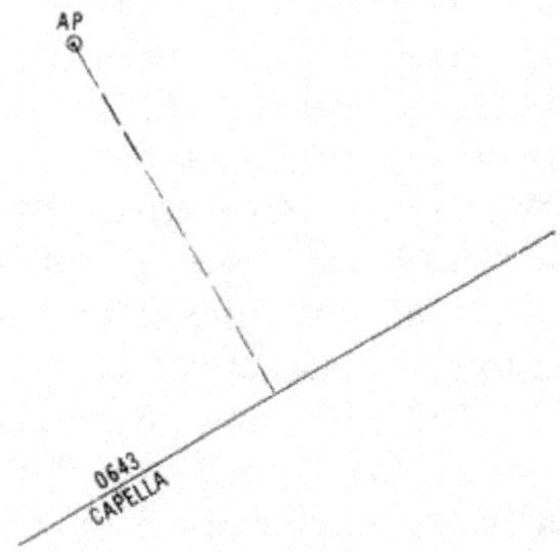

FIGURE 1703b.—A line of position from observation of the star Capella at 0643.

degree of latitude to the DR or EP; and the **assumed longitude ($a\lambda$)** is selected so that the local hour angle is a whole degree. The location of the line of position is independent of the location of the AP (within reasonable limits), assuming only that the altitude intercept is measured from the AP used for determining Hc. That is, each AP has its own altitude intercept, depending upon its distance from the line of position.

The altitude intercept, the numerical difference between Hc and Ho, is customarily expressed in nautical miles (minutes of arc), and labeled **T** or **A** to indicate whether the line of position is **toward** or **away** from the GP, as measured from the AP:

Hc	37°51′.6		Hc	61°57′.3
Ho	37°43′.9		Ho	62°12′.7
a	7.7A		*a*	15.4T

The azimuth is customarily determined by computation or table at the time of determining Hc.

This method of plotting a line of position from a celestial observation was suggested by Marcq St.-Hilaire (art. 2108), and generally bears his name. It is used almost universally by modern navigators. The method is based upon knowledge of one point on the line, and the direction of the line. Another method of utilizing the same principle is to assume the latitude and compute the longitude at which the line of position crosses that parallel (the time sight method, art. 2106), or vice versa. When this method is used, the azimuth is customarily found separately, from a table or graph. A third method is to compute two points on the line of position and draw a straight line through them. This line is a chord, rather than a tangent, of the circle of position, but in most cases the difference is negligible. This third method was that originally proposed by Captain Thomas H. Sumner (art. 131), and for this reason the resulting line of position is sometimes called a **Sumner line,** although the expression may be applied to any line of position resulting from celestial observation.

When celestial navigation is used, plotting is generally done on plotting sheets (art. 323) published by the Defense Mapping Agency Hydrographic/Topographic

Center. These are less expensive than charts, and the absence of detail eliminates a possible source of confusion and error.

1704. Using lines of position from celestial observations.—Like any other line of position, one resulting from a celestial observation does not pinpoint the position of the craft, but may provide all the information needed to insure safety of the vessel. The selection of a celestial body and the time of observation to provide the desired information is based upon the fact that the line of position is perpendicular to the azimuth line. If the celestial body is on or near the celestial meridian, the line of position is a **latitude line,** indicating the latitude at the time of observation, sometimes called the **observed latitude.** Similarly, a body on or near the prime vertical provides a **longitude line,** indicating the **observed longitude.** One ahead or astern provides a **speed line,** since the line of position is perpendicular to the course, and hence is an indication of the speed made good since the last speed line or fix. Similarly, a body on the beam provides a **course line** which indicates to what extent the course is being made good. If the azimuth line is perpendicular to a coastline, shoal, or other hazard, the line of position indicates the distance of the vessel from the danger. Passage parallel to such a danger, or between two of them, might be made safely by means of a series of observations of a body on the beam during passage, without fixing the position of the vessel. This problem might be simplified by precomputing the sextant altitude at intervals during passage, and plotting this versus time on cross-section paper, so that sextant altitudes can be compared immediately with the values taken from the curve to determine any deviation from the desired track. In a perpendicular approach to a coast, the point at which landfall will be made can be predicted with considerable accuracy if a body having an azimuth parallel to the beach is observed.

During twilight, with clear skies, the selection of a celestial body to provide desired information is simply a matter of choosing the body with azimuth nearest that desired, remembering that bodies having azimuths differing by 180° should provide the same line of position. Observation of bodies in opposite directions provides a check, and a better one than two observations of the same body, or observations of two bodies having nearly the same azimuth, for any constant error in the observations, such as might be caused by abnormal dip, can be eliminated by observing bodies on opposite azimuths and using a line midway between the two plotted lines of position.

When a limited number of bodies is available for a considerable period, as during daylight, the best time to make an observation to obtain a line of position in a desired direction can be determined by means of an azimuth table or diagram, or an inspection table such as Pub. No. 229. The azimuth is located, and the corresponding local hour angle is recorded. The local hour angle can then be converted to GHA, and the time at which this GHA occurs can be determined from the almanac (art. 1904).

Lines of position can be used for determining an estimated position (art. 1705), or they can be advanced or retired (art. 1706) to obtain a fix (art. 1707) or running fix (art. 1708). If a single body is available for observation, increased accuracy can usually be obtained by making three or more observations, adjusting all lines to a common time (art. 1706), and using either the middle line, or the average position of all lines.

1705. Estimated position.—As indicated in chapter VIII, a **dead reckoning (DR) position** is determined by advancing a known position for courses and distances. In the absence of additional information, the DR position is the best estimate of the position of the vessel. However, the expression **estimated position (EP)** is generally applied to one determined by using additional but inconclusive information. If the effects of wind and current can be estimated, and these effects have not been considered in establishing the DR position, they can be applied separately to establish an EP. As each additional item of information is received, an improved estimate might be made.

A single line of position can be useful in establishing an estimated position. If an accurate line is obtained, the actual position is somewhere on this line. In the absence of better information, a perpendicular from the previous DR position or EP to the line of position establishes the new EP, as shown in figure 1705a. The foot of the perpendicular from the AP has no significance in this regard, since it is used only to locate the line of position.

The establishment of a good EP is dependent upon accurate interpretation of all information available. Generally, such ability can be acquired only by experience. If, in the judgment of the navigator or captain, the course has been made good, but the speed has been uncertain, the best estimate of the position might be at the intersection of the course line and the line of position, as shown in figure 1705b. If the speed since the last fix is considered accurate, but the course is considered uncertain, the EP might be at the intersection of the line of position and an arc centered on the previous fix and of radius equal to distance traveled, as shown in figure 1705c.

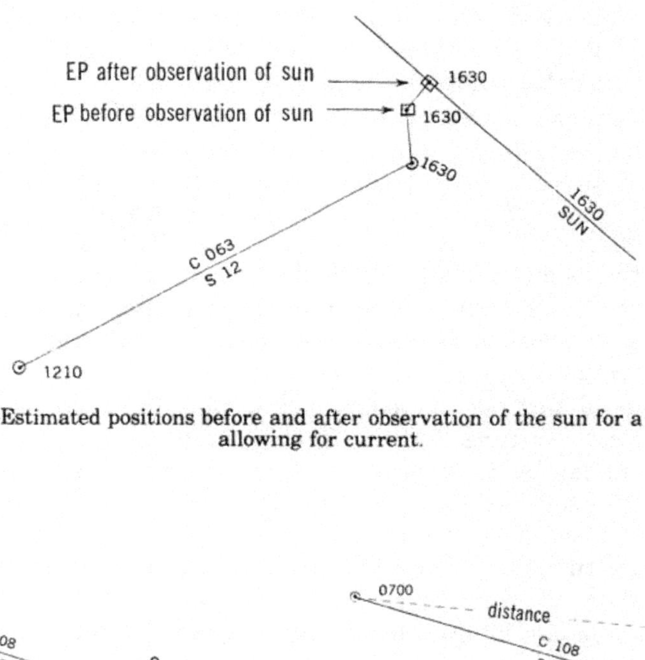

FIGURE 1705a.—Estimated positions before and after observation of the sun for a line of position, allowing for current.

FIGURE 1705b.—An estimated position when the course and a line of position are considered accurate.

FIGURE 1705c.—An estimated position when the speed and a line of position are considered accurate.

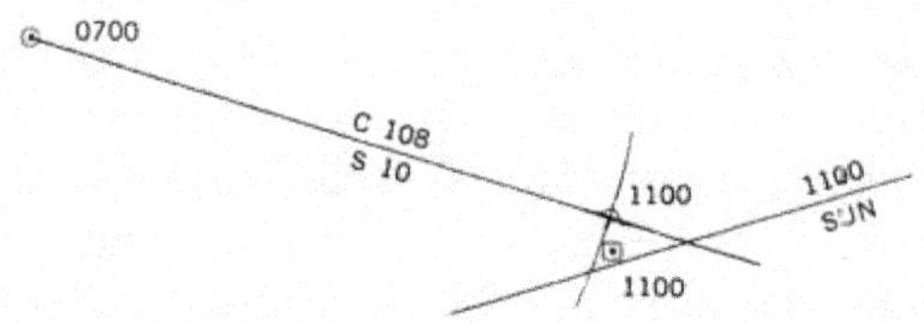

FIGURE 1705d.—An estimated position when a line of position is considered of first accuracy, speed of second accuracy, and course of third accuracy.

More often, neither course nor speed is known to be entirely accurate, but if one is considered more accurate than the other, the EP may be located accordingly. Even the line of position might properly be considered of questionable accuracy, and some estimate of its reliability established. Figure 1705d shows an EP that might be established by considering the line of position of greatest but incomplete accuracy, the speed of secondary accuracy, and the course as least accurate.

The expression **most probable position (MPP)** is sometimes used as the equivalent of *estimated position.* However, the former is of somewhat broader application, since it may apply equally well to establishment of the fix when more than two lines of position are available.

1706. Advancing and retiring lines of position.—For a stationary observer, lines of position resulting from observations made at different times are equally applicable without adjustment. However, for a moving observer, as one aboard a vessel underway at sea, any line of position (except a course line) applies only to the position at the time of observation. If lines resulting from observations made at different times are to be utilized for determining position, they should be adjusted for the motion of the observer between observations.

A line of position resulting from observation of a celestial body can be advanced or retired in the same manner as other lines of position (art. 1010), by selecting any point associated with the line of position and running it forward or backward by dead reckoning, or by estimate. For most accurate results, the best estimate of course and speed made good (over the bottom) between the time of observation and the time to which the line is to be adjusted should be used. Any error in determining these values is reflected in the adjusted line of position. However, error in speed does not affect the accuracy of an adjusted course line, nor does error in course introduce an appreciable error in the accuracy of an adjusted speed line. The time label of an adjusted line of position includes both the time of observation and the time to which the line is adjusted.

As in the case of a line of position resulting from observation of the bearing of an identifiable, charted object (art. 1004), the number of lines on the chart can be kept to a minimum, reducing the possibility of confusion, by adjusting the point from which the line is drawn. In the case of celestial navigation, this is the assumed position. This method applies equally well to all observations, and avoids some possible difficulty which might arise in advancing a line of position nearly parallel to the course line. When the AP is advanced or retired, the initial line of position need not be drawn unless it serves some useful purpose.

1707. The fix.—The common intersection of two or more lines of position constitutes a fix, regardless of the source of the position lines, provided only that the lines are based upon simultaneous observations. Celestial observations are seldom simultaneous because all sights of a group are customarily taken by a single observer, usually the navigator. If observations are made a few minutes apart (a **round of sights**), as during a twilight period, all lines are adjusted to a common time, and the position is considered a fix, rather than a running fix. Many navigators advance earlier lines to the time of the last observation, and consider the fix applicable at this time, as shown in figure 1707a. An alternative procedure, which is gaining in acceptance, is to *advance* earlier sights and *retire* later ones to an intermediate time, either the time of the mid observation or a convenient time during the period of observation, such as a whole, half, or quarter hour. This results in a more accurate and convenient time of the fix. In figure 1707b the lines of figure 1707a are adjusted to a common time at a whole hour. With any procedure, the time of the fix is the common time to which the lines of position are adjusted.

In figures 1707a and 1707b the assumed positions are typical of those which might be used with a modern method of sight reduction such as Pub. No. 229 (ch. XX). *Any* position in the vicinity might be used. If the dead reckoning (or estimated)

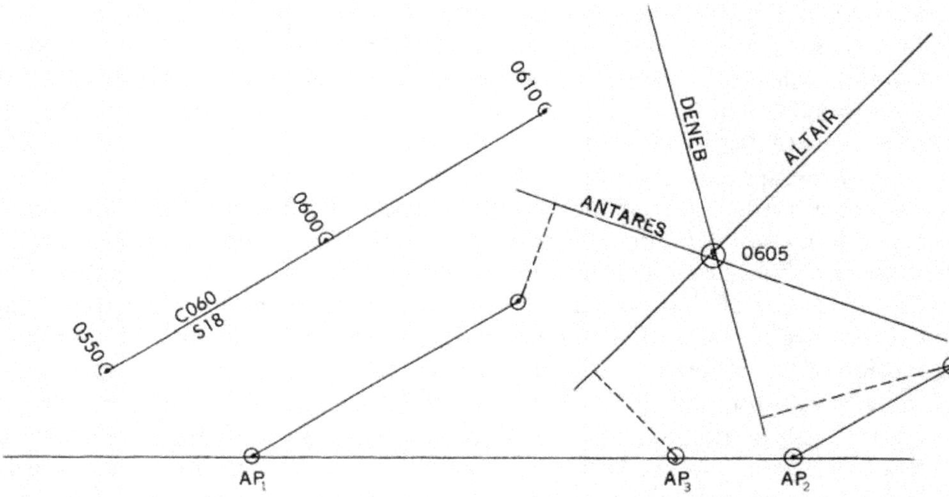

FIGURE 1707a.—A fix obtained by advancing earlier lines of position to the time of the last observation.

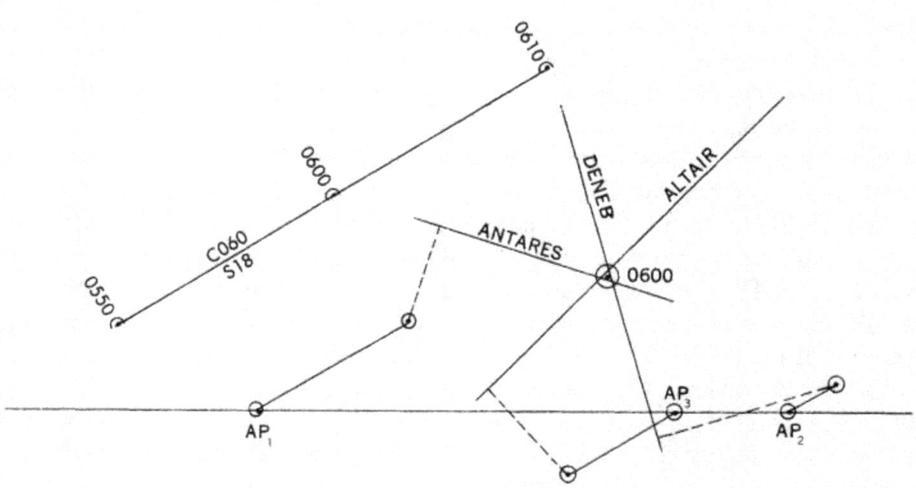

FIGURE 1707b.—A fix obtained by adjusting the lines of position of figure 1707a to a convenient time during the period of observation.

position at the time of each observation is used as the assumed position for that sight, all sights are plotted from the DR position (or EP) at the time for which the fix is desired. If the same AP is used for all sights, the advanced or retired AP's are along a straight line extending in the direction of the course line, the AP corresponding to the earliest observation being farthest advanced along this line, and others progressing along it in a direction *opposite* to that of the course. If there is any change of course or speed between observations, this should be considered in advancing or retiring a line of position, as it would in running forward the dead reckoning. Under normal conditions, lines of position adjusted for a short interval to obtain a fix are moved by dead reckoning, without separate allowance for current.

Two lines of position provide a fix, but when additional celestial bodies are available, it is good practice to observe them. Additional lines serve as a check on the accuracy of the first two, and should decrease the error of the fix. However, the increased accuracy of a fix resulting from a number of lines of position, over that resulting from only two, is not great under normal conditions, and the principal reason for the additional observations is the increased confidence the navigator has in the reliability of his fix.

In selecting bodies for observation, one should generally consider azimuth primarily, and such factors as brightness, altitude, etc., secondarily. Individual circumstances, however, may dictate departures from this procedure. During twilight, when skies are clear and the entire horizon is good, one generally has ample choice of bodies to observe. It is good practice to make several more observations than the minimum considered acceptable, so that additional lines of position will be available, if needed, to resolve possible ambiguities or confirm doubtful results.

Sights need not be solved in the order taken. During evening twilight the brightest bodies should be observed first, as soon as they can be "brought down" successfully to the horizon. During morning twilight the reverse is true, the dimmer stars being observed while they are still visible. However, with advance planning, one can include in the list of bodies to be observed those which should provide the best fix.

If all observations were precisely correct, in every detail, the resulting lines of position would meet at a point. However, this is rarely the case. Three observations generally result in lines of position forming a triangle. If this triangle is not more than two or three miles on a side under good conditions, and five to ten miles under unfavorable conditions, there is normally no reason to suppose that a mistake has been made. Even a point fix, however, is not *necessarily* accurate. An uncorrected error in time, for instance, would move the entire fix eastward if early and westward if late, at the rate of 1′ of longitude for each 4^s of error in time.

With two or four observations, the ideal is to have them crossing at angles of 90°. With three observations, the ideal is angles of 60°. With three observations it is good practice to observe bodies differing in azimuth by 120°, as nearly as possible. This provides lines of position crossing at angles of 60°, and, in addition, any constant error in altitude is eliminated, serving only to increase or decrease the size of the triangle, but not affecting the position of its center. If the azimuths differ by 60° (or the azimuth spread is less than 180°), a large constant error in altitude would result in a fix *outside* the triangle, as shown in figure 1707c. With lines of position crossing at 60°, the assumed constant error for a fix outside the triangle is three times that for a fix inside the triangle. With four bodies, azimuths differing

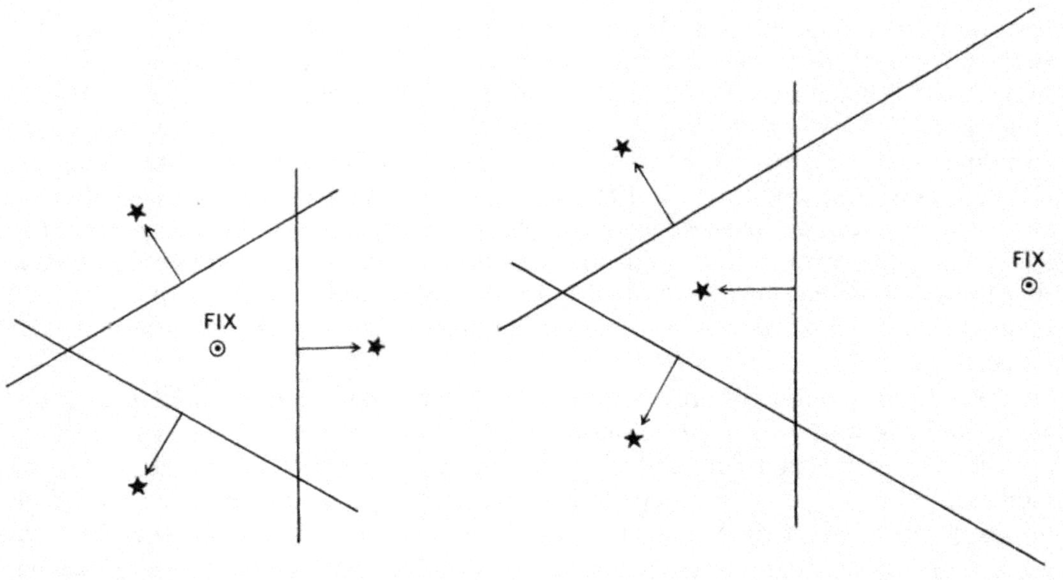

FIGURE 1707c.—A fix from three lines of position, assuming a constant error in altitude. If all lines are moved away (in this case) from the bodies observed, they would meet in a point which might be either inside (left) or outside (right) the triangle.

by 90° produce a box fix, with constant error eliminated by using the mid point as the fix. With more than four observations, the selection of the fix becomes more complex, and general rules are probably undesirable. The evaluation of each observation and the exercise of judgment become of greater importance. Whatever the number of observations, common practice, backed by logic, is to take the center of the figure formed unless there is reason for deviating from this procedure. By "center" is meant the point representing the least total error of all lines considered reliable. With three lines of position, the center is considered that point, within the triangle, which is equidistant from the three sides. It may be found by bisecting the angles, but more commonly it is located by eye. If a fix outside the triangle is to be used, and eye interpolation is not considered sufficiently reliable, the point can be found by bisecting two external angles and the internal angle at the third intersection. If a constant error is assumed, the most probable position of the fix can always be found, whether within or outside the triangle, by bisecting the angle formed by azimuth lines originating at each intersection.

1708. A running fix (R FIX), in celestial navigation, is a position obtained by observations separated by a considerable time interval, usually several hours. The usual occasion for a running fix is the availability of a single celestial body for observation, generally the sun. The delay between observations is usually to permit the azimuth to change sufficiently to provide a good angle of cut between lines of position. Thus, the sun may be observed about 0900, and again about noon.

Generally, a longer wait results in a more nearly perpendicular intersection of the two lines of position, but it may also increase the error of the advanced line. The earlier line is advanced for the course and distance made good. The ability with which these can be predicted determines the accuracy of the running fix, assuming accurate observation, sight reduction, and plotting. For this reason it is impractical to set a specific time limit upon the advancement of a line of position. This should be determined by the conditions of each situation, in the best judgment of the navigator. Experience is valuable in acquiring such judgment.

When an observation of a single body is made, with the intent of later advancing it to obtain a running fix with a second observation, the line of position should be plotted for the time of observation, regardless of the method used for advancing it, for the single line usually provides some useful information, as indicated in article 1704.

Allowance for current, when advancing a line of position, can be made by solving a vector diagram, as indicated in article 807, to determine the course and speed made good. An alternative method is to advance the AP or line without allowance for current, and then to advance it a second time in the direction of set of the current, for a distance equal to the drift multiplied by the number of hours between the time of observation and the time to which the line is advanced. This method is illustrated in figure 1708a. The distance *AB* is equal to the distance between the 0800 and 1152 DR positions. The direction *BC* is the estimated set of the current, and the length *BC* is the distance through which the current is assumed to act.

A third method provides accurate results even when a reliable estimate of the current is not available, provided (1) a good fix was obtained several hours before the time of observation, and (2) the average current between the time of the previous fix and the time of observation can be assumed to continue until the time to which the line is to be advanced. This method is illustrated in figure 1708b. The 0510 fix is shown at the left, and the DR positions at 0830 and 1215, the ship being on course 074°, speed 12 knots. The sun is observed at 0830 and again at 1215, and it is desired to advance the earlier line to obtain a running fix at 1215. The lines of position at 0830 and 1215 are plotted. To advance the 0830 line of position, the distance *AB* is assumed to increase uniformly with time interval from 0510. The interval to 0830 is 3^h20^m, and that to 1215 is 7^h05^m. Therefore,

$$A'B' = AB \times \frac{7^h05^m}{3^h20^m} = AB \times 2.1.$$

The advanced line of position is drawn through *B'*, parallel to the original line through *B*. The running fix is at the intersection of the 1215 line and the advanced 0830 line.

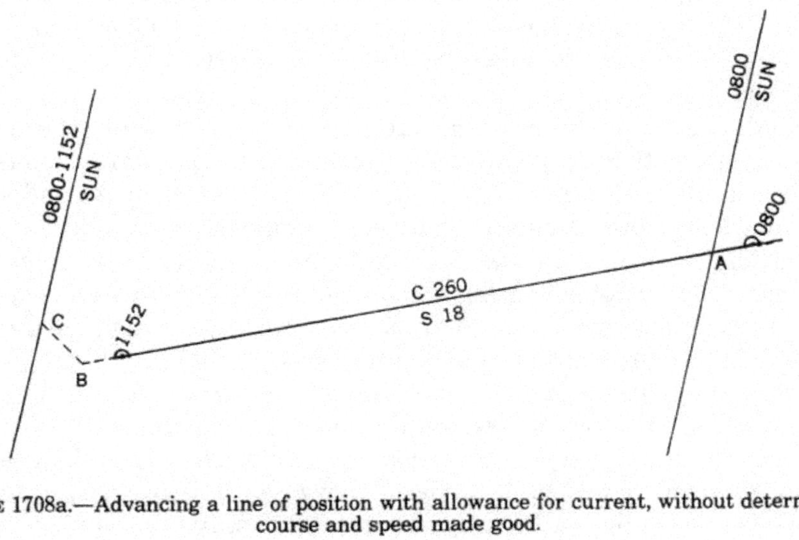

FIGURE 1708a.—Advancing a line of position with allowance for current, without determining course and speed made good.

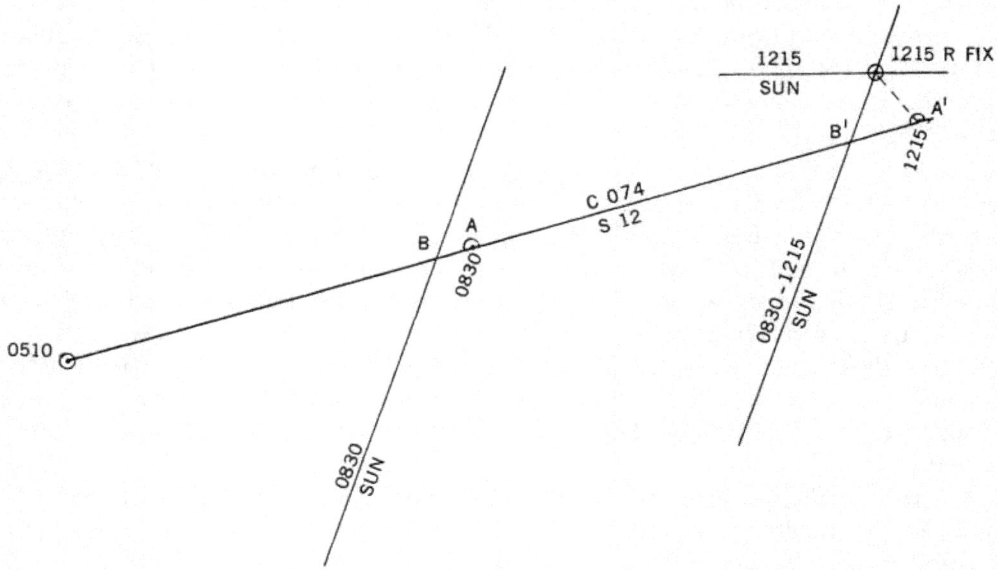

FIGURE 1708b.—Advancing a line of position without previous knowledge of the current.

The set of the average current *between 0510 and 0830* is the direction from *A'* to the 1215 running fix, and the drift is equal to this distance divided by 7^h05^m. The direction of a straight line (not shown) from the 0510 fix to the 1215 running fix is the course made good between 0510 and 0830, and the length of this line divided by the time (7^h05^m) is the speed made good to 0830.

The points *B* and *B'* need not be at the intersection of the lines of position and the course line. *Any* point on the line of position can be used, and the line *A'B'* drawn parallel to *AB*. Changes of course and speed do not affect the accuracy of the solution as long as *A'B'* is parallel to *AB*.

Several other variations are possible. A convenient one is to measure the distance from the earlier fix to point *B*, and divide this by the time to determine an "assumed" speed (based upon the assumption that point *B* represents the position of the vessel at the time of observation), and then to use this speed to advance the line of position. This variation should not be used without adjustment if a change of course or speed is involved between the earlier fix and the time to which the line is to be advanced.

This method should be used with caution. Any error in either the earlier fix or the first line of position is *increased* in proportion to the elapsed time. Thus, in figure 1708b, if *AB* is in error by one mile, *A'B'* is in error by 2.1 miles. It should not be used when there is reason to suspect a change in current between fixes.

1709. Celestial navigation and dead reckoning.—As indicated in chapter VIII, dead reckoning consists of advancing a known position for courses and speeds. Some difference in technique arises from a difference of opinion among navigators on the definition of (1) "known position" and (2) courses and speeds.

Regarding the first, no position determined by celestial navigation as commonly practiced at sea is known with perfect accuracy. An average error of two miles is realistic. Because of the varying conditions encountered, it is difficult to establish limits of a "known" position. In general, however, a reasonably reliable fix or running fix is considered sufficiently accurate to justify a new start in the dead reckoning. An estimated position or a fix or running fix of doubtful accuracy is considered an indication, but an inconclusive one, of the error in the dead reckoning. Therefore, it is considered good practice to avoid starting a new dead reckoning track from such a position unless there is a compelling reason for doing so. After long experience and the development of sound judgment, a navigator might acquire great skill in establishing a most probable position of sufficient reliability to justify more frequent breaks in the continuity of the dead reckoning, but even under these conditions any reasonable element of doubt should be given great respect.

What has been said regarding "known position" applies, also, in large measure to course and speed. The course steered and the speed at which a ship is being driven forward by its engines can be determined with relatively little error. Allowance for wind and current is a matter largely of judgment based upon experience. If the dead reckoning is to be meaningful, considerable caution should be exercised in allowing for wind and current when determining the course and speed to use for plotting. In the absence of information of a high degree of reliability, it is considered prudent to determine dead reckoning without allowing for estimated effects of wind and current.

In the absence of better information, then, it is considered good practice to start a new dead reckoning track only from a reliable fix or running fix, and to use courses and speeds without allowance for wind and current. This does not mean, however, that the navigator should not continually be aware of the possibility of error in his position as determined by dead reckoning, nor should he fail to make an estimate of the size and direction of the error. In this ability, and that of accurately interpreting all navigational information received, lies the test of a good navigator. This is largely the *art* of navigation, as distinguished from the somewhat mechanical process of making observations and computing and plotting the results, and also from the *science* of devising the aids that are used in modern navigation.

When it is desired to determine "average current," this expression being used to mean the resultant of all dead reckoning errors, the dead reckoning should be run forward from a fix (not a running fix) to the time of the next fix (or running fix if the method of art. 1708 is used). A dead reckoning position determined in any other way is not usable, unless it is adjusted to provide a "no-current" position. A straight line connecting such a dead reckoning position and the fix at the same time indicates the current. The direction of the line *from* the DR position *to* the fix is the **set** of the current, and the length of this line divided by the number of hours since the last fix is the **drift,** as in piloting.

Problems

A Position Plotting Sheet such as chart 969, covering latitudes 27° to 30° north and south is needed for most of the problems of this chapter. If this is not available, one can be constructed by means of table 5, as explained in article 307; or small area plotting sheets can be constructed as explained in article 324.

1703a. In each of the following, determine the altitude intercept, *a,* and label it T or A, as appropriate:

	Hc	*Ho*
(1)	18°21!4	18°25!9
(2)	53°02!7	52°35!5
(3)	(−)0°05!2	(−)0°12!7
(4)	(−)0°11!1	0°01!1

Answers.—(1) *a* 4.5 T, (2) *a* 27.2 A, (3) *a* 7.5 A, (4) *a* 12.2 T.

1703b. The 0930 DR position of a ship is lat. 29°20′.4N, long. 130°25′.2W. At this time the navigator observes the sun, and computes Hc and Zn for the 0930 DR position, as follows: Hc 45°42′.9, Ho 45°50′.2, Zn 157°.3. As a check, he also solves the same sight for an assumed position of lat. 29°00′.0N, long. 130°30′.0W, with the following results: Hc 46°00′.0, Zn 157°.0.

Required.—Plot the two lines of position, and account for the result.

Answer.—The two lines of position plot as approximately the same line, which is not dependent upon the assumed position, but only upon the observed altitude and the time of observation.

1705a. The 0500 fix of a ship is lat. 27°10′.0N, long. 142°55′.5W. The ship is on course 068°, speed 9 knots. At 0800 the navigator observes the sun, with the following results:

a 6.6T	*a*L 27°00′.0N
Zn 105°.0	*a*λ 142°39′.1W

The current since the morning fix is estimated to set 130°, at a drift of 1.4 knots.

Required.—(1) The 0800 DR position.

(2) The 0800 EP if there were no observation, and no current was anticipated.

(3) The 0800 EP using the current, if there were no observation.

(4) The 0800 EP using the line of position, but not the current.

(5) The 0800 EP using all available information.

Answers.—(1) 0800 DR: L 27°20′.0N, λ 142°27′.2W; (2) 0800 EP without current and line of position: L27°20′.0N, λ 142°27′.2W; (3) 0800 EP with current but no line of position: L 27°17′.4N, λ 142°23′.5W; (4) 0800 EP with line of position but no current: L 27°19′.5N, λ 142°25′.3W; (5) 0800 EP with current and line of position: L 27°18′.7N, λ 142°25′.8W.

1705b. The 0530 fix of a ship is lat. 28°55′.8N, long. 161°51′.7E. The ship is on course 060°, speed 10 knots. At 0830 the navigator observes the sun, with the following results:

a 6.7A	*a*L 29°00′.0N
Zn 110°.0	*a*λ162°28′.9E

Required.—(1) The 0830 EP if the course is believed to have been made good, and the line of position is considered accurate.

(2) The 0830 EP if the speed is believed to be correct, and the line of position is considered accurate.

(3) The 0830 EP if the course and speed are considered of equal reliability, and the line of position is considered accurate.

(4) The 0830 EP if the course is of questionable accuracy, but considered more reliable than the speed, and the line of position is considered accurate.

(5) The 0830 EP if the speed is of questionable accuracy, but considered more reliable than the course, and the line of position is considered accurate.

(6) The 0830 EP if the course is believed to have been made good, and the error contributed by the uncertainty of the line of position is believed to be twice that contributed by the uncertainty of the speed.

Answers.—(1) 0830 EP: L 29°13′.5N, λ 162°26′.3E; (2) 0830 EP: L 29°06′.5N, λ 162°23′.6E; (3) 0830 EP: L 29°09′.8N, λ 162°25′.0E; (4) 0830 EP: any place between (1) and (3); (5) 0830 EP: any place between (2) and (3); (6) 0830 EP: L 29°11′.4N, λ 162°22′.8E.

1707a. At 1740 the navigator and two assistants observe simultaneously three stars, with the following results:

	Fomalhaut	*Deneb*	*Aldebaran*
Hc	28°10′.3	34°59′.6	39°52′.8
Ho	28°05′.3	35°05′.6	39°46′.8
Zn	210°.0	308°.7	089°.3
*a*L	28°00′.0N	28°00′.0N	28°00′.0N
*a*λ	42°31′.7W	42°29′.0W	42°23′.2W

Required.—The 1740 fix.

Answer.—1740 fix: L 28°06′.6N, λ 42°30′.5W.

1707b. The 1800 DR position of a ship is lat. 27°02′.2N, long. 170°17′.0W. The ship is on course 045°, speed 14 knots. During evening twilight the navigator observes three stars, with the following results:

	Dubhe	*Altair*	*Spica*
Time	1815	1821	1830
Hc	34°45′.2	22°11′.8	47°24′.8
Ho	34°51′.3	22°15′.7	47°20′.4
Zn	331°.5	090°.3	219°.9
*a*L	27°00′.0N	27°00′.0N	27°00′.0N
*a*λ	170°10′.2W	170°05′.0W	169°54′.8W

Required.— The 1830 fix.

Answer.—1830 fix: L 27°11′.5N, λ 170°00′.5W.

1707c. The 1930 DR position of a ship is lat. 29°10′.5S, long. 122°35′.4W. The ship is on course 320°, speed 16 knots. During evening twilight the navigator observes a planet and two stars, with the following results:

	Saturn	*Regulus*	*Rigil Kent.*
Time	1931	1942	1951
Hc	46°58′.5	53°04′.0	24°19′.5
Ho	46°55′.5	53°09′.3	24°30′.0
Zn	023°.5	170°.2	297°.6
*a*L	29°00′.0S	29°00′.0S	29°00′.0S
*a*λ	122°55′.0W	122°45′.1W	122°35′.2W

Required.—The 1942 fix.

Answer.—1942 fix: L 29°05′.3S, λ 122°47′.4W.

1707d. The 0500 DR position of a ship is lat. 29°53′.9N, long. 69°32′.1W. The ship is on course 130°, speed 13 knots. During morning twilight the navigator observes a planet and two stars, with the following results:

	Mars	*Kochab*	*Spica*
Time	0451	0502	0511
Hc	17°14′.1	38°26′.2	33°35′.2
Ho	17°24′.5	38°19′.2	33°47′.8
Zn	130°.1	353°.2	237°.9
*a*L	30°00′.0N	30°00′.0N	30°00′.0N
*a*λ	69°41′.7W	69°30′.0W	69°18′.3W

Required.—The 0500 fix.

Answer.—0500 fix: L 29°54′.0N, λ 69°30′.5W.

1707e. The 0930 DR position of a ship is lat. 28°40′.4N, long. 125°30′.4E. The ship is on course 220°, speed 25 knots. The navigator observes the sun and moon, and solves each sight from the DR position at the time of sight, with the following results:

	Sun	*Moon*
Time	0936	0943
Hc	54°24′.3	37°07′.9
Ho	54°26′.3	37°14′.7
Zn	200°.2	142°.6

Required.—The 0943 fix.

Answer.—0943 fix: L 28°32′.1N, λ 125°32′.3E.

1707f. A ship is on course 314°, speed 24 knots. During evening twilight the navigator observes two stars and the moon, and solves all three sights using assumed latitude 28°00′.0S, assumed longitude 41°19′.5W as the AP, with the following results:

	Peacock	*Moon*	*Alpheratz*
Time	1855	1900	1905
Hc	57°12′.6	66°58′.2	23°00′.5
Ho	57°17′.9	67°01′.2	22°53′.7
Zn	194°.7	300°.5	038°.2

Required.—The 1900 fix.

Answer.—1900 fix: L 28°03′.5S, λ 41°26′.5W.

1707g. The 0400 DR position of a ship is lat. 27°01′.8N, long. 51°36′.0E. The ship is on course 037°, speed 20 knots. At 0545 the course is changed to 309°. During morning twilight the navigator observes two stars, with the following results:

	Vega	*Alpheratz*
Time	0537	0602
a	4.5T	7.8T
Zn	300°.5	075°.7
*a*L	27°00′.0N	27°00′.0N
*a*λ	51°45′.2E	51°50′.1E

Required.—The 0602 fix.

Answer.—0602 fix: L 27°28′.1N, λ 51°51′.1E.

1707h. The 0600 DR position of a ship is lat. 27°50′.3N, long. 20°58′.2W. The ship is on course 000°, speed 20 knots. During morning twilight the navigator observes four stars, with the following results:

	Dubhe	*Kaus Aust.*	*Spica*	*Vega*
Time	0551	0554	0558	0604
Hc	29°01′.1	21°57′.8	37°59′.4	54°33′.1
Ho	28°53′.4	22°11′.7	38°03′.5	54°28′.5
Zn	330°.0	149°.7	233°.3	057°.3
*a*L	28°00′.0N	28°00′.0N	28°00′.0N	28°00′.0N
*a*λ	20°54′.6W	21°08′.4W	20°56′.7W	20°51′.3W

Required.—The 0600 fix.

Answer.—0600 fix: L 27°53′.5N, λ 20°55′.0W.

1707i. The 1815 DR position of a ship is lat. 29°41′.5S, long. 163°52′.3W. The ship is on course 295°, speed 18 knots. During evening twilight the navigator observes three stars, with the following results:

	Regulus	*Pollux*	*Aldebaran*
Time	1810	1815	1821
Hc	45°18′.6	35°50′.7	22°50′.8
Ho	45°26′.2	36°03′.4	22°57′.7
Zn	040°.2	350°.7	300°.5
*a*L	30°00′.0S	30°00′.0S	30°00′.0S
*a*λ	163°45′.0W	163°49′.8W	163°54′.0W

Required.—(1) The 1815 fix, assuming random errors.

(2) The 1815 fix, assuming a constant error.

Answers.—(1) 1815 fix: L 29°47′.2S, λ 163°51′.2W; (2) 1815 fix: L 29°51′.4S, λ 163°50′.6W.

1708a. The 0830 DR position of a ship is lat. 29°25′.4S, long. 9°34′.7E. The ship is on course 326°, speed 22 knots. The sun is observed during the morning, and again at 1200, with the following results:

	Sun	*Sun*
Time	0830	1200
a	15.2A	28.4A
Zn	062°.3	169°.5
*a*L	29°00′.0S	29°00′.0S
*a*λ	9°37′.0E	8°52′.1E

Required.—The 1200 running fix.

Answer.—1200 R fix: L 28°31′.6S, λ 8°50′.0E.

1708b. The 0900 DR position of a ship is lat. 28°05′.6N, long. 93°44′.0W. The ship is on course 220°, speed 20 knots, and is believed to be in a current with set of 110° and a drift of 1.5 knots. The sun is observed during the morning, and again at 1200, with the following results:

	Sun	*Sun*
Time	0900	1200
a	11.2T	17.0A
Zn	103°.2	172°.0
*a*L	28°00′.0N	27°00′.0N
*a*λ	93°54′.0W	94°38′.9W

Required.—The 1200 running fix.

Answer.—1200 R fix: L 27°19′.8N, λ 94°17′.5W.

1708c. The 0715 fix of a ship is lat. 28°28′.9S, long. 81°14′.8W. The ship is on course 120°, speed 15 knots. During the morning the sun is observed twice, with the following results:

	Sun	*Sun*
Time	0945	1200
a	9.4A	0
Zn	095°.0	005°.0
*a*L	29°00′.0S	29°00′.0S
*a*λ	80°26′.1W	80°11′.2W

Required.—(1) The 1200 running fix, allowing for current.

(2) Set and drift of the current.

(3) Course made good between 0715 and 0945.

Answers.—(1) 1200 R fix: L 29°01′.0S, λ 80°00′.2W; (2) set 049°, drift 1.1 kn.; (3) course made good 116°.0.

1708d. The 0500 fix of a ship is lat. 28°36′.5N, long 143°22′.0E. The courses and speeds during the morning are as follows:

Time	*Course*	*Speed*
0500	047°	24 kn.
0600	102°	20 kn.
0715	038°	16 kn.
0845	075°	19 kn.
1000	030°	23 kn.
1045	085°	25 kn.

During the morning the sun is observed twice, with the following results:

	Sun	*Sun*
Time	0915	1200
a	8.8A	20.0A
Zn	125°.0	191°.7
*a*L	29°00'.0N	29°00'.0N
*a*λ	144°44'.8E	145°29'.8E

Required.—(1) The 1200 running fix, allowing for current.

(2) Set and drift of the current.

(3) Course and speed made good between fixes, assuming no change in current.

Answers.—(1) 1200 R fix: L 29°20'.0N, λ 145°33'.0E; (2) set 200°, drift 1.7 kn.; (3) course made good 070°, speed made good 17.7 kn.

1709a. The 0400 DR position of a ship is lat. 27°41'.8S, long. 64°54'.0E. This position has been run forward from a fix at 1715 the previous evening. The ship is on course 215°, speed 19 knots, but at 0600 the course is changed to 125°. At 0715 a fix locates the ship at lat. 28°23'.0S, long. 65°04'.3E.

Required.—Set and drift of the current between fixes.

Answers.—Set 073°, drift 1.0 kn.

1709b. The 0500 fix of a ship is lat. 27°09'.0N, long. 158°09'.5W. The ship is on course 310°, speed 14 knots. At 1155 a running fix locates the ship at lat. 28°01'.2N, long. 159°33'.2W. A new dead reckoning plot is started from this position. At 1900 a star fix is obtained, locating the ship at lat. 28°57'.8N, long. 160°54'.9W.

Required.—Set and drift of the average current between morning and evening fixes.

Answers.—Set 167°, drift 1.2 kn.

CHAPTER XVIII

TIME

1801. Introduction.—Time serves to regulate affairs aboard ship, as it does ashore. But to the navigator, it has additional significance. It is not enough to know *where* the ship is, was, or might be located in the future. The navigator wants to know *when* the various positions were or can reasonably be expected to be occupied. Time serves as a measure of progress. By considering the time at which a ship occupied various positions in the past, and by comparing the speed and various conditions it has encountered with those anticipated for the future, the skillful navigator can predict with reasonable accuracy the time of arrival at various future positions. Time can serve as a measure of safety, for it indicates when a light or other aid to navigation might be sighted, and if it is not seen by a certain time, the navigator knows he has cause for concern.

To the celestial navigator, time is of added significance, for it serves as a measure of the *phase* of the earth's rotation. That is, it indicates the position of the celestial bodies relative to meridians on the earth. Until an accurate *measure* of time became available at sea, longitude could not be found.

Very small *intervals* of time are used in certain electronic navigational aids, such as radar and Loran.

Whatever the type of navigation, a thorough mastery of the subject of time is important to the navigator.

The four independent base units of measurement currently used in science are length, mass, time, and temperature. It is true that, except for fields of science such as cosmology, geology, and astronomy, time interval is the more important time concept, and date (astronomical) is of much less importance to the rest of science. This is true because the "basic laws" of physics are differential in nature and only involve small time intervals. In essence, physical "laws" do not depend upon when (i.e., the date) they are applied.

Based on these laws and extensive experimentation, scientists have been able to demonstrate that frequency can be controlled and measured with the smallest percentage error of any physical quantity. The frequency of a periodic phenomenon is the number of cycles of this phenomenon per unit of time (i.e., per second). The name of the unit of frequency is the **hertz (Hz)** and is identical to a cycle per second (cps). Since most clocks depend upon some periodic phenomenon (e.g., a pendulum) in order to "keep time," and since one can make reliable electronic counters to count the "swings" of the periodic phenomenon, we can construct clocks with timekeeping accuracy (rate accuracy) equal to the accuracy of the frequency standard. Today's most precise and accurate clocks incorporate a cesium atomic beam as the "pendulum" of the clock.

1802. Clocks and timekeeping.—In early times, the location of the sun in the sky was the only reliable indication of the time of day. Of course, when the sun was not visible, one was unable to know the time with much precision. People developed devices (called clocks) to interpolate between checks with the sun. The sun was sort of a "master clock" that could be read with the aid of a sundial. An ordinary clock, then, was a device used to interpolate between checks with the sun. The gain in accuracy of the different clock devices over a period of years is shown in figure 1802. (Timekeeping has shown nearly 10 orders of magnitude improvement within the last 6 centuries with about 6 orders occurring within 70 years of the 20th century.) Thus, a clock could be a "primary clock" like the position of the sun in the sky, or it could be a secondary clock and only interpolate between checks with the primary clock or time standard. Historically, some people have used the word "clock" with the connotation of a secondary time reference, but today this usage would be too restrictive.

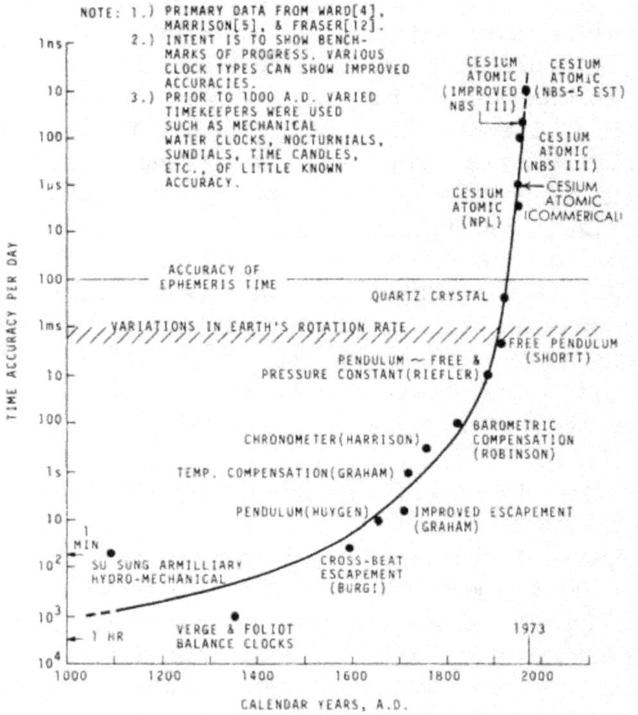

FIGURE 1802.—Progress in timekeeping accuracy.

When one thinks of a clock, it is customary to think of some kind of pendulum or balance wheel, a group of gears, and a clock face. Each time the pendulum completes a swing, the hands of the clock are moved a precise amount. In effect, the gears and hands of the clock "count" the number of swings of the pendulum. The face of the clock, of course, is not marked off in the number of swings of the pendulum but rather in hours, minutes, and seconds.

One undesirable characteristic of pendulum-type clocks is that no two clocks ever keep exactly the same time. This is one reason for looking for a more stable "pendulum" for clocks. In the past, the most stable "pendulums" were found in astronomy. Here one obtains a significant advantage because only one universe exists—at least for observation purposes, and time defined by this means is available to anyone—at least in principle. Thus, one can obtain a very reliable time scale which has the property of universal accessibility. In this chapter, time scale (art. 1804) is used to refer to a conceptually distinct method of assigning dates to events.

In a very real sense, the pendulum of ordinary, present-day electric clocks is the electric current supplied by the power company. In the United States the power utilities generally synchronize their generators to the National Bureau of Standards

(NBS) low frequency broadcast, WWVB. Thus, the right number of pendulum swings occur each day. Since all electric clocks which are powered by the same source have, in effect, the same pendulum, these clocks do not gain or lose time relative to each other; i.e., they run at the same rate. Indeed, they will remain fairly close to the time as broadcast by WWVB (± 5 seconds) and will maintain the same time difference with respect to each other (± 1 millisecond) over long periods of time.

It has been known for some time that atoms have characteristic resonances or, in a loose sense, "characteristic vibrations." The possibility therefore exists of using the "vibrations of atoms" as pendulums for clocks. Presently, microwave resonances (vibrations) of atoms are the most precisely determined and reproducible physical phenomena that man has encountered. A clock which uses "vibrating atoms" as a pendulum will generate a time scale more uniform than even its astronomical counterparts.

But due to intrinsic errors in any actual clock system, atomic clocks drift relative to other similar clocks. Of course, the rate of drift is much smaller for atomic clocks than the old pendulum clocks, but nonetheless real and important. The attribute of universal accessibility for atomic time is accomplished by coordination between laboratories generating atomic time.

1803. Basic concepts of time.—One can use the word "time" in the sense of date. (By "date" we mean a designated mark or point on a time scale.) One can also consider the concept of time interval or "length" of time between two events. The difference between these concepts of date and time interval is important and has often been confused in the single word "time."

The *date* of an event on an earth-based time scale is obtained from the number of cycles (and fractions of cycles) of the apparent sun counted from some agreed-upon origin. Similarly, atomic time scales are obtained by counting the cycles of a signal in resonance with certain kinds of atoms.

The word "epoch" is sometimes used in a similar manner to the word "date." However, dictionary definitions of epoch show gradations of meanings such as time duration, time instant, or a particular time reference point, as well as a geological period of time. Thus, epoch often simultaneously embodies concepts of both date and duration. Because of such considerable ambiguity in the word "epoch," its use in this volume will be restricted to a time instant.

Another aspect of time is that of *simultaneity;* i.e., coincidence in time of two events. For example, we might synchronize clocks upon the arrival of portable clocks at a laboratory. Here we introduce an additional term, *synchronization,* which implies that the two clocks are made to have the same reading in some frame of reference. Note that the clocks need not be synchronized to an absolute time scale. As an example, two people who wish to communicate with each other might not be critically interested in the date, they just want to be *synchronized* as to when they use their communications equipment.

1804. Time scales.—A system of assigning dates to events is called a **time scale.** The apparent motion of the sun in the sky constitutes one of the most familiar time scales but is certainly not the only time scale. Note that to completely specify a date using the motion of the sun as a time scale, one must count days (i.e., make a calendar) from some initially agreed-upon beginning. In addition (depending on accuracy needs) one measures the fractions of a day (i.e., "time of day") in hours, minutes, seconds, and maybe even fractions of seconds. That is, one counts cycles (and even fractions of cycles) of the sun's daily apparent motion around the earth.

1805. Fundamental kinds of time.—There are three fundamentally different kinds of time. These are time based on the rotation of the earth on its axis; time based on long term observations of the annual revolution of the earth around the sun; and time based on transitions in the atom.

Time based on the rotation of the earth on its axis has several forms, all of which are related to each other by rigorous formulae or by appropriate tables. These forms are the various sidereal times, mean and apparent, and solar times, mean and apparent.

Time defined by the daily rotation of the earth with respect to the equinox or first point of Aries is known as **sidereal time.** The sidereal time is numerically measured by the hour angle of the equinox, which represents the position of the equinox in the daily rotation. The period of one rotation of the equinox in hour angle, between two consecutive upper meridian transits, is a **sidereal day;** it is divided into 24 sidereal hours, reckoned from 0^h at upper transit which is known as **sidereal noon.** The true equinox is at the intersection of the true celestial equator of date with the ecliptic of date; the time measured by its daily rotation is **apparent sidereal time.** The position of the true equinox is affected by the nutation of the axis of rotation of the earth; and the nutation consequently introduces irregular periodic inequalities into the apparent sidereal time and the length of the sidereal day. The time measured by the daily motion of the mean equinox of date, which is affected only by the secular inequalities due to the precession of the axis, is **mean sidereal time.** The maximum difference between apparent and mean sidereal times is only a little over a second, and its greatest daily change is a little more than a hundredth of a second. Because of its variable rate, apparent sidereal time is used by astronomers only as a measure of epoch; it is not used for time interval. Mean sidereal time is deduced from apparent sidereal time by applying the equation of equinoxes.

Universal Time (UT) is a particular case of the measure known in general as **mean solar time** (art. 1808). Universal Time is the mean solar time on the Greenwich meridian, reckoned in days of 24 mean solar hours beginning with 0^h at midnight. Universal Time and sidereal time are rigorously related by a formula so that if one is known the other can be found. The ratio of the mean solar day to the mean sidereal day is 1.0027379093, and the equivalent measures of the length of the day are:

Mean sidereal day .... $23^h56^m04\overset{s}{.}09054$ of mean solar time
Mean solar day .... $24^h03^m56\overset{s}{.}55536$ of mean sidereal time.

Universal Time, in principle, is determined by the *average* rate of the apparent daily motion of the sun relative to the meridian of Greenwich; but in practice the numerical measure of Universal Time at any instant is computed from sidereal time.

Universal Time is the standard in the application of astronomy to navigation. Observations of Universal Times are made by observing the times of transit of stars.

The Universal Time determined directly from astronomical observations is denoted **UT0.** Since the earth's rotation is nonuniform, corrections must be applied to UT0 to obtain a more uniform time. This more uniform time is obtained by correcting for two known periodic motions.

One motion, the **polar motion** (the motion of the geographic poles), is the result of the axis of rotation continuously moving with respect to the earth's crust. The corrections for this motion are quite small (±15 milliseconds for Washington, D.C.).

On applying the correction to UT0, the result is **UT1**, which is the same as **Greenwich mean time (GMT)** used in celestial navigation.

The second known periodic motion is the variation in the earth's speed of rotation due to winds, tides, and other phenomena. As a consequence, the earth suffers an annual variation in its speed of rotation of about ±30 milliseconds. When UT1 is corrected for the mean seasonal variations in the earth's rate of rotation, the result is **UT2.**

Although UT2 was at one time believed to be a uniform time system, it was later determined that there are secular variations in the earth's rate of rotation, possibly caused by random accumulations of matter in the convection core of the earth. Such accumulations would change the earth's moment of inertia and thus its rate of rotation.

The second fundamental kind of time, **Ephemeris Time (ET)**, now known as **dynamical time**, is, by definition, a uniform time system. Ephemeris Time is the uniform measure of time defined by the law of dynamics and determined in principle from the orbital motions of the planets, specifically the orbital motion of the earth as represented by Newcomb's *Tables of the Sun.* Ephemeris Time is the measure of time in which Newcomb's *Tables of the Sun* agree with observation. Ephemeris Time is time based on the **ephemeris second** defined as 1/31556925.9747 of the tropical year for 1900 January $0^{d}12^{h}$ ET. The ephemeris day is 86,400 ephemeris seconds. The ephemeris second is a fundamental invariable unit of time.

The Ephemeris Time at any instant is obtained from observation by directly comparing observed positions of the sun, moon, and planets with gravitational ephemerides of their coordinates; observations of the moon are most effective and expeditious for this purpose. Ephemeris Time is used by astronomers in the fundamental ephemerides of the sun, moon, and planets, but is not used by navigators.

The third fundamental kind of time, **Atomic Time (AT)**, is based on transitions in the atom.

The basic principle of the **atomic clock** is that electromagnetic waves of a particular frequency are emitted when an atomic transition occurs. The frequency of the **cesium beam atomic clock** was found to be 9,192,631,770 cycles per second of Ephemeris Time in an experiment conducted jointly by the National Physical Laboratory, Teddington, England, and the U. S. Naval Observatory during 1955–1958.

In 1967 the **atomic second** was defined by the Thirteenth General Conference on Weights and Measures as the duration of 9,192,631,770 periods of the radiation corresponding to the transition between two hyperfine levels of the ground state of the cesium atom 133. This value was established to agree as closely as possible with the ephemeris second. Thus, the atomic second became the unit of time in the **International System of Units (SI).**

UT2 and A1, the atomic time scale established by the U. S. Naval Observatory in 1958, were identical on January 1, 1958. To the accuracy currently available, A1 and ET differ only by a constant such that $ET-A1=32^{s}.18$.

The advent of atomic clocks, which have accuracies better than 1 part in 10^{10}, led in 1961 to the coordination of time and frequency emissions of the U. S. Naval Observatory and the Royal Greenwich Observatory. The master oscillators controlling the signals were calibrated in terms of the cesium standard (A1) and corrections determined at the U. S. Naval Observatory and the Royal Greenwich Observatory were made simultaneously at all transmitting stations. Because of the divergence of the astronomical and atomic time scales due to the unpredictable variations in the earth's rotation, the time emissions were adjusted by applying a frequency offset to the oscillator so that the rate defined by the timing pulses was in general agreement with UT2. If, in spite of this, the departure of the time signals from UT2 became unacceptable, the epoch of the signals was adjusted by 100 millisecond steps. These adjustments kept the transmitted time synchronized with the rotation of the earth within a tolerance of $0^s.1$. This system became known as **Coordinated Universal Time (UTC)** and was accepted by many authorities following its formal recommendation in 1961 by the International Astronomical Union (IAU) and in 1963 by the International Radio Consultative Committee (CCIR), a committee of the International Telecommunications Union (ITU) which controls international coordination of time signal transmission.

In February 1970 at the Plenary Assembly of the ITU, it was agreed that commencing January 1, 1972, the use of frequency offsets would be discontinued and that all time signal transmissions would be based strictly on the internationally adopted definition of the second. The ITU also agreed that the coordinated time transmission based on the atomic second be maintained in approximate agreement with UT1 by stepping the transmitted time one whole second whenever necessary.

In accordance with the implementation resolutions of the IAU, the new system was inaugurated January 1, 1972, using the second defined in terms of an **International Atomic Time (TAI)** scale (art. 1806) as the unit of time and UT1 as the astronomical reference. Beginning at this time, UTC was then maintained in approximate agreement with UT1 by step adjustments **(leap seconds)** as directed by the Bureau Internationale de l'Heure (BIH).

At the end of 1971, before the new system was inaugurated, there was a difference of almost 10 seconds between TAI and UTC. In order that this difference be an integral number of whole seconds (in this case 10^s) a special negative adjustment of approximately 0.108^s was made in accordance with BIH directive so that the reading on the UTC scale was 1 January 1972, $00^h00^m00^s$ at the instant the TAI scale was 1 January 1972, $00^h00^m10^s$.

1806. International Atomic Time (TAI) scale.—In October 1971, the General Conference on Weights and Measures endorsed the Bureau Internationale de l'Heure (BIH) atomic time scale as the International Atomic Time (TAI) scale defined as follows:

"International Atomic Time is the time reference coordinate established by the Bureau Internationale de l'Heure on the basis of the readings of atomic clocks functioning in various establishments in accordance with the definition of the second, the SI Unit (International System of Units) of time."

The Atomic Time (AT) scales maintained in the United States by the National Bureau of Standards and the U. S. Naval Observatory constitute approximately 37½ percent of the stable reference information used in maintaining a stable TAI scale by the BIH.

1807. Time interval and time scales.—One should note sources of confusion which can exist in the measurement of time and in the use of the word "second." Suppose that two events occurred at two different dates. For example the dates of these two events were 15 December 1970, $15^h30^m00^s.000000$ UTC and 15 December 1970, $16^h30^m00^s.000000$ UTC. At first thought one would say that the time interval between these two events was exactly 1 hour = 3600.000000 seconds, but this is *not* true. (The actual interval was longer by about 0.000108 seconds [3600 seconds $\times 300 \times 10^{-10}$]. Refer to table 1807.) Recall that the UTC time scale (like all the UT scales and the ET scale) was not defined in accordance with the definition of the interval of time, the second. Thus, one cannot simply subtract the dates of two events as assigned by the UTC scale (or any UT scale or the ET scale) in order to obtain the precise time interval between these events. Historically, the reason behind this state of affairs is that navigators need to know the earth's position (i.e., UT1)—not the duration of the second. Yet, many scientists need to know an exact and reproducible time interval. Note that this might also be true of the new UTC system if the particular time interval included one or more leap seconds.

It is also confusing that the dates assigned by the UT, ET, and UTC scales involve the same word as the unit of time interval, the second. For accurate and precise measurements, this distinction can be extremely important.

Year	Offset rate of UTC in parts per 10^{10}
1960	–150
1961	–150
1962	–130
1963	–130
1964	–150
1965	–150
1966	–300
1967	–300
1968	–300
1969	–300
1970	–300
1971	–300
1972 → future	0

TABLE 1807.—Frequency offsets of UTC from 1960 to 1972.

1808. Solar Time.—The basis of time measurement in celestial navigation is the period of rotation of the earth. This period is not quite constant; it is subject to variations which may reach a few milliseconds per day. These variations will be disregarded initially; the earth will be conceived as rotating at a constant rate.

The earth's rotation causes the sun and other celestial bodies to appear to cross the sky from east to west each day. If a person located on the earth's equator measured the time interval between two successive transits overhead of a very distant star, he would thereby measure the period of the earth's rotation. If he then made similar measurements on the sun instead of a star, he would obtain a result about 4 minutes longer than before. This difference is due to the earth's motion around the sun, which continuously changes the apparent place of the sun among the stars. Thus, during the course of a day the sun appears to move a little to the east among the stars so that the earth must rotate on its axis through more than 360° in order to bring the sun overhead again. Of course this apparent eastward movement of the sun cannot be observed directly.

If the sun is on the observer's meridian when the earth is at point *A* (fig. 1808) in its orbit around the sun, it will not be on the observer's meridian after the earth has rotated through 360° because the earth will have moved along its orbit to point *B*. Before the sun is again on the observer's meridian, the earth must turn still more on its axis. The sun will be on the observer's meridian again when the earth has moved to point *C* in its orbit. Thus, during the course of a day the sun appears to move eastward with respect to the stars.

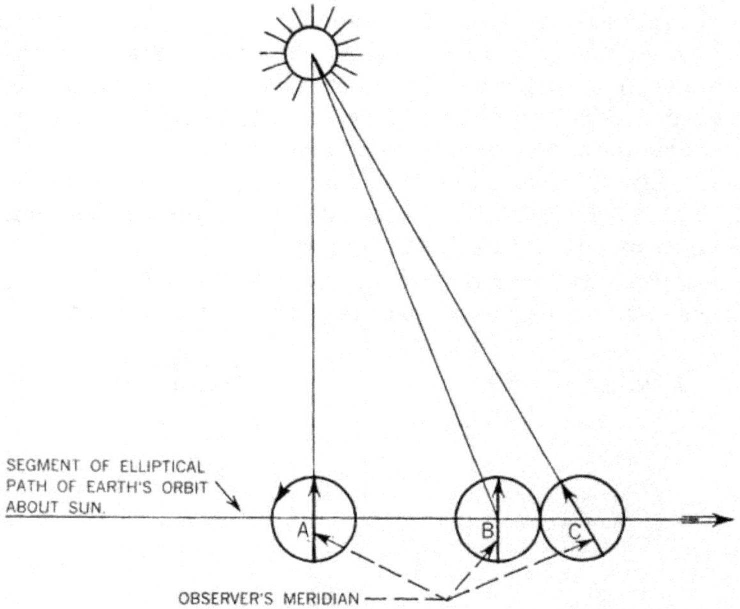

FIGURE 1808.—Apparent eastward movement of the sun with respect to the stars.

Even if the earth did not rotate on its own axis, the sun would rise and set once during the year because of the earth's orbit around it. The stars, however, are not within the earth's orbit. Since they are generally more than a million times as distant as the sun, their apparent positions are only very slightly affected by the earth's orbital motion. The apparent positions of the stars are commonly reckoned with reference to an imaginary point called the **vernal equinox**, which is the intersection of the celestial equator and the ecliptic. The sun is at the vernal equinox at the beginning of spring, when it passes over the equator on its apparent journey northward. The period of the earth's rotation measured with respect to the vernal equinox is called a **sidereal day.** The period with respect to the sun is called an **apparent solar day.**

With the sun moving eastward among the stars so that the difference between the apparent solar and sidereal day is about 4 minutes of time, on any night a star will rise about 4 minutes earlier than on the previous night. Thus, the celestial sphere appears to shift westward about 1° each night. The complete shift through 360° is associated with the **year,** the period of one revolution of the earth around the sun. By the calendar, one year is 365 days duration for a **common year** and 366 days for a **leap year.** A leap year is any given year divisible by 4, unless it is a century year, which must be divisible by 400 to be a leap year. Thus, 1900 was not a leap year, but 2000 will be. This calendar, now in general use, is called the **Gregorian calendar.**

When measuring time by the rotation of the earth, the time is **apparent solar time** if the apparent (real) sun is used as the celestial reference.

Use of the apparent sun as a celestial reference for time results in time of nonconstant rate for at least three reasons. First, revolution of the earth in its orbit is not constant. Second, motion of the apparent sun is along the ecliptic, which is tilted with respect to the celestial equator, along which time is measured. Third, rotation of the earth on its axis is not constant. The effect due to this third cause is extremely small.

For the various forms of mean solar time, the apparent sun is replaced by a fictitious **mean sun**, conceived as moving eastward along the celestial equator at a uniform speed equal to the average speed of the apparent sun along the ecliptic, thus providing a nearly uniform measure of time equal to the approximate average apparent time. The speed of the mean sun along the celestial equator is taken as 15° per hour of mean solar time.

1809. Equation of time.—Mean solar time, or **mean time** as it is commonly called, is sometimes ahead of and sometimes behind apparent solar time (**sundial time**). The difference, which never exceeds about $16^{m}.4$, is called the **equation of time (Eq. T.)**.

By one convention, the equation of time is the time interval which must be added algebraically to the mean time to obtain apparent time. This convention is used here.

In accordance with Kepler's second law (art. 1407), the speed of the earth in its elliptical orbit around the sun varies with the changing distance between the two bodies. The earth moves faster at perihelion than it does at aphelion. Consequently, as seen from the earth the sun appears to move faster in January than it does in July. Even if the earth's orbital speed were uniform, the hour angle of the sun would still change at a variable rate because the sun as observed from the earth appears to move in the plane of the ecliptic, which is inclined at an angle of about 23°27′ to the plane of the celestial equator.

In deriving the value of the equation of time it is simpler to consider the contributions of the ellipticity and obliquity of the apparent orbit of the sun about the earth separately. In considering the ellipticity and obliquity contributions separately, it is convenient to introduce a second fictitious sun. This second sun, known as the **dynamical mean sun**, is conceived to move eastward along the ecliptic at the average rate of the apparent (true) sun. The dynamical mean sun and the apparent sun occupy the same position when the earth is at perihelion (or the sun is at perigee when using the concept that the sun orbits the earth). The dynamical mean sun and the mean sun, or **astronomical mean sun** as it is sometimes called, occupy the same position at the time of the vernal equinox.

That part of the equation of time due to the ellipticity of the orbit and known as the **eccentricity component** is the difference, in mean solar time units, between the hour angles of the apparent (true) sun and the dynamical mean sun. It is also the difference in the right ascensions of these two suns. That part of the equation of time due to the obliquity of the orbit is the difference in units of mean solar time, between the hour angles of the dynamical mean sun and the astronomical mean sun. It is also the difference in the right ascensions of these two suns.

Figure 1809a illustrates the apparent orbit of the sun about the earth. In accordance with Kepler's second law the radius vector sweeps through equal areas in equal time intervals. Therefore, the angular velocity of the true sun is greatest at perigee. With the true sun T and the dynamical mean sun D occupying the same position at perigee P around 1 January, following perigee the true sun moves ahead of the dynamical mean sun which is moving eastward along the ecliptic at the average rate of the true sun. The maximum separation of about 2° (8 minutes) occurs about 1 April. Because of Kepler's second law, the dynamical mean sun and the true sun must be in coincidence again at apogee A about 1 July. The time for the true sun to move from perigee to apogee is equal to the time for the true sun to move from apogee to perigee. Since the dynamical mean sun moves at the average rate of the true sun, the time to complete the orbit of the ecliptic is equal to the time required for the true sun to complete the same orbit. Since the line of apsides bisects the sun's apparent orbit, it follows that the time required for the dynamical mean sun to complete half the orbit is the same as that required for the true sun to complete half the orbit. Therefore, the dynamical mean sun and the true sun occupy the same position at apogee.

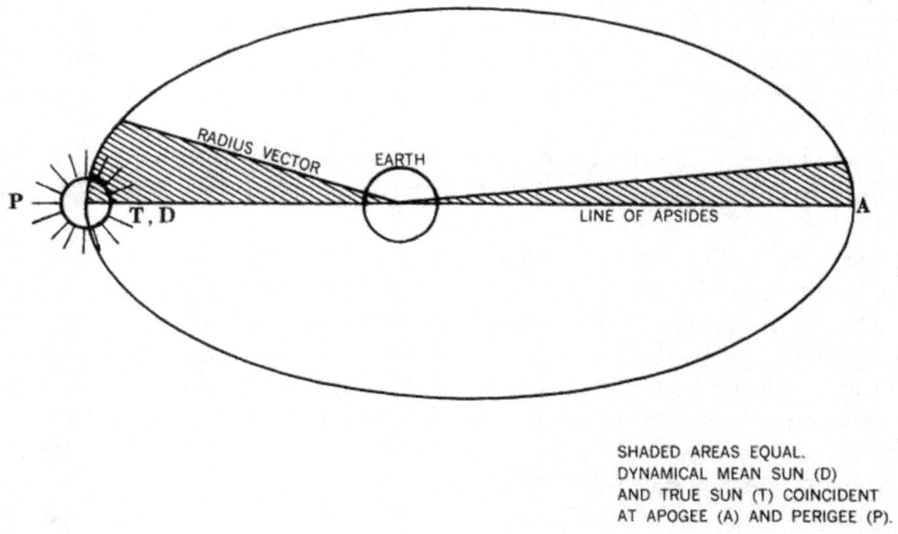

FIGURE 1809a.—Apparent orbit of the sun about the earth.

With the true sun and the dynamical mean sun occupying the same position at apogee and with the angular velocity of the true sun being least at apogee, following apogee the dynamical mean sun moves ahead of the true sun. The maximum separation of about 2° (8 minutes) occurs about 1 October. The two suns are again coincident at perigee about 1 January.

The eccentricity component of the equation of time is shown in figure 1809b. The obliquity component of the equation of time can now be found by comparing a dynamical mean sun moving uniformly along the ecliptic with an astronomical mean sun also moving uniformly at the same rate in the plane of the celestial equator.

With the dynamical mean sun and the astronomical mean sun coincident at the first point of Aries and each moving uniformly at the same rate along their respective paths, following the time of the vernal equinox the positions of the two suns are such that the celestial longitude of the dynamical mean sun equals the right ascension of the astronomical mean sun. As shown in figure 1809c, ♈D=♈M. As is also shown in this figure, following the vernal equinox the right ascension of the astronomical mean sun is greater than the right ascension of the dynamical mean sun. Thus, during this period that part of the equation of time due to the obliquity of the orbit is a negative value.

When the celestial longitude of the dynamical mean sun has increased to 90°, the right ascension of the astronomical mean sun will also be 90°. At the time of the summer solstice, the hour circles of the two suns are coincident; the elevated pole, the ecliptic pole, and the two suns all lie on the same great circle. Therefore, at the summer solstice that part of the equation of time due to the obliquity of the orbit is zero. Halfway between the time of the vernal equinox and the summer

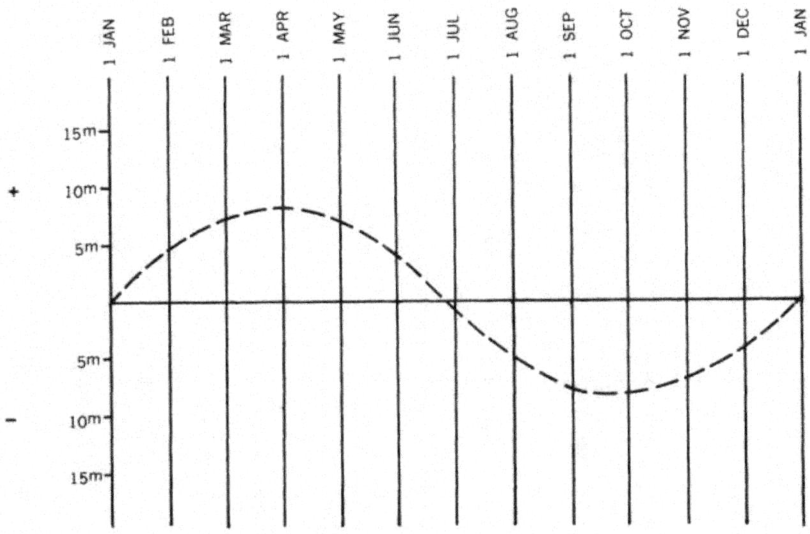

FIGURE 1809b.—Eccentricity component.

solstice that part of the equation of time due to obliquity of the orbit reaches a maximum value of about 10 minutes.

Following the summer solstice and until the time of the autumnal equinox, the right ascension of the dynamical mean sun is greater than that of the astronomical mean sun. At the autumnal equinox, the two suns are coincident. Following the autumnal equinox and until the time of the winter solstice, the right ascension of the astronomical mean sun is greater than that of the dynamical mean sun. At the winter solstice, the hour angles of the two suns are coincident; the elevated pole, the ecliptic pole, and the two suns all lie on the same great circle. Therefore, at the winter solstice that part of the equation of time due to the obliquity of the orbit is zero. Following the winter solstice and until the time of the vernal equinox, the right ascension of the dynamical mean sun is greater than the right ascension of the astronomical mean sun.

Figure 1809d illustrates that part of the equation of time due to obliquity of the orbit. Figure 1809e illustrates the combining of the two parts. From inspection of the curve it can be seen that the equation of time is zero on or about 15 April, 14 June, 1 September, and 24 December. The greatest value is about 16^m22^s in November.

1810. Expressing time.—As a measure of part of a day, time based upon the rotation of the earth can be stated in a number of different ways. At any given moment, the time depends upon (1) the point on the celestial sphere used as reference, (2) the reference meridian on the earth, and (3) the somewhat arbitrary starting point of the day.

When the sun is used as the celestial reference point, **solar time** results. If the actual sun observable in the sky is used, **apparent solar time** is involved, and if a fictitious **mean sun** is used to provide a time having an almost constant rate, **mean solar time** results. Time reckoned by use of the first point of Aries (♈) as the celestial reference point is called **sidereal time.** Use of the moon as the celestial reference point provides a variable-length **lunar day**, the basis of **lunar time,** which is useful in tide prediction and analysis. Because of its application, a lunar day is sometimes called a **tidal day.** It averages about 24^h50^m (mean solar units) in length.

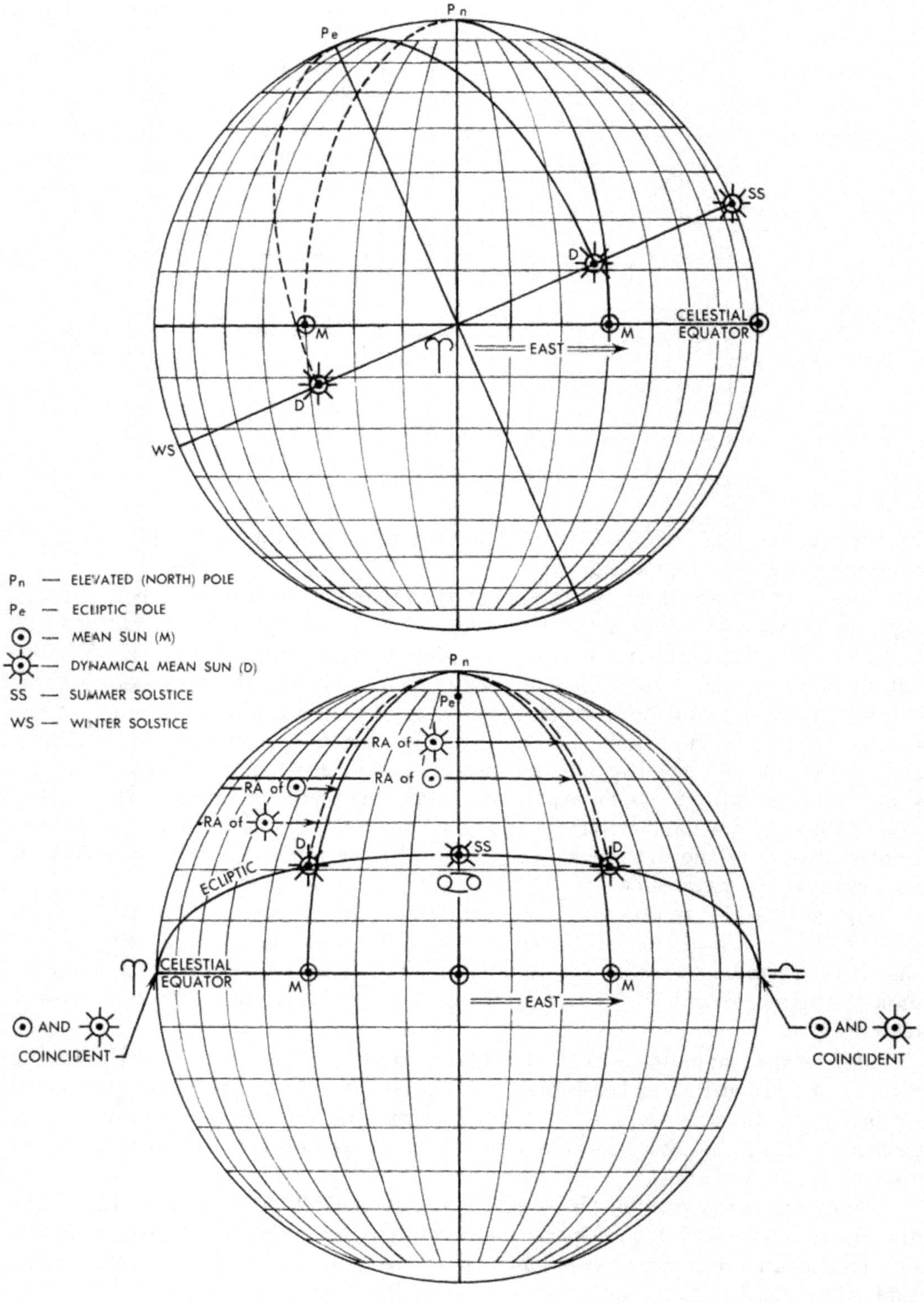

FIGURE 1809c.—Right ascensions of dynamical and astronomical mean suns.

If the meridian of the observer is used as the terrestrial reference, **local time** is involved. If a **zone** or **standard meridian** is used as the **time meridian** for mean

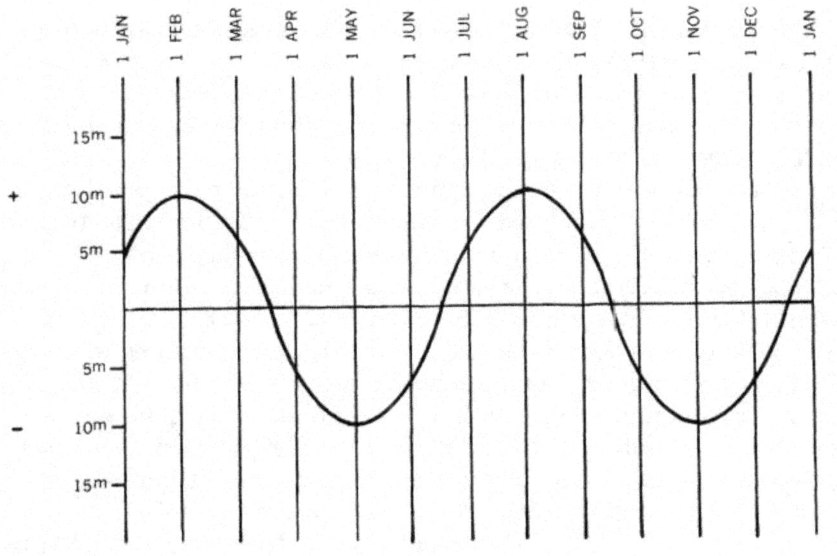

FIGURE 1809d.—Obliquity component.

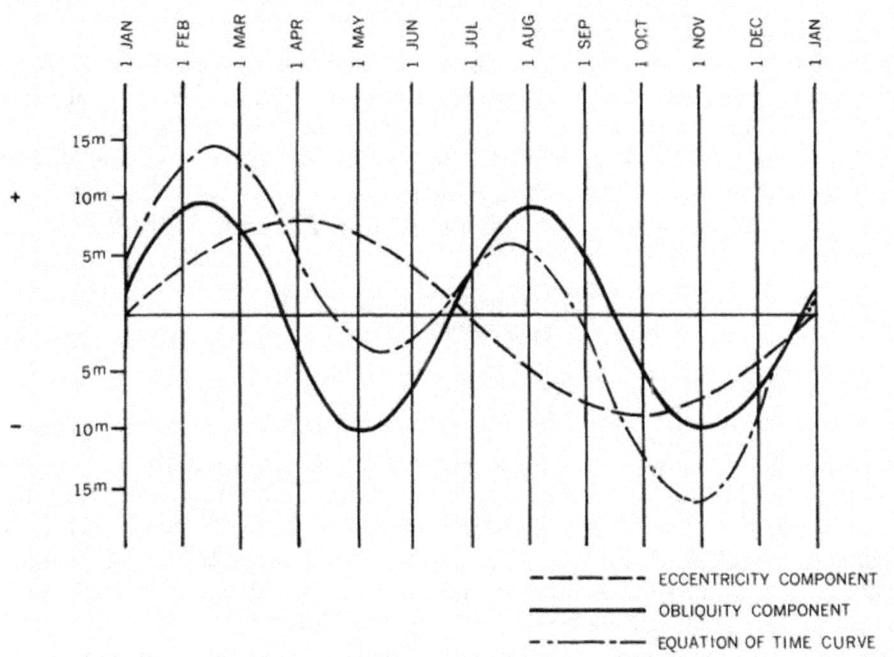

FIGURE 1809e.—Equation of time curve constructed from eccentricity and obliquity components.

solar time over an area, **zone** or **standard time** results. Use of a meridian farther east than would normally be used, so that the period of daylight is shifted later in the day produces a form of zone time called **daylight saving** or **summer time.** Time based upon the Greenwich meridian is called **Greenwich time. Greenwich mean time (GMT)** is of particular interest to a navigator because it is the principal entering argument for the almanacs.

One complete revolution of the earth with respect to a celestial reference point is called a **day.** In modern usage every kind of solar time has its zero or starting point at **midnight,** when the celestial reference point is directly over the *lower* *branch* of the terrestrial reference meridian. This has not always been so. Until January 1, 1925, the **astronomical day** began at *noon*, 12 hours *later* than the start of the calendar day of the same date. The **nautical day** began at *noon,* 12 hours *earlier* than the calendar day, or 24 hours earlier than the astronomical day of the same date. The **sidereal day** begins at **sidereal noon,** when the first point of Aries is over the *upper branch* of the reference meridian. There is no sidereal date.

Time is customarily expressed in time units, from 0^h through 24^h. To the nearest 1^m it is generally stated by navigators in a four-digit unit without punctuation. Thus, 0000 is midnight at the start of the day. One minute later the time is 0001. Half an hour after the start of the day the time is 0030, at one hour the time is 0100, at one hour and four minutes it is 0104, at 19 minutes after noon (solar time) it is 1219, at four hours and 23 minutes after (solar) noon it is 1623, etc. The term "hours" is sometimes used with the four-digit system to indicate that the number refers to the time or "hour" of the day. However, in those few occasions when any reasonable doubt may exist as to whether time is indicated, the fact can better be indicated in another way. Thus, the expression "1600 hours" to indicate "1600" or "16 hours" is not strictly correct, and is better avoided. **Watch time (WT),** indicated by a watch or clock having a 12-hour dial, and **chronometer time (C)** are expressed on a 12-hour basis, with designations AM (ante meridian) and PM (post meridian), as in ordinary civil life ashore.

In contrast, a time interval is expressed as hours and minutes, as 5^h26^m. When either the time of day or a time interval is given to seconds, this same form is used, as $21^h15^m18^s$. The kind of time may be indicated, usually by abbreviation.

When a time interval is to be added to or subtracted from a time, the solution can be arranged conveniently in tabular form.

Example 1.—What is the time and date $14^h36^m53^s$ after $21^h14^m18^s$ on July 24?

Solution.—

$$\begin{array}{rl} 21^h14^m18^s & \text{July 24} \\ \underline{14^h36^m53^s} & \\ \underline{35^h51^m11^s} & \text{July 24} \\ =11^h51^m11^s & \text{July 25} \end{array}$$

The fact that the sum of hours exceeds 24 is an indication that the date increases by one. Similarly, in *subtracting* an interval, the date is one day earlier if 24^h must be added to the time before the subtraction can be made. That is, since 2400 of one day is 0000 of the following day, one might say that 2700 on one day is $2700-2400=0300$ on the following day. In the example above, $11^h51^m11^s$ on July 25 is the same as $11^h51^m11^s+24^h00^m00^s=35^h51^m11^s$ on July 24.

Date is sometimes expressed as an additional unit of the time sequence. Thus, $21^h14^m18^s$ on July 24 might be stated $24^d21^h14^m18^s$. This system is of particular value when an interval of several days is to be added or subtracted.

Example 2.—What is the time and date $9^d16^h35^m04^s$ before $5^h11^m33^s$ on September 15?

Solution.—

$$\begin{array}{l} 15^d05^h11^m33^s \\ \underline{\ 9^d16^h35^m04^s} \\ 5^d12^h36^m29^s \text{ or } 12^h36^m29^s \text{ on Sept. 5.} \end{array}$$

By this method the month and day, if of significance, are recorded separately, or they, too, can be added to the sequence.

Example 3.—What is the time and date 3 years, 6 months, 25 days, 12 hours, 19 minutes, and 44 seconds after $7^h52^m24^s$ on November 14, 1958?

Solution.—

$$\begin{array}{r} 1958^y11^m14^d07^h52^m24^s \\ 3^y06^m25^d12^h19^m44^s \\ \hline 1962^y06^m08^d20^h12^m08^s \end{array} = 20^h12^m08^s \text{ on June 8, 1962.}$$

Since a month may contain a variable number of days, both the months and days should be solved together. Thus, in the example above, the answer would be 17 months, 39 days. If 12 months are converted to one year, this becomes five months, 39 days. Since the fifth month is May, this might be stated as May 39. Since there are 31 days in May, this is 39−31=8 days into the next month, or June 8.

A simpler method of determining the number of elapsed *days* between any two dates is to use the **Julian day** of each date, if the information is available. This also eliminates possible error due to change of calendar if long intervals are involved. The Julian day is the consecutive number of the day starting at 1200 on January 1, 4713 BC. Julian day is listed in the *Astronomical Almanac.*

1811. Time and arc.—The time of day is an indication of the interval since the day began. One day represents one complete rotation of 360° of the earth with respect to a selected celestial point. Each day is divided into 24 **hours** of 60 **minutes,** each minute having 60 **seconds.** Thus, each day has $24\times60=1{,}440$ minutes or $1{,}440\times60=86{,}400$ seconds. This is time regardless of the celestial reference point used, and since the various references are in motion with respect to each other, as "seen" from the earth, apparent solar, mean solar, and sidereal days are of different lengths. Since they all have the same number and kind of fractional parts, these parts are themselves of different length in the different kinds of time. Mean solar units are customarily used to indicate time intervals. The smallest unit normally used in celestial navigation is the second, but in some electronic equipment the **millisecond** (one-thousandth of a second), **microsecond** (one-millionth of a second), and the **millimicrosecond** or **nanosecond** (one-billionth of a second) are used.

Time of day is an indication of the *phase* of rotation of the earth. That is, it indicates how much of a day has elapsed, or what part of a rotation has been completed. Thus, at zero hours the day begins. One hour later, the earth has turned through 1/24 of a day, or 1/24 of 360°, or

$$\frac{360^\circ}{24}=15^\circ.$$

Six hours after the day begins, it has turned through $6/24=1/4$ day, or

$$\frac{360^\circ}{4}=90^\circ.$$

Twelve hours after the start of the day, the day is half gone, having turned through 180°. Smaller intervals can also be stated in angular units, for since 1 hour or 60 minutes is equivalent to 15°, 1 minute of time is equivalent to

$$\frac{15^\circ}{60}=0^\circ\!.25=15',$$

and 1 second of time is equivalent to

$$\frac{15'}{60}=0'.25=15''.$$

Thus,

Time	*Arc*
$1^d=$	$24^h=360°=1$ circle
$60^m=$	$1^h=15°$
	$4^m=1°=60'$
$60^s=$	$1^m=15'$
	$4^s=1'=60''$
	$1^s=15''=0'.25$

Any time interval can be expressed as an angle of rotation, and vice versa. Interconversion of these units can be made by the relationships indicated above.

To convert time to arc:

1. Multiply the hours by 15 to obtain degrees.
2. Divide the minutes of time by four to obtain degrees, and multiply the remainder by 15 to obtain minutes of arc.
3. Divide the seconds of time by four to obtain minutes and tenths of minutes of arc, or multiply the remainder by 15 to obtain seconds of arc.
4. Add degrees, minutes, and tenths (or seconds).

Example 1.—Convert $14^h21^m39^s$ to arc units.

Solution.—

(1) $14^h \times 15 = 210°$

(2) $21^m \div 4 = 5°15'$ (remainder $1^m \times 15 = 15'$)

(3) $39^s \div 4 = 9'45''$ (remainder $3^s \times 15 = 45''$)

(4) $14^h21^m39^s = 215°24'45'' = 215°24'.8$ (to the nearest $0'.1$).

To convert arc to time:

1. Divide the degrees by 15 to obtain hours, and multiply the remainder by four to obtain minutes of time.
2. Divide the minutes of arc by 15 to obtain minutes of time, and multiply the remainder by four to obtain seconds of time.
3. Divide the seconds of arc by 15 to obtain seconds of time.
4. Add hours, minutes, and seconds.

Example 2.—Convert $215°24'45''$ to time units.

Solution.—

(1) $215° \div 15 = 14^h20^m$ (remainder $5° \times 4 = 20^m$)

(2) $24' \div 15 = 1^m36^s$ (remainder $9' \times 4 = 36^s$)

(3) $45'' \div 15 = 3^s$

(4) $215°24'45'' = 14^h21^m39^s$

Example 3.—Convert $161°53'.7$ to time units.

Solution.—

(1) $161° \div 15 = 10^h44^m$ (remainder $11° \times 4 = 44^m$)

(2) $53'.7 \div 15 = 3^m34^s.8$ (remainder $8'.7 \times 4 = 34^s.8$)

(3) $161°53'.7 = 10^h47^m34^s.8 = 10^h47^m35^s$.

The navigator should be able to make these solutions mentally, writing only the answer. As a check, the answer can be converted back to the original value. Solution can also be made by means of arc to time tables in the almanacs. In the

Nautical Almanac the table, given near the back of the volume (app. F), is in two parts, permitting separate entries with degrees, minutes, and quarter minutes of arc. The table is arranged in this manner because the navigator is confronted with the problem of converting arc to time more often than the reverse.

Example 4.—Convert 334°18′22″ to time units, using the *Nautical Almanac* arc to time conversion table.

Solution.—

$$\begin{aligned} 334^\circ &= 22^h16^m \\ 18'.25 &= \underline{\quad\ \ 1^m13^s} \\ 334^\circ18'22'' &= 22^h17^m13^s \end{aligned}$$

The 22″ are converted to the nearest quarter minute of arc for solution to the nearest second of time. Interpolation can be used if more precise results are required, since exact relationships are tabulated in the *Nautical Almanac* conversion table.

Example 5.—Convert 83°29′.6 to time units, using the *Nautical Almanac* arc to time conversion table.

Solution.—

$$\begin{aligned} 83^\circ &= 5^h\,32^m \\ 29'.6 &= \underline{\quad\ \ 1^m58^s.4} \\ 83^\circ29'.6 &= 5^h\,33^m58^s.4 \end{aligned}$$

In this solution, 58ˢ.4 was obtained by eye interpolation in the quarter-minute part of the table.

Example 6.—Convert 17ʰ09ᵐ42ˢ to arc units, using the *Nautical Almanac* arc to time conversion table.

Solution.—

$$\begin{aligned} 17^h08^m &= 257^\circ \\ 1^m\ 42^s &= \underline{\quad\ \ 25'.5} \\ 17^h09^m\ 42^s &= 257^\circ25'.5 \end{aligned}$$

A similar table appears near the back of the *Air Almanac* (app. G); however, quarter minutes of arc are not included.

Example 7.—Convert 334°47′.2 to time units, using the *Air Almanac* arc to time conversion table.

Solution.—

$$\begin{aligned} 334^\circ - 180^\circ = 154^\circ &= 10^h16^m \\ 47'.2 &= \underline{\quad\ \ 3^m09^s} \\ 154^\circ47'.2 &= 10^h19^m09^s \\ 334^\circ47'.2 &= 22^h19^m09^s \end{aligned}$$

Example 8.—Convert 15ʰ13ᵐ18ˢ to arc units, using the *Air Almanac* arc to time conversion table.

Solution.—

$$\begin{aligned} 15^h12^m &= 228^\circ \\ 1^m\ 18^s &= \underline{\quad\ \ 19'.5} \\ 15^h13^m\ 18^s &= 228^\circ19'.5 \end{aligned}$$

Because the almanac conversion tables are exact relationships, interpolation in them can be carried to any degree of precision desired without introducing an error.

1812. Time and longitude.—As indicated in the preceding article, time is a measure of rotation of the earth, and any given time interval can be represented by a corresponding angle through which the earth turns. Suppose the celestial reference point were directly over a certain reference of the earth. An hour later the earth would have turned through 15°, and the celestial reference would be directly over a meridian 15° farther west. Any difference of longitude is a measure of the angle through which the earth must rotate for the local time at the western meridian to become what it was at the eastern meridian before the rotation took place. Therefore, places to the eastward of an observer have *later* time, and those to the westward have *earlier* time, and the difference is exactly equal to the difference in longitude, expressed in time units. When a meridian other than the local meridian is used as the time reference, the difference in time of two places is equal to the difference of longitude of their time reference meridians.

1813. The date line.—Since time becomes later toward the east, and earlier toward the west, time at the lower branch of one's meridian is 12 hours earlier or later depending upon the direction of reckoning. A traveler making a trip around the world gains or loses an entire day. To prevent the *date* from being in error, and to provide a starting place for each day, a **date line** is fixed by international agreement. This line coincides with the 180th meridian over most of its length. In crossing this line, one alters his date by one day. In effect, this changes his time 24 hours to compensate for the slow change during a trip around the world. Therefore, it is applied in the opposite direction to the change of time. Thus, if a person is traveling *eastward* from east longitude to west longitude, time is becoming *later*, and when the date line is crossed, the date becomes 1 day *earlier*. That is, at any moment the date immediately to the *west* of the date line (east longitude) is 1 day *later* than the date immediately to the *east* of the line, except at GMT 1200, when the (mean time) date is the same all over the world. At any other time two dates occur, one boundary between dates being the date line, and the other being the midnight line along the lower branch of the meridian over which the mean sun is located. At GMT 1200 these two boundaries coincide. In the solution of problems, error can sometimes be avoided by converting local time to Greenwich time, and then converting this to local time on the opposite side of the date line. Examples are given in the following articles.

1814. Zone time.—At sea, as well as ashore, watches and clocks are normally set approximately to some form of **zone time (ZT)**. At sea the *nearest* meridian exactly divisible by 15° is usually used as the **time meridian** or **zone meridian.** Thus, within a **time zone** extending 7.°5 on each side of each time meridian the time is the same, and time in consecutive zones differs by exactly one hour. The time is changed as convenient, usually at a whole hour, near the time of crossing the boundary between zones. Each time zone is identified by the number of times the longitude of its zone meridian is divisible by 15°, positive in west longitude and negative in east longitude. This number and its sign, called the **zone description (ZD)**, is the number of whole *hours* that are added to or subtracted from the zone time to obtain **Greenwich mean time (GMT)**, which is the zone time at the Greenwich (0°) meridian, and is often called **Universal Time (UT)**. The mean sun is the celestial reference point for zone time.

Example 1.—For an observer at long. 141°18.′4W the ZT is $6^h18^m24^s$.

Required.—(1) Zone description.

(2) GMT.

Solution.—(1) The nearest meridian exactly divisible by 15° is 135°W, into which 15° will go nine times. Since longitude is west, ZD is (+) 9.

ZT		$6^h18^m24^s$
ZD	(+)	9
(2) GMT		$15^h18^m24^s$

In converting GMT to ZT, a positive ZD is *subtracted,* and a negative one *added,* but its sign remains the same, being part of the description. The word "reversed" (rev.) is written to the right in the work form to indicate that the "reverse" process is to be performed.

Example 2.—The GMT is $15^h27^m09^s$.

Required.—(1) ZT at long. 156°24′.4 W.

(2) ZT at 39°04′.8 E.

Solution.—

(1) GMT		$15^h27^m09^s$		(2) GMT		$15^h27^m09^s$	
ZD	(+)	10	(rev.)	ZD	(−)	3	(rev.)
ZT		$5^h27^m09^s$		ZT		$18^h27^m09^s$	

When time at one place is converted to that at another, the date should be watched carefully. If a sum exceeds 24 hours, subtract this amount and add one day. If 24 hours are added before a subtraction is made, the date at the place is one day *earlier.*

Example 3.—At long. 73°29′.2 W the ZT is $21^h12^m53^s$ on May 14.

Required.—(1) GMT and date.

(2) ZT and date at long. 107°15′.7 W.

Solution.—

ZT		$21^h12^m53^s$	May 14
ZD	(+)	5	
(1) GMT		$2^h12^m53^s$	May 15
ZD	(+)	7	(rev.)
(2) ZT		$19^h12^m53^s$	May 14

The second part of this problem might have been solved by using the *difference* in zone description. Since the second place is two zones farther west, its time is two hours earlier. Problems involving zone times at various places generally involve nothing more than addition or subtraction of one small number, so solutions can generally be made mentally. However, when this forms part of a larger problem, or when a record of the solution is desired, the full solution should be recorded, including labels.

Example 4.—On November 30 the 1430 DR long. of a ship is 51°32′.4 W. Ten hours later the DR long. is 53°07′.2 W.

Required.—ZT and date of arrival at the second longitude.

Solution.—

ZT		1430	Nov. 30
ZD	(+)	3	
GMT		1730	Nov. 30
int.		10	
GMT		0330	Dec. 1
ZD	(+)	4	(rev.)
ZT		2330	Nov. 30

If a time zone boundary had not been crossed, there would have been no need to find GMT. It is particularly helpful to retain this step when the date line is crossed. This line is the *center* of a time zone, the western (east longitude) half being designated (−) 12, and the eastern (west longitude) half (+) 12.

Example 5.—On December 31 the 0800 DR long. of a ship is 177°23′.9E. Forty hours later the DR long. is 171°53′.9 W.

Required.—ZT and date of arrival at the second longitude.

Solution.—

ZT		0800	Dec. 31
ZD	(−)	12	
GMT		2000	Dec. 30
int.		40	
GMT		1200	Jan. 1
ZD	(+)	11	(rev.)
ZT		0100	Jan. 1

Alternative solution

ZT	$31^d08^h00^m$	
ZD	(−)12	
GMT	$30^d20^h00^m$	
int.	1^d16^h	
GMT	$1^d12^h00^m$	
ZD	(+)11	(rev.)
ZT	$1^d01^h00^m$	

For certain communication purposes it is sometimes convenient to designate a time zone by a single letter. The system used is shown in figure 1814.

Use of time zones on land began in 1883, when railroads adopted four standard zones for the continental United States. The division of the United States into time zones was not officially adopted by Congress, however, until March 19, 1918, when a fifth zone was also established for Alaska. The system of time zones is now used almost universally throughout the world, although on land the zone boundaries are generally altered somewhat for convenience. In a few places, half-hour zones are used but these are not standard time zones.

On land, normal zone time is usually called **standard time,** often with an adjective to indicate the zone, as **eastern standard time.** In some areas timepieces are *advanced* one or more hours during the summer to provide greater use of daylight. This "fast" time is called **daylight saving time** in the United States, and **summer time** elsewhere. When time is one hour fast, the zone description is (algebraically) one *less* than normal. When daylight saving or summer time is specified, an advance of one hour is understood unless a greater number is indicated.

Example 6.—What is the standard time and date at Tokyo, long. 140°E, when the daylight saving time at Washington, D.C., long. 77°W, is 1600 on Oct. 5?

Solution.—

ZT		1600	Oct. 5
ZD	(+)	4	
GMT		2000	Oct. 5
ZD	(−)	9	(rev.)
ZT		0500	Oct. 6

During hostilities daylight saving time may be kept all year long throughout a nation, and designated **war time.**

1815. Chronometer time (C) is time indicated by a chronometer. Since a chronometer is set approximately to GMT, and not reset until it is overhauled and cleaned, perhaps three years later (art. 1514), there is nearly always a **chronometer error (CE),** either *fast* (F) or *slow* (S). The change in chronometer error in 24 hours is called **chronometer rate,** or **daily rate,** and designated *gaining* or *losing.* With a

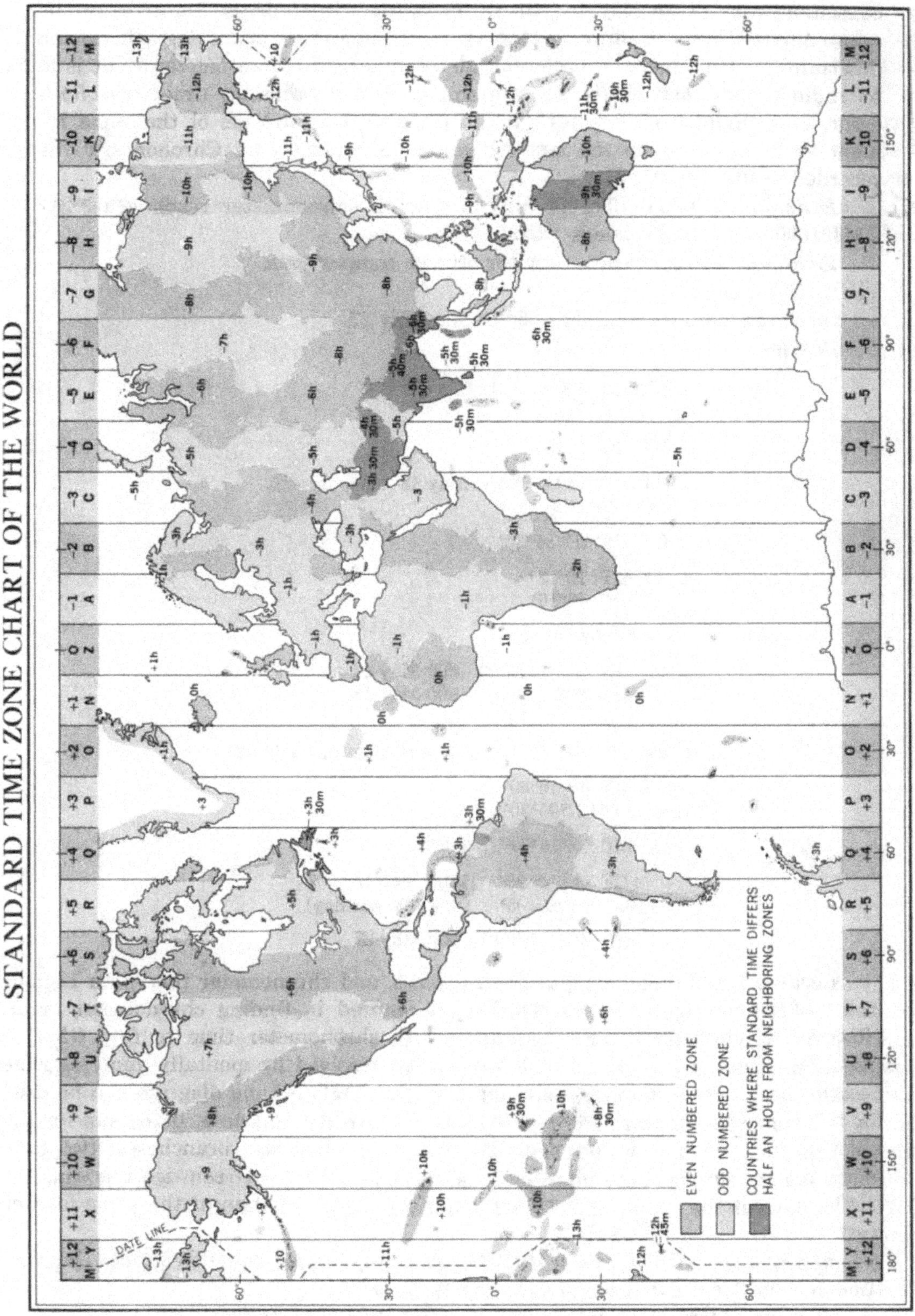

FIGURE 1814.—Time zone chart of the world.

consistent rate of 1^s per day for three years, the chronometer error would be approximately 18^m. Since chronometer error is subject to change, it should be determined from time to time, preferably daily at sea. Chronometer error is found by radio time signal (art. 1826), by comparison with another timepiece of known error, or by applying chronometer rate to previous readings of the same instrument. It is recorded to the nearest whole or half second. Chronometer rate is recorded to the nearest $0^s.1$.

Example 1.—At GMT 1200 on May 12 the chronometer reads $12^h04^m21^s$. At GMT 1600 on May 18 it reads $4^h04^m25^s$.

Required.—(1) Chronometer error at both comparisons.

(2) Chronometer rate.

(3) Chronometer error at GMT 0530 on May 27.

Solution.—

GMT	$12^h00^m00^s$	May 12
C	$12^h04^m21^s$	
(1) CE	(F) 4^m21^s	
GMT	$16^h00^m00^s$	May 18
C	$4^h04^m25^s$	
(1) CE	(F) 4^m25^s	
GMT	12^d12^h	
GMT	18^d16^h	
diff.	$6^d04^h = 6^d.2$	
CE	(F) 4^m21^s	1200 May 12
CE	(F) 4^m25^s	1600 May 18
diff.	4^s	gained
(2) daily rate	$0^s.6$	per day, gaining. ($4^s \div 6^d.2$)
GMT	$18^d16^h00^m$	
GMT	$27^d05^h30^m$	
diff.	$8^d13^h30^m$	$= 8^d.5$
CE	(F) 4^m25^s	1600 May 18
corr.	$(+)5^s$	($8^d.5 \times 0^s.6$ per day)
(3) CE	(F) 4^m30^s	0530 May 27

Because GMT is stated on a 24-hour basis, and chronometer time on a 12-hour basis, a 12-hour ambiguity exists. This is ignored in finding chronometer error. However, if chronometer error is applied to chronometer time to find GMT, a possible 12-hour error can result. This can be resolved by mentally applying zone description to local time to obtain approximate GMT. A time diagram can be used for resolving doubt as to approximate GMT and Greenwich date. If the sun for the kind of time used (mean or apparent) is between the *lower* branches of two time meridians (as the standard meridian for local time, and the Greenwich meridian for GMT), the date at the place farther *east* is one day *later* than at the place farther *west.*

Example 2.—On August 14 the DR long. of a ship is about 124°E, and the zone time is about 0500. Chronometer error is 12^m27^s slow.

Required.—GMT and date when the chronometer reads $8^h44^m22^s$.

Solution.—

approx.	ZT	0500	Aug. 14
	ZD	(−)8	
approx.	GMT	2100	Aug. 13

C	$8^h44^m22^s$	
CE	(S) 12^m27^s	
GMT	$20^h56^m49^s$	Aug. 13

The *A* chronometer, usually the best (having the most nearly uniform rate), is compared directly with the time signal (art. 1826). Other chronometers, designated *B*, *C*, etc., may then be compared with the *A* chronometer.

Example 3.—At GMT 1400 chronometer *A* is checked by time signal, and found to read $1^h57^m09^s$. A little later, when it reads $2^h05^m00^s$, chronometer *B* reads $2^h11^m38^s$.

Required.—(1) Error of chronometer *A*.

(2) Error of chronometer *B*.

Solution.—

	GMT	$14^h00^m00^s$
	C_A	$1^h57^m09^s$
(1)	CE_A	(S) 2^m51^s
	C_A	$2^h05^m00^s$
	GMT	$14^h07^m51^s$
	C_B	$2^h11^m38^s$
(2)	CE_B	(F) 3^m47^s

If time signals are not available at the chronometer, a good comparing watch (art. 1516) should be compared with the radio signal, and this watch used to determine chronometer error, as indicated in example 3, substituting the watch for chronometer *A*.

1816. Watch time (WT) is time indicated by a watch. This is usually an approximation of zone time, except that for timing celestial observations it is good practice to set a comparing watch (art. 1516) to GMT. If the watch has a second setting hand, the watch can be set exactly to ZT or GMT, and the time is so designated. If the watch is not set exactly to one of these times, the difference is known as **watch error (WE)**, labeled *fast* (F) or *slow* (S) to indicate whether the watch is ahead of or behind the correct time, respectively.

If a watch is to be set exactly to ZT or GMT, it is set to some whole minute slightly ahead of the correct time, and stopped. When the set time arrives, the watch is started. It should then be checked for accuracy.

Example 1.—A chronometer 9^m46^s fast on GMT reads approximately 7^h23^m. At the next whole five minutes of GMT a comparing watch is to be set to GMT exactly.

Required.—(1) What should the watch read at the moment of starting?

(2) What should the chronometer read?

Solution.—

	C	$7^h23^m00^s$	
	CE	(F) 9^m46^s	
	GMT	$7^h13^m14^s$	
(1)	GMT	$7^h15^m00^s$	(next whole 5^m)
	CE	(F) 9^m46^s	
(2)	C	$7^h24^m46^s$	

The GMT may be in error by 12^h, but if the watch is graduated to 12 hours, this will not be reflected. If a watch with a 24-hour dial is used, the actual GMT should be determined.

If watch error is to be determined, it is done by comparing the reading of the watch with that of the chronometer at a selected moment. This may be at some selected GMT, as in example 1.

Example 2.—If, in example 1, the watch had read $7^h14^m48^s$ at the moment the chronometer read $7^h24^m46^s$, what would be the watch error on GMT?

Solution.—

GMT	$7^h15^m00^s$
WT	$7^h14^m48^s$
WE	(S) 12^s

A more convenient chronometer time might be selected, as a whole minute.

Example 3.—A watch is set to zone time approximately. The longitude is about 48°W. The watch is compared with a chronometer which is 19^m44^s fast on GMT. When the chronometer reads $5^h22^m00^s$, the watch reads $2^h01^m53^s$.

Required.—Watch error on zone time.

Solution.—

C	$5^h22^m00^s$
CE	(F) 19^m44^s
GMT	$5^h02^m16^s$
ZD	(+)3 (rev.)
ZT	$2^h02^m16^s$
WT	$2^h01^m53^s$
WE	(S) 23^s

The possible 12^h error is not of significance. When such a watch is used for determining GMT, however, as for entering an almanac, the 12-hour ambiguity is important. Unless a watch is graduated to 24 hours, its time is designated AM before noon and PM after noon.

Example 4.—On January 3 the DR long. is 94°14.′7E. An observation of the sun is made when the watch reads $12^h16^m23^s$ PM. The watch is 22^s fast on zone time.

Required.—GMT and date.

Solution.—

WT	$12^h16^m23^s$	PM Jan. 3
WE	(F) 22^s	
ZT	$12^h16^m01^s$	
ZD	(−)6	
GMT	$6^h16^m01^s$	Jan. 3

Note that between 1200 and 1300 watch designations are PM. Between 0000 and 0100 they are AM.

Comparison of a watch and a chronometer should be made carefully. If two observers are available, one can give a warning "stand-by" a few seconds before the selected time, and a "mark" at the appointed moment, while the other notes the time of the watch. A single observer can make a satisfactory comparison by counting with the chronometer. Chronometers beat in half seconds, with an audible "tick." Ten seconds before the selected time (perhaps a whole minute), the observer starts counting with the beats, as he watches the chronometer second hand, "50, and, 1, and, 2, and, 3, and, . . . 9, and, mark." During the count the observer shifts his view from the chronometer to the second hand of the watch, continuing to count in cadence with the chronometer beats. At the "mark," the second, minute, and hour hands of the watch are read in that order, and the time recorded. A comparison of this time with the GMT or ZT corresponding to the selected chronometer time indicates the watch error.

Even though a watch is set to zone time approximately, its error on GMT can be determined and used for timing observations. In this case the 12-hour ambiguity in GMT should be resolved, and a time diagram used to avoid possible error. This method requires additional work, and presents a greater probability of error, without compensating advantages.

Still another method of determining GMT, generally used before zone time came into common use at sea, is to *subtract* watch time from chronometer time, to find C−WT. This is then *added* to the watch time of an observation to obtain chronometer time (C−WT+WT=C). Chronometer error is then applied to the result to obtain GMT. A time diagram should *always* be used with this method, to resolve the 12-hour ambiguity and to be sure of the correct Greenwich date, unless an auxiliary solution is made using approximate ZT and ZD. This method has little to recommend it.

If a watch has a **watch rate** of more than a few seconds per day, watch error should be determined both before and after a round of sights, and any difference distributed proportionally among observations.

If a stopwatch is used for timing observations, it should be started at some convenient GMT, as a whole 5^m or 10^m. The time of each observation is then this GMT *plus* the reading of the watch.

1817. Local mean time (LMT), like zone time, uses the mean sun as the celestial reference point. It differs from zone time in that the local meridian is used as the terrestrial reference, rather than a zone meridian. Thus, the local mean time at each meridian differs from that of every other meridian, the difference being equal to the difference of longitude, expressed in time units. At each zone meridian, including 0°, LMT and ZT are identical.

Example 1.—At long. 124°37.′2W the LMT is $17^h24^m18^s$ on March 21.

Required.—(1) GMT and date.

(2) ZT and date at the place.

Solution.—

LMT	$17^h24^m18^s$	Mar. 21
λ	$8^h18^m29^s$W	
(1) GMT	$1^h42^m47^s$	Mar. 22
ZD (+)	8	(rev.)
(2) ZT	$17^h42^m47^s$	Mar. 21

In navigation the principal use of LMT is in rising, setting, and twilight tables. The problem is usually one of converting the LMT taken from the table to ZT. At sea, the difference between these times is normally not more than 30^m, and the conversion is made directly, without finding GMT as an intermediate step. This is done by applying a correction equal to the difference of longitude (dλ). If the observer is *west* of his time meridian, the correction is *added*, and if *east* of it, the correction is *subtracted*. If Greenwich time is desired, it is found from ZT.

Example 2.—At long. 63°24.′4E the LMT is 0525 on January 2.

Required.—(1) ZT and date.

(2) GMT and date.

Solution.—

LMT	0525	Jan. 2
dλ	(−) 14	
(1) ZT	0511	Jan. 2
ZD	(−)4	
(2) GMT	0111	Jan. 2

On land, with an irregular zone boundary, the longitude may differ by more than 7°.5 (30^m) from the time meridian.

If LMT is to be corrected to daylight saving time, the difference in longitude between the local and time merdian can be used, or the ZT can first be found and then increased by one hour.

Conversion of ZT (including GMT) to LMT is the same as conversion in the opposite direction, except that the sign of dλ is reversed. This problem is not normally encountered in navigation.

1818. Apparent time utilizes the **apparent** (real) **sun** as its celestial reference, and a meridian as the terrestrial reference. **Local apparent time (LAT)** uses the local meridian. The LAT at the 0° meridian is called **Greenwich apparent time (GAT).**

The LAT at one meridian differs from that at any other by the difference in longitude of the two places, the place to the eastward having the later time, and conversion is the same as converting LMT at one place to LMT at another.

Use of the apparent sun as a celestial reference point for time results in time of nonconstant rate for at least three reasons. First, revolution of the earth in its orbit is not constant. Second, motion of the apparent sun is along the ecliptic, which is tilted with respect to the celestial equator, along which time is measured. Third, rotation of the earth on its axis is not constant. The effect due by this third cause is extremely small.

For the various forms of mean time, the apparent sun is replaced by a fictitious **mean sun** conceived as moving eastward along the *celestial equator* at a uniform speed equal to the average speed of the apparent sun along the ecliptic, thus providing a nearly uniform measure of time equal to the approximate average apparent time. At any moment the accumulated difference between LAT and LMT is indicated by the **equation of time (Eq. T)**, which reaches a maximum value of about $16^m.4$ in November. This quantity is tabulated at 12-hour intervals at the bottom of the right-hand daily page of the *Nautical Almanac*. In the United States, the sign is considered positive (+) if the time of sun's "Mer. Pass." is earlier than 1200, and negative (−) if later than 1200. If the "Mer. Pass." is given as 1200 (as on June 12–14, 1975), the sign is positive if the GHA at GMT 1200 is between 0° and 1°, and negative if it is greater than 359°. The sign is correct for conversion of GMT to GAT. In Great Britain, this convention is reversed. Since GMT is the entering argument for the almanacs, interconversion of apparent and mean time should preferably be made from Greenwich time, rather than from local time.

Example.—Find the LAT and date at ZT $15^h10^m40^s$ on May 31, 1975, for long. 73°18′.4W.

Solution.—

ZT	$15^h10^m40^s$	May 31
ZD (+)	5	
GMT	$20^h10^m40^s$	May 31
Eq. T (+)	2^m25^s	
GAT	$20^h13^m05^s$	May 31
λ	$4^h53^m14^s$	W
LAT	$15^h19^m51^s$	May 31

In conversion from apparent to mean time, a second solution may be needed if the equation of time is large and changing rapidly, using the GAT for entering the almanac for the first solution, and using the GMT from this solution as the almanac entry value for the second solution.

Apparent time can also be found by converting hour angle to time units, and adding or subtracting 12 hours. If LAT is required, but not GAT, conversion of arc to time should be made from LHA, rather than GHA, to avoid the need for conversion of longitude to time units. Equation of time can be found by *subtracting* mean time from apparent time at the same meridian. This method of finding apparent time and equation of time is the only one available with the *Air Almanac,* which does not tabulate equation of time.

The navigator has little or no use for apparent time, as such. However, it can be used for finding the time of **local apparent noon (LAN)**, when the apparent sun is on the celestial meridian.

The mean sun averages out the irregularities in time due to the variations of the speed of revolution of the earth in its orbit and the fact that the apparent sun moves in the ecliptic while hour angle is measured along the celestial equator. It does not eliminate the error due to slight variations in the *rotational* speed of the earth. When a correction for the accumulated error from this source is applied to mean time, **Ephemeris Time** results. This time is of interest to astronomers, but is not used directly by the navigator.

1819. Sidereal time uses the first point of Aries (vernal equinox) as the celestial reference point. Since the earth revolves around the sun, and since the direction of the earth's rotation and revolution are the same, it completes a rotation with respect to the stars in less time (about $3^m56^s.6$ of mean solar units) than with respect to the sun, and during one revolution about the sun (1 year) it makes one complete rotation more with respect to the stars than with the sun. This accounts for the daily shift of the stars nearly 1° westward each night. Hence, sidereal days are shorter than solar days, and its hours, minutes, and seconds are correspondingly shorter. Because of nutation (art. 1417) sidereal time is not quite constant in rate. Time based upon the average rate is called **mean sidereal time,** when it is to be distinguished from the slightly irregular sidereal time. The ratio of mean solar time units to mean sidereal time units is 1:1.00273791.

The sidereal *day* begins when the first point of Aries is over the *upper* branch of the meridian, and extends through 24 hours of sidereal time. The sun is at the first point of Aries at the time of the vernal equinox, about March 21. However, since the solar day begins when the sun is over the *lower* branch of the meridian, apparent solar and sidereal times differ by 12 hours at the vernal equinox. Each month thereafter, sidereal time *gains* about 2 hours on solar time. By the time of the summer solstice, about June 21, sidereal time is 18 hours *ahead* or 6 hours *behind* solar time. By the time of the autumnal equinox, about September 23, the two times are together, and by the time of the winter solstice, about December 22, the sidereal time is 6 hours *ahead* of solar time. There need be no confusion of the date, for there is no sidereal date.

Local sidereal time (LST) uses the local meridian as the terrestrial reference. At the prime meridian this is called **Greenwich sidereal time (GST).** The difference between LST at two meridians is equal to the difference of longitude between them, the place to the eastward having the later time. Local sidereal time is LHA♈ expressed in time units. To determine LST at any given moment, find GHA♈ by means of an almanac, and then apply the longitude to convert it to LHA♈. Then convert LHA♈ in arc to LST in time units.

Example.—Find LST at ZT $8^h25^m51^s$ on May 31, 1975, for long. 103°16′.3E.

Solution.—

ZT	$8^h25^m51^s$	May 31
ZD	(−)7	
GMT	$1^h25^m51^s$	May 31
1^h	262°54′.8	
25^m51^s	6°28′.8	
GHA♈	269°23′.6	
λ	103°16′.3	E
LHA♈	12°39′.9	
LST	$0^h50^m40^s$	

Unless GST is required, conversion from arc to time units should be made from LHA♈, rather than from GHA♈, to avoid the need for converting longitude from arc to time units.

Conversion of sidereal to solar time is the reverse. Local sidereal time is converted to arc (LHA♈), and the longitude is applied to find GHA♈. This is used as an argument for entering the almanac to determine GMT, which can then be converted to any other kind of time desired. This is similar to one method of finding time of meridian transit, described in article 2104. Normally, the problem is not encountered by the navigator.

Sidereal time, as such, is little used by the navigator. It is the basis of star charts (art. 2204) and star finders (art. 2210), and certain sight reduction methods (notably Pub. No. 249), but generally in the form LHA♈. This kind of time is used for these purposes because its celestial reference point remains almost fixed in relation to the stars. Sidereal time is used by astronomers to regulate mean time. Timepieces regulated to sidereal time can be purchased.

1820. Time and hour angle.—Both time and hour angle are a measure of the phase of rotation of the earth, since both indicate the angular distance of a celestial reference point *west* of a terrestrial reference meridian. Hour angle, however, applies to *any* point on the celestial sphere. Time might be used in this respect, but only the apparent sun, mean sun, the first point of Aries, and occasionally the moon are commonly used.

Hour angles are usually expressed in arc units, and are measured from the upper branch of the celestial meridian. Time is customarily expressed in time units. Sidereal time is measured from the upper branch of the celestial meridian, like hour angle, but solar time is measured from the lower branch. Thus, LMT=LHA mean sun plus or minus 180°, LAT=LHA apparent sun plus or minus 180°, and LST=LHA♈.

As with time, **local hour angle (LHA)**, based upon the local celestial meridian, at two places differs by the longitude between them, and LHA at longitude 0° is called **Greenwich hour angle (GHA)**. In addition, it is often convenient to express hour angle in terms of the *shorter* arc between the local celestial meridian and the body. This is similar to measurement of longitude from the Greenwich meridian. Local hour angle measured in this way is called **meridian angle (t)**, which is labeled *east* or *west,* like longitude, to indicate the direction of measurement. A westerly meridian angle is numerically equal to LHA, while an easterly meridian angle is equal to 360° − LHA; also, LHA = t (W), and LHA = 360° − t (E). Meridian angle is used in the solution of the navigational triangle (art. 1433).

Example 1.—Find LHA and t of the sun at GMT $3^h24^m16^s$ on June 1, 1975, for long. 118°48'.2W.

Solution.—

GMT	$3^h24^m16^s$	June 1
3^h	225°35'.7	
24^m16^s	6°04'.0	
GHA	231°39'.7	
λ	118°48'.2	W
LHA	112°51'.5	
t	112°51'.5	W

Example 2.—Find LHA and t of Kochab at ZT $18^h24^m47^s$ on May 31, 1975, for long. 55°27'.3W.

Solution.—

	Kochab	
ZT	$18^h24^m47^s$	May 31
ZD	(+)4	
GMT	$22^h24^m47^s$	May 31
22^h	218°46'.5	
24^m47^s	6°12'.8	
SHA	137°17'.7	
GHA	2°17'.0	
λ	55°27'.3	W
LHA	306°49'.7	
t	53°10'.3	E

1821. The legal basis of standard time in the United States is contained in the "Uniform Time Act of 1966" (Public Law 89–387) and the U. S. Code, Title 15. This act reiterates the policy of the United States to "promote the adoption and observance of uniform time within prescribed Standard Time Zones . . ." and establishes the annual advancement and retardation of standard time by 1 hour the last Sunday of April and October, respectively. The Department of Transportation is the agency designated for enforcement of the law.

The "Uniform Time Act" establishes eight Standard Time Zones for the United States (fig. 1821) and notes that standard time is based on the mean solar time of specified longitudes. The reference meridians are spaced 15° apart in longitude beginning with the meridian through Greenwich, England. Time zones extend 7½° in longitude on each side with considerable variation in boundaries to conform to political or geographic boundaries or both. Since the time zones are 15° apart, the time difference between two adjacent zones is 1 hour.

A comprehensive delineation of these zones is given in the **Code of Federal Regulations,** entitled "Standard Time Zone Boundaries."

This system of time zones is now used almost universally throughout the world, although on land the zone boundaries are generally altered somewhat for convenience (fig. 1814). In a few places, half-hour zones are used.

STANDARD TIME ZONES OF THE UNITED STATES

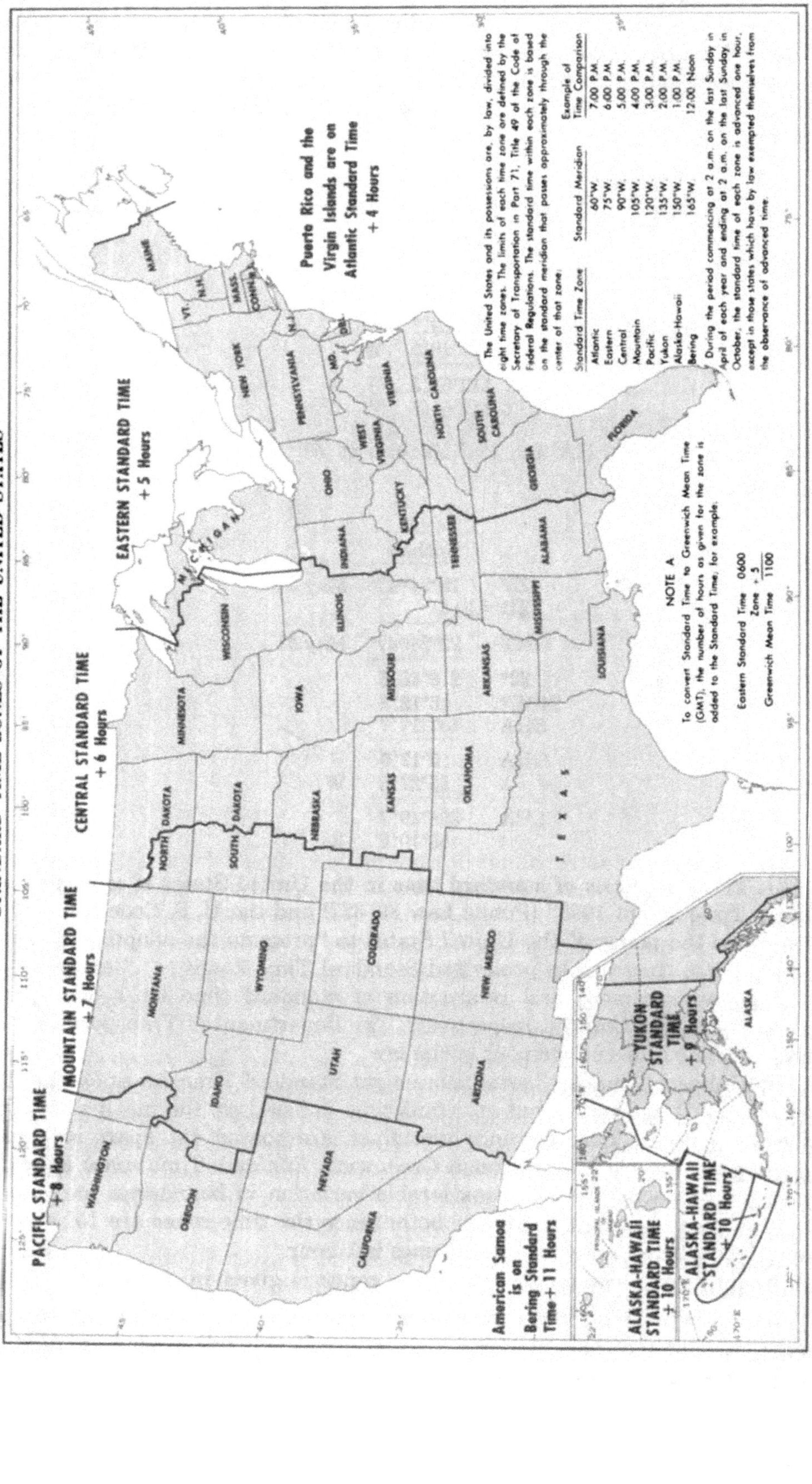

The standard times used in various countries and places are tabulated in the almanacs.

Radio Dissemination of Time Signals

1822. Dissemination systems.—Of the many systems for time and frequency dissemination, the majority employ some type of radio transmission, either in dedicated time and frequency emissions or established systems such as radionavigation systems and television. The most accurate means of time and frequency dissemination today is through on site visits or aircraft flyovers with portable atomic clocks.

Radio time signals can be used either to perform a clock function or to set clocks. When one uses a radio wave instead of a clock, however, new considerations evolve. One is the delay time of approximately 3 microseconds per kilometer it takes the radio wave to propagate and arrive at the reception point. Thus, a user 1,000 kilometers from a transmitter receives the time signal about 3 milliseconds later than the on-time transmitter signal. If time is needed to better than 3 milliseconds, correction must be made for the signal to pass through the receiver.

In most cases standard time and frequency emissions as received are more than adequate for ordinary needs. However, many systems exist for the more exacting scientific requirements.

Astronomers, geodesists, navigators, and others using time based on the earth's rotation (UT1) require that the emissions of Coordinated Universal Time (UTC) also include the difference between UTC and UT1. This difference is discussed in art. 1827.

1823. Characteristic elements of dissemination systems.—A number of common elements characterize most time and frequency dissemination systems. Among the more important elements are accuracy, ambiguity, repeatability, coverage, availability of time signal, reliability, ease of use, cost to the user, and the number of users served. There does not now appear to be any single system which incorporates all desired characteristics. The relative importance of these characteristics will vary from one user to the next, and the kind of compromise solution for one user may not be satisfactory to another. These common elements are discussed in the following examination of a possible radio signal.

Consider a very simple system consisting of an unmodulated 10-kHz signal as shown in figure 1823. A positive going zero-crossing of this signal, leaving the transmitter at 0000 UTC, will reach the receiver at a later time equivalent to the propagation delay. The user must know this delay because the accuracy of his knowledge of time can be no better than the degree to which this delay is known. (By *accuracy* is meant the degree of conformity to some specified value or definition.) Since all cycles of the signal are identical, the signal is *ambiguous* and the user must somehow decide which cycle is the "on time" cycle. This means, in the case of the hypothetical 10-kHz signal, that the user must know the time to ±50 microseconds (half the period of the signal). Further, the user may desire to use this system, say once a day, for an extended period of time to check his clock or frequency standard. However, it may be that the delay will vary from one day to the next, and if the user is unaware of this variation, his accuracy will be limited by the lack of *repeatability* of the signal arrival time.

Many users are interested in making time coordinated measurements over large geographic areas. They would like all measurements to be referenced to one time system to eliminate corrections for different time systems used at scattered or

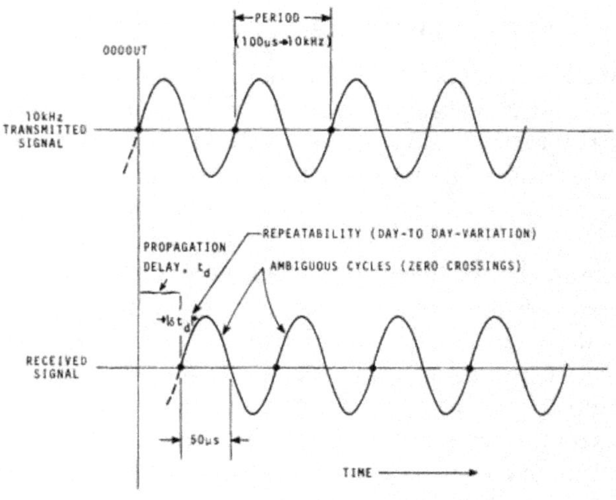

FIGURE 1823.—Single tone time dissemination.

remote locations. This is a very important practical consideration when measurements are undertaken in the field. In addition, a one reference system, such as a single time broadcast, increases confidence that all measurements can be related to each other in some known way. Thus, the *coverage* of a system is an important concept. Another important characteristic of a timing system is the *percent of time available.* The man on the street who has to keep an appointment needs to know the time perhaps to a minute or so. Although he requires only coarse time information, he wants it on demand so he carries a wristwatch that gives the time to him 24 hours a day. On the other hand, a user who needs time to a few microseconds employs a very good clock which only needs an occasional update, perhaps only once or twice a day. An additional characteristic of time and frequency dissemination is *reliability,* i.e., the likelihood that a time signal will be available when scheduled. Propagation fadeout can sometimes prevent reception of HF signals.

1824. Radio propagation factors.—Radio has offered good means of transferring standard time and frequency signals since the early 1900's. As opposed to the physical transfer of time via portable clocks, the transfer of information by radio entails propagation of electromagnetic energy through some propagation medium from a transmitter to a distant receiver.

In a typical standard frequency and time broadcast, the signals are directly related to some master clock and are transmitted with little or no degradation in accuracy. In a vacuum and noise free background, the signals should be received at a distant point essentially as transmitted, except for a constant path delay with the radio wave propagating near the speed of light (i.e., 299,773 kilometers per second). The propagation media, including the earth, atmosphere, and ionosphere, as well as physical and electrical characteristics of transmitters and receivers, influence the stability and accuracy of received radio signals, dependent upon the frequency of the transmission and length of signal path. Propagation delays are affected in varying degrees by extraneous radiations in the propagation media, solar disturbances, diurnal effects, and weather conditions, among others.

Radio dissemination systems can be classified in a number of different ways. One way is to divide those carrier frequencies low enough to be reflected by the ionosphere (below 30 MHz) from those sufficiently high to penetrate the ionosphere (above 30 MHz). The former can be observed at great distances from the transmitter but suffer from ionospheric propagation anomalies that limit accuracy; the latter are restricted to line-of-sight applications but show little or no signal deterioration caused by propagation anomalies. The most accurate systems tend to be those which use the higher, line-of-sight frequencies, while broadcasts of the lower carrier frequencies show the greatest number of users.

1825. Standard time broadcasts.—The World Administrative Radio Council (WARC) has allocated certain frequencies in five bands for standard frequency and time signal emission as shown in table 1825. For such dedicated standard frequency transmissions, the International Radio Consultative Committee (CCIR) recommends that carrier frequencies be maintained so that the average daily fractional frequency deviations from the internationally designated standard for measurement of time interval should not exceed 1×10^{-10}. The U.S. Naval Observatory *Time Service Announcement Series 1, No. 2,* gives characteristics of standard time signals that are assigned to allocated bands, as reported by the CCIR.

1826. Time signals.—The usual method of determining chronometer error and daily rate is by radio time signals, popularly called **time ticks.** Most maritime nations broadcast time signals several times daily from one or more stations, and a vessel equipped with radio receiving equipment normally has no difficulty in obtaining a time tick anywhere in the world. Normally, the time transmitted is maintained virtually uniform with respect to atomic clocks. The Coordinated Universal Time (UTC) as received by a vessel may differ from UT1 (GMT) by as much as $0\overset{s}{.}9$ (art. 1828).

The majority of radio time signals are transmitted automatically, being controlled by the standard clock of an astronomical observatory. Absolute reliance may be had in these signals, and they should be correct to 0.05 second. Some stations transmit by a combination of hand and automatic signals, and care should be exercised to differentiate between the two at the time of actual comparison of the chronometer.

Other radio stations, however, have no automatic transmission system installed, and the signals are given by hand. In this instance the operator is guided by the standard clock at the station. The clock is checked by either astronomical observations or by reliable time signals. The hand transmission should be correct to 0.25 second.

At sea the spring-driven chronometer should be checked daily by radio time signal, and in port daily checks should be maintained, or begun at least three days prior to departure, if conditions permit. Error and rate are entered in the chronometer record book (or record sheet) each time they are determined.

The various time signal systems used throughout the world are discussed in Pubs. Nos. 117A and 117B, *Radio Navigational Aids,* and volume 5 of *Admiralty List of Radio Signals.* Only the United States signals are discussed here.

Band No.	Designation	Frequency Range
4	VLF (Very Low Frequency).	20.0 kHz±50 Hz.
6	MF (Medium Frequency).	2.5 MHz±5 kHz.
7	HF (High Frequency)	5.0 MHz±5 kHz. 10.0 MHz±5 kHz. 15.0 MHz±10 kHz. 20.0 MHz±10 kHz. 25.0 MHz±10 kHz.
9	UHF (Ultra High Frequency).	400.1 MHz±25 kHz (satellite).
10	SHF (Super High Frequency).	4.202 GHz±2 MHz (satellite-space to earth). 6.427 GHz±2 MHz (satellite-earth to space).

TABLE 1825.—International standard time and frequency radio assignments.

The U.S. Naval Observatory at Washington, D.C., controls the transmissions of time signals from U. S. Naval radio stations. Beginning at 5 minutes before each even hour of GMT, dashes are transmitted on every second, except the 29th and certain others near the end of each minute, as shown in the following diagram:

Minutes	Seconds										
	50	51	52	53	54	55	56	57	58	59	60
55	—		—	—	—	—					—
56	—	—		—	—	—					—
57	—	—	—		—	—					—
58	—	—	—	—		—					—
59	—										——

The seconds marked "60" indicate the start of the next minute. The final dash, marking the hour, is considerably longer than any of the others. The number of dashes in the group near the end of any minute indicates the number of minutes before the hour. This is known as the **United States system.** In all cases the beginnings of the dashes indicate the beginning of the seconds, and the ends of the dashes are without significance.

Although the broadcasts of the National Bureau of Standards (NSB) stations WWV and WWVH are intended primarily for dissemination of frequency and time *interval* for scientific purposes, time ticks are also provided.

Station WWV broadcasts from Fort Collins, Colorado at the international allocated frequencies of 2.5, 5.0, 10.0, 15.0, 20.0, and 25.0 MHz; station WWVH transmits from Kauai, Hawaii on the same frequencies with the exclusion of 25.0 MHz. The hourly broadcast formats are shown in figure 1826. The broadcast signals include standard time and frequencies and various voice announcements. Details of these broadcasts are given in NBS Special Publication 236, *NBS Frequency and Time Broadcast Services.* Both HF emissions are directly controlled by cesium beam frequency standards with periodic reference to the NBS atomic frequency and time standards; corrections are published monthly.

The time ticks in the WWV and WWVH emissions are shown in figure 1826. The 1-second UTC markers are transmitted continuously by WWV and WWVH, except for omission of the 29th and 59th marker each minute. With the exception of

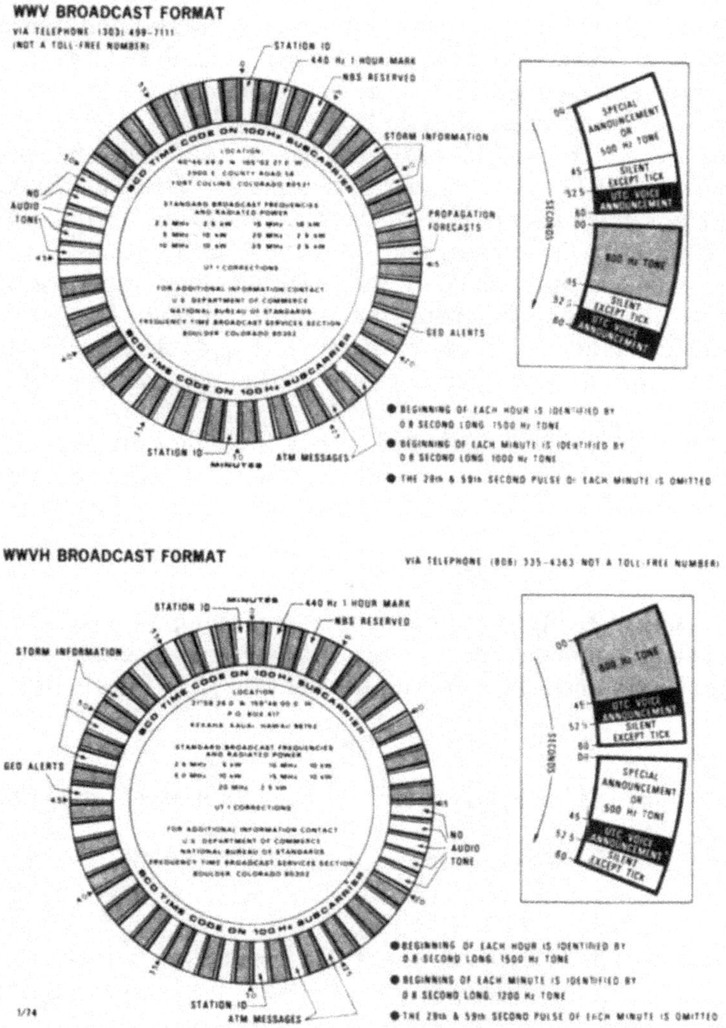

FIGURE 1826.—Broadcast format of stations WWV and WWVH.

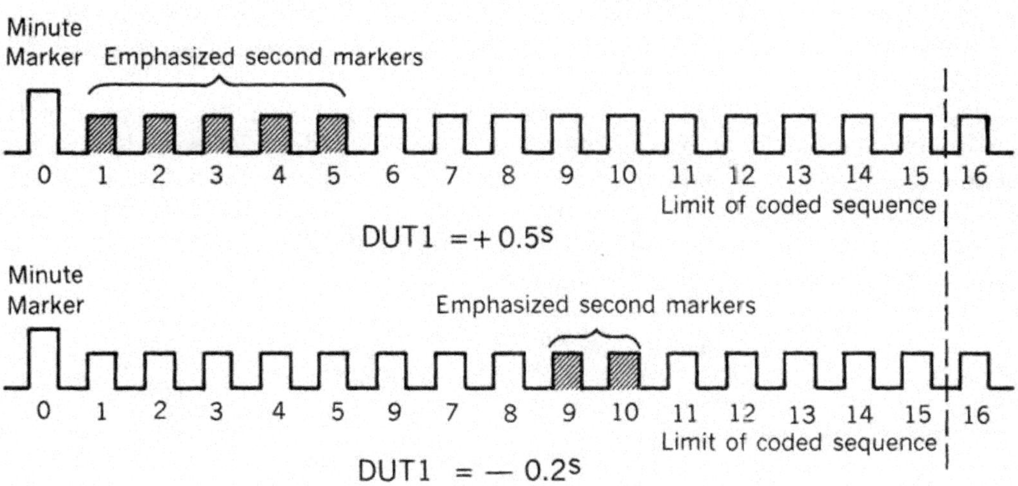

FIGURE 1827a.—CCIR code for transmission of DUT1.

the beginning tone at each minute (800 milliseconds) all 1-second markers are of 5 milliseconds duration. Each pulse is preceded by 10 milliseconds of silence and followed by 25 milliseconds of silence. Time voice announcements are given also at 1-minute intervals. All time announcements are UTC.

1827. Codes for the transmission of DUT1.—The difference between UTC and and UT1 is known as DUT1, the relationship being DUT1=UT1−UTC. By means of a coding system incorporated in the actual emissions, primary time signal sources promulgate DUT1 in integral multiples of 100 milliseconds.

The CCIR standard format is in the form of emphasized second markers utilizing the first 15 seconds following the minute marker. The emphasis of the second markers can take the form of lengthening, doubling, splitting, or tone modulation of the normal seconds markers. Each emphasized second represents a DUT1 value of $0\overset{s}{.}1$.

A positive value of DUT1 is indicated by emphasizing a number (n) of consecutive seconds markers following the minute marker from seconds markers one to seconds marker (n) inclusive; (n) is an integer from 1 to 8 inclusive (fig. 1827a).

$$\mathrm{DUT1}=(\mathrm{n}\times 0.1)\mathrm{s}.$$

A negative value of DUT1 is indicated by emphasizing a number (m) of consecutive seconds markers following the minute marker from seconds marker nine to seconds marker (8+m) inclusive; (m) is an integer from 1 to 8 inclusive.

$$\mathrm{DUT1}=-(\mathrm{m}\times 0.1)\mathrm{s}.$$

A zero value of DUT1 is indicated by the absence of emphasized seconds markers.

The National Bureau of Standards stations WWV and WWVH transmit DUT1 using the CCIR standard format. The CCIR standard format is used by most coordinated stations, including CHU, Canada.

In the **USSR extended format,** the CCIR format is followed for DUT1. In addition dUT1 is given to specify more precisely the difference UT1−UTC to multiples of $0\overset{s}{.}02$, the total value of the correction being DUT1+dUT1. Positive values of dUT1 are transmitted by the marking of p second markers between the 21st and 24th second (fig. 1827b) so that dUT1=(p×0.02)s. Negative values of dUT1 are transmitted by the marking of q second markers between the 31st and 34th second, so that dUT1=−(q×0.02)s.

DUT1 may also be given by voice announcement or in Morse code. In the **Morse code method,** U. S. Naval Radio Stations use standard Morse code (15 words per minute) between seconds 56 and 59 inclusive of each minute not used for time ticks to indicate the sign and value in tenths of a second of DUT1. Positive values are indicated by the letter "A" and the appropriate digit; negative values are indicated by the letter "S" and the appropriate digit.

Standard Morse

A ·—	S···
1 ·————	6—····
2 ··———	7——···
3 ···——	8———··
4 ····—	9————·
5 ·····	0—————

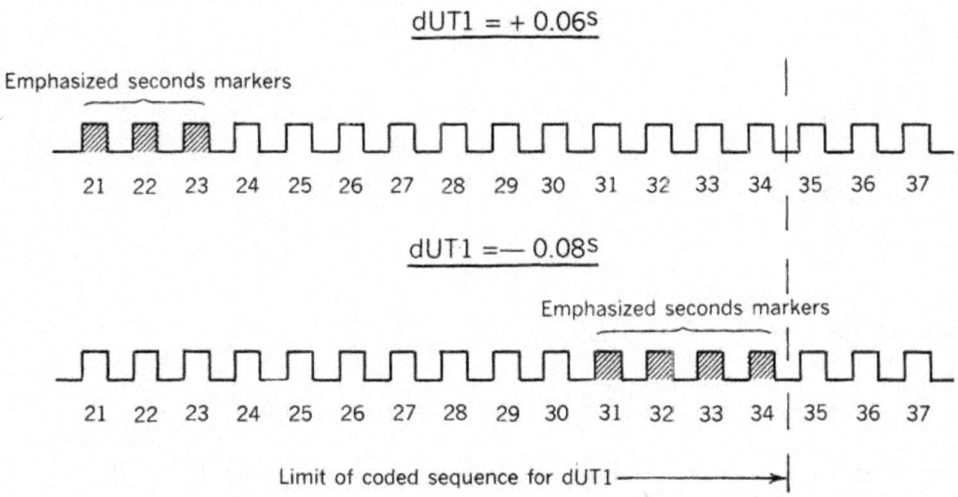

FIGURE 1827b.—USSR extended format for transmission of dUT1.

For example:

“. . .   . — — — — —” means DUT1 = $-0^{s}.1$ and UT1 = UTC $-0^{s}.1$.

Pubs. Nos. 117A and 117B, *Radio Navigational Aids,* should be referred to for up-to-date information on time signals.

1828. Leap-second adjustments.—By international agreement, UTC is maintained within about $0^{s}.9$ of the celestial navigator's time scale, UT1. The introduction of **leap seconds** allows a good clock to keep approximate step with the sun. Because of the variations in the rate of rotation of the earth, however, the occurrences of the leap seconds are not predictable in detail.

The Bureau International de l'Heure (BIH) decides upon and announces the introduction of a leap second. The BIH announces the new leap second at least several weeks in advance. A positive or negative leap second is introduced the last second of a UTC month, but first preference is given to the end of December and June, and second preference is given to the end of March and September. A positive leap second begins at $23^{h}59^{m}60^{s}$ and ends at $00^{h}00^{m}00^{s}$ of the first day of the following month. In the case of a negative leap second, $23^{h}59^{m}58^{s}$ is followed one second later by $00^{h}00^{m}00^{s}$ of the first day of the following month.

The dating of events in the vicinity of a leap second is effected in the manner indicated in figures 1828a and 1828b.

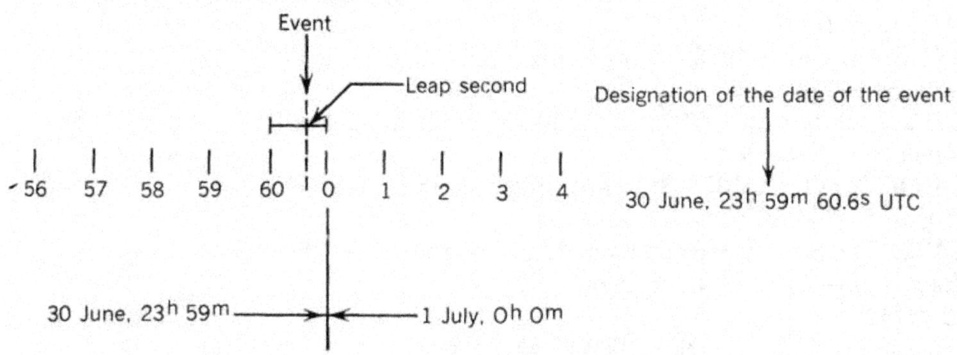

FIGURE 1828a.—Dating of event in the vicinity of a positive leap second.

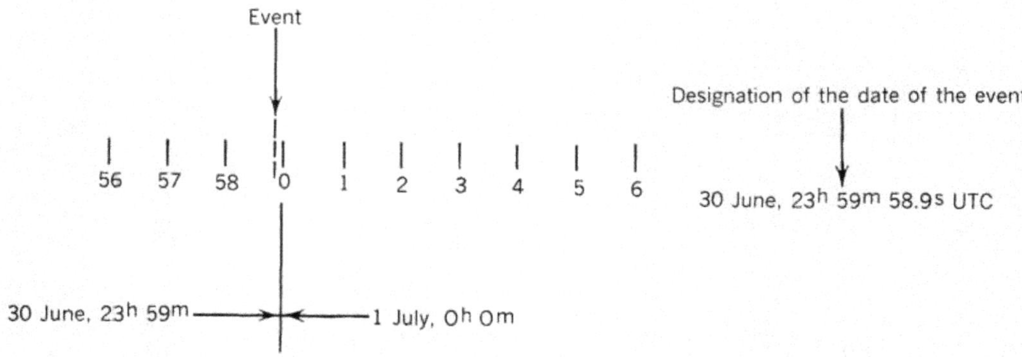

FIGURE 1828b.—Dating of event in the vicinity of a negative leap second.

Whenever leap second adjustments are to be made to UTC, mariners are advised by HYDROLANT/HYDROPAC messages originated by the Defense Mapping Agency Hydrographic/Topographic Center.

Problems

1810a. What is the time and date $9^h13^m29^s$ before $3^h16^m34^s$ May 9?

Answer.—T $18^h03^m05^s$ May 8.

1810b. What is the time and date $4^d19^h22^m50^s$ after $9^h31^m04^s$ on December 25?

Answer.—T $4^h53^m54^s$ on Dec. 30.

1810c. What is the time and date 2 years, 11 months, 16 days, 10 hours, 23 minutes, and 48 seconds before $2^h46^m17^s$ on October 4, 1958?

Answer.—T $16^h22^m29^s$ on Oct. 17, 1955.

1810d. What is the time and date 412 days, 15 hours, 6 minutes, and 56 seconds after $22^h27^m03^s$ on March 16, 1958?

Answer.—T $13^h33^m59^s$ on May 3, 1959.

1811a. Convert $6^h28^m31^s$ to arc units, without use of a conversion table.

Answer.—97°07′45″ or 97°07′.8.

1811b. Convert 217°28′.8 to time units, without use of a conversion table.

Answer.—$14^h29^m55^s.2$ or $14^h29^m55^s$.

1811c. Convert 196°21′46″ to time units, without use of a conversion table.

Answer.—$13^h05^m27^s.1$ or $13^h05^m27^s$.

1811d. Convert 107°49′44″ to time units, using appendix F.

Answer.—$7^h11^m19^s$.

1811e. Convert 211°37′.3 to time units, using appendix F.

Answer.—$14^h06^m29^s.2$.

1811f. Convert $8^h49^m33^s$ to arc units, using appendix F.

Answer.—132°23′.2.

1811g. Convert 251°09′.2 to time units, using appendix G.

Answer.—$16^h44^m37^s$.

1811h. Convert $23^h07^m38^s$ to arc units, using appendix G.

Answer.—346°54′.5.

1814a. For an observer at long. 97°24′.6E the ZT is $19^h10^m26^s$.

Required.—(1) Zone description.

(2) GMT.

Answers.—(1) ZD (−) 6, (2) GMT $13^h10^m26^s$.

1814b. The GMT is $11^h32^m07^s$.

Required.—(1) ZT at long. 133°24′.7W.

(2) ZT at long. 111°43′.9E.

Answers.—(1) ZT $2^h32^m07^s$, (2) ZT $18^h32^m07^s$.

1814c. At long. 165°18′.2E the ZT is $17^h08^m51^s$ on July 11.

Required.—(1) GMT and date.

(2) ZT and date at long. 125°36′.7W.

Answers.—(1) GMT $6^h08^m51^s$ on July 11, (2) ZT $22^h08^m51^s$ on July 10.

1814d. On January 26 the 0800 DR long. of a ship is 128°03′.2E. Twenty-six hours later the EP long. is 125°01′.4E.

Required.—ZT and date of arrival at the second longitude.

Answer.—ZT 0900 Jan. 27.

1814e. On April 1 the 1200 running fix long. of a ship is 179°55′.2W. Eight hours later the DR long. is 178°48′.9E.

Required.—ZT and date of arrival at the second longitude.

Answer.—ZT 2000 Apr. 2.

1814f. Inch'ŏn, long. 137°E, uses ZD (−) 8^h30^m for standard time. Find the standard time and date at San Francisco, long. 122°W, when the summer time at Inch'ŏn is 2000 on August 9.

Answer.—ZT 0230 Aug. 9.

1815a. At GMT 1400 on July 2 the chronometer reads $1^h42^m28^s$. At GMT 0800 on July 12 it reads $7^h42^m40^s$.

Required.—(1) Chronometer error at GMT 1400 on July 2.

(2) Chronometer error at GMT 0800 on July 12.

(3) Chronometer rate.

(4) Chronometer time at ZT 1800 July 20, at long. 153°21′.7W.

Answers.—(1) CE 17^m32^s slow, (2) CE 17^m20^s slow, (3) rate $1^s.2$ gaining, (4) C $3^h42^m51^s$.

1815b. On March 5 the DR long. of a ship is about 151 °E, and the zone time is about 1800. Chronometer error is 6^m40^s fast.

Required.—GMT and date when the chronometer reads $8^h02^m23^s$.

Answer.—GMT $7^h55^m43^s$ on Mar. 5.

1815c. On November 7 the EP long. of a ship is about 71° W, and the zone time is about 1900. Chronometer error is 1^m18^s slow.

Required.—GMT and date when the chronometer reads (1) $11^h55^m20^s$, (2) $11^h59^m50^s$.

Answers.—(1) GMT $23^h56^m38^s$ Nov. 7, (2) GMT $0^h01^m08^s$ Nov. 8.

1815d. At GMT 2200 a comparing watch is checked by time signal, and found to read $10^h00^m05^s$. The chronometer errors are then determined by means of the comparing watch. When the watch reads $10^h06^m00^s$, chronometer *A* reads $10^h11^m17^s$, and when the watch reads $10^h08^m00^s$, chronometer *B* reads $9^h59^m06^s$.

Required.—(1) Watch error.

(2) Error of chronometer *A*.

(3) Error of chronometer *B*.

Answers.—(1) WE 5^s fast on GMT, (2) CE_A 5^m22^s fast, (3) CE_B 8^m49^s slow.

1816a. A chronometer 7^m22^s slow on GMT reads approximately 3^h45^m. About 2 minutes later, when the GMT is a whole minute, a comparing watch will be set to GMT exactly.

Required.—(1) Reading of the watch at starting.

(2) Reading of the chronometer.

Answers.—(1) WT $3^h54^m00^s$, (2) C $3^h46^m38^s$.

1816b. A chronometer 5^m10^s fast on GMT reads approximately 5^h50^m. About 1 minute later, when the GMT is a whole minute, a comparing watch with a 24-hour dial will be set to GMT exactly. The ZT is approximately 1145 and the long. 94°W.

Required.—(1) Reading of the watch at starting.

(2) Reading of the chronometer.

(3) Watch error if, instead of being set to GMT, the watch setting is unchanged and the watch reads $17^h45^m32^s$ at comparison.

Answers.—(1) WT $17^h46^m00^s$, (2) C $5^h51^m10^s$, (3) WE 28^s slow on GMT.

1816c. A watch is set to zone time, approximately. The long. is about 160° E. The watch is compared with a chronometer which is 3^m16^s fast on GMT. When the chronometer reads $1^h48^m00^s$, the watch reads $12^h45^m02^s$.

Required.—Watch error on zone time.

Answer.—WE 18^s fast on ZT.

1816d. On February 14 the DR long. is 63°46.′1W. An observation of Dubhe is made when the watch reads $6^h07^m30^s$ PM. The watch is 11^s slow on zone time.

Required.—GMT and date.

Answer.—GMT $22^h07^m41^s$ Feb. 14.

1816e. On December 11 a watch is set to zone time, approximately. The long. is 137°W. The chronometer is 3^m36^s fast on GMT. When the chronometer reads $4^h40^m00^s$, the watch reads $7^h36^m06^s$ PM.

Required.—(1) Convert WT to GMT and determine watch error.

(2) GMT and date about 20 minutes later, when the uncorrected watch reads $7^h55^m52^s$.

Answers.—(1) WE 18^s slow, (2) GMT $4^h56^m10^s$ Dec. 12.

1816f. Shortly before taking morning sights on January 17 the navigator compares his watch with the chronometer. When the chronometer reads $2^h30^m00^s$, the watch reads $6^h13^m12^s$ AM. The chronometer is 17^m15^s fast on GMT. The long. is 118°W.

Required.—(1) C−WT.

(2) GMT and date a little later when Regulus is observed at W $6^h28^m47^s$ AM.

Answers.—(1) C−WT $8^h16^m48^s$, (2) GMT $14^h28^m20^s$ Jan. 17.

1817a. At long. 138°09.′3E the LMT is $0^h09^m57^s$ on April 23.

Required.—(1) GMT and date.

(2) ZT and date at the place.

Answers.—(1) GMT $14^h57^m20^s$ Apr. 22, (2) ZT $23^h57^m20^s$ Apr. 22.

1817b. At long. 157°18.′4W the LMT is 1931 on June 29.

Required.—(1) ZT and date.

(2) GMT and date.

Answers.—(1) ZT 2000 June 29, (2) GMT 0600 June 30.

1817c. At long. 99°35.′7W the daylight saving time is $21^h29^m45^s$ on August 31.

Required.—(1) Standard time and date.

(2) LMT and date.

Answers.—(1) Standard time $20^h29^m45^s$ Aug. 31, (2) LMT $20^h51^m22^s$ Aug. 31.

1818a. Find the LAT and date at ZT $5^h26^m13^s$ on June 12, 1975, for long. 9°28.′1E.

Answer.—LAT $5^h04^m21^s$ June 12.

1818b. At long. 77°15.′5W the LAT is 1500 on June 13, 1975.

Required.—(1) ZT.

(2) LMT.

Answers.—(1) ZT $15^h08^m56^s$, (2) LMT $14^h59^m54^s$.

1818c. Using the *Air Almanac,* find (1) LAT at long. 117°55′W, and (2) the Eq. T, at ZT $20^h43^m09^s$ on June 1, 1975.

Answers.—(1) LAT $20^h53^m44^s$, (2) Eq. T (+) 2^m15^s.

1819a. Find LST at ZT $19^h24^m26^s$ on June 1, 1975, for long. 87°51.′2E.

Answer.—LST $11^h53^m29^s$.

1819b. Find the ZT at LST $21^h20^m07^s$ on May 31, 1975, for long. 54°21.′3W.

Answer.—ZT $4^h24^m40^s$.

CHAPTER XIX

THE ALMANAC

1901. Introduction.—A requirement of celestial navigation is the availability of accurate predictions of the positions of the celestial bodies used. These predictions, with respect to the celestial equator system of coordinates (art. 1428), are contained in three publications of the United States Naval Observatory. The solution for a celestial line of position consists principally of the conversion of tabulated coordinates to those on the horizon system of coordinates (art. 1430).

The ***Astronomical Almanac*** replaced the ***American Ephemeris and Nautical Almanac*** in 1981. This annual publication is arranged to suit the convenience of the astronomer, for whom it is primarily intended. It contains celestial data to a high degree of precision such as various astronomical constants, details of eclipses, planetary configurations (art. 1422), and miscellaneous phenomena but is not needed for ordinary purposes of navigation. Each volume contains instructions for its use.

The ***Nautical Almanac,*** an annual publication, contains the astronomical information needed by the marine navigator. It is conveniently arranged to suit his needs, and the information is tabulated to a practical degree of precision, in general to the nearest 0.′1 of arc and 1^s of time, at hourly intervals. In 1958, the *Nautical Almanac* became a joint publication of the U.S. Naval Observatory and H. M. Nautical Almanac Office, Royal Greenwich Observatory, and incorporated a number of changes from previous editions. Extracts from the *Nautical Almanac* for 1975 are given in appendix F. These extracts, illustrating the various features of that publication, can be used in the solution of the various illustrative and sample problems of the present volume.

The ***Air Almanac,*** published two times per year, is intended primarily for air navigators. In general, the information is similar to that of the *Nautical Almanac,* but is given to a precision of 1′ of arc and 1^s of time, at intervals of 10^m (values for the sun and Aries are given to a precision of 0.′1). This publication is suitable for ordinary navigation at sea, but may lack the precision that is sometimes needed. The *Air Almanac* is a joint publication of the U.S. Naval Observatory and H. M. Nautical Almanac Office, Royal Greenwich Observatory. Extracts from the *Air Almanac* are given in appendix G.

A highly abbreviated, long-term almanac is given in appendix H. Because of the large intervals between entries, and the fact that no provision is made for nutation, information taken from this almanac may be of reduced accuracy. Although this accuracy is sufficient for most purposes of navigation, the almanac is not as convenient to use as either of those published by the U.S. Naval Observatory, and is not recommended when one of them is available. Instructions for its use are included in appendix H.

The Explanatory Supplement to The Astronomical Ephemeris and to The American Ephemeris and Nautical Almanac, contains detailed explanations of the basis and derivation of each ephemeris in the *AE* in the edition for 1960; it also contains other useful material that is relevant to positional and dynamical astronomy and to chronology. Footnotes indicate the changes that have been introduced since 1960 and it contains a reprint of *The Supplement to A.E. 1968*, which gives an account of the introduction of the IAU (1964) system of astronomical constants. It was published by H. M. Stationery Office, London, but is now out of print (1984). A new Explanatory Supplement to the Astronomical Almanac is in preparation.

1902. *Nautical Almanac*.—The major portion of the *Nautical Almanac* is devoted to hourly tabulation of Greenwich hour angle and declination, to the nearest 0'.1 of arc. On each set of facing pages, information is given for three consecutive days. On the left-hand page, successive columns give GHA of Aries and both GHA and declination of Venus, Mars, Jupiter, and Saturn, followed by the SHA and declination of 57 stars. The GHA and declination of the sun and moon, and the horizontal parallax of the moon, are given on the right-hand page. Where applicable, the quantities v and d are given to assist in interpolation. The quantity v is the difference between the actual change of GHA in 1 hour and a constant value used in the interpolation tables, while d is the change in declination in 1 hour. Both v and d are given to the nearest 0'.1. To the right of the moon data is given the LMT (art. 1817) of sunrise, sunset, and beginning and ending of nautical and civil twilight for various latitudes from 72°N to 60°S. The LMT of moonrise and moonset at the same latitudes is given for each of the three days for which other information is given, and for the following day. Magnitude (art. 1405) of each planet at GMT 1200 of the middle day is given at the top of the column. The GMT (art. 1814) of transit across the celestial meridian of Greenwich is given as "Mer. Pass." The value for the first point of Aries for the middle of the three days is given to the nearest $0^m.1$ at the bottom of the Aries column. The time of transit of the planets for the middle day is given to the nearest whole minute, with SHA (at GMT 0000 of the middle day) to the nearest 0'.1, below the list of stars. For the sun and moon, the time of transit to the nearest whole minute is given for each day. For the moon, both upper and lower transits are given. This information is tabulated below the rising, setting, and twilight information. Given there, also, are the equation of time for 0^h and 12^h, and the age and phase of the moon (art. 1423). Equation of time is given, without sign, to the nearest whole second. Age is given to the nearest whole day. Phase is given by symbol.

The main tabulation is preceded by a list of religious and civil holidays, phases of the moon, a calendar, information on eclipses occurring during the year, and notes and a diagram giving information on the planets.

The main tabulation is followed by explanation and examples. Next are four pages of standard times (zone descriptions) in use in various places in the world. Star charts are given next, followed by a list of 173 stars in order of increasing sidereal hour angle. This list includes the stars given on the daily pages. It gives the SHA and declination each month, and the magnitude. Stars are listed by Bayer's name and also by popular name where there is one. Following the star list are three pages of Polaris tables giving the azimuth and the corrections to be applied to the observed altitude to find the latitude. Next is a table for converting arc to time units. This is followed by a 30-page table called "Increments and Corrections," used for interpolation of Greenwich hour angle and declination. This table is printed on tinted paper, for quick location. Then come tables for interpolating for times of rising, setting, and twilight; followed by two indices of the 57 stars listed on the daily pages, one index being in alphabetical order, and the other in order of decreasing SHA.

Sextant altitude corrections are given at the front and back of the almanac. Tables for the sun, stars, and planets, and a dip table, are given on the inside front cover and facing page, with an additional correction for nonstandard temperature and atmospheric pressure on the following page. Tables for the moon, and an abbreviated dip table, are given on the inside back cover and facing page. Use of the altitude correction tables is explained in chapter XVI. Corrections for the sun, stars, and planets for altitudes greater than 10°, and the dip table, are repeated on one side of a loose bookmark. The star indices are repeated on the other side.

1903. ***Air Almanac.***—As in the *Nautical Almanac,* the major portion of the *Air Almanac* is devoted to a tabulation of GHA and declination. However, in the *Air Almanac* values are given at intervals of 10 minutes, to a precision of 0ʹ.1 for the sun and Aries and to a precision of 1ʹ for the moon and the planets. Values are given for the sun, first point of Aries (GHA only), the three navigational planets most favorably located for observation, and the moon. The magnitude of each planet listed is given at the top of its column, and the phase of the moon is given at the top of its column. Values for the first 12 hours of the day are given on the right-hand page, and those for the second half of the day on the back. In addition, the right-hand page has a table of the moon's parallax in altitude, and below this the semidiameter of the sun, and both the semidiameter and age of the moon (art. 1423). Each daily page includes the LMT of moonrise and moonset; and a difference column for finding the time of moonrise and moonset at any longitude.

Critical tables for interpolation for GHA are given on the inside front cover, which also has an alphabetical listing of the stars, with the number, magnitude, SHA, and declination of each. The same interpolation table and star list are printed on a flap which follows the daily pages. This flap also contains a star chart, a star index in order of decreasing SHA, and a table for interpolation of the LMT of moonrise and moonset for longitude.

Following the flap are instructions for the use of the almanac; a list of symbols and abbreviations in English, French, and Spanish; a list of time differences between Greenwich and various other places; a number of sky diagrams (art. 2212); a planet location diagram (art. 2209); star recognition diagrams for periscopic sextants; sunrise, sunset, and civil twilight tables; rising, setting, and depression graphs; semiduration graphs of sunlight, twilight, and moonlight in high latitudes; list of 173 stars by number and Bayer's name (also popular name where there is one), giving the SHA and declination each month (to a precision of 0ʹ.1), and the magnitude; tables for interpolation of GHA sun and GHA Aries; a table for converting arc to time; a single Polaris correction table; an aircraft standard dome refraction table; a refraction correction table; a Coriolis correction table; and on the inside back cover a correction table for dip of the horizon.

1904. Use of the almanacs.—The time used as an entering argument in the almanacs is 12^h + Greenwich hour angle of the mean sun and is denoted by GMT. This scale may differ from the broadcast time signals by an amount which, if ignored, will introduce an error of up to 0ʹ.2 in longitude determined from astronomical observations. The difference arises because the time argument depends on the variable rate of rotation of the earth while the broadcast time signals are now based on an atomic time-scale. Step adjustments of exactly one second are made to the time signals as required (primarily at 24^h on December 31 and June 30) so that the difference between the time signals and GMT, as used in the almanacs, may not exceed $0^s.9$. Those who require to reduce observations to a precision of better than 1^s must therefore obtain the correction to the time signals from coding in the signal, or from other sources. The correction may be applied to each of the times of observation. Alternatively, the longitude, when determined from astronomical observations, may be corrected by the corresponding amount shown in the following table:

Correction to time signals	Correction to longitude
$-0^s.7$ to $-0^s.9$	$0'.2$ to east
$-0^s.6$ to $-0^s.3$	$0'.1$ to east
$-0^s.2$ to $+0^s.2$	no correction
$+0^s.3$ to $+0^s.6$	$0'.1$ to west
$+0^s.7$ to $+0^s.9$	$0'.2$ to west

The main contents of the almanacs consist of data from which the Greenwich hour angle (GHA) and the declination (Dec.) of all the bodies used for navigation can be obtained for any instant of Greenwich mean time (GMT). The local hour angle (LHA) can then be obtained by means of the formula:

$$\text{LHA} = \text{GHA} \begin{matrix} -\text{ west} \\ \\ +\text{ east} \end{matrix} \text{ longitude.}$$

For the sun, moon, and the four navigational planets, the GHA and declination are tabulated directly in the *Nautical Almanac* for each hour of GMT throughout the year; in the *Air Almanac,* the values are tabulated for each whole 10^m of GMT. For the stars the sidereal hour angle (SHA) is given, and the GHA is obtained from:

$$\text{GHA Star} = \text{GHA Aries} + \text{SHA Star.}$$

The SHA and declination of the stars change slowly and may be regarded as constant over periods of several days or even months if lesser accuracy is required. The SHA and declination of stars tabulated in the *Air Almanac* may be considered constant to a precision of $1'.5$ to $2'$ for the period covered by each of the volumes providing the data for a whole year, with most data being closer to the smaller value. GHA Aries, or the Greenwich hour angle of the first point of Aries (the vernal equinox), is tabulated for each hour of GMT in the *Nautical Almanac* and for each whole 10^m of GMT in the *Air Almanac.* Permanent tables give the appropriate increments to the tabulated values of GHA and declination for the minutes and seconds of GMT.

In the *Nautical Almanac,* the permanent table for increments also includes corrections for v, the difference between the actual change of GHA in one hour and a constant value used in the interpolation tables and d, the change in declination in 1 hour.

In the *Nautical Almanac,* v is always positive unless a negative sign ($-$) is given. This can occur only in the case of Venus. For the sun, the tabulated values of GHA have been adjusted to reduce to a minimum the error caused by treating v as negligible; there is no v tabulated for the sun.

No sign is given for tabulated values of d, which is positive if declination is increasing, and negative if it is decreasing. The sign of a v or d value is given also to the related correction.

In the *Air Almanac,* the tabular values of the GHA of the moon are adjusted so that use of an interpolation table based on a fixed rate of change gives rise to negligible error; no such adjustment is necessary for the sun and planets. The tabulated declination values, except for the sun, are those for the *middle* of the interval between the time indicated and the next *following* time for which a value is given, making interpolation unnecessary. Thus, it is always important to take out the GHA and declination for the tabular GMT immediately before the time of observation.

In the *Air Almanac,* GHA Aries and the GHA and declination of the sun are tabulated to a precision of 0′.1. If these values are extracted with the tabular precision, the "Interpolation of GHA" table on the inside front cover (and flap) should not be used; use the "Interpolation of GHA Sun" and "Interpolation of GHA Aries" tables, as appropriate. These tables for interpolation to a precision of 0′.1 just precede the Polaris Table.

The instructions in the explanation of each volume to ignore the decimal in smaller type when extracting GHA Aries and GHA and declination of the sun to a precision of 1′ instead of rounding-off in the normal way are intended for the air navigator.

1905. Finding GHA and declination of the sun.—*Nautical Almanac.* Enter the daily-page table with the whole hour next preceding the given GMT, unless this time is itself a whole hour, and take out the tabulated GHA and declination. Record, also, the *d* value given at the bottom of the declination column. Next, enter the increments and corrections table for the number of minutes of GMT. If there are seconds, use the next *earlier* whole minute. On the line corresponding to the seconds of GMT take the value from the sun-planets column. Add this to the value of GHA from the daily page to find GHA at the given time. Next, enter the correction table for the same minute with the *d* value, and take out the correction. Give this the sign of the *d* value, and apply it to the declination from the daily page. The result is the declination at the given time.

Example 1.—Find the GHA and declination of the sun at GMT $18^h24^m37^s$ on June 1, 1975, using the *Nautical Almanac.*

Solution.—

	Sun			Sun	
GMT	$18^h24^m37^s$	June 1	GMT	$18^h24^m37^s$	June 1
18^h	90°34′.3		18^h	22°02′.5	N *d*(+)0′.3
24^m37^s	6°09′.3		*d* corr.	(+)0′.1	
GHA	96°43′.6		Dec.	22°02′.6	N

The correction table for GHA of the sun is based upon a rate of change of 15° per hour, the average rate during a year. At most times the rate differs slightly from this. The slight error thus introduced is minimized by adjustment of the tabular values.

The *d* value is the amount that the declination changes between 1200 and 1300 on the middle day of the three shown.

Air Almanac. Enter the daily page with the whole 10^m next preceding the given GMT, unless the time is itself a whole 10^m, and extract the tabulated GHA. The declination is extracted, without interpolation, from the same line as the tabulated GHA or, in the case of planets, the top line of the block of six. If the values extracted are rounded to the nearest minute, next enter the "Interpolation of GHA" table on the inside front cover (and flap), using the "Sun, etc." entry column, and take out the value for the remaining minutes and seconds of GMT. If the entry time is an exact tabulated value, use the correction given half a line *above* the entry time. Add this correction to the GHA taken from the daily page to find the GHA at the given time. No adjustment of declination is needed. If the values are extracted with a precision of 0′.1, the table for interpolating the GHA of the sun to a precision of 0′.1 must be used. No adjustment of declination is needed.

Example 2.—Find the GHA and declination of the sun at GMT $18^h24^m37^s$ on June 1, 1975, using the *Air Almanac.*

Solution.—

	Sun			Sun	
GMT	$18^h24^m37^s$	June 1	GMT	$18^h24^m37^s$	June 1
18^h20^m	95°34′		18^h20^m	95°34′.3	
4^m37^s	1°09′		4^m37^s	1°09′.3	
GHA	96°43′		GHA	96°43′.6	
Dec.	22°03′ N		Dec.	22°02′.6 N	

1906. Finding GHA and declination of the moon.—*Nautical Almanac.* Enter the daily-page table with the whole hour next preceding the given GMT, unless this time is itself a whole hour, and take out the tabulated GHA and declination. Record, also, the corresponding *v* and *d* values tabulated on the same line, and determine the sign of the *d* value. The *v* value of the moon is always positive (+), and is not marked in the almanac. Next, enter the increments and corrections table for the minutes of GMT, and on the line for the seconds of GMT take the GHA correction from the moon column. Then, enter the correction table for the same minute with the *v* value, and extract the correction. Add both of these corrections to the GHA from the daily page to obtain the GHA at the given time. Then, enter the same correction table with the *d* value, and extract the correction. Give this correction the sign of the *d* value, and apply it to the declination from the daily page to find the declination at the given time.

Example 1.—Find the GHA and declination of the moon at GMT $21^h25^m44^s$ on June 1, 1975, using the *Nautical Almanac.*

Solution.—

	Moon			Moon	
GMT	$21^h25^m44^s$	June 1	GMT	$21^h25^m44^s$	June 1
21^h	225°28′.1		21^h	3°06′.8S	*d*(−)10′.7
25^m44^s	6°08′.4	*v*(+)15′.8	*d* corr.	(−)4′.5	
v corr.	(+)6′.7		Dec.	3°02′.3S	
GHA	231°43′.2				

The correction table for GHA of the moon is based upon the *minimum* rate at which the moon's GHA increases, 14°19′.0 per hour. The *v* correction makes the adjustment for the actual rate. The *v* value itself is the difference between the minimum rate and the actual rate during the hour following the tabulated time. The *d* value is the amount that the declination changes during the hour following the tabulated time.

Air Almanac. Enter the daily page with the whole 10^m next preceding the given GMT, unless this time is itself a whole 10^m, and take out the tabulated GHA and the declination, without interpolation. Next, enter the "Interpolation of GHA" table on the inside front cover, using the "moon" entry column, and take out the value for the remaining minutes and seconds of GMT. If the entry time is an exact tabulated value, use the correction given half a line *above* the entry time. Add this correction to the GHA taken from the daily page to find the GHA at the given time. No adjustment of declination is needed.

Example 2.—Find the GHA and declination of the moon at GMT $21^h25^m44^s$ on June 1, 1975, using the *Air Almanac.*

Solution.—

	Moon	
GMT	$21^h25^m44^s$	June 1
21^h20^m	230°20′	
5^m44^s	1°23′	
GHA	231°43′	
Dec.	3°02′ S	

The declination given in the table is correct for the time *5 minutes later than tabulated,* so that it can be used for the 10-minute interval without interpolation, to an accuracy to meet most requirements. If greater accuracy is needed, it can be obtained by interpolation, remembering to allow for the 5 minutes indicated above.

1907. Finding GHA and declination of a planet.—*Nautical Almanac.* Enter the daily-page table with the whole hour next preceding the given GMT, unless the time itself is a whole hour, and take out the tabulated GHA and declination. Record, also, the *v* value given at the bottom of each of these columns. Next, enter the increments and corrections table for the minutes of GMT, and on the line for the seconds of GMT take the GHA correction from the sun-planets column. Next, enter the correction table with the *v* value and extract the correction, giving it the sign of the *v* value. Add the first correction to the GHA from the daily page, and apply the second correction in accordance with its sign, to obtain the GHA at the given time. Then, enter the correction table for the same minute with the *d* value, and extract the correction. Give this correction the sign of the *d* value, and apply it to the declination from the daily page to find the declination at the given time.

Example 1.—Find the GHA and declination of Venus at GMT $5^h24^m07^s$ on June 2, 1975, using the *Nautical Almanac.*

Solution.—

	Venus	
GMT	$5^h24^m07^s$	June 2
5^h	206°59.′4	
24^m07^s	6°01.′8	*v* (−)0.′4
v corr.	(−)0.′2	
GHA	213°01.′0	

	Venus	
GMT	$5^h24^m07^s$	June 2
5^h	23°30.′8 N	*d* (−)0.′5
d corr.	(−)0.′2	
Dec.	23°30.′6 N	

The correction table for GHA of planets is based upon the mean rate of the sun, 15° per hour. The *v* value is the difference between 15° and the change of GHA of the planet between 1200 and 1300 on the middle day of the three shown. The *d* value is the amount that the declination changes between 1200 and 1300 on the middle day.

Venus is the only body listed which ever has a negative *v* value.

Air Almanac. Enter the daily page with the whole 10^m next preceding the given GMT, unless this time is itself a whole 10^m, and extract the tabulated GHA and declination, without interpolation. The tabulated declination is correct for the time 30^m later than tabulated, so that interpolation during the hour following tabulation is not needed for most purposes. Next, enter the "Interpolation of GHA" table on the inside front cover, using the "sun, etc." column, and take out the value for the remaining minutes and seconds of GMT. If the entry time is an exact tabulated value, use the correction half a line *above* the entry time. Add this correction to the

GHA from the daily page to find the GHA at the given time. No adjustment of declination is needed.

Example 2.—Find the GHA and declination of Venus at GMT $5^h48^m45^s$ on June 2, 1975, using the *Air Almanac.*

Solution.—

	Venus	
GMT	$5^h48^m45^s$	June 2
5^h40^m	216°59′	
8^m45^s	2°11′	
GHA	219°10′	
Dec.	23°31′ N	

The declination is taken for the next *earlier* tabulated time, and is correct for GMT 5^h45^m.

1908. Finding GHA and declination of a star.—If the GHA and declination of each navigational star were tabulated separately, the almanacs would be several times their present size. But since the sidereal hour angle (art. 1426) and the declination are nearly constant over several days (to the nearest 0.′1) or months (to the nearest 1′), separate tabulations are not needed. Instead, the GHA of the first point of Aries, from which SHA is measured, is tabulated on the daily pages, and a single listing of SHA and declination is given for each double page of the *Nautical Almanac,* and for an entire volume of the *Air Almanac.* The finding of GHA ♈ is similar to finding GHA of the sun, moon, and planets.

Nautical Almanac. Enter the daily-page table with the whole hour next preceding the given GMT, unless this time is itself a whole hour, and take out the tabulated GHA ♈. Record, also, the tabulated SHA and declination of the star from the listing on the left-hand daily page. Next, enter the increments and corrections table for the minutes of GMT, and on the line for the seconds of GMT take the GHA correction from the Aries column. Add this correction and the SHA of the star to the GHA ♈ of the daily page to find the GHA of the star at the given time. No adjustment of declination is needed.

Example 1.—Find the GHA and declination of Canopus at GMT $3^h24^m33^s$ on June 2, 1975, using the *Nautical Almanac.*

Solution.—

	Canopus	
GMT	$3^h24^m33^s$	June 2
3^h	294°58.′0	
24^m33^s	6°09.′3	
SHA	264°09.′3	
GHA	205°16.′6	
Dec.	52°41.′1 S	

The SHA and declination of 173 stars, including Polaris and the 57 listed on the daily pages, are given for the middle of each month, on almanac pages 268–273. For a star not listed on the daily pages this is the only almanac source of this information. Interpolation in this table is not necessary for ordinary purposes of navigation, but is sometimes needed for precise results. Thus, if the SHA and declination of *β Crucis* (Mimosa) are desired for March 1, 1975, they are found by simple eye interpolation to by SHA 168°25′.2 and Dec. 59°33′.2S.

If GHA ♈ is desired, it is found as indicated in example 1, but omitting the addition of SHA of a star. In the example GHA ♈ is 294°58′.0+6°09′.3=301°07′.3.

Air Almanac. Enter the daily page with the whole 10^m next preceding the given GMT, unless this is itself a whole 10^m, and extract the tabulated GHA ♈. Next, enter the "Interpolation of GHA" table on the inside front cover, using the "sun, etc." entry column, and take out the value for the remaining minutes and seconds of GMT. If the entry time is an exact tabulated value, use the correction given half a line *above* the entry time. From the tabulation at the left side of the same page, extract the SHA and declination of the star. Add the GHA from the daily page and the two values taken from the inside front cover to find the GHA at the given time. No adjustment of declination is needed.

Example 2.—Find the GHA and declination of Peacock at GMT $12^h17^m58^s$ on June 1, 1975, using the *Air Almanac.*

Solution.—

	Peacock	
GMT	$12^h17^m58^s$	June 1
12^h10^m	71°52′	
7^m58^s	2°00′	
SHA	54°03′	
GHA	127°55′	
Dec.	56°49′ S	

Rising, Setting, and Twilight

1909. Rising, setting, and twilight.—In both almanacs the times of sunrise, sunset, moonrise, moonset, and twilight information at various latitudes between 72°N and 60°S are given to the nearest whole minute. By definition, rising or setting occurs when the *upper* limb of the body is on the *visible* horizon, assuming standard refraction for zero height of eye. Because of variations in refraction and height of eye, computation to a greater precision than 1^m is not justified.

In high latitudes some of the phenomena do not occur during certain periods. The symbols used to indicate this condition are:

▭ Sun or moon does not set, but remains continuously above the horizon.

▬ Sun or moon does not rise, but remains continuously below the horizon.

//// Twilight lasts all night.

The *Nautical Almanac* makes no provision for finding the times of rising, setting, or twilight in polar regions. The *Air Almanac* has graphs for this purpose.

In the *Nautical Almanac,* sunrise, sunset, and twilight tables are given only once for the middle of the three days on each page opening. For most purposes this information can be used for all three days. Both almanacs have moonrise and moonset tables for each day.

The tabulations are in local mean time (art. 1817). On the zone meridian, this is the zone time (ZT). For every 15′ of longitude that the observer's position differs from that of the zone meridian, the zone time of the phenomena differs by 1^m, being *later* if the observer is *west* of the zone meridian, and *earlier* if he is *east* of the zone meridian. The local mean time of the phenomena varies with latitude of the observer, declination of the body, and hour angle of the body relative to that of the mean sun.

Sunrise and sunset are also tabulated in the tide tables (from 76°N to 60°S) and in a supplement to the American ephemeris of 1946 entitled *Tables of Sunrise, Sunset, and Twilight* (from 75°N to 75°S). The meridian angle of any body at the time of its rising and setting can be computed by the formulas given in article 715 of volume II. The data concerning the rising and setting of the sun and moon and the duration of twilight for high southern latitudes are published as graphs in United States Naval Observatory Circular No. 147, *Sunlight, Moonlight, and Twilight for Antarctica;* these graphs are similar to the graphs in the *Air Almanac* for northern latitudes (art. 1912).

1910. Finding time of sunrise and sunset.—*Nautical Almanac.* Enter the table on the daily page, and extract the LMT for the tabulated latitude next *smaller* than the observer's latitude (unless this is an exact tabulated value). Apply a correction from table I on almanac page xxxii to interpolate for latitude, determining the sign of the correction by inspection. Then convert LMT to ZT by means of the difference in longitude (dλ) between the local and zone meridians.

Example.—Find the zone time of sunrise and sunset at lat. 43°31'.4N, long. 36°14'.3W on June 1, 1975.

Solution.—

L 43°31'.4N June 1
λ 36°14'.3W

	Sunrise		Sunset
40°	0433	40°	1922
T I	(−)11	T I	(+)11
LMT	0422	LMT	1933
dλ	(+)25	dλ	(+)25
ZT	0447	ZT	1958

Air Almanac. The procedure is the same as that for the *Nautical Almanac,* except that the LMT is extracted from the tables of sunrise and sunset instead of the daily page, and latitude correction is by linear interpolation.

The tabulated times are for the Greenwich meridian. Except in high latitudes near the times of the equinoxes, the time of sunrise and sunset varies so little from day to day that no interpolation is needed for longitude. If such an interpolation is considered justified, it can be made in the same manner as for the moon (art. 1912).

In high latitudes, interpolation is not always possible. For instance, on June 1, 1975, sunrise at latitude 66°N occurs at 0115, but at latitude 68°N the sun does not set. Between these two latitudes the time of sunrise might be found from the graphs in the *Air Almanac,* or by computation, as explained in article 715 of volume II. However, in such a marginal situation, the time of sunrise itself is uncertain, being greatly affected by a relatively small change of refraction or height of eye.

1911. Finding time of twilight.—Morning twilight *ends* at sunrise, and evening twilight *begins* at sunset. The time of the darker limit can be found from the almanacs. The *time* of the darker limits of both *civil* and *nautical* twilights (center of the sun 6° and 12°, respectively, below the celestial horizon) is given in the *Nautical Almanac.* The *Air Almanac* provides tabulations of civil twilight from 60°S to 72°N. The brightness of the sky at any given depression of the sun below the horizon may vary considerably from day to day, depending upon the amount of cloudiness and other atmospheric conditions. In general, however, the most effective period for observing stars and planets occurs when the center of the sun is between about 3° and 9° below the celestial horizon. Hence, the darker limit of civil twilight occurs at about the mid point of this period. At the darker limit of nautical twilight the horizon is generally too dark for good observations. At the darker limit of *astronomical twilight* (center of the sun 18° below the celestial horizon) full night has set in. The time of this twilight is given in the *Astronomical Almanac*. Its approximate value can be determined by extrapolation (art. 207, vol. II) in the *Nautical Almanac,* noting that the duration of the different kinds of twilight is not proportional to the number of degrees of depression at the darker limit. More precise determination of the time at which the center of the sun is any given number of degrees below the celestial horizon can be determined by a large-scale diagram on the plane of the celestial meridian (art. 1432) or by computation (art. 715, vol. II). Duration of twilight in latitudes higher than 65°N is given in a graph in the *Air Almanac.*

Nautical Almanac. The method of finding the darker limit of twilight is the same as that for sunrise and sunset (art. 1910).

Example 1.—Find the zone time of beginning of morning nautical twilight and ending of evening nautical twilight at lat. 21°54′.7S, long. 109°34′.2E on June 1, 1975.

Solution.—

L 21°54′.7S June 1
λ 109°34′.2E

	Nautical twilight		Nautical twilight
20°S	0537	20°S	1819
T I	(+)3	T I	(−)3
LMT	0540	LMT	1816
dλ	(−)18	dλ	(−)18
ZT	0522	ZT	1758

Air Almanac. The method of finding the darker limit of twilight is the same as that for sunrise and sunset as explained in article 1910.

Example 2.—Find the zone time of beginning of morning civil twilight and ending of evening civil twilight at lat. 47°18′.8S, long. 87°28′.3W on June 1, 1975.

Solution.—

L 47°18′.8S June 1
λ 87°28′.3W

	Civil twilight		Civil twilight
45°S	0654	45°S	1701
corr.	(+)7	corr.	(−)7
LMT	0701	LMT	1654
dλ	(−)10	dλ	(−)10
ZT	0651 (twilight)	ZT	1644 (twilight)

Sometimes in high latitudes the sun does not rise but twilight occurs. This is indicated in the *Air Almanac* by the symbol ■ in the sunrise and sunset column. To find the time of beginning of morning twilight, *subtract* half the duration of twilight as obtained from the duration of twilight graph from the time of meridian transit of the sun; and for the time of ending of evening twilight, *add* it to the time of meridian transit. The LMT of meridian transit never differs by more than 16^{m}4 (approximately) from 1200. The actual time on any date can be determined from the almanac.

1912. Finding time of moonrise and moonset is similar to finding time of sunrise and sunset, with one important difference. Because of the moon's rapid change of declination, and its fast eastward motion relative to the sun, the time of moonrise and moonset varies considerably from day to day. These changes of position on the celestial sphere (art. 1403) are continuous, as moonrise and moonset occur successively at various longitudes around the earth. Therefore, the change in time is distributed over all longitudes. For precise results, it would be necessary to compute the time of the phenomena at any given place, by the method described in article 715 of volume II. For ordinary purposes of navigation, however, it is sufficiently accurate to interpolate between consecutive moonrises or moonsets at the Greenwich meridian. Since apparent motion of the moon is westward, relative to an observer on the earth, interpolation in west longitude is between the phenomenon on the given date and the *following* one. In east longitude it is between the phenomenon on the given date and the *preceding* one.

Nautical Almanac. For the given date, enter the daily-page table with latitude, and extract the LMT for the tabulated latitude next *smaller* than the observer's latitude (unless this is an exact tabulated value). Apply a correction from table I of the almanac "Tables for Interpolating Sunrise, Moonrise, etc." to interpolate for latitude, determining the sign of the correction by inspection. Repeat this procedure for the day following the given date, if in west longitude; or for the day preceding, if in east longitude. Using the difference between these two times, and the longitude, enter table II of the almanac "Tables for Interpolating Sunrise, Sunset, etc." and take out the correction. Apply this correction to the LMT of moonrise or moonset at the Greenwich meridian on the given date to find the LMT at the position of the observer. The sign to be given the correction is such as to make the corrected time fall between the times for the two dates between which interpolation is being made. This is nearly always positive (+) in west longitude and negative (−) in east longitude. Convert the corrected LMT to ZT.

Example 1.—Find the zone time of moonrise and moonset at lat. 58°23′.6N, long. 144°07′.5W on June 1, 1975, using the *Nautical Almanac.*

Solution.—

L 58°23′.6N June 1
λ 144°07′.5W

	Moonrise			Moonset	
58°N	0007	June 1	58°N	1110	June 1
T I	(+)1		T I	(−)1	
LMT (G)	0008	June 1	LMT (G)	1109	June 1

58°N	0021 June 2		58°N	1221 June 2
T I	0		T I	0
LMT (G)	0021 June 2		LMT (G)	1221 June 2
LMT (G)	0008 June 1		LMT (G)	1109 June 1
diff.	13		diff.	72
T II	(+)5		T II	(+)28
LMT (G)	0008 June 1		LMT (G)	1109 June 1
LMT	0013 June 1		LMT	1137 June 1
dλ	(−)24		dλ	(−)24
ZT	2349 May 31		ZT	1113 June 1

Air Almanac. For the given date, determine LMT for the observer's latitude at the Greenwich meridian, in the same manner as with the *Nautical Almanac,* except that linear interpolation is made directly from the main tabulation, since no interpolation table is provided. Extract, also, the value from the "Diff." column to the right of the moonrise and moonset column, interpolating if necessary. This "Diff." is one-fourth of one-half of the daily difference. The error introduced by this approximation is generally not more than a few minutes, although it increases with latitude. Using this difference, and the longitude, enter the "Interpolation of Moonrise, Moonset" table on flap F4 of the *Air Almanac* and take out the correction. The *Air Almanac* recommends the taking of the correction from this table without interpolation. The results thus obtained are sufficiently accurate for ordinary purposes of navigation. If greater accuracy is desired, the correction can be taken by interpolation. However, since the "Diff." itself is an approximation, the *Nautical Almanac* or computation (art. 715, vol. II) should be used if accuracy is a consideration. Apply the correction to the LMT of moonrise or moonset at the Greenwich meridian on the given date to find the LMT at the position of the observer. The correction is positive (+) for west longitude, and negative (−) for east longitude, unless the "Diff." on the daily page is preceded by the negative sign (−), when the correction is negative (−) for west longitude, and positive (+) for east longitude. If the time is near midnight, record the date at each step, as in the *Nautical Almanac* solution.

Example 2.—Find the zone time of moonrise and moonset at lat. 58°23′.6N, long. 144°07′.5W on June 1, 1975, using the *Air Almanac.*

Solution.—

L 58°23′.6N June 1
λ 144°07′.5W

	Moonrise			Moonset
diff.	(+)07		diff.	(+)36
58°N	0007		58°N	1110
corr.	(+)1		corr.	(−)1
LMT (G)	0008		LMT (G)	1109
corr.	(+)4		corr.	(+)29
LMT	0012		LMT	1138
dλ	(−)24		dλ	(−)24
ZT	2348 May 31		ZT	1114

As with the sun, there are times in high latitudes when interpolation is inaccurate or impossible. At such periods, the times of the phenomena themselves are uncertain, but an approximate answer can be obtained by moonlight graph in the *Air Almanac* or by computation, as explained in article 715 of volume II. With the moon, this condition occurs when the moon rises or sets at one latitude, but not at the next higher tabulated latitude, as with the sun. It also occurs when the moon rises or sets on one day but not on the preceding or following day. This latter condition is indicated in the *Air Almanac* by the symbol * in the "Diff." column.

Because of the eastward revolution of the moon around the earth, there is one day each synodical month (29½ days) when the moon does not rise, and one day when it does not set. These occur near last quarter and first quarter, respectively. Since this day is not the same at all latitudes or at all longitudes, the time of moonrise or moonset found from the almanac may occasionally be the preceding or succeeding one to that desired. When interpolating near midnight, one should exercise caution to prevent an error.

Refer to the right-hand daily page of the *Nautical Almanac* for June 12, 13, 14 (app. F). On June 13 moonset occurs at 2350 at latitude 70°N, and at 0031 at latitude 72°N. These are *not* the same moonset, the one at 0031 occurring approximately 1 day *later* than the one occurring at 2350. This is indicated by the two times, which differ by nearly 24 hours. The table indicates that with increasing northerly latitude, moonset occurs *later.* Between 70°N and 72°N the time crosses midnight *to the following* day. Hence, between these latitudes interpolation should be made between 2350 on June 13 and 0007 on June 14.

The effect of the revolution of the moon around the earth is to cause the moon to rise or set *later* from day to day. The daily retardation due to this effect does not differ greatly from 50^m. The change in declination of the moon may increase or decrease this effect. The effect due to change of declination increases with latitude, and in extreme conditions it may be greater than the effect due to revolution of the moon. Hence, the interval between successive moonrises or moonsets is more erratic in high latitudes than in low latitudes. When the two effects act in the same direction, daily differences can be quite large. Thus, at latitude 72°N the moon rises at 0550 on June 13, and at 0806 on June 14. When they act in opposite directions, they are small, and when the effect due to change in declination is larger than that due to revolution, the moon sets *earlier* on succeeding days. Thus, at latitude 72°N the moon sets at 0031 on June 13, and at 0007 on June 14. This condition is reflected in the *Air Almanac* by a negative "Diff." If this happens near last quarter or first quarter, two moonrises or moonsets might occur on the same day, one a few minutes after the day begins, and the other a few minutes before it ends. On June 14, 1975, for instance, at latitude 72°N, the moon sets at 0007, rises at 0806, and sets again at 2350 the same day. On those days on which no moonrise or no moonset occurs, the next succeeding one is shown with 24^h added to the time. Thus, at latitude 68°N the moon rises at 2342 on May 25, while the next moonrise occurs 24^h45^m later, at 0027 on May 27. This is listed both as 2427 on May 26 and as 0027 on May 27 (not shown in app. F).

Interpolation for longitude is always made between *consecutive* moonrises or moonsets, regardless of the days on which they fall.

Example 3.—Find the zone time of moonset at lat. 71°38.′7N, long. 56°21.′8W during the night of June 13–14, 1975, using the *Nautical Almanac.*

Solution.—

L 71°38′.7N June 13–14
λ 56°21′.8W

	Moonset	
70°N	2350	June 13
T I	(+)15	
LMT (G)	0005	June 14
70°N	2342	June 14
T I	(+)7	
LMT (G)	2349	June 14
LMT (G)	0005	June 14
diff.	16	
T II	(−)2	
LMT (G)	0005	June 14
LMT	0003	June 14
dλ	(−)15	
ZT	2348	June 13

Interpolation for the first entry is between 2350 on June 13 (lat. 70°N) and 0007 on June 14 (lat. 72°N); for the second entry, between 2342 on June 14 and 2350 on June 14.

Beyond the northern limits of the almanacs the values can be obtained from a series of graphs given near the back of the *Air Almanac.* These graphs are shown in appendix G. For high latitudes, graphs are used instead of tables because graphs give a clearer picture of conditions, which may change radically with relatively little change in position or date. Under these conditions interpolation to practical precision is simpler by graph than by table. In those parts of the graph which are difficult to read, the times of the phenomena's occurrence are themselves uncertain, being altered considerably by a relatively small change in refraction or height of eye.

On all of these graphs any given latitude is represented by a horizontal line, and any given date by a vertical line. At the intersection of these two lines the duration is read from the curves, interpolating by eye between curves.

The "Semiduration of Sunlight" graph gives the number of hours between sunrise and meridian transit or between meridian transit and sunset. The dot scale near the top of the graph indicates the LMT of meridian transit, the time represented by the minute dot nearest the vertical dateline being used. If the intersection occurs in the area marked "sun above horizon," the sun does not set; and if in the area marked "sun below horizon," the sun does not rise.

Example 1.—Find the zone time of sunrise and sunset at lat. 71°30′.0N, long. 10°00′.0W near Jan Mayen Island, on August 25, 1975.

Solution.—

	August 25	
LMT	1202	LAN, from top of graph
dλ	(−)20	
ZT	1142	LAN
semidur.	840	from graph
ZT	0302	sunrise (−semidur.)
ZT	2022	sunset (+semidur.)

A vertical line through August 25 passes nearest the dot representing LAN 1202 on the scale near the top of the graph. This is LMT; at longitude 10°00.′0W the ZT is 20^m earlier, or at 1142. The intersection of the vertical dateline with the horizontal latitude line occurs between the 8^h and 9^h curves, at approximately 8^h40^m. Hence, sunrise occurs at this interval before LAN and sunset at this interval after LAN.

The "Duration of Twilight" graph gives the number of hours between the beginning of morning *civil* twilight (center of sun 6° below the horizon) and sunrise, or between sunset and the end of evening civil twilight. If the sun does not rise, but twilight does occur, the time taken from the graph is half the total length of the single twilight period, or the number of hours from beginning of morning twilight to LAN, or from LAN to end of evening twilight. If the intersection occurs in the area marked "continuous twilight or sunlight," the center of the sun does not get more than 6° below the horizon, and if in the area marked "no twilight nor sunlight," the sun remains more than 6° below the horizon throughout the entire day.

Example 2.—Find the zone time of beginning of morning twilight and ending of evening twilight at the place and date of example 1.

Solution.—

Twilight			Twilight		
ZT	0302	sunrise, from example 1	ZT	2022	sunset, from example 1
dur.	153	from graph	dur.	153	from graph
ZT	0109	morning twilight	ZT	2215	evening twilight

The intersection of the vertical dateline and the horizontal latitude line occurs approximately one-sixth of the distance from the 2^h line toward the 1^h20^m line; or at about 1^h53^m. Morning twilight begins at this interval before sunrise, and evening twilight ends at this interval after sunset.

The "Semiduration of Moonlight" graph gives the number of hours between moonrise and meridian transit or between meridian transit and moonset. The dot scale near the top of the graph indicates the LMT of meridian transit, each dot representing one hour. The phase symbols indicate the date on which the principal moon phases occur, the open circle indicating full moon and the dark circle indicating new moon. If the intersection of the vertical dateline and the horizontal latitude line falls in the "moon above horizon" or "moon below horizon" area, the moon remains above or below the horizon, respectively, for the entire 24 hours of the day.

If approximations of the times of moonrise and moonset are sufficient, the values of semiduration taken from the graph can be used without adjustment. For more accurate results, the times on the required date and the adjacent date (the following date in west longitude and the preceding date in east longitude) should be determined, and an interpolation made for longitude, as in any latitude, since the intervals given are for the Greenwich meridian.

Example 3.—Find the zone time of moonrise and moonset at lat. 74°00.′0N, long. 108°00.′0W on May 8, 1975, and the phase of the moon on this date.

Solution.—

	May 8		May 9	
LMT	0923	LMT	1006	meridian transit,
dλ	(+)12	dλ	(+)12	from graph
ZT	0935	ZT	1018	meridian transit
semidur.	7^h59^m	semidur.	9^h15^m	from graph
ZT	0136	ZT	0103	(moonrise−semidur.)
ZT	1734	ZT	1933	(moonset+semidur.)

	Moonrise			Moonset	
ZT	0136	May 8	ZT	1734	May 8
ZT	0103	May 9	ZT	1933	May 9
diff.	(−)33		diff.	(+)119	
33×108.0/360	(−)10		119×108.0/ 360	(+) 36	
ZT	0126		ZT	1810	

The phase is crescent, about 2 days before new moon. The LMT of meridian transits are found by noting the intersections of the vertical datelines with the dot scale near the top of the graph, interpolating by eye. At longitude 108°00′.0W the ZT is 12^m later. The semiduration is found by noting the position, with respect to the semiduration curves, of the intersection of the vertical dateline with the horizontal latitude line. This interval is subtracted from the time of meridian transit to obtain moonrise, and added to obtain moonset. These solutions are made for both May 8 and 9, and the difference determined in minutes. The adjustment to be applied to the ZT on May 8 at Greenwich is determined by multiplying this difference by the ratio λ/360. The phase is determined by noting the position of the vertical dateline with respect to the phase symbols. If the answer indicates that the phenomenon occurs on a date differing from that desired, a new solution should be made, adjusting the starting date accordingly. The phenomenon may occur twice on the same day, or it may not occur at all. In high latitudes the effect on the time of moonrise and moonset of a relatively small change in declination is considerably greater than in lower latitudes, resulting in greater differences from day to day.

Sunlight, twilight, and moonlight graphs are not given for south latitudes. Beyond latitude 65°S, the northern hemisphere graphs can be used for determining the semiduration or duration, by using the vertical dateline for a day when the declination has the same numerical value but opposite sign. The time of meridian transit and the phase of the moon are determined as explained above, using the correct date. Between latitudes 60°S and 65°S solution is made by interpolation between the tables and the graphs.

Several other methods of solution of these phenomena are available. The *Tide Tables* tabulate sunrise and sunset from latitude 76°N to 60°S. A supplement to the American Ephemeris of 1946, entitled *Tables of Sunrise, Sunset, and Twilight,* provides tabulations from latitude 75°N to 75°S and graphs for semiduration of sunlight and duration of twilight, with separate graphs for civil, nautical, and astronomical twilights. Semiduration or duration can be determined graphically by means of a diagram on the plane of the celestial meridian (art. 1432), or by computation. When computation is used, solution is made for the meridian angle at which the required negative altitude occurs. The meridian angle expressed in time units is the semiduration in the case of sunrise, sunset, moonrise, and moonset; and the semiduration of the combined sunlight and twilight, or the time from meridian transit at which morning twilight begins or evening twilight ends. For sunrise and sunset the altitude used is (−)50′. Allowance for height of eye can be made by algebraically subtracting (numerically adding) the dip correction from this altitude. The altitude used for twilight is (−)6°, (−)12°, or (−)18° for civil, nautical, or astronomical twilight, respectively. The altitude used for moonrise and moonset is −34′−SD+HP, where SD is semidiameter and HP is horizontal parallax, from the daily pages of the *Nautical Almanac.*

1913. Rising, setting, and twilight at a moving craft.—Instructions given in the preceding three articles relate to a fixed position on the earth. Aboard a moving craft the problem is complicated somewhat by the fact that time of occurrence depends upon position of the craft, and vice versa. At ship speeds, it is generally sufficiently accurate to make an approximate mental solution, and use the position of the vessel at this time to make a more accurate solution. If higher accuracy is required, the position at the time indicated in the second solution can be used for a third solution. If desired, this process can be repeated until the same answer is obtained from two consecutive solutions. However, it is generally sufficient to alter the first solution by 1^m for each 15′ of longitude that the position of the craft differs from that used in the solution, adding if west of the estimated position, and subtracting if east of it. In applying this rule, use both longitudes to the nearest 15′. The first solution is known as the first estimate; the second solution is the second estimate.

CHAPTER XX

SIGHT REDUCTION

2001. Introduction.—The process of deriving from a celestial observation the information needed for establishing a line of position is called **sight reduction.** The observation itself consists of measuring the altitude of a celestial body and noting the time. The process of finding such a line of position may be divided into six steps:

1. Correction of sextant altitude (ch. XVI).
2. Determination of GHA and declination (ch. XIX).
3. Selection of assumed position and finding local hour angle or meridian angle at that point.
4. Computation of altitude and azimuth.
5. Comparison of computed and observed altitudes.
6. Plot of the line of position.

Broadly speaking, tables which assist in any of these steps can be considered **sight reduction tables.** However, the expression is generally limited to tables intended primarily for computation of altitude and azimuth. A great variety of such tables exists. In chapter XXI various methods of sight reduction, including graphical and mechanical solutions, are contrasted. All are based, directly or indirectly, upon solution of the navigational triangle (art. 1433). Thus, the process of sight reduction, in its limited sense, is one of converting coordinates of the celestial equator system (art. 1428) to those of the horizon system (art. 1430).

The correction of the sextant altitude (hs) to find observed altitude (Ho) is not necessarily performed first. If any form of time other than GMT is used for timing the observation, it is first converted to GMT because this is the kind of time used for entering the almanacs. From the almanac, the GHA and declination are determined.

2002. Selection of the assumed position (AP).—The following variables are needed to compute the altitude and azimuth:

1. Latitude (L).
2. Declination (d or Dec.).
3. Local hour angle (LHA) or meridian angle (t).

Except for declination, these variables are dependent upon the position from which the altitude and azimuth are to be computed for the time of the observation. Although the dead reckoning or estimated position can be used, unnecessary interpolation can be avoided when using modern sight reduction tables by selecting an AP for the reduction that will result in two of the three variables being exact entry values or table arguments. In these tables altitudes and azimuth angles are given for each whole degree of latitude and each whole degree of either meridian angle or local hour angle. Since the assumed position should be within 30′ of the actual position, the whole degree of latitude *nearest* to the DR or EP at the time of the sight is selected as the **assumed latitude (aL).** The **assumed longitude ($a\lambda$)** is also selected within 30′ of the DR or EP so that no minutes of arc will remain after it is applied to GHA. This means that in west longitude the minutes of $a\lambda$ must be the same as those of GHA; while in east longitude the minutes of $a\lambda$ must be equal to 60′ minus the minutes of GHA.

2003. Finding the local hour angle and meridian angle.—Meridian angle is the angular distance that the celestial body is east or west of the celestial meridian. It is found from local hour angle (LHA), which, in turn, is found from Greenwich hour angle by adding east longitude or subtracting west longitude. A time diagram (art. 1428) is useful in visualizing this relationship.

Example 1.—The GHA is 168°42′.6.

Required.—The LHA and t at (1) long. 137°24′.6W, and (2) 158°24′.7E.

Solution.—

(1) GHA	168°42′.6	(2) GHA	168°42′.6
λ	137°24′.6W	λ	158°24′.7E
LHA	31°18′.0	LHA	327°07′.3
t	31°18′.0W	t	32°52′.7E

In west longitude, if GHA is less than longitude, add 360° to GHA before subtracting. In east longitude, if the sum exceeds 360°, subtract this amount. If LHA is less than 180°, it is numerically equal to meridian angle, which is labeled W (west). If LHA is greater than 180°, t is 360° − LHA and is labeled E (east).

Example 2.—The GHA is 168°42′.6; observations are made at (1) long. 137°24′.6W, and (2) long. 158°24′.7E.

Required.—The $a\lambda$ providing whole degrees of LHA and t.

Solution.—

(1) GHA	168°42′.6	(2) GHA	168°42′.6
$a\lambda$	137°42′.6W	$a\lambda$	158°17′.4E
LHA	31°00′.0	LHA	327°00′.0
t	31°00′.0W	t	33°00′.0E

2004. Comparison of computed and observed altitudes.—After appropriate corrections are applied to the **sextant altitude (hs)**, the **observed altitude (Ho)** is obtained. For the instant of observation, the altitude and azimuth at some convenient **assumed position (AP)** near the actual position of the observer are determined by calculation or equivalent process. The difference between this **computed altitude (Hc)** and Ho is the **altitude intercept (*a*)**, sometimes called **altitude difference.**

Since *a* is the difference in altitude at the assumed and actual positions, it is also the difference in zenith distance, and therefore the difference in radii of the circles of equal altitude at the two places. The position having the greater altitude is on the circle of smaller radius, and hence is closer to the GP of the body. In figure 2004 the AP is shown on the inner circle. Hence, Hc is greater than Ho.

The altitude intercept, the numerical difference between Hc and Ho, is customarily expressed in nautical miles (minutes of arc), and labeled **T** or **A** to indicate whether the line of position is **toward** or **away** from the GP, as measured from the AP.

Two useful aids in labeling the intercept are: Coast Guard Academy for Computed Greater Away, and **Ho Mo To** for **Ho More Toward.**

For example,

Hc	37°51′.6	Hc	61°57′.3
Ho	37°43′.9	Ho	62°12′.7
a	7.7A	*a*	15.4T

2005. Plot of the line of position.—The line of position can be plotted using part of the information within the broken circle of figure 2004, as shown in figure 2005. First, the AP is plotted. The circle of equal altitude through this position is not needed, and is not plotted. From the AP the azimuth line is measured toward or away from the GP as appropriate, and the altitude intercept is measured along this line. At the point thus located, a line is drawn perpendicular to the azimuth line. This perpendicular is the line of position.

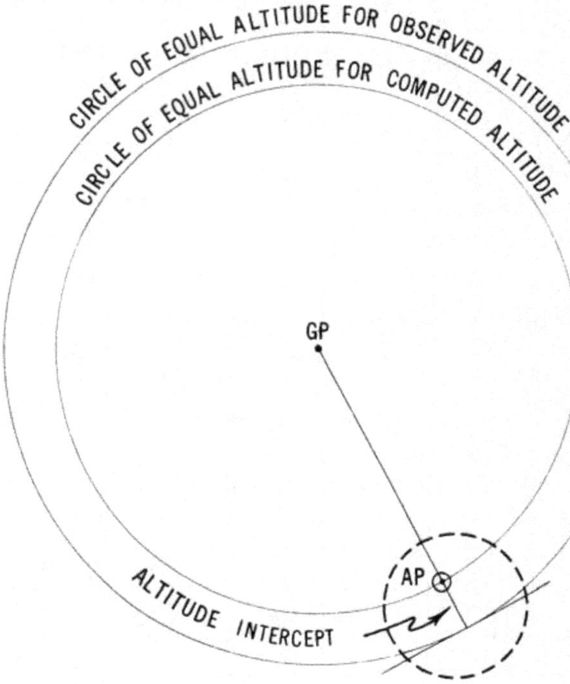

FIGURE 2004.—The basis for the line of position from a celestial observation.

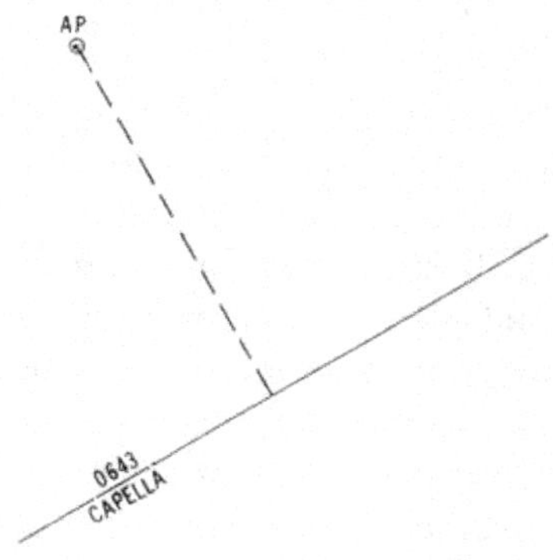

FIGURE 2005.—A line of position from observation of the star Capella at 0643.

2006. Computation of altitude and azimuth.—In modern practice, solutions of the navigational triangle for altitude and azimuth are usually effected by means of sight reduction tables of the inspection type. These inspection tables may contain tabulations of altitude and azimuth angle for arguments of local hour angle (or meridian angle), declination, and latitude, or tabulations of altitude and azimuth of selected stars for arguments of epoch, latitude, and sidereal time (LHA ♈). Values are taken directly from the tables, without need for logarithms, auxiliary functions, or mathematical solutions (except interpolation).

Mathematical solutions of altitude and azimuth angle are presented in chapter VII of volume II. Chapter IX of volume II contains examples of the use of logarithms and auxiliary functions in sight reduction.

The principal inspection tables are Pub. No. 229, *Sight Reduction Tables for Marine Navigation;* Pub. No. 249, *Sight Reduction Tables for Air Navigation;* and Pub. No. 214, *Tables of Computed Altitude and Azimuth.*

2007. *Sight Reduction Tables for Marine Navigation* (Pub. No. 229).—These tables are published by the Defense Mapping Agency Hydrographic/Topographic Center to facilitate the practice of celestial navigation at sea. A secondary purpose of the tables is to provide, within the limitations of the tabular precision and interval, a table of the solutions of the spherical triangle of which two sides and the included angle are known and it is necessary to find the values of the third side and adjacent angle.

The tables have been designed primarily for use with the Marcq St.-Hilaire or intercept method of sight reduction, utilizing a position assumed or chosen so that interpolation for latitude and local hour angle is not required. For entering arguments of integral degrees of latitude, declination, and local hour angle, altitudes and their differences are tabulated to the nearest tenth of a minute, azimuth angles to the nearest tenth of a degree. But the tables are designed for precise interpolation of altitude for declination only by means of interpolation tables which facilitate linear interpolation and provide additionally for the effect of second differences (art. 2008). The data are applicable to the solutions of sights of all celestial bodies; there are no limiting values of altitude, latitude, hour angle, or declination.

The tables are divided into six volumes, each of which includes two eight-degree zones of latitude. An overlap of 1° occurs between volumes. The six volumes cover latitude bands 0° to 15°, 15° to 30°, 30° to 45°, 45° to 60°, 60° to 75°, and 75° to 90°.

Each consecutive opening of the pages of a latitude zone differs from the preceding one by 1° of local hour angle (LHA). As shown in figures 2007a and 2007b, the values of LHA are prominently displayed at the top and bottom of each page; the horizontal argument heading each column is latitude, and the vertical argument is declination (Dec.). For each combination of arguments, the tabulations are: the tabular altitude (ht or Tab. Hc), the altitude difference (d) with its sign, and the azimuth angle (Z).

Within each opening, the data on the left-hand page are the altitudes, altitude differences, and azimuth angles of celestial bodies when the latitude of the observer has the same name as the declinations of the bodies. For any LHA tabulated on a left-hand page and any combination of the tabular latitude and declination arguments, the tabular altitude and associated azimuth angle respondents on the left-hand page are those of a body above the celestial horizon of the observer.

The LHA's tabulated on the left-hand pages are limited to the following ranges: 0° increasing to 90°, and 360° decreasing to 270°. On any left-hand page there are two tabulated LHA's, one LHA in the range 0° increasing to 90°, and the second in the range 360° decreasing to 270°.

On the right-hand page of each opening, the data above the horizontal rules are the tabular altitudes, altitude differences, and azimuth angles of celestial bodies above the celestial horizon when the latitude of the observer has a name contrary to the name of the declinations of the bodies and the LHA's of the bodies are those

60°, 300° L.H.A. **LATITUDE SAME NAME AS DECLINATION** N. Lat. {L.H.A. greater than 180°......Zn=Z; L.H.A. less than 180°..........Zn=360°−Z}

Dec.	38° Hc	d	Z	39° Hc	d	Z	40° Hc	d	Z	41° Hc	d	Z	42° Hc	d	Z	43° Hc	d	Z	44° Hc	d	Z	45° Hc	d	Z	Dec.
°	° ′	′	°	° ′	′	°	° ′	′	°	° ′	′	°	° ′	′	°	° ′	′	°	° ′	′	°	° ′	′	°	°
0	23 12.2	+40.1	109.6	22 51.9	+40.9	110.0	22 31.3	+41.6	110.4	22 10.2	+42.4	110.7	21 48.8	+43.1	111.1	21 27.0	+43.8	111.5	21 04.8	+44.6	111.9	20 42.3	+45.3	112.2	0
1	23 52.3	39.8	108.8	23 32.8	40.6	109.2	23 12.9	41.4	109.6	22 52.6	42.2	110.0	22 31.9	42.9	110.4	22 10.8	43.7	110.8	21 49.4	44.4	111.1	21 27.6	45.0	111.5	1
2	24 32.1	39.5	107.9	24 13.4	40.4	108.4	23 54.3	41.2	108.8	23 34.8	41.9	109.2	23 14.8	42.7	109.6	22 54.5	43.4	110.0	22 33.8	44.2	110.4	22 12.6	45.0	110.8	2
3	25 11.6	39.3	107.1	24 53.8	40.1	107.6	24 35.5	40.9	108.0	24 16.7	41.7	108.4	23 57.5	42.5	108.8	23 37.9	43.3	109.3	23 18.0	43.9	109.7	22 57.6	44.7	110.1	3
4	25 50.9	39.0	106.3	25 33.9	39.8	106.7	25 16.4	40.6	107.2	24 58.4	41.5	107.6	24 40.0	42.3	108.1	24 21.2	43.0	108.5	24 01.9	43.8	108.9	23 42.3	44.5	109.3	4
5	26 29.9	[illegible]	105.4	26 13.7	+39.5	105.9	25 57.0	+40.4	106.4	25 39.9	+41.1	106.8	25 22.3	+41.9	107.3	25 04.2	+42.8	107.7	24 45.7	+43.5	108.2	24 26.8	+44.3	108.6	5
6	27 08.5	38.4	104.6	26 53.2	39.2	105.1	26 37.4	40.0	105.5	26 21.0	40.9	106.0	26 04.2	41.8	106.5	25 47.0	42.5	107.0	25 29.2	43.3	107.4	25 11.1	44.0	107.9	6
7	27 46.9	[illegible]	103.7	27 32.4	38.9	104.2	27 17.4	39.8	104.7	27 01.9	40.7	105.2	26 46.0	41.4	105.7	26 29.5	42.3	106.2	26 12.5	43.1	106.7	25 55.1	43.8	107.1	7
8	28 24.9	[illegible]	102.8	28 11.3	38.6	103.3	27 57.2	39.5	103.9	27 42.6	40.3	104.4	27 27.4	41.2	104.9	27 11.8	41.9	105.4	26 55.6	42.8	105.9	26 38.9	43.6	106.4	8
9	29 02.6	[illegible]	101.9	28 49.9	38.2	102.5	28 36.7	39.1	103.0	28 22.9	40.0	103.5	28 08.6	40.8	104.1	27 53.7	41.7	104.6	27 38.4	42.5	105.1	27 22.5	43.3	105.6	9
10	29 39.9	+37.0	101.0	29 28.1	+37.9	101.6	29 15.8	+38.8	102.1	29 02.9	+39.7	102.7	28 49.4	+40.6	103.2	28 35.4	+41.4	103.8	28 20.9	+42.2	104.3	28 05.8	+43.1	104.8	10
11	30 16.9	36.5	100.1	30 06.0	37.5	100.7	29 54.6	38.4	101.3	29 42.6	39.3	101.8	29 30.0	40.2	102.4	29 16.8	41.1	102.9	29 03.1	42.0	103.5	28 48.9	42.8	104.0	11
12	30 53.4	36.2	99.2	30 43.5	37.2	99.8	30 33.0	38.1	100.4	30 21.9	39.0	101.0	30 10.2	39.9	101.5	29 57.9	40.8	102.1	29 45.1	41.6	102.7	29 31.7	42.4	103.2	12
13	31 29.6	35.8	98.3	31 20.7	36.7	98.9	31 11.1	37.7	99.5	31 00.9	38.6	100.1	30 50.1	39.6	100.7	30 38.7	40.5	101.2	30 26.7	41.3	101.8	30 14.1	42.2	102.4	13
14	32 05.4	35.3	97.3	31 57.4	36.3	97.9	31 48.8	37.3	98.6	31 39.5	38.3	99.2	31 29.7	39.1	99.8	31 19.2	40.0	100.4	31 08.0	41.0	101.0	30 56.3	41.9	101.6	14
15	32 40.7	+34.9	96.4	32 33.7	+35.9	97.0	32 26.1	+36.9	97.6	32 17.8	+37.8	98.3	32 08.8	+38.8	98.9	31 59.2	+39.8	99.5	31 49.0	+40.6	100.1	31 38.2	+41.5	100.7	15
16	33 15.6	34.5	95.4	33 09.6	35.5	96.1	33 03.0	36.4	96.7	32 55.6	37.5	97.3	32 47.6	38.4	98.0	32 39.0	39.3	98.6	32 29.6	40.3	99.3	32 19.7	41.1	99.9	16
17	33 50.1	[illegible]	94.4	33 45.1	35.0	95.1	33 39.4	36.0	95.7	33 33.1	37.0	96.4	33 26.0	38.0	97.1	33 18.3	38.9	97.7	33 09.9	39.9	98.4	33 00.8	40.8	99.0	17
18	34 24.0	[illegible]	93.4	34 20.1	34.5	94.1	34 15.4	35.6	94.8	34 10.1	36.5	95.5	34 04.0	37.6	96.1	33 57.2	38.6	96.8	33 49.8	39.5	97.5	33 41.6	40.5	98.1	18
19	34 57.5	[illegible]	92.4	34 54.6	34.0	93.1	34 51.0	35.1	93.8	34 46.6	36.1	94.5	34 41.6	37.1	95.2	34 35.8	38.1	95.9	34 29.3	39.1	96.6	34 22.1	40.0	97.2	19
20	35 30.5	[illegible]	91.4	35 28.6	+33.6	92.1	35 26.1	+34.5	92.8	35 22.7	+35.7	93.5	35 18.7	+36.6	94.2	35 13.9	+37.7	94.9	35 08.4	+38.6	95.6	35 02.1	+39.6	96.3	20
21	36 02.9	[illegible]	90.4	36 02.2	32.9	91.1	36 00.6	34.1	91.8	35 58.4	35.1	92.5	35 55.3	36.2	93.3	35 51.6	37.2	94.0	35 47.0	38.2	94.7	35 41.7	39.2	95.4	21
22	36 34.8	[illegible]	89.3	36 35.1	32.5	90.0	36 34.7	33.6	90.8	36 33.5	34.6	91.5	36 31.5	35.7	92.3	36 28.8	36.7	93.0	36 25.2	37.8	93.7	36 20.9	38.8	94.5	22
23	37 06.1	[illegible]	88.2	37 07.6	31.9	89.0	37 08.3	32.9	89.8	37 08.1	34.1	90.5	37 07.2	35.2	91.3	37 05.5	36.2	92.0	37 03.0	37.2	92.8	36 59.7	38.2	93.5	23
24	37 36.9	[illegible]	87.2	37 39.5	31.3	87.9	37 41.2	32.5	88.7	37 42.2	33.5	89.5	37 42.4	34.6	90.2	37 41.7	35.7	91.0	37 40.2	36.8	91.8	37 37.9	37.8	92.6	24
25	38 07.1	+29.5	86.1	38 10.8	+30.7	86.8	38 13.7	+31.8	87.6	38 15.7	+33.0	88.4	38 17.0	+34.0	89.2	38 17.4	+35.2	90.0	38 17.0	+36.2	90.8	38 15.7	+37.3	91.6	25
26	38 36.6	28.9	85.0	38 41.5	30.0	85.7	38 45.5	31.2	86.5	38 48.7	32.4	87.4	38 51.0	33.5	88.2	38 52.6	34.6	89.0	38 53.2	35.7	89.8	38 53.0	36.8	90.6	26
27	39 05.5	28.3	83.8	39 11.5	29.5	84.6	39 16.7	30.6	85.4	39 21.1	31.7	86.3	39 24.5	32.9	87.1	39 27.2	34.0	87.9	39 28.9	35.2	88.7	39 29.8	36.3	89.6	27
28	39 33.8	27.5	82.7	39 41.0	28.7	83.5	39 47.3	30.0	84.3	39 52.8	31.1	85.2	39 57.4	32.3	86.0	40 01.2	33.4	86.8	40 04.1	34.5	87.7	40 06.1	35.6	88.5	28
29	40 01.3	26.9	81.5	40 09.7	28.1	82.4	40 17.3	29.2	83.2	40 23.9	30.5	84.0	40 29.7	31.7	84.9	40 34.6	32.8	85.7	40 38.6	34.0	86.6	40 41.7	35.1	87.5	29
30	40 28.2	+26.1	80.4	40 37.8	+27.4	81.2	40 46.5	+28.6	82.1	40 54.4	+29.8	82.9	41 01.4	+30.9	83.8	41 07.4	+32.2	84.6	41 12.6	+33.3	85.5	41 16.8	+34.5	86.4	30
31	40 54.3	25.5	79.2	41 05.2	26.6	80.0	41 15.1	27.9	80.9	41 24.2	29.1	81.8	41 32.3	30.3	82.6	41 39.6	31.4	83.5	41 45.9	32.6	84.4	41 51.3	33.8	85.3	31
32	41 19.8	24.6	78.0	41 31.8	25.9	78.8	41 43.0	27.1	79.7	41 53.3	28.3	80.6	42 02.6	29.6	81.5	42 11.0	30.8	82.4	42 18.5	32.0	83.3	42 25.1	33.2	84.2	32
33	41 44.4	23.9	76.8	41 57.7	25.1	77.6	42 10.1	26.4	78.5	42 21.6	27.6	79.4	42 32.2	28.8	80.3	42 41.8	30.1	81.2	42 50.5	31.3	82.1	42 58.3	32.5	83.0	33
34	42 08.3	23.0	75.5	42 22.8	24.4	76.4	42 36.5	25.6	77.3	42 49.2	26.9	78.2	43 01.0	28.1	79.1	43 11.9	29.3	80.0	43 21.8	30.6	81.0	43 30.8	31.8	81.9	34
35	42 31.3	+22.3	74.3	42 47.2	+23.5	75.2	43 02.1	+24.8	76.1	43 16.1	+26.0	77.0	43 29.1	+27.3	77.9	43 41.2	+28.6	78.8	43 52.4	+29.8	79.8	44 02.6	+31.0	80.7	35
36	42 53.6	21.4	73.0	43 10.7	22.7	73.9	43 26.9	23.9	74.8	43 42.1	25.3	75.7	43 56.4	26.5	76.7	44 09.8	27.8	77.6	44 22.2	29.0	78.6	44 33.6	30.3	79.5	36
37	43 15.0	20.5	71.7	43 33.4	21.8	72.6	43 50.8	23.1	73.5	44 07.4	24.3	74.5	44 22.9	25.7	75.4	44 37.6	26.9	76.4	44 51.2	28.3	77.3	45 03.9	29.5	78.3	37
38	43 35.5	19.7	70.4	43 55.2	20.9	71.3	44 13.9	22.2	72.3	44 31.7	23.6	73.2	44 48.6	24.8	74.1	45 04.5	26.1	75.1	45 19.5	27.4	76.1	45 33.4	28.7	77.1	38
39	43 55.2	18.7	69.1	44 16.1	20.0	70.0	44 36.1	21.4	71.0	44 55.3	22.6	71.9	45 13.4	24.0	72.9	45 30.6	25.3	73.8	45 46.9	26.5	74.8	46 02.1	27.8	75.8	39
40	44 13.9	+17.8	67.8	44 36.1	+19.2	68.7	44 57.5	+20.4	69.6						71.5	45 55.9	+24.3	72.5	46 13.4	+25.6	73.5	46 29.9	+27.0	74.5	40
41	44 31.7	16.9	66.5	44 55.3	18.1	67.4	45 17.9	19.5									23.4	71.2	46 39.0	24.8	72.2	46 56.9	26.0	73.2	41
42	44 48.6	15.9	65.1	45 13.4	17.2	66.0	45 37.4	18.5									22.5	69.9	47 03.8	23.8	70.9	47 22.9	25.2	71.9	42
43	45 04.5	15.0	63.8	45 30.6	16.3	64.7	45 55.9											68.5	47 27.6	22.8	69.5	47 48.1	24.1	70.5	43
44	45 19.5	13.9	62.4	45 46.9	15.2	63.3	46 13											67.1	47 50.4	21.8	68.1	48 12.2	23.2	69.2	44
45	45 33.4	+12.9	61.0	46 02.1	+14.2	61.9													48 12.2	+20.8	66.8	48 35.4	+22.2	67.8	45
46	45 46.3	11.9	59.6	46 16.3	13.1	60.5													48 33.0	19.8	65.3	48 57.6	21.1	66.4	46
47	45 58.2	10.9	58.2	46 29.4	12.1														48 52.8	18.7	63.9	49 18.7	20.0	65.0	47
48	46 09.1	9.7	56.8	46 41.5	11.1															17.6	62.5	49 38.7	19.0	63.5	48
49	46 18.8	8.8	55.3	46 52.6	9.9															16.4	61.0	49 57.7	17.8	62.0	49
50	46 27.6	+7.6	53.9	47 02.5																+15.3	59.5	50 15.5	+16.7	60.5	50
51	46 35.[illegible]	6.5	52.5	47 11.4																14.2	58.0	50 32.2	15.4	59.0	51
52	46 41.7	5.5	51.0	47 19.1																	56.5	50 47.6	14.3	57.5	52
53	46 47.[illegible]	4.3	49.6	47 25.8																	55.0	51 01.9	13.0	56.0	53
54	46 51.5	3.3	48.1																		53.4	51 14.9	11.8	54.4	54
55	46 54.8	+2.1	46.6																		51.9	51 26.7	+10.6	52.8	55
56	46 56.9	+1.0	45.2																		50.3	51 37.3	9.2	51.3	56
57	46 57.9	−0.2	43.7																			51 46.5	7.9	49.7	57
58	46 57.7	1.2	42.3																			51 54.4	6.7	48.1	58
59	46 56.5	2.4	40.8																			52 01.1	5.3	46.4	59
60	46 54.1	−3.5	39.3																			52 06.4	+3.9	44.8	60
61	46 50.6	4.6	37.9																			52 10.3	2.6	43.2	61
62	46 46.0	5.7	36.4																			52 12.9	+1.3	41.6	62
63	46 40.3	6.8	35.0																			52 14.2	−0.1	39.9	63
64	46 3[illegible].5	7.9	33.5																			52 14.1	1.4	38.3	64
65	46 25.6	−9.0	32.1																			52 12.7	−2.8	36.7	65
66	46 1[illegible].6	10.0	30.6																			52 09.9	4.1	35.0	66
67	46 06.6	11.1	29.2																			52 05.8	5.5	33.4	67
68	45 55.5	12.1	27.8																			52 00.3	6.8	31.8	68
69	45 43.4	13.2	26.4																			51 53.5	8.1	30.2	69
70	45 30.2	−14.2	25.0																			51 45.4	−9.4	28.6	70
71	45 16.0	15.1	23.6																			51 36.0	10.7	27.0	71
72	45 00.9	16.2	22.2																		24.9	51 25.3	12.0	25.4	72
73	44 44.7	17.1	20.9																		23.4	51 13.3	13.2	23.8	73
74	44 27.6	18.0	19.5																		21.8	51 00.1	14.4	22.3	74
75	44 09.6	−19.0	18.2	45 06.5																	20.3	50 45.7	−15.6	20.8	75
76	43 50.6	19.8	16.9	44 48.0																	18.8	50 30.1	16.9	19.2	76
77	43 30.8	20.8	15.6	44 28.5																18.4	17.4	50 13.2	17.9	17.7	77
78	43 10.0	21.6	14.3	44 08.1																19.5	15.9	49 55.3	19.1	16.2	78
79	42 48.4	22.4	13.0	43 46.8																20.5	14.5	49 36.2	20.2	14.8	79
80	42 26.0	−23.3	11.8	43 24.7	−23.0															−21.6	13.1	49 16.0	−21.2	13.3	80
81	42 02.7	24.0	10.5	43 01.7	23.8															22.6	11.7	48 54.8	22.3	11.9	81
82	41 38.7	24.9	9.3	42 37.9	24.7														47 33.4	23.5	10.3	48 32.5	23.3	10.5	82
83	41 13.8	25.6	8.1	42 13.2	25.4	8.2													47 09.9	24.5	8.9	48 09.2	24.3	9.1	83
84	40 48.2	26.3	6.9	41 47.8	26.2	7.0													46 45.4	25.4	7.6	47 44.9	25.3	7.7	84
85	40 21.9	−27.0	5.7	41 21.6	−26.9	5.8												6.2	46 20.0	−26.3	6.3	47 19.6	−26.2	6.4	85
86	39 54.9	27.7	4.5	40 54.7	27.7	4.6	41 54.5										27.3	4.9	45 53.7	27.2	5.0	46 53.4	27.1	5.1	86
87	39 27.2	28.5	3.4	40 27.0	28.3	3.4	41 26.9										28.1	3.6	45 26.5	28.1	3.7	46 26.3	27.9	3.8	87
88	38 58.7	29.0	2.2	39 58.7	29.0	2.3	40 58.7	29.0								43 58.5	28.9	2.4	44 58.4	28.8	2.4	45 58.4	28.8	2.5	88
89	38 29.7	29.7	1.1	39 29.7	29.7	1.1	40 29.7	29.7	1.1						1.2	43 29.6	29.6	1.2	44 29.6	29.6	1.2	45 29.6	29.6	1.2	89
90	38 00.0	−30.3	0.0	39 00.0	−30.3	0.0	40 00.0	−30.3	0.0	41 00.0	−30.3	0.0	42 00.0	−30.3	0.0	43 00.0	−30.4	0.0	44 00.0	−30.4	0.0	45 00.0	−30.4	0.0	90
	38°			39°			40°			41°			42°			43°			44°			45°			

60°, 300° L.H.A. **LATITUDE SAME NAME AS DECLINATION**

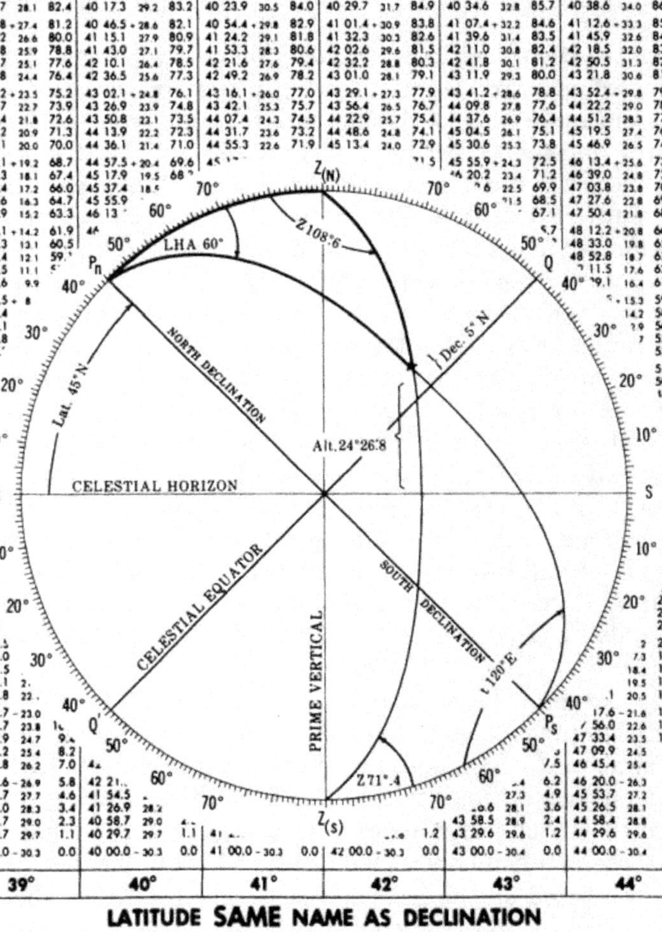

FIGURE 2007a.—Left-hand page of opening of volume 3 of Pub. No. 229.

tabulated at the top of the page. The data below the horizontal rules are the tabular altitudes, altitude differences, and azimuth angles of celestial bodies above the celestial horizon when the latitude of the observer has the same name as the

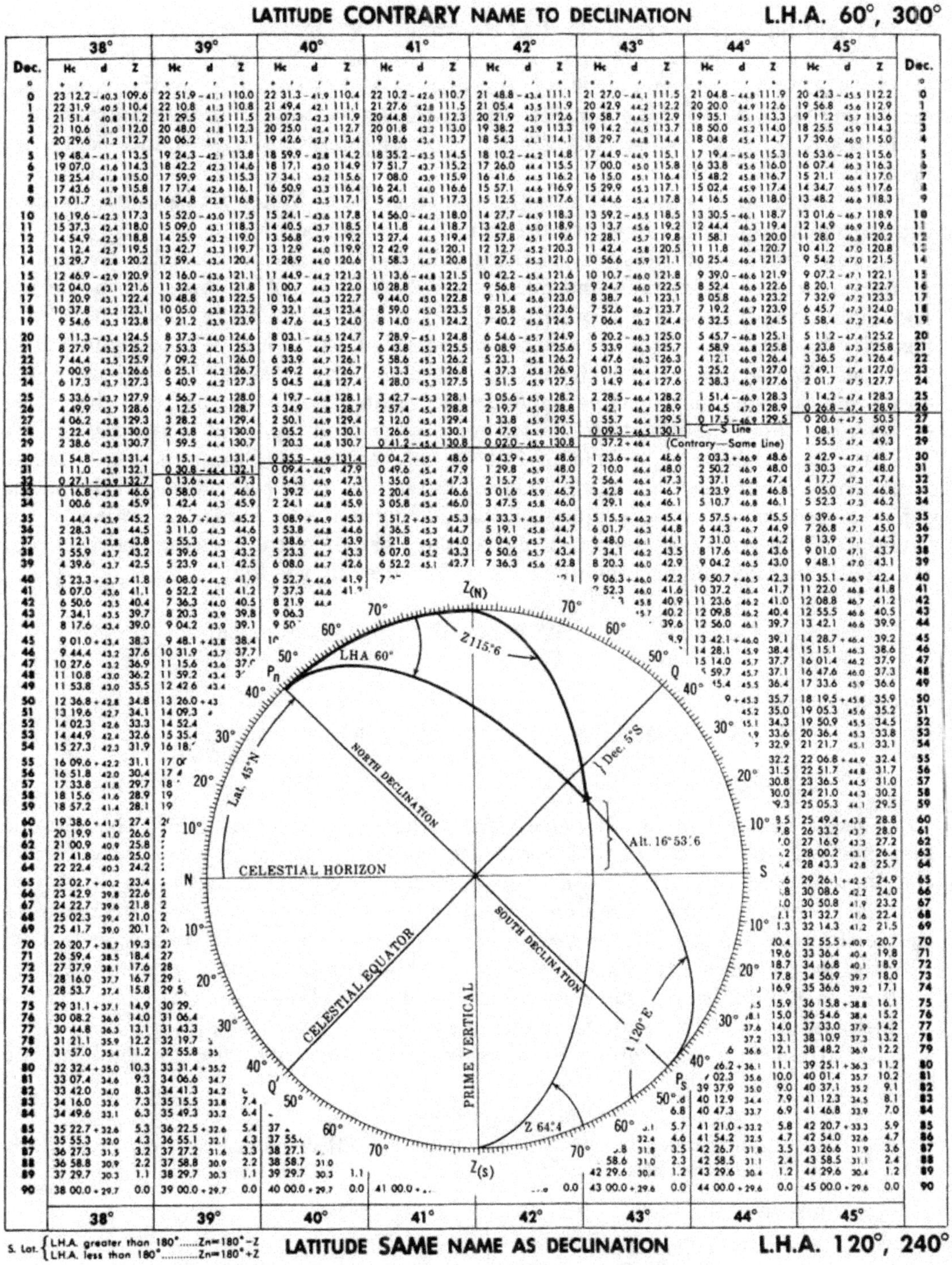

FIGURE 2007b.—Right-hand page of opening of volume 3 of Pub. No. 229.

declinations of the bodies, and the LHA's of the bodies are those tabulated at the bottom of the page.

The LHA's tabulated at the top of a right-hand page are the same as those tabulated on the left-hand page of the opening. The LHA's tabulated at the bottom of the right-hand page are limited to the range 90° increasing to 270°, one of the two LHA's at the bottom of the page is in the range 90° increasing to 180°; the other LHA is in the range 180° increasing to 270°; the LHA in the range 90° increasing to 180° is the supplement of the LHA at the top of the page in the range 0° increasing to 90°. When the LHA is 90°, the left and right-hand pages are identical.

The horizontal rules, known as the **Contrary-Same Line** or **C–S Line,** indicate the degree of declination in which the celestial horizon occurs.

Figure 2007c illustrates four of the eight possible celestial triangles for specific numerical values of latitude and declination, and the LHA's tabulated on the left and right-hand pages of an opening of the tables (figs. 2007a and 2007b). The diagram on the plane of the celestial meridian at the upper left of figure 2007c indicates that the celestial body always lies above the celestial horizon when the observer's latitude has the same name as the declination of the body and the values of LHA are those tabulated on the left-hand page of an opening of the tables. The diagram at the upper right reveals that for the various combinations of arguments on the right-hand page, including whether the name of the observer's latitude is the same as or contrary to the name of the declination, the numerical value of the declination governs whether the body is above or below the celestial horizon. For example, the following arguments are used for entering the tables:

LHA	60°
Latitude	45°N (Contrary Name to Declination)
Declination	5°S

The respondents are:

Tabular altitude,	ht (Tab. Hc)	16°53'.6
Altitude difference,	d	(–)46'.2
Azimuth angle,	Z	115°.6

As can be verified by an inspection of the upper-right diagram, the altitude respondent is for a body 16°53'.6 above the celestial horizon. Further inspection of the tabular data (fig. 2007b) and the diagram reveals that with the LHA and latitude (Contrary Name) remaining constant, the altitude of the body decreases as the declination increases in numerical value. Between values of declination 26° and 27° the body crosses the celestial horizon. When the declination reaches 35°, the altitude is 6°39'.6 below the celestial horizon; the tabular azimuth angle is the supplement of the actual azimuth angle of 134°.4.

As an additional example, the following arguments are used for entering the tables:

LHA	240°	(t 120°E)
Latitude	45°S	(Same Name as Declination)
Declination	5°S	

The respondents are:

Tabular altitude,	ht (Tab. Hc)	16°53'.6
Altitude difference,	d	(–)46'.2
Azimuth angle,	Z	115°.6

However, inspection of the diagram on the plane of the celestial meridian at the lower right of figure 2007c reveals that the altitude is 16°53'.6 *below* the celestial horizon; the tabular azimuth angle is the *supplement* of the actual azimuth angle of 64°.4. Further inspection of the tabular data and the diagram reveals that with the LHA and latitude (Same Name) remaining constant, the altitude of the body increases as the declination increases numerically. Between values of declination of 26° and 27° the body crosses the celestial horizon. When the declination reaches 35°, the altitude is 6°39.'6 above the celestial horizon; the tabular azimuth angle is the actual azimuth angle of 45.°6.

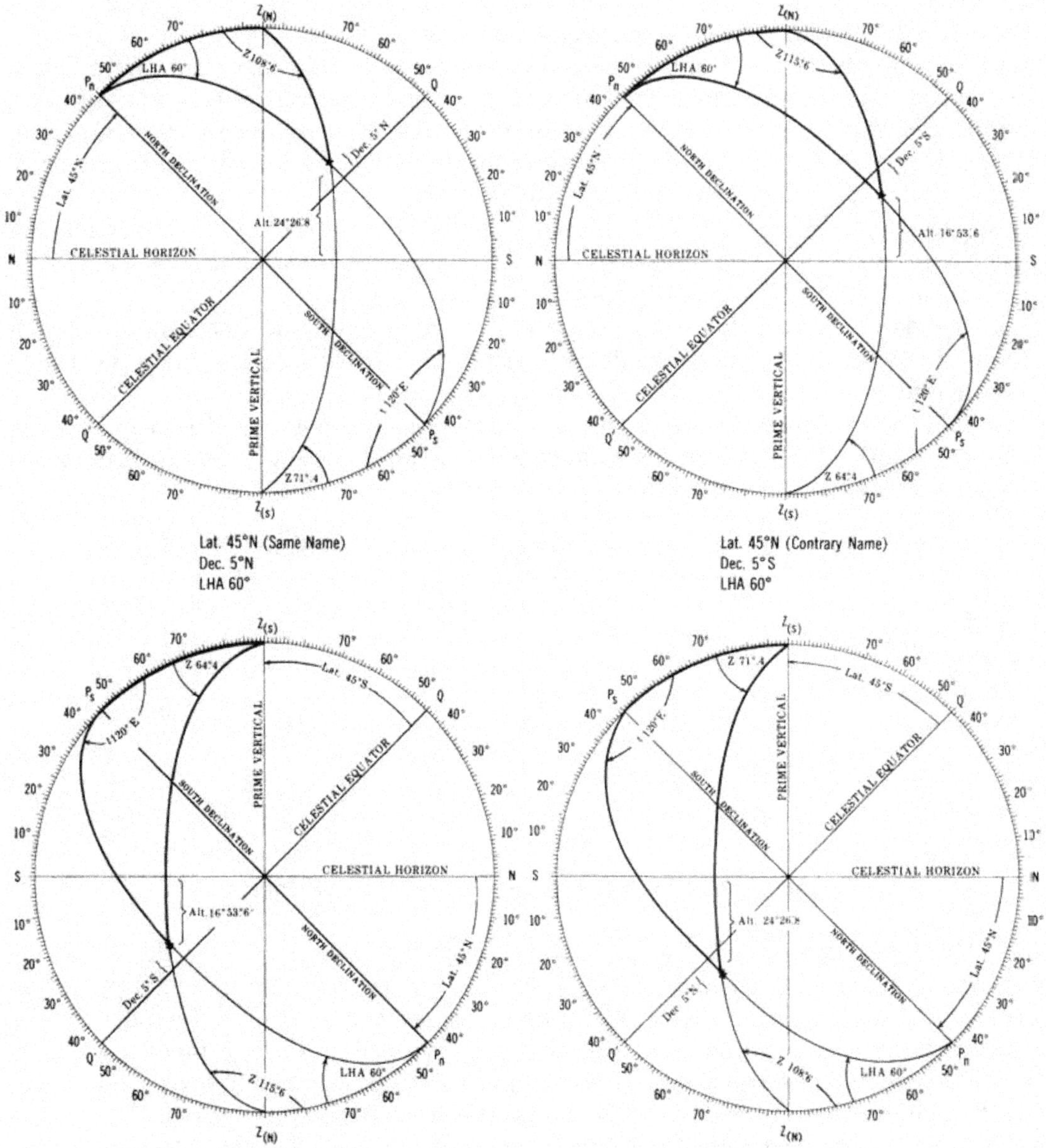

Lat. 45°N (Same Name)
Dec. 5°N
LHA 60°

Lat. 45°N (Contrary Name)
Dec. 5°S
LHA 60°

Lat. 45°S (Same Name)
Dec. 5°S
LHA 240° (t 120°E)

Lat 45°S (Contrary Name)
Dec. 5°N
LHA 240° (t 120° E)

FIGURE 2007c.—Diagrams on the plane of the celestial meridian.

Inspection of figures 2007a, 2007b, and 2007c reveals that if the left-hand page of an opening of the tables is entered with latitude of contrary name and one of the LHA's tabulated at the bottom of the facing page, the tabular altitudes are negative; the tabular azimuth angles are the supplements of the actual azimuth angles.

2008. Pub. No. 229 interpolation.—In the normal use of the tables with the Marcq St.-Hilaire method, it is only necessary to interpolate the tabular altitude and azimuth angle for the excess of the actual declination of the celestial body over the integral declination argument. When the tabular altitude is less than 60°, the required interpolation can always be effected through the use of the tabulated altitude differences. When the tabular altitude is in excess of 60°, it may be necessary to include the effects of second differences. When the tabular altitude difference is printed in italic type followed by a small dot, the effects of the second differences should be included in the interpolation. Although the effects of second differences may not be required, these effects can always be included in the interpolation whenever it is desired to obtain greater accuracy.

If the sight reduction is from a position such that interpolation for latitude and local hour angle increments is necessary, the required additional interpolation of the altitude can be effected by graphical means.

The data in the column for latitude 45° (Same Name as Declination) as contained in figure 2007a is rearranged in table 2008 to illustrate the first and second differences.

Table 2008 illustrates that the **first differences** are the differences between successive altitudes in a latitude column; the **second differences** are the differences between successive first differences.

LHA 60°, Lat. 45° (Same Name as Declination)

Dec.	ht (Tab. Hc)	First Difference	Second Difference
4°	23°42′.3		
		+44′.5	
5°	24°26′.8		−0′.2
		+44′.3	
6°	25°11′.1		−0′.3
		+44′.0	
7°	25°55′.1		

TABLE 2008.—First and second differences of tabular altitudes.

The usual case is that the change of altitude with 60′ increase in declination is nearly linear as illustrated in figure 2008. In this case, the required interpolation can be effected by multiplying the altitude difference (a first difference) by the excess of the actual declination over the integral declination argument divided by 60′. This excess of declination in minutes and tenths of minutes of arc is referred to as the **declination increment** and is abbreviated **Dec. Inc.**

Using the data of table 2008, the computed altitude when the LHA is 60°, the latitude (Same Name) is 45°, and the declination is 5°45′.5 is determined as follows:

$$\text{Correction} = \text{Altitude difference} \times \frac{\text{Dec. Inc.}}{60'} = (+)44'.3 \times \frac{45'.5}{60'} = 33'.6$$

$$\text{Hc} = \text{ht} + \text{correction} = 24°26'.8 + 33'.6 = 25°00'.4.$$

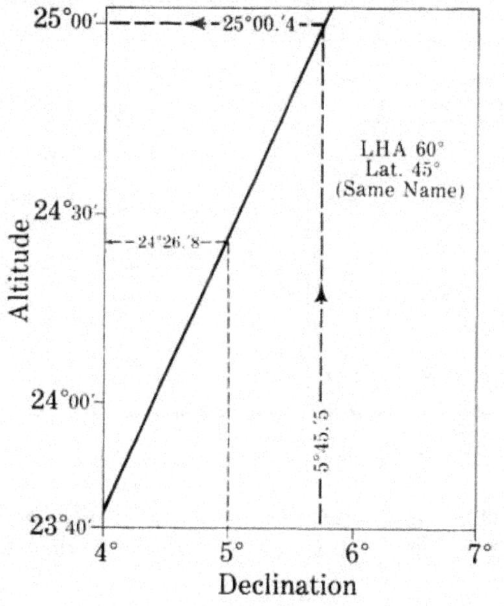

FIGURE 2008.—Linear interpolation by graph.

2009. Pub. No. 229 Interpolation Table.—The main part of the four-page Interpolation Table is basically a multiplication table providing tabulations of:

$$\text{Altitude Difference} \times \frac{\text{Declination Increment}}{60'}.$$

The design of the Interpolation Table is such that the desired product must be derived from component parts of the altitude difference. The first part is a multiple of 10′ (10′, 20′, 30′, 40′, or 50′) of the altitude difference; the second part is the remainder in the range 0.′0 to 9.′9. For example, the component parts of altitude difference 44.′3 are 40′ and 4.′3.

In the use of the first part of the altitude difference, the Interpolation Table arguments are Dec. Inc. and the integral multiple of 10′ in the altitude difference, d. As shown in figure 2009a, the respondent is:

$$\text{Tens} \times \frac{\text{Dec. Inc.}}{60'}.$$

In the use of the second part of the altitude difference, the Interpolation Table arguments are the nearest Dec. Inc. ending in 0.′5 and Units and Decimals. The respondent is:

$$\text{Units and Decimals} \times \frac{\text{Dec. Inc.}}{60'}.$$

In computing the table, the values in the Tens part of the multiplication table were modified by small quantities varying from −0.′042 to +0.′033 before rounding to the tabular precision to compensate for any difference between the actual Dec. Inc. and the nearest Dec. Inc. ending in 0.′5 when using the Units and Decimals part of the table.

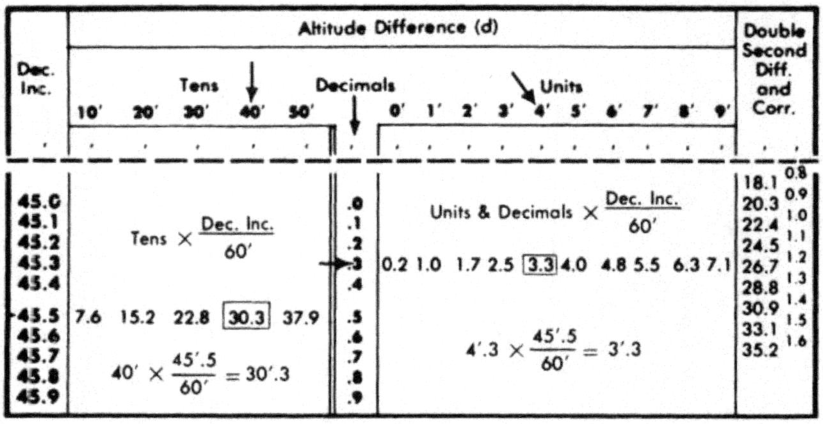

FIGURE 2009a.—Interpolation Table.

As an example of the use of the Interpolation Table, the computed altitude and true azimuth are determined for Lat. 45°N, LHA 60°, and Dec. 5°45.′5N. Data are exhibited in figure 2009b.

The respondents for the entering arguments (Lat. 45° Same Name as Declination, LHA 60°, and Dec. 5°) are:

Tabular altitude,	ht	24°26.′8
Altitude difference,	d	(+)44.′3
Tabular azimuth angle,	Z	108.°6

Note that Dec. Inc. 45.′5 is the vertical argument for entering the Interpolation Table to extract the correction for tens of minutes of altitude difference, d, and that it also indicates the subtable where the correction for minutes and tenths of minutes (Units and Decimals) of altitude difference, d, is found. Entering the Interpolation Table with Dec. Inc. 45.′5 as the vertical argument, the correction for 40′ of the altitude difference is 30.′3; the correction for 4.′3 of the altitude difference is 3.′3. Adding the two parts, the correction is (+)33.′6, the sign of the correction being in accordance with the sign of the altitude difference, d.

No special table is provided for interpolation of the azimuth angle, and the differences are not tabulated. With latitude and local hour angle constant, the successive azimuth angle differences corresponding to 1° increase in declination are less than 10.°0 for altitudes less than 84°, and can easily be found by inspection. If formal interpolation of azimuth angle is desired, the degrees and tenths of degrees of azimuth angle difference are treated as minutes and tenths of minutes in obtaining the required correction from the Units and Decimals subtable to the right of the declination increment. But for most practical applications, interpolation by inspection usually suffices. In this example of formal interpolation, using an azimuth angle difference of −0.°7 and a Dec. Inc. of 45.′5, the correction as extracted from the Units and Decimals subtable to the right of the Dec. Inc. is −0.°5. Therefore, the azimuth angle as interpolated for declination increment is 108.°1 (108.°6−0.°5). In summary,

Tabular altitude	ht	24°26.′8
Correction for 40′ of alt. diff.		(+)30.′3
Correction for 4.′3 of alt. diff		(+) 3.′3
Computed altitude	Hc	25°00.′4

(See figures 2008 and 2009b)

Tabular azimuth angle	Z	108°.6
Correction for Dec. Inc. 45′.5		(−)0°.5
Interpolated azimuth angle	Z	N108°.1W
True azimuth	Zn	251°.9

INTERPOLATION TABLE

Dec. Inc.	Altitude Difference (d) Tens 10′	20′	30′	40′	50′	Decimals	Units 0′	1′	2′	3′	4′	5′	6′	7′	8′	9′	Double Second Diff. and Corr.
45.0	7.5	15.0	22.5	30.0	37.5	.0	0.0	0.8	1.5	2.3	3.0	3.8	4.5	5.3	6.1	6.8	18.1 0.8 20.3 0.9
45.1	7.5	15.0	22.5	30.0	37.6	.1	0.1	0.8	1.6	2.4	3.1	3.9	4.6	5.4	6.1	6.9	22.4 1.0
45.2	7.5	15.0	22.6	30.1	37.6	.2	0.2	0.9	1.7	2.4	3.2	3.9	4.7	5.5	6.2	7.0	24.5 1.1
45.3	7.5	15.1	22.6	30.2	37.7	.3	0.2	1.0	1.7	2.5	3.3	4.0	4.8	5.5	6.3	7.1	26.7 1.2
45.4	7.6	15.1	22.7	30.3	37.8	.4	0.3	1.1	1.8	2.6	3.3	4.1	4.9	5.6	6.4	7.1	28.8 1.3
45.5	7.6	15.2	22.8	30.3	37.9	.5	0.4	1.1	1.9	2.7	3.4	4.2	4.9	5.7	6.4	7.2	30.9 1.4
45.6	7.6	15.2	22.8	30.4	38.0	.6	0.5	1.2	2.0	2.7	3.5	4.2	5.0	5.8	6.5	7.3	33.1 1.5 35.2 1.6
45.7	7.6	15.3	22.9	30.5	38.1	.7	0.5	1.3	2.0	2.8	3.6	4.3	5.1	5.8	6.6	7.4	
45.8	7.7	15.3	22.9	30.6	38.2	.8	0.6	1.4	2.1	2.9	3.6	4.4	5.2	5.9	6.7	7.4	
45.9	7.7	15.3	23.0	30.6	38.3	.9	0.7	1.4	2.2	3.0	3.7	4.5	5.2	6.0	6.7	7.5	

60°, 300° L.H.A. LATITUDE SAME NAME AS DECLINATION

N. Lat. {L.H.A. greater than 180° ... Zn=Z; L.H.A. less than 180° ... Zn=360°−Z}

Dec.	38° Hc	d	Z	39° Hc	d	Z	40° Hc	d	Z	41° Hc	d	Z
°	° ′	′	°	° ′	′	°	° ′	′	°	° ′	′	°
0	23 12.2	+40.1	109.6	22 51.9	+40.9	110.0	22 31.3	+41.6	110.4	22 10.2	+42.4	110.7
1	23 52.3	39.8	108.8	23 32.8	40.6	109.2	23 12.9	41.4	109.6	22 52.6	42.2	110.0
2	24 32.1	39.5	107.9	24 13.4	40.4	108.4	23 54.3	41.2	108.8	23 34.8	41.9	109.2
3	25 11.6	39.3	107.1	24 53.8	40.1	107.6	24 35.5	40.9	108.0	24 16.7	41.7	108.4
4	25 50.9	39.0	106.3	25 33.9	39.8	106.7	25 16.4	40.6	107.2	24 58.4	41.5	107.6
5	26 29.9	+38.6	105.4	26 13.7	+39.5	105.9	25 57.0	+40.4	106.4	25 39.9	+41.1	106.8
6	27 08.5	38.4	104.6	26 53.2	39.2	105.1	26 37.4	40.0	105.5	26 21.0	40.9	106.0
7	27 46.9	38.0	103.7	27 32.4	38.9	104.2	27 17.4	39.8	104.7	27 01.9	40.7	105.2
8	28 24.9	37.7	102.8	28 11.3	38.6	103.3	27 57.2	39.5	103.9	27 42.6	40.3	104.4
9	29 02.6	37.3	101.9	28 49.9	38.2	102.5	28 36.7	39.1	103.0	28 22.9	40.0	103.5

42° Hc	d	Z	43° Hc	d	Z	44° Hc	d	Z	45° Hc	d	Z	Dec.
° ′	′	°	° ′	′	°	° ′	′	°	° ′	′	°	°
21 48.8	+43.1	111.1	21 27.0	+43.8	111.5	21 04.8	+44.6	111.9	20 42.3	+45.3	112.2	0
22 31.9	42.9	110.4	22 10.8	43.7	110.8	21 49.4	44.4	111.1	21 27.6	45.0	111.5	1
23 14.8	42.7	109.6	22 54.5	43.4	110.0	22 33.8	44.2	110.4	22 12.6	45.0	110.8	2
23 57.5	42.5	108.8	23 37.9	43.3	109.3	23 18.0	43.9	109.7	22 57.6	44.7	110.1	3
24 40.0	42.3	108.1	24 21.2	43.0	108.5	24 01.9	43.8	108.9	23 42.3	44.5	109.3	4
25 22.3	+41.9	107.3	25 04.2	+42.8	107.7	24 45.7	+43.5	108.2	24 26.8	+44.3	108.6	5
26 04.2	41.8	106.5	25 47.0	42.5	107.0	25 29.2	43.3	107.4	25 11.1	44.0	107.9	6
26 46.0	41.4	105.7	26 29.5	42.3	106.2	26 12.5	43.1	106.7	25 55.1	43.8	107.1	7
27 27.4	41.2	104.9	27 11.8	41.9	105.4	26 55.6	42.8	105.9	26 38.9	43.6	106.4	8
28 08.6	40.8	104.1	27 53.7	41.7	104.6	27 38.4	42.5	105.1	27 22.5	43.3	105.6	9

FIGURE 2009b.—Data from main tables and Interpolation Table.

2010. Pub. No. 229 interpolation when second differences are required.—The accuracy of linear interpolation usually decreases as the altitude increases. At altitudes above 60° it may be necessary to include the effect of second differences in the interpolation. When the altitude difference, d, is printed in italic type followed by a small dot, the second-difference correction may exceed 0′.25, and should normally be applied. The need for a second-difference correction is illustrated by the graph of table 2010 data in figure 2010a.

Other than graphically, the required correction for the effects of second differences is obtained from the appropriate subtable of the Interpolation Table. However, before the Interpolation Table can be used for this purpose, what is known as the double-second difference must be formed. The **double-second difference (DSD)** is the sum of two successive second differences. Although second differences are not tabulated, the DSD can be formed readily by subtracting, algebraically, the tabular altitude difference immediately above the respondent altitude difference from the tabular altitude difference immediately below. The result will always be a negative value.

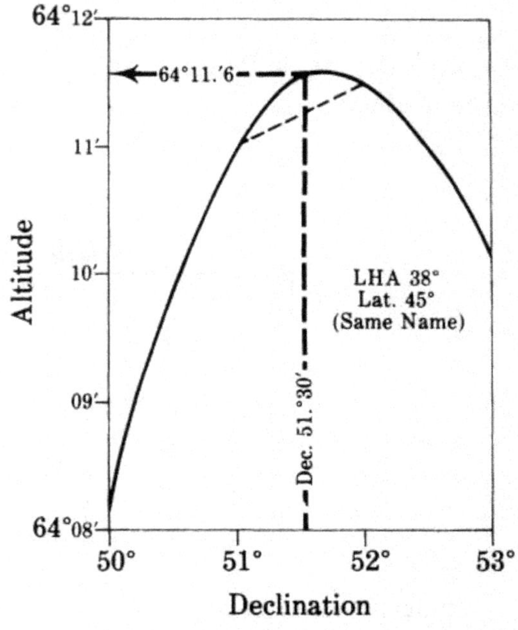

FIGURE 2010a.—Nonlinear interpolation by graph.

LHA 38°, Lat. 45° (Same Name as Declination)

Dec.	ht (Tab. Hc)	First Difference	Second Difference
50°	64°08′.2		
		+2′.8	
51°	64°11′.0		−2′.3
		+0′.5	
52°	64°11′.5		−2′.1
		−1′.6	
53°	64°09′.9		

TABLE 2010.—First and second differences of tabular altitudes.

As shown in figure 2010b, that compartment of the DSD table opposite the block in which the Dec. Inc. is found is entered with the DSD to obtain the DSD correction to the altitude. The correction is always plus. Therefore, the sign of the DSD need not be recorded. When the DSD entry corresponds to an exact tabular value, always use the upper of the two possible corrections.

As an example of the use of the double-second difference, the computed altitude and true azimuth are determined for Lat. 45°N, LHA 38°, and Dec. 51°30′.0N. Data are exhibited in figure 2010b.

The respondents for the entering arguments (Lat. 45° Same Name as Declination, LHA 38°, and Dec. 51°) are:

Tabular altitude,	ht	64°11′.0
Altitude difference,	d	(+)0′.5
Azimuth angle,	Z	62°.8

The linear interpolation correction to the tabular altitude for Dec. Inc. 30′.0 is (+)0′.3.

Hc = ht + linear correction = 64°11′.0 + 0′.3 = 64°11′.3

INTERPOLATION TABLE

Dec Inc.	Tens 10′	20′	30′	40′	50′	Decimals	Units 0′	1′	2′	3′	4′	5′	6′	7′	8′	9′
30.0	5.0	10.0	15.0	20.0	25.0	.0	00	05	10	15	20	25	30	36	41	46
30.1	5.0	10.0	15.0	20.0	25.1	.1	01	06	11	16	21	26	31	36	41	46
30.2	5.0	10.0	15.1	20.1	25.1	.2	01	06	11	16	21	26	32	37	42	47
30.3	5.0	10.1	15.1	20.2	25.2	.3	02	07	12	17	22	27	32	37	42	47
30.4	5.1	10.1	15.2	20.3	25.3	.4	02	07	12	17	22	27	33	38	43	48
30.5	5.1	10.2	15.3	20.3	25.4	.5	03	08	13	18	23	28	33	38	43	48
30.6	5.1	10.2	15.3	20.4	25.5	.6	03	08	13	18	23	28	34	39	44	49
30.7	5.1	10.3	15.4	20.5	25.6	.7	04	09	14	19	24	29	34	39	44	49
30.8	5.2	10.3	15.4	20.6	25.7	.8	04	09	14	19	24	29	35	40	45	50
30.9	5.2	10.3	15.5	20.6	25.8	.9	05	10	15	20	25	30	35	40	45	50

Double Second Diff	and Corr
0.8	
	0.1
2.4	
	0.2
4.0	
	0.3
5.6	
	0.4
7.2	
	0.5
8.8	
	0.6
10.4	
	0.7
12.0	
	0.8
13.6	
	0.9
15.2	
	1.0
16.8	

38°, 322° L.H.A. **LATITUDE SAME NAME AS DECLINATION**

N. Lat. {L.H.A. greater than 180° ...... Zn=Z; L.H.A. less than 180° ...... Zn=360°−Z}

Dec.	38° Hc	d	Z	39° Hc	d	Z	40° Hc	d	Z	41° Hc	d	Z
°	° ′	′	°	° ′	′	°	° ′	′	°	° ′	′	°
49	60 41.7	8.8	55.6	61 15.0	7.1	57.1	61 46.9	5.3	58.7	62 17.3	3.3	60.3
50	60 32.9	−10.7	53.6	61 07.9	−9.0	55.1	61 41.6	−7.2	56.6	62 14.0	−5.4	58.2
51	60 22.2	12.4	51.6	60 58.9	10.8	53.0	61 34.4	9.1	54.5	62 08.6	7.3	56.0
52	60 09.8	14.1	49.6	60 48.1	12.6	51.0	61 25.3	10.9	52.4	62 01.3	9.2	53.9
53	59 55.7	15.9	47.7	60 35.5	14.3	49.0	61 14.4	12.8	50.4	61 52.1	11.2	51.8
54	59 39.8	17.5	45.8	60 21.2	16.1	47.0	61 01.6	14.6	48.3	61 40.9	13.0	49.7
55	59 22.3	−19.1	43.9	60 05.1	−17.8	45.1	60 47.0	−16.3	46.3	61 27.9	−14.8	47.7
56	59 03.2	20.7	42.0	59 47.3	19.3	43.2	60 30.7	18.1	44.4	61 13.1	16.6	45.6
57	58 42.5	22.2	40.2	59 28.0	21.0	41.3	60 12.6	19.6	42.4	60 56.5	18.3	43.7
58	58 20.3	23.6	38.4	59 07.0	22.5	39.5	59 53.0	21.3	40.6	60 38.2	20.0	41.7
59	57 56.7	25.0	36.7	58 44.5	23.9	37.7	59 31.7	22.8	38.7	60 18.2	21.6	39.8

42° Hc	d	Z	43° Hc	d	Z	44° Hc	d	Z	45° Hc	d	Z	Dec.
° ′	′	°	° ′	′	°	° ′	′	°	° ′	′	°	°
62 46.3	−1.4	62.0	63 13.7	+0.6	63.7	63 39.4	2.7	65.5	64 03.4	4.8	67.4	49
62 44.9	−3.4	59.8	63 14.3	−1.4	61.5	63 42.1	+0.6	63.3	64 08.2	+2.8	65.1	50
62 41.5	5.5	57.6	63 12.9	3.6	59.3	63 42.7	−1.5	61.0	64 11.0	+0.5	62.8	51
62 36.0	7.4	55.5	63 09.3	5.5	57.1	63 41.2	3.6	58.8	64 11.5	−1.6	60.5	52
62 28.6	9.5	53.3	63 03.8	7.7	54.9	63 37.6	5.7	56.5	64 09.9	3.7	58.2	53
62 19.1	11.3	51.2	62 56.1	9.6	52.7	63 31.9	7.8	54.3	64 06.2	5.9	56.0	54
62 07.8	−13.3	49.1	62 46.5	−11.6	50.5	63 24.1	−9.9	52.1	64 00.3	−8.0	53.7	55
61 54.5	15.1	47.0	62 34.9	13.5	48.4	63 14.2	11.8	49.9	63 52.3	10.1	51.4	56
61 39.4	16.9	44.9	62 21.4	15.4	46.3	63 02.4	13.9	47.7	63 42.2	12.2	49.2	57
61 22.5	18.6	42.9	62 06.0	17.2	44.2	62 48.5	15.7	45.6	63 30.0	14.1	47.0	58
61 03.9	20.4	40.9	61 48.8	19.0	42.2	62 32.8	17.6	43.5	63 15.9	16.1	44.8	59

FIGURE 2010b.—Data from main tables and Interpolation Table.

However, by inspection of figure 2010a, illustrating this solution graphically, the computed altitude should be 64°11′.6. The actual change in altitude with an increase in declination is nonlinear. The altitude value lies on the curve between the points for declination 51° and declination 52° instead of the straight line connecting these points.

The DSD is formed by subtracting, algebraically, the tabular altitude difference immediately above the respondent altitude difference from the tabular altitude difference immediately below. Thus, the DSD is formed by algebraically subtracting (+)2′.8 from (−)1′.6; the result is (−)4′.4.

As shown in figure 2010b, that compartment of the DSD table opposite the block in which the Dec. Inc. (30′.0) is found is entered with the DSD (4′.4) to obtain the DSD correction to the altitude. The correction is 0′.3. The correction is always plus.

Hc = ht + linear correction + DSD correction
Hc = 64°11′.0 + 0′.3 + 0′.3 = 64°11′.6.

2011. Complete solution by Pub. No. 229 and *Nautical Almanac*.—The complete solution includes all of the parts listed in article 2001. Because of the various alternatives available for the separate parts, a large number of variations might be used in the complete solution.

It is good practice to have a standard work form. If this is not printed, or on a rubber stamp, it should be copied in its entirety before the solution is started. The first step should then be to fill in the known information. If the solution for observed altitude is made first, this value can then be copied in the main solution, so that it will be ready for comparison when Hc is determined. The best form to use is that which the individual navigator finds most logical and least likely to result in errors.

Some navigators include a time diagram (art. 1428) in the form as a check both on the time and meridian angle computation.

There is a growing tendency among navigators to keep the navigational watch set to GMT. This is particularly helpful when a number of observations are made, as during twilight, to eliminate the need for repeated application of watch error and zone description, and determination of Greenwich date.

Example.—On June 2, 1975, the 1742 dead reckoning position of a ship is lat. 41°10'S, long. 128°00'E. The ship is on course 315°, speed 20 knots. Observations are made from a height of eye of 31 feet using a sextant having an index correction of (−)1'.0 as indicated below. Determine the 1742 fix.

Body	*GMT*	*Sextant Altitude*	*SHA*	*Declination*
Spica	08ʰ24ᵐ03ˢ	32°30'.4	159°01'.1	11°02'.2S
Regulus	08ʰ29ᵐ58ˢ	36°57'.1	208°13'.9	12°05'.2N
Procyon	08ʰ35ᵐ59ˢ	35°05'.1	245°29'.8	5°17'.2N
Canopus	08ʰ41ᵐ55ˢ	52°47'.7	264°09'.3	52°41'.1S

		Spica		*Regulus*
GMT (June 2)		08ʰ24ᵐ03ˢ		08ʰ29ᵐ58ˢ
GHA ♈ for 8ʰ GMT		10°10'.3		10°10'.3
Increments	24ᵐ03ˢ	6°01'.7	29ᵐ58ˢ	7°30'.7
SHA☆		159°01'.1		208°13'.9
GHA☆		175°13'.1		225°54'.9
*a*λ		127°46'.9E		128°05'.1E
LHA☆		303°00'.0		354°00'.0
Dec		11°02'.2S		12°05'.2N
Dec. Inc		02'.2		05'.2
*a*L		41°00'.0S		41°00'.0S
ht (Tab. Hc)		31°55'.0		36°42'.6
d and correction	(+)39'.7	(+)1'.5	(−)59'.7	(−)5'.2
Hc		31°56'.5		36°37'.4
Ho		32°22'.5		36°49'.4
a		26. 0T		12. 0T
Z and Zn	S104°.1E	075°.9	S172°.7E	007°.3

		Procyon		*Canopus*
GMT (June 2)		08ʰ35ᵐ59ˢ		08ʰ41ᵐ55ˢ
GHA ♈ for 8ʰ GMT		10°10'.3		10°10'.3
Increments	35ᵐ59ˢ	9°01'.2	41ᵐ55ˢ	10°30'.5
SHA☆		245°29'.8		264°09'.3
GHA☆		264°41'.3		284°50'.1
*a*λ		128°18'.7E		128°09'.9E
LHA☆		33°00'.0		53°00'.0
Dec		5°17'.2N		52°41'.1S
Dec. Inc		17'.2		41'.1
*a*L		41°00'.0S		41°00'.0S
ht (Tab. Hc)		34°59'.1		52°48'.5
d and correction	(−)52'.0	(−)14'.9	(+)3'.8	(+)2'.6
Hc		34°44'.2		52°51'.1
Ho		34°57'.3		52°40'.6
a		13.1T		10. 5A
Z and Zn	S138°.7W	318°.7	S53°.3W	233°.3

2012. *Sight Reduction Tables for Air Navigation* (Pub. No. 249).—Although these tables are designed to satisfy the needs of the air navigator, they are frequently used for sight reduction at sea. The following description and explanation of the use of the tables is limited to the marine application.

Volume I contains tabulations of altitude (to the nearest 1′) and azimuth (to the nearest 1°) in parallel columns. For each 1° of latitude a two-page table (one-page above 69°) is given. For each 1° (2° beyond latitude 69°) of LHA ♈, altitude and azimuth are given for seven stars carefully selected with regard to azimuth, magnitude, altitude, and continuity. Stars of the first magnitude are shown in capital letters, and those of second and third magnitude in lower case with initial capital. After each 15 entries a break occurs and a new listing of stars is given, whether or not there are any changes from the previous list. Stars are listed in the order of increasing azimuth at the beginning of each period. Forty-one stars are included, 19 of which are of the first magnitude, 17 of the second magnitude, and 5 of the third magnitude. The tables are intended for use with an assumed position selected so that latitude and LHA ♈ are each the nearest whole degree (nearest *even* degree of LHA ♈ at latitudes higher than 69°).

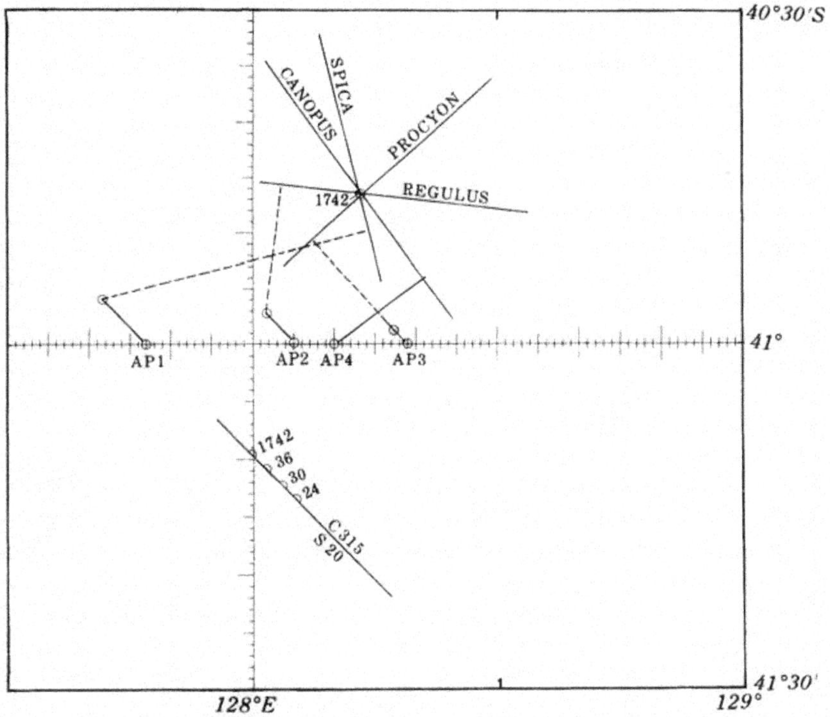

Volume I, for selected stars, is arranged for entering with latitude, LHA ♈, and the appropriate star name. This arrangement minimizes the time and effort required in sight reduction. Progressive changes in the coordinates affecting the tabulated data necessitate recomputation at approximately 5-year intervals in order to reduce the effects of this source of cumulative error. Of the seven stars selected for each 15° of LHA ♈, the three marked by the diamond symbol (♦) provide sets favorably situated in altitude and azimuth for the three-body fix normally used in *air navigation.* The volume for epoch 1980.0 will replace the volume for epoch 1975.0.

Tabulation by name of the star eliminates the need for finding the declination, but a correction for precession of the equinoxes (art. 1419) and nutation (art. 1417) may be needed. This is given in an auxiliary table near the back of the volume. The correction which may reach a value as high as 3′ is applied to the fix, not to each altitude.

Tabulation of *azimuth* (not azimuth angle) eliminates the need for conversion.

Tabulation by LHA ♈ instead of meridian angle of the star eliminates the need for finding and applying SHA. It also makes of the tables a star finder for the *seven* stars given, since all values given for any entry of LHA ♈ are for the *same time.*

An almanac giving GHA ♈ is included for use should the *Air Almanac* or *Nautical Almanac* not be available.

Example 1.—During evening twilight on June 2, 1975, the 1724 DR position of a ship is lat. 40°39′.2S, long. 128°01′.2E. At GMT $8^h24^m03^s$ the navigator observes Canopus with a marine sextant having no IC, from a height of eye of 38 feet. The hs is 55°57′.1.

Required.—The *a*, Zn, and AP, using Pub. No. 249 (epoch 1975.0), vol. I, and the *Air Almanac.*

Solution.—

June 2		Canopus		+ ☆	−
GMT	$8^h24^m03^s$ June 2		IC	−	−
8^h20^m	15°11′		D		6′
4^m03^s	1°01′		R		1′
GHA ♈	16°12′		sum	−	7′
*a*λ	127°48′E		corr.		(−)7′
LHA ♈	144°00′		hs		55°57′
*a*L	41°00′S		Ho		55°50′
Hc	55°48′				
Ho	55°50′				
a	2T	*a*L 41°00′S			
Zn	233°	*a*λ 127°48′E			

Example 2.—During evening twilight on June 2, 1975, the 1724 DR position of a ship is lat. 40°39′.2S, long. 128°01′.2E. At GMT $8^h24^m03^s$ the navigator observes Canopus with a marine sextant having no IC, from a height of eye of 38 feet. The hs is 55°57′.1.

Required.—The *a*, Zn, and AP, using Pub. No. 249 (epoch 1975.0), vol. I, and the *Nautical Almanac.*

Solution.—

June 2	Canopus	+ ☆ −
GMT 8ʰ24ᵐ03ˢ June 2		IC − −
8ʰ00ᵐ 10°10.′3		D 6.′0
24ᵐ03ˢ 6°01.′7		R 0.′7
GHA ♈ 16°12.′0		sum − 6.′7
*a*λ 127°48.′0 E		corr. (−) 6.′7
LHA ♈ 144°00.′0		hs 55°57.′1
*a*L 41°00.′0 S		Ho 55°50.′4
Hc 55°48.′0		
Ho 55°50.′4		
a 2.4 T	*a*L 41°00′S	
Zn 233°	*a*λ 127°48′E	

Volumes II and III are somewhat similar in many respects to Pub. No. 229. Altitude and azimuth angle are given in parallel columns for every whole degree of latitude (0° to 89°), every whole degree of declination (0° to 29°), and every whole degree (2° beyond lat. 69°) of LHA for all values at which the altitude is greater than several degrees *below* the celestial horizon. The values for latitude and declination contrary name are tabulated with values of meridian angle (LHA less than 180°) increasing *upward* on the page. This permits better utilization of space where same- and contrary-name tabulations are given on the same page. It also serves to emphasize the difference between the same- and contrary-name tabulations, the contrary-name tabulation being given in a "contrary" manner on the page. A more convenient arrangement of declination entries is provided by having the "top" of each page of the tables along the left side, requiring the turning of the page through 90°.

The altitude difference (labeled *d*) is tabulated between the altitude and azimuth angle to facilitate interpolation of altitude for declination. No interpolation is needed for latitude and LHA because the assumed position is selected so that these are the nearest whole degree (nearest *even* degree of LHA beyond latitude 69°). The altitude difference is the difference in minutes, with sign, between the accompanying altitude and that for declination 1° *greater,* at the same latitude and LHA. It is used for entering an auxiliary table for determining the correction to be applied to altitude for minutes of declination, Dec. Inc. Interpolation is normally made in the direction of increasing declination.

Volume II covers latitudes 0° to 39°, and volume III contains similar information for latitudes 40° to 89°. Since these tables are entered with LHA of the celestial body, they do not become inaccurate in succeeding years, and no correction is needed for precession and nutation, as in volume I. In contrast with volume I, azimuth angle is tabulated instead of azimuth. A long-term almanac giving the GHA and declination of the sun is included in both volumes for use should the *Air Almanac* or *Nautical Almanac* not be available. These volumes are intended for solution of observations of the sun, moon, planets, and any stars within the declination range.

Example 3.—During morning twilight on June 2, 1975, the 0724 DR position of a ship is lat. 40°39.′2S, long. 131°01.′2E. At GMT 22ʰ24ᵐ03ˢ (June 1) the navigator observes Alpheratz with a marine sextant having no IC, from a height of eye of 38 feet. The hs is 20°15.′3.

Required.—The *a*, Zn, and AP, using Pub. No. 249, vol. III, and the *Air Almanac.*

Solution.—

	June 2		Alpheratz		+ ☆ −
GMT	$22^h24^m03^s$	June 1		IC	– –
22^h20^m	224°46′			D	6′
4^m03^s	1°01′			R	3′
SHA	358°13′			sum –	9′
GHA	224°00′			corr.	(−)9′
*a*λ	131°00′E			hs	20°15′
LHA	355°00′			Ho	20°06′
d	28°57′N	Dec. Inc. 57′			
*a*L	41°00′S				
ht	20°51′	d (−) 60		Z	S175°E
corr.	(−)57′				
Hc	19°54′				
Ho	20°06′				
a	12T		*a*L 41°00′S		
Zn	005°		*a*λ 130°40′E		

Example 4.—During morning twilight on June 2, 1975, the 0724 DR position of a ship is lat. 40°39′.2S, long. 131°01′.2E. At GMT $22^h24^m03^s$ (June 1) the navigator observes Alpheratz with a marine sextant having no IC, from a height of eye of 38 feet. The hs is 20°15′3.

Required.—The *a*, Zn, and AP, using Pub. No. 249, vol. III, and the *Nautical Almanac.*

Solution.—

	June 2		Alpheratz		+ ☆ −
GMT	$22^h24^m03^s$	June 1		IC –	–
22^h	219°45′.7			D	6′.0
24^m03^s	6°01′.7			☆-P	2′.6
SHA	358°13′.2			sum	– 8′.6
GHA	224°00′.6			corr.	(−)8′.6
*a*λ	130°59′.4E			hs	20°15′.3
LHA	355°00′.0			Ho	20°06′.7
d	28°57′.2N	Dec. Inc. 57′			
*a*L	41°00′.0S				
ht	20°51′.0	*d* (−)60		Z	S175°E
corr.	(−)57′.0				
Hc	19°54′.0				
Ho	20°06′.7				
a	12.7T		*a*L 41°00′.0S		
Zn	005°		*a*λ 130°59′.4E		

2013. *Tables of Computed Altitude and Azimuth* (Pub. No. 214).—The publication of these sight reduction tables has been discontinued by the Defense Mapping Agency Hydrographic/Topographic Center. However, it is expected that navigators will use the volumes they possess for many years to come. The tables were published in nine volumes, each covering 10° of latitude in increments of 1°. For each degree of latitude there is a series of tables, with cutaway tabs providing quick reference to the first page of the tables for that latitude. Declination entries are given at intervals of 0°.5 from 0° to 29°. Beyond this, 37 selected declination entries are given to provide solutions for all of the stars listed on the daily pages of the almanacs, and most of the additional stars listed near the back of the *Nautical Almanac.* A total of 96 declination entries are given for each latitude, arranged eight to a page. Each declination entry is given at the top of a column. The third variable, meridian angle, is given in the column at the extreme left and right sides of each page. These columns are labeled "H.A.," the abbreviation for "hour angle," the expression formerly used for meridian angle, but replaced because of confusion with *local* hour angle, *Greenwich* hour angle, etc. Meridian angle entries are given at intervals of 1° from 0° at the top of the page to the maximum value at which the altitude is 5° or greater.

At most page openings, separate tables are given for declination having the *same* name (N or S) as the latitude and those having *contrary* name (one N, the other S). That is, declination values on the left-hand page (same name) are duplicated on the right-hand page (contrary name). A maximum of ninety-one meridian angle entries (0°–90°) are given on the left-hand page (same name). As either the declination or the latitude increases, the number of same-name entries increases, and the number of contrary-name entries decreases. When the same-name entries exceed 90° of meridian angle, the additional ones are placed on the right-hand page, below the contrary-name entries. At extreme values of declination and latitude there are no contrary-name entries, the same-name entries occupying both pages.

In each declination column there are four sets of figures. The first, given in bold type, is the altitude (labeled "Alt.") to the nearest 0'.1. Following this is Δd in small type. This is the change of altitude for a unit change of declination. Except when the value is 1.0, entries are given in hundredths of a unit, although the position of the decimal point is not shown. Following Δd, and also in small type, is Δt, the change of altitude for a unit change of meridian angle. This is given in the same form as Δd. The last set of figures in the column is the azimuth angle (labeled "Az."), to the nearest 0°.1. The current abbreviation for azimuth angle, Z, is used in example solutions.

At latitude 0° the arrangement is modified because there is no "same" or "contrary" name of declination. Here a single set of declination entries is given. Declination replaces latitude as the prefix label for azimuth angle.

Following the altitude-azimuth section of each latitude is a two-page star identification table. The use of this table is explained in article 2213.

On the inside front cover and its facing page is a speed-time-distance table, which is useful in advancing or retiring lines of position, as well as for other purposes. This table contains information similar to that in table 19, but in somewhat different form. Volume VIII and older printings of other volumes have sextant altitude correction tables on these pages.

Following the speed-time-distance table is an arc to time conversion table.

Following the title page and preface are given a description of the tables, and sample problems.

On the inside back cover and facing page is given a "multiplication table" to multiply Δd or Δt by the number of minutes between the declination or meridian angle and the value used for entering the main table. This is used in interpolating the altitude for declination or meridian angle.

On the two pages next preceding the multiplication table is given a somewhat similar table to provide easy interpolation for latitude.

The use of the various parts of Pub. No. 214 is explained in articles 2014–2016 and 2213. The primary purpose of Pub. No. 214 is to provide an easy method of sight reduction for use with the *Nautical Almanac* aboard ship. It may also be used with the *Air Almanac,* and for solution of any spherical triangle for which entry values are given. Therefore, it can be used in great-circle sailing for determining the initial course and the distance.

2014. Pub. No. 214 solution by Δd only.—If interpolation is made for all three variables—latitude, declination, and meridian angle—a triple interpolation is needed. A simpler solution, almost universally used with Pub. No. 214, is to select an assumed position that will eliminate interpolation for latitude and meridian angle, leaving a simple interpolation for declination.

Example.—Find computed altitude (Hc) and azimuth (Zn) if *a*L is 41°00.′0N, d is 22°14.′3N, and t is 36°00.′0W.

Solution.—

t	36°00.′0W		
d	22°14.′3N	d diff. 14.′3	
*a*L	41°00.′0N		
ht	54°16.′8	Δd (+) 0.65	Z N111.°0W
corr.	(+)9.′3		
Hc	54°26.′1		
Zn	249.°0		

The main table is entered with the three variables, t, d, and *a*L (being sure to note whether d and *a*L are of same or contrary name); and the values of ht (tabulated altitude, labeled "Alt." in Pub. No. 214), Δd, and Z are taken directly from the table, without interpolation. The tabulated altitude (ht) is the computed altitude (Hc) for the values used for entering the table. The designation ht is used to distinguish it from the Hc obtained by applying a correction to the value taken from the table.

The declination entry argument used should be the tabulated entry nearest the declination for which a solution is sought, normally differing by not more than half a degree. The difference between this value and the actual declination is recorded as "d diff". No sign (+ or −) is assigned to this value. It is good practice to show Δd as a decimal, even though it is not tabulated in this way. The sign of this value should be determined carefully by inspection of the main table of Pub. No. 214. Interpolation of altitude for declination is made between the base value taken from the table and the value given on the same line in the next column to the right or left. The choice of the second column depends upon the actual declination. If it is *greater* than the value used for entering the table, use the next column to the *right,* and if less, use the next column to the *left.* If the value in the second column is greater than the base value, the sign is plus (+), and if less, the sign is minus (−). The accuracy of this important step can be checked by comparing the computed altitude (Hc) with the altitudes given in the main table of Pub. No. 214. If Δd has been given the correct sign (and applied correctly), Hc should lie between the tabulated altitudes in the columns for tabulated declination next smaller and next larger than the actual declination.

The azimuth angle is given a prefix N or S to agree with the latitude, and a suffix E or W to agree with the meridian angle. For this reason it is good practice to label these values *when they are recorded.*

The next step is to multiply Δd by d diff., to interpolate between the altitude entries for consecutive declination columns. In most instances, the easiest way to do this is to use the multiplication table on the inside back cover of Pub. No. 214 and its facing page, entering separately with minutes and tenths of minutes of Δd and adding the two parts. The correction, which is recorded below ht, is given the sign of Δd. The correction is then added or subtracted, in accordance with its sign, to ht. The answer is computed altitude (Hc).

Azimuth (Zn), which is recorded below Hc, is found by converting azimuth angle (Z) in accordance with its labels, as explained in article 1430. Usually, azimuth angle is found without interpolation.

If Δd is changing rapidly, or when it changes sign (at the maximum altitude for the given meridian angle and latitude), interpolation may be somewhat less accurate than in other parts of the tables, but this should not introduce a large error unless the celestial body is near the zenith, when the method of Pub. No. 214 is not recommended.

2015. Pub. No. 214 solution by Δd and Δt is similar to that using Δd only, but with the additional step of interpolating between the altitude entries for consecutive meridian angle entries, in a similar manner to interpolation for declination.

Example.—Find computed altitude and azimuth if *a*L is 41°00′.0N, d is 20°48′.7S, and t is 22°14′.0E.

Solution.—

t 22°14′.0E	t diff. 14′.0	t corr. (−) 4′.2
d 20°48′.7S	d diff. 11′.3	d corr. (+) 10′.8
*a*L 41°00′.0N		corr. (+) 6′.6
ht 24°43′.1	Δd (+) 0.95 Δt (−) 0.30	Z N157°.4E
corr. (+)6′.6		
Hc 24°49′.7		
Zn 157°.4		

In this solution, t diff. is the difference between the meridian angle and the nearest whole degree of t used for entering the table. The t corr. is t diff.×Δt, found by using the same multiplication table used for the d corr. The sign of Δt is found by inspection of the main table. In this example interpolation is between the base altitude for t 22°, and the altitude for t 23°. Since the altitude for t 23° is less than that for t 22°, the correction should be subtracted, so that the interpolated value will lie between the two values between which interpolation is being made. The sign of Δd is found by comparing the altitude for d 21°00′ with that for d 20°30′. The total correction is the algebraic sum of the t corr. and d corr.

The principal advantage of this solution is that a round of sights can be worked and plotted from the same assumed position. However, this advantage is offset by the additional length of the solution. The method is little used.

2016. Pub. No. 214 solution by Δd, Δt, and ΔL.—If the altitude and azimuth at a particular place are desired, interpolation should be made for all three variables, t, d, and L, if needed. The change in altitude for a change of latitude of 1′ (ΔL) is not tabulated. The table on the two pages *preceding* the multiplication table of Pub. No. 214 is used for finding the correction for latitude.

Example 1.—Find computed altitude and azimuth if *a*L is 41°12′.8S, d is 21°32′.5S, and t is 8°52′.3W.

Solution.—

t	8°52'.3W	t diff. 7'.7			+	−
d	21°32'.5S	d diff. 2'.5		t corr.	2'.4	
aL	41°12'.8S	L diff. 12'.8		d corr.	2'.4	
ht	69°04'.0	Δd (+)0.94	Δt (+)0.32	L corr.		11'.7
corr.	(−)6'.9			sum	4'.8	11'.7
Hc	68°57'.1			corr.		(−)6'.9
Zn	336°.0			Z		S156°.0W

The corrections for meridian angle and declination are found as explained in articles 2014 and 2015. The L corr. is found by entering the correction table with azimuth angle and L diff., the difference between the latitude of the assumed position and the nearest whole degree used for entering the main table. It is customary to interpolate in this table, where applicable, but the error introduced by not doing so is always less than 0'.3 if the *nearest* whole degree of azimuth angle is used. The sign of the latitude correction is determined by the rules given at the bottom of the correction table. The total correction is the algebraic sum of the three individual corrections.

All of these are *altitude* corrections. The azimuth is that corresponding to the values used for entering the main table. If the exact value at the place is desired, it can be found by interpolation. If such an interpolation is made, the interpolated value should be used for entering the latitude correction table for altitude interpolation.

In sight reduction for plotting lines of position, it is not customary to interpolate for azimuth angle when Pub. No. 214 is used. However, if greater accuracy is desired, as for determining compass error, triple interpolation should be made. This is customarily accomplished by entering the main table of Pub. No. 214 with the *nearest* values of t, d, and L, and taking out the corresponding tabulated value. Simple eye interpolation is then used to determine separately the correction for each of the three variables. The algebraic sum of these is the correction applied to the base value. The Δd, Δt, and ΔL corrections are not used because these refer to the *altitude,* not the azimuth. Corrections are made to azimuth angle *before* it is converted to azimuth.

Interpolation for azimuth is discussed in more detail in article 2028.

The complete solution includes all of the parts listed in article 2001. Because of the various alternatives available for the separate parts, a large number of variations might be used in the complete solution.

Example 2.—During morning twilight on June 1, 1975, the 0624 EP of a ship is lat. 41°12'.3S, long. 178°39'.2E. At ZT $6^h24^m57^s$ the navigator observes Mars with a marine sextant having an IC of (−) 1'.0, from a height of eye of 53 feet. The hs is 41°45'.9.

Required.—The *a,* Zn, and AP, using Pub. No. 214 (Δd, Δt, ΔL) and the *Nautical Almanac.*

Solution.—

	June 1		Mars		+ M	−
ZT	$6^h24^m57^s$	18^h	1°34'.4N *d* (+)0.7	IC		1'.0
ZD (−)	12	corr.	0'.3	D		7'.1
GMT	$18^h24^m57^s$ May 31	d	1°34'.7N	☆-P		1'.0
18^h	150°48'.3			add'l.	0'.1	
24^m57^s	6°14'.3	*v*(+)	0.8	sum	0'.1	9'.1
v corr.	(+)0'.3			corr.		(−)9'.0
GHA	157°02'.9			hs		41°45'.9
*a*λ	178°39'.2E			Ho		41°36'.9
LHA	335°42'.1					
t	24°17'.9E	t diff.	17'.9		+	−
d	1°34'.7N	d diff.	4'.7	t corr.		7'.5
*a*L	41°12'.3S	L diff.	12'.3	d corr.		4'.2
ht	42°13'.1	Δd(−)	0.91 Δt(−)0.42	L corr.		10'.3
corr.	(−)22'.0			sum		22'.0
Hc	41°51'.1			corr.		(−)22'.0
Ho	41°36'.9			Z		S146°.7E
a	14.2A	*a*L	41°12'.3S			
Zn	033°.3	*a*λ	178°39'.2E			

Special Techniques

2017. Adjustment of straight line of position.—Table 4 gives the corrections to the straight line of position (LOP) as drawn on a chart or plotting sheet to provide a closer approximation to the arc of the circle of equal altitude, a small circle of radius equal to the zenith distance. As shown in figure 2017, the corrections are offsets of points on the LOP and are drawn at right angles to the LOP in the direction of the observed body. The offset points are joined to obtain the arc of the small circle. Usually the desired approximation to the arc of the small circle can be obtained by drawing a straight line through two offset points. The magnitudes of the offsets are dependent upon altitude and the distance of the offset point from the intercept.

2018. Graphical interpolation of altitude for latitude and local hour angle.—In principle the graphical method for interpolating altitude for latitude and local hour angle is the measurement of the difference of the radii of two circles of equal altitude corresponding to the altitudes of a celestial body from two positions at the same instant. One circle passes through the assumed position (integral latitude and that longitude providing an integral LHA), and the second circle passes through the dead reckoning position or other position from which the computed altitude is required.

The measurement, which is the difference in zenith distances as measured from the zenith of the assumed position and the zenith of some nearby position, is effected as follows:

1. Draw the azimuth line from the assumed position (AP) as shown in figure 2018 (the azimuth angle is interpolated for declination increment before conversion to true azimuth).

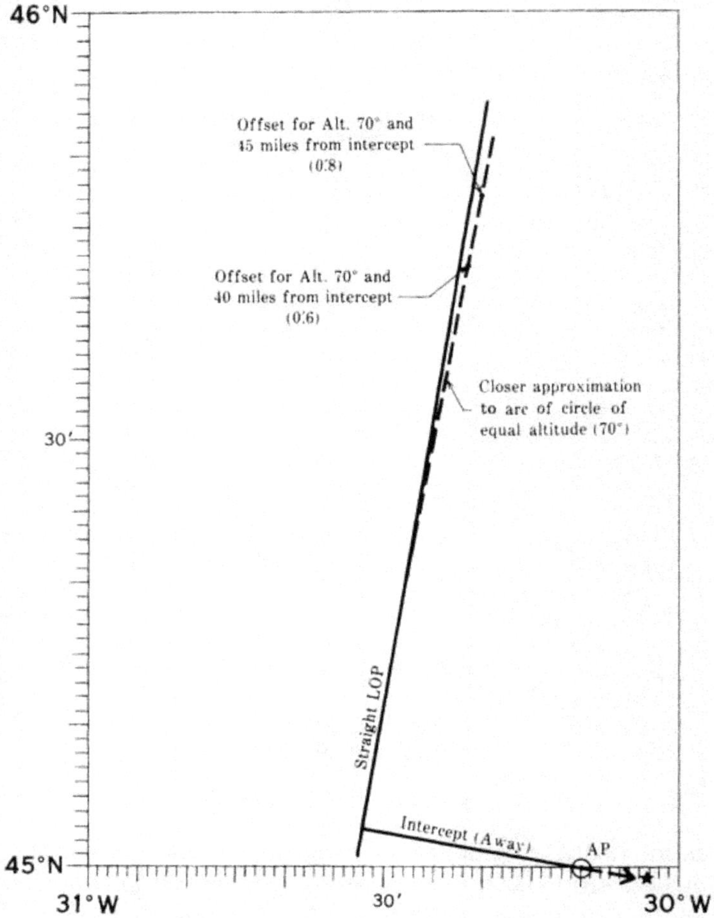

FIGURE 2017.—Adjustment of straight line of position.

2. From the position (DR) for which the computed altitude is required, draw a line perpendicular to the azimuth line or its extension. This line approximates the arc of the circle of equal altitude passing through the DR.

3. Measure the distance from the foot of the perpendicular to the DR in nautical miles.

4. Enter table 4 with the distance of the DR from the foot of the perpendicular, and the altitude of the body as interpolated for declination increment; extract the offset.

5. From the foot of the perpendicular and in a direction away from the celestial body, lay off the offset on the azimuth line or its extension.

6. As shown in figure 2018, a closer approximation to the arc of the circle of equal altitude through the DR is made by drawing a straight line from the offset point to the DR.

7. The required correction, in units of minutes of latitude, for the latitude and LHA increments is the length along the azimuth line between the AP and the arc of the circle of equal altitude through the DR.

If the arc of the circle of equal altitude through the DR crosses the azimuth line between the AP and the body, the correction is to be added to the altitude interpolated for declination increment; otherwise the correction is to be subtracted.

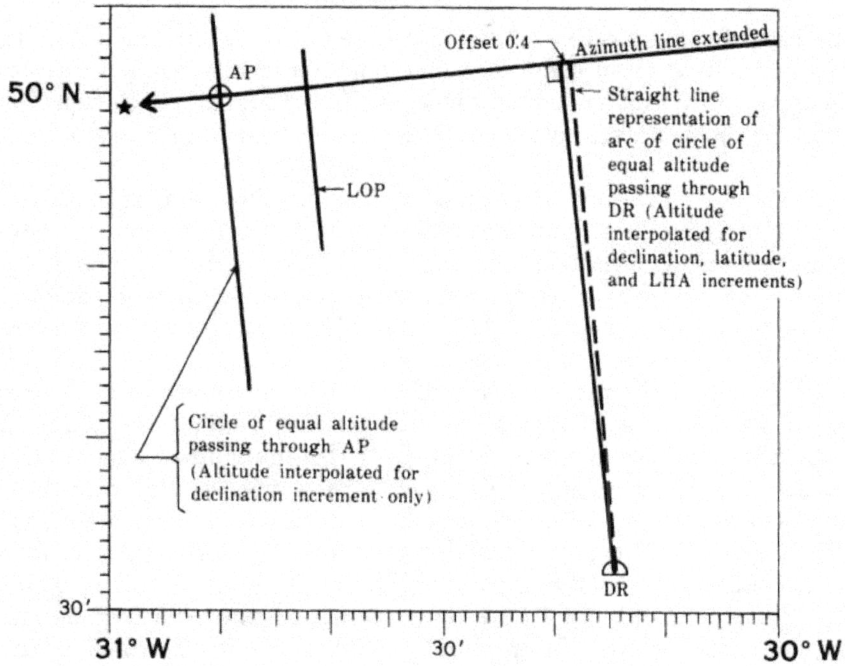

FIGURE 2018.—Graphical interpolation of altitude.

The method will give highly satisfactory results except when plotting on a Mercator chart in high latitudes.

Example.—

Computed altitude from AP	Hc	70°05.′0
Observed altitude	Ho	70°00.′0
Intercept	*a*	5.0A
Computed altitude from AP	Hc	70°05.′0
Difference of the radii		20.′4
Computed altitude from DR	Hc	69°44.′6
Computed altitude from DR	Hc	69°44.′6
Observed altitude	Ho	70°00.′0
Intercept	*a*	15.4T

The basic method should have most frequent application in great-circle solutions.

2019. Pub. No. 229 interpolation near the horizon.—This discussion is restricted to the interpolation of altitude for declination within the 1° interval containing the horizon, indicated by the horizontal segments of the C–S Line. Interpolation of altitude in the interval under consideration is accomplished by using the last tabular altitude and altitude difference appearing above the C–S Line. Since the last tabular altitude above the C–S Line indicates the body's altitude above the horizon for LHA at top of page, for the pertinent latitude, and for the last integral declination above the horizontal segment of the C–S Line pertaining to that particular latitude, interpolation resulting in positive altitudes may be carried out for increments of declination of contrary name so long as the interpolated altitude correction does not exceed the last tabular altitude above the C–S Line; for the

LHA at bottom of page, positive altitudes will result when interpolating altitude for increments of declination of same name so long as the interpolated altitude correction exceeds the last tabular value above the C–S Line. Interpolation for declinations and increments of declination in excess of the above limits results in negative altitudes.

The tabular azimuth angle pertinent to this one-degree interval of declination is that immediately above or that immediately below the C–S Line, according as the entering arguments are contrary or same name, respectively. The difference in azimuth angle for the interval is determined by taking the value of tabular azimuth angle, on the same side of the C–S Line as the LHA argument, from the supplement of that on the opposite side of the line.

2020. Pub. No. 229 interpolation of negative altitudes.—This paragraph is restricted to tabular and interpolated altitudes for declinations other than one-degree intervals of declination containing the C–S Line. For all local hour angles at the top of the right-hand page, all tabular or interpolated altitudes on that page for declinations below the C–S Line are negative; also for any local hour angle at the bottom of the right-hand page, all tabular or interpolated altitudes for declinations above the C–S Line are negative; additionally, for these same local hour angles and latitudes changed to Contrary Name, the tabular or interpolated altitudes on the left-hand page are negative. Interpolation of altitudes for declination increments within these areas of negative altitude should, however, be accomplished as if the altitudes were positive, adhering strictly to the sign given d. Then, after interpolation, regard the results as negative. In all instances involving negative altitudes, except the one-degree interval of declination which includes the C–S Line, the supplement of the pertinent tabular azimuth angle is that to be converted to true azimuth by the rules to be found on each opening of the basic tables.

2021. Interpolation near the zenith.—In the region within 4° of the zenith where normal interpolation methods are inadequate, the following method can usually be used to interpolate both altitude and azimuth angle. The Interpolation Table is employed in carrying out the desired interpolation, but the values of altitude and azimuth angle extracted from the basic tables constitute data which require independent differencing; the tabular altitude difference, d, is not used.

To carry out the altitude interpolation, the basic tables are entered with the pertinent LHA and Dec., and with the integral degree of Lat. so chosen that, when increased by the declination increment, it is within 30′ of the known or DR latitude; this practice will prevent long intercepts. For these entering arguments and for a latitude and declination one degree more than the above referenced latitude and declination, respectively, extract the tabular altitudes and azimuth angles. The altitudes and azimuth angles are then differenced and with these differences interpolation of altitude and azimuth angle for the desired declination is made, utilizing the Interpolation Table. The computed altitude is then compared with that observed to determine the intercept, which together with the interpolated azimuth angle converted to true azimuth makes possible the construction of a line of position, which is plotted from the assumed longitude, and from the latitude of the entering argument, augmented by the declination increment.

Example 1.—

LHA	Lat.	Dec.	Ho
3°27′	31°06′N	28°35′.1N	86°05′.5

Lat.	Dec.	Tab. Hc	diff.	Tab. Z	diff.
31°	28°	86°01′.4		138°.2	
			(+)1′.0		(+)0°.3
32°	29°	86°02′.4		138°.5	

Interpolate to Dec. = 28°35′.1
Dec. Inc. = 35′.1, diff. = (+)1′.0, Z diff. = (+)0°.3

Tab. Hc		86°01′.4	Tab. Z	138°.2
Correction		(+)0′.6		(+)0°.2
	Hc	86°02′.0	Z	138°.4
	Ho	86°05′.5		
Intercept		3.5T	Zn	221°.6

Plot from Lat. 31°35′.1N

Example 2.—

LHA	Lat.	Dec.	Ho
357°19′	37°58′S	36°13′.1S	87°14′.2

Lat.	Dec.	Tab. Hc	diff.	Tab. Z	diff.
38°	36°	86°52′.8		128°.9	
			(+)1′.4		(+)0°.4
39°	37°	86°54′.2		129°.3	

Interpolate to Dec. = 36°13′.1
Dec. Inc. = 13′.1, diff. = (+)1′.4, Z diff. = (+)0°.4

Tab. Hc		86°52′.8	Tab. Z	128°.9
Correction		(+)0′.3		(+)0°.1
	Hc	86°53′.1	Z	129°.0
	Ho	87°14′.2		
Intercept		21.1T	Zn	051°.0

Plot from Lat. 38°13′.1S

2022. Precomputation.—Sometimes it is desired to determine computed altitude *before* the observation, generally for the purpose of obtaining a line of position quickly after the observation has been completed. This is called **precomputation.** When it is done, sextant altitude corrections are generally applied *with reversed sign* to Hc to obtain **precomputed altitude (Hp),** which is then compared directly with hs to obtain the altitude intercept for plotting a line of position. Where altitude is needed for entering correction tables, the computed altitude (Hc) is used. The error introduced by this practice is negligible except at low altitudes, where the corrections should be adjusted by using the Hp to reenter the tables. If greater accuracy is required, limit precomputation to Hc and Zn, and apply corrections to the sextant altitude after observation.

Example.—On June 2, 1975, the 1025 DR position of a ship is lat. 42°21′.4S, long. 118°47′.1W. The navigator plans to observe the lower limb of the sun at this time with a marine sextant having an IC of (+)2′.5, from a height of eye of 25 feet.

Required.—(1) Precomputed altitude by Pub. No. 214, Δd only, (2) the *a,* Zn, and AP if hs is 22°23′.6. Use *Nautical Almanac.*

Solution.—

	June 2			Sun		+ ☉	−
ZT	$10^h25^m00^s$		18^h	22°10′.5N d(+)0′.3	IC	2′.5	
ZD	(+)8		d corr.	(+)0′.1	D		4′.9
GMT	$18^h25^m00^s$	June 2	d	22°10′.6N	☉	13′.7	
18^h	90°32′.0				sum	16′.2	4′.9
25^m00^s	6°15′.0				corr.		(+)11′.3
GHA	96°47′.0						
$a\lambda$	118°47′.0W						
LHA	338°00′.0						
t	22°00′.0E						
d	22°11′.5N		d diff.	10′.6			
aL	42°00′.0S						
ht	22°50′.5		Δd	(−)0.95		Z S157°.9E	
corr.	(−)10′.1						
Hc	22°40′.4						
corr.	(+)11′.3	(rev.)					
(1) Hp	22°29′.1						
hs	22°23′.6						
(2) a	5.5A		aL	42°00′.0S			
Zn	022°.1		$a\lambda$	118°47′.0W			

At the AP used in the calculation, Hp is correct only for the time used. However, if the observation is made early or late, the same Hp and Zn can be used by moving the AP along the parallel of latitude, eastward for early observations, or westward for late observations, a distance equal to 0′.25 of longitude for each second (15′.0 for each minute) difference between actual and predicted times. This adjustment is based upon the assumptions that the apparent motion of the body is westward at the rate of 15° per hour, and the declination is constant. Over the seconds or minutes likely to be involved, these assumptions and the possible increased length of the plotted lines do not introduce a significant error, except possibly for the moon.

2023. Low altitudes.—When Hc is determined by inspection tables such as Pub. No. 214, a minimum tabulated altitude may be available. In Pub. No. 214, altitudes below 5° are not given. These tables can be used for low-altitude observations by selection of an AP that will result in Hc being 5° or greater. To do this, proceed as follows: At the time of observation, note the approximate azimuth of the celestial body. Plot the azimuth line through the DR or EP and measure off, toward the celestial body, a distance equal to 6° − hs (or 6° − Ho). Select the AP in relation to this point as if it were the DR or EP. Occasionally it may be necessary to use 7° − hs (or 7° − Ho). The increased length of the altitude intercept line does not introduce a significant error over the distance that a rhumb line can be considered identical to a great circle. Only in high latitudes is this a problem, and here the error can be virtually eliminated by using a chart projection on which a great circle plots as a straight line or approximately so. The error introduced by using a rhumb line to represent the circle of equal altitude (the line of position) is not increased because the AP selected is near the azimuth line.

Example.—On June 1, 1975, the 1625 DR position of a ship is lat. 43°39′.7S, long. 15°07′.0W. At GMT $17^h24^m22^s$ the navigator observes the lower limb of the sun as it breaks out below an overcast, shortly before setting. He uses a marine sextant having no IC, and makes his observation from a height of eye of 52 feet. The hs is

1°01'.6, air temperature (art. 807, vol. II) 24°F, and the atmospheric pressure (art. 808, vol. II) 30.16 inches. The sun's azimuth is approximately 305°.

Required.—The *a*, Zn, and AP using Pub. No. 214 (Δd only), the *Nautical Almanac*, and tables 23 and 24.

Solution.—

	June 1		Sun		+ ⊙ −
GMT	$17^h24^m22^s$ June 1	17^h	22°02'.1N *d*(+)0'.3	IC	− −
17^h	75°34'.4	*d* corr.	(+)0'.1	D	7'.0
24^m22^s	6°05'.5	d	22°02'.2N	sum	− 7'.0
GHA	81°39'.9			corr.	(−)7'.0
*a*λ	20°39'.9W			hs	1°01'.6
LHA	61°00'.0			ha	0°54'.6
t	61°00'.0W				
d	22°02'.2N	d diff.	2'.2	⊙	9'.1
*a*L	41°00'.0S			T	1'.3
ht	5°21'.8	Δd(−)	0.75 Z S125°.5W	B	0'.3
corr.	(−)1'.7			sum	− 10'.7
Hc	5°20'.1			corr.	(−)10'.7
Ho	0°43'.9			ha	0°54'.6
a	276.2A	*a*L	41°00'.0S	Ho	0°43'.9
Zn	305°.5	*a*λ	20°39'.9W		

Refer to figure 2023. From the 1625 DR position, the approximate azimuth of 305° is plotted, as shown by the broken line. Along this line a distance of 6°00'.0−0°43'.9=5°16'.1, or 316.1 miles, is measured, locating the point labeled *A*. The AP is selected with respect to this point as if it were the DR position. The sight is plotted from this AP as in any observation. If it makes the plot easier, record *a* in the solution in degrees and minutes of arc instead of in miles (5°20'.1−0°43'.9= 4°36'.2=276'.2=276.2 miles).

Large altitude differences can be avoided by using a method of solution that provides for low altitudes. Among such methods are Pub. Nos. 229 and 249; nearly any trigonometric method such as the cosine-haversine formula, the Ageton Method (table 35); and most graphical and mechanical methods. All of these methods are discussed in chapter XXI. If a trigonometric method is used, the *signs* of the various functions (or special rules) should be used if there is a possibility of Hc being negative. The need for special care can be eliminated by using an assumed position about half a degree or more from the DR position or EP, in the direction of the celestial body, if the altitude is less than 0°30'.

By any method of solution, if *either* Hc or Ho (but not both) is negative, the altitude intercept is found by numerically *adding* the two altitudes. Thus, if Hc is (+)0°12'.6 and Ho is (−)0°03'.2 the altitude intercept, *a*, is 15'.8, or 15.8 miles. The positive altitude is greater than the negative one. Therefore, the *a* in this case is *away.* If *both* Hc and Ho are negative, the difference is found by subtraction, but in this case the one which is numerically smaller is the greater altitude. Thus, if Hc is (−)0°09'.6 and Ho is (−)0°04'.3, the altitude intercept is 5.3T.

2024. High altitudes are usually avoided for at least two reasons. First, bodies near the zenith are difficult to observe. A star or planet is difficult to "bring down" to the horizon. It is not always easy to determine the azimuth accurately, and when near the zenith, a body may be changing azimuth rapidly. On the other hand, such observations are little affected by astronomical refraction. The second reason for avoiding high altitudes is one of geometry. As the altitude increases, the radius of

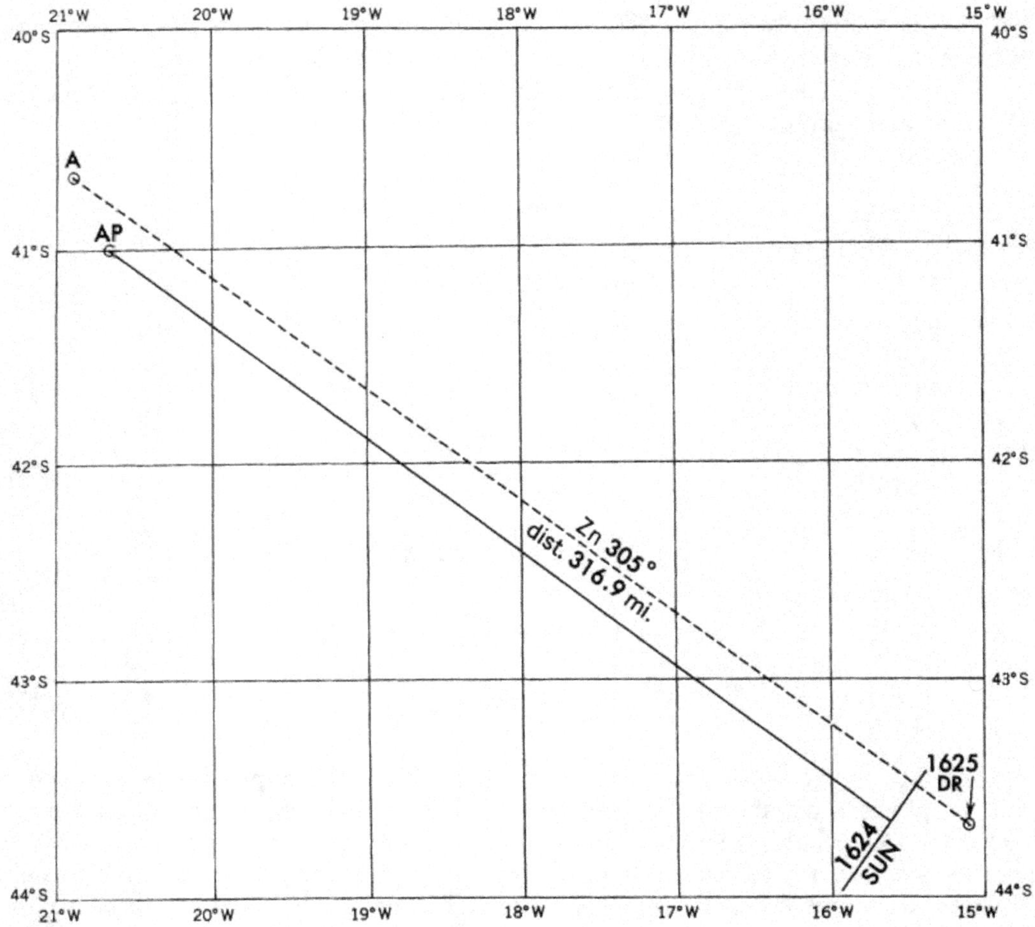

FIGURE 2023.—Selecting an AP for low-altitude solutions by Pub. No. 214.

the circle of position decreases. For a body near the zenith, the radius is so small that the use of a straight line to approximate the circle may introduce serious error.

With higher altitudes, it is good practice to avoid use of lines of position extending a considerable distance from the azimuth line. Since the decrease in radius is gradual, there is no one altitude at which the curvature becomes excessive. However, a safe general rule, if one is needed, is to use the DR position or EP as the assumed position, and interpolate for azimuth angle, for all altitudes greater than 70°. The purpose of this is not primarily to decrease the altitude intercept, as sometimes suggested, but to decrease the length of the line of position.

Within perhaps three degrees of the zenith, the curvature of the circle of position becomes so great that even for a short distance a straight line is not an adequate representation of the circle. At these altitudes, it is good practice to plot the line of position as a circle. This is done by using the geographical position (GP) of the celestial body as the center, and the zenith distance as the radius. Hence, no sight reduction tables are needed. The same body can be used for obtaining a fix from two observations separated by several minutes. In celestial navigation, as in piloting, a circle of position is advanced or retired by moving its center.

Example.—On May 31, 1975, the 1224 DR position of a ship is lat. 20°17'.4N, long. 50°07'.4W. The ship is on course 127°, speed 18 knots. Using a marine sextant having no IC, the navigator observes the lower limb of the sun twice, from a height of eye of 65 feet. The first observation is made at GMT $15^h15^m15^s$, and hs is 88°01'.1. The second observation is made at GMT $15^h24^m13^s$, and hs is 87°34'.7. The GHA of the sun at GMT $15^h15^m15^s$ is 49°25'.6. Use the same declination as at GMT $15^h24^m13^s$.

Required.—The 1224 fix.

Solution.—

	May 31		Sun		+	⊙ −
GMT	$15^h15^m15^s$ May 31		d21°53'.1N	IC	–	–
GHA	49°25'.6			D		7'.8
				⊙	15'.9	
				sum	15'.9	7'.8
GP L_1	21°53'.1N			corr.		(+)8'.1
GP λ_1	49°25'.6W			hs		88°01'.1
radius	110.8 mi.			Ho		88°09'.2
				z		1°50'.8
			Sun		+	⊙ −
GMT	$15^h24^m13^s$ May 31	15^h	21°53'.0N d (+)0'.3	IC	–	–
15^h	45°36'.8	d corr.	(+)0'.1	D		7'.8
24^m13^s	6°03'.3	d	21°53'.1N	⊙	15'.9	
GHA	51°40'.1			sum	15'.9	7'.8
				corr.		(+)8'.1
GP L_2	21°53'.1N			hs		87°34'.7
GP λ_2	51°40'.1W			Ho		87°42'.8
radius	137.2 mi.			z		2°17'.2

Answer.—1224 fix: L 20°09'.0N, λ 50°06'.0W.

The plot of this problem is shown in figure 2024. No significant error would be introduced by assuming the same declination and sextant altitude correction for both observations, and a change of GHA equal to the arc equivalent of the time difference between observations (art. 1811). In east longitude the GP longitude would be 360° − GHA.

Latitude by Meridian Transit

2025. Meridian altitudes.—The latitude of a place on the surface of the earth, being its angular distance from the equator, is measured by an arc of the meridian between the zenith and the equator, and hence is equal to the declination of the zenith; therefore, if the zenith distance of any heavenly body when on the meridian be known, together with the declination of the body, the latitude can be found.

Figure 2025a shows the celestial sphere surrounding the earth: P_nMP_s is the upper branch of a celestial meridian and LL′ a portion of the corresponding geographic meridian. The declination of a body at M (arc QM) is numerically equal to the latitude of its geographical position at GP. The zenith distance of a body is equivalent to the distance on earth between the geographical position of the body and the position of the observer. In figure 2025a the zenith distance of M is 30° and its declination is 20°N. If the body is on the meridian, the GP is also on the meridian. Since P_n, Z, and M are all on the celestial meridian, the navigational triangle flattens out to a line. The observer is 30° *north* of the GP (L 50°N) if the

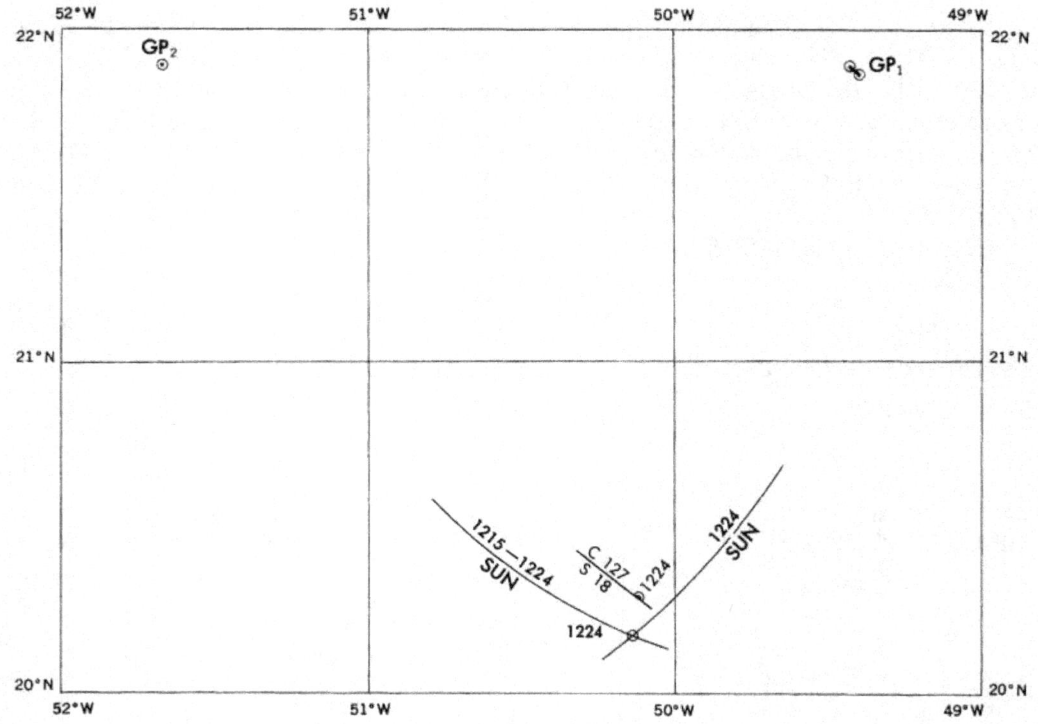

FIGURE 2024.—Plotting high-altitude observations.

body is seen to bear south, or 30° *south* of the GP (L′ 10°S) if the body is seen to bear *north.* The navigator knows whether the GP is north or south, because it is the same as the direction he faces when making his observation.

In the diagram on the plane of the celestial meridian shown in figure 2025b, M is the position of a celestial body north of the equator but south of the zenith; QM is the declination of the body; SM is the altitude (h); and MZ is the zenith distance (z).

From the diagram:

$$QZ = QM + MZ, \text{ or } L = d + z.$$

With attention to the direction of the GP and the name of the declination, the above equation may be considered general for any position of the body at upper transit, as M, M′, M″.

When the body is below the pole, as at M‴—that is, at its lower transit—the same formula may be used by substituting 180° − d for d. Another solution is given in this case by observing that:

$$NP_n = P_nM''' + NM''', \text{ or } L = p + h.$$

By drawing that half of the diagram on the plane of the celestial meridian containing the zenith, the proper combination of zenith distance and declination is made obvious, as shown in the following examples:

Example 1.—The navigator observes the sun on the meridian, bearing south. The declination of the sun is 10°00′.0N; the corrected sextant altitude (Ho) is 60°00′.0.

Required.—The latitude.

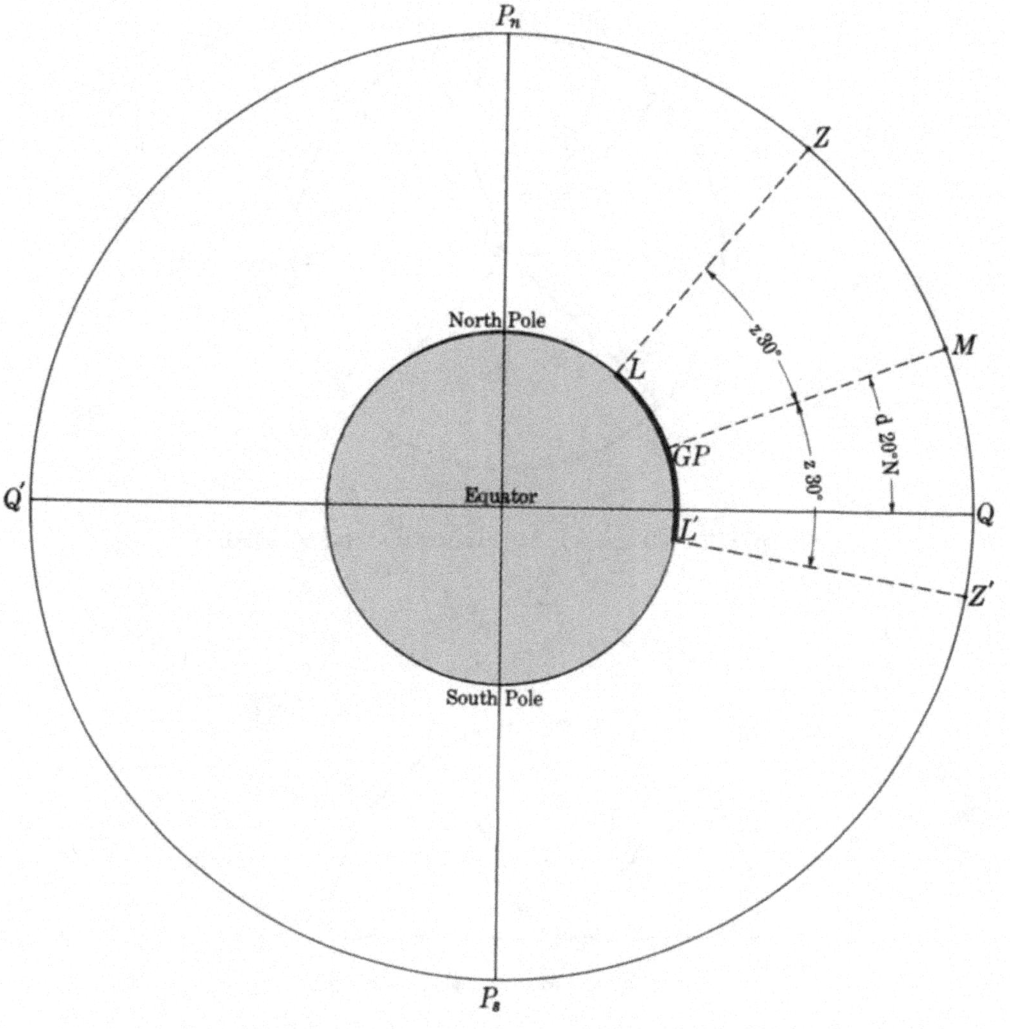

FIGURE 2025a.—Body on celestial meridian.

Solution.—L=z+d. (Fig. 2025c.)

	90°00′.0
Ho	60°00′.0
z	30°00′.0
d	10°00′.0N
L	40°00′.0N

Example 2.—The navigator observes the sun on the meridian, bearing south. The declination of the sun is 10°00′.0S; the corrected sextant altitude (Ho) is 65°00′.0.

Required.—The latitude.

Solution.—L=z−d. (Fig. 2025d.)

	90°00′.0
Ho	65°00′.0
z	25°00′.0
d	10°00′.0S
L	15°00′.0N

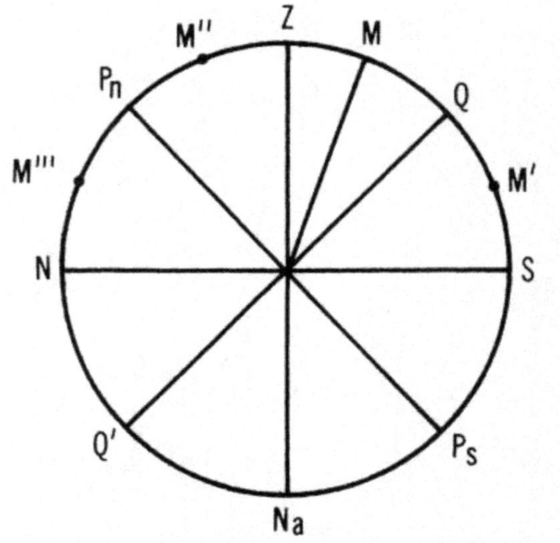

FIGURE 2025b.—Diagram on the plane of the celestial meridian.

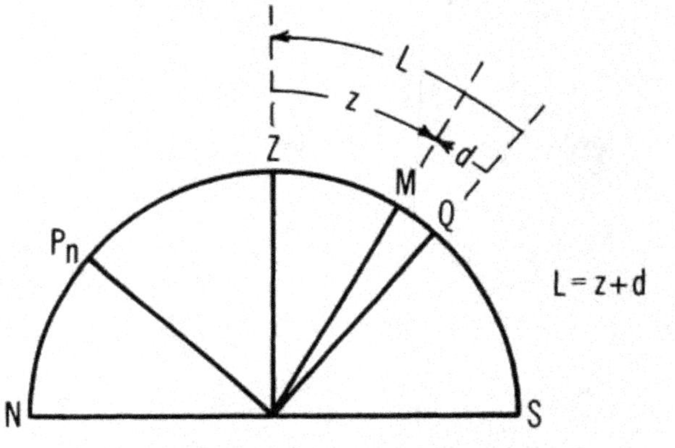

FIGURE 2025c.—Meridian altitude diagram.

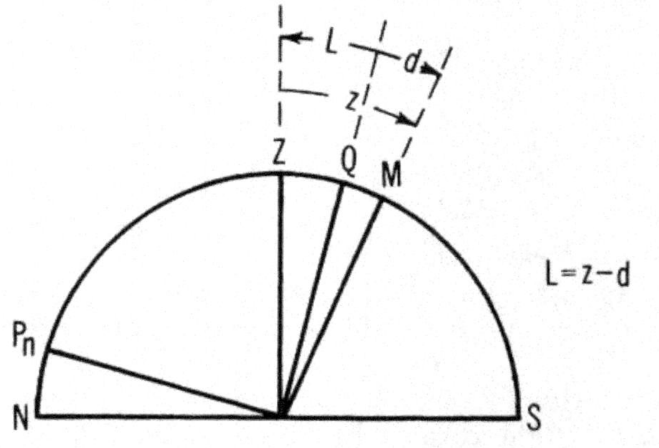

FIGURE 2025d.—Meridian altitude diagram.

Example 3.—The navigator observes the sun on the meridian, bearing north. The declination of the sun is 20°00ʹ.0S; the corrected sextant altitude (Ho) is 60°00ʹ.0.

Required.—The latitude.

Solution.—L=z+d. (Fig. 2025e.)

	90°00′.0
Ho	60°00′.0
z	30°00′.0
d	20°00′.0S
L	50°00′.0S

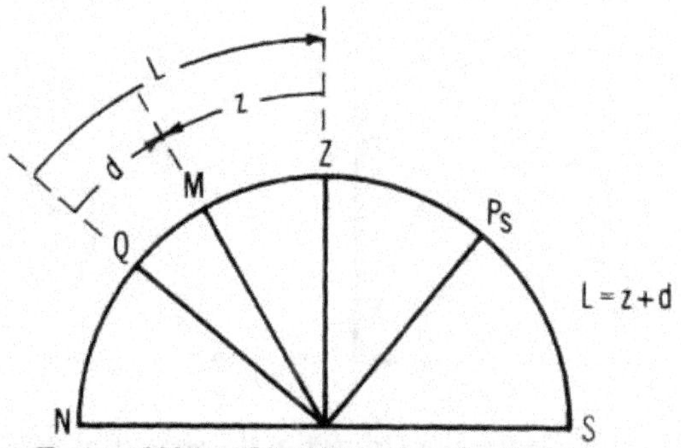

FIGURE 2025e.—Meridian altitude diagram.

Example 4.—The navigator observes the sun on the meridian, bearing north. The declination of the sun is 23°00′.0N; the corrected sextant altitude (Ho) is 72°00′.0.

Required.—The latitude.

Solution.—L=d−z. (Fig. 2025f.)

	90°00′.0
Ho	72°00′.0
z	18°00′.0
d	23°00′.0N
L	5°00′.0N

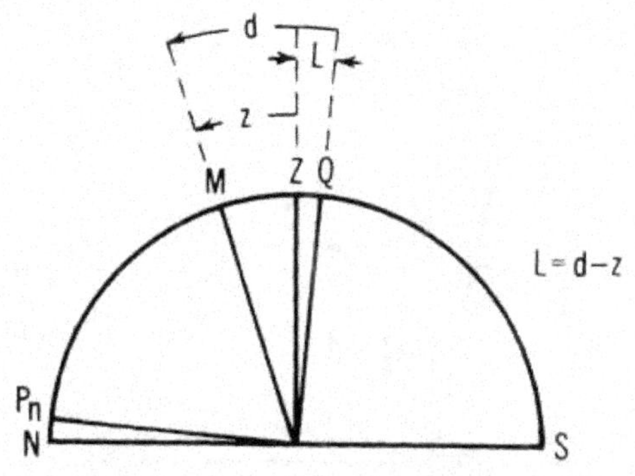

FIGURE 2025f.—Meridian altitude diagram.

Example 5.—In the vicinity of the equator, the navigator observes the sun on the meridian, bearing north. The declination of the sun is 22°05′.0N; the corrected sextant altitude (Ho) is 67°45′.0.

Required.—The latitude.

Solution.—L=z−d. (Fig. 2025g.)

	90°00′.0
Ho	67°45′.0
z	22°15′.0
d	22°05′.0N
L	0°10′.0S

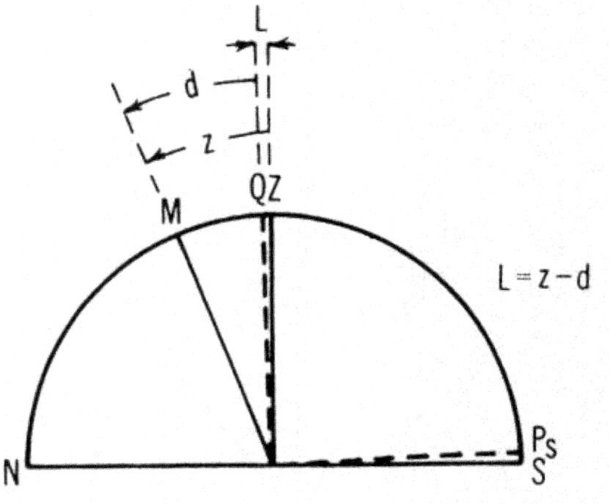

FIGURE 2025g.—Meridian transit in the vicinity of the equator.

Example 6.—The navigator in high northern latitudes observes the sun on the celestial meridian, bearing north. The declination of the sun is 18°46′.0N; the corrected sextant altitude (Ho) is 6°22′.0.

Required.—The latitude.

Solution.—L=(180° − d) − z, or
L=p+h. (Fig. 2025h.)

	90°00′.0
Ho	6°22′.0
z	83°38′.0
180° − d	161°14′.0N
L	77°36′.0N

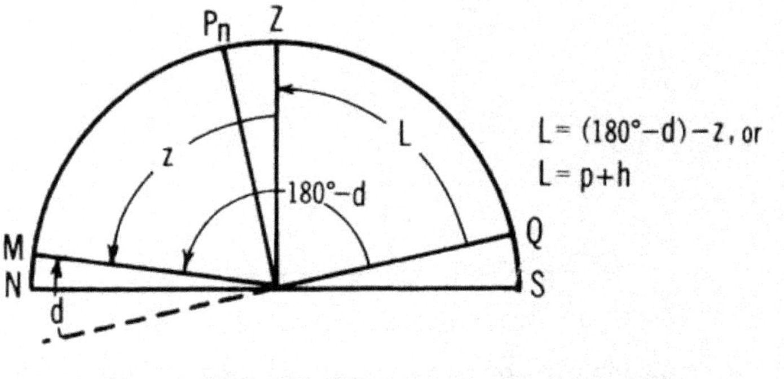

FIGURE 2025h.—Meridian altitude at lower transit.

Since the sun's GP is 83°38′.0 north of the observer in high northern latitudes, the GP is beyond the pole, or on the *lower branch* of the observer's meridian.

If an observation is made near but not exactly at meridian transit, it can be solved as a meridian altitude, with one modification. Enter table 29 with the approximate latitude of the observer and the declination of the body, and take out the **altitude factor** *(a)*. This is the difference between meridian altitude and the altitude one minute of time later (or earlier). Next, enter table 30 with the altitude factor and the difference of time between meridian transit and the time of observation, and take out the correction. *Add* this value to Ho if near upper transit, or *subtract* it from Ho if near lower transit. Then proceed as for a meridian altitude, remembering that the value obtained is the latitude at the time of observation, not at the time of meridian transit. This method should not be used beyond the limits of table 30 unless reduced accuracy is acceptable. This process is called **reduction to the meridian**, the altitude before adjustment an **ex-meridian altitude**, and the observation an **ex-meridian observation.** It requires knowledge of the meridian angle, which depends upon knowledge of longitude.

2026. Finding time of meridian transit.—If a meridian altitude is to be observed other than by chance, a knowledge of the time of transit of the body across the meridian is needed.

On a slow-moving vessel, or one traveling approximately east or west, the time need not be known with great accuracy. The right-hand daily page of the *Nautical Almanac* gives the GMT of transit of the sun and moon across the Greenwich meridian (approximately LMT of transit across the local meridian) under the heading "Mer. Pass." In the case of the moon, an interpolation should be made for longitude. This is performed in the same manner as finding the LMT of moonrise and moonset (art. 1912). In the case of planets, the tabulated accuracy is normally sufficient without interpolation. The time of transit of the navigational planets is given at the lower right-hand corner of each left-hand daily page of the *Nautical Almanac.* The tabulated values are for the middle day of the page. These times are the GMT of transit across the Greenwich meridian, but are approximately correct for the LMT of transit across the local meridian. Observations are started several minutes in advance and continued until the altitude reaches a maximum and starts to decrease (a minimum and starts to increase for lower transit). The greatest altitude occurs at upper transit (and the least at lower transit). This method is not reliable if there is a large northerly or southerly component of the vessel's motion, because the altitude at meridian transit changes slowly, particularly at low altitudes. At this time the change due to the vessel's motion may be considerably greater than that due to apparent motion of the body (rotation of the earth), so that the highest altitude occurs several minutes before or after meridian transit.

If the moment at which the azimuth is 000° or 180° can be determined accurately, the observation can be made at this time. However, this generally does not provide a high order of accuracy.

If the longitude is known with sufficient accuracy, the time of transit can be computed. A number of methods of computation have been devised, but perhaps the simplest is to consider the GHA of the body equal to the longitude if west, or $360° - \lambda$ if east, and find the time at which this occurs.

Example 1.—Find the zone time of meridian transit of the sun at longitude 156°44′.2W on May 31, 1975.

Solution.—

	May 31	
λ	156°44'.2W	
GHA	156°44'.2	
22^h	150°36'.1	
24^{m}32^s	6°08'.1	
GMT	22^{h}24^{m}32^s	May 31
ZD(+)	10	(rev.)
ZT	12^{h}24^{m}32^s	

This solution is the reverse of finding GHA. The largest tabulated value of GHA that does not exceed the desired GHA is found in the tabulation for the day, and recorded, with its time. The difference between this value and the desired GHA is then used to enter the "Increments and Corrections" table. The time interval corresponding to this value is added to the time taken from the daily page. If there is a *v* correction, it is subtracted from the GHA difference before the time interval is determined. The GMT can be converted to any other kind of time desired. If the Greenwich date differs from the local date at the time of transit (for the sun this can occur only near the 180th meridian), a second solution may be needed. This possibility can often be avoided by making an approximate mental solution in advance. As the basis for this approximate solution, it is convenient to remember that the GMT of Greenwich transit (GHA 0°) is about the same as the LMT of local transit. To find the time of transit of a star, subtract its SHA from the desired GHA to find the desired GHA♈. Determine the time corresponding to GHA♈, as explained above for the sun.

Aboard a moving vessel, the longitude at transit usually depends upon the time of transit. An approximate mental solution may provide a time sufficiently close. In the absence of better information, use ZT 1200 for the sun. Find the time of transit for the position at this time, and then make an adjustment, if necessary, for the sun between 1200 and the time found by computation. This adjustment is equal to four seconds for each minute of longitude involved. If the ship is *west* of the 1200 position at the computed time of transit, *add* the correction; and if *east, subtract* it. The result is the **first estimate** of the zone time of **local apparent noon (LAN)** or of meridian transit. For high accuracy a second adjustment may occasionally be needed, but this is seldom justified because of the uncertainty of the vessel's position. If the second adjustment is made, the result is the **second estimate.**

The time of transit of the sun can also be found by means of apparent time (art. 1818). Meridian transit occurs at LAT 12^{h}00^{m}00^s. This can be converted to any other kind of time desired.

Example 2.—Find the zone time of meridian transit of the sun as observed aboard a ship steaming at 20 knots on course 255° on May 31, 1975, using the positional data given in figure 2026.

Solution.—

	May 31	
	360°00′.0	
λ	112°55′.0E	
GHA	247°05′.0	
4^h	240°37′.7	
25^m49^s	6°27′.3	
GMT	$4^h25^m49^s$	May 31
ZD(−)	8	(rev.)
ZT	$12^h25^m49^s$	
dλ	$(+)0^m36^s$	
ZT	$12^h26^m25^s$	(first estimate)

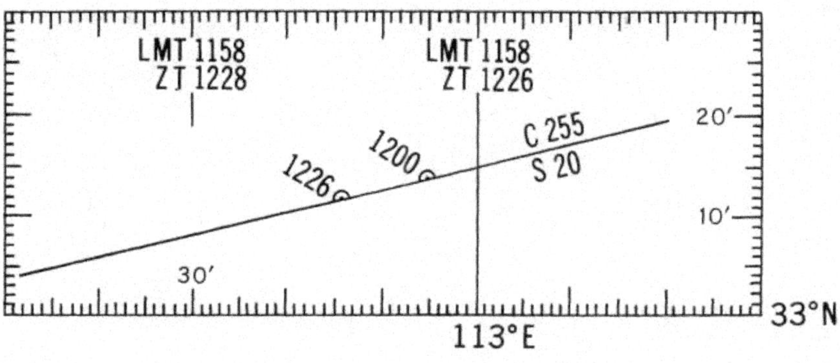

FIGURE 2026.—Time of meridian passage.

The second estimate of the zone time of meridian transit is found by plotting the DR position for the first estimate of the zone time of transit and then applying the dλ between this DR and the 1200 DR to the time found by computation.

ZT	$12^h25^m49^s$	
dλ	$(+)0^m37^s$	
ZT	$12^h26^m26^s$	(second estimate)

As shown in figure 2026, the zone times of meridian transit are noted on several successive meridians. This is accomplished by extracting the LMT of meridian transit from the daily page of the *Nautical Almanac* and converting this time to the zone time for each meridian. The time when the ship and the sun are on the same meridian can then be obtained by inspection to within approximately one-half minute.

Polaris

2027. Latitude by Polaris.—Another special method of finding latitude, available in most of the Northern Hemisphere, utilizes the fact that Polaris is less than 1° from the north celestial pole. As indicated in article 1432, the altitude of the elevated pole above the celestial horizon is equal to the latitude. Since Polaris is never far from the pole, its observed altitude (Ho), with suitable correction, is the latitude.

The nature of this correction as tabulated in the *Air Almanac* is suggested by inspection of figure 2027b in which the circle represents the path of Polaris around the north celestial pole (P_n), as seen by an observer on earth looking along the axis P_sP_n (fig. 2027a). The line *ab* represents a small portion of the observer's meridian. Polaris is at upper transit at *a* and at lower transit at *b*. This is also shown in figure 2027b, to larger scale. Latitude is equal to the altitude *minus* the polar distance (*p*) when Polaris is at *a* and *plus* the polar distance when it is at *b*. When the star is at any point *c*, the Polaris correction is polar distance times the cosine of the local hour angle (corr.$=p$ cos LHA). Thus, the correction is a function of LHA of the star, and hence also of LHA ♈, insofar as the difference between these quantities (the SHA) can be considered constant.

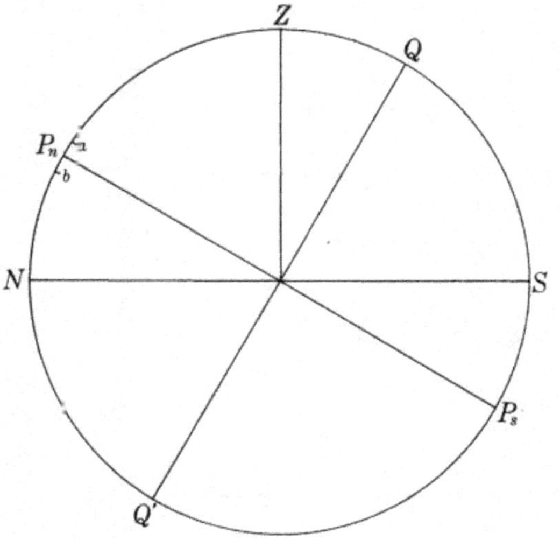

FIGURE 2027a.—Latitude is equal to the (1) declination of the zenith and (2) the altitude of the elevated pole. Compare with figure 2027b.

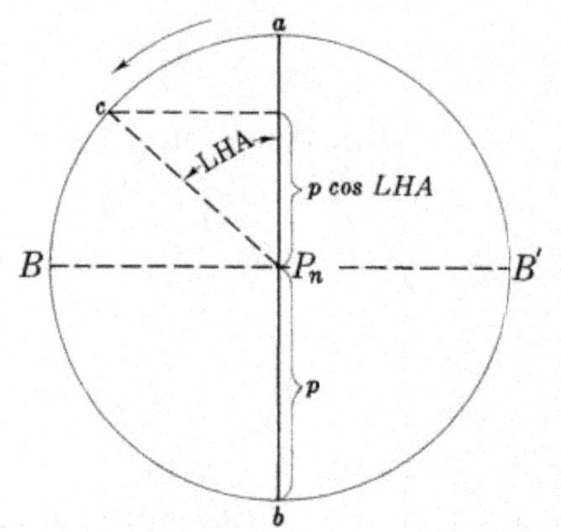

FIGURE 2027b.—The correction is (−)*aPn* when Polaris is at *a* and (+)*bPn* when at *b*. At any point *c* the correction is *p* cos LHA. At *B* or *B'* the correction is zero.

Although the above method usually provides sufficient accuracy for air navigation, a higher degree of accuracy may be obtained by use of Polaris correction tables included in the *Nautical Almanac*. These tables are based on the following formula:

$$\text{Latitude} - \text{corrected sextant altitude} = -p \cos h + \tfrac{1}{2} p \sin p \sin^2 h \tan (\text{latitude}),$$

where p=polar distance of *Polaris*$=90° -$Dec.

h=local hour angle of *Polaris*=LHA Aries+SHA.

The value a_0, which is a function of LHA Aries only, is the value of both terms of the above formula calculated for mean values of the SHA and Dec. of Polaris, for a mean latitude of 50°, and adjusted by the addition of a constant (58′.8). The value a_1, which is a function of LHA Aries and latitude, is the excess of the value of the second term over its mean value for latitude 50°, increased by a constant (0′.6) to make it always positive. The value a_2, which is a function of LHA Aries and date, is the correction to the first term for the variation of Polaris from its adopted mean position; it is increased by a constant (0′.6) to make it positive. The sum of the added constants is 1°, so that:

$$\text{Latitude} = \text{corrected sextant altitude} - 1° + a_0 + a_1 + a_2.$$

The table at the top of each Polaris correction page is entered with LHA Aries, and the first correction (a_0) is taken out by single interpolation. The second and third corrections (a_1 and a_2, respectively) are taken from the double entry tables without interpolation, using the LHA Aries column with the latitude for the second correction and with the month for the third correction.

Example 1.—During morning twilight on June 2, 1975, the 0525 DR position of a ship is lat. 15°43′.6N, long. 110°07′.3W. At watch time $5^h24^m49^s$ AM the navigator observes Polaris with a marine sextant having an IC of (−)3′.0, from a height of eye of 44 feet. The watch is 23^s slow on zone time. The hs is 16°24′.0.

Required.—The latitude.

Solution.—

	June 2			Polaris			+ ☆ −
WT	$5^h24^m49^s$	AM		+	−	IC	3′.0
WE	(S)23^s		a_0	38′.2		D	6′.4
ZT	$5^h25^m12^s$		a_1	0′.3		☆-P	3′.3
ZD (+)	7		a_2	0′.3		sum	(−)12′.7
GMT	$12^h25^m12^s$	June 2	add'l		60′.0	corr.	(−)12′.7
12^h	70°20′.2		sum	38′.8	60′.0	hs	16°24′.0
25^m12^s	6°19′.0		corr.		(−)21′.2	Ho	16°11′.3
GHA♈	76°39′.2						
λ	110°07′.3	W					
LHA♈	326°31′.9						
Ho	16°11′.3						
corr.	(−)21′.2						
L	15°50′.1	N					

Since LHA ♈ is an entering value in all three correction tables, and since this is affected by the longitude, other observations, if available, should be solved and plotted first, to obtain a good longitude for the Polaris solution. For greater accuracy, particularly in higher latitudes, and especially if considerable doubt exists as to the longitude, it is good practice to find the azimuth of Polaris and draw the line of position perpendicular to it, through the point defined by the latitude found in the computation and the longitude used in the solution. The azimuth at various latitudes to 65°N is given below the Polaris corrections. This table can be extrapolated to higher latitudes, but Polaris would not ordinarily be used much beyond latitude 65°. In the example given above the azimuth is 000°.8.

The *Air Almanac* provides only the first correction, which it designates *Q*.

Polaris observations can be solved like those of other celestial bodies, using the declination and SHA given in the tabulation near the back of the *Nautical Almanac*.

The Polaris correction table in the *Air Almanac* lists the correction *Q* to be applied to the sextant altitude of Polaris to give the latitude of the observer. The correction is given in a critical table, with argument LHA Aries (♈) obtained from the daily pages. The effect of refraction is not included and so the sextant altitude must be fully corrected before use.

Example 2.—On 1 June 1975 at GMT $02^h53^m32^s$ in longitude 49°18′W, the corrected sextant altitude of Polaris is 54°46′.

Required.—The latitude.

Solution.—

GMT	$2^h53^m32^s$ June 1	Corr. sextant altitude	54°46′
2^h50^m	291°28′	Q (LHA ♈, 243°03′)	(+) 44′
3^m32^s	0°53′	Latitude	55°30′
GHA ♈	292°21′		
λ	49°18′ W		
LHA ♈	243°03′		

2028. Azimuth by tables.—One of the more frequent applications of sight reduction tables is their use in computing the azimuth of a celestial body for comparison with an observed azimuth in order to determine the error of the compass. In computing the azimuth of a celestial body, for the time and place of observation, it is normally necessary to interpolate the tabular azimuth angle as extracted from the tables for the differences between the table arguments and the actual values of declination, latitude, and local hour angle. The required triple interpolation of the azimuth angle using Pub. No. 229, *Sight Reduction Tables for Marine Navigation,* is effected as follows:

1. Refer to figure 2028a. The main tables are entered with the nearest integral values of declination, latitude, and local hour angle. For these arguments, a base azimuth angle is extracted.

43°, 317° L.H.A. **LATITUDE SAME NAME AS DECLINATION**

	30°			31°			32°			33°			34°			35°			
Dec.	Hc	d	Z	Hc	d	Z	Hc	d	Z	Hc	d	Z	Hc	d	Z	Hc	d	Z	Dec.
°	° ′	′	°	° ′	′	°	° ′	′	°	° ′	′	°	° ′	′	°	° ′	′	°	°
20	50 00.7	+23.2	94.2	49 55.7	+24.6	95.4	49 49.4	+26.1	96.6	49 41.9	+27.5	97.8	49 33.2	+28.9	98.9	49 23.3	+30.3	100.1	20
21	50 23.9	22.0	92.8	50 20.3	23.6	94.0	50 15.5	25.0	95.2	50 09.4	26.6	96.4	50 02.1	28.0	97.6	49 53.6	29.4	98.8	21

44°, 316° L.H.A. **LATITUDE SAME NAME AS DECLINATION**

	30°			31°			32°			33°			34°			35°			
Dec.	Hc	d	Z	Hc	d	Z	Hc	d	Z	Hc	d	Z	Hc	d	Z	Hc	d	Z	Dec.
°	° ′	′	°	° ′	′	°	° ′	′	°	° ′	′	°	° ′	′	°	° ′	′	°	°
20	49 08.9	+23.0	93.7	49 04.5	+24.5	94.8	48 58.8	+26.0	96.0	48 52.0	+27.4	97.1	48 44.0	+28.8	98.2	48 34.8	+30.2	99.4	20
21	49 31.9	22.0	92.2	49 29.0	23.5	93.4	49 24.8	25.0	94.6	49 19.4	26.4	95.7	49 12.8	27.8	96.9	49 05.0	29.2	98.0	21

FIGURE 2028a.—Extracts from Pub. No. 229.

2. The tables are reentered with the same latitude and LHA arguments but with the declination argument 1° greater or less than the base declination argument depending upon whether the actual declination is greater or less than the base argument. The difference between the respondent azimuth angle and the base azimuth angle establishes the azimuth angle difference (Z Diff.) for the increment of declination.

3. The tables are reentered with the base declination and LHA arguments but with the latitude argument 1° greater or less than the base latitude argument depending upon whether the actual (usually DR) latitude is greater or less than the base argument to find the Z Diff. for the increment of latitude.

4. The tables are reentered with the base declination and latitude arguments but with the LHA argument 1° greater or less than the base LHA argument depending upon whether the actual LHA is greater or less than the base argument to find the Z Diff. for the increment of LHA.

5. The correction to the base azimuth angle for each increment is

$$\text{Z Diff.} \times \frac{\text{Inc.}}{60'}.$$

The auxiliary interpolation table (art. 206, vol. II) can normally be used for computing this value because the successive azimuth angle differences are less than 10°.0 for altitudes less than 84°.

Example 1.—In DR lat. 33°24′.0N, the azimuth of the sun is observed as 096°.5 pgc. At the time of the observation, the declination of the sun is 20°13′.8N; the local hour angle of the sun is 316°41′.2.

Required.—The gyro error.

Solution.—By Pub. No. 229:

The error of the gyrocompass is found as shown in figure 2028b.

	Actual	Base Arguments	Base Z	Tab* Z	Z Diff.	Increments	Correction (Z Diff × Inc. ÷ 60)
Dec.	20°13′.8N	20°	97°.8	96°.4	−1°.4	13′.8	−0°.3
DR Lat.	33°24′.0N	33°(Same)	97°.8	98°.9	+1°.1	24′.0	+0°.4
LHA	316°41′.2	317°	97°.8	97°.1	−0°.7	18′.8	−0°.2

Base Z	97°.8
Corr.	(−) 0°.1
Z	N 97°.7E
Zn	097°.7
Zn pgc	096°.5
Gyro Error	1°.2E

Total Corr.	−0°.1

*Respondent for two base arguments and 1° change from third base argument, in vertical order of Dec., DR Lat., and LHA.

FIGURE 2028b.—Azimuth by Pub. No. 229.

Using Pub. No. 214, *Tables of Computed Altitudes and Azimuth,* to determine the azimuth from the tabular values which are actually azimuth angles, interpolation is made for meridian angle, declination, and latitude.

Example 2.—In DR lat. 41°25′.9S, the azimuth of the sun is observed as 016°.0 pgc. At the time of the observation, the declination of the sun is 22°19′.6N; the meridian angle of the sun is 17°22′.4E.

Required.—The gyro error.

Solution.—By Pub. No. 214:

(1) Refer to appendix N. In this example the tabular azimuth angle used as base Z is compared with the tabular azimuth angle for the base declination and latitude, but meridian angle 18° to determine the Z Diff. for meridian angle. Similarly, the base Z is compared with the tabular value for the base meridian angle and latitude, but declination 22° to determine the Z Diff. for declination; and with the base value of meridian angle and declination, but latitude 42° to determine the Z Diff. for latitude.

(2) Interpolation is between whole degrees of meridian angle and latitude, but between half degrees of declination when the declination is 29° or less. For declinations of more than 29°, the interpolation interval for declination varies.

$$\text{t corr.} = \text{Z Diff.} \times \frac{\text{Increments}}{1°} = (-)1°.0 \times \frac{0°.4}{1°.0} = (-)0°.4.$$

$$\text{d corr.} = \text{Z Diff.} \times \frac{\text{Increments}}{0°.5} = (-)0°.1 \times \frac{0°.2}{0°.5} = (+)0°.0.$$

$$\text{L corr.} = \text{Z Diff.} \times \frac{\text{Increments}}{1°} = (+)0°.2 \times \frac{0°.4}{1°.0} = (+)0°.1.$$

The error of the gyrocompass is found as shown in figure 2028c.

	Actual	Base Arguments	Base Z	Tab* Z	Z Diff.	Increments	Correction +	Correction −
t	17°.4E	17°	162°.7	161°.7	(−) 1°.0	0°.4		0°.4
d	22°.3N	22°30′	162°.7	162°.6	(−) 0°.1	0°.2	0°.0	
L	41°.4S	41°(Contrary)	162°.7	162°.9	(+) 0°.2	0°.4	0°.1	
						Total Corr.		(−) 0°.3

Base Z	162°.7
Corr.	(−) 0°.3
Z	S 162°.4E
Zn	017°.6
Znpgc	016°.0
Gyro Error	1°.6E

*Respondent for two base arguments and 1° (30′ for d) change from the third base argument, in vertical order of t, d, and L.

FIGURE 2028c.—Azimuth by Pub. No. 214.

Solutions for azimuth are discussed in more detail in chapter VII of volume II.

2029. Amplitudes.—For checking the compass, an azimuth observation of a celestial body at low altitude is desirable because it can be measured easiest and most accurately. If the body is observed when its center is on the *celestial* horizon, the **amplitude (A)**, which is the arc of the horizon between the prime vertical and the body, can be taken directly from table 27.

The amplitude is given the prefix E (east) if the body is rising and W (west) if setting. It is given the suffix N if the body rises or sets north of the prime vertical (which it does if it has northerly declination) and S if it rises or sets south of the prime vertical (having southerly declination). The suffix is given to agree with the declination of the body. Interconversion of amplitude and azimuth is similar to that of azimuth angle and azimuth. Thus, if A=E15°S, the body is 15°south of east or 90°+15°=105°. For any given body, the numerical value of amplitude would be the same at rising and setting if the declination did not change.

When the center of the sun is on the celestial horizon, its *lower limb* is about two-thirds of a diameter above the visible horizon. When the center of the moon is on the celestial horizon, its *upper limb* is on the visible horizon. When planets and stars are on the celestial horizon, they are a little more than one sun diameter above the visible horizon. In high latitudes, amplitudes should be observed on the visible horizon.

If the body is observed when its center is on the *visible horizon,* the *observed* value should be corrected by the value from table 28, using the rules given with the table, before comparison with the value taken from table 27. If preferred, the correction can be applied with reversed sign to the value taken from table 27 and compared with the uncorrected observed value. This is the procedure used if amplitude or azimuth is desired when the celestial body is on the visible horizon.

CHAPTER XXI

COMPARISON OF VARIOUS METHODS OF SIGHT REDUCTION

2101. Introduction.—Before the development of a means of determining accurate time at sea (art. 127), longitude could not be found by celestial observation. Celestial bodies were used for determination of latitude, and as an indication of direction, often in a very general way. The development of the marine chronometer opened up a whole new vista to the navigator. Immediately, methods began to appear to utilize this new dimension of navigation. During the two centuries that have elapsed, many of the best minds have been directed to the problem of providing easier or more adequate methods of "reducing" the observations to a form suitable for determination of position.

2102. Kinds of methods.—Various "special" methods have been devised to take advantage of some unique relationship to provide a simplified solution. The most widely used are **latitude methods** for determination of latitude by meridian altitude or observation of Polaris, and **longitude methods** for determination of longitude by observation of a body near the prime vertical. Both latitude and longitude methods have now been largely superseded by the **altitude method,** based upon the discovery of the altitude intercept, or difference, by the Frenchman Marcq St.-Hilaire (art. 132). Most modern methods are of this type, although some latitude and longitude methods are still in use.

The most commonly used methods utilize computation for determining certain information which is then plotted as a line of position, two or more such lines being needed for a fix. The "method" might consist of one or more formulas to be solved by general mathematical tables, a set of special tables conveniently arranged for use with the formulas, or a set of tables constituting a list of computed answers. A method which determines latitude or longitude separately requires no plot. In fact, a plot would be misleading unless the celestial body were almost exactly on the celestial meridian or prime vertical, or unless the azimuth were considered. While a number of methods determine latitude and longitude by computation, without plotting, other methods substitute a graphical or mechanical solution for computation.

2103. Meridian altitudes.—If a celestial body is on the celestial meridian at the time of observation, a modification of the high-altitude method (art. 2024) can be used at any altitude, without plotting the GP. Both GP and observer are on the same meridian, and the difference of latitude between them is the zenith distance of the body (90° − Ho). The direction of the GP is the direction faced during observation (unless a backsight is made). The line of position is a latitude line.

Several hundred years ago, when longitude could not be found accurately, and logarithms had not been invented, the finding of latitude furnished the only reliable navigation available on long sea voyages. Since most of these were in a generally easterly or westerly direction, it became common practice to sail first to the latitude of destination ("run down the latitude") and then to follow this parallel until landfall was made. The meridian observation of the sun at local apparent noon was the most important navigational event of the day, and became a well-established routine. On the basis of this observation at "high noon," clocks were reset, and a new day, the **nautical day**, began. Intentional meridian altitudes of other celestial bodies were not as widely used as those of the sun.

As accurate time became available at sea, and then more convenient tables and more accurate almanacs appeared, the noon sight lost its importance. Since the modern inspection table has been available, the use of meridian altitudes has decreased rapidly, and reduction to the meridian has all but disappeared. True, the solution of a meridian altitude is simple and quick, but this is more than offset by the need for determining the time at which to make the observation (art. 2025), the dislike of many mariners for having to make an observation at a predetermined time, the inconvenience sometimes experienced when local apparent noon occurs at a time when other activities conflict with observation, and the possibility of missing the observation because of overcast conditions. The practice of observing a body when a line of position is desired, and solving those which happen to have a meridian angle of 0° or 180° in the same manner as other observations, is a growing practice that eliminates the need for remembering a separate procedure for bodies on the celestial meridian. The modern navigator thinks primarily in terms of lines of position, rather than of latitude and longitude observations.

Meridian altitudes are discussed in more detail in article 2025.

2104. Ex-meridian altitudes.—If an observation is made near but not exactly at meridian transit, it can be solved as a meridian altitude, with one modification. Enter table 29 with the approximate latitude of the observer and the declination of the body, and take out the **altitude factor** (*a*). This is the difference between meridian altitude and the altitude one minute of time later (or earlier). Next, enter table 30 with the altitude factor and the difference of time between meridian transit and the time of observation, and take out the correction. *Add* this value to Ho if near upper transit, or *subtract* it from Ho if near lower transit. Then proceed as for a meridian altitude, remembering that the value obtained is the latitude at the time of observation, not at the time of meridian transit. This method should not be used beyond the limits of table 30 unless reduced accuracy is acceptable. This process is called **reduction to the meridian,** the altitude before adjustment an **ex-meridian altitude,** and the observation an **ex-meridian observation.** It requires knowledge of the meridian angle, which depends upon knowledge of longitude. If reasonable doubt exists regarding the longitude, the azimuth of the body at the time of observation should be determined, and the line of position drawn perpendicular to it (through the point defined by the "observed" latitude and the assumed longitude), rather than as a latitude line. There are alternative methods available. A correction to latitude can be applied, using the factor f from table 26. In 1899 A. A. Vilkitskiy, a captain in the Russian Navy, developed a mechanical device for determining the correction to be applied for reduction to the meridian.

2105. Latitude by Polaris.—Another special method of finding latitude, available in most of the Northern Hemisphere, utilizes the fact that Polaris is less than 1° from the north celestial pole. As indicated in article 1432, the altitude of the elevated pole above the celestial horizon is equal to the latitude. Since Polaris is never far from the pole, its observed altitude (Ho), with suitable correction, is the latitude.

Like other special solutions, latitude by Polaris has lost much of its popularity since modern inspection tables have become available. Being of magnitude 2.1, Polaris is not a bright star. It is normally considered available to the mariner only during twilight, when the azimuths of various celestial bodies relative to each other are of more interest than an "easy" solution which is little, if at all, simpler than the usual solution by inspection table. If provision were made for solution of Polaris sights by inspection table, the special method would no longer be needed for ordinary navigation.

2106. Longitude methods.—A celestial observation for a line of position, whether reduction is to be by longitude method or by latitude method, consists of measurement of the altitude of a body with the noting of the time. If sight reduction is to be by the longitude method, the latitude must be known, or the best estimate used. With altitude, latitude, and declination (from the almanac), one is able to solve the navigational triangle (art. 1433) for meridian angle. This is converted to local hour angle. The Greenwich hour angle at the time of observation is determined by means of the almanac. The difference between the GHA and LHA is the longitude. A time diagram (art. 1428) is useful in establishing the correct relationship.

Longitude can also be determined by establishing the exact time of meridian transit, at which time the GHA (or 360° − GHA) is the longitude.

If the latitude is known accurately, the longitude method provides a direct and relatively simple solution for position. However, since latitude is rarely known to the desired accuracy, a line of position is usually needed. This is obtained by either (1) solving for longitude at two or more assumed latitudes, and drawing a straight line through the points thus found (the Sumner method), or (2) solving for longitude at one point, determining the azimuth at this point, and drawing the line of position through the single point thus found, perpendicular to the azimuth of the body.

The error introduced in the computed longitude as a result of an inaccurate latitude used in the solution increases as the celestial body departs from the prime vertical. If it is learned that an incorrect latitude has been used in the solution, a correction can be applied, using the factor F from table 26. If the body is near the celestial meridian, a small error in the latitude introduces a large error in the longitude. At any location, the azimuth of the body can be determined by observation or computation, and a line of position drawn perpendicular to it, through the position defined by the latitude used in the computation, and the calculated longitude. Alternatively, solution can be made at two or more latitudes, and the line of position drawn through the two positions. It was the use of this second method in 1837 by Captain Thomas H. Sumner, when his latitude was in doubt, that led to the discovery of the line of position from celestial observation (art. 131).

No longitude method is more accurate than the GMT used for timing the observation. Before chronometers (art. 127) and time signals (art. 127) were available, relatively few navigators attempted to determine longitude, and it was never established reliably. The search for a method of "discovering" longitude at sea was primarily a search for a means of determining time at the Greenwich meridian (arts. 126, 127).

If the longitude is to be determined, most accurate results are obtained, by observation of a body on the prime vertical. The observation having been made, sight reduction can be made by time sight or, more conveniently, by an ordinary solution for a line of position, using an inspection table such as Pub. No. 229. Any general method of sight reduction can be used, without need for a special solution.

Solution by the longitude method is usually called a **time sight.** The various longitude methods are all basically the same, differing only in choice of formulas and arrangement of tables. The basic formula is:

$$\cos t = \frac{\sin h - \sin L \sin d}{\cos L \cos d},$$

in which t is the meridian angle, h is the altitude, L is the latitude, and d is the declination. Early tables for solution of meridian angle were called **horary tables.**

The time sight came into use following the development of the marine chronometer in 1763. Solution for meridian angle is usually by the formula

$$\text{hav } t = \sec L \csc p \cos S \sin (S - h),$$

in which $p = 90° - d$ if L and d have same name, and $90° + d$ if L and d have contrary names; and $S = \frac{1}{2}(h + L + p)$.

When azimuth angle is used with the method, it is usually computed by one of the formulas

$$\sin Z = \sin t \cos d \sec h$$

or

$$\text{hav } (180° - Z) = \sec h \sec L \cos S \cos (S - p).$$

There are no rules with this method, but it is subject to possible large errors in high latitudes or if the body has a high declination. Various special tables have appeared for solution of the time sight:

Cassini. The first "inspection tables" were probably prepared by M. Cassini, a Frenchman, in 1770. These "horary tables" provided tabulated solutions for meridian angle.

Lalande. The horary tables prepared in 1793 by Jerome Lalande, a Frenchman, provided tabulated solution for meridian angle for the sun and stars for all latitudes to 61°.

Lynn Horary Tables, by Thomas Lynn, a commander of the East India Company Service, were published in 1827. These 242-page tables consisted of tabulated solutions of meridian angle computed by the time sight formula. Two years later they were followed by a volume of 364 pages of azimuth angle (*Lynn Azimuth Tables*) computed by the haversine azimuth formula. Entries are given for whole degrees of latitude to 60°, declination to 24°, and altitude to 60° (later 90°). The tables are accurate and well arranged, but the triple interpolation is tedious.

Hommey. Louis Hommey's *Table d'angles horaires* (horary tables), published in two volumes in France in 1863, contained more than 40,000 hour angles calculated for "all latitudes." These tables were an improvement on those of Cassini and Lalande.

Martelli. In 1873 a small volume of 49 pages by G. F. Martelli, an Italian, was published in New Orleans. This book, called simply *Tables of Logarithms,* provided a relatively short, fast solution for meridian angle, with very few rules and only one interpolation. Martelli abandoned the inspection table and provided five short tables for a four-place logarithmic solution by the formula

$$\text{hav } t = \frac{\cos (L \sim d) - \cos z}{2 \cos L \cos d}.$$

Solution required six book openings, six table entries, and four mathematical steps. Hour angles were given only to eight hours, and no provision was made for azimuth.

This method proved very popular, and is still used among navigators of several countries. A 1932 edition was published in Glasgow, Scotland, with explanations in French, Dutch, Italian, and Spanish, as well as in English. A 1944 edition added provision for finding azimuth angle, and for solution by the altitude method.

Thomson. A table of only nine pages by Sir William Thomson, better known as Lord Kelvin, was published in London in 1876 to provide a solution for the longitude method. This very thin volume, called *Tables for Facilitating Sumner's Method at Sea,* contains the first known solution by dividing the navigational triangle into two right triangles. In 1849 Towson (art. 2126) had divided the triangle in the same manner, but this solution was for reduction to the meridian. Lord Kelvin divided the triangle by dropping a perpendicular from the celestial body to the celestial meridian of the observer, as shown in figure 2111. He used a for the length of the perpendicular v, b' for x, and b for w (of fig. 2111). His solution uses the formulas

$$\sin \mathrm{t}=\frac{\sin a}{\cos \mathrm{d}},$$

$$\cos b=\frac{\sin \mathrm{d}}{\cos a},$$

$$\sin \mathrm{h}=\cos a \cos b',$$

$$\sin \mathrm{z}=\frac{\sin a}{\cos \mathrm{h}}.$$

The tables are entered with *half* the colatitude (using colatitude to the nearest whole degree) in column b. With a pair of dividers, search is made in the "cohypo" column for two numbers, one agreeing with the altitude, and the other with the declination. The number in column A opposite the altitude in the cohypo column is the azimuth angle, and that opposite the declination is the meridian angle, interpolation being used if needed. The line of position is adjusted for the difference between the interpolated altitude and the observed altitude.

Although the tables are among the shortest of the various methods, their manipulation is difficult. In 1880 Kortazzi, a Russian astronomer, attempted to modify the tables to provide an easier solution, but without great success.

Davis' *Chronometer Tables,* providing a solution for the longitude method, were published in 1897 in London. They are similar to Lynn's tables, using his values but providing assistance in interpolation by adding values of change with latitude, declination, and altitude. As with Lynn, a separate volume is given for azimuth angle, in which there is no interpolation. Originally Davis' tables were limited to latitude 50° and declination 24°, but later tables were published for declinations 23° to 64°. A limited number of altitude entries is given.

Blackburne. Tables by H. S. Blackburne, a New Zealander, were published in London in 1914 under the title *The Excelsior Azimuth and Position Finding Table.* The tables, providing a solution by the longitude method, are similar to *Lynn Horary Tables* and Davis' *Chronometer Tables* but with a new determination of azimuth based upon the ratio of variation of latitude to variation of meridian angle.

Azimuth angles (ten pages) are given in a separate tabulation in the same volume with meridian angles (242 pages).

Blackburne's arrangement is more modern than that of Davis. This was the first publication to include columns for variations of t for 1′ of declination, latitude, and altitude. Meridian angles are given to $0^s.1$. Latitude is limited to 30°, and declination to 23°.

Rust. In 1918 the *Practical Tables for Navigators and Aviators,* by Captain Armistead Rust, USN, were published in Philadelphia. This small volume of 37 pages of tables reverted to a logarithmic solution, as did Martelli's, using the following formula for determining meridian angle:

$$\log \text{hav } t = \log \sec L + \log \sec d + \log \tfrac{1}{2} [\cos (L \sim d) - \sin h].$$

The volume has three tables. Table *A* tabulates log secants for obtaining the first two terms of the formula. Table *B* is a double-entry table giving log ½ [cos (L∼d)−sin h]. Table *C* gives log haversines. Values are given to four places.

Azimuth angle is obtained from an original diagram computed from the well-known formula

$$\sin Z = \sin t \cos d \sec h.$$

This diagram had been given in a volume of ex-meridian tables by Rust published in 1908. In the *Practical Tables* an auxiliary diagram was added to indicate the meridian angle when the celestial body is on the prime vertical. The purpose of the diagram is to resolve possible ambiguities when the azimuth angle is near 90°. The Rust azimuth diagram was used later by Goodwin and Weems, and in the Italian *Tavole H* (art. 2110).

Goodwin. *The Alpha, Beta, Gamma Navigation Tables* of H. B. Goodwin, an Englishman, were first published in London in 1921. This is a small volume having two tables with a total of 34 pages. Meridian angle is found from the formula

$$\text{ver } t = \frac{\cos (L \sim d) - \cos z}{\cos L \cos d}.$$

Table I has two values, α being the angle in seconds of arc, and β being four-place natural cosines multiplied by 1,000 to eliminate the decimal. Table II provides γ, the logarithms for the values of versine t.

The Rust diagram is used for determining azimuth angle.

Instructions are included for use of the tables for altitude method of solution, and for reduction to the meridian.

H.O. Pubs. Nos. 203 and 204 (Littlehales), *The Sumner Line of Position of Celestial Bodies,* were published by the U.S. Navy Hydrographic Office in 1923. These tables, prepared by George W. Littlehales, provide in two large volumes (847 and 675 pages, respectively) tabulated solutions of the meridian angle and azimuth angle, using the general time sight formulas. The arrangement is similar to that of Davis and Blackburne, but t and Z are tabulated together in consecutive columns. Latitude is limited to 60°, and declination to 27° in H.O. Pub. No. 203 and to 64° in H.O. Pub. No. 204. Interpolation for latitude is avoided by using the nearest whole degree, and shifting the line of position for the difference between the altitude at this latitude and the observed altitude.

These publications are no longer in print.

Soule and Dreisonstok. In 1932 these two Americans prepared a small volume providing a logarithmic solution of the longitude method, using the formulas

$$\frac{1}{\text{hav } t}=\frac{\sec S \csc (S-h)}{\sec L \csc p}$$

and

$$\frac{1}{\text{hav } (180°-Z)}=\frac{\sec S \sec (S-p)}{\sec L \sec h},$$

where $S=\frac{1}{2}(h+L+p)$ and $p=90°\pm d$.

The "azimuth" determined in this way is the direction of the line of position ($Z\pm90°$) rather than that of the celestial body.

Weems' *Secant Time Sight* was published in 1944 by Captain P. V. H. Weems, USN (Ret.), to provide a short solution based entirely upon secants and cosecants, using the formulas

$$\csc^2 \tfrac{1}{2} t=\frac{\sec S \csc (S-h)}{\sec L \sec d}$$

and

$$\csc Z=\frac{\csc t \sec d}{\sec h},$$

where $S=\frac{1}{2}(h+L+p)$. A Rust azimuth diagram is included for those who prefer a diagrammatic solution.

2107. Finding time on prime vertical.—Best results by time sight are obtained when the celestial body is on the prime vertical. As explained in article 1432, a celestial body having a declination of opposite name to the latitude crosses the prime vertical below the horizon. Its nearest visible approach is at the time of rising and setting.

If a celestial body has a declination of the same name as the latitude, but is numerically *greater,* it does not cross the prime vertical. Its nearest approach (in azimuth) is at the point at which its azimuth angle is maximum. At this point the meridian angle is given by the formula

$$\sec t=\tan d \cot L,$$

and its altitude by the formula

$$\csc h=\sin d \csc L.$$

A celestial body having a declination of the same name as the latitude, and numerically *smaller,* crosses the prime vertical at some point before it reaches the celestial meridian, and again after meridian transit. At these two crossings of the prime vertical, the meridian angles are equal and are always less than 90°. They are given by the formula

$$\cos t=\tan d \cot L.$$

The altitudes are also equal, and are given by the formula

$$\sin h=\sin d \csc L.$$

Meridian angle and altitude of bodies on the prime vertical, and similar data for the nearest approach (in azimuth) of those bodies of same name which do not cross the prime vertical, are given in table 25 for various latitudes, and for declinations from 0° to 23°, inclusive. Similar information can be determined by means of Pub. No. 214, entering with latitude and declination, and finding the meridian angle and altitude corresponding to an azimuth angle of 90° (or the maximum azimuth angle for nearest approach). Since this information is generally not required to great accuracy, interpolation is not needed.

To find the *time* of crossing the prime vertical, convert t to LHA, and add west longitude or subtract east longitude to find GHA. The GMT at which this GHA occurs can be found and converted to any other time desired.

Example.—Determine (1) the approximate zone time, and (2) the approximate altitude of the sun when it crosses the prime vertical during the afternoon of May 30, 1975, at lat. 51°32′.3N, long. 160°21′.7W, using table 25 and the *Nautical Almanac.*

Solution.—

	May 30	
t	71°.6W	(from table 25)
LHA	71°.6	
λ	160°.4W	
GHA	232°.0	
3^h	225°.6	
26^m	6°.4	
GMT	0326	May 31
ZD	(+)11	(rev.)
(1) ZT	1626	May 30
(2) h	28°.4	(from table 25)

At the time of crossing the prime vertical, or at nearest approach (in azimuth), a celestial body is changing azimuth slowly, and therefore this is considered a good time to check compass deviation or to swing ship.

The prime vertical at any place is the celestial horizon of a point 90° away, on the same meridian. Therefore, a celestial body crosses the prime vertical at approximately the same time it rises and sets at the point 90° away. Thus, if one is at latitude 35°N, the sun crosses his prime vertical at about the same time it rises or sets at latitude 55°S. If time of sunrise and sunset are to be obtained accurately by this method, corrections must be applied for semidiameter and refraction.

2108. Altitude methods, like longitude methods, require an accurately timed observed altitude of a celestial body. Usually, in both types of solution, the navigational triangle is solved, but in the altitude method, t, d, and L are used in solving for altitude. The method is based upon the concept of circles of equal altitude explained by Commander Marcq St.-Hilaire, a Frenchman, in 1875 (art. 132). For this reason it is often called the **Marcq St.-Hilaire method.** It may also be called the **altitude intercept method** because it uses the difference between computed and observed altitudes, a value called an **altitude intercept.**

The altitude method has largely replaced the latitude and longitude methods, although some navigators still prefer the older methods. The principal advantage of the altitude method is that it provides a universal solution that is equally reliable in all latitudes, with all values of declination, and with all values of meridian angle. Even for observations of celestial bodies near the zenith the altitude method is applicable, although in this case an arc of the circle itself is plotted, without the use of the altitude intercept (art. 2004). However, the formulas selected for some of the "short methods" do impose some limitations when those methods are used.

For many years following introduction of the altitude method, the concept was termed the "new navigation," an expression now seldom heard. At first, an attempt was made to adapt existing tables to the altitude method. Some were more readily adaptable than others. In due course, various methods designed for use with the altitude method made their appearance. These methods may be grouped in six classifications: those which do not divide the triangle, those which divide it by dropping a perpendicular from each of the three vertices, those which do not use the navigational triangle, and the modern "inspection table." However, not all inspection tables are for altitude methods. In practice, the dropping of a perpendicular from the pole has not been used except in great-circle sailing (art. 903). This would result in dividing both the meridian angle and zenith distance into two parts, a condition that has not proved attractive.

2109. Altitude methods, triangle not divided.—The basic formula for solution of the undivided navigational triangle is

$$\sin h = \sin L \sin d + \cos L \cos d \cos t,$$

derived from the law of cosines. A number of special tables have been prepared for solution of the undivided triangle:

Davis' *Requisite Tables,* published in London in 1905, introduced the **cosine-haversine formula** to navigation, although it had been used previously by astronomers. This formula is

$$\text{hav } z = \text{hav } (L \sim d) + \cos L \cos d \text{ hav } t,$$

in which z is zenith distance $(90° - h)$. This is sometimes written

$$\text{hav } z = \text{hav } (L \sim d) + \text{hav } \theta,$$

in which $\text{hav } \theta = \cos L \cos d \text{ hav } t$. It might also be written entirely in haversines:

$$\text{hav } z = \text{hav } (d - L) + \text{hav } t\, [\text{hav } (180° - L - d) - \text{hav } (d - L)].$$

In this formula the sign of d is reversed if L and d are of contrary name. The haversine of an angle is positive whether the angle is positive or negative.

These tables were the first to give log haversines and natural haversines in one table. The method was little used at first, but later proved very popular, and as haversines became available from additional sources, the formula was used even more widely. Davis' original tables made no provision for azimuth.

Since the cosine-haversine formula can be used for solution with tables 33 and 34, an example is given below, with solution of azimuth by the formula

$$\sin Z = \sin t \cos d \sec h.$$

Example.—A celestial body a little to the south of west is observed, with the following results: t 80°45′.9W, *a*L 41°12′.3S, d 21°50′.7S.

Required.—The Hc and Zn by the formulas given above.

Solution.—

t	80°45′.9W	*l* hav	9.62300		*l* sin	9.99434
*a*L	41°12′.3S	*l* cos	9.87643			
d	21°50′.7S	*l* cos	9.96764		*l* cos	9.96764
θ	—	*l* hav	9.46707	*n* hav 0.29313		
L~d	19°21′.6			*n* hav 0.02827		
z	69°04′.3			*n* hav 0.32140		
Hc	20°55′.7				*l* sec	10.02964
Zn	258°.8			Z S78°47′.0W	*l* sin	9.99162

This is typical of logarithmic solutions, except that there are no "rules" for the altitude computation. As pointed out in article 2125, the formula used for azimuth angle does not indicate whether the body is north or south of the prime vertical.

Ball. In 1907 Rev. Frederick Ball's *Altitude or Position Line Tables* were published in London. There are two volumes of 244 and 240 (later 313) pages, respectively, the first volume having tables for latitude 0° to 30°, and the second, 31° to 60° (later editions 24° to 60°). These were the first inspection tables for the altitude method. The tabulated altitudes were computed by the haversine formula. The assumed position is selected so that latitude and meridian angle are the nearest whole degree, but no assistance is given for interpolation of altitude for declination.

Azimuth angle is not tabulated, being found by the altitude tables, interchanging altitude and declination, and finding azimuth angle (in hours and minutes) in the meridian angle columns. Since declination is limited to 24° in the first edition and 60° in later editions, this method is not available for azimuth if altitude is greater than this amount. In this case azimuth angle is found by the formula

$$\sin Z = \frac{\sec L \times \Delta h \text{ (for } 8^m)}{120}.$$

This formula had not previously been used in navigation.

Davis' ***Alt-Azimuth Tables*** were published in London in 1917. This volume lists both altitude and azimuth together for the first time. Latitudes included are from 30° to 64°, and declinations from 0° to 24°. In 1921 a second volume was published for latitudes 0° to 30°. Entries are for each whole degree of latitude and declination, and for each 4^m (1°) of meridian angle. However, for each meridian angle, altitude *or* azimuth angle is given alternately. Thus, azimuth angle is given for meridian angles of 0^m, 8^m, 16^m, 24^m, etc., and altitude for meridian angles of 4^m, 12^m, 20^m, 28^m, etc. Altitudes are given in bold type. All declinations entries (0° to 24°) are given on facing pages. Tables for latitude and declination of the same name are given in the first part of the book and those for contrary names in the last part, the two parts being separated by several auxiliary tables. Altitude is given to the nearest 1′, and azimuth angle to the nearest 0°.1. Altitudes are carried down to the horizon, and the local apparent times of sunrise and sunset are also given, with the azimuth angle at these times. Because of the 8^m interval between altitude entries, and the use of an assumed position to avoid interpolation for change in meridian angle, large altitude differences sometimes arise.

H.O. Pub. No. 201, *Simultaneous Altitudes and Azimuths of Celestial Bodies,* was published by the U. S. Navy Hydrographic Office in 1919. In this volume of 606 pages, altitudes and azimuths were tabulated in parallel columns for the first time. Latitude is limited to 60° and declination to 24°. Virtually all altitudes above the horizon are included. The tables are well arranged and very legible, but no assistance is given for interpolation of altitude or azimuth angle for a change of declination. Meridian angles are given at intervals of 10^m ($2\overset{\circ}{.}5$). Since the assumed position is selected to avoid interpolation of altitude or azimuth for meridian angle, large altitude differences result in some instances. This publication is no longer in print.

Yonemura. In 1920 S. Yonemura's *Tables for Calculating Altitude and Azimuth of Celestial Bodies* were published in Japan. This small table of 39 pages contains logarithms of haversines and secants, arranged for convenient solution of the formulas

$$\text{hav}\,(90^\circ-h)-\text{hav}\,(L\sim d)=\text{hav}\,\theta,$$

$$\log\frac{1}{\text{hav}\,\theta}-(\log\sec L+\log\sec d)=\log\frac{1}{\text{hav}\,t},$$

$$\log\csc Z=\log\csc t+\log\sec d-\log\sec h.$$

The method is similar to that of Davis' *Requisite Tables* but includes solution for azimuth angle. The table is included in the book of Ogura's tables (art. 2110).

Braga. The *Táboas de Alturas* by Romêo Braga, a Brazilian, were published in 1924 in Paris. This is a table of natural haversines arranged for solution of the formula in which

$$\text{hav}\,(90^\circ-h)=A+B,$$

in which $A=\text{hav}\,t-[\text{hav}\,(L+d)\,\text{hav}\,t]$

and $B=\text{hav}\,(L-d)-[\text{hav}\,(L-d)\,\text{hav}\,t].$

The first table of 108 pages is for solution for A and B. The second table of nine pages is for finding h.

The assumed latitude is selected so that (L+d) is a whole degree. The assumed longitude is selected so that t is a whole degree.

No provision is made for azimuth.

Japanese H.O. Pub. No. 601, *Celestial Navigation Computation Tables,* was published in 1942. The method is similar to that of Yonemura, the triangle not being divided and a modification of the cosine-haversine formula being used for altitude.

Waller. In 1946 George W. D. Waller, a naval officer on duty as a navigation instructor at the U. S. Naval Academy, proposed a solution by means of Gaussian logarithms, commonly called "addition and subtraction logs." The formula

$$\csc h=\frac{\csc d\csc L}{1+\dfrac{\csc d\csc L}{\sec d\sec L\sec t}}$$

was derived from the basic formula given above, and

$$\csc Z=\frac{\csc t\sec d}{\sec h}$$

was derived from the time and altitude azimuth formula given in article 2125.

A single table of 30 pages would contain in consecutive columns the following values:

$$A = \log \text{secant}$$
$$B = \log \text{cosecant}$$
$$C- = \log (\text{cosecant}+1) - \log \text{cosecant}$$
$$C+ = \log \text{cosecant} - \log (\text{cosecant}-1).$$

All values would be multiplied by 100,000 to eliminate decimals. One additional page would contain A and B values for all whole degrees from 0° to 180°. The values C+ and C− are the Gaussian logs.

The method is reasonably short and simple. Its publication as the "A, B, C Method," with suitable explanation, was prevented by the untimely death of its originator.

Hugon. *Nouvelles Tables Pour le Calcul de la Droite de Hauteur a Partir du Point Estimé,* by the French hydrographic engineer, Professor P. Hugon, were published in 1947. This logarithmic solution is based upon the fundamental formula

$$\sin h = \sin L \sin d + \cos L \cos d \cos t,$$

from which the following is derived:

$$\text{hav } z = Xy + Yx$$

in which

$$X = \text{hav } (180° - t) = \text{cohav } t$$
$$y = \text{hav } (d - L)$$
$$Y = \text{hav } t$$
$$x = \text{hav } [180° - (d + L)] = \text{cohav } (d + L).$$

The formula for z may then be written

$$\text{hav } z = A + B$$

in which

$$\log A = \log X + \log y$$
$$\log B = \log Y + \log x.$$

Solution is by means of a table of 90 pages which lists in parallel columns values of log cohav, log hav, and natural hav for every minute of arc from 0° to 180°. The solution requires six book openings, seven table entries, and five mathematical steps.

Azimuth is found from a diagram in a pocket on the inside back cover. This diagram is designed to solve the formula

$$M = \alpha X + \beta Y,$$

in which

$$M = \cos h \cos Z$$

$$\alpha = \sin (d - L)$$

$$X = \cos^2 \frac{t}{2}$$

$$\beta = \sin (d + L)$$

$$Y = \sin^2 \frac{t}{2}.$$

Chiesa. About 1948 the *Tavole nautiche e Tavole dei Semisenoversi* of the Italian Stefano Chiesa were published in Genova, Italy. These include tables for computation of altitude by the cosine-haversine formula, and "A, B, C" tables for computation of azimuth angle by the formula

$$\text{hav } Z = [\text{hav } p - \text{hav}(L - h)] \sec L \sec h.$$

Rose. In 1952 the *Nautische Tafeln* of G. Rose were published in Germany. This volume has a convenient table for computation of the altitude by the cosine-haversine formula. It also includes the "A, B, C" azimuth tables of Lecky (art. 2126). Various other tables are included, a number of them having been taken from an earlier work of the same name by Dr. Otto Fulst, published in numerous editions since 1860.

Doniol. The *Miniature Navigation Table for Altitude and Azimuth,* by R. Doniol, a Frenchman, was published in 1955. This is undoubtedly the shortest of all sight reduction tables, consisting of only two pages. The formula for altitude was derived from the basic formula given above. The formula used is

$$\sin h = n - (n + m)\, a,$$

in which $n = \cos(L - d)$, $m = \cos(L + d)$, and $a = \text{hav } t$.
The formula for Z was derived from the formula

$$\cot Z = \frac{\tan d \cos L - \cos t \sin L}{\sin t}.$$

The formula used is

$$\tan Z = \frac{\cos d}{\gamma},$$

in which $\gamma = f\Delta_M + f'\Delta_N$. In this expression,

$$f = \frac{\tan \frac{1}{2} t}{2 \sin 1'},\quad f' = \frac{\cot \frac{1}{2} t}{2 \sin 1'},$$

$\Delta_M = \sin(d + L) \sin 1'$, and $\Delta_N = \sin(d - L) \sin 1'$.

The first of the two tables gives sines and cosines for each half degree, and tangents for half degrees of 45° and more. Interpolation is performed by means of a tabulated Δ value which is the change of sine or cosine for 1′. Interpolation is minimized by selecting an assumed position so that t and either (L+d) or (L−d) are an exact half degree.

The second table gives the value of t in degrees, minutes, and seconds, and the values of f and f′ corresponding to selected values of a (natural haversines). The interval between consecutive tabulated values of haversine varies from 0.0002 to 0.005.

The solution is generally accurate to 0′.1 of altitude and 0°.1 of azimuth, but the method requires a number of relatively simple mathematical steps, making it somewhat longer than most "short" solutions.

2110. Altitude methods, perpendicular from zenith.—In figure 2110 the navigational triangle is shown in heavy lines on a diagram on the plane of the celestial meridian (art. 1432). The broken line is a perpendicular from the zenith to the hour circle of the celestial body. This perpendicular may fall outside the triangle. In figure 2110 it divides both the azimuth angle (at Z) and the codeclination side into two parts. The length of the perpendicular is designated v and the two parts of the codeclination are designated w and x. By means of Napier's rules (art. 142, vol. II), the following basic formulas can be derived:

$$\sin v = \cos L \sin t \quad (1)$$
$$\cos w = \sin L \sec v, \text{ or } \tan w = \cot L \cos t \quad (2)$$
$$\sin Z' = \sin w \sec L, \text{ or } \cot Z' = \sin L \tan t \quad (3)$$
$$\sin h = \cos v \cos x \quad (4)$$
$$\sin Z'' = \sin x \sec h, \text{ or } \cos Z'' = \tan v \tan h, \quad (5)$$

in which $x = 90° - (d + w)$.

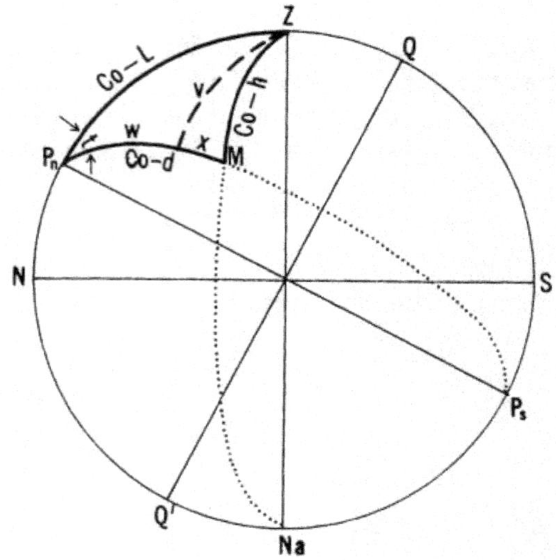

FIGURE 2110.—Navigational triangle with perpendicular from zenith to hour circle.

This basic method has been modified in a number of ways, having proved the most popular altitude method.

Souillagouet, a French professor of hydrography, was the first to divide the navigational triangle by dropping a perpendicular from the zenith. His *Tables du Point Auxiliare* were published in France in 1891. He designates various parts of the diagram of figure 2110 as follows:

v is designated a
w is designated b
x is designated $90° - (d \sim b)$.

His formulas for altitude are

$$\tan b = \cot L \cos t$$
$$\sin a = \cos L \sin t$$
$$\sin h = \cos a \sin (d \sim b).$$

For azimuth angle, the perpendicular is dropped from the celestial body to the celestial meridian, a being the perpendicular and b that part of the celestial meridian from the pole to the foot of the perpendicular. The following formulas are used:

$$\tan b = \cos t \cot d$$
$$\sin a = \sin t \cos d$$
$$\cot Z = \cos (L + b) \cot a.$$

The assumed position is selected so that latitude is the nearest 15′ and meridian angle is the nearest 1^m or 15′ (2^m or 30′ for latitudes greater than 60°). There are four separate tables with a total of 408 pages. The method requires five book openings, seven table entries, and six mathematical steps. Interpolation is not needed.

Bertin. A French professor of hydrography, Charles Bertin, devised tables similar to those of Souillagouet, which were published in Paris in 1919 under the title *Tablette de Point Sphérique.* Bertin used Souillagouet's formulas for altitude, but for azimuth angle he dropped the perpendicular from the zenith, as for altitude, and used the following formulas:

$$\sin Z_1 = \sin c \sec L$$
$$\cot Z_2 = \sin b \tan (c+d)$$
$$Z = Z_1 + Z_2.$$

In these formulas, b and c are substituted for the a and b, respectively, of Souillagouet. This is the first method in which Z is divided into two parts, found separately.

The assumed position is selected so that latitude and meridian angle are each to the nearest 20′. The tables are shorter than those of Souillagouet, having 324 pages, but still bulky for this type solution. The method has fewer steps and requires only two book openings, but interpolation is needed in two steps.

Ogura. In 1920 the *New Altitude and Azimuth Tables* by Sinkiti Ogura, Japanese hydrographic engineer, were published in Tokyo. The solution for altitude is generally similar to that of Souillagouet, a perpendicular being dropped from the zenith, but Ogura introduced secants and cosecants for the first time in this type solution. His solution for azimuth is similar to that of Blackburne (art. 2106) and Lecky (art. 2126).

The assumed position is selected so that latitude and meridian angle are each to the nearest whole degree. The altitude is determined by means of two tables (A and B–C) of a total of 27 pages, and azimuth by means of three additional tables (D, E, F) of a total of 29 pages. The altitude can be obtained to an accuracy of 0′.6 without interpolation. Latitude and declination are limited to 65°. The rules are numerous and complicated.

Both the Ogura and Yonemura (art. 2109) tables are given in the same book, the Japanese Hydrographic Office Pub. No. 225. An English edition, with slight modifications in the Ogura method, was published in 1924.

The Ogura tables have been widely copied.

Smart and Shearme's *Position Line Tables* were published in London in 1922, based upon a division of the triangle by a perpendicular from the zenith. The altitude formulas of Souillagouet were used, but the arrangement of the earlier tables was improved. It is somewhat similar to that of Ogura, but with the positions of meridian angle and latitude interchanged, providing a better arrangement when solutions of several observations are made simultaneously. Solution requires a log sine table which is not provided. There is no solution for azimuth. The assumed position is selected so that the meridian angle and latitude are each the nearest whole degree. No interpolation is needed.

Newton and Pinto. The *Navegação Moderna* of J. A. Newton and J. C. Pinto was published in Lisbon, Portugal, in 1924, providing a solution by dropping a perpendicular from the zenith. The method is based upon ideas expressed by Newton in 1912 and 1913. The formulas for altitude are almost the same as those of Souillagouet. The method of finding azimuth angle resembles somewhat the method of Bertin, but with the use of auxiliary angles. There are only two tables, the first occupying 120 pages, and the second two pages. The assumed position is selected so that latitude and meridian angle are each to the nearest 30′. No interpolation is needed, but the rules are somewhat complicated.

Weems' ***Line of Position Book,*** published in 1927, combines the Ogura altitude tables and the Rust azimuth diagram (art. 2106).

H.O. Pub. No. 208 (Dreisonstok), *Navigation Tables for Mariners and Aviators,* was published by the U. S. Navy Hydrographic Office in 1928 to provide a solution by the method of dropping a perpendicular from the zenith. The method, devised by Lieutenant Commander J. Y. Dreisonstok, USN, is similar to Ogura's. For altitude, it uses the Souillagouet formula inverted so as to be in secants and cosecants. For azimuth angle the formula is similar to that of Newton and Pinto, except that it does not use the parallactic angle at the celestial body. In the first edition the latitude was limited to 65°. There were two tables, one of 45 pages and the other of 18 pages. Later, a 23-page addition to the first table extended the coverage to the poles.

The assumed position is selected so that the latitude and meridian angle are each the nearest whole degree. The method requires four book openings, eight table entries, and six mathematical steps. Although values are usually obtained by relatively easy interpolation, altitude accuracy of 0′.5 can be obtained without interpolation.

As with H.O. Pub. No. 211 (art. 2111), the rules for this method were made on the assumption that only bodies above the celestial horizon would be observed. The rules may be restated to allow for both positive and negative altitudes, as follows:

If t is less than 90°, give b same name as latitude.

If t is greater than 90°, give b opposite name to latitude, and mark Z′ minus.

If (d+b) is numerically greater than 90°, mark Z″ minus.

If (d+b) is contrary name to latitude, the altitude is negative; use the supplement of Z″.

If Z is minus, subtract from 360° and mark plus.

The value labeled "t" in the tables is actually LHA. If t, east or west, is used, as in modern practice, the printed values greater than 180° can be ignored. The rules can be stated in abbreviated form on alternate pages, as follows:

At the top of each left-hand page of table I:

$t<90°$, b same name as L.

At the top of each right-hand page of table I:

$t>90°$, b contrary name to L, Z′ (−).

At the top of each left-hand page of table II:

$(d+b)>90°$, Z″ (−).

At the top of each right-hand page of table II:

Z(−), use 360° − Z.

At the bottom of each page of table II (if desired):

(d+b) contrary name to L, Hc (−): use 180° − Z″.

If the D+B value used for finding Z″ exceeds 10,000, it is reduced by this amount, the remainder being used for entering table II. If desired, this can be stated in abbreviated form at the bottom of alternate pages of table II, as follows:

$$(C+D)>10{,}000, \text{ use } (C+D)-10{,}000.$$

Like H.O. Pub. No. 211 (art. 2111), H.O. Pub. No. 208 has been largely superseded by those tables constituting a list of computed answers.

Gingrich. The *Aerial and Marine Navigation Tables,* by Lieutenant John E. Gingrich, USN, were published in 1931 to provide another solution by the method of dropping a perpendicular from the zenith. The formulas for altitude are similar to those of Ogura, and the formulas for azimuth are similar to those of Perrin (art. 2126). The first two tables, of 31 and seven pages, respectively, are similar to those of Ogura. A single third table of 13 pages is given for azimuth. The general arrangement is in many respects similar to that of H.O. Pub. No. 208, and as with the earlier method, the assumed position is selected so that latitude and meridian angle are each to the nearest whole degree. The precision of tabulation of K, an auxiliary function, is not consistent. Consequently, if the tables are used without interpolation, errors as great as 0′.5 can arise in the computed altitude.

Weems' *New Line of Position Tables* are sometimes called the *Manuscript Tables* because they were in manuscript form from 1932 until they were published in 1943. They are similar to his earlier tables but arranged with the position of latitude and meridian angle values interchanged so that values for several observations can be taken from the tables with a single book opening. The latitude values are extended from the 65° given in earlier tables to 90°. As in the earlier edition, the Rust azimuth diagram (art. 2106) is included, but provision is also made for computation of azimuth angle. One part is found in terms of latitude and meridian angle, using the formula of H.O. Pub. No. 208, and the other part is found in terms of altitude and the perpendicular from the zenith. If the azimuth is required to a greater precision that 0°.5, interpolation is needed. The assumed position is selected so that the latitude and meridian angle are each the nearest whole degree.

Collins. The *I. C. S. Altitude and Azimuth Tables for Air and Sea Navigation,* by Elmer B. Collins, formerly of the U. S. Navy Hydrographic Office, were published by the International Correspondence Schools in 1934. The tables and method of solution are generally similar to those of H.O. Pub. No. 208.

F-Tafel, published by the German Oberkommandos der Kriegsmarine about 1937, divides the triangle by a perpendicular from the zenith. The formulas of Souillagouet are used for altitude. Azimuth is found by the familiar formula

$$\sin Z=\sin t \cos d \sec h.$$

There are four tables. Latitude, declination, and altitude are limited to 70°. The assumed position is selected so that latitude and meridian angle are each to the nearest whole degree.

Comrie. In 1938 the *Hughes' Tables for Sea and Air Navigation,* by L. J. Comrie, former Superintendent, H. M. Nautical Almanac Office, were published in London. These tables are similar to those of H.O. Pub. No. 208, but arranged with the positions of latitude and meridian angle interchanged as in the Weems' *New Line of Position Tables.*

Myerscough and Hamilton. The *Rapid Navigation Tables,* by W. Myerscough and W. Hamilton, were published in London in 1939. A perpendicular is dropped from the zenith to the hour circle of the celestial body. With slight modification, the altitude formulas of Souillagouet and the azimuth formula of Gingrich are used. Six quantities are tabulated in a single table of 90 pages. Both declination and latitude are limited to 70°. In the 1950 edition, the limits of declination and latitude were extended to 89°. A revision of the 1950 edition was published in 1965 as *Rapid Navigation Tables for Mariners,* a table of 195 pages.

Ageton's *Manual of Celestial Navigation,* published in 1942, combines the first table of Weems' *New Line of Position Tables* as table I, and H.O. Pub. No. 211 (art. 2111) as table II. The basic formulas are restated in terms of secants and cosecants. The result is a short, easy solution without interpolation, involving four book openings, eight table entries, and four mathematical steps. Since the H.O. Pub. No. 211 table is included, the book can be used for Ageton's earlier method.

Benest and Timberlake. The *Astro-Navigation Tables for the Common Tangent Method* by two British professors, E. E. Benest and E. M. Timberlake, were published in 1945. In three tables of 61, 18, and 12 pages is given a logarithmic solution for altitude only, by dropping a perpendicular from the zenith. The formulas are slight modifications of those of Ogura.

The location of the line of position is somewhat similar to the method sometimes used in longitude method solutions such as H.O. Pubs. Nos. 203 and 204 (art. 2106). Two assumed positions are selected, usually 1° apart on the same meridian. The altitude intercept at each position is determined, and a circle, or arc of a circle, is drawn with the assumed position as the center, and the altitude intercept as the radius. The line of position is the common tangent to the two circles. Since there are four common tangents, the general direction of the body is required. Where doubt exists as to which of two or more answers is the correct one, additional solutions from other assumed positions may resolve the ambiguity. If the celestial body is near the meridian, the two assumed positions are better taken on the same parallel of latitude. Even with these precautions, there is danger of selection of the wrong line.

***Tavole H** (I. I. 3113),* published by the Istituto Idrografico della Marina of Italy in 1947, combines table I of Ogura and table II of Weems' *New Line of Position Tables,* including, also, the Rust azimuth diagram (art. 2106). This table is a modification of an earlier *Tavole F.*

Čumbelić. In 1969 Captain Petar Čumbelić of Yugoslavia published his single volume and compact *Nautičke Tablice.* This method is based upon a triangle divided by dropping a perpendicular from the zenith. The table includes an English explanation.

2111. Altitude methods, perpendicular from body.—Figure 2111 is a diagram on the plane of the celestial meridian (art. 1432), with the navigational triangle shown in heavy lines. A perpendicular from the celestial body, *M,* to the celestial meridian divides the triangle into two right spherical triangles. In figure 2111 the length of the perpendicular is designated v and the two parts of the colatitude are designated w and x. By means of Napier's rules (art. 142, vol. II), the following basic formulas can be derived:

$$\sin v = \cos d \sin t \quad (1)$$

$$\cos w = \sin d \sec v, \text{ or } \sin w = \cot t \tan v \quad (2)$$

$$\sin h = \cos v \cos x \quad (3)$$

$$\sin Z = \sin v \sec h, \text{ or } \cos Z = \tan x \tan h \quad (4)$$

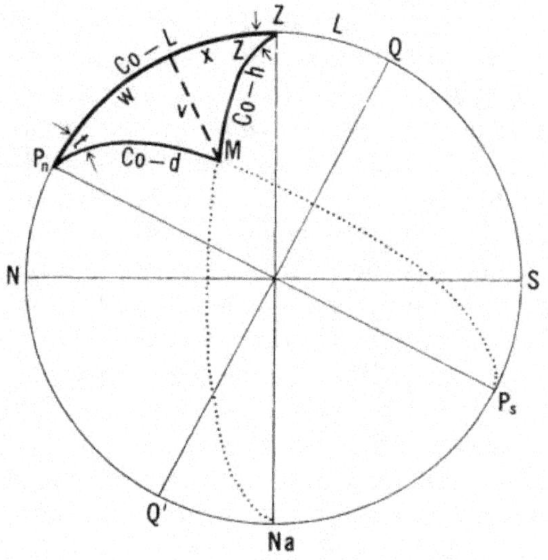

FIGURE 2111.—Navigational triangle with perpendicular from celestial body to celestial meridian.

Since $x = 90° - (w + L)$, formula (3) can be written in terms of latitude, and w found from equation (2). Thus, both h and Z can be determined by means of t, d, and L and auxiliary functions found from them.

William Thomson (Lord Kelvin) was the first to divide the navigational triangle as shown in figure 2111 for sight reduction, but his method (art. 2106) was for determination of longitude. Various later methods made such a division for determination of altitude.

Fuss. The *Tables to Find Altitudes and Azimuths,* devised by V. E. Fuss, an astronomer at the Kronstadt (Russia) Naval Observatory, were published in 1901. In these tables a perpendicular is dropped from the celestial body, the following notation being used (fig. 2111):

v is designated a
w is designated $90° - b$
x is designated $B - 90°$
$B = 90° - L + b$ (if v falls between Z and Q).

Solution is by the following formulas:

$$\sin a = \cos d \sin t$$
$$\cot b = \cot d \cos t$$
$$\sin h = \cos a \sin B$$
$$\cot Z = \cot a \cos B.$$

The assumed latitude is selected to provide the nearest 15′ value of B. The assumed longitude is selected so that t will be the nearest whole 1^m ($0\overset{\circ}{.}25$). The tables are entered twice, first with t and d to find a and b, interpolating for d, and then with B and a to find h and Z, interpolating for a. The method involves two book openings, eight table entries, four interpolations, and ten mathematical steps. There are 144 pages of tables.

Aquino. The *Altitude and Azimuth Tables* of Radler de Aquino, a Brazilian naval officer, were first published in 1909. These were followed the next year by his *Sea and Air Navigation Tables.* Several later editions of both publications appeared with some modification, principally of the auxiliary material given. Aquino dropped a perpendicular from the celestial body to the celestial meridian, and used the same formulas as Fuss and generally the same arrangement, except that longitude is assumed so as to provide a meridian angle to the nearest whole degree.

H.O. Pub. No. 209 (Pierce), *Position Tables for Aerial and Surface Navigation,* was published by the U. S. Navy Hydrographic Office in 1930. These tables were devised by Commander M. R. Pierce, USN, in 1925, when he was navigator of the dirigible USS *Los Angeles.* The method is based upon a triangle divided by a perpendicular from the celestial body. It is generally similar to those designed by Fuss and Aquino, but the arrangement is somewhat different, requiring 206 pages of tables. This method was never widely used, and is now out of print.

H.O. Pub. No. 211 (Ageton), *Dead Reckoning Altitude and Azimuth Table,* was published by the U. S. Navy Hydrographic Office in 1931. This method, designed by Lieutenant Arthur A. Ageton, USN, while a student of the Post Graduate School, then at Annapolis, Maryland, is based upon a triangle divided by dropping a perpendicular from the celestial body. It is generally similar to those of Fuss and Aquino. However, Ageton modified the formulas so as to include only secants and cosecants. In terms of figure 2111, his notation is as follows:

v is designated R
w is designated $90° - K$
x is designated $K \sim L$
$K = x + L$.

Ageton's formulas are

$$\csc R = \csc t \sec d$$

$$\csc K = \frac{\csc d}{\sec R}$$

$$\csc h = \sec R \sec (K \sim L)$$

$$\csc Z = \frac{\csc R}{\sec h}.$$

A single table of 36 pages gives five-place log cosecants (labeled A) and log secants (labeled B), both multiplied by 100,000 to eliminate the decimal. These values are given in parallel columns for each 0.′5 of angle from 0° to 180°. The table is well arranged and indexed for quick reference. The rules are relatively simple and well presented. The method can be used for solution from the dead reckoning or any other assumed position. The method is intended for use without interpolation. These features combined to make this a popular method, although solution is somewhat tedious, and large errors may be encountered if t is near 90°. The method has been largely superseded by those tables constituting a list of computed answers. However, the method is now published as table 35.

If a celestial body near the visible horizon is observed, it may be *below* the celestial horizon (zenith distance greater than 90°), because of refraction and dip. Under these conditions the *computed* altitude, Hc, is negative (art. 2023). In the solution by H.O. Pub. No. 211, Hc is negative if K is of the same name as L and greater than $(90° + L)$, *or* if K is of contrary name to L and greater than $(90° - L)$. Under the second of these conditions, Z is less than 90° and should be taken from the top of the table *if* K is greater than $(180° - L)$.

Fontoura da Costa and Penteado's *Tabuas de Altura e Azimute* were published in Lisboa, Portugal, in 1936. These consist of 26 pages of log secant and log cosecant tables similar to those of H.O. Pub. No. 211. The method and formulas are slight modifications of those of H.O. Pub. No. 211 (table 35).

Tillman. The *Altitude Tables for Mariners and Aviators,* by E. Tillman, were published in 1936 in Sweden. Solution is by three tables using the basic formulas given above.

USSR tables. About 1940 the USSR replaced the Fuss tables with a method that is similar but uses a much shorter table. However, the solution is about the same length as with the Fuss tables, requiring six book openings, nine table entries, and five mathematical steps. Visual interpolation is used.

Japanese H.O. Pub. No. 602, *Brief Celestial Navigation Table,* was published in 1942. A perpendicular is dropped from the celestial body, as in figure 2111. Side *w* is designated K, and the following formulas are derived from the basic formulas given above:

$$\log \tan K = \log \cot d + \log \cos t$$

$$\log \cot Z = \log \cot t + \log \csc K + \log \cos (K+L)$$

$$\log \cot h = \log \cot (K+L) + \log \sec Z.$$

These formulas result in a simple solution, at the expense of some duplication in the three tables of 49 pages.

Hickerson. In 1944 Thomas F. Hickerson, professor of applied mathematics at the University of North Carolina, published a small volume called *Navigational Handbook with Tables,* in which the tables of H.O. Pub. No. 211 are given with the interval between entries reduced to 0.′2. All values are given on 45 pages, by tabulating values only to 45° and interchanging the A and B values for angles between 45° and 90°. In 1947 a second edition was published under the title *Latitude, Longitude and Azimuth by the Sun or Stars.*

2112. Altitude methods without use of navigational triangle.—The navigational triangle is composed of arcs of three great circles: the celestial meridian of the observer, the hour circle of the celestial body, and the vertical circle of the celestial body. Arcs of other great circles might also be used in forming spherical triangles that can be solved to find altitude and azimuth.

Kotlarić. In 1954 Stjepo M. Kotlarić, scientific co-worker, Hydrographic Institute of the Yugoslav Navy, proposed a method and in 1958 published *Tables K1* based upon the solution of three right spherical triangles composed of arcs of great circles, as follows:

triangle 1—celestial horizon, hour circle, and celestial equator;

triangle 2—celestial horizon, hour circle, and celestial meridian (lower branch);

triangle 3—celestial horizon, hour circle, and vertical circle.

The formulas are derived from Napier's rules (F, M, and C are auxiliary parts):

$$\begin{aligned}\cos C &= \cos L \sin t\\ \tan M &= \cot L \cos t\\ \tan (Z+F) &= -\sin L \tan t\\ \tan F &= \cos C \tan (M+d)\\ \sin Hc &= \sin C \sin (M+d)\\ Z &= (Z+F)-F.\end{aligned}$$

Hc and Z are solved from tables I and II, tables III and IV being the multiplication tables for obtaining the Hc corrections. The four tables total 190 pages. Table III is not used if the assumed position is selected so that latitude and meridian angle are the nearest whole or half degree. The pages could be reduced considerably if values for half degrees were not tabulated. With an assumed position selected as indicated above, the method requires only four table entries and three mathematical steps. All signs are printed in the heading of the tables. The computation is simple. *Tables K1* were reprinted in 1963 and 1971. The tables include a comprehensive English explanation.

2113. Modern inspection tables may contain lists of altitude or azimuth, or both. Another type tabulates the information needed for finding longitudes. Values are taken directly from the tables, without the need for logarithms, auxiliary functions, or mathematical solutions (except interpolation). Inspection tables are not new, the horary tables of Cassini in 1770, Lalande in 1793, Lynn in 1827, and Hommey in 1863 (art. 2106) being of this type. Other inspection tables include Davis' *Chronometer Tables,* Blackburne, H.O. Pubs. Nos. 203 and 204, Ball, Davis' *Alt-Azimuth Tables,* and H.O. Pub. No. 201 (arts. 2106 and 2109). None of these tables is used to any extent today, largely because interpolation is difficult, and coverage is limited. A short logarithmic solution with wide coverage has often proved more popular.

In contrast, the modern inspection table, made practicable by recent developments in computation techniques, has largely replaced the trigonometric solution. The principal modern inspection tables are:

Pub. No. 214, *Tables of Computed Altitude and Azimuth,* were published by the U. S. Navy Hydrographic Office between 1936 and 1946, in nine volumes. Between 1951 and 1953 the British Admiralty published identical tables (H.D. 486) in six volumes, with altered explanation to suit British practice. The first volume of an identical Spanish edition was published in Spain in 1953, and the second volume in 1956. Several volumes of an Italian edition based on Pub. No. 214 have also been published. Pub. No. 214 was superseded by Pub. No. 229, *Sight Reduction Tables for Marine Navigation.* Both tables are described in detail in chapter XX.

British Air Pub. 1618 (H.O. Pub. No. 218), *Astronomical Navigation Tables,* were published by the British Admiralty between 1938 and 1944 in 15 volumes (lat. 0°–79°). In 1941 the first 14 volumes (lat. 0°–69°) were republished by the U. S. Navy Hydrographic Office as H.O. Pub. No. 218. The tables are intended primarily for aviators.

These tables are similar to Pub. No. 249, but with several major differences. In A.P. 1618 values are given to the nearest whole minute for altitude, and the nearest whole degree for azimuth. The altitude values include allowance for refraction at a height of 5,000 feet. The minimum altitude in most cases is 10°. Provision is made for interpolation for declination only, and this always from the *next smaller* whole degree, instead of from the *nearest* whole degree. Declination is given for each whole degree from 0° to 28° only. In addition, values of altitude and azimuth angle are given for the declination (in 1940) of 22 stars. This part of the table is entered with the star name (or an arbitrarily-assigned number), so the declination of the body need not be known. An auxiliary table provides a correction for changes in declination during the years following 1940 (to the year 2000).

During World War II these tables were widely used by aviators. Some marine navigators also used them. Publication of these tables has been discontinued.

Japanese H.O. Pub. No. 351, *Celestial Navigation Observation Table,* was published in 1940–42, in seven volumes for latitudes 0°–70°. The original printing was classified "secret." The tables are similar to British Air Pub. 1618, with several differences. In H.O. Pub. No. 218 all star-name entry tables are given first, followed by all declination entry tables. In Pub. No. 351 the declination entry table for each degree of latitude is followed by the star-name entry table. Altitudes, including refraction at 4,000 meters (13,123 feet), are tabulated to a minimum value of 2°. Declination is extended to 29°. The latitude-declination contrary-name entries are inverted so that meridian angles increase *upward* on the page as in Pub. No. 260, resulting in better utilization of space on the pages having both "same name" and "contrary name" entries. Twenty stars are used, the selection differing somewhat from that of H.O. Pub. No. 218. In H.O. Pub. No. 218 the stars are listed and numbered alphabetically. In Pub. No. 351 they are given in order of declination, from Dubhe, listed as 62°03′N, to Sirius, listed as 16°38′S.

Hoehne. In October 1941 George G. Hoehne, then a navigation instructor at the Pan American Airways Navigation School, Miami, Florida, submitted a set of *Star Air Navigation Tables* to the U. S. Navy Hydrographic Office which were superior to the star section of H.O. Pub. No. 218 in basic design. His manuscripts included the tabulation of the altitudes and true azimuths of carefully selected bright stars arranged in a format such that this data could be rapidly extracted for at least ten stars from two facing pages with but one opening of the tables. The use of LHA ♈ instead of the LHA of each star as a table argument simplified the sight reduction by: (1) eliminating the need to apply the SHA of a star to the GHA ♈ to obtain the LHA of a star; (2) enabling the optimum arrangement for rapid extraction of tabular values of altitude and true azimuth; and (3) providing for the selection of the best stars for observation for a given LHA ♈ at an assumed position. The use of LHA ♈ as a table argument with the data arranged in parallel columns so that the stars would be tabulated, from left to right, in increasing numerical order of true azimuth served to make the tables a star finder. The same basic format was later used with 360° of LHA ♈ per table opening for volume I of Pub. No. 249, *Sight Reduction Tables for Air Navigation.*

In early 1942, Hoehne was originally granted permission to use, in part, tabular values of altitude, azimuth angle, and pertinent data contained in the star section of H.O. Pub. No. 218 and ingeniously constructed his preliminary volume I of *Celestial Navigation Tables.* Its success led to his publication of volume II of *Practical Celestial Air Navigation Tables* in 1943. For the period 1942–1990, two simple correction tables prevent the tabulated altitudes from becoming inaccurate because of precession of the equinoxes (art. 1419). Refraction at altitude 5,000 feet is included as in H.O. Pub. No. 218.

Figures 2113a and 2113b illustrate facing pages on which data for at least 10 stars are tabulated for a 60° range of LHA ♈ (T).

The tabulated altitudes are corrected for annual change in declination using Table I on the inside of the front cover (fig. 2113c). This correction must be added or subtracted according to the sign prefixed to (t). The year at the bottom of each star column is the year before which this correction cannot exceed 1′.

GHA ♈ as obtained from the almanac is adjusted for the annual change in right ascension using Table VII at the back of the book (fig. 2113d). The correction obtained from this table can be readily combined with the increment to be added to the tabular GHA ♈ extracted from the almanac.

30°N																30°N
T	CAPH			ANTARES			ALPHERATZ			ARCTURUS			FOMALHAUT			T
				CAPELLA						ALDEBARAN						
	Alt.		Az	Alt.		Az	Alt.		Az	Alt.		Az	Alt.		Az	
	° ′	t	°	° ′	t	°	° ′	t	°	° ′	t	°	° ′	t	°	
301	40 11−	3	36	13 38−	5	228	37 31+	6	75	11 20−	9	286	17 28+	15	142	301
302	40 42	3	36	12 59	5	229	38 21	6	75	10 30	9	287	17 59	15	142	302
303	41 13	3	36	12 20	5	230	39 11	6	76	9 41	9	287	18 31	15	143	303
304	41 44	4	36	11 41	5	230	40 01	6	76	8 42	9	287	19 02	15	144	304
305	42 14	4	36	11 01	5	231	40 52	6	76	8 03	9	288	19 32	16	145	305
306	42 44−	4	36	10 21−	5	231	41 42+	5	77	7 14−	9	288	20 02+	16	145	306
307	43 15	5	36	9 41	5	232	42 33	5	77	6 25	9	289	20 31	16	146	307
308	43 45	5	35	9 00	5	233	43 24	5	77	5 37−	9	289	20 59	16	147	308
309	44 15	5	35	8 19	5	233	44 14	5	78				21 27	16	148	309
310	44 45	6	35	7 38	5	234	45 05	5	78				21 55	16	149	310
311	45 14−	6	35	6 56−	5	234	45 56+	5	78				22 21+	16	149	311
312	45 44	6	34	6 14	5	235	46 47	5	79				22 48	17	150	312
313	46 14	7	34	5 32−	5	235	47 38	5	79				23 13	17	151	313
314	46 43	7	34				48 29	5	79				23 38	17	152	314
315	47 12	7	34				49 20	5	80				24 02	17	153	315
316	47 41−	8	33				50 11+	5	80				24 26+	17	153	316
317	48 09	8	33				51 02	5	80				24 49	17	154	317
318	48 37	8	33				51 53	4	81				25 11	17	155	318
319	49 06	9	32				52 45	4	81				25 32	17	156	319
320	49 33	9	32	CAPELLA			53 36	4	81				25 53	18	157	320
321	50 01−	9	32				54 27+	4	82				26 13+	18	158	321
322	50 28	10	31	5 35+	2	39	55 19	4	82				26 32	18	159	322
323	50 55	10	31	6 08	2	39	56 10	4	82				26 50	18	160	323
324	51 22	11	31	6 40	2	40	57 02	4	82				27 08	18	161	324
325	51 48	11	30	7 13	2	40	57 53	4	83				27 25	18	161	325
326	52 14−	11	30	7 46+	2	41	58 45+	4	83				27 41+	18	162	326
327	52 40	12	29	8 20	2	41	59 36	4	83				27 56	18	163	327
328	53 05	12	29	8 54	2	41	60 28	4	84				28 11	18	164	328
329	53 30	12	28	9 29	2	42	61 19	4	84				28 24	18	165	329
330	53 54	13	28	10 03	2	42	62 11	4	84				28 37	19	166	330
331	54 18−	13	27	10 38+	2	43	63 03+	4	85				28 49+	19	167	331
332	54 42	13	26	11 13	2	43	63 55	3	85	ALDEBARAN			29 00	19	168	332
333	55 05	14	26	11 48	2	43	64 46	3	86				29 11	19	169	333
334	55 27	14	25	12 24	2	44	65 38	3	86				29 20	19	170	334
335	55 49	14	25	13 00	2	44	66 30	3	86	5 43+	4	74	29 29	19	171	335
336	56 11−	15	24	13 36+	2	44	67 22+	3	87	6 33+	4	75	29 36+	19	172	336
337	56 31	15	23	14 12	2	45	68 14	3	87	7 22	4	75	29 43	19	173	337
338	56 52	15	23	14 49	2	45	69 06	3	87	8 12	3	76	29 49	19	174	338
339	57 12	16	22	15 25	2	45	69 58	3	88	9 02	3	76	29 54	19	175	339
340	57 31	16	21	16 02	2	46	70 49	3	88	9 52	3	77	29 58	19	176	340
341	57 49−	16	20	16 40+	2	46	71 41+	3	89	10 43+	3	77	30 01+	19	177	341
342	58 07	17	19	17 17	2	46	72 33	3	89	11 33	3	78	30 03	19	178	342
343	58 24	17	19	17 55	1	47	73 25	3	90	12 23	3	78	30 05	19	179	343
344	58 40	17	18	18 32	1	47	74 17	3	90	13 14	3	78	30 05	19	180	344
345	58 56	17	17	19 10	1	47	75 09	3	91	14 05	3	79	30 05	19	181	345
346	59 11−	18	16	19 49+	1	48	76 01+	3	91	14 56+	3	79	30 03+	19	182	346
347	59 25	18	15	20 27	1	48	76 53	3	92	15 47	3	80	30 01	19	183	347
348	59 38	18	14	21 06	1	48	77 45	3	92	16 38	3	80	29 58	19	184	348
349	59 50	18	13	21 44	1	48	78 37	3	93	17 29	3	81	29 54	19	185	349
350	60 02	19	12	22 23	1	49	79 29+	3	94	18 20	3	81	29 49	19	186	350
351	60 13−	19	11	23 02+	1	49				19 11+	3	82	29 43+	19	187	351
352	60 22	19	10	23 41	1	49				20 02	3	82	29 36	19	188	352
353	60 32	19	9	24 21	1	49				20 54	3	83	29 29	19	189	353
354	60 40	19	8	25 00	1	50				21 45	3	83	29 20	19	190	354
355	60 47	20	7	25 40	1	50				22 37	3	84	29 11	19	191	355
356	60 53−	20	6	26 20+	1	50				23 28+	3	84	29 00+	19	192	356
357	60 58	20	5	27 00	1	50				24 20	3	84	28 49	19	193	357
358	61 03	20	4	27 40	1	51				25 12	3	85	28 37	19	194	358
359	61 06	20	3	28 20	1	51				26 03	3	85	28 24	18	195	359
360	61 09−	20	2	29 00+	1	51				26 55+	3	86	28 11+	18	196	360
Year	1944			1948			1944			1944			1944			Year
				1957						1949						

FIGURE 2113a.—Left-hand page of opening of *Practical Celestial Air Navigation Tables.*

30°N 30°N

T	ALTAIR Alt.		Az	RASALAGUE Alt.		Az	VEGA Alt.		Az	ETAMIN Alt.		Az	DENEB Alt.		Az	T
	° ′	t	°	° ′	t	°	° ′	t	°	° ′	t	°	° ′	t	°	
301	68 23+	9	191	50 49−	1	252	69 56−	1	302	58 03	0	321	73 21−	11	23	301
302	68 12	9	193	50 00	1	253	69 11	1	301	57 31	0	321	73 41	12	20	302
303	67 59	9	196	49 10	1	253	68 27	1	300	56 58	0	320	73 58	12	18	303
304	67 44	9	199	48 20	1	254	67 42	1	300	56 25	0	320	74 13	12	16	304
305	67 26	9	201	47 30	1	255	66 56	1	299	55 51	0	319	74 26	12	13	305
306	67 07+	9	203	46 40−	1	256	66 11−	1	299	55 17	0	319	74 37−	13	11	306
307	66 45	9	206	45 49	1	256	65 25	1	298	54 43	0	318	74 45	13	8	307
308	66 21	9	208	44 59	1	257	64 39	1	298	54 08	0	318	74 51	13	5	308
309	65 56	9	210	44 08	1	258	63 53	1	298	53 33	0	318	74 55	13	3	309
310	65 29	8	212	43 17	1	258	63 07	1	297	52 58	0	317	74 56	13	0	310
311	65 01+	8	214	42 26−	1	259	62 21−	1	297	52 23	0	317	74 55−	13	357	311
312	64 30	8	216	41 35	1	260	61 35	0	297	51 51	0	317	74 51	13	355	312
313	63 59	8	218	40 44	1	260	60 48	0	297	51 11	0	316	74 45	13	352	313
314	63 26	8	220	39 53	1	261	60 02	0	296	50 36	0	316	74 37	13	349	314
315	62 52	8	222	39 01	1	262	59 15	0	296	49 59	0	316	74 26	12	347	315
316	62 16+	8	224	38 10−	1	262	58 29	0	296	49 23	0	315	74 13−	12	344	316
317	61 40	7	225	37 18	1	263	57 42	0	296	48 47	0	315	73 58	12	342	317
318	61 02	7	227	36 27	1	263	56 55	0	296	48 10	0	315	73 41	12	340	318
319	60 24	7	229	35 35	1	264	56 09	0	296	47 34	0	315	73 21	11	337	319
320	59 45	7	230	34 44	1	264	55 22	0	296	46 57	0	315	73 01	11	335	320
321	59 04+	7	231	33 52−	1	265	54 35	0	296	46 20	0	315	72 38−	11	333	321
322	58 23	7	233	33 00	1	266	53 48	0	296	45 43	0	315	72 14	10	331	322
323	57 42	7	234	32 08	1	266	53 02	0	296	45 06	0	314	71 48	10	329	323
324	56 59	7	235	31 17	1	267	52 15	0	296	44 29	0	314	71 21	10	328	324
325	56 16	7	237	30 25	1	267	51 28	0	296	43 52	0	314	70 53	9	326	325
326	55 32+	6	238	29 33−	1	268	50 41	0	296	43 15	0	314	70 23−	9	325	326
327	54 48	6	239	28 41	1	268	49 54	0	296	42 37	0	314	69 52	9	323	327
328	54 03	6	240	27 49	1	269	49 08	0	296	42 00	0	314	69 21	8	322	328
329	53 18	6	241	26 57	1	269	48 21	0	296	41 23	0	314	68 48	8	321	329
330	52 32	6	242	26 05	1	270	47 34	0	296	40 46	0	314	68 15	8	319	330
331	51 46+	6	243	25 13−	1	270	46 48	0	296	40 08	0	314	67 40−	7	318	331
332	51 00	6	244	24 22	1	271	46 01	0	296	39 31	0	314	67 05	7	317	332
333	50 13	6	245	23 30	1	271	45 15	0	296	38 54	0	314	66 30	7	316	333
334	49 25	6	246	22 38	1	272	44 28	0	296	38 16	0	314	65 54	7	315	334
335	48 38	6	247	21 46	1	272	43 42	0	297	37 39	0	314	65 17	6	314	335
336	47 50+	6	248	20 54−	1	273	42 55	0	297	37 02	0	314	64 39−	6	314	336
337	47 01	6	249	20 02	1	273	42 09	0	297	36 24	0	314	64 02	6	313	337
338	46 13	5	250	19 11	1	274	41 22+	1	297	35 47	0	314	63 23	5	312	338
339	45 24	5	250	18 19	1	274	40 36	1	297	35 10	0	314	62 45	5	312	339
340	44 35	5	251	17 27	1	275	39 50	1	297	34 33	0	314	62 06	5	311	340
341	43 16+	5	252	16 36−	1	275	39 04+	1	298	33 56	0	314	61 26−	5	310	341
342	42 56	5	253	15 44	1	276	38 18	1	298	33 19	0	315	60 47	4	310	342
343	42 07	5	253	14 53	1	276	37 32	1	298	32 42	0	315	60 07	4	309	343
344	41 17	5	254	14 01	1	277	36 46	1	298	32 05	0	315	59 27	4	309	344
345	40 27	5	255	13 10	1	277	36 00	1	298	31 28	0	315	58 46	4	309	345
346	39 37+	5	255	12 19	1	277	35 15+	1	298	30 52	0	315	58 05−	3	308	346
347	38 46	5	256	11 27	1	278	34 29	1	299	30 15	0	315	57 25	3	308	347
348	37 56	5	257	10 36	1	278	33 44	1	299	29 38	0	315	56 43	3	308	348
349	37 05	5	257	9 45	1	279	32 58	1	299	29 02	0	315	56 02	3	307	349
350	36 14	5	258	8 54	1	279	32 13	1	299	28 26	0	316	55 21	3	307	350
351	35 23+	5	259	8 04−	1	280	31 28−	1	300	27 49	0	316	54 39−	2	307	351
352	34 32	5	259	7 13	1	280	30 43	1	300	27 13	0	316	53 58	2	307	352
353	33 41	5	260	6 23−	1	281	29 58	1	300	26 37	0	316	53 16	2	306	353
354	32 50	5	261				29 13	1	300	26 01	0	316	52 34	2	306	354
355	31 59	5	261				28 28	1	301	25 26	0	317	51 52	2	306	355
356	31 08+	5	262				27 43+	1	301	24 50	0	317	51 10−	1	306	356
357	30 16	5	262				26 59	1	301	24 15	0	317	50 28	1	306	357
358	29 25	5	263				26 14	1	301	23 39	0	317	49 46	1	306	358
359	28 33	5	263				25 30	1	302	23 04	0	317	49 04	1	306	359
360	27 42+	5	264				24 46+	1	302	22 29	0	318	48 21−	1	306	360
Year	1947			1963			1958						1945			Year

From Practical Celestial Air Navigation Tables, Volume II.

Figure 2113b.—Right-hand page of opening of *Practical Celestial Air Navigation Tables.*

TABLE I.

t / Year	1	2	3	4	5	6	7	8	9	10	11	12	13	14	15	16	17	18	19	20	t / Year
1942	0	0	0	0	0	0	0	0	0	0	0	0	0	0	0	1	1	1	1	1	**1942**
1943	0	0	0	0	0	0	0	0	0	0	1	1	1	1	1	1	1	1	1	1	**1943**
1944	0	0	0	0	0	0	0	1	1	1	1	1	1	1	1	1	1	1	1	1	**1944**
1945	0	0	0	0	0	0	1	1	1	1	1	1	1	1	1	1	1	2	2	2	**1945**
1946	0	0	0	0	0	1	1	1	1	1	1	1	1	1	2	2	2	2	2	2	**1946**
1947	0	0	0	0	1	1	1	1	1	1	1	1	2	2	2	2	2	2	2	2	**1947**
1948	0	0	0	1	1	1	1	1	1	1	1	2	2	2	2	2	2	2	3	3	**1948**
1949	0	0	0	1	1	1	1	1	1	2	2	2	2	2	2	2	3	3	3	3	**1949**
1950	0	0	0	1	1	1	1	1	2	2	2	2	2	2	2	3	3	3	3	3	**1950**
1951	0	0	1	1	1	1	1	1	2	2	2	2	2	3	3	3	3	3	3	4	**1951**
1952	0	0	1	1	1	1	1	2	2	2	2	2	3	3	3	3	3	4	4	4	**1952**
1953	0	0	1	1	1	1	2	2	2	2	2	3	3	3	3	3	4	4	4	4	**1953**
1954	0	0	1	1	1	1	2	2	2	2	3	3	3	3	4	4	4	4	4	5	**1954**
1955	0	0	1	1	1	2	2	2	2	2	3	3	3	4	4	4	4	4	5	5	**1955**
1956	0	1	1	1	1	2	2	2	2	3	3	3	3	4	4	4	5	5	5	5	**1956**
1957	0	1	1	1	1	2	2	2	3	3	3	3	4	4	4	5	5	5	5	6	**1957**
1958	0	1	1	1	2	2	2	2	3	3	3	4	4	4	4	5	5	5	6	6	**1958**
1959	0	1	1	1	2	2	2	3	3	3	3	4	4	4	5	5	5	6	6	6	**1959**
1960	0	1	1	1	2	2	2	3	3	3	4	4	4	5	5	5	6	6	6	7	**1960**
1961	0	1	1	1	2	2	2	3	3	4	4	4	5	5	5	6	6	6	7	7	**1961**
1962	0	1	1	1	2	2	3	3	3	4	4	4	5	5	6	6	6	7	7	7	**1962**
1963	0	1	1	2	2	2	3	3	3	4	4	5	5	5	6	6	7	7	7	8	**1963**
1964	0	1	1	2	2	2	3	3	4	4	4	5	5	6	6	6	7	7	8	8	**1964**
1965	0	1	1	2	2	2	3	3	4	4	5	5	5	6	6	7	7	8	8	8	**1965**
1966	0	1	1	2	2	3	3	3	4	4	5	5	6	6	6	7	7	8	8	9	**1966**
1967	0	1	1	2	2	3	3	4	4	4	5	5	6	6	7	7	8	8	9	9	**1967**
1968	0	1	1	2	2	3	3	4	4	5	5	6	6	7	7	7	8	8	9	9	**1968**
1969	0	1	1	2	2	3	3	4	4	5	5	6	6	7	7	8	8	9	9	10	**1969**
1970	0	1	2	2	2	3	4	4	4	5	6	6	6	7	8	8	8	9	10	10	**1970**
1971	1	1	2	2	3	3	4	4	5	5	6	6	7	7	8	8	9	9	10	10	**1971**
1972	1	1	2	2	3	3	4	4	5	5	6	6	7	7	8	9	9	10	10	11	**1972**
1973	1	1	2	2	3	3	4	4	5	6	6	7	7	8	8	9	9	10	10	11	**1973**
1974	1	1	2	2	3	3	4	5	5	6	6	7	7	8	8	9	10	10	11	11	**1974**
1975	1	1	2	2	3	4	4	5	5	6	6	7	8	8	9	9	10	10	11	12	**1975**
1976	1	1	2	2	3	4	4	5	5	6	7	7	8	8	9	10	10	11	11	12	**1976**
1977	1	1	2	2	3	4	4	5	6	6	7	7	8	9	9	10	10	11	12	12	**1977**
1978	1	1	2	3	3	4	4	5	6	6	7	8	8	9	10	10	11	11	12	13	**1978**
1979	1	1	2	3	3	4	5	5	6	6	7	8	8	9	10	10	11	12	12	13	**1979**
1980	1	1	2	3	3	4	5	5	6	7	7	8	9	9	10	11	11	12	13	13	**1980**
1981	1	1	2	3	3	4	5	5	6	7	8	8	9	10	10	11	12	12	13	14	**1981**
1982	1	1	2	3	4	4	5	6	6	7	8	8	9	10	10	11	12	13	13	14	**1982**
1983	1	1	2	3	4	4	5	6	6	7	8	9	9	10	11	11	12	13	14	14	**1983**
1984	1	1	2	3	4	4	5	6	7	7	8	9	10	10	11	12	12	13	14	15	**1994**
1985	1	2	2	3	4	4	5	6	7	8	8	9	10	10	11	12	13	14	14	15	**1985**
1986	1	2	2	3	4	5	5	6	7	8	8	9	10	11	12	12	13	14	15	15	**1986**
1987	1	2	2	3	4	5	5	6	7	8	9	9	10	11	12	13	13	14	15	16	**1987**
1988	1	2	2	3	4	5	6	6	7	8	9	10	10	11	12	13	14	14	15	16	**1988**
1989	1	2	2	3	4	5	6	7	7	8	9	10	11	11	12	13	14	15	16	16	**1989**
1990	1	2	2	3	4	5	6	7	8	8	9	10	11	12	12	13	14	15	16	17	**1990**

FIGURE 2113c.—Correction to tabulated altitude due to annual change in declination of stars.

TABLE VII.

CORRECTION TO GHA ARIES DUE TO ANNUAL CHANGE IN RIGHT ASCENSION OF STARS

Name of Star						Name of Star
	1. ALDEBARAN 2. ALPHERATZ 3. ALPHECCA 4. ALTAIR 5. ANTARES	6. ARCTURUS 7. BETELGEUX 8. CAPELLA 9. CAPH 10. CANOPUS	11. DENEB 12. DENEBOLA 13. DUBHE 14. ETAMIN 15. FOMALHAUT	16. POLLUX 17. PROCYON 18. RASALAGUE 19. REGULUS 20. RIGEL	21. SIRIUS 22. SPICA 23. VEGA	

No. / Year	1	2	3	4	5	6	7	8	9	10	11	12	13	14	15	16	17	18	19	20	21	22	23	No. / Year
1942	-09	+39	-04	+01	+32	-15	-01	-06	+29	+20	+08	+28	-03	+11	+23	+33	-04	-04	-19	+04	+21	+28	+15	**1942**
1943	10	38	04	00	31	16	01	07	28	20	08	28	03	11	22	33	05	04	20	03	21	27	15	**1943**
1944	11	38	05	-01	30	17	02	08	27	19	07	27	04	10	22	32	06	05	21	02	20	26	14	**1944**
1945	11	37	05	01	30	17	03	09	27	19	07	26	05	10	21	31	06	06	22	02	19	26	14	**1945**
1946	-12	+36	-06	-02	+29	-18	-04	-10	+26	+19	+06	+25	-06	+10	+20	+30	-07	-06	-22	+01	+18	+25	+13	**1946**
1947	13	35	07	03	28	19	05	12	25	18	06	25	07	09	19	29	08	07	23	00	18	24	13	**1947**
1948	14	35	07	04	27	19	05	13	24	18	05	24	08	09	18	28	09	08	24	-01	17	23	12	**1948**
1949	15	34	08	04	26	20	06	14	23	18	05	23	09	09	17	27	10	08	25	01	16	22	12	**1949**
1950	16	33	09	05	25	21	07	15	23	17	04	22	10	08	17	26	10	09	26	02	16	22	11	**1950**
1951	-17	+32	-09	-06	+24	-21	-08	-16	+22	+17	+04	+22	-11	+08	+16	+25	-11	-10	-26	-03	+15	+21	+11	**1951**
1952	17	31	10	07	23	22	09	17	21	17	03	21	12	08	15	24	12	11	27	03	15	20	10	**1952**
1953	18	31	10	07	22	23	09	18	20	16	03	20	13	07	14	23	13	11	28	04	14	19	10	**1953**
1954	19	30	11	08	21	23	10	19	19	16	02	19	14	07	13	22	14	12	29	05	13	19	09	**1954**
1955	20	29	12	09	20	24	11	20	19	16	02	19	15	07	12	22	14	13	30	06	13	18	09	**1955**
1956	-21	+28	-12	-10	+19	-25	-12	-22	+18	+15	+01	+18	-15	+06	+12	+21	-15	-13	-30	-06	+12	+17	+08	**1956**
1957	22	28	13	10	19	25	13	23	17	15	01	17	16	06	11	20	16	14	31	07	11	16	08	**1957**
1958	22	27	14	11	18	26	13	24	16	15	00	16	17	06	10	19	17	15	32	08	11	15	07	**1958**
1959	23	26	14	12	17	27	14	25	15	14	00	15	18	05	09	18	17	15	33	08	10	15	07	**1959**
1960	24	25	15	12	16	28	15	26	15	14	-01	15	19	05	08	17	18	16	34	09	09	14	06	**1960**
1961	-25	+25	-15	-13	+15	-28	-16	-27	+14	+14	-01	+14	-20	+05	+08	+16	-19	-17	-34	-10	+09	+13	+06	**1961**
1962	26	24	16	14	14	29	17	28	13	13	02	13	21	04	07	15	20	18	35	11	08	12	05	**1962**
1963	26	23	17	15	13	30	18	29	12	13	02	12	22	04	06	14	20	19	36	11	07	11	05	**1963**
1964	27	22	17	15	12	30	18	30	11	13	03	12	23	03	05	13	21	20	37	12	07	11	04	**1964**
1965	28	21	18	16	11	31	19	32	11	12	04	11	24	03	04	12	22	20	38	13	06	10	04	**1965**
1966	-29	+20	-19	-17	+10	-32	-20	-33	+10	+12	-04	+10	-25	+03	+03	+11	-23	-21	-38	-14	+05	+09	+03	**1966**
1967	30	20	19	18	09	32	21	34	09	12	05	09	26	02	03	11	24	22	39	14	05	08	03	**1967**
1968	31	19	20	18	08	33	22	35	08	11	05	09	27	02	02	10	24	22	40	15	04	07	02	**1968**
1969	32	18	21	19	07	34	22	36	07	11	06	08	27	02	+01	09	25	22	41	16	03	07	02	**1969**
1970	33	18	21	20	07	34	23	37	07	11	06	07	28	01	00	08	26	23	42	16	03	06	01	**1970**
1971	-34	+17	-22	-21	+06	-35	-24	-38	+06	+10	-07	+06	-29	+01	-01	+07	-27	-24	-42	-17	+02	+05	+01	**1971**
1972	34	16	22	21	05	36	25	39	05	10	07	06	30	01	02	06	28	24	43	18	01	04	00	**1972**
1973	35	15	23	22	04	36	26	40	04	10	08	05	31	00	02	05	28	25	44	19	01	03	00	**1973**
1974	36	14	24	23	03	37	26	41	03	09	08	04	32	00	03	04	29	26	45	19	00	03	-01	**1974**
1975	37	14	24	23	02	38	27	43	03	09	09	03	33	00	04	03	30	27	46	20	-01	02	01	**1975**
1976	-38	+13	-25	-24	+01	-39	-28	-44	+02	+09	-09	+02	-34	-01	-05	+02	-31	-27	-46	-21	-01	+01	-02	**1976**
1977	39	12	26	25	00	39	29	45	01	08	10	02	35	01	06	01	31	28	47	21	02	00	02	**1977**
1978	40	11	26	26	-01	40	30	46	00	08	10	01	36	01	07	00	32	29	48	22	03	00	03	**1978**
1979	40	11	27	26	02	41	31	47	-01	08	11	00	37	02	07	00	33	29	49	23	03	-01	03	**1979**
1980	41	10	28	27	03	41	31	48	01	07	11	-01	38	02	08	-01	34	30	50	24	04	02	04	**1980**
1981	-42	+09	-28	-28	-04	-42	-32	-49	-02	+07	-12	-01	-39	-02	-09	-02	-35	-31	-50	-24	-05	-03	-04	**1981**
1982	43	08	29	29	04	43	33	50	03	07	12	02	39	03	10	03	36	31	51	25	05	04	05	**1982**
1983	44	07	29	29	05	43	34	51	04	06	13	03	40	03	11	04	36	32	52	26	06	04	05	**1983**
1984	45	07	30	30	06	44	35	53	05	06	13	04	41	03	11	05	37	33	53	26	07	05	06	**1984**
1985	45	06	31	31	07	45	35	54	05	06	14	04	42	04	12	06	38	33	54	27	07	06	06	**1985**
1986	-46	+05	-31	-32	-08	-45	-36	-55	-06	+05	-14	-05	-43	-04	-13	-07	-39	-34	-54	-28	-08	-07	-07	**1986**
1987	47	04	32	32	09	46	37	56	07	05	15	06	44	05	14	08	39	35	55	29	09	08	07	**1987**
1988	48	04	33	33	10	47	38	57	08	05	15	07	45	05	15	09	40	36	56	29	09	08	08	**1988**
1989	49	03	33	34	11	47	39	58	09	04	16	07	46	05	16	10	41	36	57	30	10	09	09	**1989**
1990	-50	+02	-34	-34	-12	-48	-39	-59	-09	+04	-16	-08	-47	-06	-16	-11	-42	-37	-58	-31	-11	-10	-09	**1990**

Correction applied to GHA Aries in accordance with prefixed sign.

From Practical Celestial Air Navigation Tables, Volume II. Copyright 1942, 1943, 1969 & 1970, by Navigation Publishing Co., and by George G. Hoehne, Author. Used by permission.

FIGURE 2113d.—Correction to GHA ♈ due to annual change in right ascension of stars.

Example.—During morning twilight on June 1, 1975, the 0430 DR position of a ship is lat. 30°10′N, long. 45°02′W. At GMT $4^h32^m30^s$ the navigator observes Fomalhaut with a marine sextant having no IC, from a height of eye of 38 feet. The hs is 24°58′.

Required.—The *a*, Zn, and AP, using volume II of *Practical Celestial Air Navigation Tables* and the *Air Almanac*.

Solution.—

	June 1	
GMT	$7^h32^m30^s$	June 1
7^h30^m	1°40′	
2^m30^s	0°38′	
GHA ♈	362°18′	
1975 corr.	(—) 04′	
adj. GHA ♈	362°14′	
*a*λ	45°14′	W
LHA ♈	317°00′	
*a*L	30°00′N	
ht	24°49′	t (+)17
1975 corr.	(+) 10′	
Hc	24°59′	
Ho	24°52′	
a	7A	
Zn	154°	

Fomalhaut	+	☆	−
IC	—		—
D			6′
sum	—		6′
corr.			(—) 6′
hs			24°58′
Ho			24°52′

*a*L 30°00′N
*a*λ 45°14′W

Since refraction at 5,000 feet is included in the tabulated altitudes, the sextant altitude is not corrected for refraction. An auxiliary table provides an additional refraction correction to be applied to low-altitude observations. For marine observations there is no correction for altitudes above 15°; the correction is (−)1′ for altitudes 6° to 10° and (−)2′ for altitude 4°. The correction is applied to the sextant altitude.

Japanese H.O. Pub. No. 603, *Simplified Celestial Observation Table,* was published in 1943. This publication is virtually the same as Pub. No. 351, except that eight additional stars are given, all farther south than those of Pub. No. 351. This extends the list to α *Crucis* (Acrux), given as declination 62°48′S.

Altitude and Azimuth Almanac was published by the Japanese Hydrographic Office, beginning in 1944. Originally, this was a secret publication. Several different versions were printed, and there were some modifications after the first editions. In each, however, the functions of almanac and sight reduction tables were combined. For each of several specific locations, the altitude and azimuth of one or more celestial bodies are tabulated for the date and time, usually at ten-minute intervals. In the earlier editions, the locations selected were important points in the western Pacific. From this practice, these publications are sometimes called "destination tables." Later editions used positions differing in latitude by 5°. These tables provided a quick solution for observations made at the tabulated times. On a worldwide basis such a system would involve a very voluminous tabulation each year, or cumbersome corrections. The *Altitude and Azimuth Almanac* is no longer published.

Hohentafeln nach Sternzeit, an official German table, was published in 1944 as an experimental edition with a very limited range of latitude. The tables were similar to those of Hoehne, but with six stars listed for each minute of local sidereal time.

Ménéclier and Chevalier. The *Cálculo del Punto* of Victor Ménéclier and Roberto Chevalier was published in 1945–49 by Aeronáutica Argentina. There are six volumes for latitudes 0° to 59° *south.* At intervals of 4^mLST (or 1° LHA♈) the altitude, correction factor, and azimuth (not azimuth angle) of selected stars are tabulated. Twelve columns are provided, but a number of blank areas appear, resulting in an average of about nine altitude-azimuth entries for each time entry. In most cases altitudes are carried to a minimum value of 5°, and azimuth to the horizon. These tables are similar to those of Hoehne and volume I of Pub. No. 249.

H.O. Pub. No. 230 (Goetz), *High Latitude Celestial Navigation Tables,* designed in 1945 by Roy F. Goetz, was published by the U. S. Navy Hydrographic Office in 1946.

The first section, called "Star Tables," is entered with the latitude to the nearest 1° from 70°N to 89°N, the name of the star (for ten selected stars), and LHA♈ at intervals of 2° for latitude 70° to 79°, 5° for 80° to 84°, and 10° for 85° to 89°. Altitude is tabulated to the nearest 1′ and azimuth (not azimuth angle) to the nearest 0°.1. A "ΔH" value is given for use with an auxiliary table to interpolate for precession of the equinoxes (art. 1419).

In the second section, called "Declination Tables," declination is substituted for the name of the star. A separate table is given for each 1° declination from 0° to 28°. For each degree a "same name" section is given first, followed by a "contrary name" section (to declination 19°). The minimum altitude is 1°. The declination tables give "d" in place of "ΔH" for use with an auxiliary table to interpolate for declination.

Only 400 of these tables were published. They were intended only for use in military aircraft operating beyond the latitude range of H.O. Pub. No. 218. After Pub. No. 249 became available, H.O. Pub. No. 230 was canceled.

Pub. No. 249, *Sight Reduction Tables for Air Navigation,* in three volumes, are published by the Defense Mapping Agency Hydrographic/Topographic Center. A preliminary edition of volume I for selected stars was published in 1947 under the title *Star Tables for Air Navigation,* using the principles and features of tables proposed previously by George G. Hoehne, Commander C. H. Hutchings, USN, and others. The altitudes of this edition were adjusted for refraction at a height of 10,000 feet. By the time the "first" edition was printed in 1951, for epoch 1955.0, more than 20,000 copies of the preliminary edition had been distributed. The 1951 edition dropped the refraction adjustment feature from the altitudes, and had an improved selection of stars. It was followed in 1952 with two volumes for declination entry at 1° intervals from 0° to 29°. In 1952 and 1953 a British edition was published with identical tables (A. P. 3270) but altered explanation. The tables have been accepted as standard by the air forces of Great Britain, Canada, Australia, New Zealand, and the United States.

Pub. No. 249 is described in detail in Chapter XX.

Experimental Air Navigation Tables. During the early part of World War II the British Royal Air Force felt the need for an inspection table that would be faster than Air Pub. 1618 (H.O. Pub. No. 218), but free from the limitations of the astrograph (art. 2123). Wing Commander R. C. Alabaster suggested the addition of SHA to the hour angle (measured eastward) of the stars given in Air Pub. 1618, converted to time at the sidereal rate of 15°02′.5 per hour. This would give the time interval until the next meridian transit of the vernal equinox. Before observation, the time of passage of the vernal equinox across a convenient meridian would be marked on the chart or plotting sheet. After observation, the tables would be entered with assumed latitude and the nearest tabulated altitude. The (SHA+HA) corresponding to this altitude would be added to GMT at the time of observation. The result should be close to the time marked on the chart. The difference would be converted to arc units (or a time scale would be marked on the chart or plotting sheet) and the corresponding longitude determined. This point would serve as the assumed position. The difference between the observed altitude and that used for entering the table would be the altitude difference to be used with the azimuth for plotting the line of position.

Squadron Leaders A. Potter and A. J. Hagger suggested a method of printing a time scale on the chart or plotting sheet with an auxiliary table to assist in locating the assumed position.

Various modifications and conventions were later added to avoid negative values and other complications. As the method finally emerged, a quantity known as "scale time" was adopted. This value, designated T, would be equal to 26 hours plus the GMT of the next transit of the vernal equinox occurring after 0600 during the night of the flight. The GMT of observation would be designated t. The quantity $T-t$ would be the value tabulated.

Further attempts were made to simplify the conversion of mean to sidereal time so that the single setting might be used during an entire flight. One of these, called the "Astro-Scales," was suggested by Wing Commander E. W. Anderson in 1945. In 1953 he and D. H. Sadler suggested an improved version.

Although a considerable amount of thought was given to this method, and experimental tables were published for a limited band of latitude, the limitations of a longitude method and the inconvenience of converting mean time to sidereal time resulted in the method being discarded in favor of the less restrictive Pub No. 249 method.

Ashton. In 1943 Philip Ashton proposed a set of tables called *Astrograph-time Star Tables for Air Navigation,* based upon the principle of the *Experimental Air Navigation Tables.* A permanent table would be entered with the name of the star, latitude, and "astrograph mean time" (art. 2123), and altitude and azimuth would be taken from the table. A set of tables issued each year would list the values to be used with GMT each night to determine the astrograph mean time. Before take-off, the chart or plotting sheet would be marked to agree with the astrograph mean time, and a metal tape would then be used to convert mean time to sidereal time for finding the assumed position.

Heard. About 1950 John F. Heard, associate professor of astronomy at the University of Toronto, prepared a modification of the *Experimental Air Navigation Tables.* The tabulation would be altered so that altitude would be given in the left-hand column at intervals of 20′. A delta ("diff.") value would be tabulated and this used with the difference between entering and observed altitudes to enter an auxiliary table to determine a correction to be applied to $T-t$ so that the altitude difference need not be plotted. A correction for 60 minus seconds of T would also be applied. The "bearing" of the line of position (azimuth plus or minus 90°) would also be tabulated. The line of position would be plotted through the assumed position, in the direction indicated by the "bearing." For any given time three stars differing in azimuth by approximately 120° would be given. The part of the table to use would be determined by a rough computation of $T-$t.

Kotlarić. In 1976 Dr. Stjepo M. Kotlarić, (art. 2112 and 2116), then Assistant Director, Hydrographic Institute of the Yugoslav Navy, conceived and designed a set of inspection tables generally similar to Pub. No. 214 in format but differing considerably from the latter tables in the manner of tabulation.

Tables K21, Computed Results of Altitudes and Azimuths for all Latitudes and all Celestial Bodies, are in three volumes, each covering a 30° band of latitude. The table arguments are latitude, meridian angle, and declination. The respondents are tabular altitude, index for altitude correction due to the excess of actual declination over the declination argument, and azimuth angle. Unlike Pub. No. 214 but like Pub. No. 229, the sign of the altitude correction is tabulated.

The altitude and azimuth angle data for each degree of latitude are arranged on nine successive openings of the volume. Meridian angle is the vertical argument and declination is the horizontal argument. As in Pub. No. 214, tabular altitudes and corresponding azimuth angles are limited to those of bodies approximately 5° and more above the horizon. Data for a body having declination of same name as latitude are tabulated on the left-hand page and continued on the right-hand page as necessary. Data for a body having declination of contrary name to the latitude are tabulated below on the right-hand page beginning with meridian angle 0° and continuing until the 5° altitude limitation is reached.

At each opening declination is tabulated in eight columns at 1° intervals from 0° to 30°; from 31°30′ to 89° the declinations and intervals are selected to provide those declinations required for the reduction of the selected stars. The declination in the right-hand column of each page is repeated in the first column of the pages of the next opening. This repetition of data is used as an aid to determining whether interpolation of azimuth angle for declination increment is required. There is no need to turn pages to determine the change in azimuth angle for 1° increase in declination.

The tables are intended for use with an assumed position selected so that interpolation for latitude and meridian angle is not required. Entering the tables with integral values of latitude, declination, and meridian angle, the respondents are found in a form which is a unique feature of *Tables K21.* The degrees and minutes of the tabulated altitude are printed as a two to four digit group without spacing between digits if four digits are used. For example: Tabular altitude 32°45′ is printed as 3245. Although the digital tabulation is only to the minute for tabular altitude and the index for altitude correction, and to the degree for azimuth angle, the addition of a decimal point following these values enables their extraction to a greater precision than that tabulated. The decimal point following the tabular altitude or the index for altitude correction means 0.′5; following the tabular azimuth angle, the decimal point means 0.°5. The table designer added this feature to include the maximum amount of data in the smallest space practicable and to give the user an option with respect to the precision of the data extracted. When using the decimal point option, the extracted values of altitude and index correction do not differ from values computed to the nearest 0.′1 by more than 0.′2; the extracted azimuth angle values do not differ from azimuth angles computed to the nearest 0.°1 by more than 0.°2.

The multiplication table is simple to use. It is entered with the index correction as the vertical argument and the declination increment (1′ to 59′ followed by decimal parts from 0.′ to 0.′9) as the horizontal argument.

The tables include a comprehensive English explanation.

Davies. In 1980 Rear Admiral Thomas D. Davies, USN (Ret.), published *Star Sight Reduction Tables for 42 Stars: Assumed Altitude Method of Celestial Navigation.* The tables list for each degree of latitude the LHA♈ and azimuth of stars when their altitudes are integral degrees. The table entering arguments are assumed latitude and assumed altitude. A list of about 15 or 20 stars under the altitude heading approximating the observed altitude is examined to find the combination of LHA♈ and azimuth approximating the time and azimuth of the star observation.

The altitude intercept is plotted from an adjusted assumed position: the closest integral degree of latitude and the DR longitude offset by the difference between the tabulated LHA♈ and LHA♈ at the time of the observation.

The identity of the star observed need not be known.

The tabulated data are solutions of the conventional celestial triangle in which the declination and sidereal hour angle are those of 42 selected stars. Consequently the *Nautical Almanac* is not required. Values of LHA and azimuth have been generated by a computer program using nine significant figures and based on the equations:

$$\cos t = \sin h \sec d \sec L - \tan d \tan L$$
$$\cos Z = \sin d \sec L \sec h - \tan h \tan L,$$

where t=hour angle, h=altitude (assumed), d=declination, L=latitude, and Z=azimuth. The first equation is obviously a rearrangement of the fundamental formula of spherical trigonometry:

$$\sin h = \sin L \sin d + \cos L \cos d \cos t.$$

The expression for azimuth is simply derived from that for hour angle, by interchanging h and d, so that the angle being solved for is that between the arcs of h and L. For a given value of altitude the computer program solves the two equations and establishes the LHA and azimuth of a rising and setting value for each star. The totality of stars intersecting that altitude circle is then sorted to tabulate them by ascending values of LHA for ease of location.

In 1982, Admiral Davies expanded the tables described above and published *Sight Reduction Tables for Sun, Moon and Planets: Assumed Altitude Method of Celestial Navigation.*

The method utilizes the observed altitude to simplify the reduction of the sight. This is accomplished by generating an assumed altitude by rounding the observed altitude to the nearest whole degree. The method provides accurate and simple solutions for most latitudes and altitudes. There are more complicated techniques required for some limited areas of the tables. These are referred to in the instruction section, but are treated in detail only in the table's appendix. Complications have been further minimized by limiting the tables to the most useful altitudes (15° to 70°) and latitudes (0° to 60°).

The tabulated data are solutions of the conventional celestial triangle in which the declination is provided for integral degrees covering the navigational bodies in or near the plane of the ecliptic. Provision is made for adjusting the results for the residual minutes of declination as appropriate. Values of the meridian angle t, the azimuth angle Z, and the difference d have been generated by a computer program based upon the following equations:

$$\cos t = \sin h \sec d \sec L - \tan L \tan d$$
$$\cos Z = \sin d \sec h \sec L - \tan L \tan h$$
$$d = 60 (\sin L \sec h \sec d - \tan h \tan d),$$

where t=meridian angle, h=altitude, d=declination, L=latitude, Z=azimuth angle, and d = rate of change of altitude with respect to declination.

2114. Azimuth methods.—Nearly all methods proposed for obtaining a line of position are based upon the use of altitudes. The azimuth might also be used if an instrument becomes available for measuring it to the required accuracy. The accuracy needed would depend upon the acceptable error of the line of position. The error would be proportional to the cosine of the altitude. For a celestial body on the celestial horizon an error of 1′ in the azimuth would introduce an error of 1 mile in the line of position, the same as it does in an altitude observation. For any altitude greater than 0°, the error would be less.

Each method of determining a line of position by altitude has its counterpart in the azimuth problem. Thus, if it can be determined that a celestial body is exactly on the celestial meridian, the west longitude is the same as the GHA of the body. If the body is exactly on the prime vertical, the latitude can be computed. As a more general case, two points on a given azimuth line can be computed and joined by a straight line, by assuming two latitudes or two longitudes. However, if one such position is known, the azimuth line of position can be drawn through it in the direction of the azimuth. If the celestial body is sufficiently high, or if a small scale is acceptable and allowance is made for chart distortion, the azimuth line can be plotted directly, just as the circle of position can be drawn if the altitude is known. The difference between the observed azimuth and that computed for an assumed position can be used in a manner similar to the altitude difference. The azimuth difference in minutes multiplied by the cosine of the altitude would be the "intercept" measured off from the assumed position in a direction perpendicular to the computed azimuth. Through the point thus determined, a line would be drawn in the direction of the observed azimuth. For small differences, the line could be drawn perpendicular to the line from the assumed position. The relative values of the observed and computed azimuths would indicate the direction (right or left) to draw the line from the assumed position.

If the altitude and azimuth were both known to sufficient accuracy, a single celestial body would suffice for determining position by any combination of altitude and azimuth methods or by direct computation of latitude and longitude. The two lines of position would always be perpendicular to each other.

Double altitudes. For a stationary observer the longitude can be determined by observing the altitude shortly before meridian transit (either upper or lower), and noting the time when the altitude has returned to exactly the same value after meridian transit. If there has been no change in declination between observations, the mid time represents the moment of meridian transit, at which time the azimuth is 000° or 180°. The GHA (or 360°−GHA for east longitude) is the longitude of the observer. This method might be considered as either a longitude or an azimuth method. A variation is to observe a number of altitudes shortly before and after meridian transit. These are then plotted against time on cross-section paper and a smooth curve plotted through them. The time corresponding to the maximum altitude (minimum altitude for lower transit) is the moment of meridian transit.

2115. Determination of latitude and longitude.—Most methods provide information needed for plotting a line of position. The fix is at the common intersection of two or more such lines. A line of position might be plotted in one of several ways. In the latitude and longitude methods, the lines are plotted at the computed coordinate. When one coordinate has been determined, the other can be computed without plotting. Thus, the longitude determined by time sight is generally correct only for the latitude used in its solution, and the plotting of a longitude line is misleading, unless the celestial body is on the prime vertical. A better procedure is to compute two points, using different latitudes (or longitudes, if latitude is being computed). These two points are on the line of position. A straight line connecting them is a good approximation of the circle of equal altitude. This was the method used by Captain Sumner when he discovered the line of position (art. 131), and the method of H.O. Pubs. Nos. 203 and 204 (art. 2106).

Another method is to compute one point and the azimuth (or Zn±90°), and plot the line of position through the point. This is the method used by Soule and Dreisonstok (art. 2106).

If only the altitude difference is computed for two points, the line of position is a common tangent of circles of radius equal to the altitude difference at these two points. This is the method of Benest and Timberlake (art. 2110).

The most common modern method of plotting the line of position is by means of the assumed position, altitude intercept, and azimuth.

It is possible, too, to plot circles of position by using the geographical position of each body as the center of its circle, and the zenith distance as its radius. This is the method used for high-altitude observations (art. 2024), but is generally not practical for ordinary altitudes because of the small scale that would be needed, and the error that would be introduced by chart distortion, unless plotting were done on the surface of a sphere (art. 2124).

The use of a circle of equal altitude is similar to the use of a circle of position around a landmark of known range. The bearing of such a landmark also furnishes a line of position. Similarly, a line of position can be obtained by plotting the azimuth line of a celestial body, and a fix by plotting two such lines. This is generally not done because of the scale and chart limitations mentioned above, and also because the needed accuracy in observation is beyond the capability of equipment generally available to the navigator. Errors in both compass and measurement of azimuth are involved.

Various methods of determining position by computation from observations of two or more celestial bodies or four observations of a single celestial body are discussed in articles 2116 and 2117.

2116. Computed position from observation of two or more bodies.—Several methods have been proposed for computing the position directly from the observation of two or more celestial bodies. These generally consist of some combination of latitude and time sight methods. One form of automatic celestial navigation, proposed by Collins Radio Company, uses the principle of the planetarium in reverse, two bodies serving to position a horizontal-stabilized sphere (in principle) for latitude and local sidereal time. If the device is accurately set to Greenwich sidereal time, longitude is indicated.

de Jonge. In 1945 Joost H. Kiewiet de Jonge, a lieutenant in the Netherlands East Indies Army Air Force, proposed a method of determining position from the observation of three stars. The U. S. Navy Hydrographic Office published experimental tables for several star pairs for latitudes 20° to 30° under the title *Three Star Position Tables for Aerial Navigation.* It was anticipated that if the method proved popular, all possible three-star combinations (of the stars in the main tabulation of the almanacs) would be given, so that the navigator would not be limited in his selection.

No assumed position is needed with the method. Three stars are observed at intervals of three minutes, the stars being observed in the order of listing in the main table. Table I is entered with the three altitudes, h_1, h_2, and h_3, and for each a value is taken from the table. These values are labeled H_1, H_2, and H_3, respectively. They are combined to form $H_1+H_2=H_{12}$, and $H_2+H_3=H_{23}$. These combined values, H_{12} and H_{23}, are then used to enter the main table, from which local sidereal time (in arc units) and latitude are obtained. Greenwich sidereal time minus local sidereal time equals longitude (measured westward). Delta values and auxiliary tables provide corrections for motion of the observer and observation intervals differing from three minutes. Mean corrections for both atmospheric refraction and Coriolis are included in the tables, which are limited to altitudes between 20° and 75°, and azimuth difference between consecutive stars to 165°.

Dozier. In 1949 Charles T. Dozier proposed a method based upon the simultaneous observation of two celestial bodies and the solution of two spherical triangles, with vertices as follows:

triangle 1—the two celestial bodies and the elevated pole,
triangle 2—the two celestial bodies and the zenith.

The method involves the successive solution of seven formulas:

$$\cos D = \sin d_1 \sin d_2 + \cos d_1 \cos d_2 \cos S \tag{1}$$

$$\sin (X_1 \pm A_1) = \frac{\sin S \cos d_2}{\sin D} \tag{2}$$

$$\cos A_1 = \frac{\sin h_2}{\cos h_1 \sin D} \pm \frac{\tan h_1}{\tan D} \tag{3}$$

$$X_1 = (X_1 \pm A_1) \pm A_1 \tag{4}$$

$$\sin L = \sin d_1 \sin h_1 + \cos d_1 \cos h_1 \cos X_1 \tag{5}$$

$$\sin t_1 = \frac{\sin X_1 \cos h_1}{\cos L} \tag{6}$$

$$\lambda = GHA_1 \pm t_1 \tag{7}$$

in which D is the great-circle distance (angular) between the two celestial bodies, d_1 is the declination of the first body, d_2 is the declination of the second body, S is the difference of SHA of the two bodies, X_1 is the parallactic angle of the first body, A_1 is the angle at the first body between its vertical circle and the great circle between it and the second body, h_1 is the altitude of the first body (Ho is used), h_2 is the altitude (Ho) of the second body, L is the latitude of the observer, t_1 is the meridian angle of the first body, λ is the longitude of the observer, and GHA is the Greenwich hour angle of the first body.

If the great circle joining the two celestial bodies is on that side of the zenith opposite the elevated pole (if Z is within the angle formed by the vertical circle and hour circle of the first body), (X_1+A_1) is used in formulas (2) and (4), the sign in formula (3) is positive (+), and the sign of A_1 in formula (4) is negative (−). These signs are all reversed if the line adjoining the celestial bodies is on the opposite side of the zenith (Z outside the angle). If the great circle joining the two bodies passes almost through the zenith, an error might be made in the selection of the sign, and it is well to select another star pair. In formula (7) the sign is positive if the first celestial body is east of the observer's celestial meridian, and negative if it is west. The answer is in longitude measured westward from the Greenwich meridian. If the value exceeds 180°, it is subtracted from 360°, and the longitude is east.

If the quadrant of angle $(X_1 \pm A_1)$ or if t_1 is in doubt, the following formulas are suggested to replace (2) or (6):

$$\cot (X_1 \pm A_1) = \frac{\cos d_1 \tan d_2 - \sin d_1 \cos S}{\sin S} \tag{2A}$$

$$\cot t_1 = \frac{\cos d_1 \tan h_1 - \sin d_1 \cos X_1}{\sin X_1}. \tag{6A}$$

In the presentation of the method it was suggested that simultaneous observations be obtained by a two-star tracker mounted on a stable platform, or by a double sextant. Several such sextants have been proposed, but none is in common use. Other possibilities would be to have two observers or to adjust the value of one observation for the change in altitude due to its apparent motion and the motion of the observer between observations.

It was proposed that values obtained by solution of formula (1) be published in a permanent table, since these values for various star pairs would be constant except for the very slight change due to proper motion (art. 1418). Since the values obtained by formula (2) change slowly with precession of the equinoxes (art. 1419), it was proposed that the angle $(X_1 + A_1)$ for a number of star pairs be published annually, perhaps in the almanacs. The other formulas would be solved after observation of the celestial bodies.

Kotlarić. In 1954 Stjepo M. Kotlarić (arts. 2112 and 2113), of Yugoslavia, proposed a method based upon the simultaneous observation of two selected stars and the solution of three spherical triangles with vertices as follows:

triangle 1—the two stars and the elevated pole,
triangle 2—the two stars and the zenith, and
triangle 3—the elevated pole, the zenith, and the second star.

The parallactic angle (X) of the second star is determined by solutions of triangles 1 and 2, using haversine formulas. The latitude and the local hour angle of the second star are then computed by solution of triangle 3, using observed altitude, parallactic angle, and declination of the second star as the known parts. By subtracting the GHA of the second star from its LHA, the longitude is obtained.

From triangle $P_nM_1M_2$ (fig. 2116), angle A and the interstellar distance $(90° - V_x)$ are computed by means of the difference of sidereal hour angles (SHA_1 and SHA_2) of the two stars observed and their declinations (d_1 and d_2) from the formulas:

$$\sin^2 \frac{90° - V_x}{2} = \sin^2 \frac{d_2 - d_1}{2} + \sin^2 \frac{SHA_1 - SHA}{2} \cos d_2 \cos d_1 \tag{1}$$

$$\sin^2 \frac{A}{2} = \csc (90° - V_x) \sec d_2 \cos R \sin (R - d_1) \tag{2}$$

$$R = \frac{d_1 + (90° - V_x) + d_2}{2}.$$

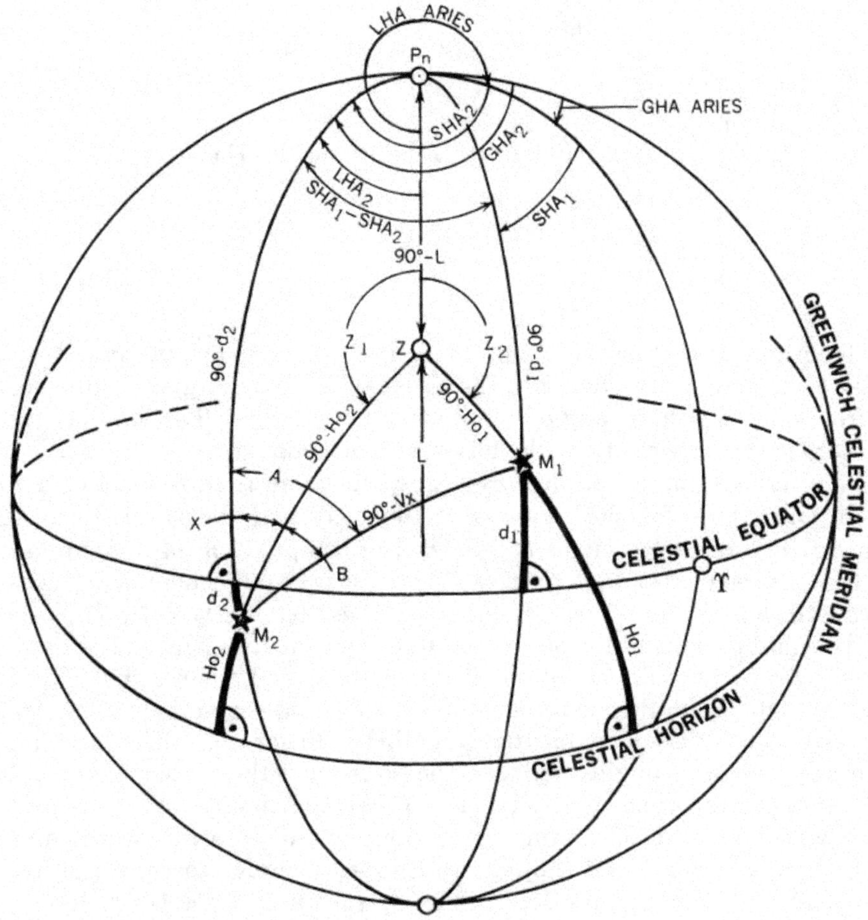

FIGURE 2116.—The spherical triangles on the celestial sphere used in Tables K11.

From triangle ZM_1M_2, angle B is computed, using the two observed altitudes (Ho_1 and Ho_2) and the interstellar distance $(90°-V_x)$ as the known parts, from the formula:

$$\sin^2\frac{B}{2} = \csc(90°-V_x)\sec Ho_2 \cos R \sin(R-Ho_1) \tag{3}$$

$$R=\frac{Ho_1+(90°-V_x)+Ho_2}{2}.$$

Angle B, combined with angle A, gives the parallactic angle (X) from one of the following relations, depending upon the relative positions of the zenith and the two stars:

$$X= A+B,\ X=A-B,\ X=360°-(A+B). \tag{4}$$

From triangle $PnZM_2$ the local hour angle of the second star (LHA_2) and the latitude of the observer are computed, using d_2, Ho_2, and X as the known parts, from the formulas:

$$\sin^2\frac{90^\circ - L}{2} = \sin^2\frac{Ho_2 - d_2}{2} + \sin^2\frac{X}{2}\cos Ho_2 \cos d_2 \tag{5}$$

$$\sin^2\frac{LHA_2}{2} = \csc (90^\circ - d_2) \sec L \cos R \sin (R - Ho_2) \tag{6}$$

$$R = \frac{Ho_2 + (90^\circ - d_2) + L}{2}.$$

The longitude of the observer and its sign (− for west longitude and + for east longitude) are found by subtracting the Greenwich hour angle of the second star (GHA_2) from the local hour angle of the same star (LHA_2). The second star should not be observed in the vicinity of the observer's meridian.

Nonsimultaneous observations, preferably not more than 4 minutes apart, may be used if corrections for the motions of the body and the observer are applied.

A modified version of this direct method of computation of latitude and longitude of the observer's position proposed in 1954 was published in 1971 as *Tables K11, Two-Star Fix Without Use of Altitude Difference Method.* In *Tables K11* the developed formulas were not used for tabulation of latitude and LHA_2 as suggested in 1954, but for the tabulation of correction indices and their signs. These indices are used through the multiplication table in obtaining corrections to the tabulated entering arguments (assumed latitude and LHA ♈) for the difference in declinations and sidereal hour angles between the constant values used in the tabulation and the values at the time of observation. Correction indices are also provided in the main table for use in obtaining corrections to the entering arguments for the difference between actual and tabulated altitudes in order to obtain accurate latitude and LHA ♈. The longitude is obtained by subtracting the GHA_2 from the sum of LHA ♈ and tabulated SHA_2.

Tables K11 are unique among all other tabular methods as they give tabulated corrections to be applied to the assumed position in order to obtain the observed position without plotting lines of position. The method is simple; there are no rules for signs nor mental interpolations. The tabulated corrections are given on one page of the main tables at the opening for assumed latitude and LHA ♈.

The tables are published in five volumes: Volume I N (lat. 0° to 9°30′N), in press; Volume II N (lat. 10° to 19°30′N), published 1975; Volume III N (lat. 20° to 29°30′N), published 1974; Volume IV N (lat. 30° to 39°30′N), published 1972; and Volume V N (lat. 40° to 49°30′N), published 1971.

Data for 37 stars in 155 star combinations are included in the tables. Each of the 360 pages of the main table of each volume contains tabulations for two star pairs for arguments of a single latitude (whole degrees or whole degrees and half degree) and a 20° range of LHA ♈ at half-degree intervals.

In 1974 Dr. Kotlarić simplified the correction procedure required when the star observations are nonsimultaneous; eliminating the need to observe the compass azimuth of the first star. Corrections for the elapsed time between observations are applied directly to the tabulated latitude and LHA ♈.

The modification is designed for the new edition in a way to double the number of star pairs without increasing the number of pages of each volume. To reach this goal four different star pairs will be tabulated on each page within the same 20° range of LHA ♈ but divided separately for each 10° range of LHA ♈. Eight different star pairs are given for each assumed position found at each opening of the tables. This is deemed quite satisfactory for the practice of navigation.

The tables include a comprehensive English explanation.

2117. Position from observation of single body.—If azimuth could be determined and plotted to sufficient accuracy, the altitude and azimuth of a single body could be used for establishing a fix. Any combination of altitude and azimuth methods (arts. 2108 and 2114) might be used, or the position could be computed without plotting. The following formulas might be used:

$$\sin t = \sin Z \cos h \sec d$$
$$\tan K_1 = \cos t \cot d$$
$$\tan K_2 = \cos Z \cot h$$
$$L = 90° - (K_1 + K_2) \quad \text{(Approximate latitude must be known).}$$

A single body can be used for a running fix, of course, and if the body is near the zenith, a relatively short time might be needed. This is the case for high-altitude observations (art. 2024) and has been used by a submarine measuring azimuth through its periscope when the sun is near the zenith.

Willis. Another method of determining position by a single body is by the use of altitude and rate of change of altitude. Three methods of doing this were suggested by Edward J. Willis in 1928.

Prime vertical observation. It can be shown by the use of differential calculus (art. 144, vol. II) that

$$\cos L = \frac{dh}{dt} \csc Z \tag{1}$$

when

$$\frac{dh}{dt}$$

is the rate of change of altitude with respect to time, specifically the change of altitude in minutes of arc during a one-minute-of-arc (four-seconds-of-time) change of hour angle of the body. However, to obtain latitude accurately in this way it is necessary to determine

$$\frac{dh}{dt}$$

to an accuracy of perhaps four decimal places, and Z to an accuracy of perhaps one minute of arc. Two possible methods of obtaining

$$\frac{dh}{dt}$$

are given below, but present instrument limitations do not permit measurement of azimuth to the required accuracy. However, the cosecant of 90° is unity, so that if the observation is made when the celestial body is on the prime vertical, the formula becomes

$$\cos L = \frac{dh}{dt}. \tag{2}$$

Relatively little error is introduced if the body is within 1° of the prime vertical. The determination of position consists of the following steps:

1. Observe the altitude (h) and rate of change of altitude

$$\left(\frac{dh}{dt}\right)$$

when the celestial body is within 1° of the prime vertical.

2. Compute latitude (L) by formula (2).

3. Determine longitude by any standard method, such as Pub. No. 214 or other line of position method, or by time sight (art. 2106).

Perpendicular lines of position. The great circle through the zenith and the celestial body (the vertical circle or azimuth line) furnishes an azimuth line of position that can be established if rate of change of altitude can be accurately determined. This line is perpendicular to the circle of equal altitude and therefore nearly perpendicular to the line of position determined in the usual manner. The intersection of the two lines is the position of the observer. The method involves the following steps:

1. Observe the altitude (h) and rate of change of altitude

$$\left(\frac{dh}{dt}\right).$$

2. Compute the direction of the great circle through the zenith and the celestial body (the vertical circle) at the point where the great circle crosses the celestial equator. This is the complement of the latitude of the vertex and so can be found from a modification of formula (2), which gives the latitude of the vertex:

$$\sin Z_o = \frac{dh}{dt}. \tag{3}$$

3. Compute the longitude (λ_o) at which the vertical circle crosses the celestial equator, using the formula

$$\sin (\lambda_o \sim \lambda_b) = \tan Z_o \tan d. \tag{4}$$

The value λ_b is the longitude of the geographical position of the celestial body.

4. Solve for the latitude (L) at which the azimuth line of position crosses the meridian of the dead reckoning position, or for longitude (λ) at which the line crosses the parallel of latitude of the dead reckoning position, using one of the following formulas:

$$\tan L = \cot Z_o \sin (\lambda_o \sim \lambda_{DR}) \tag{5}$$

or

$$\sin (\lambda_o \sim \lambda) = \tan Z_o \tan L_{DR} \tag{6}$$

in which L_{DR} and λ_{DR} are the DR latitude and longitude, respectively. Any assumed position in the vicinity can be used in place of the DR. In general, it is preferable to use (5) if azimuth angle is between 45° and 135°, and (6) if it is outside these limits.

5. Solve for the direction (Z) of the azimuth line of position at the point determined in step (4), using the formula

$$\sin Z = \sin Z_o \sec L. \tag{7}$$

If the DR position or the AP is near the actual position, the azimuth can be considered the same at both without appreciable error.

6. Plot the azimuth line of position through the point found in step (4), in the direction found in step (5).

7. Compute *a* and Zn by any method and plot the resulting line of position. The intersection of the two lines of position is the fix.

Latitude and longitude by computation. This method is independent of a dead reckoning position, and requires no plotting. It is free from limitations except that observations near meridian transit should be avoided. At this time the rate of change of altitude decreases to zero and then reverses, introducing a possible error. The steps by this method are:

1. Observe the altitude (h) and rate of change of altitude

$$\left(\frac{dh}{dt}\right).$$

2. Compute Z_0, using formula (3).

3. Compute the latitude (L) of the observer by the formula

$$\sin L = \cos Z_0 \cos\left[h \pm \sin^{-1}\left(\frac{\sin d}{\cos Z_0}\right)\right]. \qquad (8)$$

In the solution of this equation, the angle whose sine is

$$\frac{\sin d}{\cos Z_0}$$

is added to or subtracted from h. The cosine of this angle is then multiplied by cos Z_0, and the result is the sine of the latitude of the observer. The sign is positive (+) unless L is greater than d and has the same name, when it is negative (−). However, if d is of the same name and greater, the angle to be added may be greater than 90°.

4. Compute the meridian angle of the observer by the formula

$$\sin t = \sin Z_0 \cos h \sec d \sec L. \qquad (9)$$

5. Determine GHA for the time of observation.

6. Convert t to LHA, and compute longitude (λ) by the formula

$$\lambda = GHA - LHA. \qquad (10)$$

If λ is greater than 180°, subtract it from 360° and label it E (east).

Formulas (8) and (10) yield a position on the circle of equal altitude regardless of the value of Z_0 used. The correct position is given only if the correct value of Z_0 is used.

Any of the three methods requires determination of

$$\frac{dh}{dt}.$$

Two methods are proposed:

In the first, the time needed for the sun (or moon) to change altitude an amount equal to its own diameter is measured. If the body is rising, the upper limb of the reflected image is brought a short distance below the horizon. As it makes contact with the horizon, a stopwatch is started. When the lower limb makes contact with the horizon (usually between 127.8 seconds, the minimum for a stationary observer, and ten minutes after the first contact) the watch is stopped, and the time is read to the nearest tenth of a second, if possible. If the body is setting, the lower limb of the reflected image is brought a short distance above the horizon and the watch started when the lower limb makes contact and stopped when the upper limb makes contact with the horizon. At sunrise or sunset no sextant is needed. Any lag in starting or stopping the watch will not affect the result if it is the same at both ends of the period. The diameter of the body, in minutes of arc, divided by one-fourth the number of seconds is

$$\frac{dh}{dt}.$$

Since semidiameter is tabulated, the most convenient procedure for determining

$$\frac{dh}{dt}$$

is probably to solve the equation

$$\frac{dh}{dt}=\frac{8\ SD}{T},$$

where SD is the semidiameter of the body in minutes and T is the time interval in seconds. The semidiameter is given to the nearest 0.′1 in the *Nautical Almanac.* More accurate results will be obtained if the value is taken from the *Astronomical Almanac,* where semidiameter is given to the nearest 0.″01.

The motion of the observer introduces an error which can be corrected as follows: multiply *half* the run of the vessel between upper and lower limb contacts, expressed in nautical miles, by the cosine of the angle between the course of the vessel and the azimuth of the celestial body at the mid time of observation. If this angle is *less* than 90°, the correction is *added* to the tabulated semidiameter if the body is setting, and *subtracted* if it is rising. If the angle is *greater* than 90°, the correction is *added* if the body is rising and *subtracted* if it is setting.

Some practice may be needed to obtain an accurate measurement of the time interval. This practice might be obtained by making a number of observations at a known position and comparing these with values obtained by computation, using the formula

$$T=8\ SD \cos h \sec d \sec L \csc t,$$

using Hc for h.

The time of an observation is at the middle of the interval between contacts. In correcting hs, the reading of the sextant, to obtain Ho, omit the correction for semidiameter. This might be done by correcting in the usual manner, with an additional correction equal to the semidiameter. The additional correction is negative (−) if the lower limb correction is applied, and positive (+) if the upper limb correction is applied. Another way is to apply neither the lower nor upper limb correction, but a value equal to the algebraic average of both.

The second method of determining

$$\frac{dh}{dt}$$

is given as the more accurate of the two.

It consists of observing three altitudes of the celestial body at exactly equal intervals of from 15 to 30 minutes. A shorter interval may result in too great an error in rate, while a longer one increases the time without advantage. If h_1, h_2, and h_3 are the three altitudes and t_1 and t_3 are the meridian angles at the times of the first and third observations, respectively,

$$\frac{dh}{dt}$$

can be computed by means of the formula

$$\frac{dh}{dt} = \sin \tfrac{1}{2}(h_1 - h_3) \cos \tfrac{1}{2}(h_1 + h_3) \csc \tfrac{1}{2}(t_1 - t_3) \sec h_2.$$

If difficulty is experienced in making an accurate observation at a given time, better results might be obtained by computing the time for the third observation, by adding the interval between the first two observations to the time of the second observation, and then making several observations starting shortly before the computed time. These can then be plotted on cross-section paper with altitude as one coordinate and time as the other. The altitude indicated by the intersection of the line representing the required time and a line faired through the plotted points is used as the third altitude. A similar procedure might increase the accuracy of the first two observations. A quicker but less accurate way of determining the third altitude is to take one observation shortly before the required time and another shortly after it, and interpolating to find the altitude at the required time. Another variation is to take an altitude at *about* the required time and adjust the second altitude to the corresponding value midway between the first and third observations, using the mean value found by interpolating from the first or third observation and extrapolating (art. 207, vol. II) from the other. The time and altitude are those of the second observation.

This method assumes no change of declination between observations, and no change in the position of the observer. When the observer is not stationary, a correction is applied to h_1 and h_3 to convert them to the equivalent values at the position of the second observation. Assuming constant course and speed, this correction in minutes of arc is equal to the vessel's run between consecutive observations multiplied by the cosine of the angle between the course of the vessel and the average azimuth of the body. If the angle is *less* than 90°, the correction is *added* to h_1 and *subtracted* from h_3. If the angle is *greater* than 90°, the correction is *subtracted* from h_1 and *added* to h_3.

A possible variation of either method of determining

$$\frac{dh}{dt}$$

would be to make a comparatively large number of observations (10 to 15) at short intervals and plot the altitudes versus time on cross-section paper. A point near each end of the line faired through the plotted points would then be corrected for the run of the vessel, as in the second method. Two points might then be selected, one near each end of the altitude-time line. The change in altitude, in minutes, divided by one-fourth the number of seconds between the two points is

$$\frac{dh}{dt}.$$

If preferred, three points might be selected at equal intervals and the formula of the second method used.

Rate determined by two individual observations a few minutes apart would not be sufficiently accurate for practical navigation.

None of the methods employing rate of change of altitude have proved popular, probably because of the difficulty of obtaining an accurate value of

$$\frac{dh}{dt}.$$

The use of azimuth and rate of change of azimuth, altitude and rate of change of azimuth, or azimuth and rate of change of altitude have been even less attractive because of the even greater difficulty of obtaining accurate measurements of azimuth or rate of change of azimuth. With the further development of automatic devices for continuously measuring altitude or azimuth, with allowance for motion of the observer, such methods might prove more attractive.

2118. Use of unique situations.—Various unique situations might be used for determining position or a line of position. As a general rule these have not been attractive because they could be used only when the conditions were met. As an example, if a celestial body of known coordinates were known to be in the zenith, the declination of the body would be the same as the latitude of the observer. His longitude would be the same as GHA of the body (360° − GHA in east longitude).

Near the geographical poles, the poles can be used as the assumed position. Here the declination of the body is the same as the computed altitude, and GHA replaces azimuth.

Meridian altitudes (art. 2103) and latitude by Polaris (art. 2105) are examples of methods depending upon unique situations. These have both been used extensively, but are decreasing in popularity because of their reliance upon unique conditions, without adequately compensating advantages.

Shchetkin. In 1899 N. O. Shchetkin proposed a method of computing latitude and meridian angle from measurement of the times at which two or more pairs of stars have the same altitude. Each star pair would provide, in effect, a single great-circle line of position. Variations of the method were proposed by Zinger, Pewzow, and W. W. Kawraisky, a Russian. The necessary tables for latitude 60°N to 80°N were published by the Astronomical Institute of Russia in 1936. A similar method was prepared by Simon Swahn in 1943.

Collins. In 1946 Oliver C. Collins, an astronomer at the University of Nebraska, proposed a variation of the method of Shchetkin, and extended it to include observations when two celestial bodies have the same azimuth.

McKee. In 1951 Lieutenant Merlin A. McKee, USMS, proposed a graphical solution of the same-altitude method of Collins.

Pierce. About 1951 Rear Admiral M. R. Pierce, USN (Ret.), suggested a method of establishing a line of position perpendicular to the course line when the altitude of a celestial body is observed at the moment it crosses the great circle through the observer and his destination.

2119. Graphical and mechanical solutions.—All of the methods described above require tables, either for a mathematical solution or to extract computed values of altitude and azimuth. The total number of possible tabular solutions must be very great. The number of graphical and mechanical solutions is almost endless. The ones selected for mention below are representative of the types that have been prepared or made available.

Graphical solutions are almost as old as tabular ones, having existed at least since 1790, when *Margetts' Horary Tables* appeared in graphical form. These were intended "for shewing by Inspection the Apparent Diurnal Motion of the Sun, Moon, and Stars, the Latitude of a Ship and the Azimuth, Time, or Altitude corresponding with any Celestial Object." They were intended primarily for use with the longitude method of laying down a line of position.

In general, graphical and mechanical solutions have not proved popular, for several reasons: First, they generally involve a small scale, yielding results of less accuracy than desired, even with careful work. Second, some of the methods must be used as a whole, and cannot be divided into parts to increase the scale. Third, such methods usually do not provide a record of the solution, and it is often difficult to check the results. Fourth, solutions requiring instruments are subject to errors due to lack of proper adjustment or mechanical damage which may not be apparent. Fifth, the required diagrams or instruments may be quite bulky, requiring considerable space for stowage and manipulation. Finally, in some cases the necessary instruments are expensive.

2120. Altitude and azimuth angle by graph.—One type of graphical solution is by means of a diagram that solves an equation.

d'Ocagne. Typical of such diagrams is that prepared by Maurice d'Ocagne, a Frenchman. Both altitude and azimuth angle can be found by means of this diagram, which is based upon the following formulas:

$$\text{hav } z = \text{hav } (L-d) + \{\text{hav } [180^\circ - (L+d)] - \text{hav } (L-d)\} \text{ hav } t,$$
$$\text{hav } (90^\circ \pm d) = \text{hav } (L-h) + \{\text{hav } [180^\circ - (L+h)] - \text{hav } (L-h)\} \text{ hav } Z,$$

in which $z = 90^\circ - h$.

The sides of a square are divided according to the haversines of angles, from 0° to 180°, and the corresponding graduations of opposite sides are connected with straight lines, forming a diagram as shown in figure 2120a. The graduations on the two sides run in opposite directions. To find the zenith distance, locate the value corresponding to (L−d) along the left of the diagram, and the value corresponding to (L+d) along the right of the diagram. Draw a straight line through these points. Locate the intersection of this line with the vertical line corresponding to meridian angle. A horizontal line from this intersection to the left edge indicates the zenith distance.

To find azimuth angle, draw a straight line between (L−h) at the left and (L+h) at the right. Locate the intersection of this line and the *horizontal* line corresponding to (90°−d). A *vertical* line from this intersection to the top of the diagram indicates the azimuth angle.

If the altitude, latitude, and declination are known, the first solution can be made in reverse for meridian angle, for a longitude method solution.

The diagram was first published in 1899 in *Traité de Nomographie* by d'Ocagne. Similar diagrams have since been published under the name *Spherical Triangle Nomogram* by Wimperis, and under the title *Altitude, Azimuth, and Hour Angle Diagram* by Littlehales in 1906, and by the U. S. Navy Hydrographic Office in 1917.

Favé and Rollet de l'Isle.—If a perpendicular is dropped from the celestial body to the celestial meridian, a diagram can be prepared to solve the basic formulas given in article 2111, or others derived from these. Such a diagram is shown in

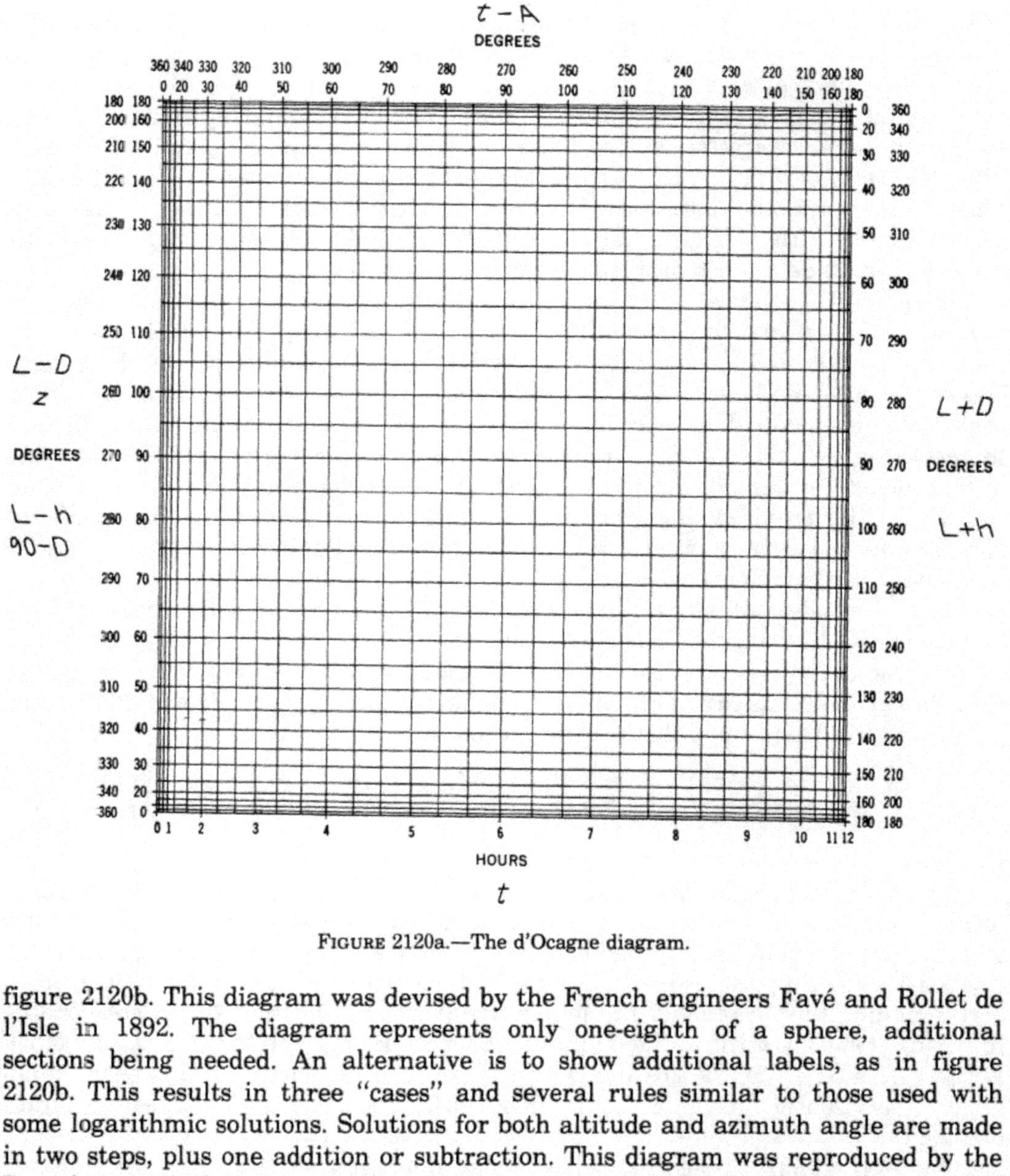

FIGURE 2120a.—The d'Ocagne diagram.

figure 2120b. This diagram was devised by the French engineers Favé and Rollet de l'Isle in 1892. The diagram represents only one-eighth of a sphere, additional sections being needed. An alternative is to show additional labels, as in figure 2120b. This results in three "cases" and several rules similar to those used with some logarithmic solutions. Solutions for both altitude and azimuth angle are made in two steps, plus one addition or subtraction. This diagram was reproduced by the Frenchman M. E. Pereire in 1894 and by another Frenchman, P. Constan, in 1906 as a method of finding azimuth

Jernæs.—In 1953 Leiv Jernæs, a Norwegian, invented a device he called a "Nauticator," which consists of various scales in a semicircle with radial scales on a plastic arm pivoted at the center of curvature of the semicircle. The device is used with a pair of dividers to solve various problems of spherical trigonometry to an accuracy of about 15′.

Bertin. In 1955 Rev. Maurice Bertin, a Frenchman, devised a graphical solution for the longitude method, using the formulas:

$$\tan^2 \tfrac{1}{2}\, t = \tan \tfrac{1}{2}\,(90^\circ - \alpha) \tan \tfrac{1}{2}\,(90^\circ - \beta) \tag{1}$$

and

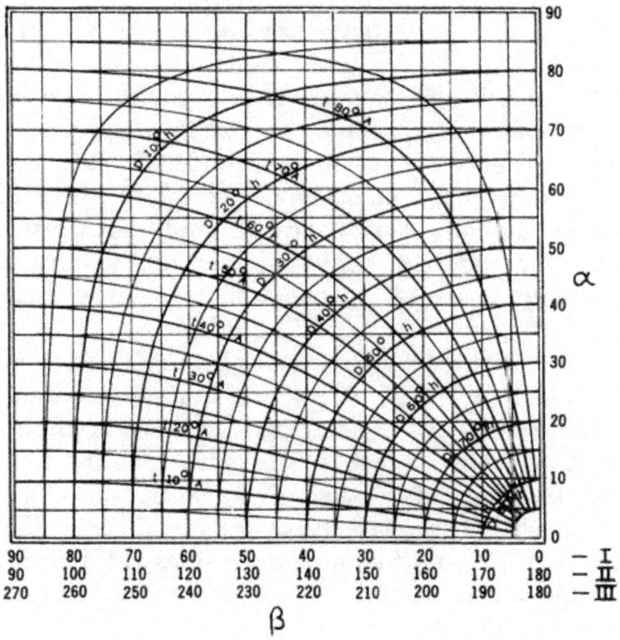

$\beta' = \beta + (90° - L)$

Case 1. L and D same Name —t<90°
Read β on scale II
Azimuth from upper pole, E or W as star is E or W of meridian

Case 2. L and D same name —t>90°
Read (180° —t) instead of t
Read β on scale I
Azimuth from upper pole, E or W as star is E or W of meridian

Case 3. L and D opposite names
Read β on scale I
Azimuth from lower pole, E or W as star is E or W of meridian

FIGURE 2120b.—The Favé diagram.

$$\tan^2 \tfrac{1}{2} Z = \frac{\tan \tfrac{1}{2} (90° - \alpha)}{\tan \tfrac{1}{2} (90° - 2\beta)} \qquad (2)$$

in which

$$\tan \tfrac{1}{2} \alpha = \tan \tfrac{1}{2} (h + d) \tan \tfrac{1}{2} (90° - L)$$

and

$$\tan \tfrac{1}{2} \beta = \frac{\tan \tfrac{1}{2} (h - d)}{\tan \tfrac{1}{2} (90° - L)}.$$

The diagram consists of three families of straight lines, one vertical, one horizontal, and the third at an angle of 45° to the others. The accuracy depends upon the scale of the diagram, but a large one is needed for navigational accuracy.

2121. Altitude and azimuth angle by computer.—Slide rules, like diagrams, have been devised to solve formulas. In the case of the navigational triangle, both suffer from the need for a scale that can be read to a subdivision at least as small as 1′. A number of such slide rules have been devised for use in reducing celestial observations.

Richer. In 1791 Jean Francisco Richer, a Frenchman, constructed a device composed of six arms, some hinged and some sliding, which won a prize offered by the Paris Academy of Science for a simple method of "clearing" lunar distances (art. 131) in the solution for longitude. The device solved a formula devised by the French mathematician Joseph Louis Lagrange, and was capable also of solving other problems involving spherical triangles, such as those related to time sight solution (art. 2106), computation of altitude, and great-circle sailing problems (art. 903).

Poor. A slide rule invented by Professor Charles L. Poor is shown in figure 2121a. This device, called the "Line of Position Computer," was designed to solve the cosine-haversine formula (art. 2109). Eight concentric circular scales are engraved on a metal disk about 15 inches in diameter. A plastic arm and circular sheet are pivoted at the center of the disk. The arm may be clamped to the plastic sheet. The seven outer scales are used in solving for altitude. The altitude scale is graduated at intervals of 10′, and further subdivisions can be estimated. The inner scale is used for determining azimuth angle. Several rules are needed, and the number of scales adds to the possibility of error.

Bygrave. A cylindrical slide rule was designed by the Englishman Bygrave to solve the navigational triangle divided by dropping a perpendicular from the celestial body to the celestial meridian (fig. 2111). This device, shown in figure 2121b, consists of three concentric tubes. The inner one has a spiral scale of logarithmic tangents, the middle one a spiral scale of logarithmic cosines, and the outer one a pointer for each scale. Solution is simple and relatively fast, but altered procedures are required if the azimuth angle is near 90°, or the meridian angle or declination is very small. The overall dimensions are about 2½ inches in diameter by nine inches long. An accuracy of about 1′ or 2′ is generally attainable.

Bertin. In 1955 Rev. Maurice Bertin devised an 18-inch slide rule to provide a solution of the longitude method to an accuracy of about 1°, using the formulas upon which his graphical solution (art. 2120) is based. He also devised a solution of the same formulas by a circular slide rule consisting essentially of two spirals. The inner one is on a disk 23 centimeters (9.2 inches) in diameter, and the outer one is on an annular ring 39 centimeters (15.6 inches) in outside diameter. The graduations are proportional to the log cotangents of half-angles. A window on a cover is provided with a radial line to serve as an index. Solution is facilitated if an approximation of the answer is known in advance. An accuracy of better than 3′ is claimed for this device. Still another solution proposed at the same time is by a computer consisting of a strip four centimeters (1.6 inches) wide and 12 meters (nearly 40 feet) long, wound on two rollers and engraved with three sets of graduations. An accuracy of better than 1′ is claimed, but several arithmetical steps are required.

LeSort. A computing device based upon solution of formulas for a divided navigational triangle was designed by Commander LeSort of the French Navy. Logarithmic scales are placed on eight films wound on rollers. The films operate in pairs so arranged that the two films of any pair can be locked together at any point. Alternate films carry log cosine and log tangent scales. Although an accuracy of about 0′.2 can be obtained, the method is comparatively long and has no apparent advantage over modern inspection tables.

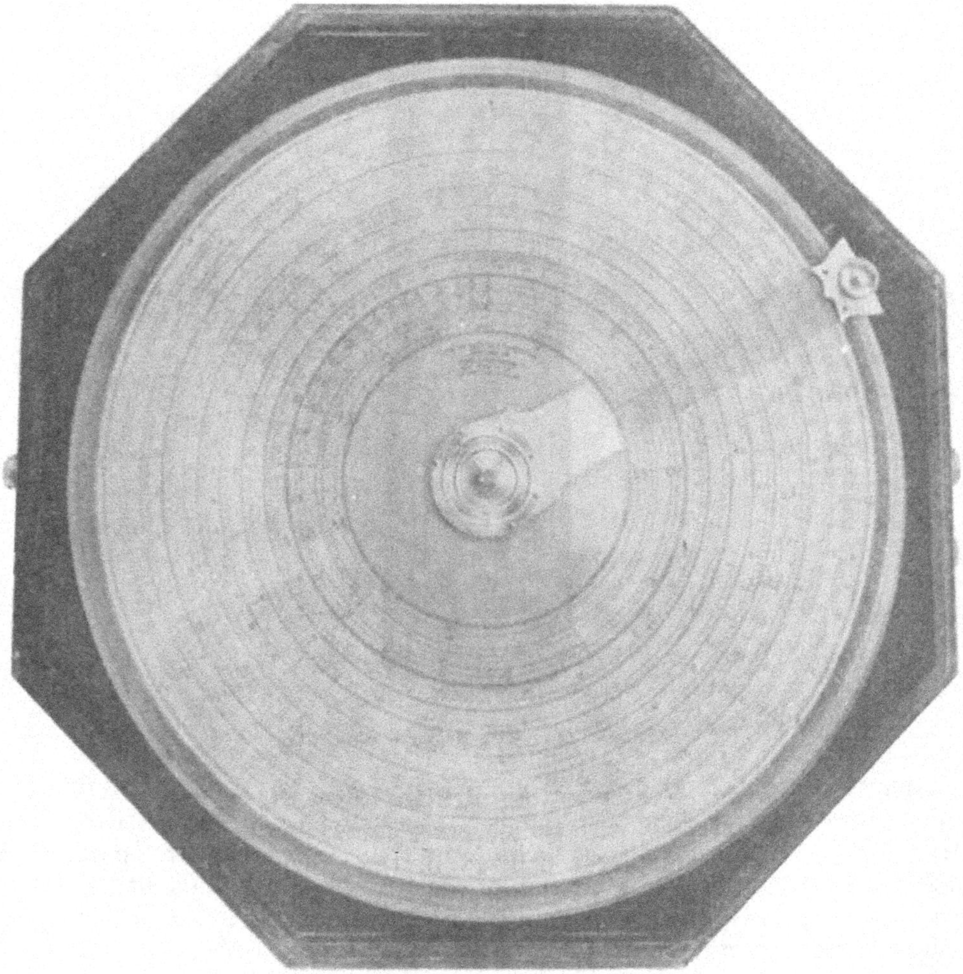

FIGURE 2121a.—The Poor Line of Position Computer.

Desk computers. Several desk-type computers have been designed to solve the navigational triangle.

2122. Altitude and azimuth angle by map projection.—If the observer were to move along his meridian to the nearer pole, and the navigational triangle were to move with him without its proportions being changed, his zenith would coincide with the pole, and the vertical circle would coincide with some celestial meridian. Zenith distance or altitude could be read directly. Since both great circles forming the azimuth angle would now coincide with celestial meridians, the azimuth angle could also be determined directly.

Littlehales. To accomplish this with a sphere, to a useful accuracy, would require a sphere of impractical size for use by the navigator. However, the solution can be made by means of a map projection. George Littlehales, of the U. S. Navy Hydrographic Office, used the stereographic projection (art. 318) and a 12-foot sphere for this purpose. The projection is divided into 368 overlapping sheets which, with a key diagram, are bound together. An accuracy of about 1′ or 2′ can be obtained by a rapid and simple process, but the volume is bulky and not particularly convenient.

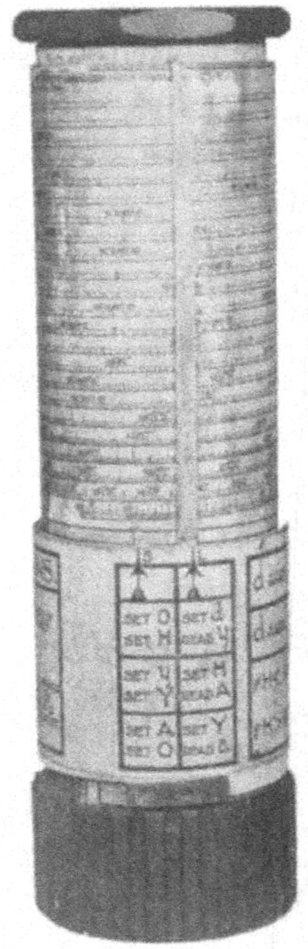

FIGURE 2121b.—The Bygrave slide rule.

Veater. Commander Veater of the British Royal Navy used the transverse Mercator projection (art. 309), with the observer's meridian as the fictitious equator.

Hyatt. A similar principle is utilized in the diagram on the plane of the celestial meridian (art. 1432). A mechanical device based upon this diagram can be made by drawing a hemisphere by equatorial orthographic (art. 319) or stereographic projection and pivoting at its center an identical hemisphere on transparent material. If the top hemisphere is rotated until the arc between poles of the two hemispheres is equal to the colatitude of the observer, the lines of one hemisphere represent coordinates of the celestial equator system (art. 1428), and those of the other, coordinates of the horizon system (art. 1430). Thus, if a body is located by meridian angle and declination on one set of lines, its altitude and azimuth angle can be read from the other set. If altitude and declination are used to locate the body, meridian angle can be read from the diagram. In the United States such a device, on both the orthographic and stereographic projections, has been prepared by Commander Delwyn Hyatt, USN, under the titles "Celestial Coordinator" and "Coordinate Transformer." It has also been produced in other countries, notably in Germany, France, and Russia, where, in addition to such a device, precision instruments based upon the same principle have been constructed. The scale of the German instrument is so small that an accuracy of about 5′ is about the best that can be expected. The Bastien-Morin (French) and Kavroyskyy (Russian) instruments might yield results of slightly greater accuracy. The plastic device, if carefully made, might be generally accurate to half a degree. It has been used primarily for instructional purposes.

Brown-Nassau. The Brown-Nassau "Navigational Computer" utilizes the same principle, but uses the azimuthal equidistant projection (art. 320) and increases the scale by limiting the device to an octant of the sphere, with separate solutions for altitude and azimuth, and various rules.

True. In his *Celestial Navigator for Aviators,* printed about 1943, Clarence H. True, of the Canal Zone, uses a single diagram on the orthographic projection. This serves as the basis for a solution by construction, claimed to be of sufficient accuracy for use in lifeboats. Various rules are needed.

Pierce. A series of diagrams on the azimuthal equidistant projection have been devised by Rear Admiral M. R. Pierce, USN (Ret.). The method is based upon the principle that angles are correctly represented at the point of tangency of this projection, and radial lines from this point represent great circles along which distances are represented by a uniform scale. A protractor is used for measuring the azimuth angle. Attached to the protractor is an arm with a linear scale graduated so that altitude can be read directly. The whole device is called a "Cadameter." The method is easy to use, and about as fast as modern inspection tables. With great care an accuracy of 1′ can be obtained. The method suffers from the need for a number of diagrams which are somewhat bulky and more susceptible to damage than a book.

2123. Latitude and longitude by diagram.—A number of graphical and mechanical solutions have been devised to yield latitude and longitude directly.

Beij. One proposed in 1924 by K. Hilding Beij, of the U. S. Bureau of Standards, was based upon the fact that latitude and local sidereal time are completely defined by the simultaneous altitudes of two celestial bodies whose declination and SHA are known. A page of his proposed diagrams is shown in figure 2123a, in which latitude is the abscissa, and LST is the ordinate. Position on the graph is located by the intersection of the curves representing the altitude of the two celestial bodies observed. The vertical line through the intersection indicates the latitude, and the horizontal line the LST. The difference between GST and LST is the longitude. If a timepiece keeping GST is available, not even an almanac is needed.

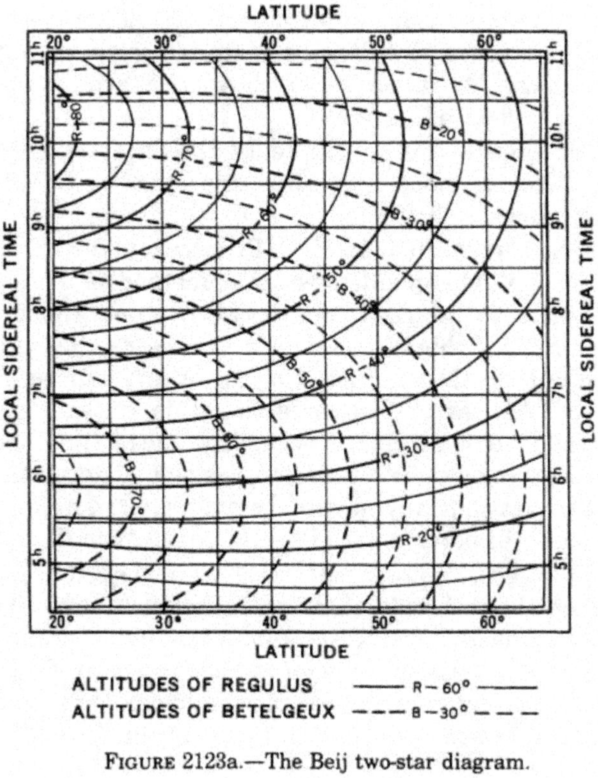

FIGURE 2123a.—The Beij two-star diagram.

The method is accurate, fast, and direct. The individual sheets can be drawn to any scale and cut to any size desired. For a large scale with sheets of a convenient size, a great many diagrams would be needed, but these might be bound together in convenient-size volumes, or placed on a tape wound around rollers, as originally proposed. A weakness of the method is the requirement for simultaneous observations. For nonsimultaneous observations a table might be provided to indicate the change in altitude during the interval between observations. Since the positions of the curves depend upon the declination and SHA of the body, the method is limited to celestial bodies whose coordinates are nearly constant, unless the curves are intended only for a particular time. Even for stars, the diagrams become out-of-date in a few years. The method is limited to the particular bodies for which curves are shown, although the number of curves need not be limited to two. This is a form of precomputation, since the computation is performed in locating the curves, rather than by the navigator. In a sense, it might be considered a graphical form of Pub. No. 249 (art. 2113).

Weems. If the Beij diagram is rotated through 90°, the parallels of latitude become horizontal, as customary on a chart. If they are spaced according to the Mercator projection, azimuth is indicated by the normal to a curve. This is the arrangement used by Captain P. V. H. Weems, USN (Ret.), in his *Star Altitude Curves,* the first volume of which was published in 1928. Later he added a third star, using a different color for each star, and included a correction for refraction at sea level. A separate volume is used for each 10° of latitude, and a correction is provided for precession of the equinoxes. Coverage extends from latitude 50°S to 70°N, with a separate volume for latitude 70°–90°N. The curves for 80°–90°N are on the polar stereographic projection. Any orthomorphic projection (art. 302) could be used at any latitude.

Lines representing observations at different times can be advanced or retired as on any chart of the same projection. In addition to the adjustment due to motion of the craft between observations, the lines are shifted right or left for the elapsed time between observations. An accuracy of about 1′ is attainable by interpolation between curves for each 10′ of altitude.

The star altitude curves are undoubtedly the most widely used of all the graphical and mechanical methods. Two-star curves similar to Weems' first edition were published in Germany in 1940.

Pritchard and Lamplough. In 1940 H. C. Pritchard and F. E. Lamplough, of the British Royal Aircraft Establishment, devised a method of reducing the work involved in the adjustment for elapsed time between observations. They placed the star altitude curves on film which is used in a projector called an **astrograph.** The curves are projected onto a Mercator plotting sheet and can be moved across it to allow for rotation of the earth. The adjustment is critical, the setting of the projector somewhat involved (a special "astrograph mean time" being needed), and a bulky and expensive projector is needed to prevent distortion. Because of these disadvantages and the fact that any advantage over short tabular methods is slight, the astrograph decreased in popularity following World War II.

Longley. In 1943 Flight Lieutenant C. D. N. Longley, RAF, suggested a "Star Computer" based upon the principle of the astrograph. A circular disk serving as a base plate would have a mean time scale around its circumference. Altitude curves of a limited number of stars would be printed on a template for each latitude. The circumference of each template would also carry a mean time scale. A radial cursor would aid in reading the device, which is set by means of the GMT at which LHA ♈

is 0° at some convenient longitude, the time of observation, and observed altitude. Longitude is determined within a 10° band, the ambiguity being resolved by means of the dead reckoning position. With a modification of the procedure, the device can be used with the altitude method.

Baker. As early as 1919 Commander T. Y. Baker, RN, prepared altitude curves and their orthogonals (normals) on transparent tape which is wound on rollers in the "Baker Navigating Machine" (fig. 2123b). The transparent tape is moved across a Mercator plotting sheet, being oriented by means of a time scale set with respect

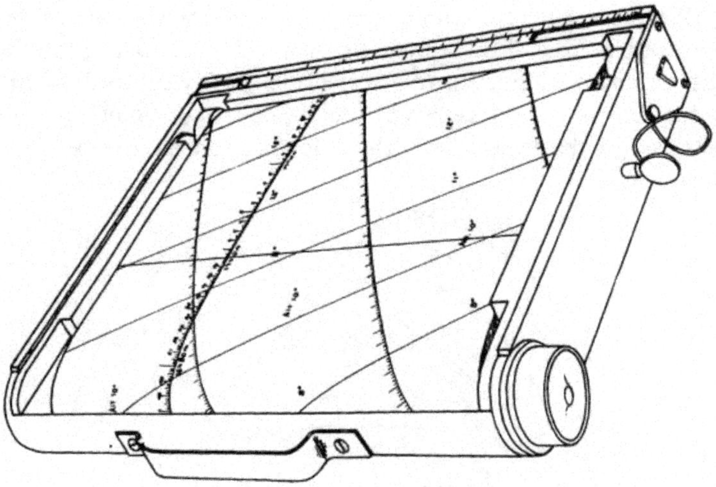

FIGURE 2123b.—The Baker Navigating Machine.

to a meridian. The line of position is transferred to the plotting sheet by means of carbon paper. A single tape has curves for several stars, and a separate tape for each 4° of declination from 24°N to 24°S permits use of the device with the sun and other bodies of the solar system. A rule attached to the machine (shown at the top of fig. 2123b) provides a correction for declination differing from that of the curves.

Weems. In 1955 Captain P. V. H. Weems, USN (Ret.), prepared a somewhat similar device called a "Polar Computer," using his star altitude curves.

Leick. In 1911 Dr. A. Leick, a German, prepared a diagram by which latitude and LST could be obtained by altitudes of Polaris and one other star. The diagram can be used for finding the correction to apply to the altitude of Polaris to determine the latitude, and then to find the LST in a second step.

Favé. In 1901 Favé devised a graphical solution based upon the Marcq St.-Hilaire principle (art. 2108). A chart on the stereographic projection (art. 318) is used. Tables of computed altitude and azimuth for the point of tangency are needed. The chart is on transparent material. An additional sheet has a set of arcs of circles, with a straight azimuth line drawn normal to them. The chart is placed over the curves with the straight azimuth line through the point of tangency and oriented in the direction of the celestial body. A large circle on the chart assists in this orientation. The chart is then moved along the azimuth line until the curve representing the computed altitude at the point of tangency is under that point. The curve representing the observed altitude is then correctly placed and a segment of it can be traced on the chart. However, due to chart distortion, error is introduced in this way. It can be removed by means of a nomogram which indicates the correct curve to use. A mark is placed on the chart at the intersection of the azimuth line and the curve representing the observed altitude. The chart is then moved along the azimuth line a second time until the correct curve is in place, and the arc is traced. This process is repeated for each celestial body observed. For stars, a one-page set of curves can be used instead of tables for determining altitude and azimuth at the point of tangency. Favé recommended use of five separate charts with points of tangency at 0°, 30°, 45°, 75°, and 90°, respectively. Each chart could be used as a plotting sheet for any longitude at the same latitude, requiring computed altitude and azimuth for only five places. Favé later put his method into instrumental form and used a special protractor and curved ruler.

Brill. In 1909 Dr. Alfred Brill, a German, invented a device based upon the same principle used by Favé, as shown in figure 2123c. In this device the plotting sheet is on the azimuthal equidistant projection (art. 320) and covers about 10° of latitude. Two sets of curves on separate sheets of tracing cloth are mounted below the plotting sheet. A handle turns the plotting sheet to the correct azimuth.

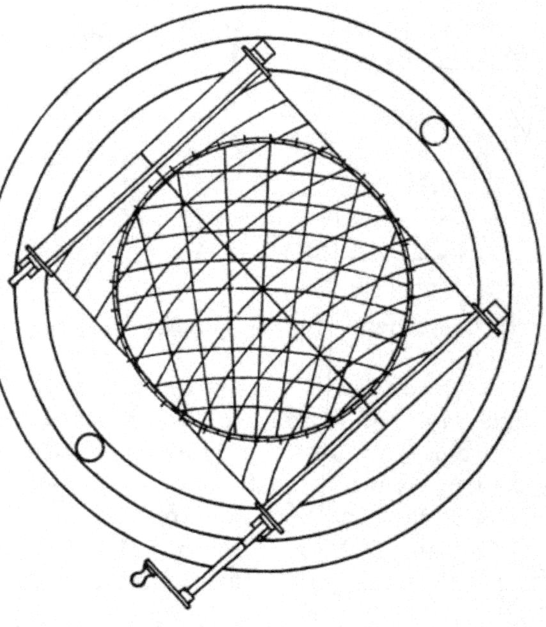

FIGURE 2123c.—The Brill device.

Voigt. The same principle used by Favé and Brill was used in the Voigt "Orion" instrument constructed in Germany in 1911. A plotting sheet on the azimuthal equidistant projection is engraved on aluminum. Each of the three plotting sheets, centered on latitudes 42°, 50°, and 55°, respectively, covers a spread of 10° of latitude. The line of position is drawn by means of a flexible ruler mounted on a bridge that can be clamped at any position over the plotting sheet. The curvature is controlled by means of gears, a scale being provided to indicate the correct value.

Vucetic. In 1921 a device called a "Toposcope" was prepared by Vucetic, a Frenchman. The device is identical with the Brill instrument except that a single set of curves is prepared and these are cut through the material as slots, and placed over the top of the plotting sheet.

Littlehales in 1918 suggested a method similar to that of Favé, but with a polyconic projection (art. 315).

Kahn. In 1928 Louis Kahn, a French naval architect, proposed that a set of navigational charts be prepared on the oblique Mercator projection (art. 310), a separate chart being provided for the great circle between various places on the earth. On each chart the small circles on the earth directly below the parallels of declination (that is, the daily paths of the geographical positions) of various navigational stars would be shown. These circles would be graduated in Greenwich sidereal time, so that the GP at any GST would be indicated. The distance from any assumed position to the GP at the instant of observation would be the zenith distance, and the direction of the line would be the azimuth. By comparing the observed zenith distance with that at the assumed position, the navigator could obtain the altitude difference, and plot the line of position. The common intersection of two or more such lines of position, advanced or retired to a common time if necessary, would define the position of the observer. The method would be limited to zenith distance within the range of the chart. A later version would produce greater accuracy, but with a little more trouble in making the measurements, by substituting the gnomonic projection (art. 317) for the oblique Mercator projection.

Dusinberre. In 1944 Lieutenant Commander H. W. Dusinberre, USN, suggested a method using star diagrams. A diagram for each 1° of latitude and 1° of LHA ♈ would be provided. Each diagram would consist of a series of radial lines extending in the directions of the prominent stars favorable for observation. The 22 stars of H.O. Pub. No. 218 (art. 2113) were suggested. Until changed by precession of the equinoxes (art. 1419) the common origin of these lines would represent a definite altitude for each star. The altitude at the next higher whole degree or half degree, adjusted for refraction, would be indicated by a tick on the appropriate azimuth line. After observation, a transparent plotting board would be properly oriented over the appropriate star diagram, using LHA ♈ and adjusting for the run between observations. The line of position would then be drawn at the correct point, perpendicular to the azimuth line, using the tick as a guide. An LHA ♈ computer was proposed for determining LHA ♈ at the time of each observation from a single LHA ♈ for a time near the start of each set of observations. When all lines of position were plotted, the fix would be transferred to the chart or plotting sheet.

2124. Solution by sphere.—Solution of a spherical triangle directly on a spherical surface, or by means of arcs representing great circles on the surface of an imaginary sphere, must have occurred to man quite early. Pictures of ancient navigators surrounded by their instruments and accessories invariably show a sphere. Solution by sphere is still suggested from time to time. Although this method is relatively simple and easy, the problem of scale is even more acute than in the graphical solutions.

Spherical methods can be classified in three groups: (1) those which solve the navigational triangle for a single line of position, (2) those which solve two or more observations for a fix, and (3) those which combine observation and solution for a fix.

The first group constructs the navigational triangle with arcs of great circles. Essentially, such a device consists of three arcs. The one representing the celestial meridian is usually fixed and a part of the frame. The base to which it is attached usually carries the azimuth scale. Movable arcs are provided for the vertical circle and the hour circle. If the latitude, meridian angle, and declination are properly set, the three arcs form the navigational triangle, and altitude and azimuth angle can be read from their scales. If altitude is used for constructing the triangle, meridian angle can be read from the instrument for a longitude solution.

Willis. A large number of teaching aids has been based upon this design or one of the many possible variations of it. Several precision instruments have been proposed or actually constructed. In 1932 such an instrument designed by Edward J. Willis, an American engineer, was constructed in Scotland. The marine version, weighing about 27 pounds, is graduated to 1′; and the aeronautical version, weighing between seven and eight pounds, is graduated to 5′. The longest dimension of either version is 11 inches.

Japanese Navy. During World War II, the Japanese Navy used an instrument virtually in the form described above. Results were accurate to approximately 1′.

McMillen. Of the various methods of determining a fix by sphere, the most obvious is that of providing an actual sphere as a plotting surface, with provision for striking arcs equal to the zenith distances, using the geographical positions of the celestial bodies as centers. In 1943 such a method was proposed by D. A. McMillen, a United States businessman in São Paulo, Brazil. His sphere, of a little more than 14 inches in diameter, had a scale of 8° (480 nautical miles) per inch along a great circle.

Hiltner. In 1945 Dr. W. F. Hiltner, a professor at Lehigh University, suggested a similar method using arcs of spheres and a billiard ball. This in effect, sets up two navigational triangles, locating the observer at the common zenith of both triangles. Simultaneous observations are needed.

U. S. Navy Training Device Center. About the same time, the Training Device Center of the U. S. Navy prepared a device called the "Sphereman Craft Positioner," combining the functions of the devices of both McMillen and Hiltner, and providing a plotting surface for dead reckoning. A line of position from a single observation can be drawn on the 17-inch aluminum globe, or the triangle of position from the observation of three stars can be mechanically set up. Provision is made for advancement or retirement of lines due to motion of the craft. The device was intended for training purposes.

Zerbee. In 1951 Louis J. Zerbee, of Bellfontaine, Ohio, proposed a device similar to that of Hiltner, but without the billiard ball. His instrument was called the "Zerbee Celestial Fix Finder." Like the Hiltner device, that of Zerbee makes no provision for nonsimultaneous observations (unless one of them is corrected to the value it would have if observed simultaneously with the other) or for a check by observation of additional bodies. Observations of bodies near the meridian or taken from high latitudes cannot be accommodated.

Combined sextant and computer. At least as early as 1895 an attempt was made to combine in a single instrument the functions of sextant and computer. Such instruments are fundamentally the same as those described above, except that they are set by alignment with one or more celestial bodies. If the instrument is level and accurately aligned with the meridian at the time of observation, the miniature sphere is oriented to the celestial sphere and the earth. If both the altitude and azimuth are used, a fix can be obtained by means of a single celestial body. If two bodies are observed simultaneously, accurate directional reference by compass is not needed.

The weakness of such methods is the need for a stable platform and either accurate directional reference or the need for observing two bodies simultaneously.

Beehler. In 1895 Lieutenant W. H. Beehler, USN, invented an instrument he called the "Solarometer," which was designed to furnish a position from observation of the sun. It requires a heavy cast iron base rigidly attached to the ship, with a bowl set in gimbals and filled with mercury. A float resting on the mercury carries the sighting instrument.

Hagner. In 1936 Fred Hagner, of San Antonio, Tex., invented a similar instrument he called the "Hagner Position Finder." This is a portable instrument operating on the same principle as the Solarometer, but obtaining the vertical by being hung from a suitable support, and therefore acting as a pendulum. This is reminiscent of the ancient astrolabe (art. 124).

Bedell. In 1953 A. L. Bedell, of St. Louis, Mo., proposed an instrument based upon simultaneous observation of two celestial bodies. The horizontal would be defined by spirit level.

2125. Azimuth.—Most of the methods described above provide for determination of both altitude and azimuth angle. Several provide only for altitude. The number of tables, diagrams, and devices providing solution for azimuth only is very great, approaching the number providing solution for both altitude and azimuth. The reason for this is that azimuth is needed for other purposes than sight reduction. One common use is for checking the compass. Since modern inspection tables have provided parallel columns of computed altitude and azimuth or azimuth angle, separate azimuth tables have decreased in popularity.

Azimuth can be determined by computation or by amplitudes (tab. 27, 28), as well as by azimuth table. The method of computation depends somewhat upon the information available. There are three general approaches:

Time azimuth is the name given an azimuth or azimuth angle computed with meridian angle (a function of time), latitude, and polar distance (or declination) as the known quantities.

Altitude azimuth is an azimuth or azimuth angle computed with altitude, latitude, and polar distance as the known quantities.

Time and altitude azimuth is computed with meridian angle, declination, and altitude as the known quantities, the most common formula being

$$\sin Z = \sin t \cos d \sec h.$$

The weakness of this method is that it does not indicate whether the celestial body is north or south of the prime vertical. Usually there is no question on this point, but if Z is near 90°, the quadrant may be in doubt. If this occurs, either the meridian angle or altitude when on the prime vertical can be determined from table 25 or by computation, using the formula

$$\cos t = \tan d \cot L$$

or

$$\sin h = \sin d \csc L.$$

If the altitude is *less*, or the meridian angle is *greater* than the value when the body is on the prime vertical, the azimuth angle should be labeled N or S to agree with the latitude. If h is *greater* or t is *less* than when on the prime vertical, Z should be given the contrary name (N or S) to that of the latitude.

Amplitudes. For checking the compass, a low altitude is desirable because it can be measured easiest and most accurately. If a celestial body is observed when its center is on the *celestial* horizon, the amplitude (art. 2029), which is the arc on the horizon between the prime vertical and the body, can be taken directly from table 27.

2126. Azimuth tables are numerous. Originally, they were designed primarily for use in determining compass error. Since the sun was the celestial body customarily used for this purpose, most of the tables were designed with the sun in mind.

Meridian angle is commonly expressed in terms of local apparent time, in intervals varying from about one to 20 minutes. In many of the tables, meridian angle increases *upward* from the bottom of the page.

The following are some of the principal azimuth tables:

Wakeley. The first known azimuth tables for use of the navigator were *The Regiment of the Pole Star* by Andrew Wakeley. These tables were part of the author's *The Mariner's Compass Rectified,* published in London in 1665. These tables show the "true hour of the day" at which the sun is at the various points of the compass.

Lynn Azimuth Tables, by Thomas Lynn (art. 2106), were published in 1829. This 364-page table gives azimuth angle computed by the haversine formula of article 2106.

Towson and Atherton. The *Tables to Facilitate the Practice of Great Circle Sailing,* by the Englishmen John Thomas Towson and J. W. Atherton, were designed primarily for great-circle sailing, but since they indicate the course, they were easily adapted to finding azimuth angle. They were published in England in 1847.

Burdwood. The *Tables of Sun's True Bearing or Azimuth,* by Staff Commander John Burdwood, RN, were first published in 1852, with additional parts being added in 1858, 1862, 1864, and 1866. Captain John E. Davis, RN, and Percy L. H. Davis, of the British Nautical Almanac Office, later added to the tables, making them complete for all values of altitude and for declination between 64°N and 64°S. These tables were standard in Great Britain for more than a century. They have now been largely replaced by H.D. 486 (Pub. No. 214) for mariners and A.P. 3270 (Pub. No. 249) for aviators. Burdwood used modifications of the time azimuth formula.

Labrosse. Azimuth tables by the Frenchman F. Labrosse were published in London in 1868, and later in Paris. In 275 pages this *Table des Azimuts du Soleil* covers latitudes from 61°N to 61°S, and declinations from 0° to 30°N or S. The following formula was used:

$$\cot Z = \frac{\tan d \cos L}{\sin t} - \sin L \cot t.$$

Fifteen editions had been published by 1920.

Shortrede. In 1869 Captain Robert Shortrede's *Azimuth and Hour Angle for Latitude and Declination* and *Tables for Finding Azimuth at Sea* were published in London.

John E. Davis. The first azimuth tables by Captain John E. Davis were published in 1875. These were published as an extension of the Burdwood tables.

Perrin. In Paris the *Nouvelles Tables Destinées à Abréger les Calculs Nautiques,* by Ensign de Vaisseau E. Perrin, French Navy, were published first in 1876. These consist of three tables of nine, seven, and six pages, respectively, providing elements for determination of azimuth by a short computation. Several editions were published.

Kortazzi, a Russian, produced a volume appropriately called *Modification des Tables d'Azimuth de Thomson* (art. 2106). These were published in Paris in 1880.

H.O. Pub. No. 66 (Schroeder and Wainwright), *Arctic Azimuth Tables.* Lieutenants Seaton Schroeder and Richard Wainwright, USN, prepared azimuth tables for use of the USS *Rodgers* in her search for the arctic steamer *Jeanette.* These were published in 1881. Azimuths to the nearest 1′ are given for each 10^m meridian angle between 4^h and 7^h, for latitudes between 70° and 88°, declination 0° to 23°, same name.

Decante. In 1882 Lieutenant de Vaisseau E. Decante, of the French Navy, prepared *Table du Cadran Solaire Azimutal,* which was published in 1904, in eight volumes for latitudes 1° to 66° and declinations 0° to 48°.

Pub. No. 260 (Schroeder and Southerland). The *Azimuths of the Sun* were prepared in 1882 by Lieutenant Seaton Schroeder, USN, and Master W. H. H. Southerland, USN. These are popularly called "Red Azimuth Tables," because of the red binding used for most printings. This designation distinguishes them from the "Blue Azimuth Tables" (Pub. No. 261). Azimuth angles are given to the nearest 1′, at 10^m intervals of local apparent time from "sunrise" to "sunset" (middle of the sun on the *celestial* horizon), with the LAT and the azimuth angle of these phenomena given at the bottom of each column. A separate table is given for each 1° of latitude from 0° to 70°. The first part of the book is a table for latitude 0°. The second part is devoted to tables of latitude and declination "same name." The third part gives "contrary name" tables. Declination entries are given at 1° intervals from 0° to 23°, with the approximate dates on which this is the declination of the sun. Extracts from these tables are given in volume II. Values are customarily taken by triple interpolation, using the right-hand "PM" LAT column as meridian angle. Pub. Nos. 260 and 261 are no longer published by DMAHTC.

Blackburne. The New Zealand nautical almanac for 1883 carried the 177-page "A and B" azimuth tables, by H. S. Blackburne. By 1911, after several modifications, these emerged as *"A, B, C" Tables for Azimuth, Great Circle Sailing, and Reduction to the Meridian.* The range of both the latitude and declination is from 90°N to 90°S.

Lecky. In 1892 Captain S. T. S. Lecky, an Englishman, modified the Blackburne tables and produced another set of "A, B, C" tables which have been widely used.

Ebsen. The *Azimut-Tabellen* of Julius Ebsen, published in Germany in 1896, uses the same formula as Labrosse, and is arranged like Pub. No. 260, except that azimuth angles are given to the nearest 0°1, and the time and azimuth angle of sunrise and sunset are given at the top of the table, in place of the dates of Pub. No. 260. In two volumes, coverage is for latitudes 72°N to 72°S, and declinations 0° to 29°. Tables are for same name only, contrary-name situations being handled by using the supplement of meridian angle, and using the supplement of the value taken from the table, as in Pub. No. 261.

Johnson. A *Combined Time and Altitude Azimuth Table* for latitudes and declinations from 0° to 80°, by A. C. Johnson of the British Royal Navy, was published in London in 1900. In the same year, his *Short, Accurate, and Comprehensive Altitude-Azimuth Tables* were published. This publication consists of three tables for computation of azimuth for each degree of latitude and altitude from 0° to 75°, and each degree of declination from 30°N to 30°S.

Zhdanko. The Russian *Tables of Azimuth of the Sun,* by M. Zhdanko, published in 1900, supplied computed azimuth angles for latitudes between 61° and 75°. These were later expanded by Yustchenko.

Percy L. H. Davis. In 1900 Percy L. H. Davis took over the work previously done by Burdwood and John E. Davis, continuing to improve and extend the tables.

Pub. No. 261, *Azimuths of Celestial Bodies,* published by the U.S. Navy Hydrographic Office in 1902, extend the Pub. No. 260 tables by providing information in similar form (but with meridian angle increasing *downward* on the page) for declinations 24° to 70°. These are popularly called "Blue Azimuth Tables," from their blue binding. Tables for "same name" only are given. If latitude and declination are of contrary name, the tables are entered with the supplement of the meridian angle. The value taken from the table is then the supplement of the azimuth angle, which is labeled N or S to agree with the latitude and E or W to agree with the meridian angle. Pub. No. 261 is no longer published by DMAHTC.

Symonds. The *Nautical Astronomy, with New Tables,* by W. P. Symonds, British Survey Commissioner, Bombay, includes azimuth tables. It was published in 1912.

Goodwin. *An Equatorial Azimuth-Table,* by H. B. Goodwin, was published in 1921.

Purey-Cust. *Azimuth by Logs,* by Admiral Sir H. E. Purey-Cust, RN, was published in England in 1929. It consists of a three-page table of the logarithms of the six principal trigonometric functions at 10′ intervals (5′ below 10°) for solution of the time azimuth and altitude azimuth formulas.

Yustchenko. In 1935 A. Yustchenko, a Russian, extended the Zhdanko tables to all latitudes, in the work entitled *Azimuty Svetil* (Azimuths of Celestial Bodies). For each 10° of latitude (5°, 15°, 25°, etc., to 85°) complete azimuth tables (to the nearest 0.°1) are given for each 1^m of meridian angle and each 30′ of declination from 0° to 30°. At the bottom of each page are given corrections for 1° of latitude. This value is multiplied by the number of degrees between the actual latitude and the latitude for which the table was computed.

Cugle. *Cugle's Two-Minute Azimuths,* by Charles H. Cugle, were printed in 1935 in two large volumes. Coverage is for latitude 0° to 65° and declination 0° to 23°. The arrangement is almost identical with that of Pub. No. 260, except that meridian angle increases *downward* on the page. The number of entries is multiplied by *five,* values being given for each 2^m of meridian angle.

Table 902. *Azimuts,* published in Paris in 1953, with the concurrence of the Marine Hydrographic Service, contains azimuth angles to the nearest 0.°1 for each whole degree of latitude from 70°N to 70°S, each whole degree of declination from 0° to 30°, and each 10^m of meridian angle. The arrangement is similar to that of Pub. No. 260, except that meridian angle increases *downward* on the page.

2127. Azimuth diagrams have appeared in various forms, in addition to the general graphical and mechanical solutions discussed above. A graphical solution is generally more acceptable for azimuth than for altitude, because the accuracy requirement for azimuth is usually less.

Godfrey. A graphical solution has been available at least since 1858 when the *Time Azimuth Diagram* of Hugh Godfrey was published in London.

Weir. The *Azimuth Diagram* devised by Captain Patrick Weir, of the British Merchant Navy, was published in London in 1890, and by the U.S. Navy Hydrographic Office in 1891, under the title *Time Azimuth Diagram.*

Molfino. In 1901 the *Nomograma degli Azimut del Sole* of Molfino was published.

Constan. In 1906 P. Constan's *Tables Graphiques d'Azimut* were published in Paris. This was a reproduction of the graph of Favé and Rollet de l'Isle (art. 2120).

Alessio. The *Diagrammi Altazimutali* of A. Alessio was published in 1908 in Italy.

Rust. In 1908 the diagram of Lieutenant Commander Armistead Rust, USN, (art. 2106) was published. This diagram was later used by Goodwin (art. 2106) and Weems (arts. 2106 and 2110), and in the Italian *Tavole H* (art. 2110).

Cornet. The *Graphique d'Azimut* of Cornet was published in 1927.

Romanovsky. About 1933 A. A. Romanovsky, a Russian, devised a simple nomogram for determining azimuth.

German Oberkommandos der Kriegsmarine. A large volume called *Azimutdiagramme,* containing sets of diagrams for each whole degree of latitude (2° beyond 80°) for all azimuth angles and for all altitudes to 80°, was published by the German Oberkommandos der Kriegsmarine in 1944.

Hugon. The azimuth diagram of Professor P. Hugon (art. 2109) was published in 1947.

Hilsenrath. About 1948 Joseph Hilsenrath, of the University of Maryland, produced a mechanical device for solving azimuth angle by the method of Weir's diagram.

2128. Summary.—The methods of sight reduction discussed in this chapter are undoubtedly only a small fraction of the number of methods that have been proposed. They are considered representative of the effort that has been made to reduce the work of the navigator. Individual preferences have largely dictated the use of the various methods. Presentation and description of a method have been important factors in the relative popularity of various methods.

There is no single "best" method for all circumstances and all navigators. The one which produces the desired results easiest and with least possibility of mistake is the one that should be selected. However, two practical precautions should be observed. First, one should be thoroughly familiar with the *limitations* or *weaknesses* of the method he selects. Second, a prudent navigator will never limit himself to a single method, particularly one requiring a special table that might some day be unavailable, or a device that is subject to mechanical damage or loss. The slight bending of an arc might be too insignificant to be noticed, yet might introduce intolerably large errors in the result. A wise practice is to memorize, or write on something always carried, fundamental formulas that can be used when no "special" tables are available.

CHAPTER XXII

IDENTIFICATION OF CELESTIAL BODIES

2201. Introduction.—A basic requirement of celestial navigation is the ability to identify the bodies observed. This is not difficult because relatively few celestial bodies are commonly used for navigation, and various aids are available to assist in their identification, as explained in this chapter.

Many navigators consider it a matter of professional pride to have a more extensive acquaintance with the heavens than required by the relatively simple demands of navigation.

2202. Bodies of the solar system.—No problem is encountered in the identification of the sun and moon. However, the planets can be mistaken for stars. A person working continually with the night sky recognizes a planet by its changing position among the relatively fixed stars. He identifies the planets by noting their positions relative to each other, the sun, the moon, and the stars. He knows that they remain within the narrow limits of the zodiac (art. 1420) but are in almost constant motion relative to the stars. The magnitude and color may be helpful. The information he needs is found in the *Nautical Almanac.* The "Planet Notes" near the front of that volume are particularly useful.

Sometimes the light from a planet seems steadier than that from a star. This is because fluctuation of the unsteady atmosphere causes **scintillation** or **twinkling** of a star, which has no measurable diameter with even the most powerful telescopes. The navigational planets are less susceptible to the twinkling because of the broader apparent area giving light.

Planets can also be identified by planet diagram (art. 2209), star finder (art. 2210), sky diagram (art. 2212), or by computation (art. 2213).

2203. Stars.—The average navigator regularly uses not more than perhaps 20 or 30 stars. The *Nautical Almanac* gives full navigational information on 19 first magnitude stars and 38 second magnitude stars, in addition to Polaris. Abbreviated information is given for 115 more. Additional stars are listed in *The Astronomical Almanac* and in various star catalogs. About 6,000 stars of the sixth magnitude or brighter (on the entire celestial sphere) are visible to the unaided eye on a clear, dark night.

Stars are designated by one or more of the following:

Name. Most names of stars, as now used, were given by the ancient Arabs and some by the Greeks or Romans. One of the stars of the *Nautical Almanac,* Nunki, was named by the Babylonians. Only a relatively few stars have names. Several of the stars on the daily pages of the almanacs had no name prior to the 1953 edition, and were given coined names so that all stars listed on the daily pages might have names. The pronunciation, meaning, and other information of general interest regarding Polaris and the 57 stars listed on the daily pages of the *Nautical Almanac* are given in appendix J.

Bayer's name. Most bright stars, including those with names, have been given a designation consisting of a Greek letter followed by the possessive form of the name of the constellation, as α *Cygni* (Deneb, the brightest star in the constellation

Cygnus, the swan). Roman letters are used when there are not enough Greek letters. Usually, the letters are assigned in order of brightness within the constellation, but in some cases the letters are assigned in another order, where it seems logical to do so. An example is the Big Dipper, where the letters are assigned in order from the outer rim of the bowl to the end of the handle. This system of star designation was suggested by John Bayer of Augsburg, Germany, in 1603. All of the 173 stars included in the list near the back of the *Nautical Almanac* are given by Bayer's name as well as regular name, where there is one.

Flamsteed's number. A similar system, accommodating more stars, numbers them in each constellation, from west to east, the order in which they cross the celestial meridian. An example is 95 *Leonis,* the 95th star in the constellation *Leo,* the lion. This system was suggested by John Flamsteed (1646–1719), who was the first British Astronomer Royal.

Catalog number. Stars are sometimes designated by the name of a star catalog and the number of the star as given in that catalog, as A. G. Washington 632. In these catalogs stars are listed in order from west to east, without regard to constellation, starting with the hour circle of the vernal equinox. This system is used primarily for dimmer stars having no other designation. Navigators seldom have occasion to use this system.

The ability to identify stars by position relative to each other is useful to the navigator. A tabulation of the relative positions of the 57 stars given on the daily pages of the *Nautical Almanac,* and Polaris, is given in appendix I. A star chart (fig. 2204) is helpful in locating these relationships and others which may be useful. This method is limited to periods of relatively clear, dark skies with little or no overcast. Stars can also be identified by the *Air Almanac* sky diagram (art. 2212), star finder (art. 2210), Pub. No. 249 (art. 2211), or by computation (art. 2213).

2204. Star charts are based upon the celestial equator system of coordinates, using declination and sidereal hour angle (or right ascension). The zenith of the observer is at the intersection of the parallel of declination equal to his latitude, and the hour circle coinciding with his celestial meridian. This hour circle has an SHA equal to $360° - \text{LHA}\ \Upsilon$ (or $\text{RA} = \text{LHA}\ \Upsilon$). The horizon is everywhere 90° from the zenith. A **star globe** is similar to a terrestrial sphere, but with stars (and often constellations) shown instead of geographical positions. Star globes are used by British navigators, but not customarily by Americans. The *Nautical Almanac* includes adequate instructions for using this device. On a star globe the celestial sphere is shown as it would appear to an observer *outside* the sphere. Constellations appear reversed. Star charts may show a similar view, but more often they are based upon the view from *inside* the sphere, as seen from the earth. On these charts, north is at the top, as with maps, but east is to the *left* and west to the *right.* The directions seem correct when the chart is held overhead, with the top toward the north, so that the relationship is similar to that in the sky. Any map projection (ch. III) can be used, but some are more suitable than others.

The *Nautical Almanac* has four star charts. The two principal ones are on the polar azimuthal equidistant projection (art. 320), one centered on each celestial pole. Each chart extends from its pole to declination 10° (same name as pole). Below each polar chart is an auxiliary chart on the Mercator projection, from 30°N to 30°S. On any of these charts, the zenith can be located as indicated above, to determine which stars are overhead. The horizon is 90° from the zenith. The charts can also be used to determine the location of a star relative to surrounding stars. The *Air Almanac* has a fold-in chart near the back, on the rectangular projection (art. 311).

STAR CHARTS

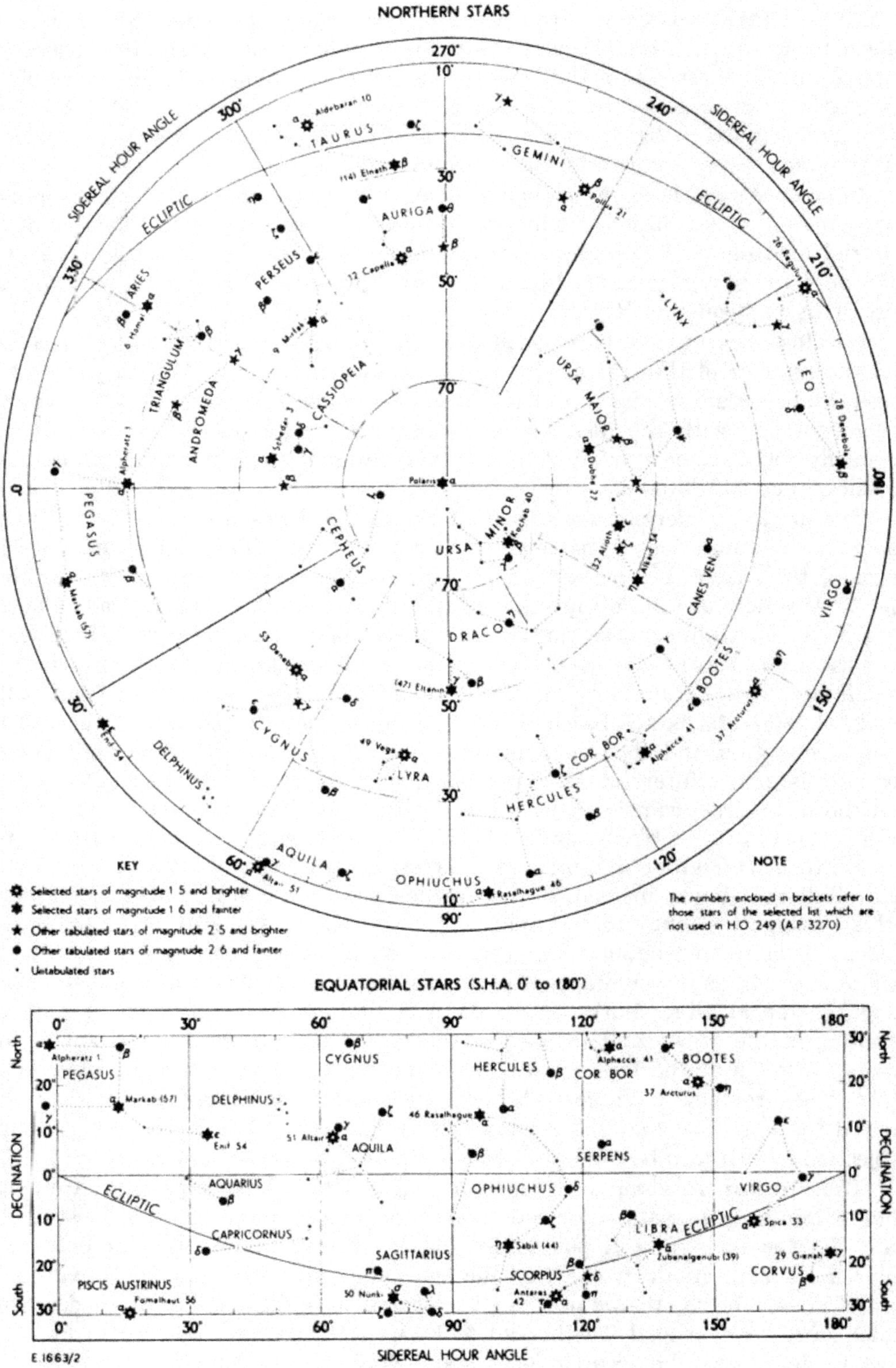

FIGURE 2204.—Star chart.

This projection is suitable for indicating the coordinates of the stars, but excessive distortion occurs in regions of high declination. The celestial poles are represented by the top and bottom horizontal *lines* the same length as the celestial equator. To locate the horizon on this chart, first locate the zenith as indicated above, and then locate the four cardinal points. The north and south points are 90° from the zenith, along the celestial meridian. The distance to the elevated pole (having the same name as the latitude) is equal to the colatitude of the observer. The remainder of the 90° (the latitude) is measured *from* the same pole, along the *lower branch* of the celestial meridian, 180° from the upper branch containing the zenith. The east and west points are on the celestial equator at the hour circle 90° east and west (or 90° and 270° in the same direction) from the celestial meridian. The horizon is a sine curve (fig. 140b, vol. II) through the four cardinal points. Directions on this projection are distorted.

The star charts shown in figures 2205–2208, on the transverse Mercator projection (art. 309), are designed to assist one in learning the stars listed on the daily pages of the *Nautical Almanac,* and Polaris. Each chart extends about 20° beyond each celestial pole, and about 60° (four hours) each side of the central hour circle (at the celestial equator). Therefore, they do not coincide exactly with that half of the celestial sphere above the horizon at any one time or place. The zenith, and hence the horizon, varies with the position of the observer on the earth, and also with the rotation of the earth (apparent rotation of the celestial sphere). The charts show all stars of fifth magnitude and brighter as they appear in the sky, but with some distortion toward the right and left edges.

The overprinted lines add certain information of use in locating the stars. Only Polaris and the 57 stars listed on the daily pages of the *Nautical Almanac* are named on the charts. The almanac star charts should be used for locating the additional stars given near the back of the *Nautical Almanac* and the *Air Almanac.* The broken lines connect stars of some of the more prominent constellations. The solid lines indicate the celestial equator and certain useful relationships among stars in different constellations. The celestial poles are marked by crosses, and labeled. By means of the celestial equator and the poles, one can locate his zenith approximately along the mid hour circle, when this coincides with his celestial meridian, as shown in the table below. At any time earlier than those shown in the table the zenith is to the *right* of center, and at a later time it is to the *left,* approximately one-quarter of the distance from the center to the outer edge (at the celestial equator) for each hour that the time differs from that shown. The stars in the vicinity of the North Pole can be seen in proper perspective by inverting the chart, so that the zenith of an observer in the Northern Hemisphere is *up* from the pole.

	Fig. 2205	*Fig. 2206*	*Fig. 2207*	*Fig. 2208*
Local sidereal time	0000	0600	1200	1800
LMT 1800	Dec. 21	Mar. 22	June 22	Sept. 21
LMT 2000	Nov. 21	Feb. 20	May 22	Aug. 21
LMT 2200	Oct. 21	Jan. 20	Apr. 22	July 22
LMT 0000	Sept. 22	Dec. 22	Mar. 23	June 22
LMT 0200	Aug. 22	Nov. 22	Feb. 21	May 23
LMT 0400	July 23	Oct. 22	Jan. 21	Apr. 22
LMT 0600	June 22	Sept. 21	Dec. 22	Mar. 23

2205. Stars in the vicinity of *Pegasus* (fig. 2205).—In autumn the evening sky has few first magnitude stars. Most of these are near the southern horizon of an observer in the latitudes of the United States. A relatively large number of second and third magnitude stars seem conspicuous, perhaps because of the small number of brighter stars. High in the southern sky three third magnitude stars and one second magnitude star form a square with sides nearly 15° of arc in length. This is *Pegasus,* the winged horse, although to many modern men it more nearly resembles a baseball diamond, complete with catcher, pitcher, batter, umpire, base umpire near second base, infield and outfield; although there does seem to be a large number of outfielders. One may even see the next batter, bat boy, and coach.

Only Markab at the southwestern corner (third base) and Alpheratz at the northeastern corner (first base) are listed on the daily pages of the *Nautical Almanac.* Alpheratz is part of the constellation *Andromeda,* the princess, extending in an arc toward the northeast and terminating at Mirfak in *Perseus,* legendary rescuer of *Andromeda.*

A line extending northward through the eastern side (first-second base line) of the square of *Pegasus* passes through the leading (western) star of M-shaped (or W-shaped) *Cassiopeia,* the legendary mother of the princess *Andromeda.* The only star of this constellation listed on the daily pages of the *Nautical Almanac* is Schedar, the second star from the leading one as the configuration circles the pole in a counterclockwise direction. If the line through the eastern side of the square of *Pegasus* is continued on toward the north, it leads to second magnitude Polaris, the North Star (less than 1° from the north celestial pole) and brightest star of *Ursa Minor,* the Little Bear. Kochab, a second magnitude star at the other end of the Little Dipper, is also listed in the almanacs. At this season the Big Dipper is low in the northern sky, below the celestial pole. A line extending from Kochab through Polaris leads to Mirfak, assisting in its identification when *Pegasus* and *Andromeda* are near or below the horizon.

Deneb, in *Cygnus,* the swan, and Vega are bright, first magnitude stars in the northwestern sky. They are discussed in article 2208. Capella, a bright star in the northeastern sky, is discussed in article 2206.

The line through the eastern side of the square of *Pegasus* (first-second base line) approximates the hour circle of the vernal equinox, shown at ♈ on the celestial equator to the south. The sun is at ♈ on or about March 21, when it crosses the celestial equator from south to north. If the line through the eastern side of *Pegasus* is extended southward and curved slightly toward the east, it leads to second magnitude Diphda. A longer and straighter line southward through the western side (home plate-third base line) of *Pegasus* leads to first magnitude Fomalhaut. A line extending northeasterly from Fomalhaut through Diphda leads to Menkar, a third magnitude star, but the brightest in its vicinity. Ankaa, Diphda, and Fomalhaut form an isosceles triangle, with the apex at Diphda. Ankaa is near or below the southern horizon of observers in latitudes of the United States. Four stars farther south than Ankaa may be visible when on the celestial meridian, just above the horizon of observers in latitudes of the extreme southern part of the United States. These are Acamar, Achernar, Al Na'ir, and Peacock. These stars, with each other and with Ankaa, Fomalhaut, and Diphda, form a series of triangles as shown in figure 2205. Almanac stars near the bottom of figure 2205 are discussed in succeeding articles.

Two other almanac stars can be located by their positions relative to *Pegasus.* These are Hamal in the constellation *Aries,* the ram, east of *Pegasus,* and Enif, west of the southern part of the square, identified as shown in figure 2205. The line leading to Hamal, if continued, leads to the *Pleiades,* not used by navigators for

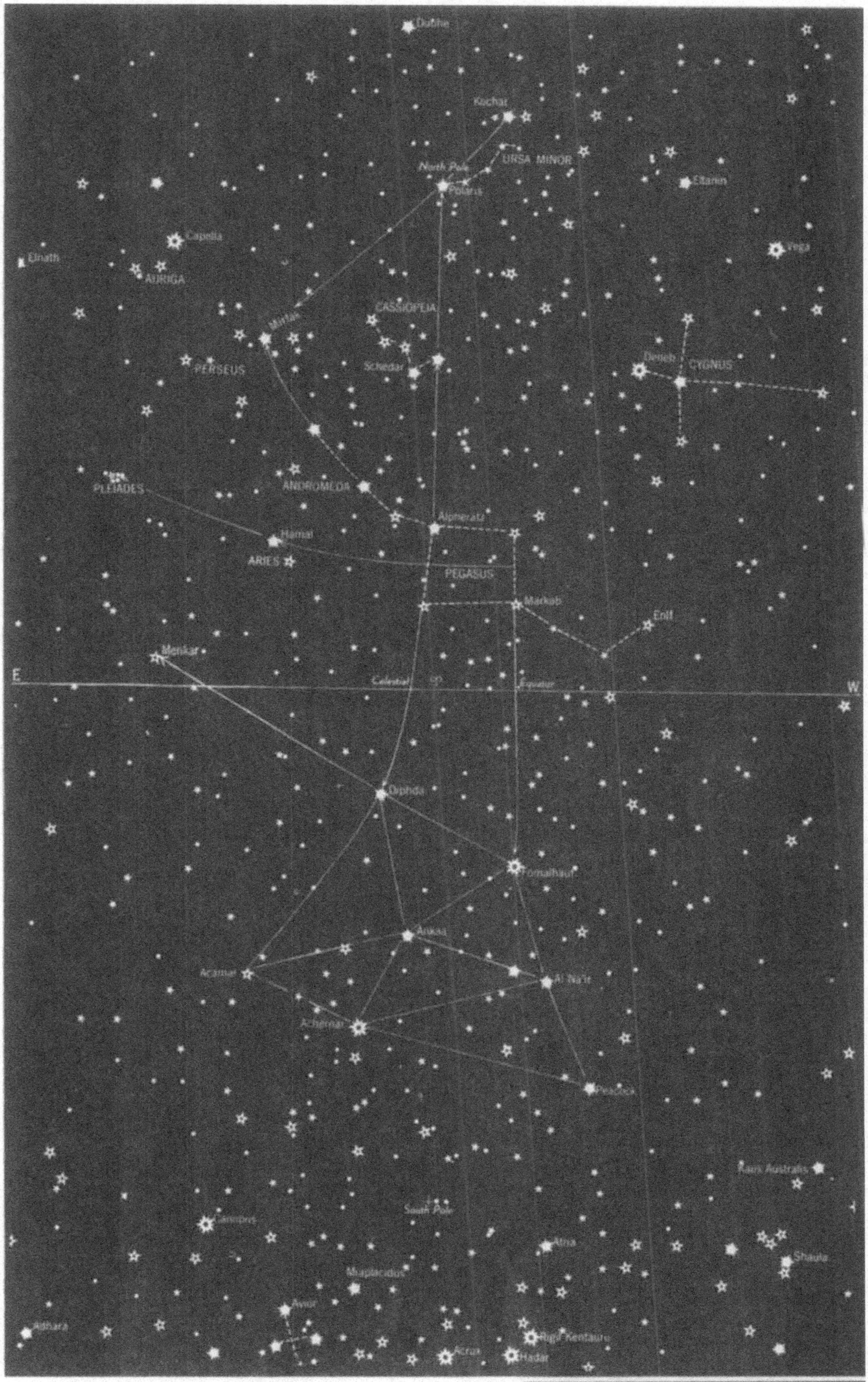

FIGURE 2205.—Stars in the vicinity of *Pegasus*.

Scale of magnitudes: 1st 2nd 3rd 4th 5th

celestial observations, but a prominent figure in the sky, heralding the approach of the many conspicuous stars of the winter evening sky, figure 2206.

2206. Stars in the vicinity of Orion (fig. 2206).—As *Pegasus* leaves the meridian and moves into the western sky, *Orion,* the mighty hunter, rises in the east. With the possible exception of the Big Dipper, no other configuration of stars in the entire sky is as well known as *Orion* and its immediate surroundings. In no other part are there so many first magnitude stars.

The belt of *Orion,* being nearly on the celestial equator, is visible by an observer in virtually any latitude, rising and setting almost on the prime vertical, and dividing equally its time above and below the horizon. Of the three second magnitude stars forming the belt, only Alnilam, the middle one, is listed on the daily pages of the *Nautical Almanac.*

Four conspicuous stars form a box around the belt. To the south is Rigel, one of the hottest and bluest of the stars, in contrast with relatively cool, red, variable Betelgeuse, at approximately an equal distance to the north. Bellatrix, bright for a second magnitude star but overshadowed by its more brilliant neighbors, is a few degrees west of Betelgeuse. Neither the second magnitude star forming the southeastern corner of the box, nor any star of the dagger, is listed on the daily pages of the *Nautical Almanac.*

A line extending eastward from the belt of *Orion* and curving toward the south leads to Sirius, the brightest star in the entire heavens, having a magnitude of (−) 1.6. Only Mars and Jupiter at or near their greatest brilliance, and the sun, moon, and Venus are brighter than Sirius. This is part of the constellation *Canis Major,* the large hunting dog of *Orion.* Starting at Sirius a curved line extends northward through first magnitude Procyon, in *Canis Minor,* the small hunting dog; first magnitude Pollux and second magnitude Castor (not listed on the daily pages of the *Nautical Almanac*), the twins of *Gemini;* brilliant Capella in *Auriga,* the charioteer; and back down to first magnitude Aldebaran, the follower, which trails the *Pleiades,* the seven sisters. Aldebaran, brightest star in the head of *Taurus,* the bull, may also be found by a curved line extending northwestward from the belt of *Orion.* The V-shaped figure forming the outline of the head and horns of *Taurus* points toward third magnitude Menkar. At the summer solstice the sun is between Pollux and Aldebaran.

If the curved line from *Orion's* belt southeastward to Sirius is continued, it leads to a conspicuous, small, nearly equilateral triangle of three bright second magnitude stars of nearly equal brilliancy. This is part of *Canis Major.* Only Adhara, the westernmost of the three stars, is listed on the daily pages of the *Nautical Almanac.* Continuing on with somewhat less curvature, the line leads to Canopus, second brightest star in the heavens and one of the two stars having a negative magnitude (−0.9). With Suhail and Miaplacidus, Canopus forms a large, equilateral triangle which partly encloses the false Southern Cross. The brightest star within this triangle is Avior, near its center. Canopus is also at one apex of a triangle formed with Adhara to the north and Suhail to the east, another triangle with Acamar to the west and Achernar to the southwest, and another with Achernar and Miaplacidus. Acamar, Achernar, and Ankaa form still another triangle toward the west. Because of chart distortion, these triangles do not appear in the sky in exactly the relationship shown on the star chart. Other daily-page almanac stars near the bottom of figure 2206 are discussed in succeeding articles.

During the winter evening sky the Big Dipper is east of Polaris, the Little Dipper is nearly below it, and *Cassiopeia* is west of it. Mirfak is northwest of

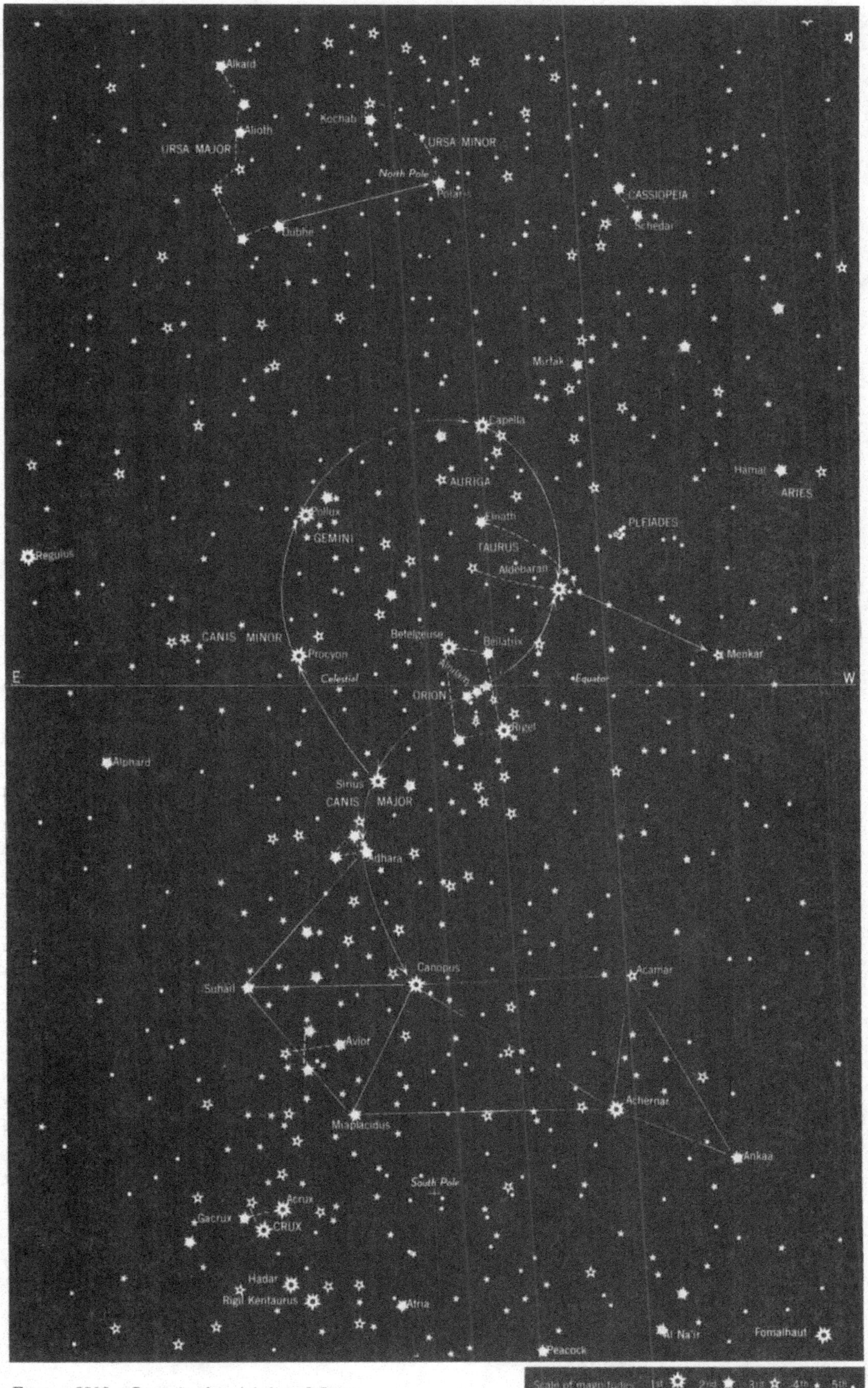

FIGURE 2206.—Stars in the vicinity of *Orion*.

Capella, nearly midway between it and *Cassiopeia*. Hamal is in the western sky. Regulus and Alphard are low in the eastern sky, heralding the approach of the configurations associated with the evening skies of spring.

2207. Stars in the vicinity of *Ursa Major* (fig. 2207).—As if to enhance the splendor of the sky in the vicinity of *Orion,* the region toward the east, like that toward the west, has few bright stars, except in the vicinity of the south celestial pole. However, as *Orion* sets in the west, leaving Capella and Pollux in the northwestern sky, a number of good navigational stars move into favorable positions for observation.

The Big Dipper, part of *Ursa Major,* the great bear, appears prominently *above* the north celestial pole, directly opposite *Cassiopeia* (only partly shown in fig. 2207), which appears as a W just above the northern horizon of most observers in latitudes of the United States. Of the seven stars forming the Big Dipper, only Dubhe, Alioth, and Alkaid are listed on the daily pages of the *Nautical Almanac.*

The two second magnitude stars forming the outer part of the bowl of the Big Dipper are often called the *pointers* because a line extending northward (*down* in spring evenings) through them points to Polaris. The Little Dipper, with Polaris at one end and Kochab at the other, is part of *Ursa Minor,* the Little Bear. Relative to its bowl, the handle of the Little Dipper curves in the opposite direction to that of the Big Dipper. Other almanac stars near the top of figure 2207 are discussed elsewhere.

A line extending southward through the pointers, and curving somewhat toward the west, leads to first magnitude Regulus, brightest star in *Leo,* the lion. The head, shoulders, and front legs of this constellation form a sickle, with Regulus at the end of the handle. Toward the east is second magnitude Denebola, the tail of the lion. On toward the southwest from Regulus is second magnitude Alphard, brightest star in *Hydra,* the sea serpent. A dark sky and considerable imagination are needed to trace the long, winding body of this figure.

A curved line extending the arc of the handle of the Big Dipper leads to first magnitude Arcturus. With Alkaid and Alphecca, brightest star in *Corona Borealis,* the Northern Crown, Arcturus forms a large, inconspicuous triangle. If the arc through Arcturus is continued, it leads next to first magnitude Spica and then to *Corvus,* the crow, which appears most like a gaff mainsail of a schooner. The brightest star in this constellation is Gienah, but three others are nearly as bright. At autumnal equinox the sun is on the celestial equator, about midway between Regulus and Spica.

A long, slightly curved line from Regulus east-southeasterly through Spica leads to Zubenelgenubi (zōō·bĕn′ĕl·jĕ·nu′bē) at the southwestern corner of an inconspicuous box-like figure called *Libra,* the (weighing) scales.

Returning to *Corvus,* a line from Gienah, extending diagonally across the figure and then curving somewhat toward the east, leads to Menkent, just beyond *Hydra.*

Far to the south, below the horizon of most northern hemisphere observers, a group of bright stars is a prominent feature of the spring sky of the Southern Hemisphere. *Crux,* the Southern Cross, is about 40° south of *Corvus.* This is a small figure and a poor cross, and hence disappointing to many who view it for the first time. The "false cross" to the west is a better but less conspicuous cross. Acrux at the southern end of the Southern Cross, and Gacrux at the northern end, are listed on the daily pages of the *Nautical Almanac.*

The triangles formed by Suhail, Miaplacidus, and Canopus, and by Suhail, Adhara, and Canopus, are west of the Southern Cross, Suhail being in line with the

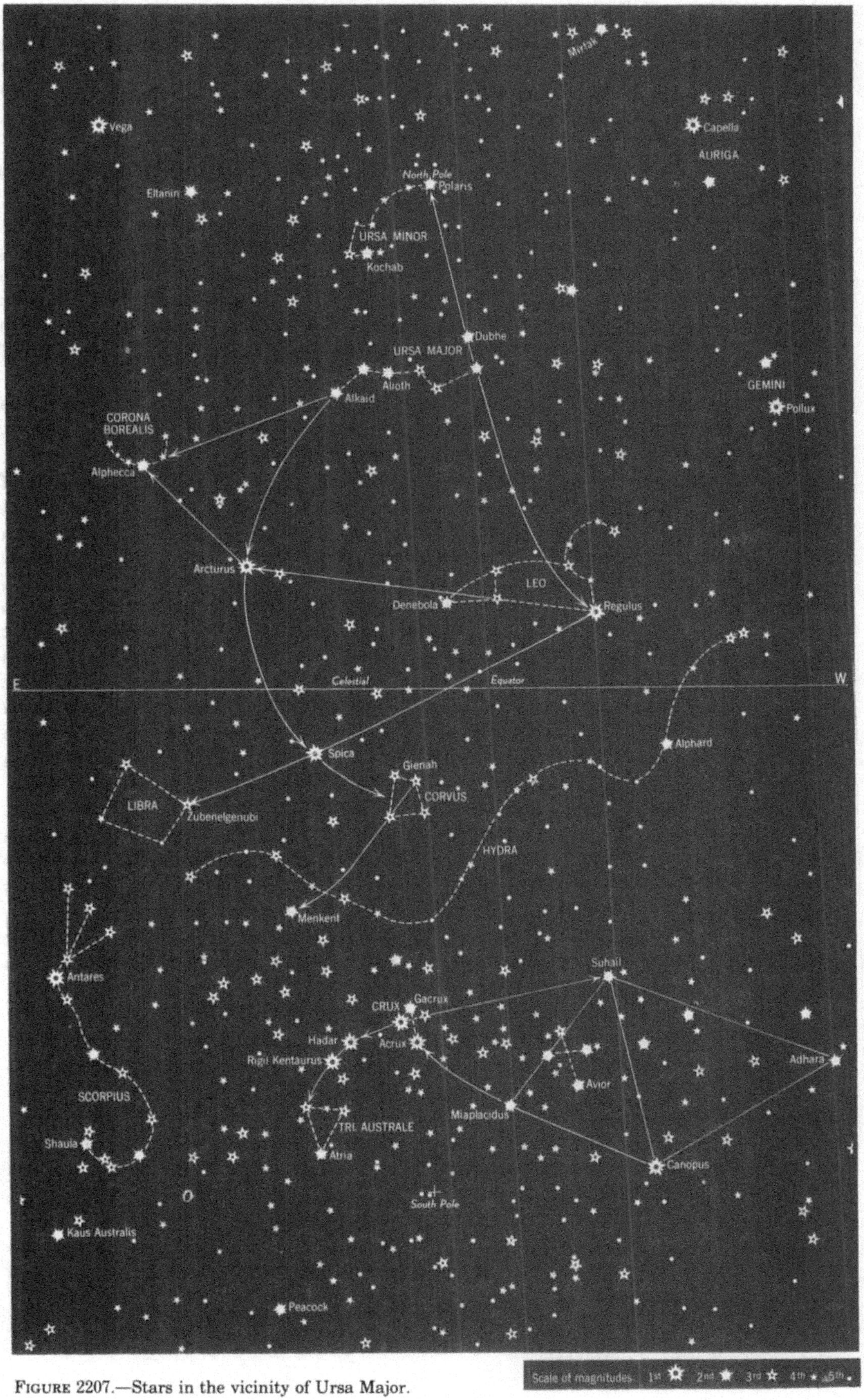

FIGURE 2207.—Stars in the vicinity of Ursa Major.

horizontal arm of the Southern Cross at this time. A line from Canopus, through Miaplacidus, curved slightly toward the north, leads to Acrux. A line through the east-west arm of *Crux,* eastward and then curving toward the south, leads first to Hadar and then to Rigil Kentaurus, two very bright stars. Continuing on, the curved line leads to small *Triangulum Australe,* the Southern Triangle, the easternmost star of which is Atria.

Scorpius, the scorpion, Kaus Australis, and Peacock, in the southeastern sky of the Southern Hemisphere, are discussed in article 2208.

2208. Stars in the vicinity of *Cygnus* (fig. 2208).—As the celestial sphere continues in its apparent westward rotation, the stars familiar to a spring evening observer sink low in the western sky. By midsummer the Big Dipper has moved to a position to the left of the north celestial pole, and the line from the pointers to Polaris is nearly horizontal. The Little Dipper is standing on its handle, with Kochab above and to the left of the celestial pole. *Cassiopeia* is at the right of Polaris, opposite the handle of the Big Dipper.

The only first magnitude star in the western sky is Arcturus, which forms a large, inconspicuous triangle with Alkaid, the end of the handle of the Big Dipper, and Alphecca, the brightest star in *Corona Borealis,* the Northern Crown.

The eastern sky is dominated by three very bright stars. The westernmost of these is Vega, the brightest star north of the celestial equator, and third brightest star in the heavens. Its magnitude is 0.1. Having a declination of a little less than 39°N, this star passes through the zenith along a path across the central part of the United States, from Washington in the east to San Francisco on the Pacific coast. Vega forms a large but conspicuous triangle with its two bright neighbors, Deneb to the northeast and Altair to the southeast. The angle at Vega is nearly a right angle. Deneb is at the end of the tail of *Cygnus,* the swan. This configuration is sometimes called the Northern Cross, with Deneb at the head. To modern youth it more nearly resembles a dive bomber while it is still well toward the east, with Deneb at the nose of the fuselage. Altair has two fainter stars close by, on opposite sides. The line formed by Altair and its two fainter companions, if extended in a northwesterly direction, passes through Vega, and on to second magnitude Eltanin. The angular distance from Vega to Eltanin is about half that from Altair to Vega. Vega and Altair, with second magnitude Rasalhague to the west, form a large equilateral triangle. This is less conspicuous than the Vega-Deneb-Altair triangle because the brilliance of Rasalhague is much less than that of the three first magnitude stars, and the triangle is overshadowed by the brighter one.

Far to the south of Rasalhague, and a little toward the west, is a striking configuration called *Scorpius,* the scorpion. The brightest star, forming the head, is red Antares. At the tail is Shaula.

Antares is at the southwestern corner of an approximate parallelogram formed by Antares, Sabik, Nunki, and Kaus Australis. With the exception of Antares, these stars are only slightly brighter than a number of others nearby, and so this parallelogram is not a striking figure. At winter solstice the sun is a short distance northwest of Nunki.

Northwest of *Scorpius* is the box-like *Libra,* the (weighing) scales, in which Zubenelgenubi marks the southwest corner.

With Menkent and Rigil Kentaurus to the southwest, Antares forms a large but unimpressive triangle. For most observers in the latitudes of the United States, Antares is low in the southern sky, and the other two stars of the triangle are below the horizon. To an observer in the Southern Hemisphere *Crux,* the Southern

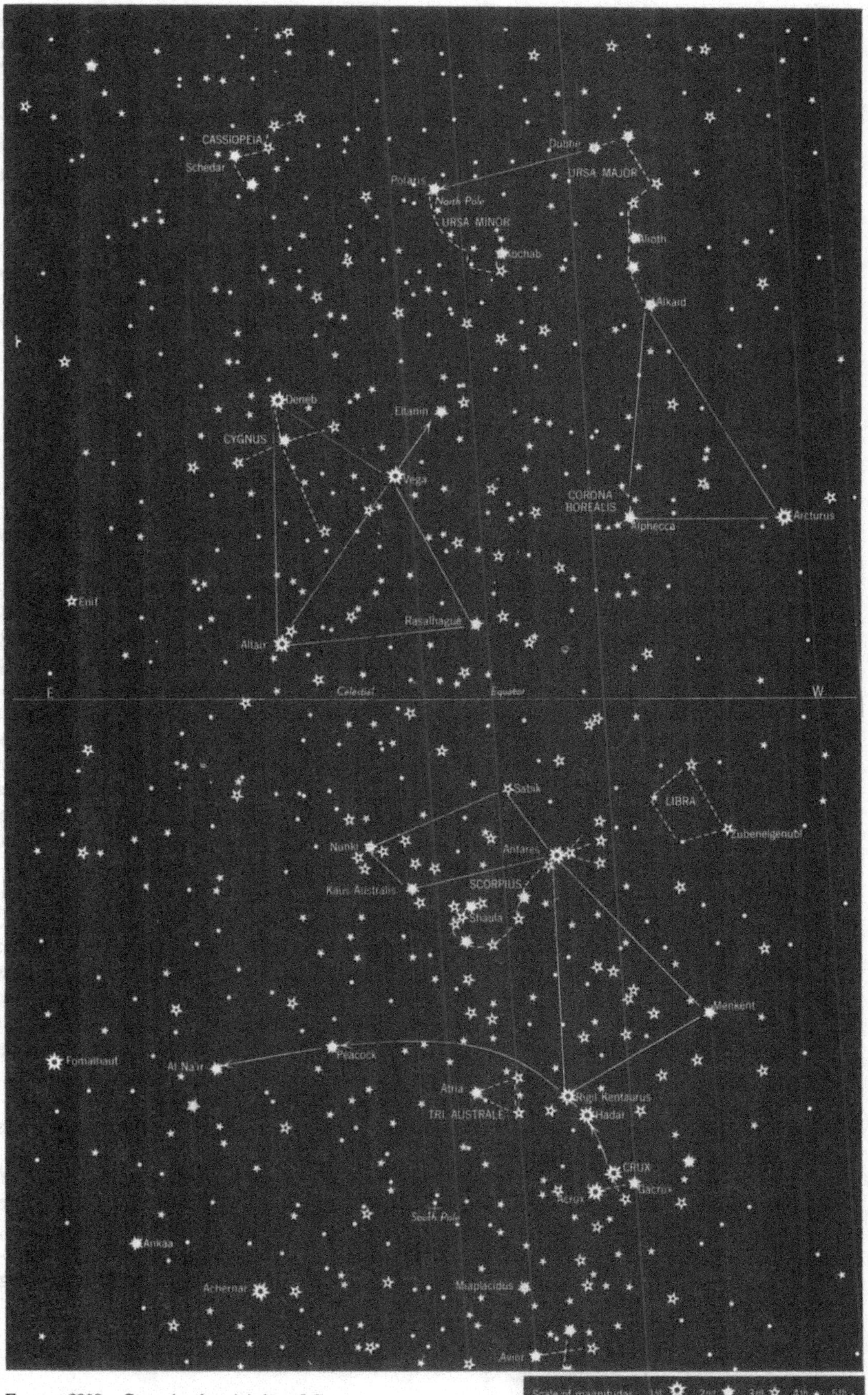

FIGURE 2208.—Stars in the vicinity of *Cygnus*.

Cross, is to the right of the south celestial pole, which is not marked by a conspicuous star. A long, curved line starting with the now-vertical arm of the Southern Cross and extending northward and then eastward passes successively through Hadar, Rigil Kentaurus, Peacock, and Al Na'ir.

Fomalhaut is low in the southeastern sky of the southern hemisphere observer, and Enif is low in the eastern sky at nearly any latitude. With the appearance of these stars it is not long before *Pegasus* will appear over the eastern horizon during the evening, and as the winged horse climbs evening by evening to a position higher in the sky, a new annual cycle approaches.

2209. Planet diagram.—The planet diagram in the *Nautical Almanac* shows, in graphical form for any date during the year, the LMT of meridian passage of the sun, of the five planets Mercury, Venus, Mars, Jupiter, and Saturn, and of each 30° of SHA. The diagram provides a general picture of the availability of planets and stars for observation, and thus the following information:

1. whether a planet or star is too close to the sun for observation;
2. whether a planet is a morning or evening star;
3. some indication of the planet's position during twilight;
4. the proximity of other planets; and
5. whether a planet is visible from evening to morning twilight.

A band 45^m wide is shaded on each side of the curve marking the LMT of meridian passage of the sun. Any planet and most stars lying within the shaded area are too close to the sun for observation.

When the meridian passage occurs at midnight, the body is in opposition (art. 1422) to the sun and is visible all night; planets may be observable in both morning and evening twilights. As the time of meridian passage decreases, the body ceases to be observable in the morning, but its altitude above the eastern horizon during evening twilight gradually increases; this continues until the body is on the meridian at twilight. From then onwards the body is observable above the western horizon and its altitude at evening twilight gradually decreases; eventually the body comes too close to the sun for observation. When the body again becomes visible, it is seen as a morning star low in the east; its altitude at twilight increases until meridian passage occurs at the time of morning twilight. Then, as the time of meridian passage decreases to 0^h, the body is observable in the west in the morning twilight with a gradually decreasing altitude, until it once again reaches opposition.

Only about one half the region of the sky along the ecliptic as shown on the diagram is above the horizon at one time. At sunrise (LMT about 6^h) the sun and, hence, the region near the middle of the diagram are rising in the east; the region at the bottom of the diagram is setting in the west. The region half way between is on the meridian. At sunset (LMT about 18^h) the sun is setting in the west; the region at the top of the diagram is rising in the east. Marking the planet diagram of the *Nautical Almanac* so that east is at the top of the diagram and west is at the bottom can be useful to interpretation.

If the curve for a planet intersects the vertical line connecting the date graduations below the shaded area, the planet is a morning star; if the intersection is above the shaded area, the planet is an evening star.

A similar **planet location diagram** in the *Air Almanac* represents the region of the sky along the ecliptic within which the sun, moon, and planets always move; it shows, for each date, the sun in the center and the relative positions of the moon, the five planets Mercury, Venus, Mars, Jupiter, Saturn and the four first magnitude stars *Aldebaran, Antares, Spica,* and *Regulus,* and also the position on the ecliptic which is north of Sirius (i.e. Sirius is 40° south of this point). The first point of Aries is also shown for reference. The magnitudes of the planets are given at suitable intervals along the curves. The moon symbol shows the correct phase. A straight line joining the date on the left-hand side with the same date of the right-hand side represents a complete circle around the sky, the two ends of the line representing the point 180° from the sun; the intersections with the curves show the spacing of the bodies along the ecliptic on the date. The time scale indicates very approximately the local mean time at which an object will be on the observer's meridian.

At any time only about half the region on the diagram is above the horizon. At sunrise the sun (and hence the region near the middle of the diagram) is rising in the east and the region at the end marked "West" is setting in the west; the region half-way between these extremes is on the meridian, as will be indicated by the local time (about 6^h). At the time of sunset (local time about 18^h) the sun is setting in the west, and the region at the end marked "East" is rising in the east.

The diagram should be used in conjunction with the *Sky Diagrams.*

2210. Star finders.—Various devices have been invented to help an observer locate and identify individual stars. The most widely used is the Star Finder and Identifier formerly published by the U. S. Navy Hydrographic Office and now published commercially. The current model, No. 2102-D, as well as the previous 2102-C model patented by E. B. Collins, employs the same basic principle as that used in the Rude Star Finder, which was patented by Captain G. T. Rude, USC&GS, and later sold to the Hydrographic Office. Successive models reflect various modifications to meet changing conditions and requirements.

The *star base* of No. 2102-D consists of a thin, white, opaque, plastic disk about 8½ inches in diameter, with a small peg in the center. On one side the north celestial pole is shown at the center, and on the opposite side the south celestial pole is at the center. All of the stars listed on the daily pages of the *Nautical Almanac* are shown on a polar azimuthal equidistant projection (art. 320) extending to the opposite pole. The south pole side is shown in figure 2210a. Many copies of an older edition, No. 2102-C, showing the stars listed in the almanacs prior to 1953, and having other minor differences, are still in use. These are not rendered obsolete by the newer edition, but should be corrected by means of the current almanac. The rim of each side is graduated to half a degree of LHA ♈ (or 360°−SHA).

Ten transparent *templates* of the same diameter as the star base are provided. There is one template for each 10° of latitude, labeled 5°, 15°, 25°, etc., plus a 10th (printed in red) showing meridian angle and declination. The older edition (No. 2102-C) did not have the red meridian angle-declination template. Each template can be used on either side of the star base, being centered by placing a small center hole in the template over the center peg of the star base. Each latitude template has a family of altitude curves at 5° intervals from the horizon (from altitude 10° on the older No. 2102-C) to 80°. A second family of curves, also at 5° intervals, indicates azimuth. The north-south azimuth line is the celestial meridian. The star base, templates, and a set of instructions are housed in a circular leatherette container.

Since the sun, moon, and planets continually change apparent position relative to the "fixed" stars, they are not shown on the star base. However, their positions at anytime, as well as the positions of additional stars, can be plotted. To do this, determine 360°−SHA of the body. For the stars and planets, SHA is listed in the *Nautical Almanac.* For the sun and moon, 360°−SHA is found by subtracting GHA

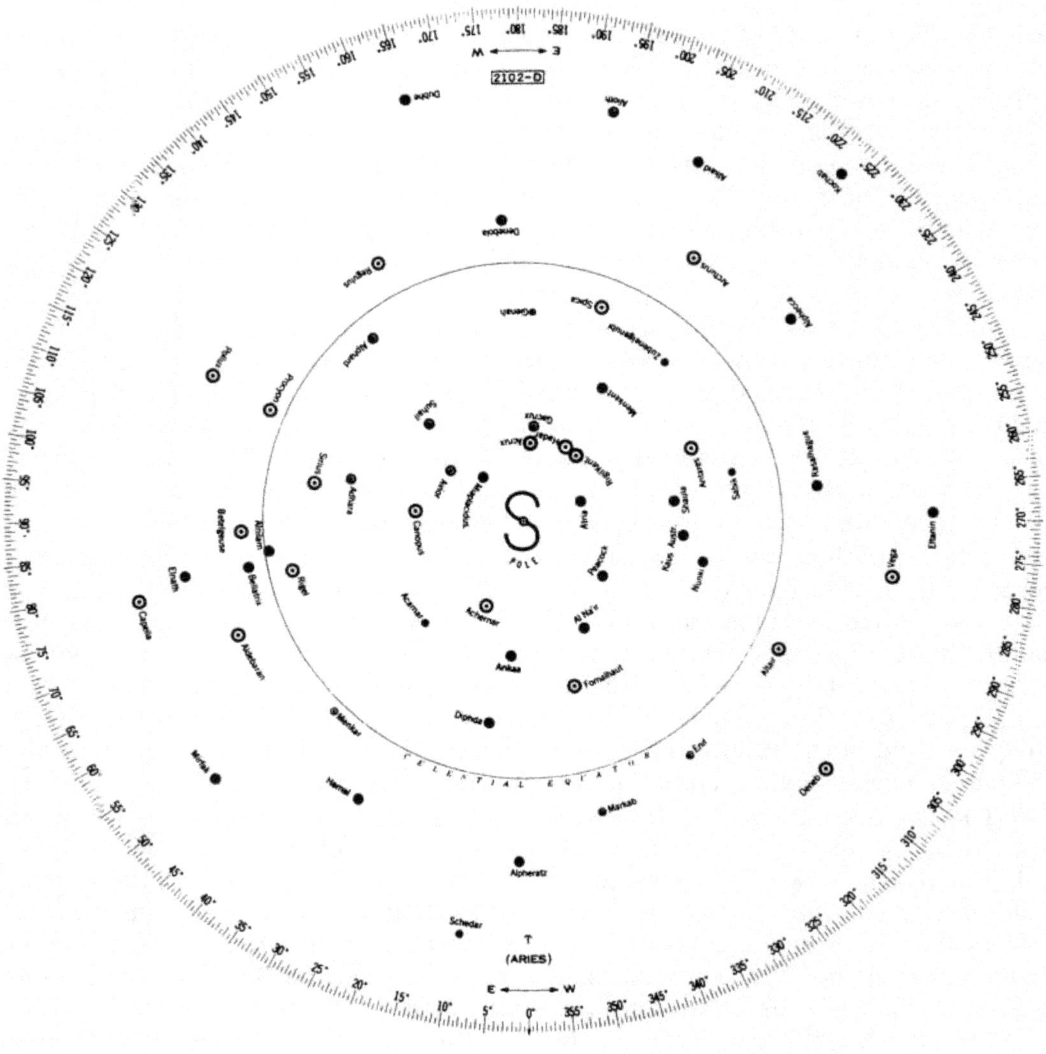

FIGURE 2210a.—The south pole side of the star base of No. 2102-D.

of the body from GHA ♈ at the same time. Locate 360° − SHA on the scale around the rim of the star base. A straight line from this point to the center represents the hour circle of the body. From the celestial equator, shown as a circle midway between the center and the outer edge, measure the declination (from the almanac) of the body *toward* the center if the pole and declination have the *same* name (*both* N or *both* S), and *away* from the center if they are of *contrary* name. Use the scale along the north-south azimuth line of any template as a declination scale. The meridian angle-declination template (the latitude 5° template of No. 2102-C) has an open slot with declination graduations along one side, to assist in plotting positions, as shown in figure 2210b. In the illustration the celestial body being located has a 360° − SHA of 285°, and a declination of 14°.5S. It is not practical to attempt to plot to greater precision than the nearest 0°.1. Positions of Venus, Mars, Jupiter, and Saturn on June 1, 1975, are shown plotted on the star base in figure 2210c. It is sometimes desirable to plot positions of the sun and moon, to assist in planning. Plotted positions of stars need not be changed. Plotted positions of bodies of the solar system should be replotted from time to time, the more rapidly moving ones oftener than others. The satisfactory interval for each body can be determined by experience. It is good practice to record the date of each plotted position of a body of the solar system, to serve later as an indication of the interval since it was plotted.

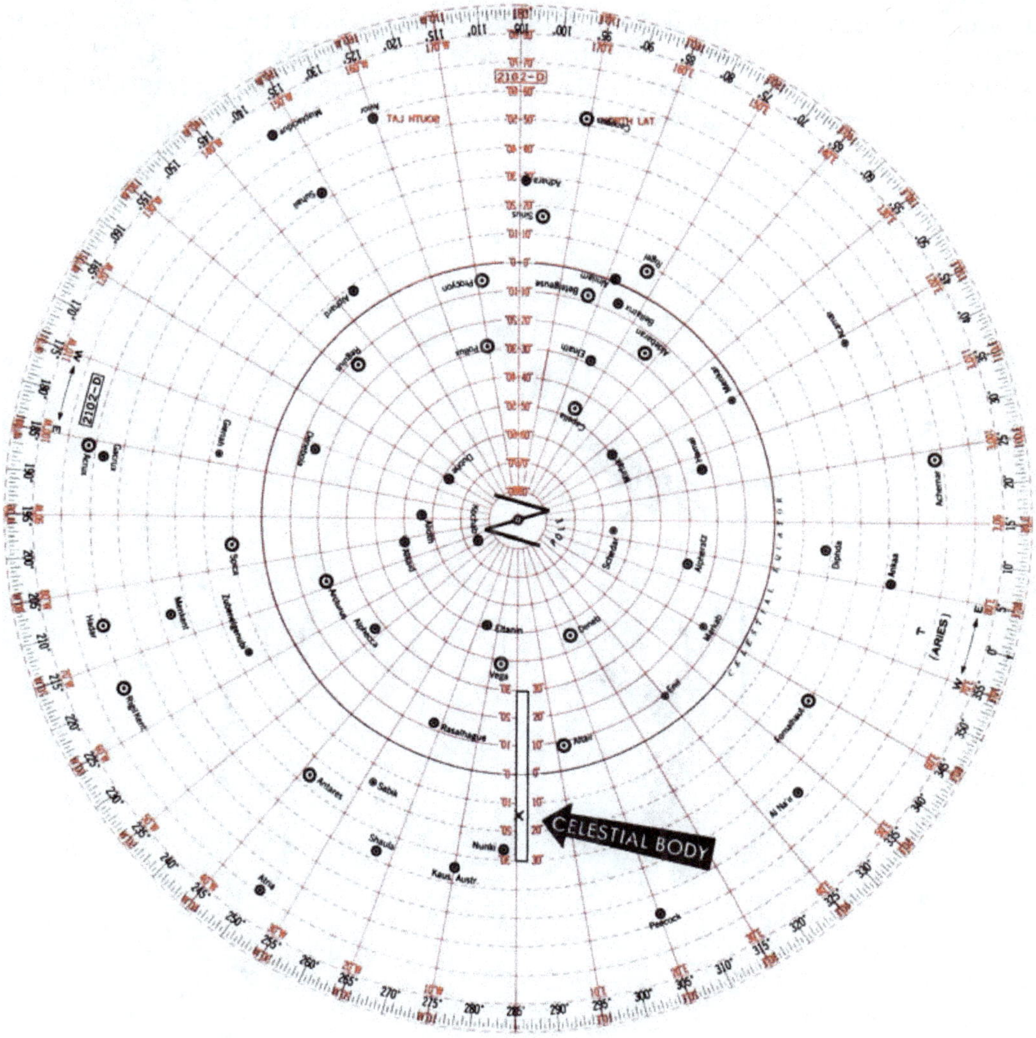

FIGURE 2210b.—Plotting a celestial body on the star base of No. 2102-D.

To orient the template properly for any given time, proceed as follows: enter the almanac with GMT, and determine GHA ♈ at this time. Apply the longitude to GHA ♈, subtracting if west or adding if east, to determine LHA ♈. If LMT is substituted for GMT in entering the almanac, LHA ♈ can be taken directly from the almanac, to sufficient accuracy for orienting the star finder template. Select the template for the latitude nearest that of the observer, and center it over the star base, being careful that the correct sides (north or south to agree with the latitude) of both template and star base are used. Rotate the template relative to the star base until the arrow on the celestial meridian (the north-south azimuth line) is over LHA ♈ on the star based graduations. The small cross at the origin of both families of curves now represents the zenith of the observer. The approximate altitude and azimuth of the celestial bodies above the horizon can be read directly from the star finder, using eye interpolation. Consider Polaris, not shown as at the north celestial pole. For more accurate results, the template can be lifted clear of the center peg of the star base, and shifted along the celestial meridian until the latitude, on the altitude scale, is over the pole. This refinement is not needed for normal use of the device. It should not be used for a latitude differing more than 5° from that for which the curves were drawn. If the altitude and azimuth of an identified body shown on the star base are known, the template can be oriented by rotating it until it is in correct position relative to that body.

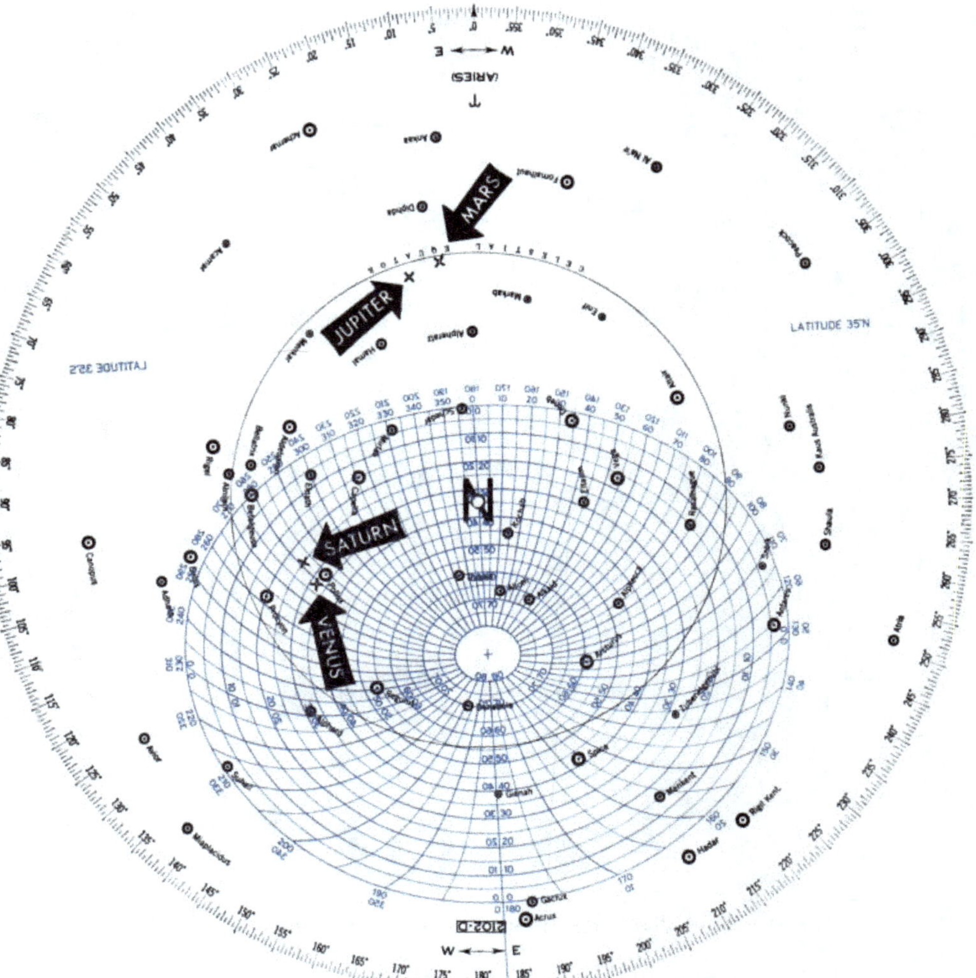

FIGURE 2210c.—A template in place over the star base of No. 2102-D.

Customarily, No. 2102-D is used in either of two ways:

1. To make an advance list of celestial bodies available for observation at a given time.

2. To identify an unknown celestial body which has been observed.

Example 1.—During evening twilight on June 1, 1975, the GMT 2324 DR position of a ship is lat. 34°12′.5N, long. 57°40′.0W.

Required.—The approximate altitude (h) and azimuth of each first magnitude star, and any planets, between altitudes 15° and 75°.

Solution (fig. 2210c).—(1) Plot the positions of the planets, as shown. The values used are those for GMT 0000 on June 1, as follows:

Venus	116°7	23°7N
Mars	8°0	1°8N
Jupiter	15°9	5°5N
Saturn	108°4	22°3N

(2) Determine LHA ♈ by means of the *Nautical Almanac,* as follows:

GMT	2324	June 1
23^h	234°48′2	
24^m	6°01′0	
GHA♈	240°49′2	
λ	57°40′0W	
LHA ♈	183°09′2	

(3) Select the template for latitude 35°, place it over the north side of the star base with "LATITUDE 35° N" appearing correctly, and orient it to 183°2. It is customary to list the bodies in order of increasing azimuth, as follows:

Vega	17°	054°
Arcturus	59°	111°
Spica	42°	157°
Regulus	53°	240°
Procyon	20°	262°
Venus	32°	278°
Saturn	25°	282°
Pollux	33°	284°
Capella	15°	316°

Example 2.—At the time and place of example 1, an unidentified celestial body is observed through a break in the clouds. Its sextant altitude is 15°27′8, and its azimuth is 085°.

Required.—Identify the celestial body.

Solution (fig. 2210c).—Orient the template as in example 1. By means of its altitude and azimuth, identify the star as Rasalhague.

If no body appears at the measured altitude and azimuth, place the red meridian angle-declination template over the altitude-azimuth template and read off, by inspection, the declination and the 360° − SHA value of the body and from this, determine its SHA. Using the SHA and declination, enter the list of stars near the back of the *Nautical Almanac,* and identify the body. If it is not found in this list, and no error has been made, one of the stars not listed in the almanac, or possibly the planet Mercury, has been observed. Unless a copy of *The Astronomical Almanac* or another book containing the required information is available, the observation cannot be used. If right ascension (art. 1426) of the body is available, but not its SHA, the value taken from the star finder (360° − SHA) is converted to time units (art. 1811) and used directly, since RA = 360° − SHA.

Example 3.—At the time and place of example 1 an unidentified celestial body is observed through a break in the clouds. Its altitude is 52°58′9, and its azimuth is 170°.

Required.—Identify the celestial body.

Solution (fig. 2210c).—Orient the template as in example 1. Since no celestial body appears at the place indicated by its altitude and azimuth, the red meridian angle-declination template is placed over the altitude-azimuth template. The declination is found to be about 1°S. The 360° − SHA value is about 190°, and SHA is therefore about 170°. From the star list near the back of the *Nautical Almanac,* the star is identified as γ *Virginis.*

Kotlarić's Star Finder and Identifier, designed and patented by Dr. Stjepo M. Kotlarić, Assistant Director of the Hydrographic Institute of the Yugoslav Navy, is actually a booklet of 18 pairs of star charts with a plastic template enclosed. It depicts the 57 selected stars plus 125 other stars, for a total of 182 stars. All of the 173 stars listed in the *Nautical Almanac* are depicted. This star finder provides greater reliability in identification than No. 2102–D.

Each pair of star charts shows the Western and Eastern Hemispheres separately. The Western Hemisphere is shown using the stereographic projection (art. 318) on a plane tangent to the celestial equator at the west point; the Eastern Hemisphere is shown using the same projection on a plane tangent to the celestial equator at the east point.

The stars are plotted with different symbols, according to their magnitude. Selected stars are in black, and other stars in green. The stars in constellations are connected by a broken yellow line, while the "star-chasing" alignments are plotted with solid yellow lines. The names of the selected stars are shown in black capital letters, while all other stars are shown in green, using capital letters for popular names.

Since the list of stars near the back of the *Nautical Almanac* is not compiled in alphabetical order, an alphabetical index of stars with their rounded values of SHA and declination is included with the star finder to facilitate location of a star in the list.

Two 20° intervals of LHA ♈, differing by 180°, are used for constructing each pair of star charts. The circle bordering the star chart (fig. 2210d) represents the observer's celestial meridian. The circle is graduated to permit orientation of the plastic template to the star chart, according to the observer's latitude, in order to portray the visible hemisphere in the horizon system of coordinates.

The latitude scale on the right-hand half is black, and that on the left half is red. By selecting the appropriate pair of star charts according to the values of LHA ♈ and the azimuth, printed on the star charts, and by placing the template over the selected star chart in such a way that their centers coincide and the zenith of the template is placed on the proper latitude value on the black border scale when the azimuth figures are in black type (or on the red border scale when the azimuth figures are in red type), the altitude and azimuth of a star can be determined. Also, with the observed altitude and azimuth the coordinates of the star in the celestial equator system can be determined.

Star symbols on all the star charts are plotted for odd tens of degrees of LHA ♈. A part of the star's path is plotted to a distance of 10° to the left and to the right of the star symbol's position. A 10° increase of LHA ♈ from the star symbol's position is plotted as a solid line path, and a 10° decrease of LHA ♈ is plotted as a dotted line path. These portions of the star paths facilitate the identification of stars and make for more certainty in identification. The position of a star, determined by means of altitude and azimuth on the template superimposed on this star chart, may be found from either the star symbol or the appropriate point on the star path when LHA ♈ at the time of the observation differs from the value of odd tens of degrees for which the star symbol is printed on the star chart.

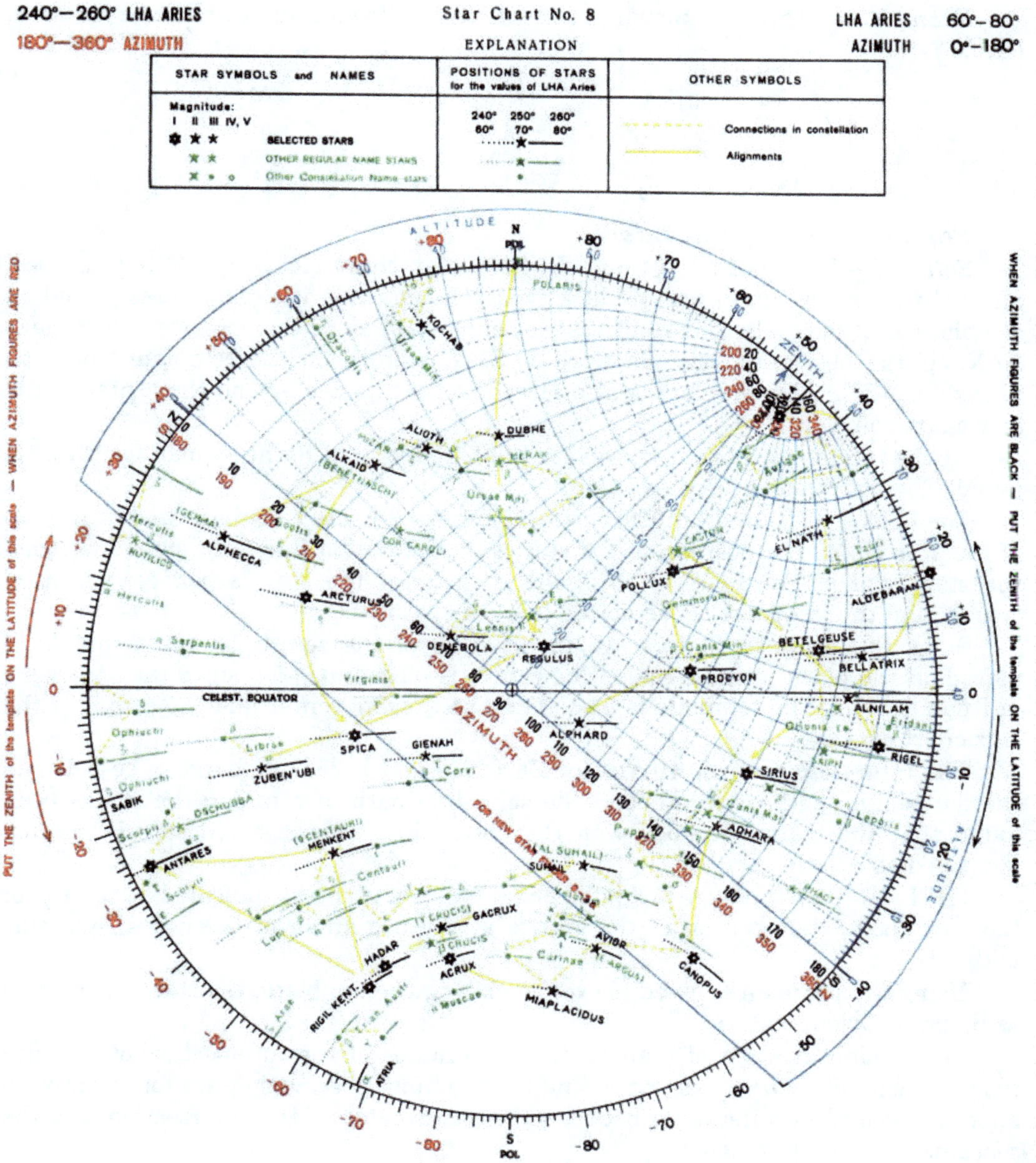

FIGURE 2210d.—Kotlarić's Star Finder and Identifier.

Example 4.—Three unknown stars with the following data are observed from lat. 50°N:

Zn	*h*	*LHA* ♈
028°	22°48′	70°
029°.5	25°00′	66°.5
152°	26°16′	61°

Required.—Identify the stars.

Solution.—For LHA ♈ 70° and Zn 028°, star chart No. 8 (fig. 2210d) is used. Since the azimuth figures are in black type, the zenith on the template is placed on the black latitude scale. The zenith is placed at +50° since the observer's latitude is 50°N. (If the observer's latitude were 50°S, the zenith on the template would be placed at −50°.) The template is also placed so that its center and the center of the star chart coincide.

At the intersection of the curves for Zn 028° and h 22°.8, the symbol for Mizar is found. The body is identified as Mizar.

For LHA ♈ 66°.5 and Zn 029°.5, star chart No. 8 is used. As indicated at the top of the star chart, the position of the star is on the dotted line path extending from the star symbol. For LHA ♈ 66°.5, the position of the star is between the star symbol and the midpoint of the dotted line star path.

Since the azimuth figures are in black type, the zenith on the template is placed on the black latitude scale. The zenith is placed at +50° since the observer's latitude is 50°N. The template is also placed so that its center and the center of the star chart coincide.

The intersection of the curves for Zn 029°.5 and h 25°.0 is found to be between the dotted line path of Alioth and the solid line path of Mizar. Since it has been found that the star's position is on the dotted line, the observed star is reliably identified as Alioth.

For LHA ♈ 61° and Zn 152°, star chart No. 8 is used. As indicated at the top of the star chart, the position of the star is at the beginning of the dotted line star path.

Using the procedures given above for placing the template, the star is identified as Saiph (κ Orionis).

In the identification of planets, the coordinates of the intersection of the altitude and azimuth curves corresponding to the planet observation are found through approximating the differences between the coordinates of the intersection and the coordinates of a nearby star.

In practice, however, the procedure is considerably simpler. For example, if the intersection of the altitude and azimuth curves is found near Regulus, the daily page of the *Nautical Almanac* is scanned to find which of the four navigational planets has coordinates closest to those of Regulus. This simple procedure is achieved through the use of the large number of stars.

2211. *Sight Reduction Tables for Air Navigation* (Pub. No. 249).—Volume I of Pub. No. 249 can be used as a star finder for the stars tabulated at any given time. For these bodies the altitude and azimuth are tabulated for each 1° of latitude and 1° of LHA ♈ (2° beyond latitude 69°). The principal limitation is the small number of stars listed.

2212. Sky diagram.—Near the back of the *Air Almanac* are a number of **sky diagrams.** These are azimuthal equidistant projections (art. 320) of the celestial sphere on the plane of the horizon, at latitudes 75°N, 50°N, 25°N, 0°, 25°S, and 50°S, at intervals of 2 hours of local mean time each month. A number of the brighter stars, the visible planets, and several positions of the moon are shown at their correct altitude and azimuth. These are of limited value because of their small scale; the large increments of latitude, time, and date; and the limited number of bodies shown. However, in the absence of other methods, particularly a star finder, these diagrams can be useful. Allowance can be made for variations from the conditions for which each diagram is constructed. Instructions for use of the diagrams are included in the *Air Almanac.*

2213. Identification by computation.—If the altitude and azimuth of the celestial body, and the approximate latitude of the observer, are known, the navigational triangle (art. 1433) can be solved for meridian angle and declination. The meridian angle can be converted to LHA, and this to GHA. With this and GHA of Aries at the time of observation, the SHA of the body can be determined. With SHA and declination, one can identify the body by reference to an almanac. Any method of solving a spherical triangle, with two sides and the included angle being given, is suitable for this purpose. A large-scale, carefully-drawn diagram on the plane of the celestial meridian, using the refinement shown in figure 1432f, should yield satisfactory results. A simple method of computation is by Pub. No. 214. Following the tables of computed altitude and azimuth for each latitude, a two-page star identification table is given, as shown in appendix N. The example given below is based upon this extract.

The steps in solution by Pub. No. 214 are:

1. Convert Zn to Z.
2. With Z and h (usually the approximate value taken from the sextant, without correction) enter the Pub. No. 214 star identification pages for the nearest whole degree of latitude, and extract the declination and meridian angle, t (given as H.A. in the table). If the declination is given in roman type, above the heavy line, it has the *same* name as the latitude. If the declination is given in *italics,* below the heavy line, it has the *contrary* name to that of the latitude. When interpolating between roman and italic declinations, consider the italic value negative, using the arithmetical *sum* as the algebraic *difference* needed for interpolation. Extract values to the nearest whole degree.
3. Convert to LHA.
4. Apply the longitude to LHA to find GHA, adding if in west longitude, and subtracting if in east longitude.
5. Enter the *Nautical Almanac* with GMT, and determine GHA ♈.
6. Subtract GHA ♈ from GHA ☆ to find SHA (since GHA☆=GHA ♈+SHA).
7. With the approximate SHA and d enter the *Nautical Almanac* star list and identify the body, checking first the SHA and then the declination. Do not overlook the possibility of having observed a planet or a star not listed in the almanac. For a planet, check first the declination. If this is approximately correct, check the GHA. It is not necessary to find the SHA of a planet.

Example.—On May 31, 1975, the 0425 DR position of a ship is lat. 41°13.′6N, long. 140°41.′7W. About this time the navigator observes an unknown star through a break in the clouds, as follows: GMT $13^h24^m46^s$, hs 15°01.′5, Zn 232°.

Required.—Identify the unknown celestial body, using Pub. No. 214.

Solution.—

	May 31		
GMT	$13^h24^m46^s$ May 31	Zn	232°
13^h	83°24'.4	Z	N128°W
24^m46^s	6°12'.5	h	15°
GHA ♈	89°36'.9 (subtract)	d	16°S (from Pub. No. 214)
GHA☆	193°	t	52°W (from Pub. No. 214)
SHA☆	103°	LHA	52°
d	16°S	λ	141°W
Body	Sabik	GHA	193°

Although no formal star identification tables are included in Pub. No. 229, a simple approach to star identification is to scan the pages of the appropriate latitudes and observe the combination of arguments which give the altitude and azimuth angle of the observation. Thus the declination and LHA☆ are determined directly. The star's SHA is found from, SHA☆=LHA☆−LHA♈. From these quantities the star can be identified from the *Nautical Almanac.*

Another solution is available through an interchange of arguments using the nearest integral values. The procedure consists of entering Pub. No. 229 with the observer's latitude (same name as declination), with the observed azimuth angle (converted from observed true azimuth as required) as LHA and the observed altitude as declination, and extracting from the tables the altitude and azimuth angle respondents. The extracted altitude becomes the body's declination; the extracted azimuth angle (or its supplement) is the meridian angle of the body. Note that the tables are always entered with latitude of same name as declination. In north latitudes the tables can be entered with true azimuth as LHA.

If the respondents are extracted from above the C-S Line on a right-hand page, the name of the latitude is actually contrary to that of the declination. Otherwise, the declination of the body has the same name as the latitude. If the azimuth angle respondent is extracted from above the C-S Line, the supplement of the tabular value is the meridian angle, t, of the body. If the body is east of the observer's meridian, LHA=360°−t; if the body is west of the meridian, LHA=t.

Problems

2210a. During morning twilight on June 3, 1975, the GMT 1825 (June 2) DR position of a ship is lat. 26°21'.4N, long. 157°17'.2E.

Required.—The approximate altitude and azimuth of each first magnitude star, and any planets, between altitudes 10° and 80°, using No. 2102-D.

Answer.—

Body	*h*	*Zn*
Jupiter	35°	101°
Mars	41°	111°
Fomalhaut	32°	160°
Altair	59°	242°
Vega	50°	301°
Deneb	67°	334°

2210b. At the time and place of problem 2210a an unidentified celestial body is observed through a break in the clouds. Its sextant altitude is 21°04'.1 and its azimuth is 044°.

Required.—Identify the celestial body, using No. 2102-D.

Answer.—Mirfak.

2210c. The dead reckoning latitude of a ship is 25°06.′4S. Two stars are observed in quick succession, as follows:

Star	*h*	*Zn*
Antares	57°	100°
Unidentified	52°	337°

Required.—Identify the unknown celestial body, using No. 2102-D.

Answer.—*ε Virginis.*

2213. On June 2, 1975, the 1725 DR position of a ship is lat. 41°27.′3S, long. 158°36.′9E. About this time the navigator observes two unknown celestial bodies through breaks in the clouds, as follows: (1) GMT $6^h24^m15^s$, hs 16°34.′9, Zn 334°; (2) GMT $6^h25^m53^s$, hs 20°26.′0, Zn 334°. The second body appears to be of the first magnitude.

Required.—Identify the unknown celestial bodies, using Pub. No. 214.

Answers.—(1) Pollux, (2) Venus.

www.ingramcontent.com/pod-product-compliance
Lightning Source LLC
LaVergne TN
LVHW061750200225
804197LV00006B/67